Developments in Clay Science

Volume 7

Nanosized Tubular Clay Minerals

Developments in Clay Science
Volume 7

Nanosized Tubular Clay Minerals

Halloysite and Imogolite

Edited by

Peng Yuan
CAS Key Laboratory of Mineralogy and Metallogeny / Guangdong Provincial Key Laboratory of Mineral Physics and Materials, Guangzhou Institute of Geochemistry, Chinese Academy of Sciences (CAS), Guangzhou, China

Antoine Thill
Laboratoire Interdisciplinaire sur l'Organisation Nanométrique et Supramoléculaire, NIMBE, CEA, CNRS, University Paris-Saclay, Paris, France

Faïza Bergaya
CNRS—ICMN (Interfaces, Confinement, Matériaux et Nanostructures), Orléans Cedex 2, France

ELSEVIER
AMSTERDAM • BOSTON • HEIDELBERG • LONDON • NEW YORK • OXFORD
PARIS • SAN DIEGO • SAN FRANCISCO • SINGAPORE • SYDNEY • TOKYO

Elsevier
Radarweg 29, PO Box 211, 1000 AE Amsterdam, Netherlands
The Boulevard, Langford Lane, Kidlington, Oxford OX5 1GB, UK
50 Hampshire Street, 5th Floor, Cambridge, MA 02139, USA

Notices

Knowledge and best practice in this field are constantly changing. As new research and experience broaden our understanding, changes in research methods, professional practices, or medical treatment may become necessary.

Practitioners and researchers must always rely on their own experience and knowledge in evaluating and using any information, methods, compounds, or experiments described herein. In using such information or methods they should be mindful of their own safety and the safety of others, including parties for whom they have a professional responsibility.

To the fullest extent of the law, neither the Publisher nor the authors, contributors, or editors, assume any liability for any injury and/or damage to persons or property as a matter of products liability, negligence or otherwise, or from any use or operation of any methods, products, instructions, or ideas contained in the material herein.

ISBN: 978-0-08-100293-3
ISSN: 1572-4352

Publisher: Candice Janco
Acquisition Editor: Amy Shapiro
Editorial Project Manager: Tasha Frank
Production Project Manager: Vijayaraj Purushothaman
Designer: Greg Harris

Typeset by SPi Global, India

Dedication

To the memory of my father, Xueqian Yuan, for the 10th anniversary of his passing away.

Peng Yuan

Contents

Part IV
Applications of Nanosized Tubular Clay Minerals

Contributors

Numbers in Parentheses indicate the pages on which the author's contributions begin.

E. Abdullayev (554), Ennis-Flint Science and Technology Center, Thomasville, NC, United States

C. Aguzzi (708), University of Granada, Granada, Spain

M.S. Amara (254), Laboratoire de Physique des Solides, CNRS, Univ. Paris-Sud, Université Paris-Saclay, Orsay Cedex, France

N. Arancibia-Miranda (458), Center for the Development of Nanoscience and Nanotechnology, CEDENNA; Facultad de Química y Biología, Universidad de Santiago de Chile, Santiago, Chile

I. Basile-Doelsch (49), Aix-Marseille Université, CNRS, IRD, CEREGE UM34, USC INRA, Aix en Provence, France

L. Belloni (361), LIONS, NIMBE, CEA, CNRS, Université Paris-Saclay, CEA/Saclay, Gif-sur-Yvette, France

F. Bergaya (1, 735), CNRS—ICMN (Interfaces, Confinement, Matériaux et Nanostructures), Orléans Cedex 2, France

B. Bonelli (279, 672), INSTM Unit of Torino Politecnico, Politecnico di Torino, Turin, Italy

E. Carazo (708), University of Granada, Granada, Spain

P. Cerezo (708), University of Granada, Granada, Spain

P. Davidson (308), Laboratoire de Physique des Solides, UMR8502, CNRS, Univ. Paris-Sud, Université Paris-Saclay, Orsay Cedex, France

I. Dozov (308), Laboratoire de Physique des Solides, UMR8502, CNRS, Univ. Paris-Sud, Université Paris-Saclay, Orsay Cedex, France

P. Du (167), CAS Key Laboratory of Mineralogy and Metallogeny / Guangdong Provincial Key Laboratory of Mineral Physics and Materials, Guangzhou Institute of Geochemistry, Chinese Academy of Sciences (CAS), Guangzhou, China

H.A. Duarte (331), Grupo de Pesquisa em Química Inorgânica Teórica (GPQIT), Instituto de Ciências Exatas, Universidade Federal de Minas Gerais (UFMG), Belo Horizonte, MG, Brazil

M. Escudey (458), Center for the Development of Nanoscience and Nanotechnology, CEDENNA; Facultad de Química y Biología, Universidad de Santiago de Chile, Santiago, Chile

A. Fernandez-Martinez (202), ISTerre, CNRS; ISTerre, Univ. Grenoble Alpes, Grenoble, France

E. Garrone (672), INSTM Unit of Torino Politecnico, Politecnico di Torino, Turin, Italy

B.C. Guo (509), South China University of Technology, Guangzhou, China

Y. Higaki (628), Institute for Materials Chemistry and Engineering, and International Institute for Carbon-Neutral Energy Research (WPI-I2CNER), Kyushu University, Fukuoka, Japan

J. Huang (509), South China University of Technology, Guangzhou, China

M.-C. Jaurand (485), Institut National de la Santé et de la Recherche Médicale (INSERM), UMR-S 116, 27 rue Juliette Dodu; Université Paris Descartes, Labex Immuno-Oncology, Sorbonne Paris Cité, Faculté de Médecine, Paris, France

E. Joussein (12), Université de Limoges, FST, GRESE 'Groupement de Recherche Eau Sol Environnement', Limoges, France

J.T. Kloprogge (115), School of Earth Sciences, The University of Queensland, St. Lucia, QLD, Australia

T. Kogure (92), Graduate School of Science, The University of Tokyo, Tokyo, Japan

P. Launois (254), Laboratoire de Physique des Solides, CNRS, Univ. Paris-Sud, Université Paris-Saclay, Orsay Cedex, France

C. Levard (49), Aix-Marseille Université, CNRS, IRD, CEREGE UM34, USC INRA, Aix en Provence, France

D. Liu (167), CAS Key Laboratory of Mineralogy and Metallogeny / Guangdong Provincial Key Laboratory of Mineral Physics and Materials, Guangzhou Institute of Geochemistry, Chinese Academy of Sciences (CAS), Guangzhou, China

Y. Lvov (554), Institute for Micromanufacturing, Louisiana Tech University, Ruston, LA, United States

W. Ma (628), Institute for Materials Chemistry and Engineering, and International Institute for Carbon-Neutral Energy Research (WPI-I2CNER), Kyushu University, Fukuoka, Japan

J. Matusik (409, 606), Faculty of Geology, Geophysics and Environmental Protection, AGH University of Science and Technology, Krakow, Poland

L.J. Michot (202), Sorbonne Universités, UPMC Univ. Paris 06, CNRS, Laboratoire PHENIX, Case 51, 4 place Jussieu, Paris, France

J. Niu (387), School of Materials Science and Engineering, China University of Mining and Technology, Xuzhou, PR China

J. Ouyang (67), Centre for Mineral Materials, School of Minerals Processing and Bioengineering; Key Laboratory for Mineral Materials and Application of Hunan Province, Central South University, Changsha, China

E. Paineau (254), Laboratoire de Physique des Solides, CNRS, Univ. Paris-Sud, Université Paris-Saclay, Orsay Cedex, France

O. Poncelet (726), CEA DRT/LITEN/DTMN, Grenoble Cedex 9, France

S. Rouzière (254), Laboratoire de Physique des Solides, CNRS, Univ. Paris-Sud, Université Paris-Saclay, Orsay Cedex, France

G. Sandri (708), University of Pavia, Pavia, Italy

J. Skrzypski (726), CEA DRT/LITEN/DTMN, Grenoble Cedex 9, France

A. Takahara (628), Institute for Materials Chemistry and Engineering, and International Institute for Carbon-Neutral Energy Research (WPI-I2CNER), Kyushu University, Fukuoka, Japan

D. Tan (167), Key Laboratory of Solid Waste Treatment and Resource Recycle, Ministry of Education, Southwest University of Science and Technology, Mianyang; CAS Key Laboratory of Mineralogy and Metallogeny / Guangdong Provincial Key Laboratory of Mineral Physics and Materials, Guangzhou Institute of Geochemistry, Chinese Academy of Sciences (CAS), Guangzhou, China

Z.H. Tang (509), South China University of Technology, Guangzhou, China

A. Thill (1, 223, 361, 429, 735), Laboratoire Interdisciplinaire sur l'Organisation Nanométrique et Supramoléculaire, NIMBE, CEA, CNRS, University Paris-Saclay, Paris, France

C. Viseras (708), University of Granada, Granada, Spain

H. Yang (67), Centre for Mineral Materials, School of Minerals Processing and Bioengineering; Key Laboratory for Mineral Materials and Application of Hunan Province, Central South University, Changsha, China

P. Yuan (1, 137, 167, 735), CAS Key Laboratory of Mineralogy and Metallogeny / Guangdong Provincial Key Laboratory of Mineral Physics and Materials, Guangzhou Institute of Geochemistry, Chinese Academy of Sciences (CAS), Guangzhou, China

X.H. Zhang (509), South China University of Technology, Guangzhou, China

Y. Zhang (67), Centre for Mineral Materials, School of Minerals Processing and Bioengineering; Key Laboratory for Mineral Materials and Application of Hunan Province, Central South University, Changsha, China

Acknowledgements

The editors, Peng Yuan, Antoine Thill and Faiza Bergaya, would like to acknowledge all the authors of this volume, who have summarized the vast amount of information in their respective fields of expertise. They also thank the anonymous reviewers who helped to improve the quality of the chapters.

Peng Yuan would like to express great gratitude to Faiza Bergaya, the investigator of this volume, for her invitation to act as first editor of this book. He genuinely thanks her for their long-term, close scientific collaboration focused on clay science, especially for the collaborative research on tubular clay minerals performed in recent years. Peng Yuan also appreciates the kind hospitality of the family of Faiza Bergaya, her husband, Badreddine, and her two daughters, Sonia and Rym, during his stay for 3 months in Orleans. Thanks to them, his trip to France was really enjoyable and memorable. Peng Yuan deeply appreciates the kind collaboration of Antoine Thill for coediting this volume. In addition, he cherishes the happy memory of a coffee break at sunny noontime in Orleans, when the three coeditors brainstormed together to share great inspirations. Finally, he is grateful to the enthusiastic encouragement of his colleagues, students and friends, who gave much assistance since this project began two-and-a-half years ago.

Antoine Thill expresses his gratitude to Faiza Bergaya, who invited him to act as a coeditor of this volume in July 2013 in Brazil. This foolish task was accepted with pleasure (maybe with the help of some piña coladas that eased the stress from the forthcoming workload). He greatly appreciated this collaboration and learned a lot from the experience of working on this edition. I want also to thank Peng Yuan, who was a brilliant and efficient first editor for this book. I wish I could have spent more time with him, and I hope that the sunny brainstorming afternoon in Orleans will soon lead to common projects.

Faiza Bergaya is deeply grateful to her husband for his patience during the long hours she spent in front of the computer. She wishes also to acknowledge Peng Yuan and Antoine Thill for accepting her proposal to share the task of editing this book, and also CNRS for giving her the opportunity as Director Emeritus, to ensure continuity of her work as Series Editor of the *Developments in Clay Science*.

The author(s) of some chapters wish to present some acknowledgements in the following:

- The author of Chapter 2 expresses acknowledgement to the main halloysite producers for provided samples: Imerys Tableware NZ Ltd. and Imerys Ceramics Centre, Limoges France for the Matauri Bay sample, PTH Intermark for the Dunino one, Esan Eczacibasi for the Turkish one, and finally Applied Minerals Inc. for the Dragonite samples. Characterization data were performed using the CarMaLim platforms in the European Ceramic Center, Limoges, France.
- The authors of Chapter 4 thank the National Science Fund for Distinguished Young Scholars (51225403), the National Natural Science Foundation of China (51304242), the Postdoctoral Science Foundation of Central South University (155219) and the Hunan Provincial Co-Innovation Centre for Clean and Efficient Utilization of Strategic Metal Mineral Resources for their support.
- The authors of Chapters 7 and 8 are grateful to the financial support from the Natural Scientific Foundation of China (Grant Nos. 41472045, 41072032 and 41502027), the National Key Technology Research and Development Program of the Ministry of Science and Technology of China (Grant No. 2013BAC01B02) and the Team Project of Natural Science Foundation of Guangdong Province, China (S2013030014241).
- One of the authors of Chapter 11, Mohamed Salah Amara, is grateful to C'Nano Ile de France for financial support of his PhD work. Germanium imogolite nanotubes were elaborated at LIONS/CEA, under the supervision of Antoine Thill.
- The authors of Chapter 13 are deeply indebted to M.S. Amara, M. Bacia-Verloop, L. Belloni, D. Constantin, J.C. Gabriel, M.E. Krapf, P. Launois, C. Levard, P. Levitz, N. Matskova, E. Paineau, O. Poncelet, S. Rouzière, J. Rose, A. Thill and M. Zinsmeister for their help with the experiments and for helpful discussions.
- The author of Chapter 14 would like to thank the Brazilian agencies Fundação de Amparo à Pesquisa de Minas Gerais (FAPEMIG), Coordenação de Aperfeiçoamento de Pessoal de Nível Superior (CAPES) and Conselho Nacional para o Desenvolvimento Científico e Tecnológico (CNPq) for the continuing support of their research. Acknowledgment is also due to the Brazilian initiative National Institute of Science and Technology for Mineral Resources, Water and Biodiversity, INCT-ACQUA for their support.
- The author of Chapter 16 acknowledges the financial support from the National Natural Science Foundation of China (Grant No. 41502032), the Natural Science Foundation of Jiangsu Province, China (BK20140212) and the Fundamental Research Funds for the Central Universities (2012QNA08).

- The author of Chapters 17 and 23 is thankful to all coworkers for help in analyses and their interpretation. The research regarding morphology transformation of Polish kaolin-group minerals shown in Chapter 17 was conducted under supervision of Professor Krzysztof Bahranowski (AGH-UST) and involved several scientists from Kraków (Poland): AGH University of Science and Technology (AGH-UST) and Institute of Catalysis and Surface Chemistry (Polish Academy of Sciences). The research on adsorption properties of Polish halloysite from Dunino deposits reported in Chapter 23 resulted from teamwork which gathered several scientists from Kraków (Poland): AGH-UST and Polish Academy of Sciences. The author thanks Dr Józef Sołtys, Managing Director of the Intermark Company, for supplying the mineral samples from the Dunino deposits. Finally, he is especially grateful to students who participated in the research: Anna Koteja, Lucyna Matykowska, Anna Śmietana (Wścisło), Paulina Maziarz and Anna Prokop. Support by the Polish National Science Centre (Grant No. DEC-2011/01/ST10/06814) is also acknowledged.
- The authors of Chapter 21 gratefully acknowledge the National Basic Research Program of China (Grant No. 2015CB654703), and National Natural Science Foundation (Grant No. 51222301, 51333003, 51473050 and U1462116) for financial supports.
- The authors of Chapter 24 acknowledges the financial support of a Grant-in-Aid for Scientific Research (A) (No. 19205031) and (No. 26248053) from the Japan Society for the Promotion of Science. The synchrotron radiation experiments were performed at BL02B2 in SPring-8 with the approval of the JASRI (Proposal No. 2010A1454).
- The authors of Chapter 26 thank the Andalusian project RNM1897 and Andalusian group CTS-946 for the financial support.

P. Yuan, A. Thill and F. Bergaya
December 2015

Chapter 1

General Introduction

P. Yuan[a,*], F. Bergaya[b] and A. Thill[c]
[a]*CAS Key Laboratory of Mineralogy and Metallogeny / Guangdong Provincial Key Laboratory of Mineral Physics and Materials, Guangzhou Institute of Geochemistry, Chinese Academy of Sciences (CAS), Guangzhou, China*
[b]*CNRS - ICMN (Interfaces, Confinement, Matériaux et Nanostructures), Orléans Cedex 2, France*
[c]*Laboratoire Interdisciplinaire sur l'Organisation Nanométrique et Supramoléculaire, NIMBE, CEA, CNRS, University Paris-Saclay, Paris, France*
[*]*Corresponding author: e-mail: yuanpeng@gig.ac.cn*

The development of clay science in recent decades has benefited substantially from its intermingling with a number of related scientific disciplines, which have opened up new frontiers in the multidisciplinary and interdisciplinary research of this science. The development of modern characterisation techniques and computational methods enabled advances in our knowledge of the structure and properties of clay minerals at the molecular and even atomic scales. With the revelation of the variety and versatility of various clay minerals, our understanding of the distribution and roles of clay minerals in Earth systems has become deeper. Furthermore, the applications of clay minerals in a variety of fields can be extended, and novel clay-based materials developed.

The research on and applications of nanosized tubular clay minerals perfectly exemplifies the above-mentioned multidisciplinary features. Nanosized tubular clay minerals, as the name implies, are clay minerals with tubular nanostructures, which means that they have at least one dimension between 1 and 100 nm (Annabi-Bergaya, 2008; Schoonheydt and Bergaya, 2011) and a hollow tubular structure. Attention was paid to the very small tubular structure of minerals long before the concept of 'nanotechnology' was first defined and used in the 1970s (Hochella et al., 2008), and the concept of 'nano' and the term 'nanotube' in the context of materials science have been popular since the 1990s (Iijima, 1991). Already, Pauling (1930) had proposed the existence of cylindrical structures formed by minerals in nature. Based on his observations of the structure of asbestos-related minerals, he suggested that a structural mismatch will exist between the layers, leading to structural deformation and curvature if the two faces of a mineral are not symmetrical. Subsequently, beginning in the 1950s, the structures of nanosized tubular clay

Developments in Clay Science, Vol. 7. http://dx.doi.org/10.1016/B978-0-08-100293-3.00001-7

minerals, such as chrysotile, halloysite and imogolite (Bates et al., 1950a,b; Cradwick et al., 1972), have been identified using microscopic and spectroscopic methods. However, for a long time thereafter, there was very little research on nanosized tubular clay minerals (Yuan et al., 2015), and relevant papers were published only occasionally.

The renewed interest in nanosized tubular clay minerals can be partially attributed to the boom in studies on nanostructured materials and related applications since the 1990s, when the first observation of carbon nanotubes (CNT) was reported (Iijima, 1991) and several nanotube materials, eg, MoS_2, WS_2 and BN (Feldman et al, 1995), were synthesized. In fact, it was proposed that many inorganic compounds (eg, the dichalcogenides of many group IV and V metals) possess layered structures comparable to that of graphite and could possibly form nanotubes, and these nanotubes have found many important applications, such as in catalysis, inorganic polymer nanocomposites and adsorption (Tenne, 2014). Various synthetic strategies have been developed for the synthesis of such inorganic nanotubes. However, the synthesis of nanotubes is somewhat more difficult than that of porous compounds. For example, when facile self-assembly methods are not straightforward with respect to the desired nanotube product (particularly its morphology), template methods have been employed using CNT as consumptive templates to obtain the nanotube structures of metal oxides (Ajayan et al., 1995). The time-consuming preparation and production of small amounts of nanotubes result in high material costs and, accordingly, limit their actual applications. By contrast, the naturally occurring analogs of manufactured nanotubes (eg, nanosized tubular clay minerals) are distinct in their economic viability and environmental benefits in comparison with artificially synthesized tubular materials such as CNT and metal oxide nanotubes. Therefore, it is not surprising that nanosized tubular clay minerals are again attracting the attention of researchers.

The interest in nanosized tubular clay minerals is driven not only by the increasing demand for advanced nanotube materials, but also by the desire to understand the geological and environmental effects of nanominerals on the local, regional and even global scale. According to the definition of nanogeoscience that has developed over the past 20 years, nanosized tubular clay minerals should be classified as typical nanominerals, based on a classification scheme proposed by Hochella et al. (2008). In this classification, nanominerals exist as one of three types of nanoscale minerals (nanorods, nanosheets and nanoparticles); that is, there are no bulk equivalents of nanominerals. By contrast, mineral nanoparticles that exist in the nano-range can also exist as larger materials. Because of the high specific surface area and strong surface reactivity of nanominerals and mineral nanoparticles, measuring and elucidating their roles and behaviours—ie, their origin (biotic or abiotic, natural or anthropogenic), geographic distribution, relevant nanochemistry and overall influence and impact within the complex chemical and physical framework of Earth systems—are important and critical challenges for the future (Hochella et al., 2008). In the context of environmental science and technology, interest has been focused not only on the role of nanosized clay minerals as the primary mineralogical components of rocks and soils involved in the occurrence, adsorption, migration and transformation of pollutants, but also on using these low-cost and environmentally friendly natural materials as adsorbents for pollution remediation. This application is made possible by the high surface activity and amenability to structural post-modifications for versatile types of pollution treatment.

Because of the continuing and increasing research interest in the above-mentioned fields, from both fundamental and applied perspectives, the body of knowledge on the occurrence, structure, properties and applications of nanotubular clay minerals has expanded significantly. According to a statistic presented by Abdullayev and Lvov in Chapter 22 of this book, research activity in the area of natural nanotubular clay minerals has increased over the past 10 years. Approximately 80% of the research papers on halloysite and imogolite applications have been published in the past 5 years, compared with 20–40% of the papers on other clay materials with wide industrial applications and other porous minerals, such as zeolite and diatomite. Therefore, it seems likely that additional active research on nanotubular clay minerals will be continuously conducted in the near future to enhance our understanding of their properties and to extend their practical applications.

Halloysite, imogolite and chrysotile are typical examples of naturally occurring nanosized clay minerals with a tubular structure. Halloysite is a dioctahedral 1:1 clay mineral of the kaolin group. It is chemically similar to kaolinite, but the unit layers in halloysite are separated by a monolayer of water molecules; thus, halloysite has a structural formula of $Al_2(OH)_4Si_2O_5{\cdot}nH_2O$. As described by Joussein et al. (2005), halloysite is abundant in both weathered rocks and soils. It has been identified as having been formed by the alteration of a wide variety of rocks (Churchman, 1999). It is also often a major component of soil derived from volcanic materials in wet tropical and subtropical regions (Chadwick et al., 2003; Joussein et al., 2005). The multilayer tubular structure of halloysite appears as curled 1:1 clay mineral layers, in which a tetrahedral Si–O sheet forms the outer surface of the nanotube, and a gibbsitelike octahedral sheet ($Al(OH)_3$) constitutes the inner surface. Normally, tubular halloysites vary in length from the submicron scale to several microns—occasionally even >30 μm (Norrish, 1995; Joussein et al., 2005), with external diameters ranging from approximately 20 to 190 nm and internal diameters from approximately 10 to 100 nm (Yuan et al., 2008, 2013; Pasbakhsh et al., 2013). The maximum reported diameter is approximately 0.5 μm for long, thick tubes (de Oliveira et al., 2007).

Imogolite occurs naturally in the soils of some weathered volcanic rocks. Typically, it is composed of single-walled aluminosilicate nanotubes with an inner diameter of approximately 1.0 nm. The tube walls consist of a curved gibbsitelike sheet ($Al(OH)_3$), where the inner hydroxyl surface of the gibbsite is replaced by $(SiO_3)OH$ groups (Guimarães et al., 2007).

Chrysotile is a clay mineral that belongs to the serpentine-mineral group and is generally transformed from ultrabasic rock-forming minerals, such as olivine, through a hydrothermal alteration process. It is a hydrous magnesium phyllosilicate, approximately represented by the formula $Mg_3(OH)_4Si_2O_5$ (Bailey, 1980). The chrysotile structure is composed of tridymite (SiO_2) and brucite ($Mg(OH)_2$) layers. The brucite octahedral sheet forms the outer surface of the tube, and SiO_4 groups are located on the inner side (Piperno et al., 2007). The diameters of chrysotile nanotubes vary, but typical values are an inner diameter of 7–8 nm and an outer diameter of 22–27 nm (Yada, 1971).

Wide industrial use of chrysotile dates to the early decades of the 20th century. As is well known, because of its peculiar crystalline character, chrysotile appears as long, ultrathin and durable fibers. This character, together with its chemical composition, determines its unique thermal, electrical and mechanical properties, which make it useful in several industrial fields, such as serving as filler for plastics and as friction materials (Ross and Virta, 2001). In industry, chrysotile was considered a type of asbestos, which is often associated with health risk issues. In fact, 'asbestos' is a poorly attributed term because it refers to two very different mineral groups with different properties: the serpentine-group minerals, of which chrysotile is the most common, and the amphibole asbestos. Worldwide production of chrysotile asbestos in 2014 was approximately 2 kt, with the main producers being Russia (53.0%), China (20.2%), Brazil (14.7%) and Kazakhstan (12.1%) (US Geological Survey, 2015). Developed countries, such as the United States, Canada and some European countries, have banned the extraction of asbestos minerals (including chrysotile) because all forms of asbestos have been deemed carcinogenic by the World Health Organization (WHO, 1986) and the International Agency for Research on Cancer (IARC, 1987). This judgment was restated in an updated version of the risk-evaluation report issued by WHO in 2014. The health risk of chrysotile is controversial; some researchers argue that it is less dangerous than amphiboles and that low levels of inhalation do not necessarily present a health risk (Sporn and Roggli, 2004; Bernstein, 2014), although prolonged exposure can lead to disease. However, in the aforementioned report, WHO stated clear conclusions from the IARC: with respect to cancer of the lungs and mesothelioma, there is sufficient evidence of carcinogenicity in humans for all forms of asbestos, including chrysotile. This conclusion was based on key new studies, and the statement itself represents the highest IARC category for describing the strength of this evidence (IARC, 2012). In addition to lung cancer and mesothelioma, the International Programme on Chemical Safety (IPCS) of the IARC concluded that occupational exposure to chrysotile causes nonmalignant lung diseases that result in deterioration in lung function, particularly a form of lung fibrosis described by the term 'asbestosis' (WHO, 1998).

Unlike chrysotile, for which serious health risks have been clearly proven, halloysite has high biocompatibility and low cytotoxicity (Vergaro et al., 2010). Regarding imogolite, although there are insufficient studies to draw conclusions regarding its potential impact on human health, its health risk is apparently less than that of chrysotile because the imogolite nanotube is very short. The unsafe nature of chrysotile severely limits its potential applications as a nanotube material. Moreover, there are numerous publications on the structure, properties and applications of chrysotile from various perspectives (Schreier, 1989; Skinner et al., 1998; Oury et al., 2004), whereas publications about halloysite and imogolite are rare. Therefore, this book will focus only on halloysite and imogolite, which are safer nanosized tubular clay minerals, although they are much less studied than chrysotile.

In addition, it is noteworthy that typical fibrous clay minerals (eg, sepiolite and palygorskite) have been incorrectly defined as nanotubular clay minerals in some publications. In fact, sepiolite and palygorskite are easily identified by their fibrous macromorphologies and micromorphologies: they vary in length from approximately 1 to 10 μm and are approximately 10 nm wide (Nolan et al., 1991; Torres-Ruíz et al., 1992); exhibit a fibrous habit, with microporous channels running parallel to the fiber length; and in the cross section of a single fiber, the channels exhibit a honeycomblike structure. This structure differs significantly from the single-channel structure of nanotubes, according to the latter's definition. In other words, sepiolite and palygorskite can be defined as nanofibers instead of nanotubes, although halloysite specimen with large nanotube length, eg, the aforementioned specimen with a length of more than about 30 μm (Norrish, 1995), can be classified as both nanotube and nanofiber.

There is increasing interest in employing halloysite and imogolite in a variety of applications, such as clay polymer nanocomposites, catalysis and adsorption. These nanosized tubular clay minerals are highly desirable in such applications because of their unique, one-dimensional tubular structure and properties that can be adjusted through modifications of both the internal and external surfaces. Thus, tubular clay minerals may serve as advanced functional materials. In addition, the structural differences between halloysite and imogolite (eg, their pore dimensions and the types of surface groups) allow them to be used for various applications. All of these applications rely on a thorough understanding of the structures and properties of the tubular clay minerals, as well as on the development of techniques for their modification and functionalization. The past 10–20 years have witnessed tremendous progress in the understanding and characterisation of these two minerals. However, the substantial reports of recent research advances of halloysite and imoglite in various aspects are scattered in the literature, and a single book focused on this topic has not been published.

This book, therefore, aims to provide a comprehensive view of the fundamental properties of halloysite and imogolite, as well as the methods used to modify their structures and properties to satisfy the requirements of target applications. The text clearly summarizes the existing knowledge about nanosized tubular clay minerals and highlights recent important advances in their practical application, as well as the challenges that remain.

The book comprises four parts, totaling 28 chapters. Following this introductory chapter, the book starts with Part I, 'Geology and Mineralogy of Nanosized Tubular Clay Minerals', which consists of two chapters focusing on halloysite (Chapter 2) and imogolite (Chapter 3). Chapter 2 provides an overview of the studies on geological genesis and occurrence, research history and nomenclature of halloysite. Also, the general mineralogical characteristics of halloysite are discussed, focusing on the crystal structure, chemical composition, morphology and hydration–dehydration behaviour because the latter is an important fingerprint of the properties of halloysite. Chapter 3 provides the geology and mineralogy of natural imogolite and imogolite-type materials. In particular, the mineralogy and formation of imogolite-type materials are elucidated with regard to their occurrence in the geological and soil environments. The effects of the reactivity of imogolite-type materials on soil properties are also discussed, including phosphate and heavy metal retention and organic matter storage.

Part II presents detailed descriptions of the 'Structure and Properties of Nanosized Tubular Clay Minerals'. Chapters 4–8 focus on halloysite and Chapters 9–13 on imogolite. Chapter 4 summarizes the surface and colloidal properties, mechanical properties and chemical stability of halloysite under acid and alkaline treatment. Chapters 5 and 6 discuss the key structural features of halloysite as identified by microscopy and spectroscopy, respectively. The use of methods such as transmission electron microscopy (TEM), scanning electron microscopy, infrared (IR) and Raman spectroscopy, X-ray photoelectron spectroscopy (XPS) and nuclear magnetic resonance (NMR) to determine the structure and related characteristics of halloysite are explained in detail. Chapter 7 describes the series of substantial changes that occur in halloysite subjected to thermal treatment, including dehydration behaviour at low-temperature heating and dehydroxylation behaviour, as well as crystalline structure changes and phase transformations during calcination. The calcination-induced changes in the texture, morphology and surface reactivity of halloysite and the utility of thermal modification for related applications are also covered. In Chapter 8, the various surface chemical modification mechanisms corresponding to three types of halloysite surfaces (the external siloxane surface, the interior alumina surface and the interlayer surface) are discussed. In addition, their selective modification by differentiating the surface chemical reactivity of a given surface is also described. This chapter also lists various modification reactions, such as intercalation, covalent grafting, ionic exchange and electrostatic attraction for several compounds.

The organization of the chapters on imogolite is similar to that of the chapters on halloysite. Chapter 9 describes key physicochemical properties of imogolite, including its colloidal properties, surface charge, hydrophilicity/hydrophobicity, acid/base properties and chemisorption and physisorption properties. Chapter 10 summarizes the microscopic and spectroscopic characterisations of the imogolite structure and some of its properties. The relevant methods include atomic force microscopy, scanning tunneling microscopy, TEM (including Cryo-TEM), IR, NMR, XPS and X-ray absorption. Scattering methods such as light-scattering and X-ray-scattering (XRS) and neutron-based analyses, as well as chemical analyses and mass analyses, are also addressed in this chapter. In Chapter 11, the deformation phenomena of imogolite (ovalization and hexagonalization), which occur during the organization of the imogolite nanotubes into bundles, are discussed. These studies are based on the XRS technique. The thermally induced structural transformations of imogolite, from the dehydroxylation process to the lamellar phase and high-temperature mullite phase transformations, are summarized as well. Chapter 12 examines the three types of surfaces occurring within imogolite bundles: pores related to imogolite inner pores, pores among three aligned nanotubes in a bundle and larger slit pores between bundles. This chapter addresses the issues of pore hydrophilicity, thermal stability, accessibility by different molecules and the corresponding modification techniques. Chapter 13 discusses the formation mechanism of liquid-crystalline phases of nanotubular clay minerals in liquid dispersions and their characteristics, mainly in regard to imogolite, but with some attention to halloysite as well. Unlike the previous chapters, which focus on experimental studies, the last chapter of Part II, Chapter 14, presents research advances in elucidating the structural, electronic and mechanical properties of nanosized tubular clay minerals and their chemical modification at a molecular level using various computational chemistry methodologies, such as classical molecular dynamics, density functional theory and approximate methods.

Part III focuses on the formation mechanism of the nanosized tubular structure of halloysite and imogolite and the corresponding synthetic routes from both the theoretical and experimental perspectives. Chapter 15 presents a theoretical discussion on the origin of the curvature of tubular clay minerals and the factors affecting their structural and morphological parameters. Chapter 16 summarizes three mechanisms that may influence the rolling of halloysite: (i) the misfit between octahedral and tetrahedral sheets, (ii) the attraction between interlayer hydroxyl groups in octahedrons and (iii) the surface tension of water. The effects of octahedral and tetrahedral substitutions on the rolling of the kaolinite layer are also discussed. Chapter 17 presents the current knowledge about synthetic routes that lead to the transformation of kaolinite morphology via intercalation and deintercalation and delamination procedures to halloysite-like nanotubes without destroying the initial aluminum silicate structure. Unlike the case with halloysite, whose direct synthesis from chemical precursors (Si and Al sources) has been rarely reported, synthetic methods have been developed to synthesize imogolite nanotubes from molecular precursors. Chapter 18 summarizes imogolite formation and growth mechanism through this synthetic route. Chapter 19 describes related materials with structures similar to that of imogolite, such as germanium-imogolite, iron-containing imogolite and oxy-imogolite. In addition, the chemical synthesis and characterisation of imogolite-like materials are detailed.

Part IV summarizes the applications of halloysite and imogolite. The first chapter of this last part of the book, Chapter 20, provides an overview of the safety of tubular clay minerals and, in particular, evaluates the safety and potential health effects of halloysite and imogolite compared with those of asbestos and CNT, whose effects on human health are known. Each of the subsequent seven chapters is devoted to discussing a specific application of halloysite or imogolite: clay polymer nanocomposites for halloysite (Chapter 21) and for imogolite (Chapter 24); controllable loading and release, in which halloysite is employed as a support material (Chapter 22); the use of halloysite for adsorption and pollution remediation (Chapter 23); the use of imogolite for catalysis and adsorption (Chapter 25); health and medical applications (Chapter 26); and real examples of tubular clay minerals' technological applications (Chapter 27) highlighting some industrial uses of these clay minerals, largely described in patents.

The book concludes with Chapter 28, which presents a broad editorial review and summary of the book, with emphasis on the key current research trends and future prospects regarding studies of nanosized tubular clay minerals.

In summary, the contents of this book aim to give useful information concerning the fundamental science and practical applications of halloysite and imogolite. This includes information on scientific topics related to tubular clay minerals, such as structure, properties, characterisations, molecular simulation, surface modifications and applications as functional materials or in clay polymer nanocomposites.

Finally, this book targets a broad audience (students and researchers) in a variety of disciplines, such as clay mineralogy, geology, chemistry, physics, materials science and environmental science, as well as participants in industries who are interested in tubular clay minerals and Clay Science.

REFERENCES

Ajayan, P., Stephan, O., Redlich, P., Colliex, C., 1995. Carbon nanotubes as removable templates for metal oxide nanocomposites and nanostructures. Nature 375, 564–567.

Annabi-Bergaya, F., 2008. Layered clay minerals. Basic research and innovative composite applications. Micropor. Mesopor. Mater. 107, 141–148.

Bailey, S., 1980. Structures of layer silicates. In: Brindley, G., Brown, G. (Eds.), Crystal Structures of Clay Minerals and Their X-ray Identification. Oxford University Press, London, pp. 1–123.

Bates, T.F., Hildebrand, F.A., Swineford, A., 1950a. Morphology and structure of endellite and halloysite. Am. Mineral. 35, 463–484.

Bates, T.F., Sand, L.B., Mink, J.F., 1950b. Tubular crystals of chrysotile asbestos. Science 111, 512–513.

Bernstein, D.M., 2014. The health risk of chrysotile asbestos. Curr. Opin. Pulm. Med. 20, 366–370.

Chadwick, O.A., Gavenda, R.T., Kelly, E.F., Ziegler, K., Olson, C.G., Elliott, W.C., Hendricks, D.M., 2003. The impact of climate on the biogeochemical functioning of volcanic soils. Chem. Geol. 202, 195–223.

Churchman, G., 1999. The alteration and formation of soil minerals by weathering. In: Sumner, M.E. (Ed.), Handbook of Soil Science. CRC Press, Boca Raton, Florida, pp. F3–F76.

Cradwick, P., Farmer, V., Russell, J., Masson, C., Wada, K., Yoshinaga, N., 1972. Imogolite, a hydrated aluminium silicate of tubular structure. Nature 240, 187–189.

de Oliveira, M.T., Furtado, S., Formoso, M.L., Eggleton, R.A., Dani, N., 2007. Coexistence of halloysite and kaolinite: a study on the genesis of kaolin clays of Campo Alegre Basin, Santa Catarina State Brazil. An. Acad. Bras. Cienc. 79, 665–681.

Feldman, Y., Wasserman, E., Srolovitz, D.J., Tenne, R., 1995. High-rate, gas-phase growth of MoS_2 nested inorganic fullerenes and nanotubes. Science 267, 222–225.

Guimarães, L., Enyashin, A.N., Frenzel, J., Heine, T., Duarte, H.A., Seifert, G., 2007. Imogolite nanotubes: stability, electronic, and mechanical properties. ACS Nano 1, 362–368.

Hochella, M.F., Lower, S.K., Maurice, P.A., Penn, R.L., Sahai, N., Sparks, D.L., Twining, B.S., 2008. Nanominerals, mineral nanoparticles, and earth systems. Science 319, 1631–1635.

IARC, 1987. Overall evaluations of carcinogenicity: an updating of IARC Monographs volumes 1 to 42. IARC Monogr. Eval. Carcinog. Risks Hum. 7, 1–440.

IARC, 2012. Asbestos (chrysotile, amosite, crocidolite, tremolite, actinolite, and anthophyllite). IARC Monogr. Eval. Carcinog. Risks Hum. 100C, 219–309.

Iijima, S., 1991. Helical microtubules of graphitic carbon. Nature 354, 56–58.

Joussein, E., Petit, S., Churchman, J., Theng, B., Righi, D., Delvaux, B., 2005. Halloysite clay minerals—a review. Clay Miner. 40, 383–426.

Nolan, R., Langer, A., Herson, G., 1991. Characterisation of palygorskite specimens from different geological locales for health hazard evaluation. Br. J. Ind. Med. 48, 463–475.

Norrish, K., 1995. An unusual fibrous halloysite. In: Churchman, G.J., Fitzpatrick, R.W., Eggleton, R.A. (Eds.), Clays Control the Environment—Proceedings of the 10th International Clay Conference 1993, Adelaide, Australia, pp. 275–284.

Oury, T.D., Roggli, V.L., Sporn, T.A. (Eds.), 2004. Pathology of Asbestos-Associated Diseases, third ed. Springer, Berlin.

Pasbakhsh, P., Churchman, G.J., Keeling, J.L., 2013. Characterisation of properties of various halloysites relevant to their use as nanotubes and microfibre fillers. Appl. Clay Sci. 74, 47–57.

Pauling, L., 1930. The structure of the chlorites. Proc. Natl. Acad. Sci. U.S.A. 16, 578.

Piperno, S., Kaplan-Ashiri, I., Cohen, S.R., Popovitz-Biro, R., Wagner, H.D., Tenne, R., Foresti, E., Lesci, I.G., Roveri, N., 2007. Characterization of geoinspired and synthetic chrysotile nanotubes by atomic force microscopy and transmission electron microscopy. Adv. Funct. Mater. 17, 3332–3338.

Ross, M., Virta, R.L., 2001. Occurrence, production and uses of asbestos. Can. Mineral. Spec. Publ. 5, 79–88.

Schoonheydt, R.A., Bergaya, F., 2011. Industrial clay minerals as nanomaterials. In: Christidis, G.E. (Ed.), Advances in the Characterization of Industrial Minerals. EMU Notes in Mineralogy, vol. 9. The Europeans Mineralogical Union and the Mineralogical Society of Great Britain, Ireland, pp. 415–440 (Chapter 10).

Schreier, H., 1989. Asbestos in the Natural Environment. Elsevier, Amsterdam.

Skinner, H.C.W., Ross, M., Frondel, C., 1998. Asbestos and Other Fibrous Materials. Oxford University Press, New York.

Sporn, T.A., Roggli, V.L., 2004. Asbestosis. In: Oury, T.D., Roggli, V.L., Sporn, T.A. (Eds.), Pathology of Asbestos-Associated Diseases, third ed. Springer, Berlin, pp. 71–103.

Tenne, R., 2014. Recent advances in the research of inorganic nanotubes and fullerene-like nanoparticles. Front. Phys. 9, 370–377.

Torres-Ruíz, J., López-Galindo, A., González-López, J., Delgado, A., 1992. Spanish fibrous clays: an approach to their geochemistry and micromorphology. In: López-Galindo, A., Rodríguez-García, M.I. (Eds.), Electron Microscopy. Materials Sciences, vol. 92. Secr. Publ. Univ., Granada, pp. 595–596.

US Geological Survey, 2015. Mineral commodity summaries 2015. US Geological Survey, Reston.

Vergaro, V., Abdullayev, E., Lvov, Y.M., Zeitoun, A., Cingolani, R., Rinaldi, R., Leporatti, S., 2010. Cytocompatibility and uptake of halloysite clay nanotubes. Biomacromolecules 11, 820–826.

WHO, 1986. Environmental Health Criteria 53: asbestos and other natural mineral fibres, World Health Organization, International Programme on Chemical Safety, Geneva. http://www.inchem.org/documents/ehc/ehc/ehc53.htm (accessed 13 March 2014).

WHO, 1998. Environmental Health Criteria 203: chrysotile asbestos, World Health Organization, International Programme on Chemical Safety, Geneva. http://www.inchem.org/documents/ehc/ehc/ehc203.htm (accessed 11 March 2014).

Yada, K., 1971. Study of microstructure of chrysotile asbestos by high-resolution electron microscopy. Acta Crystallogr. A 27, 659–664.

Yuan, P., Southon, P.D., Liu, Z., Green, M.E., Hook, J.M., Antill, S.J., Kepert, C.J., 2008. Functionalization of halloysite clay nanotubes by grafting with γ-aminopropyltriethoxysilane. J. Phys. Chem. C 112, 15742–15751.

Yuan, P., Tan, D., Annabi-Bergaya, F., Yan, W., Liu, D., Liu, Z., 2013. From platy kaolinite to aluminosilicate nanoroll via one-step delamination of kaolinite: effect of the temperature of intercalation. Appl. Clay Sci. 83, 68–76.

Yuan, P., Tan, D., Annabi-Bergaya, F., 2015. Properties and applications of halloysite nanotubes: recent research advances and future prospects. Appl. Clay Sci. 112–113, 75–93.

Part I

Geology and Mineralogy of Nanosized Tubular Clay Minerals

Chapter 2

Geology and Mineralogy of Nanosized Tubular Halloysite

E. Joussein*

Université de Limoges, FST, GRESE 'Groupement de Recherche Eau Sol Environnement', Limoges, France

*Corresponding author: e-mail: emmanuel.joussein@unilim.fr

2.1 INTRODUCTION

Halloysites are clay minerals ubiquitous in soil and weathered rock, where they occur in a variety of particle shapes and hydration states. Since the first review paper on the topic of halloysite (Joussein et al., 2005) the mineral has long fascinated clay scientists, ceramists and mineralogists with questions about its mode of formation. Its main use in the 1990s concentrated on ceramics applications, such as thin-walled porcelain and crucibles. The halloysite revival occurred with halloysite nanotubes (abbreviated HNT in some papers), which offer an attractive natural cylindrical material that could be obtained at a fraction of the cost of synthesised nanomaterials such as carbon nanotubes. Hydration properties of the mineral likely have played a major role in its exchange properties up until today. Clear trends seem to relate particle morphology and structural iron. This chapter gives a critical, but up-to-date assessment of the extensive literature on halloysite, including genesis, crystal structure and morphological diversity of the various methods of differentiating halloysite from kaolinite. Moreover, the main deposits in the world and the properties of such halloysites are mentioned to foster a better understanding of their potential applications.

2.2 BACKGROUND HISTORY AND NOMENCLATURE

The name 'halloysite' was coined by Berthier (1826) to honour its discoverer, Omalius d'Halloy, who studied a tubular mineral identified in a sample from Liège, Belgium. In the 19th century, this material was originally distinguished from kaolinite on the basis of its higher proportion of water. The two major questions about the nomenclature of halloysite are: (i) What is the distinction between the halloysite mineral species and the kaolinite species in terms of

Developments in Clay Science, Vol. 7. http://dx.doi.org/10.1016/B978-0-08-100293-3.00002-9

nomenclature? and (ii) What is the nomenclature consensus of halloysite? The first issue was summarised by Churchman and Carr (1975): 'those minerals with a kaolin layer structure which either contain interlayer water in their natural state or for which there is unequivocal evidence of their formation by dehydration from kaolin minerals containing interlayer water'. In particular, the characteristic feature of halloysite gleaned from research data, sufficient for distinguishing halloysite from kaolinite, can be indisputably established by the intercalation, or the history of intercalation, of water between its layers. Hart et al. (2002) suggested the name 'kaolinite' be used for 1:1 dioctahedral clay minerals with a platy particle morphology, and 'halloysite' be used when the particles are cylindrical or tubular.

There have been much debate over the halloysite nomenclature and the distinction between different forms of this clay mineral. Different terms such as 'halloysite', 'metahalloysite', 'hydrated halloysite', 'endellite', 'embryonic halloysite' and 'proto-halloysite' were used for halloysite minerals, causing a great deal of confusion in the literature. The various propositions are reported in Table 2.1. Before the 1930s, the different forms had been considered to be

TABLE 2.1 History and Various Suggestions About the Nomenclature of Halloysite

Mineral nomenclature according to chemical composition		
$Al_2Si_2O_5(OH)_4$	$Al_2Si_2O_5(OH)_4{\cdot}2H_2O$	**References**
Halloysite (considered as single mineral)		All authors between 1826 and 1934
Halloysite	–	Ross and Kerr (1934)
Kaolinite	Halloysite	Hofmann et al. (1934)
Metahalloysite	Halloysite	Mehmel (1935)
Halloysite	Hydrated halloysite	Hendricks (1938)
Halloysite	Endellite	Alexander et al. (1943)
Halloysite (used as a group mineral)		MacEwan suggestion (1947)
Metahalloysite/'partly hydrated halloysite'/ hydrated halloysite		MacEwan (1947)
Halloysite (no specific term for all forms)		Brindley (1951)
Halloysite (7 Å)	Halloysite (10 Å)	Brindley (1961)
Nonhydrated halloysite	Hydrated halloysite	Churchman and Carr (1975)
Halloysite (7 Å)	Halloysite (10 Å)	Brindley and Pedro (1976)

a single mineral and simply named 'halloysite', but since 1935, there has been considerable disagreement over these terms. For example, fully hydrated halloysite was named 'halloysite' by Mehmel (1935), 'hydrated halloysite' by Hendricks (1938) and 'endellite' (after one of its discoverers) by Alexander et al. (1943) and Faust (1955). This latter name is favoured by Bradley (1945). Mehmel (1935) gave the name 'metahalloysite' to the dehydrated phase, while this phase was designated 'halloysite' by Alexander et al. (1943). Therefore, the literature reveals a long-standing disagreement over what to call the different forms of halloysite. Alternative classification schemes for halloysite minerals were established, which were related to (i) the hydration state, (ii) the particle morphology (platy vs tubular), (iii) to a lesser extent, the degree of crystalline order or environmental conditions of formation and (iv) to the relationships among all these factors (Bates et al., 1950; Nagasawa and Miyazaki, 1976; Wada and Kakuto, 1985; Noro, 1986; Bailey, 1990; Ziegler et al., 2003; West et al., 2004; Inoue et al., 2012). A fifth factor, however, was that they were poorly ordered materials with a very small particle size, close to evolved allophane.

MacEwan (1947) and Churchman and Carr (1975) suggested that the hydrated and dehydrated forms of halloysite were end-members of a series, differing in their degree of hydration. According to these researchers, the name 'halloysite' was proposed for the material with a d_{001} value of 10 Å, and 'metahalloysite' for the dehydrated end-member (with a 7 Å value). MacEwan (1947) proposed that the term 'metahalloysite' was unnecessary. Finally, the actual terminology endorsed by the Nomenclature Committee of AIPEA (Association internationale pour l'Etude des Argiles) during the 5th ICC meeting at Mexico City in 1975, was the one previously proposed by Brindley (1961), which was based on the state of hydration. This author recommended employing suffixes to indicate the state of hydration of the mineral, such as halloysite (10 Å) and halloysite (7 Å). This well-accepted nomenclature was then published by Brindley and Pedro (1976). It was stated that the term 'endellite' should not be used. However, the term 'embryonic halloysite' or 'proto-halloysite' (eg, Wada and Kakuto, 1985; Farmer et al., 1991) remains, and it has been used to refer to poorly ordered materials with a halloysite (10 Å) structure, but with a very small particle size. This name has been possibly attributed to 1:1–2:1 mixed layered clay minerals (Delvaux et al., 1992).

More recently, new abbreviations such as HNT (which stands for 'halloysite nanotube') can be found in some publications. This appellation is due to specific characteristics of halloysite, such as nanoscale lumens, high length-to-diameter ratio, low hydroxyl group density on the silica surface and other factors. However, this book will follow the accepted nomenclature of 'halloysite (7 Å)' and 'halloysite (10 Å)' when the hydration state is known, and, the general nomenclature of 'halloysite' when there is no need to differentiate the hydrated and dehydrated phases.

2.3 GENESIS AND OCCURRENCE

2.3.1 Geological Processes of Main Halloysite Ores and Soils

Halloysite results from weathering, pedogenesis or hydrothermal alteration of ultramafic rocks, volcanic glass and pumice, but it is uncommon in sedimentary deposits (Nagasawa and Noro, 1987; Bobos and Gomes, 1998; Lee and Gilkes, 2005). The main deposits around the world (discussed in Section 2.3.3) are due to hydrothermal alteration. In New Zealand, the main halloysite deposits (eg, Matauri Bay deposits) are preferentially the result of the weathering of rhyolites, kaolinite being linked to hydrothermal alteration at low temperatures (typically less than 80°C) but rapid alteration of volcanic rocks (Harvey and Murray, 1997). The halloysite in Nevada (Eureka deposit, Dragon Mine) is due to the hydrothermal supergene alteration (Drews-Armitage et al., 1996), hydrothermal alteration being the main process for ore deposits (see halloysite deposits from Polish, turkish, Brazil and so on). More specifically, halloysites from hydrothermal alteration can be formed by sulphate-rich solutions associated with sulphur deposits in volcanic rocks or pumiceous tuff (for example, in Japan; Watanabe and Sudo, 1969) or also formed in karst, paleokarst or cave sediment due to acid weathering or acid low-hydrothermal speleogenesis (eg, Dupuis and Ertus, 1995; Perruchot et al., 1997; Kempe et al., 2003). In this case, the acidic fluid due to oxidation of pyrite reduces the solubility of Al, which can react with Si to precipitate halloysite. Generally, the halloysite formed in these contexts presents well-developed and regular tubes, such as halloysites coming from the Camel Lake site (Eucla Basin, South Australia).

The occurrence of halloysite as a neoformed product may be observed in marine environments: seafloor alteration by hydrothermal activity with cold seawater (Marumo and Hattori, 1999) or early-stage alteration of volcanic glass or rock (Karpoff, 1992). The formation of halloysite is related to relatively Si-rich and Ca-low environments. These conditions allowed the nucleation of halloysite. However, it is rarely reported in the literature, which is probably due to the classical problems of (i) characterisation due to low amounts of halloysite and safeguarding, and (ii) alteration during transport. The main classical occurrences of halloysite (except ore deposit) are in weathered, altered rock, saprolites and soil. Halloysite can be derived from the alteration of a large number of primary silicates (eg, feldspars, amphiboles and micas) located in many rocks such as granite, gneiss, dolerite, schist, rhyolite, pyroclastics, greywacke, greenstone, granodiorite, shale and amphibolite (eg, Romero et al., 1992; Churchman, 2000; Joussein et al., 2005; Proust et al., 2006; Pasbakhsh and Churchman, 2015). Galán (2006) explained that kaolin formation is favoured with acidic pH values of 4–6 during warm and wet conditions (precipitation $>$ 1000 mm per year) and fluids that maintain bases in solution along with some leaching of SiO_2. The rapid nucleation of halloysite is preferred, whereas kaolinite formation is restricted. Halloysite is a dominant mineral in newly formed volcanic ash soil (Parfitt et al., 1983; Delvaux et al., 1989a; Chadwick et al., 2003; West et al., 2004; Joussein et al., 2005; Pasbakhsh and Churchman, 2015) and in the early weathering product of lateritic soil (Robert and Herbillon, 1990). The occurrence of halloysite in soil is directly related to the various climatic conditions, and then environmental and leaching conditions. A prolonged persistence of water in weathering profiles may explain a greater tendency towards the formation of halloysite in deeper horizons than kaolinite in the upper one (Churchman and Gilkes, 1989; Churchman, 1990, 2000). In the case of volcanic ash soil, the pedogenic formation of halloysite is favoured in humid tropical climates with a moderate rainfall regime, which induces less desilication (see Chadwick et al., 1994, 2003 on the climatic toposequence in Hawaii) and a short dry season of up to 3 months, whereas allophane predominates under high rainfall (eg, Dahlgren et al., 1994; Chadwick et al., 2003; Sedov et al., 2003). However, for the same hydric conditions, soil derived from rhyolitic tephra tends to yield halloysite, while that from basaltic parent materials is dominated by allophane and imogolite. Finally, the local environment conditions also play an important role in the morphology of halloysite (Churchman et al., 2010).

In tropical and subtropical areas, halloysite-rich soil represents an intermediate weathering stage just between recent allophanic-rich soil and more differentiated soil rich in kaolinite and iron oxides (feralsol). Based on this finding, the relation between allophane and halloysite has been much discussed in the literature since the transformation of allophane into halloysite requires a rearrangement of atomic structures that includes dissolution and reprecipitation (Parfitt, 1990; Takahashi et al., 1993; Dahlgren et al., 1994; Churchman, 2000). Globally, from the transformation point of view, the conversion of allophane into halloysite takes more than a million years following the complete sequence: volcanic ash → allophane → halloysite (10 Å) (→) halloysite (7 Å) → kaolinite (Chadwick and Chorover, 2001; Ziegler et al., 2003), whereas for the opposing view, both allophane and halloysite form directly from the dissolution of primary minerals, depending on silica activity as a driving factor (Churchman, 2000).

2.3.2 Genetic Relation Between Halloysite and Kaolinite

As already mentioned, one of the major questions for the last 70 years or so has been the relation between polymorph halloysite and kaolinite: that is, how to distinguish between halloysite (7 Å) and kaolinite (see, for example, Fieldes, 1955; Churchman, 1990; Joussein et al., 2005). The genetic relationship between these two polymorphs, halloysite and kaolinite, is quite difficult to demonstrate because of their structural similarity, particularly when halloysite is not really tubular. Finally, the question remains, especially in the case of soil samples (halloysite ore deposit being relatively easy to characterise). In some clay samples, the classical mixture between halloysite and kaolinite and the various hydration state of samples (natural dehydration or due to preparation of samples before investigations) may have dramatic results. Many authors have reported that kaolinite tends to concentrate in the coarsest particle-size fractions, while the fine fractions are enriched in halloysite (de Souza Santos et al., 1964; Wada and Kakuto, 1985; Delvaux et al., 1992). Generally, the relative abundance of halloysite with respect to kaolinite decreases as the weathering stage increases (Nagasawa, 1992; Singh and Gilkes, 1992; de Oliveira et al., 1997). However, some older research papers (Sand, 1956; Kirkman, 1977) stated that halloysite and kaolinite appear to form independently because no transition phases (between these two minerals) occur as ageing progresses. Halloysite can rapidly convert or evolve into kaolinite by an unrolled mechanism (Churchman and Gilkes, 1989; Inoue et al., 2012), by precipitation (Singh, 1996) or by drying (de Souza Santos et al., 1965). However, some doubt remains in the latter case since the transformation of the final product has been described as 'tubular kaolinite'. The reverse transformation of kaolinite to halloysite (7–10 Å) was also reported (eg, Loughnan and Roberts, 1981; Robertson and Eggleton, 1991); however, the reasons why kaolinite becomes hydrated in a natural environment are not clear. The process was accompanied by an increase in hydration state and exchange capacity, which indicates a surface-chemical reaction. To explain this fact, the authors suggested that spiral halloysite had formed around a preferential crystallographic direction (*b*-axis) as the result of a loss of structural rigidity, due to hydration, at points along the kaolinite crystal. Singh and Gilkes (1992) explained this transformation in terms of fragmentation into laths that roll or fold to form halloysite tubes since platy kaolinite particles could roll upon hydration (Singh, 1996; Singh and Mackinnon, 1996). A possible explanation of the natural transformation of kaolinite to halloysite may be changes in the microenvironment around minerals (high reactivity at the solid–liquid interface in the microsystems) that led to the dissolution of kaolinite and the precipitation of halloysite (Churchman, 2000). Finally, it is possible experimentally to induce rolled 1:1 layers from kaolinite using a delamination process, as detailed in Chapter 17.

2.3.3 Main Halloysite Ore Deposit in the World

Halloysite is widespread throughout the world. Compared to kaolinite ore deposits, the high purity and economically interesting deposits of halloysite are relatively rare, even if halloysite can be found all over the world. The halloysite deposits are classically found near kaolinite deposits. Except for the two biggest deposits (Matauri Bay from the Northland Region of New Zealand and Tintic district, in Utah in the United States), halloysite is classically extracted from a vein or pocket in altered rock and requires selective mining. Some deposits are also only hand-extracted due to the size of the interesting pockets. Such deposits are also composed of a mixture of halloysite and kaolinite. The chemical analyses of main halloysites from ore deposits in the world are listed in Table 2.2.

The rhyolite-hosted halloysite is from Northland, New Zealand, deposit namely Matauri Bay. Halloysite deposits are classically formed from hydrothermal alteration and from the weathering of alkaline rhyolite domes. Matauri Bay halloysite, reputed to be 'the world's whitest clay', is produced from mines at Matauri Bay and nearby Mahimahi by Imerys Tableware NZ Ltd (Townsend et al., 2006). The halloysite is formed by hydrothermal alteration and the subtropical weathering of Pliocene to Pleistocene rhyolite domes and made into material comprising approximately 50% halloysite and 50% quartz, cristobalite and minor feldspar. The high-purity halloysite product possesses an overall fine particle size, coupled with low Fe_2O_3 (0.28%) and TiO_2 content (0.08%). Samples were purified by wet process methods (Wilson et al., 2006) in order to obtain different purity clay samples that were up to 97% halloysite. Clays were then extruded at 37% moisture content or shredded and dried to produce granules or powders (dehydrated halloysite; Townsend and Masters, 2002). Matauri Bay samples show a low cation exchange capacity (CEC) value and specific surface area (SSA) (Joussein et al., 2006; Pasbakhsh et al., 2013), with a heterogeneous distribution of tubes up to 3 μm, including platy kaolinite (see Fig. 2.1 and Table 2.2). The Matauri Bay halloysite deposits have been worked since 1969, and about 80,000 tpa of raw material has been mined. The plant capacity is about 25,000 tpa of processed halloysite. In the last decade, the mine is projected to be viable for another 50 years or more. The halloysite is exported to more than 20 countries for the manufacture of high-quality ceramics (mainly porcelain, but also fine bone china and technical ceramics).

Even if New Zealand Matauri Bay is the best-known halloysite, other New Zealand samples (namely Te Puke and Opotiki) have also been reported in the classical research literature, but the deposits are relatively smaller (Fig. 2.1 and Table 2.2). Opotiki halloysite (from Opotiki, Bay of Plenty), formed from rhyolitic Pahoia Tuff (Kirkman, 1977), is spheroidal dominant, whereas Te Puke halloysite (found 10 km west of Te Puke) is a water-sorted deposit that probably formed by weathering, hydrothermal action or both in rhyolite and andesite (Hughes, 1966), producing a tubular and platy morphology.

In the main Tintic mining district of Juab County in central Utah (ie, the Eureka Dragon mine deposit), the halloysite sample (namely, Dragonite) is an unusual hydrothermal replacement for susceptible dolomite beds in the adjoining Upper Cambrian Opex Formation (Pampeyan, 1989; Boden et al., 2012). Historically, the Dragon mine produced approximately 1.2 million tonnes of halloysite. Applied Minerals Incorporated owns the Dragon property and is using it to produce pure halloysite and possibly an iron-oxide pigment by-product. Recent drill results in the Dragon pit indicate a measured resource of about 501,200 tonnes of 64% halloysite (Boden et al., 2012). Due to the large deposit potential in a district, a new exploitation pit (namely, North Star) looks to be effective and promising. Dragonite is very pure halloysite

TABLE 2.2 Chemistry, Mineralogy and Physical Chemical Properties of Main Halloysite Ore Deposit Samples

Country	New Zealand	United States	New Zealand	Australia	China	Poland	Turkey
Name	*Matauri Bay*	*Dragon Mine*	*Te Puke*	*Camel Lake*	*Longyan*	*Dunino processed*	*Turkish*
SiO_2	50.40	43.50	44.82	44.96	48.00	43.30	46.00
Al_2O_3	35.50	38.80	36.70	37.57	38.00	34.50	37.01
Fe_2O_3	0.25	0.33	3.40	1.21	0.29	2.60	0.70
MgO	tr	0.12	0.01	0.19	0.30	0.08	0.45
Na_2O	tr	0.07	0.01	0.09	0.10	0.19	0.10
K_2O	tr	0.07	0.05	0.31	1.71	0.05	0.30
CaO	tr	0.26	0.01	0.28	0.16	0.26	0.15
TiO_2	0.05	0.02	0.37	0.15	0.02	1.18	0.30
MnO	0.01	0.01	0.01	0.01	tr	tr	tr
P_2O_5	0.06	0.83	0.02	0.01	nd	0.05	nd
SO_3	0.06	0.26	0.02	0.63	nd	nd	nd
LOI	13.80	15.70	14.66	14.53	12.40	15.39	15.00
Halloysite (%)	96	84	98	95	79 (Hal/Kaol)	70-80	95
Accessory minerals	Quartz/ cristobalite, anatase	Kaolinite, quartz, gibbsite, alunite	Quartz/ cristobalite, anatase	Quartz, alunite, anatase, Fe-oxides	Micaceous minerals, quartz, anatase	Quartz, anatase	Quartz, anatase, alunite, feldspar

Continued

TABLE 2.2 Chemistry, Mineralogy and Physical Chemical Properties of Main Halloysite Ore Deposit Samples—Cont'd

Country	**New Zealand**	**United States**	**New Zealand**	**Australia**	**China**	**Poland**	**Turkey**
Name	*Matauri Bay*	*Dragon Mine*	*Te Puke*	*Camel Lake*	*Longyan*	*Dunino processed*	*Turkish*
Morphology	Tube/platy	Tube	Blocky and shorty tubes	Tube	Tube/platy	Tube/platy	Tube
CEC ($cmol_c/kg$)	2.5	2.1	5.2	18.8	nd	9.1	4.2
BET (m^2/g)	22.1	57.3	33.31	74.6	nd	nd	72.2

tr and *nd* mean 'traces' and 'not determined', respectively.
Data were extracted from Wilson (2004a) for Longyan China sample, and from Pasbakhsh et al. (2013) for Camel Lake and Dragon Mine samples.

FIG. 2.1 TEM images of the main morphologies of halloysite. *Left to right*: Large- and long-tubes (Matauri Bay), short- and blocky-tubes (Te Puke), long tubes morphology (PATCH), and spheroidal (Opotiki). *PATCH reprinted with permission from Yuan et al. (2008). Copyright 2008.*

(Table 2.2) and has a shorter tubular morphology (50–1500 nm), with thinner inner diameters that ranged from 5 to 30 nm. Samples were 84% halloysite, with other minerals being kaolinite (8%), quartz, gibbsite and sulphate/phosphate minerals. The SSA is about 57 m^2/g (Pasbakhsh et al., 2013).

The Dunino mine deposit is located in the Lower Silesia, Poland, and comes from basalt weathering. Halloysite is produced by Intermark/Kopalnia Haloizytu Dunino, established in 1998 as a surface mining operation near Krotoszyce, in the Legnica province. The deposit contains over 10 million tonnes of a homogeneous raw material mined using an open-pit method. The thickness of the halloysite seam is up to 20 m, and the exploitable resources prepared so far are around 500,000 tonnes. Raw halloysite from Dunino is characterised by a high porosity and is easy to modify (Lutynski et al., 2014). The Dunino halloysite consists of a mixture of nanotubes and nanoplatelets that are up to 80% halloysite. These samples also contain mainly iron and titanium oxides in the form of small, homogenous, dispersed grains (Table 2.2).

In Turkey, Esan Eczacibasi has kaolin and halloysite mines at Balikesir and Canakkale, north of Izmir in the extreme west of the country (Biga Peninsula). High-grade halloysite is produced due to the hydrothermal alteration of neogene alkaline volcanic lava and pyroclastic deposits by geothermal fluids (Ece et al., 2008, 2013). Deposits from Turkey are among the most important in the world (Yörükoğlu and Delibaş, 2012); they are relatively pure (Table 2.2). Ece et al. (2008, 2013) explained the genetic model of halloysite deposits as being due to the alteration of native clay by hydrothermal acidic fluids, which favoured the precipitation of halloysite. In some deposits, the presence of alunite is effective, and in this case, these authors stated that halloysite has more tubular forms. Some tubes of about 5 μm long were also reported by Saklar et al. (2012) in the Taban high-grade deposit. Finally, halloysite reserves in the Balikesir region were stated to be about 50,000 tonnes (Ece et al., 2008).

In Argentina, halloysite deposits are located southwest of the province of Rio Negro, in Mamil Choique and Buitrera (Cravero et al., 2012). The deposits developed from the weathering of pyroclastic rocks, rhyolitic tuffs and ignimbrites. The particle morphology depends on the original texture of the rock. Spheroidal halloysite is related to rock with low porosity, and tubular particles are related to rock with open spaces.

Some halloysite deposits also have been reported in Brazil (Wilson et al., 2006). The kaolin resources, which are widespread throughout the country and are varied in origin, are a mixture of kaolinite and halloysite samples derived from weathered pegmatites (Minas Gerais), weathered granitic (near Sao Paulo) basement or weathered anorthositic (Encruzilhada at the east of Porto Alegre). The amount of iron and titania is low, but samples from granitic generally have greater Fe content. Halloysite in mixtures in these deposits is classically tubular, as described in de Souza Santos (1993) and de Souza Santos et al. (2009).

China is a significant producer of kaolin, including halloysite (Wilson, 2004a,b). Chinese kaolin deposits are derived from the alteration of granitic- and volcanic-type rocks. The Chinese kaolin deposits exhibit varying morphologies due to the mixture of kaolinite (platy and stacks) and tubular halloysite (Longyan deposits from kaolinitised granites in Fujian Province). Wilson (2004a) also reported some high-grade deposits, such as Mengson in Yunnan Province, formed from the weathering of granite. Samples are low-iron- and -titanium-bearing minerals (Table 2.2). Moreover, other high-grade deposits are reported in the Guizhou Province, with halloysite deposits in the Dafang, Qingxi, Zunyi and Shijin areas derived from volcanic rock. Dafang halloysites show a tubular morphology with fine-grained tubes from 0.2 to 4 μm long with very low iron and titania, and often 9 mass% < 2 μm. However, these deposits are very small and were seen as many individual pocket deposits (ie, ore deposits are presented in discrete form).

The Eucla Basin in southern Australia is one of the world's largest onshore sites of Cenozoic marine deposits. These provide a huge store of sulphidic (acidic) material, as well as alunite and kaolin, which accumulated in the vicinity of these discharge sites. Halloysite in these deposits near Playa Lake, informally referred to as the 'Camel Lake sites', can show pure samples (up to 95% halloysitic-rich) and very regular tube morphology of up to 1500 nm long (Keeling et al., 2012; Pasbakhsh et al., 2013). Other samples from Australia have been reported by the authors: halloysite from Jarrahdale (in Western Australia) from the lower pallid zone of a deeply weathered profile in dolerite, and Patch Clay (nickel mine in Siberia, 85 km NW of Kalgoorlie, Western Australia; see Fig. 2.1 and Table 2.2), which is a thin halloysite with very long, regular tubes of up to 3000 nm in length (Norrish, 1995).

Other deposits are also reported in Japan and the Sancheong district of South Korea since halloysites are widespread and subject to weathering or hydrothermal alteration of volcanic deposits. Halloysitic samples present a dominant tubular morphology. Such deposits are also reported from granitic weathering, where halloysite can exceed 5 μm in length (Nagasawa and Miyazaki, 1976).

Halloysite has now become important in Thailand. The deposit from Thung Yai district (Nakhon Si Thammarat Province) in southern Thailand is found in an alluvial and flood plain of the Tapi River, consisting mainly of hollow microtubules and plates with typical dimensions of 80–200 nm in diameter and 500–450 nm long (Bordeepong et al., 2012). Kaolin samples (approximately 70% halloysite) contained some quartz and anatase as well. Moreover, the Ranong and Narathiwat provinces in southern Thailand also produce a representative crude kaolin derived from the alteration of granite. Samples consist predominantly of tubular halloysite (70%), poorly crystallised kaolinite and quartz, with minor amounts of mica and K-feldspars. The suitable reserves of Ranong are considerable and are largely exploited.

With the extraction process of halloysite, the flowsheet is quite similar to that of kaolinite process one. Briefly, halloysite is mined with classical mechanical shovels (or mined by hand in specific deposits), loaded into trucks and then transported to the plant. Halloysite samples are blunged, dispersed and degritted (see, for example, Wilson et al., 2006). Processing of Matauri Bay samples from New Zealand has been described by Pasbakhsh and Churchman (2015); here, halloysites are centrifuged to isolate the coarse fraction and then separated using wet process methods to purify the halloysite samples using hydrocyclones. A magnetic separation then can be applied onto the slurry to remove magnetic minerals (eg, hematite, magnetite, biotite) and bleached. In addition, samples can be submitted to acidic leaching treatment using organic and inorganic acids to remove iron impurities (Saklar et al., 2012). After purification, samples were centrifuged (which also potentially reduced the particle size) and dewatered by filter pressing or rotary filters. At this point, halloysite was already in hydrated form (halloysite (10 Å) or (7–10 Å), depending on the deposit). Next, the samples were shredded and spray-dried to produce powder with 3–6% moisture content. Halloysite was in dehydrated form. The latter route saves on the price of water mass, but at the expense of hydration properties. Note that some manufacturers allow the obtaining of slurry, which can be an effective option to conserve these properties, but the slurry must be reprocessed in the laboratory.

2.4 MINERALOGICAL CHARACTERISATION

2.4.1 Crystal Structure

2.4.1.1 Crystal Structure and Related Characterisations

Halloysite is a polymorph of the 1:1 dioctahedral kaolin group, which includes kaolinite, nacrite and dickite. The ability of halloysite to hydrate has always been what distinguishes halloysite from other kaolin polymorphs. The interlayer water content of halloysite (10 Å) is about 12.3 mass% (eg, Kohyama et al., 1978), corresponding to two water molecules per formula unit. In the fully hydrated structure, the theoretical halloysite formula is $Si_2Al_2O_5(OH)_4{\cdot}2H_2O$, where each layer is composed of silica tetrahedral and alumina octahedral sheets similar to those of kaolinite. The unit-cell parameters for halloysite (10 Å) is $a \sim 5.1$ Å, $b \sim 8.9$ Å and $c \sim 10.2$ Å (basal spacing, $c\sin\beta \sim 10.1$ Å and $\beta \sim 100$ Å), whereas $a \sim 5.1$ Å, $b \sim 8.9$ Å and $c \sim 7.3$ Å (basal spacing, $c\sin\beta \sim 7.2$ Å and $\beta \sim 96.5$ Å for halloysite (7 Å)). The structure of halloysite often features a main tubular morphology with the Al-OH layer forming the inside of the tube and the Si-O the outside (Fig. 2.2). The structure of halloysite is well known, but the hydrated form remains to be explored somewhat since it is not easy to avoid loss of water during characterisation. The first papers about the structure of halloysite were published in the 1950s, lack of understanding just begins to decrease from the

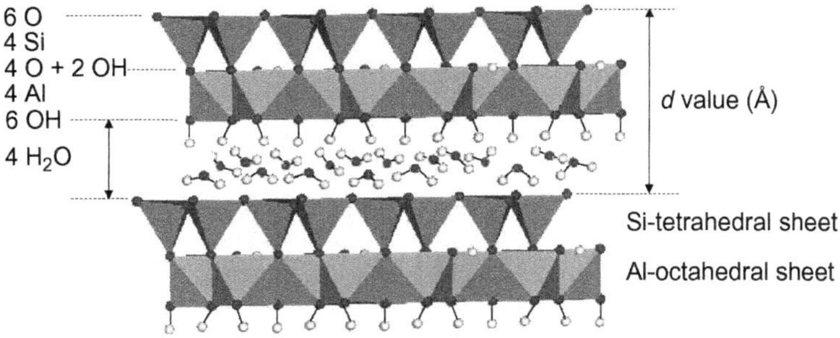

FIG. 2.2 Halloysite (10 Å) crystalline structure. The position of water molecules is only representative. *Adapted and reproduced from Joussein et al. (2005) with the kind permission of the Mineralogical Society of Great Britain and Ireland.*

development of high-performance machines. The structure of halloysite (7 Å) was deduced from selected area electron diffractions. Briefly, the structure was defined as a two-layer monoclinic structure by Kohyama et al. (1978), whereas Honjo et al. (1954) proposed a triclinic layer arrangement. Indeed, Honjo et al. (1954), and then Chukhrov and Zvyagin (1966), considered the stacking structure to form a two-layer periodicity. However, Honjo et al. (1954) proposed a stacking sequence with a triclinic symmetry, whereas Chukhrov and Zvyagin (1966) suggested a more realistic monoclinic *Cc* symmetry ($\beta_{\text{ideal}} = 97$ degrees) and a layered structure with idealised intralayer and interlayer displacements. This structure corresponds to the 2M1 polytypes of the trioctahedral polytypes (Bailey, 1969). Kohyama et al. (1978) suggested a two-layer monoclinic structure in accordance with the Chukhrov and Zvyagin structure for both halloysite (7 Å) and halloysite (10 Å). Bookin et al. (1989) investigated the nature of stacking faults in a defect-free 1Tc kaolinite from powder X-ray diffraction (XRD). These authors showed that a 1:1 layer structure, for a regular alternation of translations along the *b*-axis, leads to a structure similar to that described by Chukhrov and Zvyagin for halloysite. They concluded that the halloysite-like structure may be the end result of such defective kaolinite, despite the quite different mutual arrangement of the adjacent layers in these structures.

Recently, Kogure et al. (2011) succeeded in recording clear two-dimensional, high-resolution transmission electron microscopy (HRTEM) from tubular halloysite. The authors showed no symptoms of a two-layer periodicity, but more likely a one-layer periodicity with high density of stacking disorder. Moreover, a stacking sequence never observed in kaolinite was identified. Later, Kogure et al. (2013) investigated the structure of prismatic halloysite. The model proposed there suggests that the hydrogen bonds in the interlayer of halloysite are generated after the formation of concentric or spiralled kaolinite layers. According to this concept, tubular halloysite initially takes a hydrated form, and then dehydrates and transforms to a prismatic one consisting of sectored flat layers with the complete hydrogen-bonded interlayers (Kogure et al., 2013). Chukhrov and Zvyagin (1966) explained that the halloysite-like structure may be the end result of such defective kaolinite. However, the results of Kogure et al. (2013) seem to contradict this point, since the structural difference between kaolinite and halloysite may be related to their formation processes promoting the tendency to form a regular interstratification of the different layer displacements. This latter point (ie, stacking with the same layer displacement to the adjacent layer) is thermodynamically more favourable than stacking with a different layer displacement (Kogure et al., 2013). More details concerning the HRTEM study on the structure of halloysite are given in Chapter 5.

Halloysites can occur with various morphologies (listed in Table 2.3), the main classical one being tubes. In this case, the long axis is frequently coincident with the crystallographic *b*-axis (eg, Anand et al., 1985; Bailey, 1990). The formation of halloysite tubes has been hotly debated in the literature, without any real consensus (Pasbakhsh and Churchman, 2015): in the formation towards progressive alteration of platy kaolinite, with HNT being attached to a kaolinite plate curled smoothly rolling up part of the plate, lateral misfits of the smaller octahedral sheet compared to a larger tetrahedral one in terms of iron content, direct precipitation and other characteristics. For example, Nabarro (1967), using the theory of a continuous distribution of dislocations, explained that a continuum of such dislocations would produce lattice rotations, but not long-range lattice strains. Thus, this type of rotation may apply to hydrated and dehydrated halloysite and may contribute to the *a*- and *b*-axis disorder characteristic of halloysites. These dislocations are inherent in the crystallisation process and are not induced by external stress (Kirkman, 1981). Later, Bailey (1990) explained that the tubes are formed by layer rolling due to the dimensional misfit between the octahedral and tetrahedral sheets and weak interlayer bonding. Finally, this author assumed the important role of Al^{3+} substitution for Fe^{3+} in the octahedral position for platy halloysite (the more iron-rich variety) by increasing the overall size of the octahedral sheet and limiting the curvature effects (like the planar shape of kaolinite). Singh (1996) explained from the theoretical model that the rolling mechanism is energetically more favourable than tetrahedral rotation for halloysite compared to kaolinite to accommodate the lateral misfit. The rolling mechanism encounters significantly less resistance from Si–Si repulsion than from tetrahedral rotation. The octahedral sheet probably provides only negligible resistance to rolling. However the model may be efficient for a lower number of staking layers. A general trend is that HNT initially takes the hydrated form, and then dehydrates. Indeed, it seems that initially, halloysite grows as a hydrated form without hydrogen bonds between the basal oxygens and hydroxyls in the interlayer space. This fact induces rolling of the layer. Then, halloysite (10 Å) dehydrates, inducing

TABLE 2.3 Various Morphologies of Halloysite Reported in the Research Literature and Their Occurrence in the World

Morphologies	Occurrences	References
Tubular, long and thin, short and stubby tubes	Cryptokarstic sediment, volcanic glass and pumice, feldspar and micas alteration	(1), (2), (3), (4), (5), (6), (7), (8), (9), (10), (11), (12), (13), (14), (15), (16), (17), (18), (19), (20), (21), (22), (23), (24), (25), (26), (27), (28), (50)
Pseudospherical and spheroidal	Weathered volcanic ash and pumices, volcanic glass in marine environments (eg, Guatemala soil, New Zealand)	(5), (6), (7), (8), (11), (12), (14), (16), (18), (20), (28), (29), (30), (31), (32), (33), (34), (35), (36), (37), (38), (39)
Platy or tabular	Volcanic ash soil, weathered pyroclastic, lateritic profiles, fissures within granite, hydrothermal alteration, tuff bed (eg, Texas, Brazil, Guatemala)	(8), (16), (34), (40), (41), (42), (43), (44), (45)
Fibre	Lateritic soil, weathered granite (eg, Australia, Brazil)	(46), (47)
Prismatic, rolled, crinkly, walnut meat	Volcanic ash soil, weathered granite (Japan)	(34)
Cylindrical, disk	Rhyolitic tephra (New Zealand)	(20), (36)
Spherulic, irregular lath with rolling edge	Weathered granite/gabbros (Scotland)	(48)
Crumpled lamellar	Weathered pumices (Japan)	(44)
Lath, scroll	Altered volcanic glass (Japan)	(8), (10)
Glomerular or 'onion like'	Volcanic ash (Cameroon)	(49)

(1) Bates et al. (1950); (2) Honjo et al. (1954); (3) de Souza Santos et al. (1964); (4) Chukhrov and Zvyagin (1966); (5) Parham (1969); (6) Askenasy et al. (1973); (7) Dixon and McKee (1974); (8) Nagasawa and Miyazaki (1976); (9) Kohyama et al. (1978); (10) Nagasawa (1978); (11) Saigusa et al. (1978); (12) Loughnan and Roberts (1981); (13) Noro et al. (1981); (14) Churchman and Theng (1984); (15) Nagasawa and Noro (1987); (16) Noro (1986); (17) Delvaux et al. (1992); (18) Romero et al. (1992); (19) Singh and Gilkes (1992); (20) Jeong and Kim (1993); (21) Churchman et al. (1995); (22) Dupuis and Ertus (1995); (23) Polyak and Guven (2000); (24) Perruchot et al. (1997); (25) Adamo et al. (2001); (26) de Putter et al. (2002); (27) Baioumy and Hassan (2004); (28) Singer et al. (2004); (29) Sudo et al. (1981); (30) Tomura et al. (1983); (31) Ward and Roberts (1990); (32) Sudo (1953); (33) Sudo and Yotsumoto (1977); (34) Tazaki (1979, 1982); (35) Askenasy et al. (1973); (36) Kirkman (1977); (37) Dixon and McKee (1974); (38) Imbert and Desprairies (1987); (39) Kawano and Tomita (2001); (40) de Souza Santos et al. (1966); (41) Kunze and Bradley (1964); (42) Wilke et al. (1978); (43) Carson and Kunze (1970); (44) Wada and Mizota (1982); (45) Quantin et al. (1984); (46) de Souza Santos et al. (1965); (47) Norrish (1995); (48) Wilson and Tait (1977); (49) Sieffermann and Millot (1968); (50) Wilson (2004a).
Reproduced from Joussein et al. (2005) with the kind permission of the Mineralogical Society of Great Britain and Ireland.

the removal of water molecules from the interlayer space. During the ripening pathway of halloysite, each layer moves relative to the adjacent layer and adjusts itself by $\pm b/3$ or $\pm b/6$ shifts along the tube-axis to form an interlayer configuration with hydrogen bonding similar to the kaolinite one.

The crystallinity (comprising stacking order and internal variability) of halloysites can be evaluated by various methods (Churchman and Theng, 1984). However, most crystallinity indexes can be used only to compare halloysite samples. For example, the crystallinity index measured by Smykatz-Kloss (1974) by using differential thermal analysis (DTA) or a low-frequency 'tail' in the ^{27}Al NMR spectrum, as suggested by Newman et al. (1994) for measuring crystallinity. However, the application of the Hinckley index is impossible because XRD reflection is not evident in halloysite as opposed to kaolinite. Finally, the better potential crystallinity index can be relative to infrared (IR) spectroscopy using the Lietard index or the R2 one. A quite important point is that in a large variety of halloysites, the XRD, DTA and IR crystallinity indices were closely related to the structural iron content (proportion of Fe^{3+} in a particular location) of the samples (Churchman and Theng, 1984), and the morphology (crystallinity) of tubular halloysite particles was higher than that of spheroidal particles (Takahashi, 1958). The same results were obtained for kaolinite.

The powder XRD patterns of the main halloysites are given in Fig. 2.3. The differences between these patterns and the kaolinite XRD pattern are due to the structural disorders and the curvature of the HNT, inducing the absence of *hkl* reflection (eg, 20–25°2θ region) and the presence of scattering bands typical of a layered structure with two-dimensional order. Classically, HNT can be characterised from oriented amounts after oriented deposit onto ceramics, glass or Si-buffer slide. From oriented patterns (Fig. 2.3), the XRD identification of halloysite (10 Å) is straightforward because it gives a basal reflection of classically wide (d_{001} spacing) at 10.1 Å, representing the thickness of a single 1:1 layer (7.1 Å) and that of a monolayer of water molecules (see Section 2.4.1.1). As halloysite presents water molecules only in the interlayer position, and there is no (or almost no) substitution, there is no difference in basal spacing with the saturated state of samples even if Joussein et al. (2006) have shown that cations can be interlayered. When halloysite (10 Å) is submitted to ethylene glycol treatment, the 10.1-Å reflection shifts to 10.8 Å due to the intercalation of one glycol layer. The main problem is relative to the rapid conversion of halloysite (10 Å) into halloysite (7 Å) under ambient conditions (of temperature and humidity) or by moderate heating at 35°C (Joussein et al., 2006). In this case, it is quite difficult to determine one of the kaolin polymorphs (kaolinite or halloysite (7 Å)) and the use of organic molecules such as ethylene glycol or formamide is necessary for further distinction. Finally, dehydration of halloysite (10 Å) during XRD analysis may be prevented, or at least inhibited, by using water-saturated ceramic tiles as sample holders. The identification of halloysite (7 Å) by XRD is ambiguous because its diffraction pattern is almost identical to that of disordered

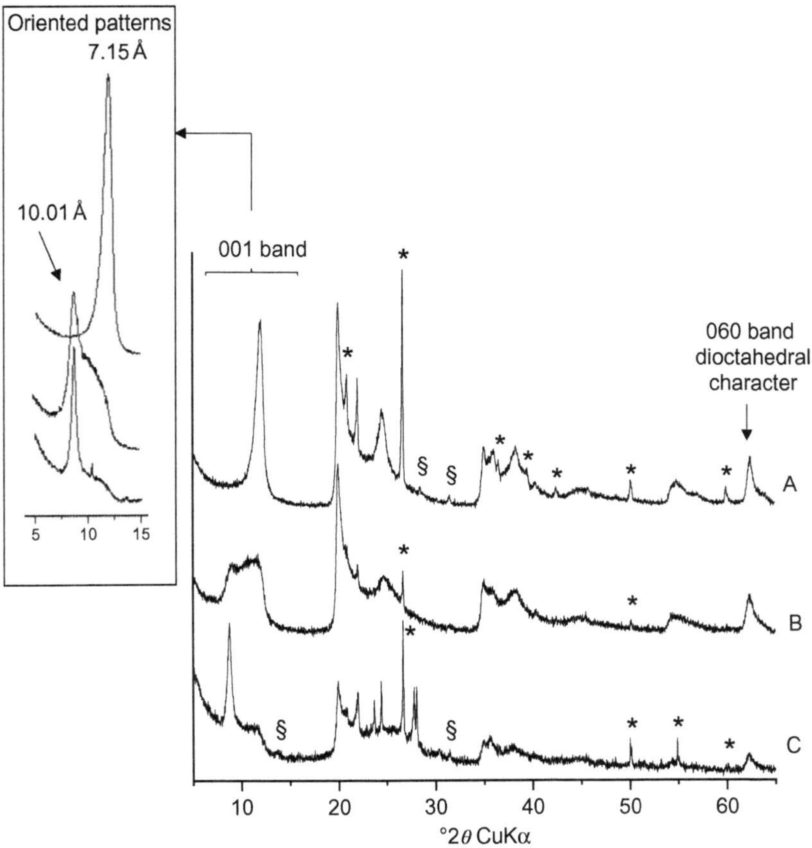

FIG. 2.3 XRD powder patterns and the corresponding oriented patterns of three New Zealand halloysites; A: Matauri Bay, B: Te Puke and C: Opotiki. * and § refer to quartz and feldspars, respectively.

kaolinite (Brindley, 1980). Halloysite (7 Å) usually shows a very broad or weak basal (d_{001}) reflection between 7.2 and 7.6 Å due to the tubular or spheroidal morphology, high degree of disorder, small crystal size and possible interstratification of layers with various hydration states (Brindley, 1980; Joussein et al., 2006). Some authors reported that preheating at 100–350°C sharpens the basal reflection and reduces the spacing to about 7.2 Å, but never to as low as 7.14 Å, which is characteristic of kaolinite but well-settled XRD instruments are needed, assuming that kaolinite is well ordered. The differentiation between halloysite (7 Å) and kaolinite will be discussed further later in this chapter.

The DTA curve of halloysite presents three thermal events: (i) an endothermic peak between 50°C and 150°C, corresponding to the loss of adsorbed water (surface and interlayer); (ii) a second endothermic peak in the 450–600°C range due to structural dehydroxylation; and (iii) a third exothermic peak between 885°C and 1000°C, which must be attributed to mullite formation. The DTA curve for halloysite (7 Å) is identical with that for kaolinite and halloysite (10 Å) except for the absence of the first endothermic peak. The main differences can be shown in the dehydroxylation temperature zone: a poorly ordered halloysite sample is lower than that of kaolinite (450°C vs 550°C). However, there is no difference in terms of the dehydroxylation peak shift between halloysite and disordered kaolinite since the temperature is affected by particle-size distribution, crystallinity, type of isomorphous substitution, mechanical treatment and other minerals. According to this, the method based on the estimation of the halloysite content relative to kaolinite in mixtures, by measuring the relative intensity (or area) of their respective dehydroxylation peaks (Hewitt and Churchman, 1982), is subject to a large uncertainties. More details about the dehydration and dehydroxylation of halloysite are discussed in Chapter 7.

IR and Raman spectroscopies are efficient and sensitive tools for identifying minerals of the kaolin group since each mineral exhibits a specific IR spectrum (Farmer, 1974). Further, IR spectroscopy can show the presence of very small quantities of kaolin minerals that are not detectable by XRD (Joussein et al., 2001). For more information on this subject and representative IR spectra, see Chapter 6. Briefly, halloysite exhibits only two Al_2OH-stretching bands (ν_1 and ν_4) at 3695 and 3620 cm^{-1}, each OH being linked to two Al atoms, whereas kaolinite gives three or four bands (depending on the monoclinic character). Halloysite generally shows a single Al_2OH-bending band at about 920 cm^{-1}, but in kaolinite, this band has a shoulder at about 938 cm^{-1}. Occasionally, the presence of a shoulder at 3600 cm^{-1}, coupled with a weak band at 875 cm^{-1}, is effective in Fe-rich halloysite. This band is ascribed to $AlFe^{3+}OH$ vibrations (de Oliveira et al., 1997) and is in accordance with the presence of structural iron. Finally, because high-defect kaolinite and halloysite exhibit very similar IR spectra, the relative content of halloysite and kaolinite in a mixture of kaolin minerals cannot be precisely estimated. Raman spectroscopy can be also a good method to investigate HNT, particularly micro-Raman allowed orientation-dependant along the different crystal axes. For example, Frost (1998) showed the difference of hydroxyl deformation between high- or low-defect kaolinites and low-defect halloysite. However, a limiting factor currently associated with Raman spectroscopy is related to fluorescence problems, which largely minimises its suitability for use on soil samples.

2.4.1.2 *Qualitative and Quantitative Differentiation of Halloysite and Kaolinite*

If one of the main questions is related to the relation between polymorph halloysite and kaolinite, the second question is how to differentiate in practice by XRD halloysite (7 Å) from kaolinite in mixtures. It is accepted that the use of organic test intercalation seems to be a better option. As indicated previously, a coupled XRD and spectroscopy investigation can allow halloysite and kaolinite to be distinguished, but only qualitatively. By revisiting MacEwan (1948), Hillier and Ryan (2002) demonstrated that ethylene glycol-solvated and heated samples, such as are taken routinely for many studies, are sufficient to determine whether halloysite (7 Å) is present in a sample. However, ethylene glycol solvation is also influenced by the crystallinity of both halloysite and kaolinite. This method is routinely used since 15 years ago at the James Hutton Institute (a well known place of clay mineralogy located in Scotland) with sensitivity from 20% halloysite in mixtures with kaolinite. Even if a large number of empirical tests have been developed based on differences in chemical reactivity of kaolin polymorphs (eg, K-acetate/ethylene glycol, hydrazine/water/glycerol, DMSO; see the discussions of various tests in Theng et al., 1984; Churchman, 1990; Joussein et al., 2005 and references therein), the general trend is based on the formamide test (Churchman et al., 1984). This is mainly due to (i) the simplicity of use, (ii) the nontoxic character of this product, as well as (iii) the well-done quantitative results obtained for a large number of mixtures with or without other phases. After oriented tile preparation, the samples were sprayed with formamide and observed by XRD after 5–10 min because the formamide intercalation occurred for halloysite < 1 h, whereas it requires at least 4 h for kaolinite; and even then, the process was often incomplete (an example is given in Fig. 2.4, showing the tendency for the three main New Zealand halloysites). The proportion of halloysite in the mixture is obtained by computing the $I_{10}/(I_7+I_{10})$ ratio, where I_{10} and I_7 denote the intensity of the XRD reflections near 10 and 7 Å, respectively. Churchman and Theng (1984) reported that the intercalation of formamide by halloysite is favoured by a large particle size, a high degree of crystallinity, and low iron content (halloysite ore deposit samples). However, air-drying and mild heating tend to inhibit layer expansion, as reported by Joussein et al. (2007). Samples from soil that has been air-dried and stored for 10 years do not respond in the same way to formamide, inducing an important underestimation after storage. Other authors have also reported the fact that some tubular forms of halloysite (7 Å) do not expand with formamide (Hart et al., 2002; Kautz and Ryan, 2003). So the proportion of halloysite in a sample may be estimated by microscopy by assuming that the particles occur only as tubes, spherules or laths. However, the information obtained by TEM and SEM is often ambiguous since halloysite can adopt a large variety of particle morphologies, some of which are similar to those of kaolinite. The question about the difference between tubular halloysite and tubular kaolinite remains. However Joussein et al. (2007) have used simultaneously formamide test, physicochemical and microscopy data to quantify the proportion of halloysite and kaolinite in soil samples. Finally, the quantitative analysis of kaolin minerals in mixtures would require the application of a combination of chemical and instrumental techniques.

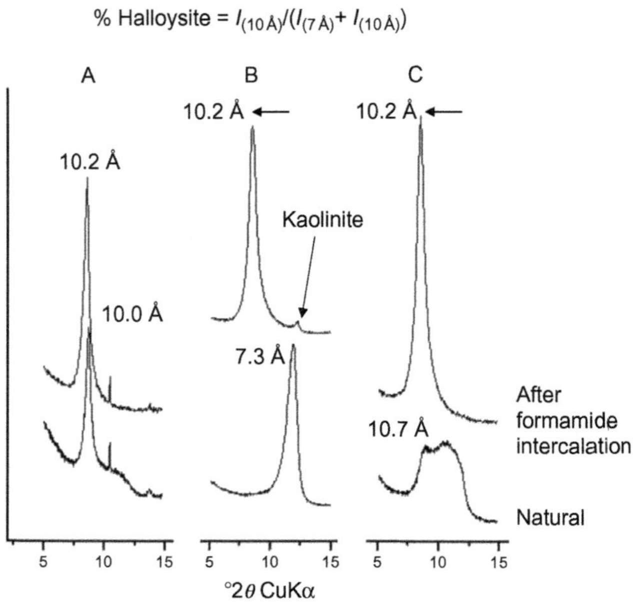

FIG. 2.4 XRD patterns of oriented preparation before (*bottom*) and after (*top*) formamide treatment for three New Zealand halloysites; A: Opotiki, B: Matauri Bay and C: Te Puke. The degree of intercalation (ie, % halloysite) was estimated from the ratio $I_{(10Å)}/(I_{(7Å)}+I_{(10Å)})$, where $I_{(10Å)}$ and $I_{(7Å)}$ represent the reflection intensity of the formamide complex at 10.2 Å and of the unexpanded kaolin component (halloysite plus kaolinite) at about 7 Å, respectively.

2.4.2 Chemical Composition and Affecting Factors

Halloysite has the same theoretical chemical composition as kaolin polymorphs except for its water content (if present). The ideal unit formula for halloysite (7 Å) and halloysite (10 Å) is $Si_2Al_2O_5(OH)_4{\cdot}nH_2O$, where $n=0$ (the 7 Å) up to 2 (the 10 Å). The chemical analyses of main halloysites from ore deposits in the world are reported in Table 2.2. The main oxides are SiO_2 and Al_2O_3, with the presence of some samples of Fe_2O_3 (iron in raw samples assumed as Fe^{3+}). The presence of impurities in such halloysite ore deposit is classical and induces difficult to assess the chemical composition (associated clay minerals, iron oxides or poorly organised minerals). This difficulty is already important in the case of impurities localised inside the hollow halloysite tubes. The main phases can be easily determined by XRD or TEM investigation (with Ti assumed as anatase, or Fe as hematite, eg, Schroeder and Shiflet, 2000). Finally, the presence of iron remains problematic since it can be ascribed partly to associated iron oxides up to 12.8 mass% of Fe_2O_3 (such as hematite or maghemite) and partly to isomorphous substitution of Fe^{3+} for Al^{3+} in the octahedral sheet (Hart et al., 2002). The isomorphous substitution of Fe^{3+} for Si^{4+} in the tetrahedral sheet was never reported for halloysite.

The presence of iron in structural position may also play an important role in the tubular halloysites (see Section 2.4.3). Some authors assumed also that some irregular replacement of Al^{3+} by Fe^{3+} in octahedral positions, as well as of Si^{4+} by Al^{3+} in the tetrahedral sheet, is effective for halloysite (Soma et al., 1992). Joussein et al. (2005) showed that a large number of halloysite particle analyses in the research literature fall close to the line joining the Al-halloysite end-member and the Fe-halloysite theoretical end-member in the Si–Al–Fe/O ternary diagram. The authors suggested that the Fe^{3+} for Al^{3+} substitution is stoichiometric. However, it should be kept in mind that Iriarte et al. (2005) showed that the possible incorporation of Fe^{3+} in synthetic kaolinite is limited, suggesting that the same behaviours may be possible for halloysite.

The chemical composition follows directly the main physicochemical properties of halloysites, described in more detail in Chapter 4. Briefly, the physicochemical properties play an important role in the potential uses of HNT. The CEC of halloysites ranges from 2 to 60 $cmol_c\,kg^{-1}$ (Table 2.2). The fact that CEC is higher for halloysite than for kaolinite is not really surprising due to its structure, which induces different configurations in the hydrogen bonds, the presence of negative charges and the various hydration states (Wilson, 2015). The net negative charge arises from the substitution of Al^{3+} for Si^{4+} in the tetrahedral sheet or Fe^{2+} for Al^{3+} in the octahedral one. This charge is balanced by hydrated exchangeable cations that are associated more with the tetrahedral than the octahedral side of the interlayer surface. Joussein et al. (2006) showed by XRD that the interlayer water in halloysite can be accompanied by charge-balancing cations. This is in accordance with White and Dixon (2002), who claimed that halloysite may contain up to 0.15 Al(IV) for two tetrahedral sites or with Komarneni et al. (1985) and Newman et al. (1994), who found less than 1% of Al(IV) using ^{27}Al MAS NMR spectroscopy onto six halloysites. The CEC is influenced by sample purity (presence of gypsum for example), particle size, cation localised in the microporosity or mesoporosity (halloysite tubes) and possibly by particle morphology (Bailey, 1990). It was generally admitted that misidentification of so-called halloysites that are potentially XRD-undetectable, especially as interstratified halloysite–smectite (Sakharov and Drits, 1973; Delvaux et al., 1989b), may account for possible wrong statements ascribing high CEC values to halloysites. Moreover, the reactivity of halloysite is relative to its hydrated state, with hydrated halloysite having a larger CEC than the corresponding dehydrated form. The SSA of halloysites is low up to 80 m^2/g (Table 2.2). However, it is important to handle the data with care since the measurements mainly depend on the experimental conditions; some samples present higher CEC values, probably due to the presence of impurities, and in the sample preparation, lower SSA due to higher heating (eg, above 250°C) before analyses inducing change in halloysite porosity. Pasbakhsh et al. (2013) showed that variation in halloysite morphology and the levels of impurities had the most effect on SSA and internal pore volume. Hart et al. (2002) also observed that the SSA of halloysitic soil from Indonesia tended to decrease with increasing Fe content, while that of Australia soil remained constant with increasing Fe content.

2.4.3 Morphology and Origin of Its Diversity

The morphology of halloysite appears to be related to crystallisation conditions, degree of alteration, chemical composition and geological occurrences. The main classical morphology of halloysite is tubular one even if spheroidal or probably platy/stacks of platelets are observed (Fig. 2.1). The different morphologies of halloysite, collected from the research literature, and their occurrence in the world are listed in Table 2.3. Up to 10 different morphologies have been reported, such as platy or tabular (like kaolinite), prismatic (in highly ordered halloysite prism form in dehydrated samples), rolled (tubes not closed), crinkly, walnut meat, cylindrical, disk (one-dimensional spherule), spherulic, irregular lath with rolling edge, crumpled lamellar (the edges of the lamellae are thin and gently curved), lath, scroll, glomerular or 'onion-like morphology', and finally tubular, long and thin, short and stubby tubes, fibres or pseudospherical and spheroidal. Overall, a large number of halloysite morphologies can be described in the literature but are not really representative of the main morphology (which is only representative of a small number of published papers). Finally, the main morphologies can be summarised as platy, tubular, spheroidal and other types of mixtures like prismatic ones. Since there may be some doubt in the case of the platy or tabular morphology of halloysite (limited between halloysite and low-crystallised kaolinite), this text focuses only on tubular and spheroidal morphologies. Indeed, SEM and TEM are the most useful techniques for morphological studies, even if the evacuation in the microscope chamber may induce some structural alterations. Actually, environmental-, cryo-TEM/SEM, and computer-assisted minimal dose systems on TEM are the most promising techniques, reflecting the real morphology.

2.4.3.1 Tubular Halloysite

As already discussed, the dominant morphology of halloysite is tubular, which is the shape most used in nanotechnology science applications. The tubes can be long and thin (Matauri Bay and PATCH in Fig. 2.1), short and stubby (Te Puke) or emerging from other tubes. The length of halloysite tubules varies widely in shape and size, ranging from 20 to 4000 nm in length, and from 20 to 200 nm for the outer diameter (Pasbakhsh et al., 2013 and references therein). The inner diameter (lumen pores) range from 5 to 30 nm, with a high concentration of small pores. In terms of the structural point of view, the greater curvature of the core of the halloysite particles seems to be linked to a smaller number of stacked layers, which sometimes show a rolled internal configuration (Dixon and McKee, 1974).

2.4.3.2 Spheroidal Halloysite

Spheroidal forms of halloysite occur commonly, particularly with the alteration of volcanic glass, ashes and pumices. The diameter of the spherules ranges from 50 to 500 nm (eg, see Churchman and Theng, 1984; Joussein et al., 2006; Fig. 2.1). The particles consist of different domains and the structure looks like a cabbage or an onion. Askenasy et al. (1973) suggest that the internal part of the particle, which is generally more irregular, consists of amorphous or poorly ordered materials such as allophanelike phase, but no evidence was clearly present. Finally, the spheroidal morphology may be related to the state of solutions, which is likely to be highly supersaturated in contact with the glass (Tomura et al., 1985).

2.4.3.3 Relation Between Morphology and Iron Content

As already shown in Section 2.4.2, the majority of halloysites are iron-rich. The influence of iron content on particle morphology has been well documented. Joussein et al. (2005) showed clear trends of the relationship between the three main morphologies of halloysite (tubular, spheroidal and platy) and the iron content: tubular particles are relatively Fe-poor, platy/platelets always contain relatively high amounts of iron and spheroidal halloysites exhibit a wide range of iron content. As the extent of isomorphous substitution of Fe^{3+} for Al^{3+} increases, the layer curvature decreases (Bailey, 1990). However, if the substitution is nonstoichiometric, the presence of octahedral vacancies might be expected to enhance the layer curvature. In this case, large numbers of structural defects may be effective probably linked to high exchange capacities. Iron content in HNT is probably linked to the tube length, with higher iron content inducing smaller tubes. Churchman and Theng (1984) suggested that iron inhibits crystallisation, leading to short tubes. Nevertheless, there is no relation between the amount of Fe content and the size of the lumen.

2.4.4 Hydration and Dehydration: An Important Fingerprint of Their Properties

Halloysite is the only 1:1 clay mineral found as a natural hydrate. This specificity allows important effects for a variety of industrial or environmental applications playing an important role in the physical properties of this mineral (Joussein et al., 2006), its exchange properties (Norrish, 1995), its reactivity in contact with organic species (Churchman and Theng, 1984; Joussein et al., 2007) and their rheological behaviours. The main problem is relative to the irreversibility of the dehydration. The hydration–dehydration of interlayer water in halloysite is strongly affected by the relative humidity (RH), temperature, drying history and finally the physicochemical and structural characteristics of the halloysite sample.

2.4.4.1 Dehydration–Rehydration Behaviour upon RH

The dehydration process of halloysite has been heavily debated in the research literature and many hypotheses have been reported. The fact is that the interlayer water in halloysite (10 Å) is easily and irreversibly lost on standing in dry air, under vacuum or on mild heating (Giese, 1988). This is why a large part of halloysite deposits is composed of halloysite (7 Å). The process involved is apparently controlled by kinetic factors for which rates are very low when there is no desiccation (Kautz and Ryan, 2003). Since the potential rehydration of halloysite is really important in order to keep its specific properties, many researchers have tried to rehydrate halloysite (7 Å) but without total success. In another study, de Souza Santos et al. (1966) observed partial rehydration in water after drying in the open air for a few hours of platy halloysite from Brazil, and Joussein et al. (2006) partially rehydrated halloysite after freeze-drying it either at 95% RH or by immersing the sample in water for 3 months. The saturation state of halloysite affects the rehydration since calcium saturation promotes the rehydration process. Finally, the only potential way is indirect, through intercalation of organic compounds by treating halloysite (7 Å) with K-acetate and then water (Wada, 1961).

The dehydration of halloysite started below 90% RH but can be still incomplete about 35–40°C at 40% RH and even at about 0% RH depending on the sample (Harrison and Greenberg, 1962; Kohyama et al., 1978; Joussein et al., 2006). Therefore, the dehydration behaviour of halloysite (10 Å) is dependent on drying history, RH and sample origin and morphology (Hughes, 1966). The rate of dehydration decreased in the order: long tubular halloysite > hexagonal platy halloysite > spheroidal halloysite (Noro, 1986; Joussein et al., 2006). Moreover, the dehydration processes of spheroidal and tubular samples are different. A complete study of dehydration behaviour with decreasing RH of two halloysites with different morphologies (tubular Te Puke and spheroidal Opotiki) was conducted by Joussein et al. (2006). The XRD pattern acquired at 27°C under various RH decreasing from 95% to 0% for two cationic saturation state (Ca^{2+} or K^+) was reported in Fig. 2.5. These authors clearly show that the two halloysite samples start to dehydrate below 70% RH, with dehydration increasing when RH falls below 10–20%. However, the dehydration process is quite different for the two samples. In the case of Te Puke halloysite, an intermediate reflection appears between the 7- and the 10 Å reflections during dehydration, whatever the cationic saturation state (K^+ or Ca^{2+}) is. Below 30% RH, the 7-Å reflection becomes dominant and the 10-Å reflection disappears for RH <20%. In the case of Opotiki halloysite sample, the hydrated phase seems to dehydrate (from 40% RH) without the formation of intermediates. However, at 0% RH, a weak reflection near 10 Å is still present, with the intensity being greater for the Ca-saturated sample than for the K one, which shows no complete dehydration. Moreover, the basal spacing of the dehydrated Opotiki

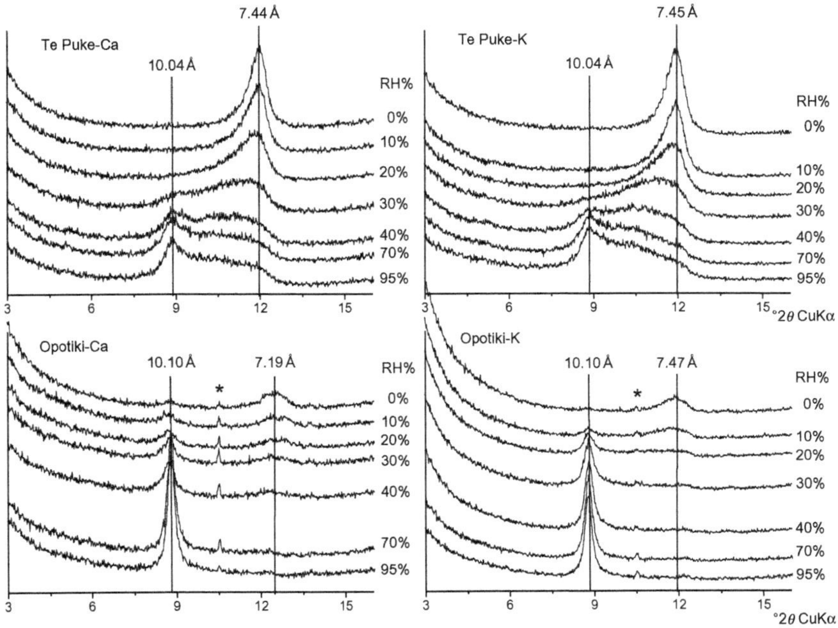

FIG. 2.5 XRD patterns of the Ca- and K-saturated Te Puke and Opotiki halloysite samples recorded under various relative humidity (RH) values ranging from 95 down to about 0% RH (T=27°C). * refers to the 8.4-Å reflection due to amphibole. Te Puke and Opotiki are tube and spheroidal dominant, respectively. *From Joussein et al. (2005), reproduced with the kind permission of The Clay Minerals Society, publisher of Clays and Clay Minerals.*

halloysite differs for K- and Ca-saturated samples (7.47 and 7.19 Å, respectively) suggesting the presence of these cations into the interlayer space of the spheroidal halloysite. Finally, the dehydration is complete for the Te Puke sample, whereas the Opotiki sample is not totally dehydrated, indicating that the two halloysites do not hold water similarly. Therefore, the dehydration of tubular particles proceeds with one intermediate hydration state (at about 8.6 Å) reacting as a separate phase due to the presence of 'hole' water molecules. The dehydration of the spheroidal halloysite gives a dehydrated phase and no 8.6 Å phase. The results suggest the presence of different types of water molecule, the 'associated' and the 'hole' water, which control the dehydration behaviour of halloysites as detailed by Giese (1988) for synthesised hydrated kaolinite. The dehydration behaviour of halloysite sample is highly heterogeneous and cannot be generalised a priori.

The halloysite (10 Å) is very susceptible to dehydrating irreversibly, producing the halloysite (7 Å) form. Joussein et al. (2006) studied the potential rehydration of the Te Puke and Opotiki halloysite samples, which were previously Ca- or K-saturated samples. Experiments were conducted after complete

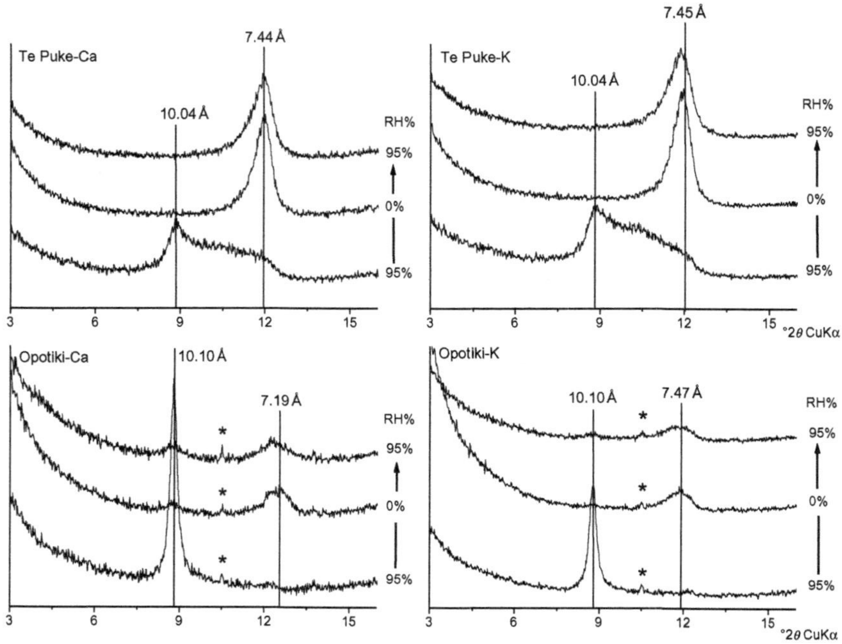

FIG. 2.6 XRD patterns obtained after dehydration from 95% RH down to 0% RH, then rehydration at 95% RH for Te Puke and Opotiki samples that previously were Ca- or K-saturated. * refers to the 8.4-Å reflection due to amphibole. Te Puke and Opotiki are tubes and spheroidal dominant, respectively. *From Joussein et al. (2005), reproduced with the kind permission of The Clay Minerals Society, publisher of Clays and Clay Minerals.*

dehydration at 0% RH. Samples were then rehydrated for a few hours at 95% RH. The study was followed by XRD (equipped with a RH control), and the results are presented in Fig. 2.6. Briefly, these authors have shown that no clear rehydration was effective for the Te Puke sample whatever the saturated cation, even if a slight decrease of the 7 Å reflection intensity suggests a partial rehydration. For Opotiki halloysite samples, a weak reflection at 10 Å is effective after a few hours at 95% RH. So a partial rehydration may occur, but it is never complete. The authors explained this latter point by the probable presence of cations with great hydration energies (such as Ca^{2+}) associated with the strongly held 'hole' water in interlayer space playing a significant role in rehydration properties.

Due to their high tendency to dehydrate, halloysite samples should be handled with great care and stored under controlled conditions of humidity to avoid dehydration even during transportation. The storage temperature does not exceed 40°C (Joussein et al., 2006). Only by storing in a sealed container in contact with water or in a water-saturated atmosphere could the material be kept fully hydrated (Giese, 1988). The use of freeze-drying is forbidden

(Hillier and Ryan, 2002; Joussein et al., 2006). Since the transportation of halloysites from extractive companies is cost dependent, samples are classically processed and dehydrated before transportation (decreasing the water cost). It will be interesting that companies could allow to sell or otherwise dispose of hydrated halloysite to favour their reactivity for nanotechnology applications.

2.4.4.2 *Interlayer Water Content and Status*

The presence of interlayer water in halloysite, with the atomic position of interlayer water being actually unknown, may be explained in terms of water activity as proposed for hydrated talc (namely, kerolite), as detailed by Joussein et al. (2005). Kerolite may be quite analogous to halloysite minerals since it is characterised by extreme layer stacking disorder, a small, if any, negative charge, a great ability to hold water molecules, and slight swelling in ethylene glycol or formamide. Most of the water associated with kerolite is weakly bound to layer surfaces, with only a minor amount being more strongly bound within the interlayer spaces (Miller et al., 1991). Halloysite can present a permanent charge due to substitutions in tetrahedral or octahedral positions which can provide the attraction of water in the interlayer space. This statement is different from the Bailey (1990) and Singh (1996) point of view. These authors consider that the driving force of water into the interlayer space was attributed to its activity and structural disorder of 1:1 layers. Indeed, the water molecules that enter into the interlayer space increase the layer-to-layer distance, as well as the stability of the mineral with the increase of hydration. This is the well-established driving force of water to make it enter interlayer space. However, the driving forces become less important, whereas the RH (assuming water activity) becomes more important as layer stacking becomes more turbostratic and less regular. Moreover, Churchman et al. (2015) explained that octahedral sheets can be positively charged up to a 8.5 pH value (assuming octahedral OH variable charge), whereas the tetrahedral sheets remain negatively charged. These octahedral positive charges can provide a driving force for the uptake of water into the interlayer space. Hendricks and Jefferson (1938) proposed an icelike H_2O model in which the interlayer water molecules in halloysite (10 Å) form hydrogen bonds with (i) another molecule, (ii) oxygens of the siloxane plane and (iii) hydroxyls of the opposite aluminol plane. Other investigations by Tarasevich and Bribina (1972), Jemai et al. (1999, 2000), Smirnov and Bougeard (1999), and Joussein et al. (2006) reported that two types of water in the interlayer space of halloysite can be described: the 'hole water' and the 'associated water'. The 'hole water' molecules are adsorbed on the tetrahedral sheet surface and characterised by two main different orientations forming hydrogen bonds with the basal oxygens, whereas the 'associated water' or icelike configuration forms an intermediate water layer in the halloysite interlayer space and has a high degree of mobility at ambient temperatures. Hole water is lost more slowly than associated water (Joussein et al., 2006). This model would explain the dynamic properties of interlayer water in halloysite (10 Å). However since dehydration takes place quite differently in spheroidal and tubular items, the model is clearly morphological dependant. Bailey (1990) suggested the potential presence of cation in the interlayer space associated with the water molecules. Joussein et al. (2006) showed by XRD the presence of interlayer Ca^{2+} cations associated with the strongly held 'hole' water in the spheroidal halloysite sample, which was previously Ca^{2+} saturated (Fig. 2.5). This latter point is crucial in the exchange properties of halloysite, as well as in their behaviour in response to intercalation.

2.5 CONCLUDING REMARKS

This chapter clearly discusses the presence of a number of deposits throughout the world. However, only a few deposits have relatively pure samples with interesting tube morphology for the potential use in nano-applications. This latter point can be (or is being) counterbalanced by the development of industrial processes to purify samples. Even if up-to-date research shows a great advance on the structure knowledge of halloysite (ie, why it is a curved lamellar mineral and how it rolls compared to other kaolin polymorphs), more investigations of the structure (as well as of the substance in relation to chemical composition) are needed. However, the understanding of the structure and properties is advancing, specifically due to the increase in nanotechnology research. However, the properties and reactivity of halloysites are hampered by the lack of knowledge about the hydration and dehydration behaviour of these minerals. In total, the following main points emerge for using HNT such as:

(i) The lack of knowledge in the synthesis pathway. Since it is very difficult to obtain pure samples in natural form with controlled morphology, it will be interesting to search for an alternative source by synthesis, but that process remains difficult due to the metastable character of halloysite (10 or 7 Å) compared to other kaolin polymorphs. A possible way is to use tubular kaolinite that can be more easily synthesised (see Chapter 17), but halloysite exhibits specific properties that are not observed for other kaolinite (mainly intercalation and charge properties).

(ii) The differentiation of kaolinite and halloysite is still problematic and requires special attention.

(iii) The reactivity of halloysite seems to vary from sample to sample, and is greatly influenced by the hydration properties of halloysite. This is one of the main factors affecting the physicochemical properties of halloysite, such as exchange capacity or intercalation of organic molecules and drugs. Further study of the interaction between water molecules and halloysite layers are still necessary. Halloysite samples should be handled with great care and stored in humid conditions.

(iv) Future research is required to understand the possible mechanism linking the morphological, chemical, hydration and charge properties of halloysite as a way to control the release of active agents or microfibre reinforcement in polymer systems.

REFERENCES

Adamo, P., Violante, P., Wilson, M.J., 2001. Tubular and spheroidal halloysite in pyroclastic deposits in the area of the Roccamonfina volcano (Southern Italy). Geoderma 99, 295–316.

Alexander, L.T., Faust, G.T., Hendricks, S.B., Insley, H., McMurdie, H.F., 1943. Relationship of the clay minerals halloysite and endellite. Am. Mineral. 28, 1–18.

Anand, R.R., Gilkes, R.J., Armitage, T.M., Hillyer, J.W., 1985. Feldspar weathering in a lateritic saprolite. Clay Clay Miner. 33, 31–43.

Askenasy, P.E., Dixon, J.B., McKee, T.R., 1973. Spheroidal halloysite in Guatemalan soil. Soil Sci. Soc. Am. Pro. 37, 799–803.

Bailey, S.W., 1969. Polytypism of trioctahedral 1:1 layer silicates. Clay Clay Miner. 17, 355–371.

Bailey, S.W., 1990. Halloysite—a critical assessment. In: Farmer, V.C., Tardy, Y. (Eds.), Surface Chemistry Structure and Mixed Layering of Clays. Proceedings of the 9th International Clay Conference 1989, Strasbourg, France. Sciences Géologiques, vol. 86, pp. 89–98.

Baioumy, H.M., Hassan, M.S., 2004. Authigenic halloysite from El-Gideda iron ore, Bahria Oasis, Egypt: characterization and origin. Clay Miner. 39, 207–217.

Bates, T.F., Hildebrand, F.A., Swineford, A., 1950. Morphology and structure of endellite and halloysite. Am. Mineral. 35, 463–484.

Berthier, P., 1826. Analyse de l'halloysite. Ann. Chim. Phys. 32, 332–335.

Bobos, I., Gomes, C., 1998. Greisen and post-greisen alteration in the Sao Vicente de pereira kaolinite deposit, Portugal. Can. Mineral. 36, 1615–1624.

Boden, T., Vanden Berg, M., Krahulec, K., Tabet, D., 2012. Utah's extractive resource industries 2012: Utah Geological Survey. Circular 116, Natural Resources Map and Bookstore, p. 17.

Bookin, A.S., Drits, V.A., Plançon, A., Tchoubar, C., 1989. Stacking faults in kaolin-group minerals in the light of real structural features. Clay Clay Miner. 37, 297–307.

Bordeepong, S., Bhongsuwan, D., Pungrassami, T., Bhongsuwan, T., 2012. Mineralogy, chemical composition and ceramic properties of clay deposits in southern Thailand. Kasetsart J. (Nat. Sci.) 46, 485–500.

Bradley, W.F., 1945. Diagnostic criteria for clay minerals. Am. Mineral. 30, 704–713.

Brindley, G.W., 1951. The nomenclature of clay minerals. Am. Mineral. 36, 370–371.

Brindley, G.W., 1961. Kaolin, serpentine and kindred minerals. In: Brown, G. (Ed.), The X-ray Identification and Crystal Structures of Clay Minerals. Mineralogical Society of London, London, pp. 51–131.

Brindley, G.W., 1980. Order-disorder in the clay mineral structures. In: Brindley, G.W., Brown, G. (Eds.), Crystal Structures of Clay Minerals and Their X-ray Identification. Mineralogical Society, London, pp. 125–196.

Brindley, G.W., Pedro, G., 1976. Meeting of the Nomenclature Committee of A.I.P.E.A., Mexico City, July 21, 1975. AIPEA Newsletter No. 12, pp. 5–6.

Carson, C.D., Kunze, G.W., 1970. New occurrences of tabular halloysite. Soil Sci. Soc. Am. Proc. 34, 538–540.

Chadwick, O.A., Chorover, J., 2001. The chemistry of pedogenic thresholds. Geoderma 100, 321–353.

Chadwick, O.A., Olson, C.G., Hendricks, D.M., Kelly, E.F., Gavenda, R.T., 1994. Quantifying climatic effects on mineral weathering and neoformation in Hawaii. In: Sanchez, P.A., Van Houten, H. (Eds.), Proceedings of the 15th International Soil Science Congress, Acapulco, Mexico, pp. 94–105.

Chadwick, O.A., Gavenda, R.T., Kelly, E.F., Ziegler, K., Olson, C.G., Elliott, W.C., Hendricks, D.M., 2003. The impact of climate on the biogeochemical functioning of volcanic soils. Chem. Geol. 202, 195–223.

Chukhrov, F.V., Zvyagin, B.B., 1966. Halloysite, a crystallochemically and mineralogically distinct species. In: Heller, L., Weiss, A. (Eds.), Proceedings of the International Clay Conference. Israel Program for Scientific Translation, Jerusalem, Israel, pp. 11–25.

Churchman, G.J., 1990. Relevance of different intercalation tests for distinguishing halloysite from kaolinite in soils. Clay Clay Miner. 38, 591–599.

Churchman, G.J., 2000. The alteration and formation of soil minerals by weathering. In: Sumner, M.E. (Ed.), Handbook of Soil Science. CRC Press, Boca Raton, FL, pp. 3–76.

Churchman, G.J., Carr, R.M., 1975. The definition and nomenclature of halloysites. Clay Clay Miner. 23, 382–388.

Churchman, G.J., Gilkes, R.J., 1989. Recognition of intermediates in the possible transformation of halloysites to kaolinites in weathering profiles. Clay Miner. 24, 579–590.

Churchman, G.J., Theng, B.K.G., 1984. Interactions of halloysites with amides: mineralogical factors affecting complex formation. Clay Miner. 19, 161–175.

Churchman, G.J., Whitton, J.S., Claridge, G.G.C., Theng, B.K.G., 1984. Intercalation method using formamide for differentiating halloysite from kaolinite. Clay Clay Miner. 32, 241–248.

Churchman, G.J., Davy, T.J., Aylmore, L.A.G., Gilkes, R.J., Self, P.G., 1995. Characteristics of fine pores in some halloysites. Clay Miner. 30, 89–98.

Churchman, G.J., Pontifex, I.R., McClure, S.G., 2010. Factors influencing the formation and characteristics of halloysites or kaolinites in granitic and tuffaceous saprolites in Hong Kong. Clay Clay Miner. 58, 220–237.

Churchman, J., Pasbakhsh, P., Lowe, D., Theng, B.K.G., 2015. Water, pH and iron: are these the keys to selecting halloysites for applications as nanotubes? Book of abstract, Euroclay 2015, 5th-10th July, Edinburgh, 2015, 68.

Cravero, F., Maiza, P.J., Marfil, S.A., 2012. Halloysite in Argentinian deposits: origin and textural constraints. Clay Miner. 47, 329–340.

Dahlgren, R.A., Shoji, S., Nanzyo, M., 1994. Mineralogical characteristics of volcanic ash soils. In: Shoji, S., Nanzyo, M., Dahlgren, R.A. (Eds.), Volcanic Ash Soils—Genesis, Properties and Utilization. Elsevier, Amsterdam, pp. 101–143.

Delvaux, B., Herbillon, A.J., Vielvoye, L., 1989a. Characterization of a weathering sequence of soils derived from volcanic ash in Cameroon. Taxonomic, mineralogical and agronomic implication. Geoderma 45, 375–388.

Delvaux, B., Mestdagh, M.M., Vielvoye, L., Herbillon, A.J., 1989b. XRD, IR and ESR study of experimental alteration of Al-nontronite into mixed-layer kaolinite/smectite. Clay Miner. 24, 617–630.

Delvaux, B., Tessier, D., Herbillon, A.J., Burtin, G., Jaunet, A.M., Vielvoye, L., 1992. Morphology, texture, and microstructure of halloysitic soil clays as related to weathering and exchangeable cation. Clay Clay Miner. 40, 446–456.

de Oliveira, M., Petit, S., Grauby, O., Formoso, M.L.L., Trescases, J.J., 1997. Characterization and distribution of halloysitic clay minerals in weathered basalts (southern Parana basin, Brazil). An. Acad. Bras. Cien. 69, 179–192.

de Putter, T., André, L., Bernard, A., Dupuis, C., Jedwab, J., Nicaise, D., Perruchot, A., 2002. Trace element (Th, U, Pb, REE) behaviour in a crytokarstic halloysite and kaolinite deposit from Southern Belgium: importance of "accessory" mineral formation for radioactive pollutant trapping. Appl. Geochem. 17, 1313–1328.

de Souza Santos, P., 1993. The use of particle clay morphology studies to characterise industrial clay deposits: examples from Brazil. Clay Miner. 28, 539–553.

de Souza Santos, P., de Souza Santos, H., Brindley, G.W., 1964. Mineralogical studies of kaolinite-halloysite clays: part II. Some Brazilian kaolins. Am. Mineral. 49, 1543–1548.

de Souza Santos, P., Brindley, G.W., de Souza Santos, H., 1965. Mineralogical studies of kaolinite-halloysite clays: part III. A fibrous kaolin mineral from piedade, São Paulo, Brazil. Am. Mineral. 50, 619–628.

de Souza Santos, P., de Souza Santos, H., Brindley, G.W., 1966. Mineralogical studies of kaolinite-halloysite clays: part IV. A platy mineral with structural swelling and shrinking characteristics. Am. Mineral. 51, 1640–1648.

de Souza Santos, P., Perche Toledo, S., de Souza Santos, H., 2009. Caulins Haloisíticos das Regiões Sudeste e Sul do Brasil. Ceram. Ind. 14, 14–20.

Dixon, J.B., McKee, T.R., 1974. Internal and external morphology of tubular and spheroidal halloysite particles. Clay Clay Miner. 22, 127–137.

Drews-Armitage, S.P., Romberger, S.B., Whitney, C.G., 1996. Clay alteration and gold deposition in the genesis and blue star deposits, Eureka County, Nevada. Econ. Geol. 91, 1383–1393.

Dupuis, C., Ertus, R., 1995. The karstic origin of the type halloysite in Belgium. In: Churchman, G.J., Fidzpatrick, R.W., Eggleton, R.A. (Eds.), Clays Control the Environment, Proceeding of the 10th International Clay Conference, 1993, Adelaide, Australia, pp. 262–366.

Ece, O.I., Schroeder, P.A., Smilley, M.J., Wampler, J.M., 2008. Acid-sulphate hydrothermal alteration of andesitic tuffs and genesis of halloysite and alunite deposits in the Biga Peninsula, Turkey. Clay Miner. 43, 281–315.

Ece, Ö.I., Ekinci, B., Schroeder, P.A., Crowe, D., Esenli, F., 2013. Origin of the Düvertepe kaolin-alunite deposits in Simav Graben, Turkey: timing and styles of hydrothermal mineralization. J. Volcanol. Geoth. Res. 255, 57–78.

Farmer, V.C., 1974. The layer silicates. In: Farmer, V.C. (Ed.), The Infrared Spectra of Minerals. Mineralogical Society, London, UK, pp. 331–363.

Farmer, V.C., McHardy, W.J., Palmeli, F., Violante, A., Violante, P., 1991. Synthetic allophanes formed incalcareous environments: nature, conditions of formation and transformation. Soil Sci. Soc. Am. J. 55, 1162–1166.

Faust, G.T., 1955. The endellite-halloysite nomenclature. Am. Mineral. 40, 1110–1118.

Fieldes, M., 1955. Clay mineralogy of New Zealand soils. Allophane and related mineral colloids. New Zeal. J. Sci. Tech. B36, 140–154.

Frost, R.L., 1998. Hydroxyl deformation in kaolins. Clay Clay Miner. 46, 280–289.

Galán, E., 2006. Genesis of clay minerals. In: Bergaya, F., Theng, B.K.G., Lagaly, G. (Eds.), Handbook of Clay Science. Elsevier, Amsterdam, pp. 1129–1162.

Giese, R.F., 1988. Kaolin minerals: structures and stabilities. In: Bailey, S.W. (Ed.), Hydrous Phyllosilicates (Exclusive of Micas). Review in Mineralogy, vol. 19. Mineralogical Society of America, Chelsea, MI, pp. 29–66.

Harrison, J.L., Greenberg, S.S., 1962. Dehydration of fully hydrated halloysite from Lawrence County, Indiana. In: Swineford, A. (Ed.), Proceeding of the 9th National Conference on Clays and Clay Minerals, 1960. Purdue University, Lafayette, IN, pp. 374–377.

Hart, R.D., Gilkes, R.J., Siradz, S., Singh, B., 2002. The nature of soil kaolins from Indonesia and Western Australia. Clay Clay Miner. 50, 198–207.

Harvey, C.C., Murray, H.H., 1997. Industrial clays in the 21st century: a perspective of exploration, technology and utilization. Appl. Clay Sci. 11, 285–310.

Hendricks, S.B., 1938. On the crystal structure of the clay minerals: dickite, halloysite and hydrated halloysite. Am. Mineral. 23, 295–301.

Hendricks, S.B., Jefferson, M.E., 1938. Structure of kaolin and talc-pyrophillite hydrates and their bearing on water sorption of the clays. Am. Mineral. 23, 863–875.

Hewitt, A.E., Churchman, G.J., 1982. Formation, chemistry and mineralogy of soils from weathered schist, Eastern Otago, New Zealand. New Zeal. J. Sci. 25, 253–269.

Hillier, S., Ryan, P.C., 2002. Identification of halloysite (7 Å) by ethylene glycol solvation: the 'MacEwan effect'. Clay Miner. 37, 487–496.

Hofmann, U., Endell, K., Wilm, D., 1934. Röntgenographische und kolloidchemische Untersuchungen über Ton. Angew. Chem. 47, 539–547.

Honjo, G., Kitamura, N., Mihama, K., 1954. A study of clay minerals by means of single-crystal electron diffraction diagrams—the structure of tubular kaolin. Clay Miner. Bull. 2, 133–141.

Hughes, I.R., 1966. Mineral changes of halloysite on drying. New Zeal. J. Sci. 9, 103–113.

Imbert, T., Desprairies, A., 1987. Neoformation of halloysite on volcanic glass in a marine environment. Clay Miner. 22, 179–185.

Inoue, A., Utada, M., Hatta, T., 2012. Halloysite-to-kaolinite transformation by dissolution and recrystallization during weathering of crystalline rocks. Clay Miner. 47, 373–390.

Iriarte, I., Petit, S., Huertas, F.J., Fiore, S., Grauby, O., Decarreau, A., Linares, J., 2005. Synthesis of kaolinite with a high level of Fe^{3+} for A1 substitution. Clay Clay Miner. 53, 1–10.

Jemai, S., Ben Haj Amara, A., Ben Brahim, J., Plançon, A., 1999. Etude structurale par diffraction des RX et spectroscopie IR des hydrates 10 et 8.4 Å de kaolinite. J. Appl. Crystallogr. 32, 968–976.

Jemai, S., Ben Haj Amara, A., Ben Brahim, J., Plançon, A., 2000. Etude structurale d'un hydrate 10 Å instable de kaolinite. J. Appl. Crystallogr. 33, 1075–1081.

Jeong, G.Y., Kim, S.J., 1993. Boxwork fabric of halloysite-rich kaolin formed by weathering of anorthosite in the Sancheong area, Korea. Clay Clay Miner. 41, 56–65.

Joussein, E., Petit, S., Decarreau, A., 2001. Une nouvelle méthode de dosage des minéraux argileux en mélange par spectroscopie IR—a new method to determine the ratio of clay minerals in mixtures by IR spectroscopy. C. R. Acad. Sci. IIA 332, 83–89.

Joussein, E., Petit, S., Churchman, J., Theng, B., Righi, D., Delvaux, B., 2005. Halloysite clay minerals—a review. Clay Miner. 40, 383–426.

Joussein, E., Petit, S., Fialips, C.-I., Vieillard, P., Righi, D., 2006. Differences in the dehydration-rehydration behavior of halloysites: new evidence and interpretations. Clay Clay Miner. 54, 473–484.

Joussein, E., Petit, S., Delvaux, B., 2007. Behavior of halloysite clay under formamide treatment. Appl. Clay Sci. 35, 17–24.

Karpoff, A., 1992. Cenozoic and mesozoic sediment from the Pigafetta Basin. In: Larson, R., Lancelot, Y. et al., (Eds.), Proceedings of the Ocean Drilling Program, ODP, College Station, TX. Scientific Results, vol. 129, pp. 3–30.

Kautz, C.Q., Ryan, P.C., 2003. The 10 Å to 7 Å halloysite transition in a tropical soil sequence, Costa Rica. Clay Clay Miner. 51, 252–263.
Kawano, M., Tomita, K., 2001. TEM-EDX study of weathered layers on the surface of volcanic glass, bytownite, and hypersthene in volcanic ash from Sakurajima volcano, Japan. Am. Mineral. 86, 284–292.
Keeling, J.L., Pasbakhsh, P., Churchman, G.J., 2012. Halloysite from the Eucla Basin, South Australia—comparison of physical properties and potential new uses. In: Broekmans, M. (Ed.), Proceeding of the 10th International Congress for Applied Mineralogy (ICAM), 2011, Trondheim, Norway, pp. 351–359.
Kempe, S., Bauer, I., Henschel, H.V., 2003. Paauhau civil defense cave on Mauna Kea, Hawaii, a lava tube modified by water erosion. J. Cave Karst Stud. 65, 76–85.
Kirkman, J.H., 1977. Possible structure of halloysite disks and cylinders observed in some New Zealand rhyolitic tephras. Clay Miner. 12, 199–216.
Kirkman, J.H., 1981. Morphology and structure of halloysite in New Zealand tephras. Clay Clay Miner. 29, 1–9.
Kogure, T., Mori, K., Kimura, Y., Takai, Y., 2011. Unraveling the stacking structure in tubular halloysite using a new TEM with computer-assisted minimal-dose system. Am. Mineral. 96, 1776–1780.
Kogure, T., Mori, K., Drits, V.A., Takai, Y., 2013. Structure of prismatic halloysite. Am. Mineral. 98, 1008–1016.
Kohyama, N., Fukushima, K., Fukami, A., 1978. Observation of the hydrated form of tubular halloysite by an electron microscope equipped with an environmental cell. Clay Clay Miner. 26, 25–40.
Komarneni, S., Fyfe, C.A., Kennedy, G.J., 1985. Order-disorder in 1:1 type clay minerals by solid-state ^{27}Al and ^{29}Si magic-angle-spinning NMR spectroscopy. Clay Miner. 20, 327–334.
Kunze, G.W., Bradley, W.F., 1964. Occurrence of a tabular halloysite in a Texas soil. Clay Clay Miner. 12, 523–527.
Lee, S.Y., Gilkes, R., 2005. Groundwater geochemistry and composition of hardpans in southwestern Australian regolith. Geoderma 126, 59–84.
Loughnan, F.C., Roberts, F.I., 1981. The natural conversion of ordered kaolinite to halloysite (10 Å) at Burning Mountain near Wingen, New South Wales. Am. Mineral. 66, 997–1005.
Lutynski, M., Sakiewicz, P., Gonzalez Gonzalez, M.A., 2014. Halloysite as mineral adsorbent of CO_2—kinetics and adsorption capacity. Inzy. Min. 33, 111–117.
MacEwan, D.M.C., 1947. The nomenclature of the halloysite minerals. Mineral. Mag. 28, 36–44.
MacEwan, D.M.C., 1948. Complexes of clay with organic compounds. I. Complex formation between montmorillonite and halloysite and certain organic liquids. J. Chem. Soc. Faraday Trans. 44, 349–367.
Marumo, K., Hattori, K.H., 1999. Seafloor hydrothermal clay alteration at Jade in the back-arc Okinawa Tough: mineralogy, geochemistry and isotope characteristics. Geochim. Cosmochim. Acta 63, 2785–2804.
Mehmel, M., 1935. Uber die struktur von halloysit und metahalloysit? Zeis. Krist. 90, 35–42 (in Dutch).
Miller, A.K., Guggenheim, S., Koster van Groos, A.F., 1991. Bond energy of adsorbed and interlayer water: kerolite dehydration at elevated pressures. Clay Clay Miner. 39, 127–130.
Nabarro, F.R.N., 1967. Theory of crystal dislocations. In: Mott, N.F., Bullard, E.C., Wilkinson, D.H. (Eds.), The International Series of Monographs on Physics. Oxford University Press, Oxford, pp. 39–42.

Nagasawa, K., 1978. Weathering of volcanic ash and other pyroclastic materials. In: Sudo, T., Shimoda, S. (Eds.), Clays and Clay Minerals of Japan. Development in Sedimentology, vol. 26. Kodansha Ltd and Elsevier, Amsterdam, pp. 189–219.

Nagasawa, K., 1992. Geology and mineralogy of kaolinitic clay deposits around Nagoya. In: Nagasawa, K. (Ed.), Clay Minerals, Their Natural Resources and Uses. Hamamatsu, Japan, pp. 1–15.

Nagasawa, K., Miyazaki, S., 1976. Mineralogical properties of halloysite as related to its genesis. In: Bailey, S.W. (Ed.), Proceedings of the International Clay Conference, 1975. Applied Publishing Ltd, Wilmette, IL, pp. 257–265.

Nagasawa, K., Noro, H., 1987. Mineralogical properties of halloysites of weathering origin. Chem. Geol. 60, 145–149.

Newman, R.H., Childs, C.W., Churchman, G.J., 1994. Aluminium coordination and structural disorder in halloysite and kaolinite by ^{27}Al NMR spectroscopy. Clay Miner. 29, 305–312.

Noro, H., 1986. Hexagonal platy halloysite in an altered tuff bed, Komaki city, Aichi prefecture, Central Japan. Clay Miner. 21, 401–415.

Noro, H., Yamada, K., Suzuki, K., 1981. An application of electron probe microanalysis for clay minerals. Kobut. Zas. 15, 42–54 (in Japanese).

Norrish, K., 1995. An unusual fibrous halloysite. In: Churchman, G.J., Fitzpatrick, R.W., Eggleton, R.A. (Eds.), Clays Control the Environment, Proceedings of the 10th International Clay Conference, 1993, Adelaide, Australia, pp. 275–284.

Pampeyan, E.H., 1989. Geologic Map of the Lynndal 30- by 60-Minutes QuadrangleUS Geological Survey. Department of the Interior, US Government Printing Office, West-central Utah. 0-234-086.

Parfitt, R.L., 1990. Allophane in New Zealand: a review. Aust. J. Soil Res. 28, 343–360.

Parfitt, R.L., Russell, M., Orbell, G.E., 1983. Weathering sequence of soils from volcanic ash involving allophane and halloysite, New Zealand. Geoderma 29, 41–57.

Parham, W.E., 1969. Formation of halloysite from feldspar: low temperature artificial weathering versus natural weathering. Clay Clay Miner. 17, 13–22.

Pasbakhsh, P., Churchman, G.J., 2015. Natural Mineral Nanotubes: Properties and Applications. Apple Academic Press, Canada, 498 pp.

Pasbakhsh, P., Churchman, G.J., Keeling, J.L., 2013. Characterisation of properties of various halloysites relevant to their use as nanotubes and microfibre fillers. Appl. Clay Sci. 74, 47–57.

Perruchot, A., Dupuis, C., Brouard, E., Nicaise, D., Ertus, R., 1997. L'halloysite karstique: comparaison des gisements types de Wallonie (Belgique) et du Périgord (France). Clay Miner. 32, 271–287.

Polyak, V.J., Guven, N., 2000. Clays in caves of the Guadalupe mountains, New Mexico. J. Cave Karst Stud. 62, 120–126.

Proust, D., Caillaud, J., Fontaine, C., 2006. Clay minerals in early amphibole weathering: tri- to dioctahedral sequence as a function of crystallization sites in the amphibole. Clay Clay Miner. 54, 351–362.

Quantin, P., Herbillon, A.J., Janot, C., Sieffermann, G., 1984. L'halloysite blanche riche en fer de Vate (Vanuatu)—hypothèse d'un édifice interstratifié halloysite-hisengérite. Clay Miner. 19, 629–643.

Robert, M., Herbillon, A.J., 1990. Application aux argiles des sols. Génèse, nature et rôle des constituants argileux dans les principaux types de sols des environnements volcaniques insulaires. In: Decarreau, A. (Ed.), Matériaux argileux. Structure, Propriétés et Applications. Société Française de Minéralogie et de Cristallographie, Paris, France, pp. 539–576.

Robertson, I.D.M., Eggleton, R.A., 1991. Weathering of granitic muscovite to kaolinite and halloysite and plagioclase-derived kaolinite to halloysite. Clay Clay Miner. 39, 113–126.

Romero, R., Robert, M., Elsass, F., Garcia, C., 1992. Abundance of halloysite neoformation in soils developed from crystalline rocks. Contribution of transmission electron microscopy. Clay Miner. 27, 35–46.

Ross, T.J., Kerr, P.F., 1934. Halloysite and allophoneUS Geological Survey. Department of the Interior. 185-G, pp. 135–148.
Saigusa, M., Shoji, S., Kato, T., 1978. Origin and nature of halloysite in ando soils from Towada tephra, Japan. Geoderma 20, 115–129.
Sakharov, B.A., Drits, V., 1973. Mixed-layer kaolinite-montmorillonite: a comparison of observed and calculated diffraction patterns. Clay Clay Miner. 21, 15–17.
Saklar, S., Agrili, H., Zimitoglu, O., Başara, B., Kaan, U., 2012. The characterization studies of the Northwest Anatolian halloysites/kaolinites. Bull. Mineral. Res. Explor. 145, 48–61.
Sand, L.B., 1956. On the genesis of residual kaolin. Am. Mineral. 41, 28–40.
Schroeder, P.A., Shiflet, J., 2000. Ti-bearing phases in the Huber Formation: an east Georgia kaolin deposit. Clay Clay Miner. 48, 151–158.
Sedov, S.N., Solleiro-Rebolledo, E., Gama-Castro, J.E., 2003. Andosol to Luvisol evolution in Central Mexico: timing, mechanisms and environmental setting. Catena 54, 495–513.
Sieffermann, G., Millot, G., 1968. L'halloysite des sols jeunes sur basaltes récents du centre Cameroun. Bull. Groupe Français des Argiles 20, 25–38.
Singer, A., Zarei, M., Lange, F.M., Stahr, K., 2004. Halloysite characteristics and formation in the northern Golan Heights. Geoderma 123, 279–295.
Singh, B., 1996. Why does halloysite roll? A new model. Clays Clay Miner. 44, 191–196.
Singh, B., Gilkes, R.J., 1992. An electron optical investigation of the alteration of kaolinite to halloysite. Clay Clay Miner. 40, 212–229.
Singh, B., Mackinnon, I.D.R., 1996. Experimental transformation of kaolinite to halloysite. Clay Clay Miner. 44, 825–834.
Smirnov, K.S., Bougeard, D., 1999. A molecular dynamics study of structure and short-time dynamics of water in kaolinite. J. Phys. Chem. B103, 5266–5273.
Smykatz-Kloss, W., 1974. The determination of the degree of (dis-)order of kaolinites by means of differential thermal analysis. Chem. Erde 33, 358–364.
Soma, M., Churchman, G.J., Theng, B.K.G., 1992. X-ray photoelectron spectroscopic analysis of halloysites with different composition and particle morphology. Clay Miner. 27, 413–421.
Sudo, T., 1953. Particle shape of a certain clay of hydrated halloysite, as revealed by electron microscope. Min. J. 1, 66–68.
Sudo, T., Yotsumoto, H., 1977. The formation of halloysite tubes from spherulitic halloysite. Clay Clay Miner. 25, 155–159.
Sudo, T., Shimoda, S., Yotsumoto, H., Aita, S., 1981. Electron micrographs of clay minerals. In: Sudo, T., Shimoda, S., Yotsumoto, H., Aita, S. (Eds.), Developments in Sedimentology, vol. 31. Elsevier Sci. Publ. Co. Kodansha Ltd, Tokyo, pp. 141–142.
Takahashi, H., 1958. Structural variations of kaolin minerals. Bull. Chem. Soc. Jpn. 31, 275–282.
Takahashi, T., Dahlgren, R.A., Van Susteren, P., 1993. Clay mineralogy and chemistry of soils formed in volcanic materials in the xeric moisture regime of northern California. Geoderma 59, 131–150.
Tarasevich, Y.I., Bribina, I.A., 1972. Infrared spectroscopy study of the state of water in halloysite. Kolloidn. Zh. 34, 405–411 (in Russian).
Tazaki, K., 1979. Micromorphology of halloysites produced by weathering of plagioclase in volcanic ash. In: Mortland, M.M., Farmer, V.C. (Eds.), Proceedings of the 6th International Clay Conference, 1978. Developments in Sedimentology, vol. 27. Elsevier, London, pp. 415–422.
Tazaki, K., 1982. Analytical electron microscopic studies of halloysite formation processes: morphology and composition of halloysite. In: Van Olphen, H., Veniale, F. (Eds.), Proceedings of the 7th International Clay Conference, 1981. Developments in Sedimentology, vol. 35. Elsevier, Amsterdam, pp. 573–584.

Theng, B.K.G., Churchman, G.J., Whitton, J.S., Claridge, G.G.C., 1984. Comparison of intercalation methods for differentiating halloysite from kaolinite. Clay Clay Miner. 32, 249–258.

Tomura, S., Shibasaki, Y., Mizuta, H., 1983. Spherical kaolinite: synthesis and mineralogical properties. Clay Clay Miner. 31, 413–421.

Tomura, S., Shibasaki, Y., Mizuta, H., Kitamura, M., 1985. Growth conditions and genesis of spherical and platy kaolinite. Clay Clay Miner. 33, 200–206.

Townsend, M.G., Masters, S., 2002. Northland halloysite—past, present and future. In: Proceedings of the AusIMM 2002 Annual Conference. Australasian Institute of Mining and Metallurgy Publication Series 6/02, pp. 139–142.

Townsend, M.G., Luke, K.A., Evans, R.B., 2006. Recent developments in the exploration and uses of halloysite clay deposits in Northland. In: Christie, A.B., Brathwaite, R.L. (Eds.), Geology and Exploration of New Zealand Mineral Deposits. The Australasian Institute of Mining and Metallurgy Monograph, vol. 25, pp. 59–64.

Wada, K., 1961. Lattice expansion of kaolin minerals by treatment with potassium acetate. Am. Mineral. 46, 78–91.

Wada, K., Kakuto, Y., 1985. Embryonic halloysites in Ecuadorian soils derived from volcanic ash. Soil Sci. Soc. Am. J. 49, 1309–1318.

Wada, S.I., Mizota, C., 1982. Iron-rich halloysite (10 Å) with crumpled lamellar morphology from Hokkaido, Japan. Clay Clay Miner. 30, 315–317.

Ward, C., Roberts, F.I., 1990. Occurrence of spherical halloysite ion bituminous coals of the Sydney basin, Australia. Clay Clay Miner. 38, 501–506.

Watanabe, T., Sudo, T., 1969. Study on small-angle scattering of some clay minerals. In: Heller, L. (Ed.), Proceedings of the International Clay Conference, 1969. Israel University Press, Jerusalem, pp. 173–181.

West, S.L., White, G.N., Deng, Y., McInnes, K.J., Juo, A.S.R., Dixon, J.B., 2004. Kaolinite, halloysite, and iron oxide influence on physical behavior of formulated soils. Soil Sci. Soc. Am. J. 68, 1452–1460.

White, G.N., Dixon, J.B., 2002. Kaolins and serpentines. In: Dixon, J.B., Schulze, D.G. (Eds.), Soil Mineralogy with Environmental Applications. SSSA Book, vol. 7, Soil Science Society of America, Madison, WI, pp. 389–411.

Wilke, B.M., Schwertmann, U., Murad, E., 1978. An occurrence of polymorphic halloysite in granite saprolite of the Bayerischer Wald, Germany. Clay Miner. 13, 66–67.

Wilson, I.R., 2004a. Kaolin and halloysite deposits of China. Clay Miner. 39, 1–15.

Wilson, I.R., 2004b. Special clays. Ind. Mineral. Mag. 54–61.

Wilson, I.R., 2015. Worldwide halloysite deposits, production and market developments. Industrial Minerals Magazine. IM Staff Published, 28p.

Wilson, M.J., Tait, J.M., 1977. Halloysite in some soils from north-east Scotland. Clay Miner. 12, 59–66.

Wilson, I.R., de Souza Santos, H., de Souza Santos, P., 2006. Kaolin and halloysite deposits of Brazil. Clay Miner. 41, 697–716.

Yörükoğlu, A., Delibaş, O., 2012. Mineral potential of Turkey. Mining Turkey 2, 18–23.

Yuan, P., Southon, P.D., Liu, Z., Green, M.E.R., Hook, J.M., Antill, S.J., Kepert, C.J., 2008. Functionalization of halloysite clay nanotubes by grafting with γ-aminopropyltriethoxysilane. J. Phys. Chem. C 112, 15742–15751.

Ziegler, K., Hsieh, J.C.C., Chadwick, O.A., Kelly, E.F., Hendricks, D.M., Savin, S.M., 2003. Halloysite as a kinetically controlled end product of arid-zone basalt weathering. Chem. Geol. 202, 461–478.

Chapter 3

Geology and Mineralogy of Imogolite-Type Materials

C. Levard* and I. Basile-Doelsch
Aix-Marseille Université, CNRS, IRD, CEREGE UM34, USC INRA, Aix en Provence, France
**Corresponding author: e-mail: levard@cerege.fr*

3.1 INTRODUCTION

While humans have controlled the synthesis of nanotubes only for the last few decades, nature has been growing them since the dawn of time. This is the case for imogolite, an aluminosilicate nanotube that forms by the weathering of volcanic fragmented rocks. Imogolite was first observed by Yoshinaga and Aomine (1962) in a clay fraction of a glassy volcanic ash soil in Japan. These authors identified a fibrous structure, although characterisation was made difficult because of the presence of other mostly amorphous phases. Miyauchi and Aomine (1966) characterised an almost pure imogolite sample by X-ray diffraction (XRD), infrared spectroscopy (IR) and transmission electron microscopy (TEM) and confirmed the fibrous aspect of imogolite. Wada and Yoshinaga (1969) estimated the diameter of the fibres to be around 10–30 nm by TEM. Additional TEM observations shows that those fibres were in fact composed of nanotubes with a diameter of about 2 nm. The first model was proposed by Cradwick et al. (1972), and the first synthesis was performed by Farmer et al. (1977). To the best of our knowledge, it is the first nanotube that was synthesised.

With a similar local structure but with a spherical shape, allophanes are also widely described, especially those of andosols. Andosols are typical dark soils of volcanic landscapes [the name is derived from the Japanese words 'an' (dark) and 'do' (soil)] that develop in glass-rich volcanic ejecta in almost any climate (WRB, 2014). The structure, as well as the formation mechanisms, of allophane are still poorly understood and are usually a catchall topic that will be discussed in this chapter, including literature data. This chapter will first review the structural properties of naturally occurring short-range ordered aluminosilicates, including imogolite and allophane. Then, the occurrence and formation of imogolite in natural environments (ie, 'geological' vs 'soil' environments) will be discussed. Finally, their reactivity and their effects on soil properties will be examined.

Developments in Clay Science, Vol. 7. http://dx.doi.org/10.1016/B978-0-08-100293-3.00003-0

3.2 STRUCTURAL PROPERTIES OF IMOGOLITE-TYPE MATERIALS

3.2.1 Imogolite

Cradwick et al. (1972) proposed the first model of the imogolite structure, which consisted of aluminosilicate ($Al_2SiO_7H_4$) single-walled nanotubes (Figs 3.1 and 3.2) with 2–3 nm external diameter, and a few being hundreds of nanometers long (Cradwick et al., 1972; Farmer et al., 1977). Imogolite is composed of a curved gibbsite layer on the outer surface and Si monomers linked to six Al inside the tube. The presence of an isolated silica tetrahedra environment (Barron et al., 1982) is very specific to the imogolite-type structure, as opposed to other clay minerals, in which Si tetrahedra are linked to form a continuous layer. This specific Si configuration [Q_0 or Q_3(3Al)] is characterised by a chemical shift at −78 ppm by ^{29}Si nuclear magnetic resonance (NMR). This specific local silicon configuration is reported as an 'imogolite local structure' (ILS) (Fig. 3.1A). More generally, all short-range ordered aluminosilicates having (in part or totally) an ILS, even with a more or less amorphous structure, are defined as imogolite-type materials (Levard et al., 2012).

3.2.2 Allophanes

The allophane structure is much more ambiguous than the imogolite structure because of its variable Al/Si atomic ratio. The most common type is Al-rich

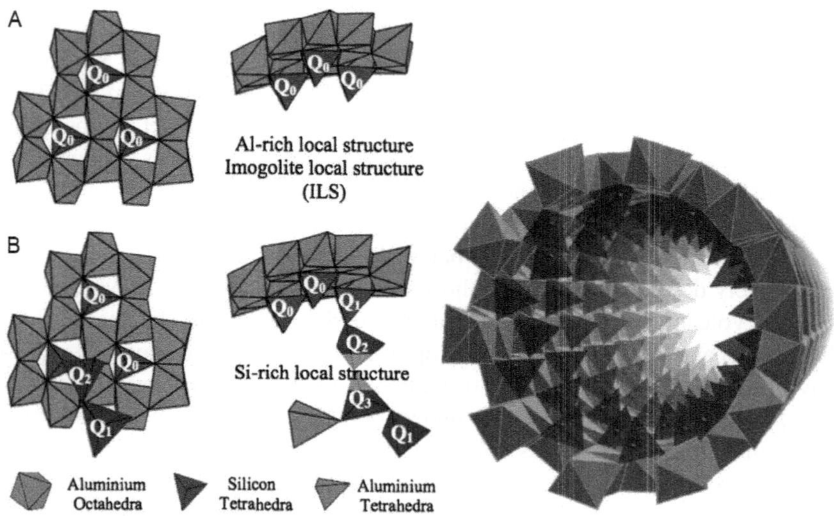

FIG. 3.1 Illustration of the structure of an imogolite nanotube. Local structures of (A) Al-rich and (B) Si-rich imogolite-type materials. Silicon sites are denoted with Q_n, where n (ranging from 0 to 4) stands for the number of directly linked silicon tetrahedrons (Levard et al., 2012).

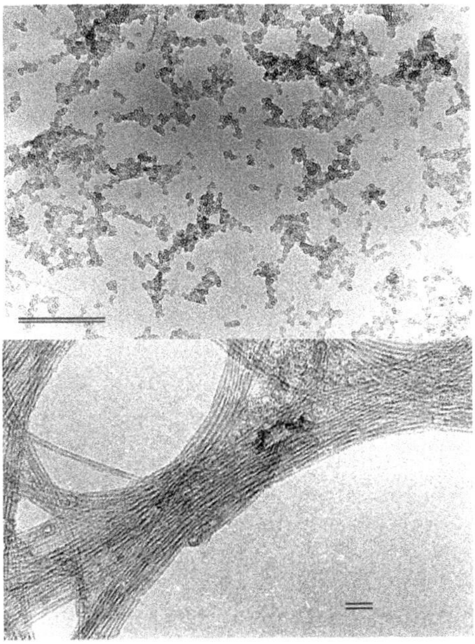

FIG. 3.2 Electron micrograph of allophane (*top*, scale bar = 100 nm) and imogolite (*bottom*, scale bar = 10 nm) from the weathered pumice from the Kitakami pumice bed in Iwate, Japan (provided generously by Wada).

allophanes (Al/Si ~ 2), sometimes named 'proto-imogolite allophanes' (Parfitt and Henmi, 1980) because their structure on a local scale is identical to the ILS (Fig. 3.1A). Allophanes with a lower Al/Si ratio, sometimes named 'Si-rich allophanes' (Al/Si ~ 1), exhibit a framework similar to that of Al-rich allophanes (ILS), with other polymerised species of Si (Q_1–Q_4) in addition to isolated Si tetrahedra (Q_0) (Fig. 3.1B). The amount of polymerized Si species increases as the Al/Si ratio decreases. As a general rule, Al-rich allophane seems to be more frequent and more stable and are readily formed in soils than Si-rich allophane (Parfitt and Kimble, 1989). On a larger scale, Al- and Si-rich allophanes are usually described as 3–5-nm spheres (Fig. 3.2; Kitagawa, 1971; Wada and Wada, 2014). Structural models have been proposed and consist of an ILS which is tentatively closed as a hollow nanosphere containing six pores due to structural constraints (Abidin et al., 2006; Creton et al., 2008).

Another type of allophanes, which are seldom observed because of their very specific location (precipitate in rivers and streams), are silica spring allophanes. They were observed for the first time in the area of Mount Ruapehu in New Zealand (Wells et al., 1977). These allophanes have an Al/Si ratio ranging from 0.9 to 1.8, and were described as being more or less complete 2–3-nm diameter spherules (Childs et al., 1990, 1999). The proposed structure consists of a polymerised SiO_4 tetrahedra framework linked to an incomplete Al octahedral layer. This specific structure does not belong to the imogolite-type material family because Si is not on a Q_0 configuration.

As stated in this introduction, the structural variability, combined with the poor crystallinity of these short-range ordered aluminosilicates, make their characterisation a real challenge. This is particularly true in andosols, where observations are complicated due to the presence of numerous other phases (organic matter (OM), iron oxides, etc.). The distinction between imogolite and allophane in andosols is mainly based on a combination of chemical extraction, NMR, XRD and Fourier-transformed infrared spectroscopy (FTIR) results (Henmi et al., 1981; Ruiz Cruz and Moreno Real, 1991; Gustafsson et al., 1995; Mossin et al., 2002; Ndayiragije and Delvaux, 2003; Hiradate et al., 2006; Gérard et al., 2007).

3.2.3 Typical Methodology Applied to Differentiate Imogolite and Allophanes

Because of their fragility under an electron beam, the typical methodology to characterise allophanes and imogolite includes a combination of chemical extraction, XRD, NMR and FTIR. The chemical method using acidified ammonium oxalate to extract Al and Si has become the standard soil science procedure for both identifying and quantifying allophanes (Parfitt, 1990; Harsh et al., 2002; Churchman and Lowe, 2012). The limits of chemical extractions are discussed in Kaufhold et al. (2009). NMR is well suited for differentiating an ILS (imogolite or Al-rich allophane) from a Si-rich local structure (Si-rich allophanes) (Table 3.1). XRD is commonly used to differentiate an imogolite structure from an allophane structure (Al- and Si-rich allophanes). Despite a poor understanding of the different reflections, reflections at around 20 and 8 Å were attributed to the tubular structure of imogolite (Cradwick et al., 1972; Table 3.1). Finally, FTIR is the only technique that can differentiate imogolite by its characteristic band at 690 cm^{-1} and doublet at 930 and 990 cm^{-1} due to the tubular structure (Table 3.1). However, the additional presence of imogolite and allophane precursors in soils is not taken into account.

3.2.4 Imogolite, Allophane, Proto-Imogolite and Proto-Allophane: Multiple Names for a Single Material?

Surprisingly, imogolite and allophane are widely described in natural systems, but their precursors (proto-imogolite and proto-allophanes) are rarely reported. In the podsolisation process, the presence of proto-imogolite was assumed as a temporary step that led to the imogolite structure (Farmer, 1987). The structure of proto-imogolite on a local scale is very similar to that reported for Al-rich allophane and imogolite (ie, ILS) (Table 3.1) in terms

TABLE 3.1 Representative Nomenclature, Occurrence and Structure of Imogolite … 2012)

	Imogolite		Allophanes			
	Imogolite	*Precursors*	*Silica-spring*	*Si-rich allo*	*Al-rich allo*	*Precursors*
Other name		Proto-imogolite	Feldspathoit allophanes		Proto-imogolite allophanes	
Natural occurrence	Pumice and andosols	Information is not available in the literature	Spring river bed	Pumice and andosols	pumice and andosols	Information is not available in the literature
Shape	Nanotubes	Tile shape particles	Spherules fragment	Spheres ovoids?	Spheres ovoids?	Information is not available in the literature

Characteristics features for commonly used techniques

	Imogolite		Allophanes			
	Imogolite	*Precursors*	*Silica-spring*	*Si-rich allo*	*Al-rich allo*	**Identification**
Chemical comp. (Al/Si)			0.9–1.8	≈1		Si-rich
	≈2	≈2			≈2 (up to 4)	Al-rich
^{29}Si NMR (ppm)	−78	−78	−78	−78	−78	Q_0
			−80 to −110	−80 to −110		Q_1 to Q_4
^{27}Al NMR (ppm)	5–6	2–6	2–6	2–6	2–6	Al^{VI}
			55	55		Al^{IV}
	930 and 995	970		1000–1100	970	Si–O stretching
FTIR (cm^{-1})	425, 500, 570	425, 500, 570	425, 500, 570	425, 500, 570	425, 500, 570	ILS
	325 and 690					Tubular
XRD (Å)	2.2 and 3.4	2.2 and 3.4	2.2 and 3.4	2.2 and 3.4	2.2 and 3.4	ILS
	8.4 and 20					(001) and (100)
TEM	Nanotubes	Amorphous	Information is not available in the literature	Spheres, rings?	Spheres, rings?	

of the silicon local environment (Q_0, NMR results), bonding vibrations (FTIR results), crystallinity (XRD results) and chemical composition (Al/Si ~ 2). Based on these four techniques, Al-rich allophanes and proto-imogolite are not dissociable. Despite these structural similarities, many studies drew conclusions about the presence of Al-rich allophanes without further characterisation data (Henmi et al., 1981; Ruiz Cruz and Moreno Real, 1991; Gustafsson et al., 1995; Mossin et al., 2002; Ndayiragije and Delvaux, 2003; Hiradate et al., 2006; Gérard et al., 2007). No studies have mentioned proto-allophane, since the allophane formation mechanisms are not clearly understood. The nature of aluminosilicates observed in soils, particularly the distinction between proto-imogolite and Al-rich allophanes, is thus still ambiguous.

3.3 OCCURRENCE AND FORMATION OF IMOGOLITE-TYPE MATERIALS IN GEOLOGIC AND PEDOLOGIC ENVIRONMENTS

Formation of imogolite and allophane are mainly presented in the research literature as the result of volcanic glass weathering. Most of the knowledge of this formation derives from studies of largely inorganic processes taking place in relatively 'clean' environments (Churchman, 2010; Churchman and Lowe, 2012). By contrast, the soil, a heterogeneous media, is a dynamic part of the biosphere, which is continually changing in response to climatic variations over time and scale spaces. The secondary minerals in soils (imogolite/allophane) can be quite different in their chemical and physical characteristics from those formed in geological environments (Churchman and Lowe, 2012).

3.3.1 Occurrence and Formation in the Geological Environment

Natural imogolite and allophane are most commonly found in tephra layers in humid climates, where volcanic glasses dissolve (Parfitt, 2009). For example, pure imogolite occurs as gel films over the surface of volcanoclastic rocks (Yoshinaga and Aomine, 1962). Using ^{27}Al and ^{29}Si NMR, the weathering process of volcanic glass to allophane was proposed by (Hiradate and Wada, 2005) as follows: (i) dissolution of Al from volcanic glass, accompanied by the transformation of Al^{IV} to Al^{VI}; (ii) formation of a gibbsite-like sheet resulting from the hydrolysis of the dissolved Al; (iii) dissolution of silica gel-like polymer in volcanic glass, resulting in the formation of monosilicic acid; and (iv) formation of $Si(OH)(OAl^{VI})_3$ structure as a result of the reaction between the gibbsite-like sheet and the monosilicic acid. These imogolite-type materials' formation reactions and their crystalline growth into either imogolite or allophane can occur in solution, as well as on the surface of volcanic glass.

Drainage conditions, as well as Si activity in solution, are the main factors controlling imogolite and allophane formation. Optimal conditions for the formation of imogolite and allophanes (Al and Si concentrations between 0.1

and 4 mM; Harsh et al., 2002) are between the optimal conditions for smectite formation (low rate of water through-flow, high Si concentration) and for gibbsite formation (high rate of water through-flow, low Si concentration) (Churchman and Lowe, 2012). The nature of the parent volcanic rock may also control the rate of imogolite-type material formation. Glass dissolution rates of basaltic (low Si content, $SiO_2 < 52\%$), andesitic (intermediate Si content, $52 < SiO_2 < 63\%$) and rhyolitic (high Si content, $SiO_2 > 63\%$) lava decrease as SiO_2 content increases. Moreover, the porosity of the lava deposit also exerts a strong control on weathering rates. Compared with hard rocks, the fragmental tephra components, especially vesicular glass and pumice, have a much greater porosity, SSA and permeability, increasing the volcanic glass weathering rate and imogolite-type material formation (Wolff-Boenisch et al., 2004).

Finally, weathering thick volcanic ash layers may lead to the formation of thick imogolite or allophane deposits. As an example, one can cite the deep 5–15-m-thick imogolite-type material layer (locally up to 80%), covering an area of more than 4000 km^2 (Santo Domingo de los Colorado, Ecuador, Kaufhold et al., 2009). Besides the volcanic glass, biotite and feldspars have been cited as minerals susceptible to forming imogolite-type materials by weathering (Parfitt and Kimble, 1989).

3.3.2 Occurrence and Formation in Soil

3.3.2.1 Andosols: An Imogolite-Type Material Mine

Imogolite-type materials have mainly been associated with andosols derived from volcanic ash. Andosols cover approximately 124 million ha, corresponding to 0.84% of the Earth's ice-free surface (McDaniel et al., 2012). They are closely associated with areas of active volcanism. As a result of intermittent volcanic activity, these soils often consist of a modern surface layer overlying a series of buried andic horizons (Fig. 3.3). The greatest concentration of andosols is found along the Pacific Ring of Fire (where Pacific Plate is subducted and associated to explosive volcanism). Numerous hotspot volcanism also forms islands in the Pacific, Atlantic and Indian Oceans where andosols are developed. The global distribution of andosols encompasses a wide variety of climatic conditions but the majority is found in higher-rainfall regions (Shoji et al., 1993).

These soils have a high potential for agricultural production and represent a very important land resource due to the high human populations living in these regions (Shoji et al., 1993). Andosols are generally fertile soils, particularly in intermediate or basic volcanic ash not exposed to excessive leaching. These soils have favourable properties for cultivation, plant roots and water storage. Andosols are planted with a wide variety of crops, including sugarcane, tobacco, sweet potato (tolerant of low phosphate levels), tea, vegetables, wheat, rice and orchard crops. Andosols on steep slopes are best kept under forest. However, in strongly hydrated andosols, the use of heavy agricultural

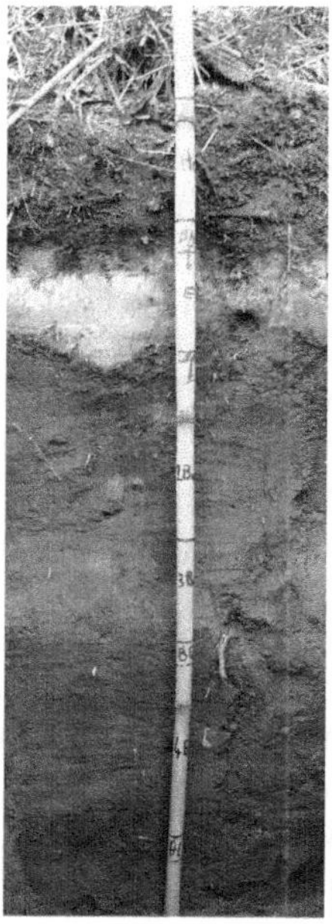

FIG. 3.3 Picture of a 200-cm-deep polyphased andosol profile of La Réunion Island (55° 21.53′E, 21°0.70′S), on the west side of the Piton des Neiges shield volcano in the Indian Ocean). Below the A_O horizon, the albic horizon (E) is composed of an accumulation of phytoliths (biogenic SiO_2), that is locally called 'Mascareignite'. The brown (grey in the print version) horizons are andic horizons (Basile-Doelsch et al., 2005, 2007).

vehicles can be complicated by the low bearing capacity and the stickiness of these soils. Moreover, the strong phosphate fixation of andosols (caused by active Al and Fe) remains a problem (Harsh et al., 2002; WRB, 2014).

A soil horizon is defined as andic according to the following diagnostic criteria (WRB, 2014): (i) the $Al_{ox} + 1/2Fe_{ox}$ is 2% or more (where 'ox' means oxalate extracted metal); (ii) bulk density is 0.9 kg/dm^3 or less; and (iii) phosphate retention is 85% or more. Imogolite-type materials, however, were identified in silandic horizons, which have an acid oxalate (pH 3) extractable silica (Si_{ox}) of 0.6% or more (or an Al_{py}/Al_{ox} of <0.5, where 'py' means pyrophosphate extracted metal). On the contrary, soils dominated by Al-humus are aluandic horizons (Si_{ox} of less than 0.6% or an Al_{py}/Al_{ox} of $\geq$0.5) and are not andosols rich in imogolite-type materials. Silandic and aluandic horizons differ in field pH: silandic horizons generally have a pH (H_2O) of 5 or higher, while aluandic horizons mainly have pH (H_2O) of less than 4.5. Once again, these chemical criteria help at predicting the presence of imogolite-type materials but do not permit any differentiation between proto-imogolite, imogolite and allophane. Most important, what makes this soil very different from tephra layers is the presence of OM in andic horizons

(5%) (WRB, 2014) that can potentially have a strong effect on the growth and structure of imogolite-type materials.

Besides andosols, imogolite-type materials occur in any environment where sufficient Si and Al exist in solution to precipitate these rapidly forming minerals. Imogolite-type materials have been identified in some horizons of podzols (soils with an eluvial horizon that has the appearance of ash, from the Russian 'pod', (underneath) and 'zola' (ash), which is directly underlaid by an alluvial horizon of black OM and reddish Fe oxide accumulation). Whether proto-imogolite forms at the soil surface and subsequently migrates into the alluvial horizon or alternatively forms in situ in the alluvial horizons has been debated (Ugolini and Dahlgren, 1991; Farmer and Lumsdon, 2001). In podzols, as in andosols, either mechanisms due to OM adsorption to imogolite-type materials or mechanisms of saturation of Al–OM complexes may occur depending on the soil conditions (Sauer et al., 2007). Imogolite-type materials have also been identified in soils developed from igneous and sedimentary rock and loess in a variety of climates and geographical regions throughout the world (Harsh et al., 2002).

3.3.2.2 *Imogolite-Type Material Formation in Soils*

Crystalline growth of imogolite and allophane have been experimentally investigated in the presence of organic compounds at around 100°C (Inoue and Huang, 1984, 1986, 1990) and in conditions relevant to soils (25°C) (Rouff et al. 2012). In most of the conditions studied, the ILS was almost systematically observed through the use of local probes (^{29}Si NMR, ^{27}Al NMR and FTIR). At low concentrations, OM would even promote the formation of imogolite-type materials (Rouff et al., 2012). Inoue and Huang (1984, 1990), however, showed by TEM that for high concentrations of humic acid (HA), the proto-imogolite–HA associations do not crystallise into imogolite.

In the andosol horizons of La Réunion island, Levard et al. (2012) suggested that poorly ordered proto-imogolite (or proto-allophane) that had reacted with organic compounds were present, rather than well-defined Al-rich allophanes and imogolite. OM thus acts as an inhibitor of the growth of allophane and imogolite. This hypothesis was also proposed based on andosol ^{27}Al NMR results that show a polymerisation state of the Al that is closer to a proto-imogolite than well-defined imogolite or allophanes (Basile-Doelsch et al., 2005, 2007).

Thus, both laboratory and soil studies suggest that organic compounds do not prevent the proto-imogolite formation; rather, they impede Al polymerisation and strongly inhibit imogolite and allophane crystalline growth (Fig. 3.4). When high OM concentrations are present, enhanced Al-humic complexation limits the ILS formation and is called the 'antiallophanic effect' (Churchman and Lowe, 2012). This effect was observed by Poulenard et al. (2003) in laboratory experiments, as well as by Rouff et al. (2012) in andosols.

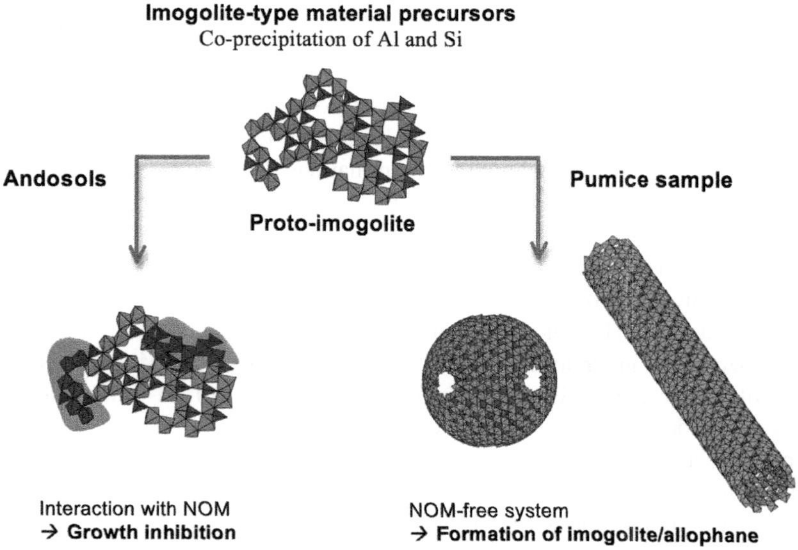

FIG. 3.4 Structures of imogolite-type materials that can be found in andosols and pumice. (*NOM*, natural organic matter). *Adapted from Levard et al. (2012).*

Soil OM content has significant control over the crystallisation state of imogolite-type materials. The allophanic andosols are characterised by the presence of proto-imogolites instead of imogolite nanotubes and allophane spheres. In most soil, the investigative methods (chemical extractions, local probes or both) do not allow differentiation between proto-imogolite and imogolite or allophane.

3.4 REACTIVITY AND EFFECT ON SOIL PROPERTIES

Imogolite-type materials are known to play a significant role in the physicochemical properties of volcanic soil and control many biogeochemical (Parfitt, 2009) processes. For example, the high hydraulic conductivity and water retention of volcanic ash soil were attributable to its highly porous structure, partly due to the presence of imogolite-type materials (Karube and Abe, 1998). A review of the properties of imogolite-type materials in term of phosphate, heavy metal retention and OM storage will be presented in the following sections.

3.4.1 Phosphate

Allophanic soil used in agriculture usually required larger amounts of phosphate fertiliser application than other types of soil because of the strong affinity of phosphate for imogolite-type materials. The interaction between phosphate- and imogolite-type materials induced the displacement of the silicon, potentially leading to the destruction of the allophanic structure and the precipitation of aluminium phosphate at high concentrations of phosphate (Rajan, 1975; Parfitt, 1989). This is why allophanic soil used in agriculture usually required larger amounts of phosphate fertiliser than other types of soil (Edmeades et al., 2006). The strong affinity of allophanes for phosphates has been used to treat effluent and sewage. Indeed, allophanes obtained from a deposit were shown to be very efficient in removing phosphorus from phosphate-rich effluent and sewage (Yuan and Wu, 2007).

3.4.2 Metals

Andosols generally exhibit high heavy metal concentrations due to their natural pedogeochemical background. Because of their high SSA, imogolite-type materials are potentially important mineral phases controlling the fate of heavy metals in those soils. Reactivity of three allophanic clay fractions with varying Al/Si ratios towards Cu(II) and Co(II) was studied, showing that both metals were chemisorbed with a higher affinity for Cu than Co (Clark and McBride, 1984). The affinity of andosols for both synthetic allophane and a natural allophanic fraction was higher than for synthetic imogolite, presumably because there are fewer sites available for chemisorption on the tubular structure of synthetic imogolite. A similar experiment with Zn was performed on both Al-rich and Si-rich allophane fractions. The amount of Zn adsorbed on Si-rich allophanes was higher than that of Al-rich allophanes. The authors assumed that this difference could be due to higher amounts of Si–OH groups in Si-rich allophanes, which causes a increase in the amount of negative charges (Ghoneim et al., 2001). The authors described the formation of both monodentate and bidentate complexes. Also, Levard et al. (2009) showed that 75% of the total Ni of an andosol in La Réunion was chemically linked to the imogolite-type materials, forming a stable complex (Fig. 3.5).

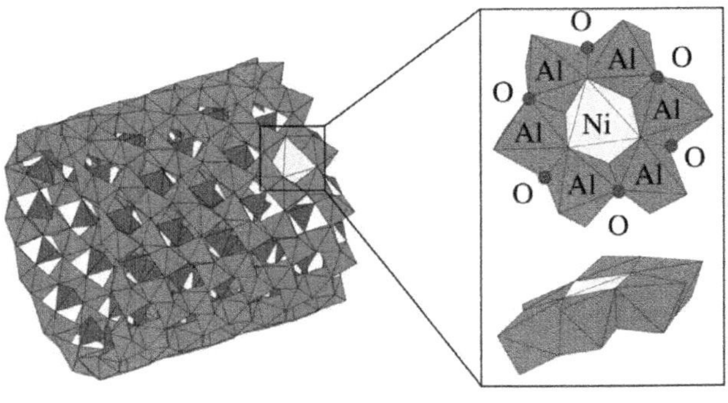

FIG. 3.5 Proposed mechanism of Ni adsorption onto an imogolite-type material (Levard et al., 2009).

Since the Fukushima nuclear accident in 2011, a new field of investigation concerns the reactivity of imogolite-type materials with radioelements such as Cesium (Cs). Due to the high surfaces occupied by andosols in Japan, it might play important roles in long-term Cs dynamics through adsorption (Fujii et al., 2014). Imogolite-type materials associated with ferrihydrite were indeed found to be the major clay mineral species of some of the upper soil layers contaminated by radioactive Cs in the Fukushima area. This surface soil layer was found to retain ^{137}Cs during at least the initial stage of contamination.

3.4.3 Organic Matter

The turnover of OM in soils is of important environmental concern, as it partially controls soil quality. In this context, the interaction of OM with minerals is an important process to be considered, as it may stabilise soil OM through strong covalent bonding, potentially enhancing soil quality. Carbon concentration in andosols can reach values as high as 20% (Wada, 1985). As an illustration, allophanic soils represent 0.8% of the global land area and contain 5% of the global soil carbon (Dahlgren et al., 2004). Allophanic soils are indeed known to accumulate significantly more OM than nonallophanic soils (Wada and Inoue, 1967). Basile-Doelsch et al. (2005) demonstrated that even though imogolite-type materials are very effective at stabilising OM, carbon associated with imogolite-type materials is still in dynamic equilibrium with its environment. Based on $\Delta^{14}C$ values, the residence time of organic carbon was estimated at about 163,000 years in an imogolite-type material-rich horizon of a volcanic ash soil from La Réunion. This value is comparable to that for organic carbon stabilised in Hawaiian volcanic soils (Torn et al., 1997).

Saggar et al. (1994) measured the carbon turnover in a range of allophanic soils spiked with ^{14}C-labelled glucose for 28 days. They found that carbon biomass was highest in soil with the highest content (13%) of imogolite-type materials and lowest in soil with less imogolite-type materials (<2%). In contrast, $^{14}CO_2$ production from the ^{14}C-labelled glucose was highest (63%) in soil with less imogolite-type material content and lowest (54%) in the soil with the most imogolite-type materials. Along the same line, carbon mineralisation rates were shown to be lower in andosols than in numerous other acid soils (Boudot et al., 1986). Furthermore, an addition of purified allophanic clay minerals to nonallophanic soils was shown to lower the carbon mineralisation rate (Martin et al., 1982; Zunino et al., 1982).

Parfitt et al. (1997) reported that after cropping for 20 years, carbon content decreased by 23 tonnes C/ha in a recent soil sample formed in alluvium (initially containing 73 tonnes C/ha), while only 10 tonnes C/ha was lost in allophanic soil (initially containing 144 tonnes C/ha). The same trend was observed in soil under dairy farming, illustrating the important role of imogolite-type materials and other Al phases in the stabilisation of OM (Schipper et al., 2007; Parfitt, 2009).

All studies agree on the fact that imogolite-type materials retard the turnover of OM by stabilising it. However, little is known about the mechanisms involved. Parfitt et al. (1999) investigated the nature of the interactions between OM and imogolite-type materials in two acidic soils from New Zealand using ^{13}C NMR. They found that imogolite-type materials interact with carboxylic groups of OM. These authors stated that this interaction is partly responsible for the stabilisation of the OM in allophanic soil. Basile-Doelsch et al. (2005) also observed a complexation of natural OM with the external Al layer of imogolite-type materials. These strong interactions have two main consequences: they stabilise the natural OM, and they inhibit the polymerisation of imogolite-type materials, leading to the formation of poorly ordered proto-imogolite natural OM complexes. More recently, density fractionation methods showed that imogolite-type materials can effectively store large amounts of OM, but at least part of the organic compounds associated with imogolite-type materials has a quite rapid turnover. These results led to consider a new model involving two types of organomineral interactions: (i) OM stabilised by strong bonds to proto-imogolite surfaces, leading to slow OM turnover; and (ii) OM retained within the porosity of the three-dimensional (3D) structure formed by imogolite-type materials (similar to a gel structure), leading to faster OM turnover (Basile-Doelsch et al., 2014).

3.5 CONCLUDING REMARKS

The mineralogy of imogolite-type materials is complex. In particular, the great similarity in terms of local structure, crystallinity and chemical composition of proto-imogolite and Al-rich allophane complicates their differentiation. However, silandic soil horizons are characterised by the presence of proto-imogolites rather than imogolite and allophanes due to their OM content. Interaction with OM is of particular concern since it controls, at least in part, the turnover of carbon in those soils. Because of their small size and high SSA, imogolite-type materials also play a significant role on water, metals and nutrient retention. Nanoscale mechanisms observed in andosols represent a typical case that highlights the role that weathering silicate phases plays in stabilising soil OM. But processes of mineral weathering also occur in other types of soils. With the aim of addressing both food security and climate change issues, understanding the mechanisms of the interactions between OM and imogolite-type materials in andosols may help to open new avenues for research geared towards gaining insight into the mechanisms of carbon storage in any type of soil (Basile-Doelsch et al., 2015). More generally, the interactions of OM with imogolite-type materials have now sparked innovative research projects in various domains, as described in chapters 12, 13, 24, and 25 of this book.

REFERENCES

Abidin, A., Matsue, N., Henmi, T., 2006. Validity of proposed model for the chemical structure of allophane with nano-ball morphology. Clay Sci. 12, 267–269.

Barron, P.F., Wilson, M.A., Campbell, A.S., Frost, R.L., 1982. Detection of imogolite in soils using solid-state Si-29 NMR. Nature 299, 616–618.

Basile-Doelsch, I., Amundson, R., Stone, W., Masiello, C., Bottero, J.Y., Colin, F., Masin, F., Borschneck, D., Meunier, J.D., 2005. Mineralogical control of organic carbon dynamics in a volcanic ash soil on La Réunion. Eur. J. Soil Sci. 56, 689–703.

Basile-Doelsch, I., Amundson, R., Stone, W.E.E., Borschneck, D., Bottero, J.Y., Moustier, S., Masin, F., Colin, F., 2007. Mineral control of carbon pools in a volcanic soil horizon. Geoderma 137, 477–489.

Basile-Doelsch, I., Derrien, D., Amundson, R., Balesdentd, J., Borschneck, D., 2014. Dynamics of organic compounds associated with non-crystalline minerals in andosols: a two-pool model. In: Sixth International Workshop on Soil Organic Matter Stabilization and Destabilization (SOM6), Charleston, South Carolina USA, on October 5–9th.

Basile-Doelsch, I., Balesdent, J., Rose, J., 2015. Are interactions between organic compounds and nanoscale weathering minerals the key drivers of carbon storage in soils? Environ. Sci. Technol. 49, 3997–3998.

Boudot, J.P., Belhadj, B.A., Chone, T., 1986. Carbon mineralization in andosols and aluminum-rich highland soils. Soil Biol. Biochem. 18, 457–461.

Childs, C.W., Parfitt, R.L., Newman, R.H., 1990. Structural studies of silica springs allophane. Clay Miner. 25, 329–341.

Childs, C.W., Hayashi, N., Newman, R.H., 1999. Five-coordinate aluminum in allophane. Clay Clay Miner. 47, 64–69.

Churchman, G.J., 2010. Is the geological concept of clay minerals appropriate for soil science? A literature-based and philosophical analysis. Phys. Chem. Earth 35, 927–940.

Churchman, G.J., Lowe, D.J., 2012. Alteration, formation, and occurrence of minerals in soils. In: Huang, P.M., Li, Y., Sumner, M.E. (Eds.), Handbook of Soil Sciences. CRC Press (Taylor & Francis), Boca Raton, FL, pp. 20.21–20.72.

Clark, C.J., McBride, M.B., 1984. Chemisorption of Cu(Ii) and Co(Ii) on allophane and imogolite. Clay Clay Miner. 32, 300–310.

Cradwick, P.D.G., Farmer, V.C., Russel, J.D., Masson, C.R., Wada, K., Yoshinaga, N., 1972. Imogolite, a hydrated aluminium silicate of tubular structure. Nature Phys. Sci. 240, 187–189.

Creton, B., Bougeard, D., Smirnov, K.S., Guilment, J., Poncelet, O., 2008. Structural model and computer modeling study of allophane. J. Phys. Chem. C 112, 358–364.

Dahlgren, R.A., Saigusa, M., Ugolini, F.C., 2004. The nature, properties and management of volcanic soils. Advances in Agronomy 82, 113–182.

Edmeades, D.C., Metherell, A.K., Waller, J.E., Roberts, A.H.C., Morton, J.D., 2006. Defining the relationships between pasture production and soil P and the development of a dynamic P model for New Zealand pastures: a review of recent developments. N. Z. J. Agric. Res. 49, 207–222.

Farmer, V.C., 1987. The role of inorganic species in the transport of aluminium in podzols. In: Righi, D., Chauvel, A. (Eds.), Podzols et Podzolization. Association française pour l'étude du sol, Plaisir, pp. 187–194.

Farmer, V.C., Lumsdon, D.G., 2001. Interactions of fulvic acid with aluminium and a proto-imogolite sol: the contribution of E-horizon eluates to podzolization. Eur. J. Soil Sci. 52, 177–188.

Farmer, V.C., Fraser, A.R., Tait, J.M., 1977. Synthesis of imogolite: a tubular aluminium silicate polymer. J. Chem. Soc. Chem. Commun, 462–463.

Fujii, K., Ikeda, S., Akama, A., Komatsu, M., Takahashi, M., Kaneko, S., 2014. Vertical migration of radiocesium and clay mineral composition in five forest soils contaminated by the Fukushima nuclear accident. Soil Sci. Plant Nutr. 60, 751–764.

Gérard, M., Caquineau, S., Pinheiro, J., Stoops, G., 2007. Weathering and allophane neoformation in soils developed on volcanic ash in the Azores. Eur. J. Soil Sci. 58, 496–515.

Ghoneim, A.M., Matsue, N., Henmi, T., 2001. Zinc adsorption on nano-ball allophanes with different Si/Al ratios. Clay Sci. 11, 337–348.

Gustafsson, J.P., Bhattacharya, P., Bain, D.C., Fraser, A.R., McHardy, W.J., 1995. Podzolisation mechanisms and the synthesis of imogolite in northern Scandinavia. Geoderma 66, 167–184.

Harsh, J., Chorover, J., Nizeyimana, E., 2002. Allophane and imogolite. In: Dixon, J.B., Schulze, D.G. (Eds.), Soil Mineralogy with Environmental Applications. SSSA Book Series, vol. 7. Soil Science Society of America, Madison, pp. 291–322.

Henmi, T., Tange, K., Minagawa, T., Yoshinaga, N., 1981. Effect of SiO2-Al2O3 ratio on the thermal-reactions of allophane.2. Infrared and X-ray-powder diffraction data. Clay Clay Miner. 29, 124–128.

Hiradate, S., Wada, S.I., 2005. Weathering process of volcanic glass to allophane determined by Al-27 and Si-29 solid-state NMR. Clay Clay Miner. 53, 401–408.

Hiradate, S., Hirai, H., Hashimoto, H., 2006. Characterization of allophanic Andisols by solid-state ^{13}C, ^{27}Al, and ^{29}Si NMR and by C stable isotopic ratio, $\Delta^{13}C$. Geoderma 136, 696–707.

Inoue, K., Huang, P.M., 1984. Influence of citric-acid on the natural formation of imogolite. Nature 308, 58–60.

Inoue, K., Huang, P.M., 1986. Influence of selected organic-ligands on the formation of allophane and imogolite. Soil Sci. Soc. Am. J. 50, 1623–1633.

Inoue, K., Huang, P.M., 1990. Perturbation of imogolite formation by humic substances. Soil Sci. Soc. Am. J. 54, 1490–1497.

IUSS Working Group WRB, 2014. World reference base for soil resources 2014: International soil classification system for naming soils and creating legends for soil maps. World Soil Resources Reports No. 106, FAO, Rome.

Karube, J., Abe, Y., 1998. Water retention by colloidal allophane and imogolite with different charges. Clay Clay Miner. 46, 322–329.

Kaufhold, S., Kaufhold, A., Jahn, R., Brito, S., Dohrmann, R., Hoffmann, R., Gliemann, H., Weidler, P., Frechen, M., 2009. A new massive deposit of allophane raw material in ecuador. Clay Clay Miner. 57, 72–81.

Kitagawa, Y., 1971. The “unit particle” of allophane. Am. Mineral. 56, 465–475.

Levard, C., Doelsch, E., Rose, J., Masion, A., Basile-Doelsch, I., Proux, O., Hazemann, J.-L., Borschneck, D., Bottero, J.-Y., 2009. Role of natural nanoparticles on the speciation of Ni in andosols of La Reunion. Geochim. Cosmochim. Acta 73, 4750–4760.

Levard, C., Doelsch, E., Basile-Doelsch, I., Abidin, Z., Miche, H., Masion, A., Rose, J., Borschneck, D., Bottero, J.Y., 2012. Structure and distribution of allophanes, imogolite and proto-imogolite in volcanic soils. Geoderma 183, 100–108.

Martin, J.P., Zunino, H., Peirano, P., Caiozzi, M., Haider, K., 1982. Decomposition of C-14-labeled lignins, model humic-acid polymers, and fungal melanins in allophanic soils. Soil Biol. Biochem. 14, 289–293.

McDaniel, P.A., Lowe, D.J., Arnalds, O., Ping, C.-L., 2012. Andisols. In: Huang, P.M., Li, Y., Sumner, M.E. (Eds.), Handbook of Soil Sciences, second ed. Properties and Processes, vol. 1. CRC Press (Taylor & Francis), Boca Raton, FL, pp. 33.29–33.48.

Miyauchi, N., Aomine, S., 1966. Mineralogy of gel-like substance in the pumice bed in kanuma and kitakami districts. Soil Sci. Plant Nutr. 12, 19–22.

Mossin, L., Mortensen, M., Nornberg, P., 2002. Imogolite related to podzolization processes in Danish podzols. Geoderma 109, 103–116.

Ndayiragije, S., Delvaux, B., 2003. Coexistence of allophane, gibbsite, kaolinite and hydroxy-Al-interlayered 2:1 clay minerals in a perudic andosol. Geoderma 117, 203–214.

Parfitt, R.L., 1989. Phosphate reactions with natural allophane, ferrihydrite and goethite. J. Soil Sci. 40, 359–369.

Parfitt, R.L., 1990. Allophane in New Zealand—a review. Aust. J. Soil Res. 28, 343–360.

Parfitt, R.L., 2009. Allophane and imogolite: role in soil biogeochemical processes. Clay Miner. 44, 135–155.

Parfitt, R.L., Henmi, T., 1980. Structure of some allophanes from New Zeland. Clay Clay Miner. 28, 285–294.

Parfitt, R.L., Kimble, J.M., 1989. Conditions for formation of allophane in soils. Soil Sci. Soc. Am. J. 53, 971–977.

Parfitt, R.L., Theng, B.K.G., Whitton, J.S., Shepherd, T.G., 1997. Effects of clay minerals and land use on organic matter pools. Geoderma 75, 1–12.

Parfitt, R.L., Yuan, G., Theng, B.K.G., 1999. A ^{13}C-NMR study of the interactions of soil organic matter with aluminium and allophane in podzols. Eur. J. Soil Sci. 50, 696–700.

Poulenard, J., Podwojewski, P., Herbillon, A.J., 2003. Characteristics of non-allophanic Andisols with hydric properties from the Ecuadorian páramos. Geoderma 117, 267–281.

Rajan, S.S.S., 1975. Phosphate adsorption and displacement of structural silicon in an allophane clay. J. Soil Sci. 26, 250–256.

Rouff, A.A., Phillips, B.L., Cochiara, S.G., Nagy, K.L., 2012. The effect of dissolved humic acids on aluminosilicate formation and associated carbon sequestration. Appl. Environ. Soil Sci. 2012, 12.

Ruiz Cruz, M.D., Moreno Real, L., 1991. Practical determination of allophane and synthetic alumina and iron oxide gels by X-ray diffraction. Clay Miner. 26, 377–387.

Saggar, S., Tate, K.R., Feltham, C.W., Childs, C.W., Parshotam, A., 1994. Carbon turnover in a range of allophanic soils amended with C-14-labeled glucose. Soil Biol. Biochem. 26, 1263–1271.

Sauer, D., Sponagel, H., Sommer, M., Giani, L., Jahn, R., Stahr, K., 2007. Podzol: soil of the year 2007. A review on its genesis, occurrence, and functions. J. Plant Nutr. Sci. 170, 581–597.

Schipper, L.A., Baisden, W.T., Parfitt, R.L., Ross, C., Claydon, J.J., Arnold, G., 2007. Large losses of soil C and N from soil profiles under pasture in New Zealand during the past 20 years. Glob. Chang. Biol. 13, 1138–1144.

Shoji, S., Nanzyo, M., Dahlgren, R., 1993. Volcanic Ash Soils: Genesis, Properties and Utilization. Elsevier, Amsterdam.

Torn, M.S., Trumbore, S.E., Chadwick, O.A., Vitousek, P.M., Hendricks, D.M., 1997. Mineral control of soil organic carbon storage and turnover. Nature 389, 170–173.

Ugolini, F.C., Dahlgren, R.A., 1991. Weathering environments and occurrence of imogolite/allophane in selected Andisols and Spodosols. Soil Sci. Soc. Am. J. 55, 1166–1171.

Wada, K., 1985. The distinctive properties of andosols. In: Stewart, B.A. (Ed.), Advances in Soil Science. Springer, New York, pp. 173–229.

Wada, K., Inoue, T., 1967. Retention of humic substances derived from rotted clover leaves in soils containing montmorillonite and allophane. Soil Sci. Plant Nutr. 13, 9–16.

Wada, S.-I., Wada, K., 2014. Visualization of the hollowness in unit particles of allophane and imogolite. J. Fac. Agric. Kyushu Univ. 59, 369–372.
Wada, K., Yoshinaga, N., 1969. The structure of imogolite. Am. Mineral. 54, 50–71.
Wells, N., Childs, C.W., Downes, C.J., 1977. Silica-Springs, Tongariro-national-park, New-Zealand—analyses of spring water and characterization of alumino-silicate deposit. Geochim. Cosmochim. Acta 41, 1497–1506.
Wolff-Boenisch, D., Gislason, S.R., Oelkers, E.H., Putnis, C.V., 2004. The dissolution rates of natural glasses as a function of their composition at pH 4 and 10.6, and temperatures from 25 to 74 degrees C. Geochim. Cosmochim. Acta 68, 4843–4858.
Yoshinaga, N., Aomine, S., 1962. Imogolite in some Ando soils. Soil Sci. Plant Nutr. 8, 22–29.
Yuan, G., Wu, L., 2007. Allophane nanoclay for the removal of phosphorus in water and waste-water. Sci. Technol. Adv. Mater. 8, 60–62.
Zunino, H., Borie, F., Aguilera, S., Martin, J.P., Haider, K., 1982. Decomposition of C-14-labeled glucose, plant and microbial products and phenols in volcanic ash-derived soils of Chile. Soil Biol. Biochem. 14, 37–43.

Part II

Structure and Properties of Nanosized Tubular Clay Minerals

Chapter 4

Physicochemical Properties of Halloysite

H. Yang[a,b,*], Y. Zhang[a,b] and J. Ouyang[a,b]
[a]*Centre for Mineral Materials, School of Minerals Processing and Bioengineering, Central South University, Changsha, China*
[b]*Key Laboratory for Mineral Materials and Application of Hunan Province, Central South University, Changsha, China*
[*]*Corresponding author: e-mail: hmyang@csu.edu.cn*

4.1 INTRODUCTION

Recent developments in nanoscience and nanotechnology opened a number of fundamental and applicable new frontiers in science and materials engineering. Advanced materials composed of natural resources with unique chemical and physical properties have shown some excellent performance. This development is also relevant in the case of halloysite-based nanocomposite materials (Du et al., 2010), due to its tubular morphology. Halloysite is highly abundant in the deposits found in many countries, and is readily obtained at a low cost, so halloysite-based materials have been used in various applications, such as fillers for rubber (Ye et al., 2007; Guo et al., 2008; Du et al., 2010; Pasbakhsh et al., 2010). Halloysite was also used in antibacterial (Levis and Deasy, 2002; Yang et al., 2010; Cavallaro et al., 2011; Wei et al., 2012), catalytic (Tae et al., 2004; Ouyang et al., 2014), optical (Giese and Van Oss, 2002), electrical (Lecouvet et al., 2013), magnetic (Zhang and Yang, 2012) and energy storage (Yang et al., 2010; Mei et al., 2011) applications. In addition, halloysite can be employed to improve the mechanical properties of cements (Wei et al., 2012) and polymers (Du et al., 2007, 2010; Pan et al., 2011). Many of these applications are conditioned by the size, shape, lumen volume and elementary composition of halloysite (Delvaux et al., 1987; Yuan et al., 2012). For instance, halloysite is a viable container for entrapping and releasing anticorrosion agents, herbicides and fungicides (Price et al., 2001; Shchukin and Möhwald, 2007; Lvov et al., 2008; Shchukin et al., 2008; Fix et al., 2009). Substances such as metal (Zhang et al., 2014b), metal oxide (Zhang and Yang, 2012; Zhang et al., 2014a) and metallocomplexes (Machado et al., 2008, 2013) can be readily

Developments in Clay Science, Vol. 7. http://dx.doi.org/10.1016/B978-0-08-100293-3.00004-2

bound to the external surfaces of the halloysite. As can be gleaned from the research literature, knowledge of the physicochemical properties of halloysite appears to be extremely important for identifying its potential application areas. This chapter summarises the typical physicochemical properties of halloysite with respect to the application aspects.

4.2 SURFACE AND COLLOIDAL PROPERTIES OF HALLOYSITE

The physicochemical characteristics of halloysites, such as impurities, cation exchange capacity (CEC), specific surface area (SSA) and pore volume, differed with regard to the sites where the raw halloysite was formed and the altered environmental conditions. Variations in halloysite morphology and the levels of impurities had the most obvious effect on SSA and internal pore volume. The different chemical compositions of the internal and external surfaces would be expected to affect the surface charges at different pH in aqueous solution. The presence of water in the interlayer space reduces the interaction between adjacent layers and facilitates curvature of the layers during halloysite formation, accommodating the mismatch in the dimensions between the octahedral and tetrahedral sheets (Pasbakhsh et al., 2013). All these factors will affect the suitability of halloysite for new applications, such as nanocontainers, nanotemplates and nanofillers. In this section, the surface and colloid chemistry of halloysite is highlighted to explain some characteristics of halloysite.

4.2.1 Cation Exchange Capacity

The ion exchange ability is possessed by many kinds of clay minerals. Thompson and several other authors (Thompson, 1850; Dixon and Mckee, 1974; Okamura and Wada, 1984; Giese and Van Oss, 2002) reported the exchange reactions that happened in clay minerals in soil. The cation exchange occurs (i) at the broken bonds around the crystal edges, (ii) as substitutions within the lattice and (iii) by replacing the hydrogen of exposed surface hydroxyls.

The CEC values of some halloysites have been determined to be as much as 20–60 cmol/kg (Table 4.1) (Joussein et al., 2005), and the hydrated halloysite has a larger CEC than the corresponding dehydrated form (Grim, 1968; Norrish, 1995). In addition, the CEC of halloysite is higher than that of kaolinite, due to its structure inducing different configurations in the hydrogen bonds, the presence of negative charges and the various hydration states of halloysite (Wilson, 2004). The CEC of halloysite is influenced by sample purity, particle size and possibly by particle morphology (Bailey, 1990). The increase in CEC with increasing Fe content and decreasing Al content has been ascribed to the substitution of Al^{3+} by Fe^{2+} in the octahedral sheet of halloysite (Wada and Mizota, 1982; Singh and Gilkes, 1992), or

TABLE 4.1 Cation Exchange Capacity (CEC) of Halloysite from Data in the Research Literature

Sample sources	Basal spacing	CEC (cmol/kg)	References
–	10 Å	40–50	Grim (1968)
–	7 Å	5–10	Grim (1968)
Yoake (Japan)	7–10 Å	20.9	Aomine and Higashi (1955)
Yoake (Japan)	–	12–13	Yoshida (1957) and Wada (1958, 1959)
–	–	58.1	Wada and Mizota (1982)
Katy soil (Texas)	7.56 Å	60	Kunze and Bradley (1964)
Komaki (Japan)	10 Å	2–22	Noro (1986)
Cameroon	10 Å	30–51	Sieffermann and Millot (1968)
Australia	10 Å	12	Norrish (1995)
Matauri Bay (New Zealand)	7 Å	2	Joussein et al. (2007)
Opotiki (New Zealand)	10 Å	7	Joussein et al. (2007)
Te Puke (New Zealand)	10 Å	5	Joussein et al. (2007)
Indonesian	7 Å	6–10	Hart et al. (2002)
Cr-halloysites (Hungary)	7–10 Å	21–35	Maksimovic and White (1973)
Cameroon	10 Å	24–35	Delvaux et al. (1990)
Camel Lake	–	18.8	Pasbakhsh et al. (2013)
Jarrahdale	–	<0.5	Pasbakhsh et al. (2013)
Matauri Bay	–	<0.4	Pasbakhsh et al. (2013)
Dragonite	–	2.1	Pasbakhsh et al. (2013)
Te Puke	–	1.3	Pasbakhsh et al. (2013)
Patch clay	–	15.3	Pasbakhsh et al. (2013)

nonstoichiometric Fe^{3+} for Al^{3+} substitution (Soma et al., 1992). Delvaux et al., (1990) classified halloysite as a clay mineral at the frontier between high- and low-activity clay minerals. However, several halloysite-rich clays show a high selectivity for poorly hydrated cations, such as NH_4^+, K^+ and Cs^+, and there was no unique mechanism to explain both the high-charge and K^+ selectivity (Fontaine et al., 1989; Joussein et al., 2004; Ndayiragije and Delvaux, 2004). Ma and Eggleton (1999) reported that the CEC of tubular halloysites being slightly higher than that of the other forms is due to the presence of some cations in the form of soluble salts in the mesoporous halloysite tubes.

The variable or pH-dependent charge is associated with broken bonds at particle edges. The adsorption of cations and anions by halloysite is influenced by the final concentration of the electrolyte solution (Wada and Ataka, 1958). With the increase of the salt concentration in solution, both CEC and anion exchange capacity increased, as predicted by classical anion and cation adsorption models for variable charges. Noncoulombic (Van der Waals) attraction may operate between the ions and the edge surfaces of halloysite nanotubes. The CEC of halloysite increases with the decrease of particle sizes due to edge and surface effects (Delvaux et al., 1992; Singh and Gilkes, 1992; Chorover and Sposito, 1995; Hart et al., 2002; Wu et al., 2013).

However, cation adsorption is assumed to be largely associated with the permanent negative charge since the CEC of halloysite soil was −65 cmol/kg at pH 7, and it only decreased to about 58 cmol/kg at pH 4 (Quantin et al., 1988). In a critical review, Bailey (1990) concluded that halloysite has a small net negative charge arising from Al^{3+} substitution to Si^{4+} sites in the tetrahedral sheet. This charge is balanced by hydrated exchangeable cations that are associated more with the tetrahedral than the octahedral side of the interlayer surface. In other words, the interlayer water in halloysite forms the hydration shell of the charge-balancing cations. Recently, Pasbakhsh et al. (2013) investigated the zeta (ζ) potential of the six halloysites throughout the pH range and concluded that the halloysite had a significantly negative surface charge in the main pH range. As a result, halloysite tends to have a polyanionic surface, except at very low pH values, and readily attracts cationic molecules from aqueous solution.

The observation that the Al:Si ratio for most halloysite is greater than unity is consistent with some Al^{3+} substitution to Si^{4+} sites (Robertson et al., 1954; Bates, 1959; Watanabe et al., 1969; Weaver and Pollard, 1973; Newman and Brown, 1987; Merino et al., 1989). Similarly, the transition from kaolinite to halloysite (7 Å) indicated an increase in permanent negative charge (from −30 to −41 mV at pH 7) (Dixon, 1989; Tarì et al., 1999; Bobos et al., 2001). For a synthetic proto-halloysite composed of hollow spheres, Wada and Kakuto (1985) and Wada (1988) calculated a structural formula with 0.5 Al^{IV} for 1.5 Si, using chemical and ^{29}Si and ^{27}Al nuclear magnetic resonance (NMR) data. White and Dixon (2002) claimed that halloysite can contain up to 0.15 Al^{IV} for every two tetrahedral sites. However, the evidence for either Al^{IV} or a permanent negative charge in halloysite is not clear. Thus, Komarneni et al. (1985) and Newman and Brown (1987) found <1% of Al^{IV} in six halloysites and two kaolinites by recording ^{27}Al MAS NMR spectroscopy of these clay minerals, and similar results were obtained by Takahashi et al. (2001) for halloysite-rich soils from northern California.

4.2.2 Specific Surface Area and Porosity

The SSA of halloysite is determined by sample purity, particle size and acid or base treatment. Pasbakhsh et al. (2013) measured the SSA and porosity of halloysite from different geographical locations (ie, Australia, New Zealand and the United States) and geological environments. The highest pore volumes associated with the cylindrical cavity or lumen in halloysite nanotubes are reported for samples with low levels of impurities and regular thin-walled nanotubes are also reported. The proportion of pore space associated with the lumen ranged from 11% to 39%, and the SSA from 22 to 81 m^2/g were greater than those of kaolinite ranged from 5 to 39 m^2/g (Theng and Wells, 1995; Levis and Deasy, 2002; Joussein et al., 2005). Previously, Birrell et al. (1955) reported a value as high as 137 m^2/g for New Zealand halloysite derived from volcanic ash, containing <5 mass% Fe_2O_3. Hart et al. (2002) measured a value of 88 m^2/g for an Indonesian kaolin sample and observed that the SSA of halloysite soil tended to decrease with increasing Fe content, while the SSA of soil from Australia remained constant with increasing Fe content.

Joo et al. (2013) analysed the SSA and pore volume of halloysite powder recovered from halloysite dispersion in acid (pH 2), neutral (pH 7) and basic (pH 11) conditions. The pore volume and pore diameter of halloysite powder increased with the decrease of the solution pH. However, the total surface area and mesopore surface area decreased with increased pH. However, micropore surface area remained constant. In acidic solution, halloysite nanotubes aggregated and many of the mesopores between halloysite nanotubes were blocked, leading to smaller pore volumes and pore size, and pore area show decreasing values with increasing pH. In basic solution, halloysite nanotubes were well dispersed and the inner-space surface area increased due to the opening of mesopores of halloysite nanotubes.

4.2.3 Dispersion Behaviour in Water

The dispersion and aggregation behaviours of halloysite nanotubes are highly related to the surface charge of halloysite (eg, the value of ζ-potential (Ouyang et al., 2014)). Dispersion behaviours of halloysite dependent on the pH values of solution were comprehensively investigated by Joo et al. (2013), and a typical ζ-potential–pH curve of halloysite with respect to solution pH is shown in Fig. 4.1A. From the plot, the surface charge of the mineral appears only slightly negative at very low pH values, and as the pH value increases from 2 to 5, the surface charge falls sharply to reach a plateau at −27 mV, before dropping further at pH values greater than 11. Similar results were reported by Tarì et al. (1999), who examined a sample of halloysite

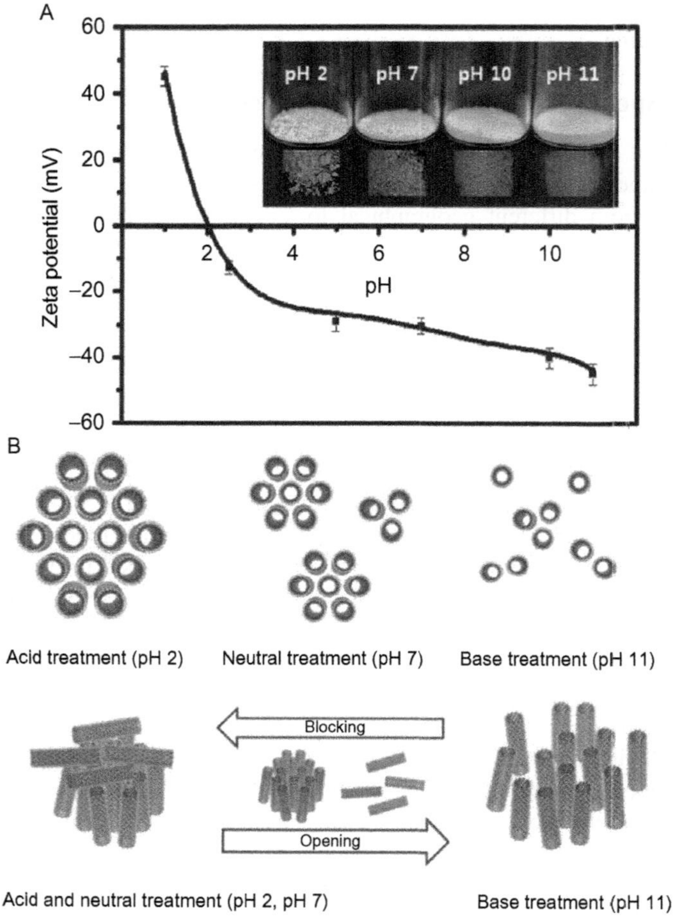

FIG. 4.1 (A) Zeta (ζ) potential plot for halloysite; the inset shows a photograph of halloysites after acid (pH 2), neutral (pH 7) and base treatment (pH 10 and 11). (B) Schematic illustration of the interspace between halloysite nanotubes after treatment with different pH solutions. *Reprinted with permission from Joo et al. (2013). Copyright 2013 The Royal Society of Chemistry.*

mined in Portugal. The reason for the generation of the negative surface charge with increasing pH is that in the curved structure, the tetrahedral (Si–O) is mainly located on the outer surfaces of the nanotubes, whereas the octahedral (Al–OH) is presented mainly on the inner surface and edges. Exposed hydroxyl groups located on the lumen surface, on the edges and on the defects of external surface could ionise. The degree of ionisation depends on the pH of the medium and the acid/base reactions occurring at a particle/solution interface, which may result in surface charge development. Both inner and outer surfaces have strongly pH-dependent surface charges, being positive or negative at pH values less or greater than the pH corresponding to their isoelectric points (pH_{IEP}), respectively. For stoichiometric structures, and in the absence of specific adsorption of other ionic species, the pH_{iep} is coincident with the point of zero charge (pH_{pzc}). The charge balance is positive for $pH < pH_{pzc}$ and negative for $pH > pH_{pzc}$ (Tarì et al., 1999). Since the outer surface of halloysite is mainly tetrahedral (Si–O), the surface charge will be negative over a wide range of pH (Fig. 4.1A). The ζ-potential at pH 2 is significantly close to zero; thus, the dispersion of halloysite is unstable. Hence, halloysite nanotubes aggregated and formed large bundles at low pH. However, in higher pH (ie, >pH 8) solution, halloysite nanotubes provided significantly low ζ-potentials (about −40 mV), which suggests that agglomeration and precipitation of particles did not occur due to Van der Waals interactions between halloysite nanotubes. At pH 11, the degree of halloysite nanotube dispersion was the highest among the tested pH. Joo et al. (2012, 2013) found that the aggregation and dispersion behaviour of halloysite nanotubes in different pH solutions apparently influence the pore diameter, pore surface area and pore volume of the recovered halloysite powder. As shown in Fig. 4.1B in basic solution, halloysite nanotubes were well dispersed, and the end of each halloysite nanotube was separated from other halloysite nanotubes (opening the inner pore of the halloysite nanotube), whereas, in acidic solution, halloysite nanotube bundles could be located adjacent to the ends of other halloysite nanotubes (blocking the inner pore of the halloysite nanotube).

The development of the concept that considers individual clay mineral particles as colloidal micelles has given the real basis for the understanding of the reactions that control the flocculation, deflocculation and dispersion of clay minerals (Hauser, 1945). Deflocculated dispersions have the advantage of a slow sedimentation rate, thereby enabling an almost uniform dose to be taken from the container, but when settling occurs, the sediment is compacted and becomes difficult to redisperse. Flocculated dispersions form loose sediments which are easily redispersible, but the sedimentation rate is fast and there is a danger of an inaccurate dose being administered. From the previous discussion, it is evident that those phenomena depend on the ion distribution in the diffuse double layer surrounding every particle or, to state it more concretely, on the ζ-potential of the clay mineral micelle. Thus, any ion that will cause a reduction in electrokinetic potential will tend to decrease the stability of the dispersion, whereas ions that aid in building up the diffuse ion double layer will act as stabilising or dispersing agents (Ford et al., 1940). Sodium phosphates, tannates and several other organic sodium compounds have proved to be very effective deflocculating or dispersing agents. However, inorganic dispersants (eg, Na_2CO_3) are difficult to eliminate, while the long-chain polysaccharide molecules of polymeric dispersants (eg, carboxyl methyl cellulose) are too large to disperse the clay mineral nanotubes. On the contrary, anionic and cationic dispersants are not only easy to eliminate by calcination, but if they are the appropriate size, they are also easy to adsorb on the clay mineral nanotubes. SDS ($C_{12}H_{25}SO_4Na$), an anionic surfactant with a tail of 12 carbon atoms attached to a sulphate group, gives the molecules amphiphilic properties. It is essential for raw halloysite to be dispersed, as it is aggregated with varying diameters and lengths under natural conditions. However, there are very few reports concerning the well dispersion of raw natural halloysite nanotubes in aqueous solution. Lun et al. (2014) demonstrated a route to disperse the raw halloysite nanotubes using SDS as a dispersant and to produce scattered and uniform halloysite nanotubes in an emulsifying machine, which helps maintain the tubular structure during the process of dispersion. The results indicated that SDS might be adsorbed and scattered on the halloysite nanotubes, which could enhance the stability and dispersibility of dispersion by the electrostatic effect (Fig. 4.2). The ζ-potential of SDS/halloysite nanotubes is negative and becomes larger than that of halloysite nanotubes within pH values ranging from 2.5 to 11.9, which reached −31 mV at a pH of around 8 when the concentration of SDS was 1.22 g/L. The diameter of dispersed halloysite nanotubes ranged from 0.6 to 1.2 μm.

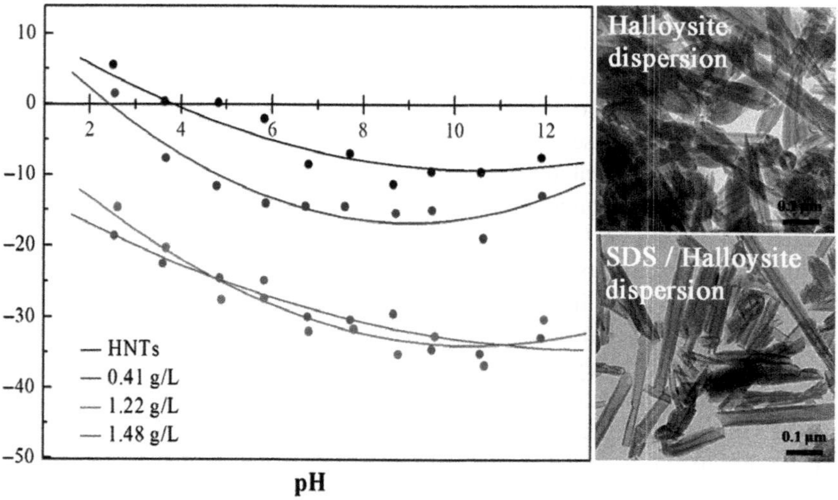

FIG. 4.2 ζ-potential vs pH for halloysite nanotube and SDS/halloysite dispersion at concentration of SDS of 0.41, 1.22 and 1.48 g/L, respectively, and transmission electron microscope (TEM) images of halloysite particles taken from halloysite dispersion (*top right*) and SDS/halloysite dispersion (*bottom right*). *Reprinted with permission from Lun et al. (2014). Copyright 2014 Springer.*

4.2.4 Hydrophilicity and Hydrophobicity

Hydrophilicity is introduced by isomorphous substitution, inducing the presence of exchangeable cations, which are hydrophilic and polarise the surface oxygen atoms. Hydrophilicity may also arise from the presence of hydroxyl groups at the surface, such as in the inner surface of halloysite and from the surface defect sites. The neutral siloxane surfaces found in halloysite have an overall hydrophobic character because the external oxygen atoms on the siloxane surface are relatively weak electron donors (Lewis bases) and are not capable of having strong interactions with water molecules. These surfaces are nonpolar and are not capable of forming hydrogen bonds with water molecules. To evaluate the hydrophilicity/hydrophobicity of halloysite, the water contact angle (CA) was measured by placing water drops (5 μL) on the surface of the moulded powder (Wu et al., 2013). The powder surfaces were prepared by gently pressing the powder onto an adhesive paste on a silicon wafer. The CA of the original halloysite was less than 10 degrees (Fig. 4.3A), indicating that the original halloysite was hydrophilic (Lvov et al., 2008). On the other hand, after the original halloysite was modified by octadecyltrimethoxy silane (ODTMS), the apparent CA increased to 164.1 ± 2.7 degrees (Fig. 4.3B), indicating that roughness-induced superhydrophobicity was achieved on the surface of the ODTMS-modified halloysite. In this case, the water droplets on the surface exhibited strong adhesion and high hysteresis.

Various authors have reported that the selective modification of the inner and outer surfaces of halloysite nanotubes with organic molecular layers containing different functional groups and hydrophobicity/hydrophilicity chemical nature, via covalent binding, to give nanotubular structures with dual chemical properties (Lvov et al., 2008; Liu et al., 2010; Cavallaro et al., 2013; Hui et al., 2013). Alteration of the halloysite nanotube surface with various organic molecules (siloxane, silanol and aluminol) could be quite

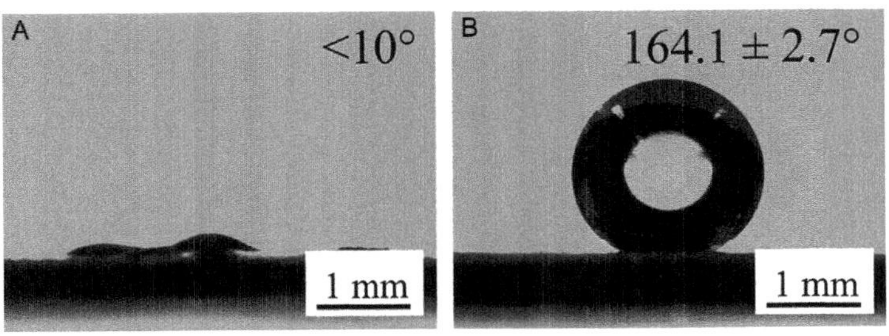

FIG. 4.3 Micrographs of 5 μL water droplets on the surfaces of (A) original halloysite and (B) ODTMS-modified halloysite held horizontally. *Reprinted with permission from Wu et al. (2013). Copyright 2013 American Chemical Society.*

useful. Halloysite was easily dispersed or aggregated without the use of mechanical methods such as sonication, ultrasonication and centrifugation, thus increasing the mechanical properties of resultant clay polymer nanocomposites (CPNs) or entrapping agents within the lumens of nanotubes.

4.3 MECHANICAL PROPERTIES OF HALLOYSITE

Halloysite can be used as an excellent nanoreinforcement for cements (Wei et al., 2012), polymers (Du et al., 2007, 2010; Pan et al., 2011) and drug delivery (Vergaro et al., 2010, 2012), due to the high length–diameter aspect ratio. Compared to montmorillonite nanoplatelets, which are often used for CPN, halloysite nanotubes have a higher intrinsic stiffness (Hedicke-Höchstötter et al., 2009). Compared to organic ones, halloysite-based drug delivery systems are not toxic, higher drug concentrations and longer- lasting time. concentrations at the surgery site, which gives a stronger therapeutic effect (Xue et al., 2015). In this section, the mechanical properties of halloysite are reported with regard to its importance for practical applications.

In order to understand the roles of halloysite nanotubes during the deformation and fracture of CPN, Lu et al. (2011) used an in situ transmission electron microscopy (TEM) with a bending stage to measure the Young's modulus of individual halloysite nanotubes. The mechanical model of the bending test is presented in Fig. 4.4A, and the Young's modulus of the halloysite nanotube can be calculated using Eq. (4.1) based on the bending and measuring method, where r_1 and r_2 are, respectively, the inner and the outer diameters of the halloysite nanotubes; P is the force applied on the length (b) of the halloysite nanotubes; L is the halloysite nanotubes total length; I represents the moment of inertia of the halloysite nanotubes and δ_B is the vertical displacement of the tip of halloysite nanotubes:

$$E = \frac{q(3L^4 - 4a^3L + a^4)}{24\delta_B I} \tag{4.1}$$

where $I = (\pi/4)(r_2^4 - r_1^4)$, $q = (p/b)$.

A single halloysite nanotube was bent a relatively large amount to test its flexibility. TEM micrographs (Fig. 4.5 A–H; Lu et al., 2011) showed that the

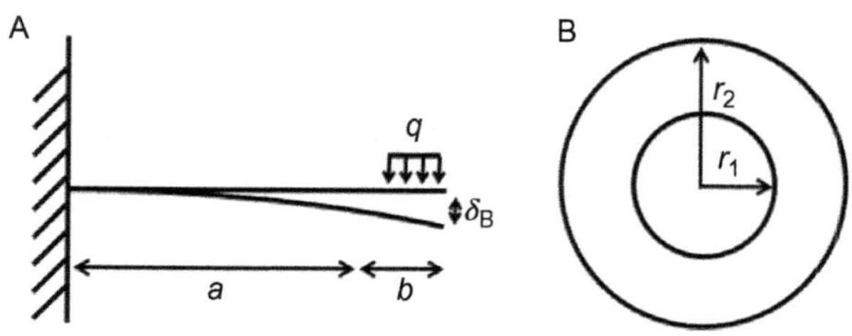

FIG. 4.4 (A) Mechanical modelling of bending test. (B) Cross section of the cantilever beam.

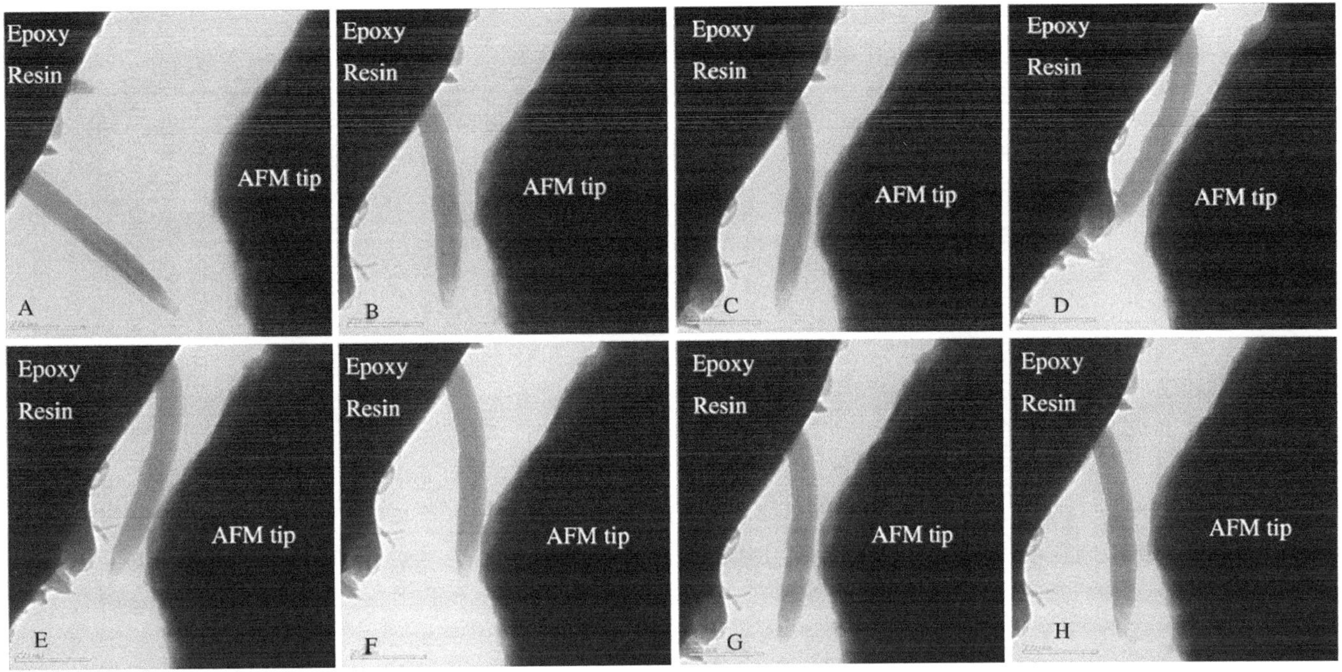

FIG. 4.5 TEM micrographs of bending tests of a single halloysite nanotube. *TEM images A to H show the gradual processes of a single nanotube is bended and then released by the ADM tip. Reprinted with permission from Lu et al. (2011). Copyright 2011 Ingentaconnect.*

halloysite nanotubes were surprisingly flexible and could be bent to almost 90 degrees without fracturing. Furthermore, based on the TEM micrographs, there was no obvious reduction in the cross-sectional area during the experiment. This phenomenon suggests that the halloysite nanotubes possess remarkable flexibility against mechanical loading. When the bending angle was larger than 5 degrees, plastic deformation may occur and the halloysite nanotubes could not return to the original state. The plastic deformation of a huge number of halloysite nanotubes in CPN may consume considerable impact energy. Accordingly, the impact strength of halloysite polymer nanocomposites may be improved by this hypothesis. These results showed that halloysite nanotubes have a high Young's modulus and remarkable flexibility. The Young's modulus of a single halloysite nanotube was about 130 ± 24 GPa, and the elastic modulus showed a tendency to decrease with increasing diameter, probably because of the higher density of structural defects in larger nanotubes. The results suggest that halloysite nanotubes have good potential to be used as nanofillers in high-performance structural materials, especially polymer-based nanocomposites. The Young's modulus of the halloysite nanotube is a key factor in the field of computer simulation for studying the mechanical behaviour of CPN.

Guimarães et al. (2010) used a self-consistent-charged density-functional tight-binding (DFTB) method to predict the Young's modulus of single-walled halloysite nanotubes and found values in the range of 230–340 GPa. Finally, the rule of mixtures has been selected to predict a Young's modulus of around 72 GPa for a single halloysite nanotube in chitosan nanocomposite membrane. De Silva et al. (2013) carried out a series of mechanical performance tests, and the tensile test data revealed that the membranes reinforced with 5 (*mass/mass%*) halloysite nanotubes yielded the highest Young's modulus (0.52 ± 0.01 GPa) and strength (81.6 ± 4.4 MPa). Lecouvet et al. (2013) reported that the measurements of the elastic modulus of halloysite nanotubes and nanoscale three-point bending tests were performed on individual nanotubes using an atomic force microscope. The experimental results showed that an average elastic modulus of 140 GPa is obtained for a set of nanotubes with outer diameters ranging between 50 and 160 nm. Moreover, the elastic modulus increases with decreasing outer diameter, with a steep jump below 50 nm. The size dependence of the elastic modulus may be attributed to the surface tension effects for thinner tubes and a nonnegligible contribution of shear deformations to the total deflection for larger tubes. The nanomechanical behaviour of halloysite nanotubes is essential to identify their potential reinforcing effects in polymeric materials. The predicted limits of the respective mechanical and structural parameters are summarised in Table 4.2.

Landau and Lifshitz (1986) described a nanoscale three-point bending test, which was performed in contact mode. Fig. 4.6 presented the force (F) vs nanotube defection (δ) curves obtained on a silicon wafer, a nanotube lying on the PC membrane and a nanotube crossing a pore (Piperno et al., 2007).

TABLE 4.2 Summary of the Predicted Limits of the Mechanical and Structural Parameters of Halloysite Nanotube

Mechanical parameter	Symbol	LB, UB
Young modulus of halloysite	E_H	1 GPa, 1000 GPa
Shear modulus of halloysite	G_H	0.4 GPa, 416.7 GPa
Effective axial stress in halloysite nanotubes, as applied to tensile-loaded halloysite nanotubes	σ_H	0.03 GPa, 4 GPa
Yield stress of halloysite	σ_y	10^3 MPa[a]
Effective residual axial strain in halloysite nanotubes, as illustrated for the case of tensile loading	ε_{res}	4×10^{-6}, 4×10^{-3}
Absolute shear stress (maximum) in halloysite nanotubes	τ_{max}^{abs}	1×10^{-1} GPa, 10 GPa
Elastic torque	T_e	15×10^{-5} pN m, 16 pN m
Elastic strain energy	U_e	9.0×10^{-24} J, 1.1×10^{-4} J
Yield torque	T_y	15×10^{-3} pN m, 16 pN m
Plastic torque	T_p	18×10^{-3} pN m, 20 pN m
Moment at yield point	M_y	31×10^{-4} pN m, 33×10^{-1} pN m
Plastic moment	M_p	24×10^{-4} pN m, 28×10^{-1} pN m
Applied force on composite	F_e	120 pN, 1 N
Axial stress within the wall of halloysite nanotubes, lumen-loaded case	σ_H	2.6×10^7 Pa, 20×10^8 Pa
Circumferential stress within the wall of halloysite nanotubes, lumen-loaded case	σ_θ	2.7×10^{10} Pa, 7.0×10^{10} Pa
Structural parameter	**Symbol**	**LB, UB**
Inner diameter of halloysite nanotubes	D_i	10 nm, 70 nm[b]
Outer diameter of halloysite nanotubes	D_o	20 nm, 200 nm[b]
Length of halloysite nanotubes	L	50 nm, 5000 nm[b]
Aspect ratio of halloysite nanotubes	q	3, 50[c]
Cross-section area of halloysite nanotubes	A_H	236 nm^2, 31,337 nm^2
Polar moment of inertia of halloysite nanotubes	J_H	1.5×10^4 nm^4, 1.6×10^8 nm^4

Continued

TABLE 4.2 Summary of the Predicted Limits of the Mechanical and Structural Parameters of Halloysite Nanotube—Cont'd

Structural parameter	Symbol	LB, UB
Bending moment of inertia of halloysite nanotubes	I_H	7.4×10^3 nm^4, 7.7×10^{-7} nm^4
Change in volume of lumen, lumen-loaded case	ΔV	3.3×10^{16} nm^3, 5.5×10^{20} nm^3

LB, lower bound; *UB*, upper bound.
[a]*Only a single value was calculated.*
[b]*These values are estimated based on data derived from experiments reported elsewhere.*
[c]*Pasbakhsh et al. (2013).*
Reprinted with permission from Lecouvet et al. (2013). Copyright 2013 IOP Science.

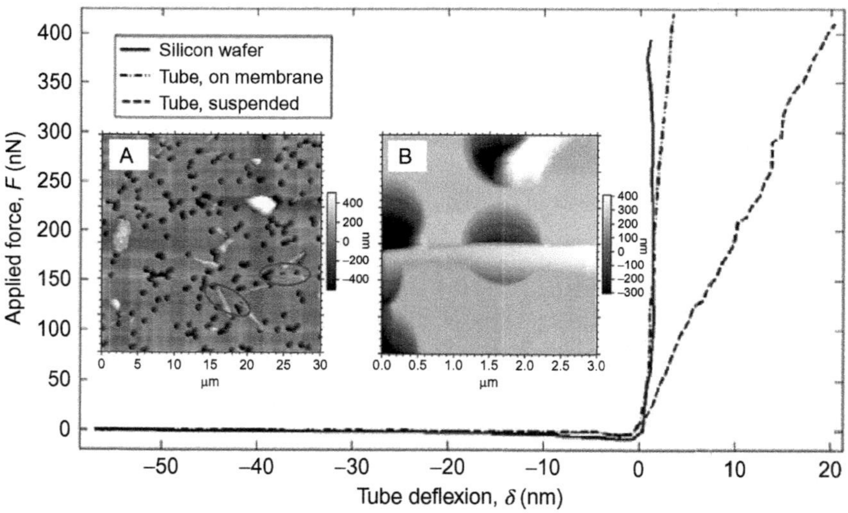

FIG. 4.6 *F* vs *δ* curves obtained on a silicon wafer (*solid line*), a nanotube lying on the PC membrane (*dashed* and *dotted line*) and a nanotube crossing a pore (*dashed line*). *Reprinted with permission from Lecouvet et al. (2013). Copyright 2013 IOP Science.*

The linear elastic behaviour of halloysite nanotubes in the range of applied forces was confirmed by the linearity and reversibility of approach and retraction curves. In addition, the force curve measured on a tube lying on the membrane indicated that there was no penetration of the atomic force microscopy tip into the clay mineral nanotube, as revealed by the infinite slope of the curve in the contact zone (Fig. 4.6). Finally, to ensure that the nanostructure does not roll during the measurements, images were taken before and after each test. Their superimposition confirmed that the nanotube was in a stable configuration during the nanoscale three-point bending tests.

Piperno et al. (2007) reported that the halloysite nanotubes exhibited elastic behaviour at small deformations (below ca. 20 nm). Young's modulus values of (159 ± 125) GPa and (279 ± 260) GPa were obtained from the force-deflection curves using the bending equation for a clamped beam under a concentrated loading. The unexpected steep jump of the elastic modulus for the small diameters (<50 nm) has already been reported and may have two main reasons:

(i) The concentration of structural defects in the multilayer structure is expected to increase with the number of constituting layers of the wall. The elastic modulus is fundamentally related to the chemical bonding between the constituent atoms and is less sensitive to the defect concentration.
(ii) The surface properties of the nanomaterials may substantially affect their mechanical properties.

4.4 CHEMICAL STABILITY OF HALLOYSITE UNDER ACID AND ALKALINE TREATMENTS

As nanostructured materials become more widely used, there is a growing need to understand the mechanism of morphological changes and physical degradation of these nanomaterials under various operational conditions (Gleiter, 2000). Natural untreated clay minerals possess a very low ability to catalyse reactions in either polar or nonpolar media (Kuroda et al., 2011). Furthermore, the structural properties of these materials can be modified by various activation methods in order to produce catalysts with high acidity, SSA, porosity and thermal stability. Therefore, knowledge of the conditions under which halloysite retains its tubular shape, as well as how it is modified by different solution environments, is vital in practical applications such as wear-resistant CPN or catalysts. The stability and transformation of halloysite in different solvent conditions are summarised in the next section.

4.4.1 Effect of Acid Treatment on Halloysite

Acid activation of clay minerals probably is one of the most effective methods that has been used to produce active materials for adsorption and catalysis purposes. Many applications of halloysite are conditioned by its lumen size, and the lumen volume is ca. 10 Vol% of the nanotube. Due to the different chemical properties of the tubes inner and outer surfaces, the lumen diameter can be enlarged by acid treatment of aluminium oxides sheets (Table 4.3). The enlarged pores allow higher amounts of active loading of chemicals with a target capacity of 40–50% (Abdullayev et al., 2012).

Recently, Zhang et al. (2012) used the acid leaching of halloysite to synthesise the SiO_2 nanotubular structure. The pretreated halloysite reacted

TABLE 4.3 Change in Halloysite Properties Including SSA, Total Pore Volume (V_{pores}), Total Concentration of Aluminium (III) $C_{Al(III)}$ and Silicon (IV) $C_{Si(IV)}$ in Aqueous Dispersions, Mole Fractions of Aluminium in Liquid X_L^{Al} and Solid X_S^{Al} Phases and Mole Fractions of Silicon in Liquid X_L^{Si} and Solid X_S^{Si} Phases Under Slow Transformation in Acid or Base Solvents

Solution composition	Time of treatment (day)	SSA (m^2/g)	V_{pores} (cm^3/g)	Al(III) $C_{Al(III)}$ ($mmol\ dm^{-3}$)	Al(III) X_L^{Al}	Al(III) X_S^{Al}	Si(IV) $C_{Si(IV)}$ ($mmol\ dm^{-3}$)	Si(IV) X_L^{Si}	Si(IV) X_S^{Si}	Solid phase composition	Morphology
Initial halloysite	0	24.3	0.090	0	0	0.47	0	0	0.53	$Al_2Si_2O_5(OH)_4$	Tubes
H_2O	28	30.5	0.108							$Al_2Si_2O_5(OH)_4+Al(OH)_3$	Tubes +nanosheets
H_2O	84	31.3	0.108	0.01	0.01	0.47	1.75	0.99	0.53		
1 mol dm^{-3} HCl	28	54.2	0.144								
1 mol dm^{-3} HCl	84	91.2	0.245	34.4	0.73	–	12.76	0.27	–	$Al_2Si_2O_5(OH)_4+SiO_2$	Tubes +nanoparticles
1 mol dm^{-3} H_2SO_4	28	70.7	0.184								
1 mol dm^{-3} H_2SO_4	84	102.2	0.322	27.5	0.71	0.31	11.5	0.29	0.60	$Al_2Si_2O_5(OH)_4+SiO_2$	Tubes +nanoparticles
1 mol dm^{-3} NaOH	28	45.0	0.165								
1 mol dm^{-3} NaOH	84	47.0	0.168	12.6	0.35	0.82	22.71	0.65	0.18	$Al_2Si_2O_5(OH)_4+Al(OH)_3$	Tubes +nanoparticles

with 6 M HCl solution at 373 K in a water bath. The process occurs by acid leaching of aluminium oxide layers and the remained product is amorphous nanotubular silica, with three times higher SSA and pore volume than those of the initial halloysite. The ^{29}Si NMR spectroscopy and DFTB calculations demonstrated the disappearance of Q^2 structure $Si(OSi)_2(OH)_2$ and Q^4 framework structure $Si(SiO)_4$, as well as the partial transformation of Q^3 structure $Si(SiO)_3(OH)$ from halloysite to nanotubular silica (Fig. 4.7). These structures are not stable and may easily collapse to form silica polycrystals. This treatment enables the possibility to functionalise the nanotubular silica surface.

White et al. (2012) investigated the long-term stability of natural halloysite nanotubes in 1 mol/dm^3 H_2SO_4 solution at room temperature. The aqueous dispersion of natural halloysite is kinetically stable in water. In low-concentrated strong acid solutions, the extensive dissolution is initiated by dissolution of the inner Al–OH surface, leading to progressive thinning of the tubes from the inside. In high-concentrated strong acid solutions, the greater solubility of Al(III) vs Si(IV) leads to the formation of small SiO_2 nanoparticles inside the tubes, yielding the enlarged SSA and pore volume in the product. It is indicated that the rate of removal of Al(III) species from the halloysite framework is affected by the H^+ ions concentration. In addition, complexing agents that form a highly stable complex with Al(III) should also increase the dissolution rate by decreasing $\Delta_f G$, the free energy of formation

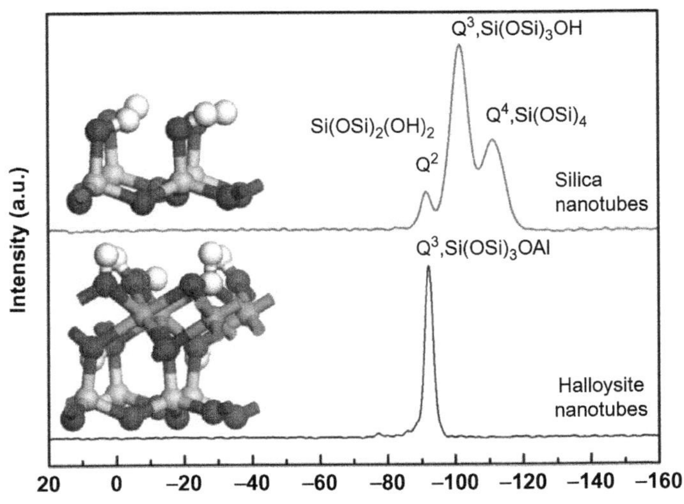

FIG. 4.7 ^{29}Si NMR spectra of halloysite nanotubes and the as-synthesised silica nanotubes (insets are corresponding atomic structures; *white* atoms represent H, *red* (*dark grey* in the print version) O, *pink* (*grey* in the print version) Al and *yellow* (*light grey* in the print version) Si). *Reprinted with permission from Zhang et al. (2012). Copyright 2012 Elsevier.*

for the solvated ionic species. On the other side, the hydrolysis of the silica framework would most likely begin from defects or edge sites.

In-depth kinetic analysis of halloysite lumen etching at various temperatures is important for obtaining nanotubes with optimal loading and release properties. Abdullayev et al. (2012) performed kinetic studies in different conditions (temperature and concentrations of acid). They revealed that chemical interaction is the rate-determining step with the activation energy of 68.0 kJ/mol and pseudo-zeroth order below 70°C. Al–OH sheet removal from the nanotubes takes place in three steps (Fig. 4.8): (i) diffusion of hydrogen ions into the inner lumen, (ii) chemical reaction with an octahedral (Al–OH) sheet on the tube inner wall and (iii) transport of the reaction products out of the lumen. In the Al–OH sheet removal process, significant changes in halloysite morphology was observed, holes in the halloysite walls were observed, free silica nanoparticles were formed, and, finally, the tubular halloysite transformed into porous silica nanorods. The result showed that upon acid treatment, the SSA of the tubes increased over six times, from 40 to 250 m^2/g. The tube lumen capacity for loading and further sustained release of active chemical agents could be increased to two to three times. In particular, halloysite loading efficiency for the benzotriazole (corrosion inhibitor) increased four times by selective etching of 60% Al–OH sheet within the tubes' lumens.

Furthermore, acid activation of halloysite has been used to produce active materials for adsorption and catalysis purposes with high acidity, SSA, porosity and thermal stability (Fujii et al., 2014; Saklar and Yorukoglu, 2015; Shu et al., 2015). The acid-activated clays have found applications in industrial processes such as alkylation of phenols, polymerisation of unsaturated hydrocarbons, clarification of edible oils and carbonless copying paper. The performance of

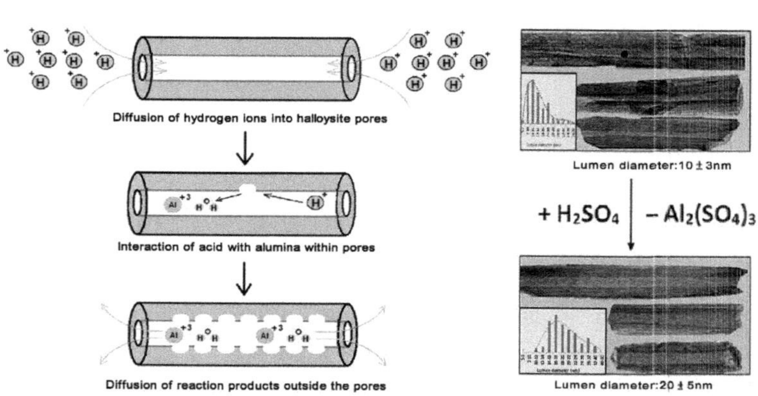

FIG. 4.8 The scheme of acid etching of an Al–OH sheet from halloysite lumen. *Reprinted with permission from Abdullayev et al. (2012). Copyright 2012 American Chemical Society.*

acid-treated halloysite catalysts in the degradation of polystyrene (PS) was investigated (Tae et al., 2004). The degradation was carried out in a semibatch reactor with a mixture of PS and catalyst at 400°C–450°C. The acid-treated halloysite showed good catalytic activity for the degradation of PS with very high selectivity over 99% to aromatic liquids. In addition, Al-containing mesoporous silica nanotubes with varying inner diameters were synthesised from acid-leached halloysite via the hydrothermal method (Xie et al., 2014). During the process of synthesis, acid-leached halloysite provided a convenient silica source and a well-defined tubular template. The results showed that the enlargement of the inner diameter of mesoporous silica nanotubes was conveniently achieved through regulating the acid-leaching time of halloysite nanotubes, while the diameters of the mesopores on the wall were uniform. In particular, compared with halloysite nanotubes, mesoporous silica nanotubes with SSA of 423.9 m^2/g (a sevenfold increase) exhibited a significant high aspirin loading of 157 mg/g (a fivefold increase).

Through the abovementioned acid treatment-assisted pore reforming, the enlarged pores will allow the loading of more active chemical components such as metals, corrosion inhibitors and drugs. In addition, more effective anchoring of polymer chains within the tube openings will be possible, which might improve the mechanical properties of the CPN.

4.4.2 Effect of Alkali Treatment on Halloysite

The alkaline treatment of clays was much less reported on than acid treatment in the literature. The progressive leaching of Al and Si sheets might be predicted, as both elements are soluble in alkaline conditions to yield aluminates and silicates. The alkaline treatment of smectites and fibrous clay minerals has received little attention, but the modifications of metakaolinite have been studied (Elert et al., 2008). Usually, the clay mineral was used as a source to synthesise zeolites for catalytic applications. Recently, the selective leaching of the Si sheets to product alumina with interesting properties was reported. For example, White et al. (2012) investigated the long-term stability of halloysite in 0.001–1 mol/dm^3 solutions of different bases (NaOH) at room temperature. In concentrated (0.01–1 mol/dm^3) strong basic solutions, extensive dissolution is initiated by the dissolution of the inner Al–OH surface catalysed by basic species, leading to progressive thinning of the tubes from the inside (Table 4.3). The better solubility of Si(IV) vs Al(III) leads to the formation of fragmented flaky particles containing layers of $Al(OH)_3$ in NaOH solutions (Fig. 4.9). Xie et al. (2014) reported the synthesis of ordered Al-MCM-41 molecular materials with uniform pore sizes via hydrothermal treatment in basic circumstances using natural halloysite as the source material, without addition of silica or aluminium reagents. The product exhibits SSA of 509.4 m^2/g, pore volume of 0.489 mL/g, and narrow pore size distribution centred at 3.2 nm. Calcinations temperature plays a significant role in

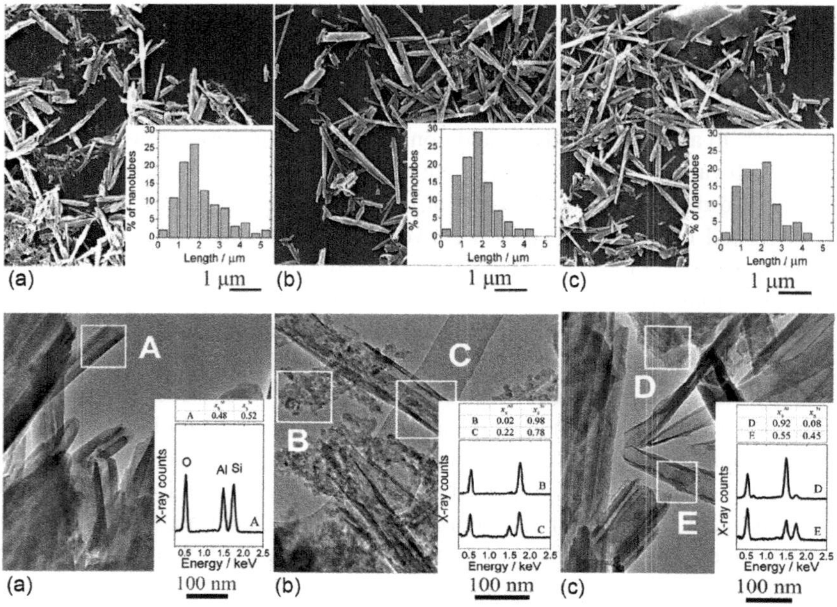

FIG. 4.9 Scanning electron microscopy (SEM) images (insets: histograms of length distribution) and high-resolution TEM (HRTEM) images (insets: energy-dispersive spectra from highlighted and labled A–E squares) of (a) untreated halloysite, halloysite treated for 84 days with (b) 1 mol/dm^3 H_2SO_4 or (c) 1 mol/dm^3 NaOH. *Reprinted with permission from White et al. (2012). Copyright 2012 Igentaconnect.*

regulating SiO_2/Al_2O_3 molar ratio for the preparation of Al-MCM-41. Furthermore, an atomic-level structural evolution from pristine halloysite to mesoporous materials Al-MCM-41 has been clarified.

4.5 CONCLUDING REMARKS

Naturally occurring halloysite is a low-cost industrial mineral resource with great potential for being used as functional material, such as in a nanoscale container. The knowledge on the physicochemical properties of halloysite is vital for the understanding of the performance of halloysite in various applications. This chapter summarises some basic aspects of the physicochemical properties of halloysite such as tubular structures, CEC, SSA, dispersion behaviour and hydrophilicity and hydrophobicity. Further studies are still needed to better understand halloysite properties, such as how the surface and structural modification of the halloysite affect the CEC and dispersion behaviour and the details about the transformation of tubular structure of halloysite in chemical treatments. In addition, the mechanical analysis combined with the theoretical modelling on halloysite may be expected to yield some interesting results.

REFERENCES

Abdullayev, E., Joshi, A., Wei, W., Zhao, Y., Lvov, Y., 2012. Enlargement of halloysite clay nanotube lumen by selective etching of aluminum oxide. ACS Nano 6, 7216–7226.

Aomine, S., Higashi, T., 1955. Clay minerals of decomposed andesitic agglomeratic lava at Yoake. Mineral. J. 1, 278–289.

Bailey, S.W., 1990. Halloysite- a critical assessment. Sciences Geologiques-Memoires 86, 89–98.

Bates, T.F., 1959. Morphology and crystal chemistry of 1:1 layer lattice silicates. Am. Mineral. 44, 78–114.

Birrell, K.S., Fieldes, M., Williamson, K.I., 1955. Unusual forms of halloysite. Am. Mineral. 40, 122–124.

Bobos, I., Duplay, J., Rocha, J., Gomes, C., 2001. Kaolinite to halloysite-7 Å transformation in the kaolin deposit of São Vicente de Pereira, Portugal. Clay Clay Miner. 49, 596–607.

Cavallaro, G., Donato, D., Lazzara, G., Milioto, S., 2011. Films of halloysite nanotubes sandwiched between two layers of biopolymer: from the morphology to the dielectric, thermal, transparency, and wettability properties. J. Phys. Chem. C 115, 20491–20498.

Cavallaro, G., Lazzara, G., Milioto, S., Parisi, F., Sanzillo, V., 2013. Modified halloysite nanotubes: nanoarchitectures for enhancing the capture of oils from vapor and liquid phases. ACS Appl. Mater. Interfaces 6, 606–612.

Chorover, J., Sposito, G., 1995. Surface charge characteristics of kaolinitic tropical soils. Geochim. Cosmochim. Acta 59, 875–884.

De Silva, R.T., Pasbakhsh, P., Goh, K.L., Chai, S.-P., Ismail, H., 2013. Physico-chemical characterisation of chitosan/halloysite composite membranes. Polym. Test. 32, 265–271.

Delvaux, B., Herbillon, A., Duffey, J., Burtin, G., Vielvoye, L., 1987. Adsorption sélective du potassium par certaines halloysites (10 Å) de sols tropicaux développés sur roches volcaniques: signification minéralogique. Comptes rendus de l'Académie des sciences. Série 2, mécanique, physique, chimie, sciences de l'univers. Sci. Terre 307, 311–317.

Delvaux, B., Herbillon, A., Dufey, J., Vielvoye, L., 1990. Surface properties and clay mineralogy of hydrated halloysitic soil clays. I: existence of interlayer K specific sites. Clay Miner. 25, 129–139.

Delvaux, B., Tessier, D., Herbillon, A., Burtin, G., Jaunet, A.-M., Vielvoye, L., 1992. Morphology, texture, and microstructure of halloysitic soil clays as related to weathering and exchangeable cation. Clay Clay Miner. 40, 446–456.

Dixon, J., 1989. Kaolin and serpentine group minerals. In: Dixon, J., Weed, S.B. (Eds.), Minerals in Soil Environments, second ed. Soil Science Society of America, Madison, pp. 467–526.

Dixon, J.B., Mckee, T.R., 1974. Spherical halloysite formation in a volcanic soil of Mexico. Trans. Int. Congr. Soil Sci. 7, 115–124.

Du, M.L., Guo, B.C., Liu, M.X., Jia, D.M., 2007. Thermal decomposition and oxidation ageing behaviour of polypropylene/halloysite nanotube nanocomposites. Polym. Polym. Compos. 15, 321–328.

Du, M.L., Guo, B.C., Jia, D.M., 2010. Newly emerging applications of halloysite nanotubes: a review. Polym. Int. 59, 574–582.

Elert, K., Sebastian, E., Valverde, I., Rodriguez-Navarro, C., 2008. Alkaline treatment of clay minerals from the Alhambra Formation: implications for the conservation of earthen architecture. Appl. Clay Sci. 39, 122–132.

Fix, D., Andreeva, D.V., Lvov, Y.M., Shchukin, D.G., Möhwald, H., 2009. Application of inhibitor-loaded halloysite nanotubes in active anti-corrosive coatings. Adv. Funct. Mater. 19, 1720–1727.

Fontaine, S., Delvaux, B., Dufey, J., Herbillon, A., 1989. Potassium exchange behaviour in Carribean volcanic ash soils under banana cultivation. Plant Soil 120, 283–290.

Ford, T., Loomis, A., Fidiam, J., 1940. The colloidal behaviour of clays as related to their crystal structure. J. Phys. Chem. 44, 1–12.

Fujii, K., Nakagaito, A.N., Takagi, H., Yonekura, D., 2014. Sulfuric acid treatment of halloysite nanoclay to improve the mechanical properties of PVA/halloysite transparent composite films. Compos. Interfaces 21, 319–327.

Giese, R.F., Van Oss, C.J., 2002. Colloid and Surface Properties of Clays and Related Minerals. Marcel Dekker Inc., New York, USA.

Gleiter, H., 2000. Nanostructured materials: basic concepts and microstructure. Acta Mater. 48, 1–29.

Grim, R.E., 1968. Clay Mineralogy. McGraw-Hill, New York.

Guimarães, L., Enyashin, A.N., Seifert, G., Duarte, H.A., 2010. Structural, electronic, and mechanical properties of single-walled halloysite nanotube models. J. Phys. Chem. C 114, 11358–11363.

Guo, B., Lei, Y., Chen, F., Liu, X., Du, M., Jia, D., 2008. Styrene butadiene rubber/halloysite nanotube nanocomposites modified by metacrylic acid. Appl. Surf. Sci. 255, 7329–7336.

Hart, R.D., Gilkes, R.J., Siradz, S., Singh, B., 2002. The nature of soil kaolins from Indonesia and Western Australia. Clay Clay Miner. 50, 198–207.

Hauser, E.A., 1945. Colloid chemistry of clays. Chem. Rev. 37, 287–321.

Hedicke-Höchstötter, K., Lim, G.T., Altstädt, V., 2009. Novel polyamide nanocomposites based on silicate nanotubes of the mineral halloysite. Compos. Sci. Technol. 69, 330–334.

Hui, W., Watanabe, H., Wei, M., Fujimoto, A., Higuchi, T., Uesugi, K., Takeuchi, A., Suzuki, Y., Jinnai, H., Takahara, A., 2013. Robust liquid marbles stabilized with surface-modified halloysite nanotubes. Langmuir 29, 14971–14975.

Joo, Y., Jeon, Y., Lee, S.U., Sim, J.H., Ryu, J., Lee, S., Lee, H., Sohn, D., 2012. Aggregation and stabilization of carboxylic acid functionalized halloysite nanotubes (HNT–COOH). J. Phys. Chem. C 116, 18230–18235.

Joo, Y., Sim, J.H., Jeon, Y., Lee, S.U., Sohn, D., 2013. Opening and blocking the inner-pores of halloysite. Chem. Commun. 49, 4519–4521.

Joussein, E., Kruyts, N., Righi, D., Petit, S., Delvaux, B., 2004. Specific retention of radiocesium in volcanic ash soils devoid of micaceous clay minerals. Soil Sci. Soc. Am. J. 68, 313–319.

Joussein, E., Petit, S., Churchman, J., Theng, B., Righi, D., Delvaux, B., 2005. Halloysite clay minerals—a review. Clay Miner. 40, 383–426.

Joussein, E., Petit, S., Delvaux, B., 2007. Behaviour of halloysite clay under formamide treatment. Appl. Clay Sci. 35, 17–24.

Komarneni, S., Fyfe, C., Kennedy, G., 1985. Order–disorder in 1:1 type clay minerals by solid-state ^{27}Al and ^{29}Si magic-angle-spinning NMR spectroscopy. Clay Miner. 20, 327–334.

Kunze, G., Bradley, W., 1964. Occurrence of a tabular halloysite in a Texas soil. Clay Clay Miner. 12, 523–527.

Kuroda, Y., Ito, K., Itabashi, K., Kuroda, K., 2011. One-step exfoliation of kaolinites and their transformation into nanoscrolls. Langmuir 27, 2028–2035.

Landau, L.D., Lifshitz, E.M., 1986. Theory of Elasticity. Pergamon, Oxford.

Lecouvet, B., Horion, J., D'Haese, C., Bailly, C., Nysten, B., 2013. Elastic modulus of halloysite nanotubes. Nanotechnology 24, 105701–105708.

Levis, S.R., Deasy, P.B., 2002. Characterisation of halloysite for use as a microtubular drug delivery system. Int. J. Pharm. 243, 125–134.

Liu, M.X., Jia, Z.X., Fang, L., Jia, D.M., Guo, B.C., 2010. Tailoring the wettability of polypropylene surfaces with halloysite nanotubes. J. Colloid Interface Sci. 350, 186–193.

Lu, D., Chen, H., Wu, J., Chan, C.M., 2011. Direct measurements of the Young's modulus of a single halloysite nanotube using a transmission electron microscope with a bending stage. J. Nanosci. Nanotechnol. 11, 7789–7793.

Lun, H.L., Ouyang, J., Yang, H.M., 2014. Enhancing dispersion of halloysite nanotubes via chemical modification. Phys. Chem. Miner. 41, 281–288.

Lvov, Y.M., Shchukin, D.G., Möhwald, H., Price, R.R., 2008. Halloysite clay nanotubes for controlled release of protective agents. ACS Nano 2, 814–820.

Ma, C., Eggleton, R.A., 1999. Cation exchange capacity of kaolinite. Clay Clay Miner. 47, 174–180.

Machado, G.S., Wypych, F., Nakagaki, S., 2008. Immobilization of metalloporphyrins into nanotubes of natural halloysite toward selective catalysts for oxidation reactions. J. Mol. Catal. A 283, 99–107.

Machado, G.S., Ucoski, G.M., Jos, O., de Lima, O.J., Ciuffi, K.J., Wypych, F., Nakagaki, S., 2013. Cationic and anionic metalloporphyrins simultaneously immobilized onto raw halloysite nanoscrolls catalyze oxidation reactions. Appl. Catal. A 460–461, 124–131.

Maksimovic, Z., White, J., 1973. Infrared study of chromium-bearing halloysites. In: Proceedings of International Clay Conference, Madrid, pp. 61–73.

Mei, D.D., Zhang, B., Liu, R.C., Zhang, H.Q., Liu, J.D., 2011. Preparation of stearic acid/halloysite nanotube composite as form-stable PCM for thermal energy storage. Int. J. Energy Res. 35, 828–834.

Merino, E., Harvey, C., Murray, H.H., 1989. Aqueous-chemical control of the tetrahedral-aluminum content of quartz, halloysite, and other low-temperature silicates. Clay Clay Miner. 37, 135–142.

Ndayiragije, S., Delvaux, B., 2004. Selective sorption of potassium in a weathering sequence of volcanic ash soils from Guadeloupe, French West Indies. Catena 56, 185–198.

Newman, A.C.D., Brown, G., 1987. The chemical constitution of clays. Monograph, Mineralogical Society, Longman Scientific and Technical, Harlow, UK.

Noro, H., 1986. Hexagonal platy halloysite in an altered tuff bed, Komaki city, Aichi prefecture, Central Japan. Clay Miner. 21, 401–415.

Norrish, K., 1995. An unusual fibrous halloysite. Agron. Monogr. 9, 275–284.

Okamura, Y., Wada, K., 1984. Ammonium-calcium exchange equilibria in soils and weathered pumices that differ in cation exchange materials. J. Soil Sci. 35, 387–396.

Ouyang, J., Zhou, Z., Zhang, Y., Yang, H., 2014. High morphological stability and structural transition of halloysite (Hunan, China) in heat treatment. Appl. Clay Sci. 101, 16–22.

Pan, J.M., Yao, H., Xu, L.C., Ou, H.X., Huo, P.W., Li, X.X., Yan, Y.S., 2011. Selective recognition of 2,4,6-trichlorophenol by molecularly imprinted polymers based on magnetic halloysite nanotubes composites. J. Phys. Chem. C 115, 5440–5449.

Pasbakhsh, P., Ismail, H., Fauzi, M.N.A., Bakar, A.A., 2010. EPDM/modified halloysite nanocomposites. Appl. Clay Sci. 48, 405–413.

Pasbakhsh, P., Churchman, G.J., Keeling, J.L., 2013. Characterisation of properties of various halloysites relevant to their use as nanotubes and microfibre fillers. Appl. Clay Sci. 74, 47–57.

Piperno, S., Kaplan-Ashiri, I., Cohen, S.R., Popovitz-Biro, R., Wagner, H.D., Tenne, R., Foresti, E., Lesci, I.G., Roveri, N., 2007. Characterization of geoinspired and synthetic chrysotile nanotubes by atomic force microscopy and transmission electron microscopy. Adv. Funct. Mater. 17, 3332–3338.

Price, R.R., Gaber, B.P., Lvov, Y., 2001. In-vitro release characteristics of tetracycline HCl, khellin and nicotinamide adenine dinucleotide from halloysite; a cylindrical mineral. J. Microencapsul. 18, 713–722.

Quantin, P., Gautheyrou, J., Lorenzoni, P., 1988. Halloysite formation through in situ weathering of volcanic glass from trachytic pumices, Vico's volcano, Italy. Clay Miner. 23, 423–437.

Robertson, R., Brindley, G., Mackenzie, R., 1954. Mineralogy of kaolin clays from Pugu, Tanganyika. Am. Mineral. 39, 118–138.

Saklar, S., Yorukoglu, A., 2015. Effects of acid leaching on halloysite. Physicochem. Probl. Miner. Process 51, 83–94.

Shchukin, D.G., Möhwald, H., 2007. Surface-engineered nanocontainers for entrapment of corrosion inhibitors. Adv. Funct. Mater. 17, 1451–1458.

Shchukin, D.G., Lamaka, S.V., Yasakau, K.A., Zheludkevich, M.L., Ferreira, M.G.S., Möhwald, H., 2008. Active anticorrosion coatings with halloysite nanocontainers. J. Phys. Chem. C 112, 958–964.

Shu, Z., Chen, Y., Zhou, J., Li, T.T., Yu, D.X., Wang, Y.X., 2015. Nanoporous-walled silica and alumina nanotubes derived from halloysite: controllable preparation and their dye adsorption applications. Appl. Clay Sci. 112, 17–24.

Sieffermann, G., Millot, G., 1968. l'Halloysite des sols jeunes sur basaltes récents du Centre Cameroun. Bulletin du Groupe Français des Argiles 20, 25–38.

Singh, B., Gilkes, R., 1992. An electron optical investigation of the alteration of kaolinite to halloysite. Clay Clay Miner. 40, 212–229.

Soma, M., Churchman, G.J., Theng, B.K.G., 1992. X-ray photoelectron spectroscopic analysis of halloysites with different composition and particle morphology. Clay Miner. 27, 413–421.

Tae, J.W., Jang, B.S., Kim, J.R., Kim, I., Park, D.W., 2004. Catalytic degradation of polystyrene using acid-treated halloysite clays. Solid State Ion. 172, 129–133.

Takahashi, T., Dahlgren, R.A., Theng, B.K.G., Whitton, J.S., Soma, M., 2001. Potassium-selective, halloysite-rich soils formed in volcanic materials from northern California. Soil Sci. Soc. Am. J. 65, 516–526.

Tarì, G., Bobos, I., Gomes, C.S., Ferreira, J.M., 1999. Modification of surface charge properties during kaolinite to halloysite-7 Å transformation. J. Colloid Interface Sci. 210, 360–366.

Theng, B.K.G., Wells, N., 1995. The flow characteristics of halloysite suspensions. Clay Miner. 30, 99–106.

Thompson, H., 1850. On the absorbent power of soils. J. Roy. Agr. Soc. Eng. 11, 68–74.

Vergaro, V., Abdullayev, E., Lvov, Y.M., 2010. Cytocompatibility and uptake of halloysite clay nanotubes. Biomacromolecules 11, 820–826.

Vergaro, V., Lvov, Y.M., Leporatti, S., 2012. Halloysite clay nanotubes for resveratrol delivery to cancer cells. Macromol. Biosci. 12, 1265–1271.

Wada, K., 1958. Adsorption of alkali chloride and ammonium halide on halloysite. Soil Sci. Plant Nutr. 4, 137–144.

Wada, K., 1959. An interlayer complex of halloysite with ammonium chloride. Am. Mineral. 44, 1237–1247.

Wada, K., 1988. Synthesis and characterization of a hollow spherical form of monolayer aluminosilicate. Clay Clay Miner. 36, 11–18.

Wada, K., Ataka, H., 1958. The ion uptake mechanism of allophane. Soil Sci. Plant Nutr. 4, 12–18.

Wada, K., Kakuto, Y., 1985. Embryonic halloysites in Ecuadorian soils derived from volcanic ash. Soil Sci. Soc. Am. J. 49, 1309–1318.

Wada, S.-I., Mizota, C., 1982. Iron-rich halloysite (10 Å) with crumpled lamellar morphology from Hokkaido, Japan. Clay Clay Miner. 30, 315–317.

Watanabe, Y., Kitagawa, Y., Sugo, S., 1969. Structural aspects of clay minerals occurring in some soils. In: Heller, L. (Ed.), Proceedings of the International Clay Conference 1969. Israel Universities Press, Jerusalem, pp. 129–140.

Weaver, C.E., Pollard, L.D., 1973. The Chemistry of Clay Minerals. Elsevier, Amsterdam.

Wei, W., Abdullayev, E., Hollister, A., Mills, D., Lvov, Y., 2012. Clay nanotube/poly(methyl methacrylate) bone cement composite with sustained antibiotic release. Macromol. Mater. Eng. 297, 645–653.

White, G.N., Dixon, J.B., 2002. Kaolin-serpentine minerals. In: Schulze, D., Dixon, J.B. (Eds.), Soil Mineralogy with Environmental Applications. Soil Science Society of America, Madison, USA, pp. 389–414.

White, R.D., Bavykin, D.V., Walsh, F.C., 2012. The stability of halloysite nanotubes in acidic and alkaline aqueous suspensions. Nanotechnology 23, 065701–065710.

Wilson, I.R., 2004. Kaolin and halloysite deposits of China. Clay Miner. 39, 1–15.

Wu, H., Watanabe, H., Ma, W., Fujimoto, A., Higuchi, T., Uesugi, K., Takeuchi, A., Suzuki, Y., Jinnai, H., Takahara, A., 2013. Robust liquid marbles stabilized with surface-modified halloysite nanotubes. Langmuir 29, 14971–14975.

Xie, Y.L., Zhang, Y., Ouyang, J., Yang, H.M., 2014. Mesoporous material Al-MCM-41 from natural halloysite. Phys. Chem. Miner. 41, 497–503.

Xue, J., Niu, Y., Gong, M., Shi, R., Chen, D., Zhang, L., Lvov, Y., 2015. Electrospun microfiber membranes embedded with drug-loaded clay nanotubes for sustained antimicrobial protection. ACS Nano 9, 1600–1612.

Yang, C., Liu, P., Zhao, Y.Q., 2010. Preparation and characterization of coaxial halloysite/polypyrrole tubular nanocomposites for electrochemical energy storage. Electrochim. Acta 55, 6857–6864.

Ye, Y., Chen, H., Wu, J., Ye, L., 2007. High impact strength epoxy nanocomposites with natural nanotubes. Polym. Test. 48, 6426–6433.

Yoshida, M., 1957. Studies on adsorptive ability of the soil. Part 3. Classification of the base exchange positions and their determination. J. Sci. Soil Manure 28, 195–198.

Yuan, P., Tan, D., Annabi-Bergaya, F., Yan, W.C., Fan, M.D., Liu, D., He, H.P., 2012. Changes in structure, morphology, porosity, and surface activity of mesoporous halloysite nanotubes under heating. Clay Clay Miner. 60, 561–573.

Zhang, Y., Yang, H.M., 2012. Halloysite nanotubes coated with magnetic nanoparticles. Appl. Clay Sci. 56, 97–102.

Zhang, Y., Fu, L.J., Yang, H.M., 2012. Insights into the physicochemical aspects from natural halloysite to silica nanotubes. Colloids Surf. A 414, 115–119.

Zhang, Y., Ouyang, J., Yang, H.M., 2014a. Metal oxide nanoparticles deposited onto carbon-coated halloysite nanotubes. Appl. Clay Sci. 95, 252–259.

Zhang, Y., Xie, Y.L., Tang, A.D., Zhou, Y.H., Ouyang, J., Yang, H.M., 2014b. Precious-metal nanoparticles anchored onto functionalized halloysite nanotubes. Ind. Eng. Chem. Res. 53, 5507–5514.

Chapter 5

Characterisation of Halloysite by Electron Microscopy

T. Kogure*
Graduate School of Science, The University of Tokyo, Tokyo, Japan
**Corresponding author: e-mail: kogure@eps.s.u-tokyo.ac.jp*

5.1 INTRODUCTION

This chapter reviews the investigation of halloysite using electron microscopy and introduces recent results with respect to its structure, mainly by the author's group. Because individual grains of halloysite are so minute, microscopic analyses of halloysite are virtually limited to electron microscopy. This is easily understood, for instance, because the distinct tubular form of halloysite can be recognised only by electron microscopy. Beside such morphological analysis, electron microscopy, especially transmission electron microscopy (TEM), is an indispensable tool to investigate the crystal structure of halloysite. Historically, electron diffraction (ED) in TEM has been used for the structural analysis of halloysite. Recently, direct imaging of the crystal structure successfully brought new insights into the structure and formation process of halloysite, although the results are not enough to achieve a complete understanding of them. Before describing these studies, basic facts about electron microscopy are introduced for the benefit of readers who are not familiar with these techniques.

5.2 BACKGROUND OF ELECTRON MICROSCOPY

According to the imaging principles and specimen forms, there are two main types of electron microscopy: scanning electron microscopy (SEM) and TEM. There is another electron microscopy called *scanning transmission electron microscopy* (*STEM*), which is just an eclectic combination of SEM and TEM. STEM is a rapidly developing technique for materials analysis, but it has not yet been actively used for the research of halloysite. In the following text, the background of SEM and TEM is described very briefly, focusing on their applications to the research of halloysite. Readers with more interest

Developments in Clay Science, Vol. 7. http://dx.doi.org/10.1016/B978-0-08-100293-3.00005-4

in these methods should refer to some of many textbooks about electron microscopy (eg, Goldstein et al., 2003; Williams and Carter, 2009), or Kogure (2013), which offers more detail about electron microscopy for clay minerals and related materials.

In SEM, a fine electron probe is scanned on the surface of the specimen and magnified images are obtained using signals emitted from the surfaces. The information obtained from the SEM images depends on the types of signal that originate the contrast of the images. The signals commonly used are secondary electrons (SE), backscattered electrons (BSE), characteristic X-rays, and cathode luminescence, among others. Of them, secondary electron images (SEI) give the surface configuration of the specimens with the highest resolution, which is suitable to observe the morphologies of fine halloysite tubes. Imaging via BSE or X-ray is also very important in the application of SEM to reveal chemical inhomogeneity in the specimens, but it is not discussed in this chapter. The resolution of SEI has been greatly improved in the end of the last century by the commercialisation of SEM with a new electron source, called 'field-emission (FE)' electron guns, owing to their extremely high brightness and small energy spread, compared to conventional 'thermal-emission (TE)' electron guns, which work by heating filaments. The resolution of SEI is primarily determined by the size of the electron probe scanning on the specimen, and the size is strongly dependent on the performance of the electron gun and objective lens. More specifically, brightness, source size and energy spread of the electron source are the important parameters for the determination of the probe size, and FE guns are far superior to TE guns with respect to them. Another merit of FE guns is that the resolution is less degraded if low-energy electrons are used, owing to their small energy spread, which realises surface-sensitive imaging, which is effective for materials consisting of relatively light elements like halloysite. Readers can understand this performance of FE-SEM by looking at several images presented in Figs. 5.1 and 5.3.

The principle of TEM is basically identical to that of optical microscopes with transmitted light, just replacing the visible light to the electron beam with a shorter wavelength, and the optical glass lenses to electron lenses using a magnetic field. However, in addition to its resolution being higher than that of optical microscopes, the advantage of TEM is its intimate collaboration with imaging and diffractometry, especially for crystalline materials. Diffracted electrons from the specimen converge to form a spot, or ED pattern, on the back-focal plane of the objective lens. If the focus length of the next imaging lens (intermediate lens) is adjusted to magnify this back-focal plane instead of the imaging plane, a magnified ED pattern can be observed on the phosphor screen. The area to form the ED pattern can be confined by an aperture (selected-area aperture) located at the imaging plane of the objective lens. Hence, the area of interest for diffractometry can be selected by observing a magnified image of the specimen and masking the outside of the area of interest by the shadow of the aperture. Although the area is primarily confined by this aperture, there is a limitation of such confinement by the aperture due to the spherical aberration of the objective lens. The minimum area for such diffraction, selected-area electron diffraction (SAED), can be close to a few hundred nanometers in diameter. Owing to this technique, it is possible to obtain ED from individual halloysite tubes. This application is described in Section 5.4.

Moreover, recent high performance in the resolution of TEM enables direct observation of atomic columns in the crystal structure of the specimens. Such imaging is generally called *high-resolution transmission electron microscopy* (*HRTEM*). The principal (or strict definition) of HRTEM is that it forms magnified images using 'phase contrast', which is generated by the interference between transmitted electrons and diffracted ones deflected back to the optical axis by the objected lens. If the specimen is thin enough and the focusing condition is optimised, the phase contrast corresponds to the projection of electrostatic potential in the crystal or atomic columns along the beam direction. In this case, the resolution to distinguish the structure is primarily determined by the wavelength of electrons and spherical aberration of the objective lens (see, for example, Fig. 5.8). HRTEM allows a significant contribution in the investigation for the structures of layered materials, particularly for their stacking features. HRTEM imaging, with sufficient resolution with the incident electrons along the constituting layers, enables direct determination of the stacking feature of individual layers. This technique is particularly irreplaceable for the analyses of nonperiodic structures such as for stacking disorder or faults or for interstratifications of different unit layers, etc. After several pioneering studies (eg, Iijima and Buseck, 1978), plenty of HRTEM research has revealed the stacking features of various clay minerals. Moreover, recent aberration-correction techniques for magnetic lenses drastically improved the resolution of TEM (Haider et al., 1998) and STEM (Krivanek et al., 1999), which has raised the significance of HRTEM/HRSTEM in the investigation of the structure of clay minerals (eg, Kogure and Okunishi, 2010). This must be also true for halloysite. However, the application of HRTEM to halloysite is less straightforward because it is so electron-beam-sensitive. These features are described in more detail in Section 5.5.

5.3 MORPHOLOGICAL ANALYSIS

Because common halloysite generally appears with a distinct tubular form—an important criterion to distinguish halloysite from the other kaolin group minerals—electron microscopy has been an irreplaceable technique for the investigation and identification of this material. The importance of TEM and SEM in the morphological investigation of halloysite has changed over time. Until around 1970, the resolution of SEM was poor, and the morphological features of halloysite were investigated only by TEM. Besides simple acicular or fibrous 'shadow' images of halloysite, which were taken from dispersed tubes on electron-transparent thin carbon film, the replica technique (eg, Chukhrov and Zvyagin, 1966; Dixon and McKee, 1974) or thin film preparation using ultramicrotomy brought information about the detail of the tubular form (for instance, cross-sectional shapes of the tubes). As mentioned previously, invention and commercialisation of FE guns improved the resolution of SEM drastically and stereoscopic imaging by SEM (Fig. 5.1) brought a number of new findings for halloysite, which was never obtained by 'shadow pictures' in TEM.

Considering their internal structure, tubular halloysites show two types of morphology: 'cylindrical' and 'prismatic or polygonal'. In the early period of investigation, Chukhrov and Zvyagin (1966) and Dixon and Mckee (1974) showed replica TEM images in which the tubular halloysite particles had a form with facets on the tube side (namely, the 'polygonal prism'). Besides TEM replicas, cross sections of halloysite tubes prepared using ultramicrotomy or ion milling often showed a polygonal form in TEM images (eg, Singh and Gilkes, 1992; Kogure et al., 2013; Fig. 5.2). Such a feature was distinctly identified by recent high-resolution SEM (Fig. 5.1B). Furthermore, ultra-high-resolution SEM imaging revealed very subtle polygonisation

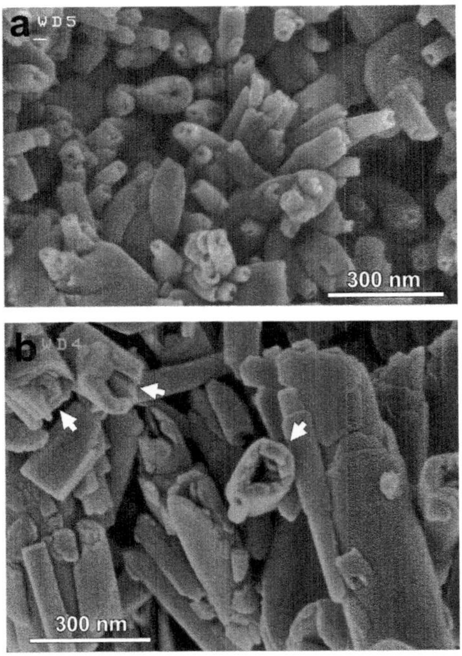

FIG. 5.1 Secondary-electron images (SEI) showing two types of tubular halloysite morphologies. (A) Cylindrical tube (Eureka, Nevada). (B) Prismatic tube showing polygonal cross sections, as indicated with the *white arrows*. (Olkhon Island, in Lake Baikal, Russia). The images were taken using a field-emission SEM (Hitachi S-4500) at an acceleration voltage of 2 kV.

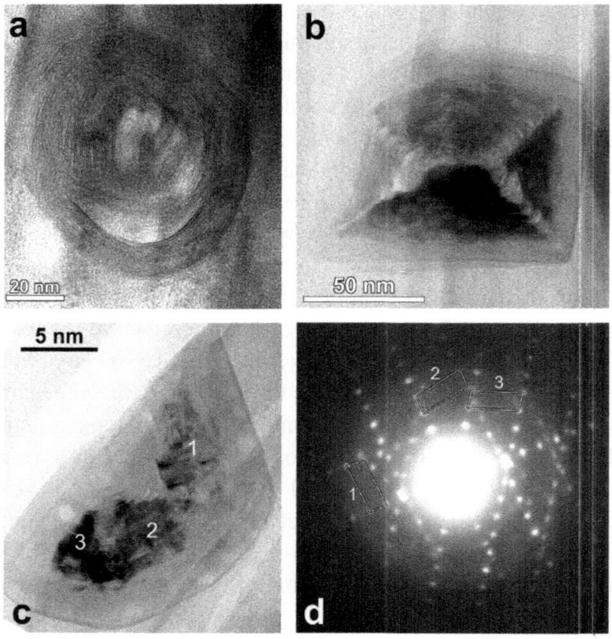

FIG. 5.2 (A–B) Cross-sectional TEM images of halloysite. (A) An almost cylindrical shape and (B) a prismatic one with a pentagonal cross section. (C) Cross section of a prismatic halloysite with three major sectors. One-half of the prism (in the *upper left*) disappeared during the ion-milling sample preparation. In addition, the crystals around the centre and outer surface amorphised before TEM recording. (D) The SAED pattern from the prism in (C), comprised of a superposition of three reciprocal lattices, which show 7-Å layer periodicity. Three sectors in the image correspond to the numbered reciprocal lattices.

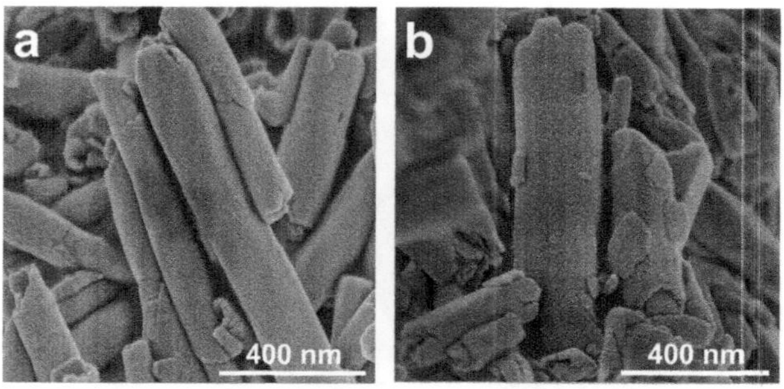

FIG. 5.3 Ultra-high-resolution SEM images of (A) cylindrical tubes and (B) prismatic ones with a number of fine facets on the tube side. Both specimens were from Olkhon Island, in Lake Baikal, Russia. The images were taken with a FEI Magellon 400 L operated at 1 kV. The fine surface contrast on the tubes is the texture of Pt-Pd film coated to prevent charging.

of the halloysite tubes with a number of facets (Fig. 5.3B). TEM images (Fig. 5.4) confirmed that such polygonal tubes adopt a 'multisectored structure' (Kogure et al., 2013) that is just like polygonal chrysotile pioneered by Yada and Wei (1987). The difference between cylindrical and polygonal

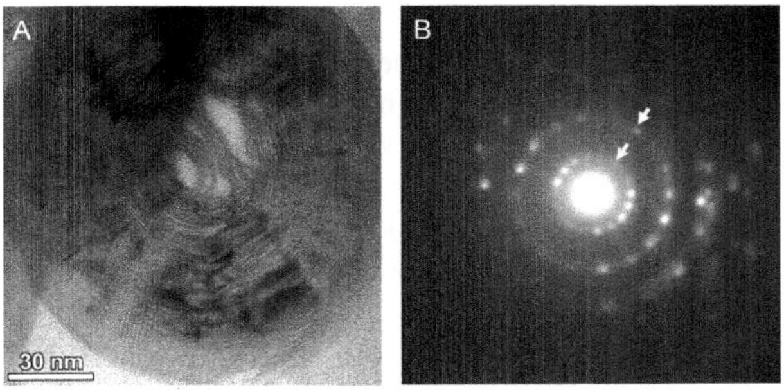

FIG. 5.4 (A) Cross-sectional TEM images of a prismatic halloysite tube, probably like that observed by the SEM image in Fig. 5.3B. (B) SAED pattern from the cross section in (A). The two *arrows* indicate the ring corresponding to 7.2 and 3.6 Å. A total of 18 intensity maximums with an almost equal interval are observed in the rings, indicating that the prism is actually a regular octadecagon with 18 sectors.

forms is important for the crystal structure of halloysite because a certain mutual position between the adjacent kaolinite layers can be established with polygonal forms, in which the layers are almost flat, but this cannot happen with cylindrical forms, in which the layers are curved.

An electron microscopy experiment conducted by Kohyama et al. (1978) showed that water molecules in the interlayer region of hydrated halloysite are rapidly removed in a vacuum environment. Hence, it was impossible to observe raw halloysite (10 Å) by normal TEM. These authors used an environmental cell with a sufficient vapour pressure around the specimen to prevent the dehydration of halloysite (10 Å). They performed an in situ observation of dehydration of halloysite. By dehydration, dense tubular halloysite (10 Å) was divided into packets of 50–100 Å in width, with slits along the tube axis formed by the collapse of the interlayer space. More details about the dehydration of halloysite are given in Chapter 7.

5.4 ELECTRON DIFFRACTION

Crystal structure is the most fundamental information about crystalline materials. A unit cell is the smallest unit of volume that contains the structural and symmetry information, and the whole crystal is described by the repeat of the cell along the three directions. Actually, the atomic arrangement in the unit cell can be determined almost routinely using X-ray diffraction (XRD). However, this procedure cannot be applied to tubular materials such as halloysite. Powder XRD patterns of halloysite generally contain little information. However, according to the results revealed by several other analytical techniques, it is confirmed that halloysite belongs to the kaolin group of minerals, with aluminium octahedra and silicon tetrahedra forming a dioctahedral 1:1 layer, as in kaolinite, dickite and nacrite. The features distinguishing halloysite from the other kaolin group minerals are (i) the rolling of kaolinite layers, resulting in a tubular morphology; and (ii) often the incorporation of water molecules in the interlayer region, allowing halloysite to be easily identified by XRD with the basal spacing close to 10 Å. Collapse of this interlayer space occurs at a high temperature around 100°C. However, how the rolled kaolinite layers are stacked in the tube is still an open question. As is well known, the major difference among the three kaolin group minerals (kaolinite, dickite and nacrite) is their stacking mode of the layers, which leads to different three-dimensional crystal structures. In other words, the three kaolin group minerals are in a polytypic relationship (generally, the same clay mineral name should be used for materials in the polytypic relationship, but the use of different names for these polytypes is allowed for historical reasons). Accordingly, the three clay minerals show different XRD patterns, though the patterns are often 'smeared' by intensive stacking disorder, especially for kaolinite. Similarly, information on the stacking mode in halloysite is of interest. For instance, the question is: Is the stacking mode of halloysite similar to the stacking mode of the other three kaolin group minerals, or is it completely unique? As mentioned previously, the tubular structure of halloysite intrinsically does not have three-dimensional periodicity, and its powder XRD generally shows only '(*hk*) diffraction bands' (Drits and Tchoubar, 1990) which correspond to the diffraction rows indexed with *hk* and intensely streaked by the disordering of layer stacking, as well as a few basal reflections that are also considerably broad (Fig. 5.5).

In such circumstances, investigation of the stacking mode in halloysite has been conducted using ED taken from individual halloysite tubes in TEM images. Because of its tubular morphology, if the tubes are dispersed in water or in ethanol and spread on a thin carbon film, which is a common technique for the simplest sample preparation for TEM, the tube axis of halloysite is oriented perpendicular to the incident electron beam. SAED patterns from a single tube revealed a periodicity of about 4.5 Å along the tube direction, indicating that the kaolinite layer in halloysite was rolled in one of the Y_i-directions (Bailey, 1988) as the tube axis (Fig. 5.6). Another finding was that the reflections on the diffraction rows beside the $00l$ rows frequently showed the interval corresponding to a two-layer periodicity, like dickite. From this result, many studies suggested that stacking mode of halloysite has a two-layer structure (Honjo and Mihara, 1954; Honjo et al., 1954; Chukhrov and Zvyagin, 1966; Kohyama et al., 1978; Singh and Gilkes, 1992). Kohyama et al. (1978) investigated the change in diffraction when hydrated halloysite was dehydrated using an environmental cell as mentioned previously, and they reported that both hydrated/dehydrated halloysites showed similar SAED patterns with a two-layer periodicity, except the change of the basal spacing from 10 to 7 Å. Beyond the periodicity of the stacking, some researchers considered the structure of the two-layer periodicity. Based on the intensity distribution in the diffraction pattern, Honjo et al. (1954) proposed a stacking sequence of $\sigma_1\tau_+\sigma_1\tau_0\sigma_1$ or $\sigma_5\tau_-\sigma_5\tau_0\sigma_5$ with a triclinic symmetry using the Zvyagin symbols (Chukhrov and Zvyagin, 1966). On the

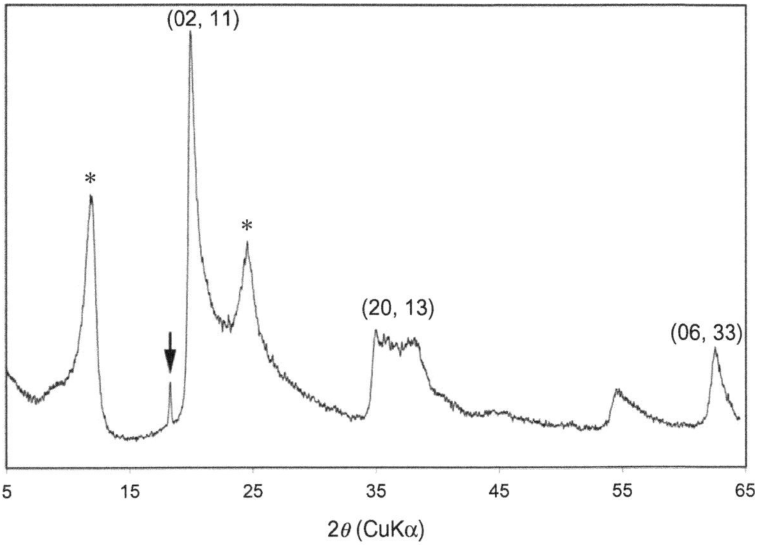

FIG. 5.5 Powder XRD pattern of halloysite from Eureka, Nevada, whose SEM image was shown in Fig. 5.1A. The indexed diffraction peaks are actually 'reflection bands'. For instance, the (02, 11) band corresponds to $02l$, $11l$ and $\overline{1}1l$ diffraction rows streaked along the c^* direction due to the missing of a finite layer-stacking sequence. In addition, the pattern consists of basal reflections, indicated with the *asterisk*, and that from an impurity phase (probably gibbsite), indicated with the *arrow*.

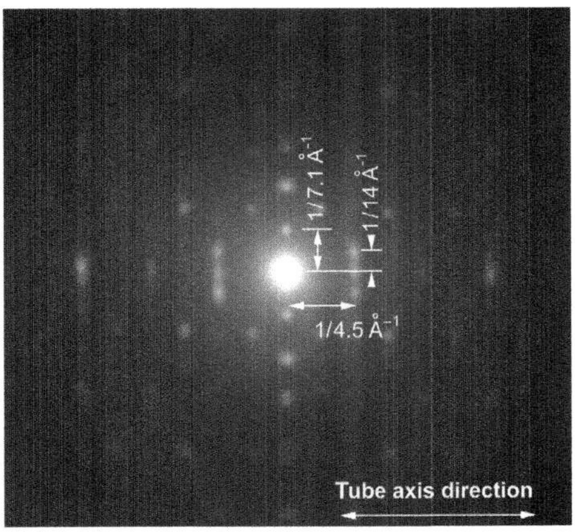

FIG. 5.6 Selected-area electron diffraction (SAED) pattern from a halloysite tube, with the incident beam direction perpendicular to the tube axis.

other hand, Chukhrov and Zvyagin (1966) suggested that the sequence is $\sigma_3\tau_+\sigma_3\tau_-\sigma_3$ with a monoclinic *Cc* symmetry, a stacking sequence that is more realistic than the other sequence, and this could lead to the intensity distribution in the oblique-texture ED pattern. In this notation, σ_3 corresponds to the lateral displacement of $a_o/3$ between the tetrahedral and octahedral sheets within a kaolinite layer, with a_o and b_o as orthogonal cell edges. On the other hand, τ_+ and τ_- correspond to the lateral displacements of $+b_o/3$ and $-b_o/3$, respectively, between the octahedral and tetrahedral sheets across the interlayer space. This sequence was also described as the regular alternation of the 'layer displacements' (Guggenheim et al., 2009) with $a_o/3+b_o/3$ ($\sigma_3\tau_+$) and $a_o/3-b_o/3$ ($\sigma_3\tau_-$). These layer displacements were redefined as t_1 and t_2, which correspond to the layer displacements of two enantiomorphs in kaolinite and also related to each other by the pseudomirror plane in the kaolinite layer, by Bookin et al. (1989) (Fig. 5.7). The stacking model for the two-layer halloysite suggested by Chukhrov and Zvyagin (1966) was just the regular interstratification of t_1 and t_2, which was also explicitly proposed by Bookin et al. (1989) (Fig. 5.7).

According to these previous studies, a two-layer periodicity in the stacking mode of halloysite is generally accepted. However, halloysite is generally regarded as being formed in a more nonequilibrium condition than kaolinite, which implies a more disordered stacking in halloysite. If halloysite adopts a two-layer periodicity, how precisely is it regulated? This question will be addressed in the next section.

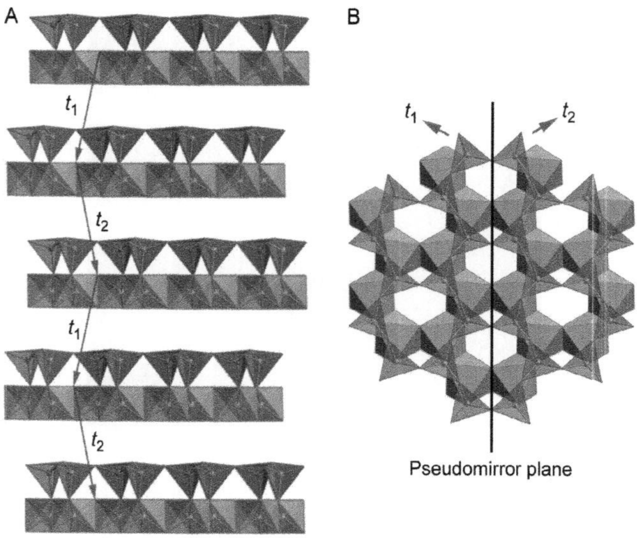

FIG. 5.7 (A) A stacking model of halloysite with a two-layer periodicity, proposed by Chukhrov and Zvyagin (1966) and Bookin et al. (1989), with regular alternation of the layer displacements t_1 and t_2 defined for the stacking of kaolinite by Bookin et al. (1989). (B) The relationship between the pseudomirror plane in the kaolinite layer, and t_1 and t_2 (Bookin et al., 1989).

5.5 HRTEM IMAGING OF THE CRYSTAL STRUCTURE

As mentioned previously in Section 5.1, HRTEM has made a significant contribution to the investigation of the structure of clay minerals and layered materials, particularly in terms of their stacking features. However, only a few studies have analysed stacking in the kaolin group of minerals (eg, Ma and Eggleton, 1999; Buatier et al., 2000). This is mainly because the kaolin group minerals are too weak to withstand electron beam radiation (Robertson and Eggleton, 1991).

In order to record the structure images, which were defined by Spence (1981) as HRTEM images from which the crystal structure is directly interpretable, a high magnification (at least 300,000 × to 400,000 ×) is necessary, and to obtain sufficient brightness for such magnified images, the electron flux to the specimen becomes very high. Due to this high flux, clay minerals are gradually damaged during recording or operation for HRTEM and converted to an amorphous state even before the recording is completed. Presumably the amorphisation proceeds from the surface, grain boundaries (Kogure and Murakami, 1998) and stacking faults (Kogure et al., 2006) of the specimens. Thin regions suitable for the structure imaging in the TEM specimens are also being damaged rapidly. The tolerance against the radiation damage is strongly dependent on the species of clay mineral. Although the reason is not clear, generally clay minerals containing iron (Fe), like biotite, chlorite and cronstedtite, are relatively electron-beam tolerant. On the other hand, kaolin minerals, like talc, pyrophyllite and illite, are beam-sensitive. In order to record structure images of these clay minerals, procedures for the recording must be completed quickly to minimise the total electron dose on the specimen. Accordingly, the experience and skill of the TEM operator are critical. In these circumstances, Kogure and Inoue (2005) first reported successful HRTEM images of diagenetic kaolinite and dickite along their X_i-directions (Bailey, 1988) with sufficient resolution to directly determine the stacking mode in the clay minerals. From their study, the nature of the stacking disorder in the diagenetic kaolinite was unambiguously confirmed; the disorder is mainly originated from the alternative of the layer displacements, t_1 and t_2 (Fig. 5.8), as suggested by Plançon et al. (1989) from XRD profile analyses. However, extension of these studies of kaolinite or dickite to the investigation of halloysite was not straightforward because halloysite is far more beam-sensitive than kaolinite. The change of the lattice image at the side of a tube with various amounts of electron doses is shown in Fig. 5.9. In the image with an electron dose of ca. 390 e/Å^2 (Fig. 5.9A), the lattice fringes of the basal plane (7.4 Å) are visible in the whole area of the tubular wall at the tube sides with no apparent evidence of the radiation damage. With an electron dose of 1950 e/Å^2 (Fig. 5.9B), the fringes became faint and only visible around the centre of the wall, suggesting that radiation-induced amorphisation proceeded from the inner and outer surface of the wall. Finally, with an electron dose of

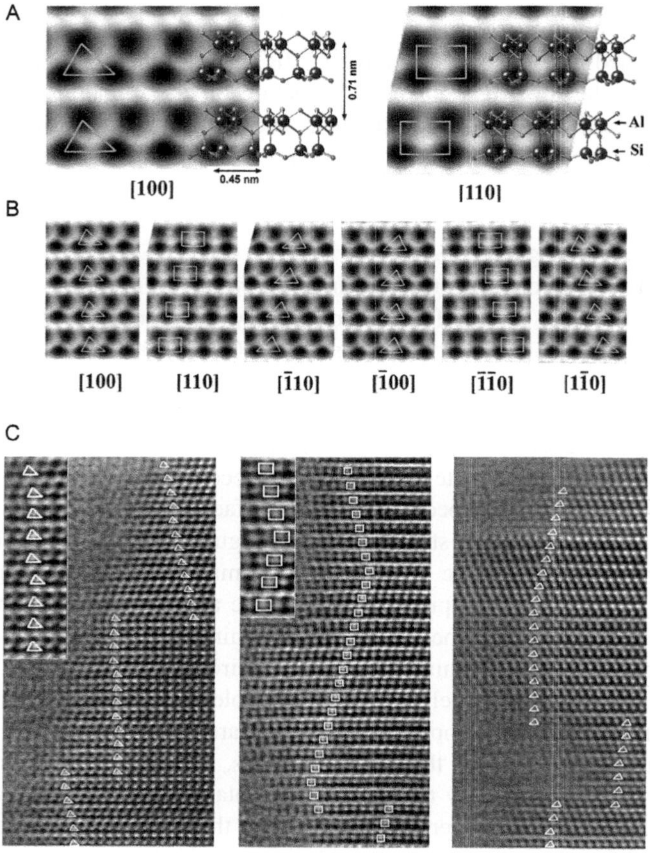

FIG. 5.8 (A) Crystal structure and corresponding simulated HRTEM image of kaolinite, viewed down [100] and [110]. In the crystal structure model, Al and Si atoms are represented as *large spheres* and oxygen (hydroxyl) atoms as *small spheres*. The *triangles* and *squares* in the left are the pattern used to identify the layer orientation in the observed images. (B) Simulated images of kaolinite viewed down the six directions. (C) HRTEM images (a noise reduction process was applied) of disordered kaolinite of diagenetic origin. The pattern (*triangle* or *square*) for each layer defined in (A) is superimposed. The insets at the *top-left* are the magnified images. See Kogure and Inoue (2005) for the detail.

8800 e/Å^2, the fringes completely disappeared (Fig. 5.9C). From this observation, it is concluded that the electron dose should be lower than several hundred e/Å^2 for taking HRTEM images of halloysite without apparent radiation damage. However, certain amounts of electron doses (at least 1 to 200 e/Å^2 for conventional films) are necessary to record near-atomic resolution structure images with sufficient magnification and without statistical noise for analysing the stacking structure. Hence, the electron dose that can be used for HRTEM recording of halloysite tubes, which involves the selection of the imaging area and adjustment of the brightness and focusing (generally

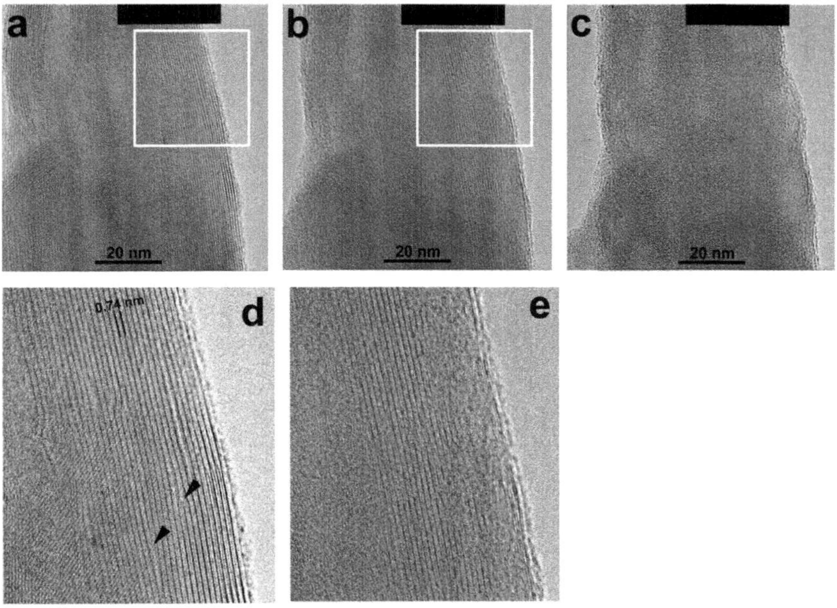

FIG. 5.9 HRTEM images of halloysite with various amounts of electron dose. (A) 390 e/Å^2. (B) 1950 e/Å^2. (C) 8800 e/Å^2. (D), (E) Magnified images of the squared areas in (A) and (B), respectively. Insertion of extra layers in the stacking is observed at the arrowed points in (D).

the setting of the crystal orientation is not necessary for tubular halloysite) before recording the final images, is very small. This is probably impossible to achieve with conventional TEM.

To overcome this problem, Kogure et al. (2011) used a TEM with a computer-assisted minimal dose system (MDS) (Hayashida et al., 2007). The MDS consisted of a charge-coupled device (CCD) camera, an electric-field-type electron-beam blanker installed beneath the condenser lens and a computer for controlling the whole TEM system, including the CCD camera and the high speed blanker. Practical TEM operations, such as brightness control, beam positioning, sample searching, magnifying, focusing and exposure, were performed using the MDS, and the TEM specimen was irradiated only when the images were taken by the CCD camera. The delay time between the blanking signal from the wave generator for the beam on/off and actual beam blanking was less than 40 μs. Moreover, high-resolution images were finally recorded using imaging plates (25 μm × 25-μm pixels).

Using this new TEM technique, Kogure et al. (2011, 2013) investigated two types of halloysite with cylindrical and prismatic morphologies. The cylindrical one was from Eureka, Nevada, USA (from the collection of the late S.W. Bailey at the University of Wisconsin-Madison). The SEM image and powder XRD pattern were already shown in Figs 5.1 and 5.5. An HRTEM image at the end of a tube is shown in Fig. 5.10, which was taken with an

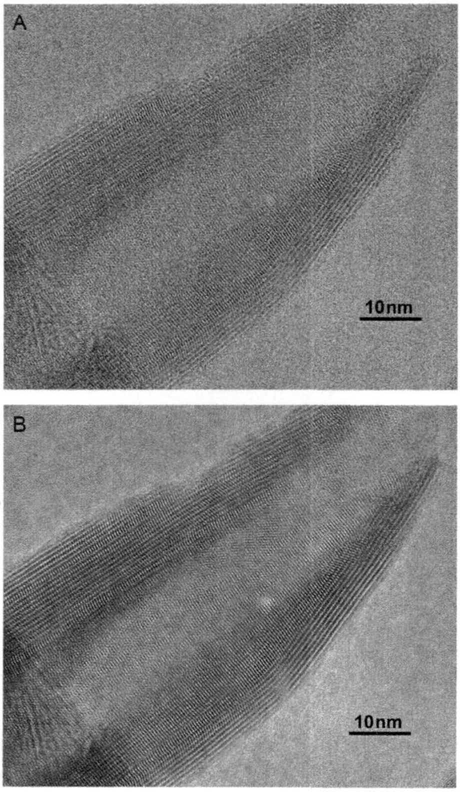

FIG. 5.10 (A) Original and (B) processed (noise-reduced) HRTEM images at the end of a cylindrical halloysite tube from Eureka, Nevada. The total electron dose was about 300 e/Å^2.

electron dose of ca. 300 e/Å^2. Two-dimensional lattice fringes are visible at both sides of the tube, where the incident electron beam was parallel to the layer and the stacking mode of halloysite should appear. Moreover, a noise reduction process (Marks 1996; Kilaas 1998; Kogure et al., 2008) was applied to the original image to make the periodic contrast more distinct (Fig. 5.10B). Although the quality of the image is still not enough to unambiguously determine the stacking sequence, stacking disorder is apparent from the lattice fringes. As mentioned previously, several studies have suggested a two-layer periodicity from ED analyses. However, no appearance of such a two-layer (ca. 14 Å) periodicity can be seen in the recorded HRTEM images. Another HRTEM image with higher magnification is presented in Fig. 5.11. The contrast for each kaolinite layer consists of two-tiered dark dots (or bright dots between the dark dots) separated by about 4.5 Å along the layer. As shown in Fig. 5.8A (Kogure and Inoue, 2005), each dark dot corresponds to a pair of SiO_4 tetrahedra (or an SiO_4 chain running parallel to the incident beam) in the tetrahedral sheet and a pair of $AlO_2(OH)_4$ octahedra in the

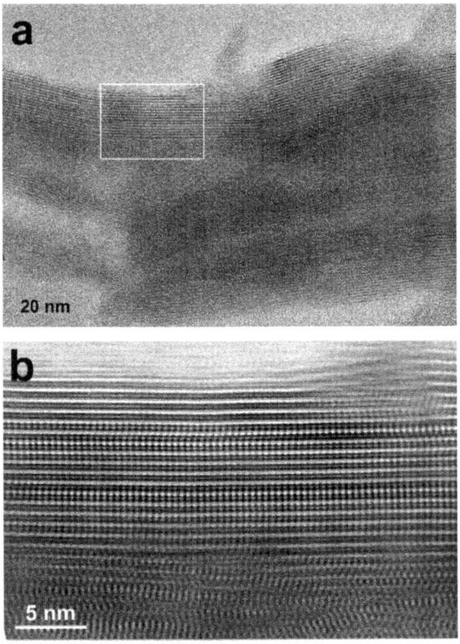

FIG. 5.11 (A) HRTEM image of a halloysite tube, and (B) magnified and noise-reduced image of the squared area in (A). Notice that the contrasts for most individual layers are two-tiered *dark dots*, as in the right of Fig. 5.8A, but the contrasts are not distinctly shifted laterally between the adjacent layers.

dioctahedral sheet. In Fig. 5.11B, the contrast of most of the layers shows vertically aligned dots with no lateral shift. This result gives an insight for the halloysite structure that the kaolinite layer rolls with the roll or tube axis perpendicular to the pseudomirror plane (Bookin et al., 1989) of the layer. Moreover, the contrasts are also not shifted laterally between the adjacent layers. This feature does not correspond to any of the simulated images for kaolinite along the six possible X_i-directions (Fig. 5.8B) indicating a different stacking sequence from the other kaolin group minerals. The contrast for each kaolinite layer with vertically aligned dots in Fig. 5.8B appears in the HRTEM images of kaolinite when viewed along [110] or $[\bar{1}\bar{1}0]$. However, the contrasts for the individual layers must be staggered between the adjacent layers if the kaolinite structure is viewed along these directions because the layer displacement (Guggenheim et al., 2009) of kaolinite is almost along $[\bar{1}00]$. As mentioned previously, a certain atomic arrangement across the interlayer is not expected with bent or rolled adjacent layers. Hence, hydrogen bonding between the adjacent layers cannot be (or can be only partly) formed across the interlayer region, unlike the platy kaolinite, which may be the origin of the unique stacking structure of the cylindrical halloysite. Kogure et al. (2013) investigated also a prismatic halloysite specimen from Olkhon Island in Lake Baikal,

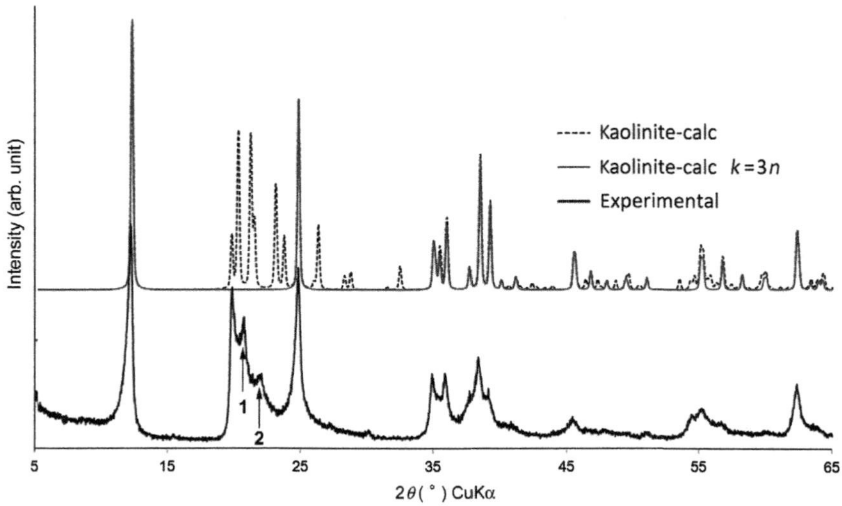

FIG. 5.12 XRD pattern of halloysite from Olkhon Island in Lake Baikal, Russia, taken at ambient conditions. The calculated patterns for kaolinite including all reflections and those with $k=3n$ are also plotted for comparison. Although they are broadened, most halloysite reflections are identified in the calculated kaolinite pattern with $k=3n$. However, those indicated by *arrows* 1 and 2 on the (02, 11) band do not correspond to kaolinite.

Russia, which was formed by the weathering of oligoclase (Chekin et al., 1972, 1976). The XRD pattern (Fig. 5.12) of the specimen indicates a basal spacing of ca. 7.2 Å, suggesting that the halloysite was almost as dehydrated as the halloysite 7 Å. Apart from the basal reflections, several reflections were distinctly identified, deviating from the typical XRD patterns of halloysite reported by other research (eg, Joussein et al., 2005). These reflections are compared with those calculated from the crystallographic parameters for kaolinite (Neder et al., 1999) in Fig. 5.12. The pattern consisting of only the reflections with $k=3n$ was drawn with a solid line, while that with all reflections is drawn with a dotted line. The experimental pattern profile higher than 2θ of 35 degrees corresponds approximately to the calculated pattern for kaolinite. In particular, calculated reflections with $k=3n$ almost correspond to those in the experimental pattern. The reflections $h\neq 0$ and $k=3n$ were called *family reflections* (Ďurovič and Weiss, 1986), and their intensity distributions are distinct for the four polytypic groups (or structure types) of 1:1 clay minerals (Zvyagin, 1967; Bailey, 1969). The rough correspondence of the family reflections between the experimental and calculated patterns indicates that the stacking sequence in the present halloysite sample belongs to group A (Bailey, 1969) with a layer displacement of approximately $\pm a_i/3$ and/or mutual rotations between adjacent layers with 0 or ±120 degrees. The experimental pattern also shows two characteristic reflections on the tail of the (02, 11) band, as indicated with arrows 1 and 2. Their d-values are 4.28 and

4.03 Å, respectively, which are in complete disagreement with the reflections for kaolinite. These *d*-values can be indexed as 021 and $\overline{1}12$ with a two-layer unit cell (Chukhrov and Zvyagin, 1966; Bailey, 1990). However, these reflections cannot be ascribed to those for dickite because those for dickite have very weak intensities, compared to other reflections such as 022 or 111. On the other hand, these reflections have the strongest intensities in the 02*l*, 11*l* and $\overline{1}1l$ reciprocal lattice rows for the stacking model proposed by Chukhrov and Zvyagin (1966). Another intense reflection, 112 with *d* being about 3.49 Å, as expected from their model, is overlapped with 004 reflections in the powder XRD pattern. Hence, the experimental XRD pattern in Fig. 5.10 is apparently consistent with the Chukhrov and Zvyagin model.

Two-dimensional HRTEM images of the specimen recorded using the TEM with a minimal dose system are shown in Fig. 5.13. The contrast of most kaolinite layers shows vertically aligned dots with no lateral shift. This is the same feature as was observed for the cylindrical halloysite (Fig. 5.11). However, the contrast for each kaolinite layer is laterally shifted to the right or left against that for the lower and upper layers in the amount of *b*/6 (only two exceptions are indicated with the white arrows in Fig. 5.13C, where no lateral shift is observed). As described by Kogure and Inoue (2005), these right and left shifts correspond to the projections of the layer displacements for t_1 and t_2, or directly correspond to the Zvyagin's interlayer shifts of τ_+ and τ_-. Hence, the two-layer stacking model proposed by Chukhrov and Zvyagin (1966) should appear as the alternating shift of the contrast for the

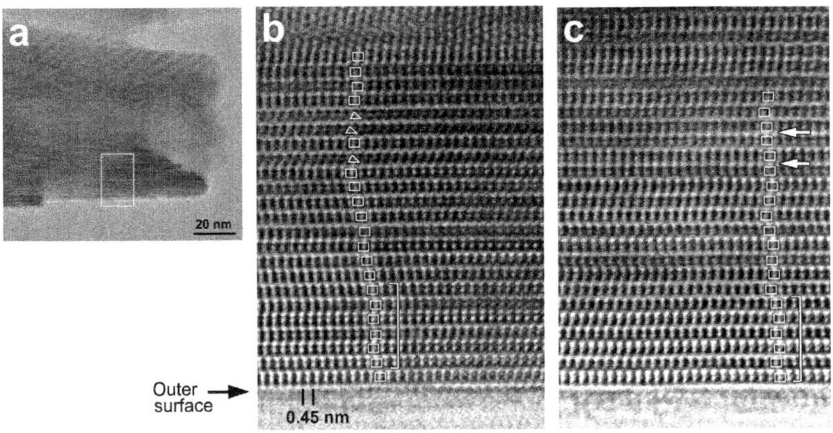

FIG. 5.13 (A) Low-magnified image of a halloysite tube. (B) Noise-reduced HRTEM image in the rectangle in (A). The *squares* and *triangles* at the individual kaolinite layers indicate the types of contrast relating to the layer orientation defined in Fig. 5.8. The square bracket indicates a position where the stacking has the two-layer periodicity in Fig. 5.7A, as proposed by Chukhrov and Zvyagin (1966). (C) Another noise-reduced HRTEM image of the side of a different halloysite tube, showing a similar contrast. The interlayers indicated with *white arrows* show no lateral shift between the adjacent layers.

kaolinite layer in Fig. 5.13. However, such alternation was only observed locally (indicated with square brackets in the figure) with six or seven layers, and not with the entire stacking. In conclusion, the interlayer configuration in the prismatic halloysite is probably identical to that in kaolinite (t_1 or t_2 with hydrogen bonding), but t_1 and t_2 occur randomly with frequent exchange and similar populations, instead of the continuity of either or the regular alternation of both of them.

5.6 RECONCILIATION OF ED AND HRTEM RESULTS

As mentioned earlier in this chapter, in previous research, the two-layer periodicity was proposed for the stacking structure of halloysite based on ED, but HRTEM results do not support the two-layer periodicity both in cylindrical and prismatic halloysite. Some explanation is necessary to account for this disagreement.

First, the ED patterns such as that in Fig. 5.6 should be reexamined. Previous ED results show the reflection at the intersection of the tube direction and the 02*l* row, which can be indexed 020. On the 02*l* row, the next reflection in both sides of 020 has a distance that nearly corresponds to the two-layer periodicity (Fig. 5.6), which is the reason why the two-layer structure was proposed. However, this reflection, which can be indexed 021 for the two-layer periodicity, is considerably streaked in the SAED patterns presented in the previous studies, except for that presented by Chukhrov and Zvyagin (1966). These patterns probably were obtained with a selected-area aperture, including the entire diameter of the halloysite tube (b in Fig. 5.14A). In this case, the intensity maximum will be inevitably formed at the 020 position by the streak in the 02*l* row, originated from the top and/or bottom parts of the tubular wall normal to the electron beam (shown by arrowheads in Fig. 5.14B). On the other hand, if only the side region of the prism is selected using the selected-area apertures (shown as 'c' in Fig. 5.14A), the reflection at the 020 position does not appear (Fig. 5.14C).

SAED patterns for the structure suggested from the HRTEM results (Fig. 5.13) have been simulated using DIFFaX (Treacy et al., 1991). This programme is dedicated to calculate diffraction patterns from layer structures containing stacking disorder. The calculation procedure is similar to that in Kogure et al. (2006, 2010) for instance. The lattice constants and atomic coordinates for the kaolinite layer were derived from the results by Neder et al. (1999) and coordinates for t_1 and t_2 layer displacements were derived from Bookin et al. (1989). To make the calculation more simple, the proportions (W_i) of t_1 and t_2 were assumed equal ($W_1=W_2=0.5$). Then the intensity distribution along the 20*l* diffraction row in the SAED pattern was calculated as a function of P_{11}, where P_{11} refers to the probability that t_1 comes next to t_1 in the stacking. The calculated XRD patterns for several P_{11} values are shown in Fig. 5.14D, with the experimental one for comparison. In Fig. 5.14D, $P_{11}=0$

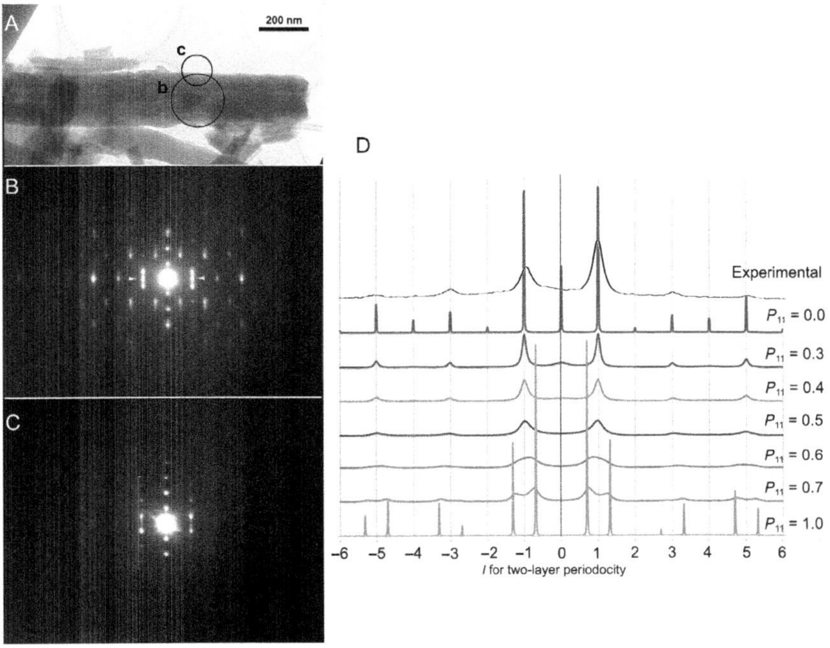

FIG. 5.14 (A) Bright-field TEM image, showing the position and size of selected-area aperture used to acquire the SAED patterns in (B) and (C). (B) SAED pattern from the area selected by the larger aperture marked 'b' in (A). (C) SAED pattern from the side of the prism, using the smaller aperture marked 'c' in (A). Note that the intensity maximums indicted with the *arrowheads* in (B) have almost disappeared in (C). (D) Calculated intensity distribution along the diffraction rows indicated with the *white bracket* in (C). The unit for the horizontal axis is the value of l for a two-layer periodicity. The experimental intensity distribution in (C) is also shown at the top.

corresponds to the stacking model by Chukhrov and Zvyagin (1966) with a two-layer periodicity, and $P_{11}=1$ to the segregated two enantiomeric kaolinite crystals with a one-layer periodicity. As P_{11} increases from zero, the reflections are broadened and those with $l=2n$ disappear, losing any evidence of a two-layer periodicity. The diffraction peaks at $l=\pm1$ correspond to those in the SAED pattern in Fig. 5.14C. As P_{11} exceeds 0.5, these peaks split and finally become those in the $11l$ diffraction row for the two enantiomeric kaolinite crystals. Accordingly, the intensity distribution of the $20l$ row in the SAED pattern (Fig. 5.14C) is corresponding to the stacking feature recorded in the HRTEM images in Fig. 5.13.

Using the same programme and parameters, powder XRD patterns were also produced for various P_{11} values (Fig. 5.15). The two characteristic peaks on the (02, 11) band in the experimental pattern are reproduced in the wide range of P_{11}, from $P_{11}=0$ to even the most disordered state of $P_{11}=0.5$. The peaks corresponding to those for kaolinite appear from $P_{11}=0.7$. The peak with $d=4.28$ Å in the XRD pattern corresponds to the peaks at $=\pm1$

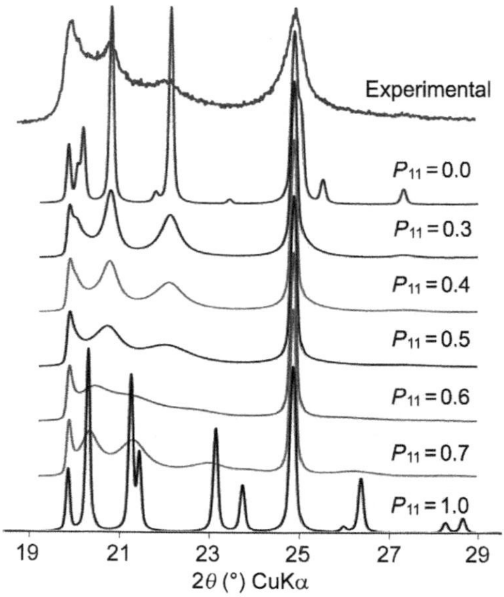

FIG. 5.15 Simulated (02, 11) band in the powder XRD patterns for prismatic halloysite as a function of P_{11}, the condition probability to continue the layer displacement of t_1 over t_1. The experimental pattern is also shown at the *top*.

in the SAED pattern in Fig. 5.14D, and that with $d=4.03$ Å in the XRD pattern corresponds to those in the other reciprocal rows.

In summary, these simulation results suggest that a two-layer periodicity of halloysite may be just due to a misunderstanding of the ED pattern using improper setting of the selected-area aperture. In the past, the aberration of the objective lens in TEM was not small, which resulted in considerable uncertainty of the defined specimen area by the selected-area aperture. Accordingly, acquisition of the diffraction pattern only from the side wall of a single halloysite tube, as shown in Fig. 5.14C, was probably difficult. Of course, Kogure et al. (2011, 2013) did not investigate the same specimen as in previous studies and the existence of a real two-layer periodicity in a halloysite specimen might still be possible.

5.7 FORMATION MECHANISM OF HALLOYSITE STRUCTURE

Two recent HRTEM studies (Kogure et al., 2011, 2013) have advanced our understanding of the halloysite structure, although it is not complete yet. The interlayer structure of prismatic halloysite is identical to that in kaolinite, but their stacking sequence is characterised by random and frequent exchange of t_1 and t_2 layer displacements whose proportion is comparable, as described previously. This stacking model can explain ED patterns from individual tubes and two peaks in the 02, 11 band in the power XRD pattern for the prismatic halloysite. On the other hand, the interlayer structure of cylindrical halloysite is different from those of other kaolin minerals. This is probably due to limitations in, if not nonexistence of, hydrogen bonding. The tube axis is perpendicular to the pseudomirror plane in the kaolinite layer for both cylindrical and prismatic halloysites. Although there is no evidence or theoretical prediction, it is suspected that the kaolinite layer can bend more easily along this direction to compensate for the lateral misfit between the tetrahedral and octahedral sheets.

Finally, a scenario for the formation process of cylindrical and prismatic halloysites with different stacking modes is proposed here (Fig. 5.16), many uncertainties remained to be solved in the future. It seems natural, and is undisputed, that initially halloysite grows as a hydrated form, halloysite (10 Å), with water molecules at the interlayer region (Fig. 5.16A). In a second step, the water molecules are removed from the interlayer space to form halloysite 7 Å (Fig. 5.16B). At this stage, because the kaolinite layers are not flat but cylindrical, a certain mutual position between the adjacent layers is not

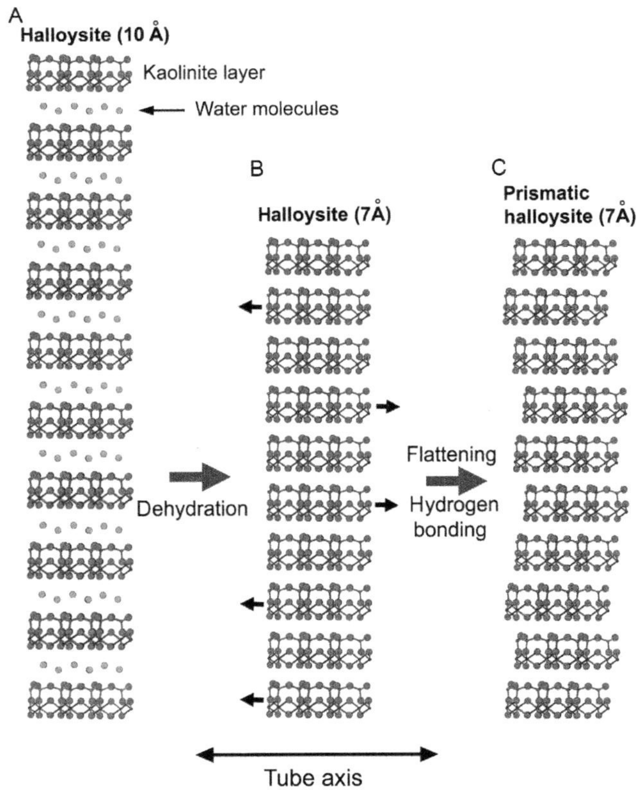

FIG. 5.16 Schematic drawing for the ripening pathway of halloysite. See text for details.

realised over the entire circumference, and hydrogen bonding across the interlayer is not formed or is very limited in a circuit. Hence, no distinct lateral shift is observed between the adjacent layers, as the HRTEM image in Fig. 5.11B demonstrates. During long-term ripening of halloysite, cylindrical forms may be converted to prismatic ones with sector domains in which the kaolinite layers are flat. In order to form hydrogen bonding and interlayer configuration like kaolinite within the sector domains, each layer must be shifted to form t_1 or t_2 layer displacement (Fig. 5.16C). At this time, in order to minimise deformation of the external form of halloysite, the proportion of t_1 or t_2 should be comparable and altered frequently, which results in the stacking sequence shown in Fig. 5.13.

5.8 CONCLUDING REMARKS

Although more observations and analyses are needed to come to a definitive conclusion, it is suggested that tubular halloysite is structurally classified into three types: hydrated-cylindrical, dehydrated-cylindrical and prismatic, as shown in Fig. 5.16. The difference between the prismatic halloysite and kaolinite is subtle—only the occurrence mode of the stacking disorder formed by the alternative of the two kinds of displacements between the adjacent layers. This difference originates in the difference of the growth mechanisms of two minerals. In the case of kaolinite, the interlayer structure with hydrogen bonding is formed during the growth of a new layer onto the preexisting layer. For prismatic halloysite, if the model presented in Fig. 5.16 is valid, the same interlayer structure is developed after the tubular structure is formed. A two-layer periodicity previously proposed from ED analyses is probably rare or very regional in halloysite. These insights will contribute to the interpretation of the analytical results obtained by macroscopic techniques such as XRD and IR spectroscopy, as well as to the understanding of the question on the genesis of this distinct clay mineral.

REFERENCES

Bailey, S.W., 1969. Polytypism of trioctahedral 1:1 layer silicates. Clays Clay Miner. 17, 355–371.

Bailey, S.W., 1988. Polytypism of 1:1 layer silicates. In: Bailey, S.W. (Ed.), Hydrous Phyllosilicates (Exclusive of Micas). Reviews in Mineralogy, Vol. 19. Mineralogical Society of America, Washington, D.C, pp. 9–27.

Bailey, S.W., 1990. Halloysite—a critical assessment. In: Farmer, V.C., Tardy, Y. (Eds.), Proceedings of the 9th International Clay Conference, Strasbourg, France. Sci. Geol. Mem. 86, pp. 89–98.

Bookin, A.S., Drits, V.A., Plançon, A., Tchoubar, C., 1989. Stacking faults in kaolin-group minerals in the light of real structural features. Clays Clay Miner. 37, 297–307.

Buatier, M.D., Deneele, D., Dubois, M., Potdevin, J.-L., Lopez, M., 2000. Nacrite in the Lodève Permian Basin: a TEM and fluid-inclusion study. Euro. J. Mineral. 12, 329–340.

Chekin, S.S., Samptoin, N.D., Finko, V.I., 1972. Formation of halloysite by weathering of oligoclase. Izvestiya Akademii Nauk SSSR, Seriya geologicheskaya N11, 88–114 (in Russian).
Chekin, S.S., Samotoin, N.D., Finko, V.I., 1976. Spiral-cylindrical shape and growth of halloysite crystallites. Izvestiya Akademii Nauk SSSR, Seriya geologicheskaya N 6, 111–124. in Russian.
Chukhrov, F.V., Zvyagin, B.B., 1966. Halloysite, a crystallochemically and mineralogically distinct species. In: Heller, L., Weiss, A. (Eds.), Israel Program for Scientific Translation, Jerusalem, Israel.
Dixon, J.B., McKee, T.R., 1974. Internal and external morphology of tubular and spheroidal halloysite particles. Clays Clay Miner. 22, 127–137.
Drits, V.A., Tchoubar, C., 1990. X-Ray Diffraction by Disordered Lamellar Structures: Theory and Applications to Microdivided Silicates and Carbons. Springer-Verlag, Berlin Heidelberg.
Ďurovič, S., Weiss, Z., 1986. OD structures and polytypes. Bull. Mineral. 109, 15–29.
Goldstein, J.I., Lyman, C.E., Newbury, D.E., Lifshin, E., Echlin, P., Sawyer, L., Joy, D.C., Michael, J.R., 2003. Scanning Electron Microscopy and X-ray Microanalysis. Springer, New York.
Guggenheim, S., Adams, J.M., Bergaya, F., Brigatti, M.F., Drits, V.A., Fromoso, M.L.L., Galan, E., Kogure, T., Stanjek, H., Stucki, J.W., 2009. Nomenclature for stacking in phyllosilicates: report of the Association Internationale pour l'Etude des Argiles (AIPEA) nomenclature committee for 2008. Clay Miner. 44, 157–159.
Haider, M., Rose, H., Uhlemann, S., Schwan, E., Kabius, B., Urban, K., 1998. A spherical-aberration-corrected 200 kV transmission electron microscope. Ultramicroscopy 75, 53–60.
Hayashida, M., Nomaguchi, T., Kimura, Y., Takai, Y., 2007. Development of computer-assisted minimal dose system with beam blanker for TEM. Micron 38, 505–512.
Honjo, G., Mihara, K., 1954. A study of clay minerals by electron-diffraction diagrams due to individual crystallites. Acta Crystallogr. 7, 511–513.
Honjo, G., Kitamura, N., Mihara, K., 1954. A study of clay minerals by means of single-crystal diagrams—the structure of tubular kaolin. Clay Miner. Bull. 2, 133–141.
Iijima, S., Buseck, P.R., 1978. Experimental study of disordered mica structures by high-resolution electron microscopy. Acta Crystallogr. A34, 709–719.
Joussein, E., Petit, S., Churchman, J., Theng, B., Righi, D., Delvaux, B., 2005. Halloysite clay minerals—a review. Clay Miner. 40, 383–426.
Kilaas, R., 1998. Optimal and near-optimal filters in high-resolution electron microscopy. J. Microsc. 190, 45–51.
Kogure, T., 2013. Electron microscopy. In: Bergaya, F., Lagaly, G. (Eds.), Handbook of Clay Science (2nd Edition) Part B: Techniques and Applications. Developments in Clay Science, vol. 5B. Elsevier, Amsterdam, pp. 275–317.
Kogure, T., Inoue, A., 2005. Determination of defect structures in kaolin minerals by high-resolution transmission electron microscopy (HRTEM). Am. Mineral. 90, 85–89.
Kogure, T., Murakami, T., 1998. Structure and formation mechanism of low-angle grain boundaries in chlorite. Am. Mineral. 83, 358–364.
Kogure, T., Okunishi, E., 2010. Cs-corrected HAADF-STEM imaging of silicate minerals. J. Electron Microsc. 59, 263–271.
Kogure, T., Jige, M., Kameda, J., Yamagishi, A., Miyawaki, R., Kitagawa, R., 2006. Stacking structures in pyrophyllite revealed by high-resolution transmission electron microscopy (HRTEM). Am. Mineral. 91, 1293–1299.
Kogure, T., Eilers, P.H.C., Ishizuka, K., 2008. Application of optimum HRTEM noise filters in mineralogy and related sciences. Microsc. Anal. 22, S11–S14.

Kogure, T., Johnston, C.T., Kogel, J.E., Bish, D., 2010. Stacking disorder in a sedimentary kaolinite. Clays Clay Miner. 58, 63–72.

Kogure, T., Mori, K., Kimura, Y., Takai, Y., 2011. Unraveling the stacking structure in tubular halloysite using a new TEM with computer-assisted minimal-dose system. Am. Mineral. 96, 1776–1780.

Kogure, T., Mori, K., Drits, V.A., Takai, Y., 2013. Structure of prismatic halloysite. Am. Mineral. 98, 1253–1260.

Kohyama, N., Fukushima, K., Fukami, A., 1978. Observation of hydrated form of tubular halloysite by an electron microscope equipped with an environmental cell. Clays Clay Miner. 26, 25–40.

Krivanek, O.L., Dellby, N., Lupini, A.R., 1999. Towards sub-angstrom electron beams. Ultramicroscopy 78, 1–11.

Ma, C., Eggleton, R.A., 1999. Surface layer types of kaolinite: a high-resolution transmission electron microscope study. Clays Clay Miner. 47, 181–191.

Marks, L.D., 1996. Wiener-filter enhancement of noisy HREM images. Ultramicroscopy 62, 43–52.

Neder, R.B., Burghammer, M., Grasl, T., Schulz, H., Bram, A., Fiedler, S., 1999. Refinement of the kaolinite structure from single-crystal synchrotron data. Clays Clay Miner. 47, 487–494.

Plançon, A., Giese, R.F., Snyder, R., Drits, V.A., Bookin, A.S., 1989. Stacking faults in the kaolin-group minerals—defect structures of kaolinite. Clays Clay Miner. 37, 203–210.

Robertson, I.D.M., Eggleton, R.A., 1991. Weathering of granitic muscovite to kaolinite and halloysite and of plagioclase-derived kaolinite and halloysite. Clays Clay Miner. 39, 113–126.

Singh, B., Gilkes, R.J., 1992. An electron optical investigation of the alteration of kaolinite to halloysite. Clays Clay Miner. 40, 212–219.

Spence, J.C.H., 1981. Experimental High-Resolution Electron Microscopy. Clarendon Press, Oxford.

Treacy, M.M.J., Newsam, J.M., Deem, M.W., 1991. A general recursion method for calculating diffracted intensities from crystals containing planar faults. Proc. R. Soc. Lond. A 433, 499–520.

Williams, D.B., Carter, C.B., 2009. Transmission Electron Microscopy: A Textbook for Material Science. Springer, New York.

Yada, K., Wei, L., 1987. Polygonal microstructures of Povlen chrysotile observed by high-resolution electron microscopy. In: Abstracts Sixth Meeting of the European Clay Groups (Seville), pp. 596–597.

Zvyagin, B.B., 1967. Electron Diffraction Analysis of Clay Mineral Structures. Plenum Press, NewYork.

Chapter 6

Characterisation of Halloysite by Spectroscopy

J.T. Kloprogge*
School of Earth Sciences, The University of Queensland, St. Lucia, QLD, Australia
*Corresponding author: e-mail: theoklop@bigpond.com

6.1 INTRODUCTION

This chapter provides an overview of various spectroscopic methods to characterise the structure of halloysite. It is obvious that the most important spectroscopic method to investigate halloysite and related kaolin group minerals is infrared (IR) spectroscopy. This is owing to the structure of halloysite, which contains at least two types of hydroxyl groups in the 1:1 layer and possible water molecules at the interlayer. The hydroxyl groups in the 1:1 layer are known as (i) the outer- or inner-surface hydroxyl groups and (ii) the inner-surface hydroxyl groups (Fig. 6.1). The inner-surface hydroxyl groups are located in the outer-upper unshared plane, whereas the inner hydroxyl groups are located in the lower shared plane of the octahedral sheet.

The hydroxyl bands are very sensitive probes for distinguishing between the kaolin minerals. In addition to the IR spectroscopy, Raman spectroscopy has proved to be very useful to study the structure of kaolin minerals (eg, Farmer, 1974; Wiewiora et al., 1979; Johnston et al., 1985; Michaelian, 1986; Frost et al., 1993). The data obtained from mid-IR (MIR) spectroscopy are not the same as those from Raman spectroscopy, as the Raman technique measures the symmetric vibrations and the physics of Raman emission are very different from those of IR absorption, where the occurrence of vibrations depend on changes in dipole moments. Raman spectroscopy can provide valuable information about the low-wavenumber region of halloysite, which is hard to obtain with IR spectroscopy. For kaolinite, the inner-surface hydroxyl groups give rise to IR bands at 3695 cm^{-1}, a band with the transition moment lying perpendicular to the *ab* plane, and two weaker bands at 3650 and 3670 cm^{-1} with transition moments in the *ab* plane. Because of the hydrogen bonding between the adjacent layers, the inner-surface hydroxyl group

Developments in Clay Science, Vol. 7. http://dx.doi.org/10.1016/B978-0-08-100293-3.00006-6

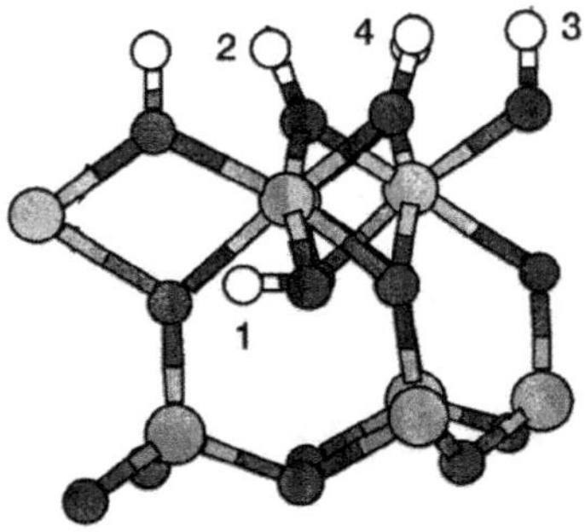

FIG. 6.1 Structure of a single halloysite layer showing inner hydroxyl group 1 and inner-surface hydroxyl groups 2–4, coordinating the Al in the octahedral sheet.

orientations are more or less fixed in comparison to halloysite, where the hydrogen bonding of these hydroxyl groups is with the interlayer water.

Solid-state magic-angle-spinning nuclear magnetic resonance (MAS-NMR) spectroscopy, Mössbauer spectroscopy and electron paramagnetic resonance (EPR) spectroscopy, also known as *electron spin resonance* (*ESR*) *spectroscopy*, are methods that use magnetic properties to obtain detailed information regarding the local environment of specific atoms. Finally, the application of X-ray photoelectron spectroscopy (XPS), NMR, ESR, Mössbauer spectroscopy and cathodeluminescence (CL) to study halloysites are briefly reviewed.

6.2 BRIEF PRESENTATION OF VARIOUS SPECTROSCOPIC METHODS

6.2.1 Infrared Spectroscopy

IR spectroscopy is a general technique based on vibrations of atoms in a molecule and has been widely used for the structural and compositional analysis of organic, organometallic, metalorganic, inorganic and polymeric materials. The IR portion of the electromagnetic spectrum is usually divided into three regions: the near-IR (NIR), MIR and far-IR, named for their relation to the visible spectrum. The higher energy NIR, approximately 14,000–4000 cm^{-1} (0.8–2.5 μm wavelength) can excite overtone or harmonic vibrations. The MIR, approximately 4000–400 cm^{-1} (2.5–25 μm) may be used to study the fundamental vibrations and associated rotational–vibrational structure. The far-IR, approximately 400–10 cm^{-1} (25–1000 μm), lying adjacent to the microwave region, has low energy and may be used for rotational spectroscopy. The names and classifications of these subregions are conventions that are only loosely based on the relative molecular or electromagnetic properties. The MIR spectrum of specifically clay minerals is sensitive to its chemical composition, isomorphous substitution and layer-stacking order. A classical mechanical model can be used as a first approximation for the portrayal of the vibrations in molecules and crystals. In this case, atomic nuclei are considered to be point masses and the interatomic interactions (both bonding and repulsive interactions) by springs. Atoms are allowed to undergo small vibrational displacements around their equilibrium positions, and their equations of motion are can be described in terms of Newtonian mechanics. If the springs are assumed to be ideal, in that the restoring force is directly proportional to the displacement (Hooke's law), then the vibrational motion is harmonic or sinusoidal in time. The proportionality constant, which relates the restoring force to the vibrational displacement, is termed the *force constant* of the spring. Solving the equations of motion for the system allows a set of vibrational frequencies to be identified. Each frequency corresponds to a particular atomic displacement pattern, known as a *normal mode of vibration*. However, although this classical model describes most of the basic features of vibrational motion, it cannot explain all the details. For a more detailed understanding, a quantum mechanical model has been developed based on Schrödinger's wave equation. In a vibrational spectroscopic experiment, the system will undergo a transition between vibrational levels with different quantum numbers and light is absorbed or emitted with an energy ($\Delta E = h\nu$) corresponding to the separation between the levels, which is generally in the order of 0–60 kJ/mol, equivalent to 0–5000 cm^{-1} (according to the relation between frequency ν, wavelength λ and wavenumber $\overline{\nu}$ in cm^{-1}: $\lambda = c/\nu$ and $\overline{\nu} = \nu/c$). To be 'IR active', a vibrational mode in a molecule must be associated with changes in the dipole. A permanent dipole is not necessary, as the rule requires only a change in dipole moment. A molecule can vibrate in many ways, and each way is called a *vibrational mode*.

6.2.2 Raman Spectroscopy

Raman spectroscopy is a relatively old technique based on theoretical grounds that had been predicted in 1923 (Kloprogge, 2005). In 1928, Sir C.V. Raman was the first to experimentally show the existence of inelastic light scattering (Kloprogge, 2005). This effect was later named the *Raman effect,* after him. The Raman principle has since been applied to fundamental chemical research, mostly in combination with IR spectroscopy. When light hits a molecule, part of the light will be dispersed with the same wavelength as the initial light. This effect is known as *elastic* or *Rayleigh scattering*. The intensity I_s of this scattering varies strongly with wavelength:

$$I_s \propto \lambda - 4 \tag{6.1}$$

A much smaller section of light, however, will show a slight shift in wavelength (approximately 10^{-6} of the initial light) due to the interaction between the vibrating molecules and the incoming photons. The Raman effect creates a shift in both positive and negative direction compared to the initial wavelength of the incoming light. These shifts are known as *Stokes* and *anti-Stokes lines*. This weak Raman scattering forms the basis of Raman spectroscopy. Raman scattering is thus mainly determined by the structure of a molecule, which can be a mineral as well. Raman spectroscopy is very sensitive for the symmetry of minerals and is very suitable for the determination of molecular vibrations of symmetric molecules. For a vibration to be allowed in the Raman spectrum, a change of polarizability will accompany the vibration. An atom or molecule in an electric field will be distorted and therefore be polarised. Thus, the induced dipole moment, which is proportional to the applied field, is calculated as follows:

$$P = \alpha \times E \tag{6.2}$$

where P and E are vectors, indicating the dipole moment and the electric field, respectively; and α is the polarisability. The polarisability is essentially a bond property, related to what might be called the 'looseness' of the electron cloud. The numerical value of the polarisability is given in volume units and is a simple number for atoms but is directional dependent for molecules. It can be completely described by three numbers and visualised by a polarisability ellipsoid. The Raman effect now is due to the changes in polarisability during intermolecular vibrations.

6.2.3 X-Ray Photoelectron Spectroscopy

XPS is a technique to probe the chemical bonding of specific elements. However, application of the XPS technique to minerals, like feldspars, can be difficult due to sample charging during the analysis. The advantage of XPS over Auger is that in XPS, the area under the photoelectron peak (which is used to calculate atom %) is easily estimated, whereas in Auger, the peak-peak (p–p) of the differential of the peak is used. The p–p of this differential is a function of the peak shape which, with Auger, can change significantly from compound to compound of the same element. Also, for Auger, the surface needs to be reasonably conducting, and Auger of insulating materials poses problems. There is also sample degradation, which can be a problem with the highly focused electron beam used for Auger emission. Degradation by the soft Al K alpha X-rays used for XPS is experienced by only a very few compounds. The only advantage for Auger is when one wants to get analysis from a very small area, as it uses a focused electron beam as in scanning electronic microscopy.

6.2.4 Solid-State Magic-Angle-Spinning Nuclear Magnetic Resonance Spectroscopy

Solid-state magic-angle-spinning nuclear magnetic resonance (MAS-NMR) is a very powerful technique capable of looking at a variety of materials. It does not require crystalline materials like diffraction techniques, and can still determine local molecular environments. The indirect magnetic interaction of the nuclear spins with the external magnetic field through the environment of electrons is called the *chemical shift*. When the magnetic field is applied, it induces currents in the electron clouds in the molecule. The circulating electrical currents in turn generate a magnetic field, and the nuclear spins sense the sum of the applied external field and the induced field generated by the molecular electrons. So the electrons in molecules cause the local magnetic fields to vary on a submolecular distance scale. The magnetic fields experienced by nuclei at two sites in the same molecule are different if the electronic environments vary. A huge variety of solid-state NMR experiments are available for the measurement of internuclear distances (dipolar recoupling), deconvolution of quadrupolar/dipolar influenced spectra, probing site symmetry and chemistry, observing solid-state dynamics, etc. In solution NMR, spectra consist of a series of very sharp transitions, due to the averaging of anisotropic NMR interactions by rapid random tumbling. By contrast, solid-state NMR spectra are very broad, as the full effects of anisotropic or orientation-dependent interactions are observed in the spectrum. High-resolution NMR spectra can provide the same type of information that is available from corresponding solution NMR spectra, but a number of special techniques/equipment are needed, including magic-angle spinning, cross-polarisation, special two-dimensional (2D) experiments and enhanced probe electronics.

6.2.5 Mössbauer Spectroscopy

Mössbauer spectroscopy is a spectroscopic technique based on the Mössbauer effect. This effect, discovered by Rudolf Mössbauer in 1957 (Mössbauer, 1958; Craig et al., 1959), consists of recoil-free, resonant absorption and emission of gamma rays in solids. Like NMR spectroscopy, Mössbauer spectroscopy probes tiny changes in the energy levels of an atomic nucleus in response to its environment. Typically, three types of nuclear interactions may be observed: an isomeric shift, also known as a chemical shift; quadrupole splitting; and magnetic or hyperfine splitting, also known as the *Zeeman effect*. Due to the high-energy and extremely narrow line widths of gamma rays, Mössbauer spectroscopy is a very sensitive technique in terms of energy (and hence frequency) resolution, capable of detecting change in just a few parts per 10^{11}. During Mössbauer absorption spectroscopy, the source is accelerated through a range of velocities using a linear motor to produce a Doppler effect and scan the gamma ray energy through a given range. A typical range of velocities for ^{57}Fe, for example, may be ± 11 mm/s (1 mm/s = 48.075 neV). In the resulting spectra, the gamma ray intensity is plotted as a function of the source velocity. At velocities corresponding to the resonant energy levels of the sample, a fraction of the gamma rays are absorbed, resulting in a drop in the measured intensity and a corresponding dip in the spectrum. The number, positions and intensities of the dips (also called *peaks*—dips in transmitted intensity are peaks in absorbance) provide information about the chemical environment of the absorbing nuclei and can be used to characterise the sample.

6.2.6 Electron Spin Resonance Spectroscopy

ESR, also called *electron proton resonance* (*EPR*), is a technique used to study chemical species with unpaired electrons. EPR spectroscopy plays an important role in the understanding of organic and inorganic radicals, transition metal complexes and some biomolecules. The basic concepts of EPR are analogous to those of NMR, but the electron spins are excited instead of the spins of atomic nuclei. ESR spectroscopy is particularly useful for studying metal complexes or organic radicals. ESR is fundamentally similar to the more widely familiar method of NMR spectroscopy, with several important distinctions. While both spectroscopies deal with the interaction of electromagnetic radiation with magnetic moments of particles, there are many differences between the two spectroscopies, to wit: (i) ESR focuses on the interactions between an external magnetic field and the unpaired electrons of whatever system it is localised to, as opposed to the nuclei of individual atoms; (ii) the electromagnetic radiation used in NMR typically is confined to the radio frequency range between 300 and 1000 MHz, whereas ESR is typically performed using microwaves in the 3–400 GHz range; (iii) in ESR, the frequency is typically held constant, while the magnetic field strength is varied, which is the reverse of how NMR experiments are typically performed, where the magnetic field is held constant, while the radio frequency is varied; (iv) due to the short relaxation times of electron spins in comparison to nuclei, ESR experiments must often be performed at very low temperatures, often below 10 K and sometimes as low as 2 K, which typically requires the use of liquid helium as a coolant; and (v) ESR spectroscopy is inherently about 1000 times more sensitive than NMR spectroscopy due to the higher frequency of electromagnetic radiation used in ESR in comparison to NMR.

6.3 INFRARED AND RAMAN SPECTROSCOPY OF HALLOYSITE AND RELATED KAOLIN MINERALS

6.3.1 The Hydroxyl Groups and Interlayer Water

In the IR spectra of kaolins, three bands were reported by Lyon and Tuddenham (1960) at 2.70 μ (≈3700 cm^{-1}), 2.75 μ (≈3640 cm^{-1}) and at 2.90 μ (≈3450 cm^{-1}). The two first bands are common to all kaolin minerals, but with different intensity ratios, and the last band was attributed to adsorbed water. The first IR spectra of halloysite were shown by Calhoun and Carlisle (1972). The hydroxyl-stretching region was characterised by three bands at 3700, 3620 and 3450 cm^{-1}. However, only two bands were reported by Madejová et al. (1997) for four Slovak halloysites at 3695 and 3622 cm^{-1}. From their published spectra, it is not completely clear whether the halloysites were dehydrated, explaining the absence of a band associated with the OH-stretching mode of water. In comparison, kaolinite shows the inner hydroxyl-stretching mode at the same position in the IR spectrum at around 3620 cm^{-1}, while the inner-surface hydroxyls commonly produce three bands around 3690, 3670 and 3650 cm^{-1}. The broad hydroxyl-stretching band associated with interlayer water below 3600 cm^{-1} is absent in the spectra of the other clay minerals (kaolinite, dickite and nacrite). Recently, Cheng et al. (2010a,b) showed more detailed bands of the hydroxyl-stretching region on coal-bearing halloysite. These authors fitted the hydroxyl-stretching region with 12 different bands, from which they tried to assign 5 bands to different types of water molecules in the interlayer space. But there is no physical or crystallographic basis for this sort of band fitting, and as Kloprogge (2015) pointed out, this studied halloysite contained more than 8% gibbsite that was not taken into account by the authors for the interpretation of the hydroxyl-stretching region.

In the Raman spectrum of halloysite (Fig. 6.2), the hydroxyl-stretching region exhibits three bands around 3630, 3623 and 3617 cm^{-1} that are attributed to the inner hydroxyl groups, while the inner-surface hydroxyl groups give rise to two bands around 3705 and 3698 cm^{-1} (Frost and Shurvell, 1997). The relative intensity of the 3630 cm^{-1} band is dependent on the orientation of the tubular halloysite. In the Raman spectrum of a hydrated tubular halloysite (10 Å) from the Weillen karst in Belgium reported by Kloprogge and Frost (1999),

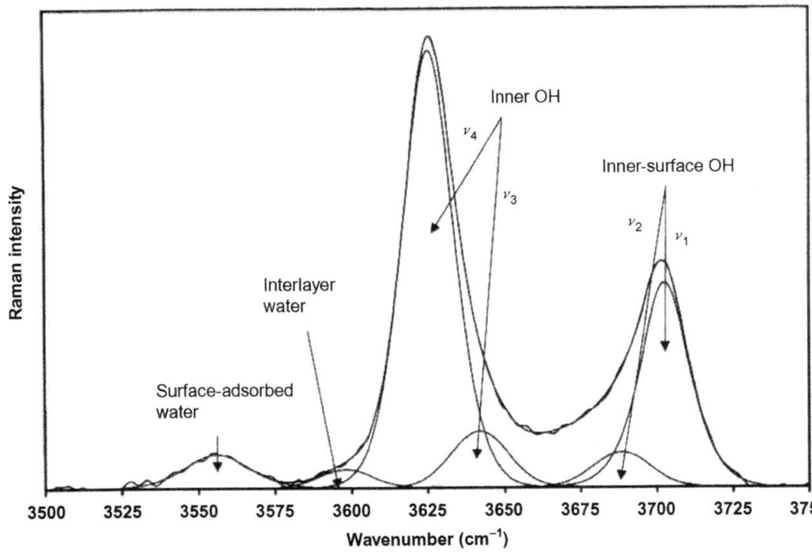

FIG. 6.2 Band component analysis of the Raman spectrum of the hydroxyl-stretching region of hydrated halloysite from the Weillen karst, Belgium. *Modifield after Kloprogge and Frost (1999).*

the hydroxyl-stretching region showed two relatively strong asymmetric bands of the halloysite hydroxyl groups. In comparison, the Raman spectrum of kaolinite is characterised by four bands at 3620, 3650, 3667 and 3682 cm^{-1} with a prominent shoulder at 3692 cm^{-1}. In highly crystalline kaolinite, the 3684 cm^{-1} band predominates, while in others, this band is weak. Similar to the IR spectra, the band at 3620 cm^{-1} has been assigned to the inner hydroxyl group, while the other bands are associated with the inner-surface hydroxyl groups. The Raman spectrum of the kaolin polytype nacrite shows comparable bands at 3618, 3628, 3634, 3650 and 3698 cm^{-1}. For the Weillen karst halloysite, in addition to these bands, two broad low-intensity bands were observed at the low-wavenumber side of the hydroxyl bands. These bands are absent in dehydrated halloysite. Therefore, it is logical to assign these bands to OH-stretching modes in water. Based on a comparison with other clay minerals, the band around 3598 cm^{-1} is probably related to the interlayer water, whereas the band around 3556 cm^{-1} represents water weakly adsorbed on the clay mineral surface (Table 6.1).

The observed wavenumbers for the inner-surface and inner hydroxyl groups indicate O–H···O bond distances of approximately 3.40–3.45 Å and 3.20–3.25 Å, respectively (Libowitzky, 1999). In the crystal structure of halloysite, however, these bonds are not straight, but bent; and therefore, the distances are overestimated. X-ray diffraction studies have indicated that those bonds are closer to around 2.9–3.0 Å instead. In kaolinite, for example, only

TABLE 6.1 Raman Spectral Band Component Analysis of the Hydroxyl-Stretching Region of Halloysite (10 Å) and Comparison with Dehydrated Halloysite

Halloysite (10 Å), Weillen Karst, Belgium (Kloprogge and Frost, 1999)	Dehydrated halloysite, New Zealand (Frost and Shurvell, 1997)	Dehydrated halloysite, Eureka5 (Frost, 1997)	Assignment
3703 cm^{-1}	3705 cm^{-1}	3712 cm^{-1}	Inner-surface OH
3688 cm^{-1}	3698 cm^{-1}	3698 cm^{-1}	Inner-surface OH
3642 cm^{-1}	3630 cm^{-1}	3631 cm^{-1}	Inner OH
3625 cm^{-1}	3623 cm^{-1}	3621 cm^{-1}	Inner OH
–	3617 cm^{-1}	–	Inner OH
3598 cm^{-1}	–	–	Interlayer H_2O
3556 cm^{-1}	–	–	Surface-adsorbed H_2O

one band is observed for the inner hydroxyl group. The folding of the octahedral layer will cause a positional disordering, resulting in the overlapping bands in the Raman spectrum. This folding and the presence of interlayer water may also explain the shift of 5–10 cm^{-1} in the band positions of the hydroxyl bands compared with the dehydrated halloysites (Table 6.1).

The IR spectrum of halloysite showed a water hydroxyl-bending mode around 1630 cm^{-1} (Calhoun and Carlisle, 1972). Two bands observed at 1633 and 1650 cm^{-1} for halloysite, were assigned to two different water-bending modes in the interlayer space (Cheng et al., 2010a,b). However, it is more likely that one mode is associated with interlayer water and the other with physically adsorbed water on the outside of the crystals, in line with the assignment of the water hydroxyl-stretching modes observed in the Raman spectra (Kloprogge and Frost, 1999).

The hydroxyl libration modes are observed in the region around 900–950 cm^{-1}. For Slovak halloysites, a single band around 913 cm^{-1} was reported in the MIR spectra with a small inflexion near 936 cm^{-1} (Madejová et al., 1997). Frost (1998) reported three Raman bands attributed to the inner-surface hydroxyl groups around 950, 938 and 923 cm^{-1} and one band at 915 cm^{-1} was ascribed to the inner hydroxyl group. In the low-wavenumber region, the corresponding Al–OH translation can be observed at 779 and 728 cm^{-1} and the O–H–O stretching and bending modes at 298, 238 and 192 cm^{-1}. In earlier work, Al–OH translation modes for dehydrated halloysite have been reported at 793, 745 and 703 cm^{-1} and 779, 728 and 693 cm^{-1}, respectively (Frost, 1995; Frost and Shurvell, 1997). Calhoun and Carlisle (1972) observed two bands in the IR spectrum at 740 and 700 cm^{-1} for halloysite. In addition to these bands, two Raman active bands at 359 and 332 cm^{-1} are observed between 300 and 400 cm^{-1}, which are absent in the other clay minerals of the kaolin group. Therefore, these vibrations are assigned to hydrogen-bonded water modes of adsorbed water and interlayer water corresponding to the two hydroxyl-stretching modes around 3598 and 3556 cm^{-1}.

IR spectroscopy has proved to be useful in the study of the thermal behaviour, such as dehydroxylation, of clay minerals (Stubičan and Roy, 1961; Pampuch, 1966; Frost and Vassallo, 1996). It is generally thought that dehydroxylation of clay minerals of the kaolin group stems from the interaction of two hydroxyl groups in a two-step process to form a water molecule by proton transfer, leaving a chemically bonded oxygen atom (in the form of a superoxide anion) in the lattice. Dehydroxylation is demonstrated by the loss of signal intensity of the hydroxyl-stretching modes between 3600 and 3700 cm^{-1} (Fig. 6.3). Below 300°C, a strong reduction in intensity is observed for the 3671 and 3625 cm^{-1} bands, in contrast to the 3693 and 3647 cm^{-1} bands, which do not show a decrease (within the experimental error). This is followed by a slow decrease in intensity of all bands up to approximately 500°C. Between approximately 500°C and 550°C, a rapid decrease to almost zero

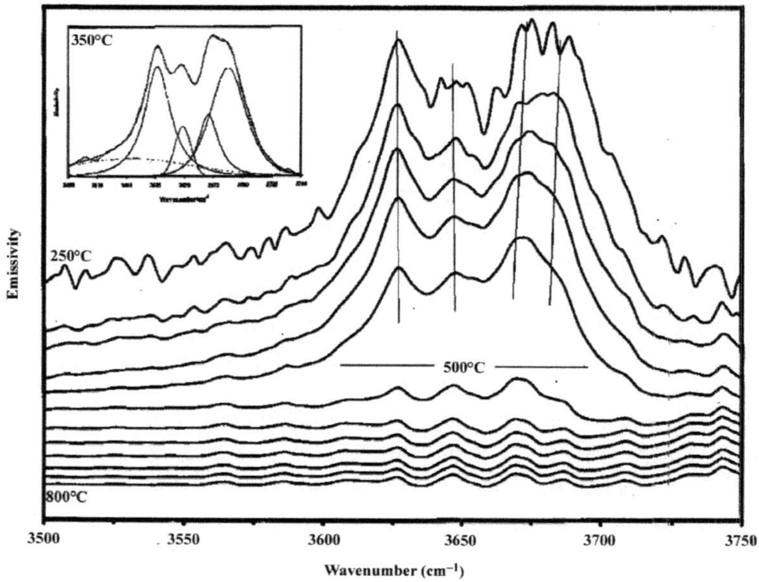

FIG. 6.3 Infrared emission spectra in the hydroxyl-stretching region between 3500 and 3750 cm^{-1} of halloysite (10 Å) from 250°C to 800°C at 50°C intervals, showing the decrease in intensity of the OH-stretching modes. Inset shows the band component analysis of the 300°C spectrum. *Modified after Kloprogge and Frost (2000).*

is observed for all the bands, and they cannot be observed at temperatures above 600°C (Kloprogge and Frost, 2000). This seems to suggest that the four different inner and inner-surface hydroxyl groups of halloysite are being removed at different temperatures, at different rates, or both, in contrast to earlier infrared emission spectroscopy (IES) observations of the dehydroxylation of kaolinite (Frost and Vassallo, 1996). The loss of the inner hydroxyl band is nearly linear over the temperature range between 300°C and 500°C. The change in the rate of intensity loss of the other three bands around 500°C supports the hypothesis that two different mechanisms occur during the dehydroxylation of the halloysite (10 Å). The corresponding inner and inner-surface Al–OH libration IES bands at 938 and 920 cm^{-1} decrease in intensity at the same rate and disappear at the same temperature as the hydroxyl-stretching bands (Kloprogge and Frost, 2000). More details about the dehydration and dehydroxylation of halloysite are reported in Chapter 7.

Recent results of the NIR region (11,000–4000 cm^{-1}) for analysing crystal chemistry have significantly extended the application of IR spectroscopy to clay mineral studies (Petit et al., 2004). However, halloysite has received very little attention. Crowley and Vergo (1988) reported a first attempt to use NIR spectroscopy to analyse mixtures of kaolinite and halloysite. Although the kaolinite–halloysite proportions could be only roughly estimated from the mixture spectra, the halloysite component was evident even when halloysite was present in only minor amounts. Cheng et al. (2010b) reported NIR spectra and detailed band fitting of a halloysite. Unfortunately, the sample contained more than 8% gibbsite. Despite the presence of gibbsite, they attempted to ascribe every band to an overtone or combination mode of halloysite (Kloprogge, 2015).

For kaolinite in the 5000–4000 cm^{-1} region, the major band at 4526 cm^{-1} and the associated one at 4620 cm^{-1} correspond to the combination of OH-stretching and deformation vibrations of Al_2OH groups. The weaker band at 4730 cm^{-1} is likely due to a combination of the internal OH-stretching vibrations (3620 cm^{-1}) with lattice vibration (Si–O stretching near 1100 cm^{-1}). Three little bands between 4100 and 4300 cm^{-1} are combinations of OH-stretching bands of kaolinite with lattice deformation vibrations (Petit et al., 1999). In the same area, Cheng et al. (2010b) reported for halloysite a major band at 4535 cm^{-1} associated with a weaker band at 4605 cm^{-1}. A band similar to that observed for kaolinite at 4730 cm^{-1} was not present in the halloysite spectrum. Similar to kaolinite, three small bands were observed between roughly 4100 and 4300 cm^{-1} at 4327, 4261 and 4207 cm^{-1}. The remaining two bands at 4450 and 4412 cm^{-1} are probably bands associated with the gibbsite in the sample (Kloprogge, 2015).

For kaolinite, two bands are clearly observed at 7176 and 7065 cm^{-1}. Between these two bands, one doublet is more or less resolved. These bands correspond to the 2υOH overtones of OH-stretching fundamental modes of Al_2OH groups (Petit et al., 1999). For halloysite, a single relatively sharp band at 7081 cm^{-1} was accompanied by a series of broader bands according to the band fitting. The band observed for kaolinite at 7065 cm^{-1} is shifted to 7031 cm^{-1} for halloysite. Here, no doublet was found between those two bands, as was the case for kaolinite. However, band fitting resulted in a number of bands at 7212, 7154, 7123, 6977, 6884, 6838, 6695 and 6464 cm^{-1}. Only the 6695 cm^{-1} band for halloysite seems to have an equivalent band for kaolinite at 6677 cm^{-1}. The relatively sharp 6677 cm^{-1} band may be due to the first overtone of the 3434 cm^{-1} vibration band, present in the MIR spectrum (Petit et al., 1999). The remaining three bands between 7000 and 6000 cm^{-1} are probably the first overtones of adsorbed water and interlayer water in halloysite. The bands around 5241, 5156 and 5017 cm^{-1} are diagnostic bands of adsorbed water and interlayer water ($\upsilon_w + \delta_w$).

6.3.2 Halloysite Layers

The lattice region of the hydrated halloysite can be interpreted on the basis of repeating units in the crystal structure. Loh (1973) has suggested that the vibrations in the lattice region of clay minerals of the kaolin group can be treated as those of distorted AlO_6 octahedra, distorted SiO_4 tetrahedra, an

O–H–O group and the bending modes of the hydroxyl groups. Based on this the region below 1200 cm^{-1} is characterised by Raman active bands associated with Al–O (910, 442, 172 and 156 cm^{-1}) and Si–O (1100, 844, 693, 540, 510, 396 and 135 cm^{-1}) bending and deformation modes. The bands around 275 and 245 cm^{-1} are due to the symmetric and antisymmetric stretching modes of the triangular O–H–O group. Halloysite is characterised by an intense band at 465 cm^{-1} accompanied by much weaker bands at 540 and 503 cm^{-1} (Fig. 6.4). These bands have two equivalent bands for dickite and four bands for kaolinite. The multiplicity is due to a reduction in site symmetry of the SiO_4 group and has been attributed to the Si–O bending vibrations, from which the 465 cm^{-1} vibration is both Raman and IR active (Farmer, 1974). The corresponding Si–O deformation mode at 396 cm^{-1} observed for the dehydrated halloysite seems to be absent in the hydrated sample, although a very small shoulder may be present at the low-wavenumber side of the 430 cm^{-1} band (Table 6.2). The Al–OH translation modes for halloysite have been reported at 793, 745 and 703 cm^{-1} and 779, 693 and 728 cm^{-1} (Frost, 1995; Frost and Shurvell, 1997). The first set of data agrees very well with the data reported for the hydrated halloysite from the Weillen karst, Belgium (Kloprogge and Frost, 1999). The difference in the band positions may indicate differences in crystal structure or folding

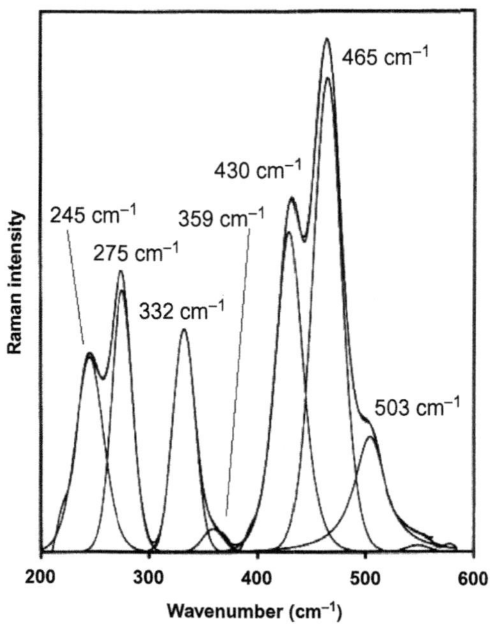

FIG. 6.4 Band component analysis of the Raman spectrum of halloysite between 200 and 600 cm^{-1}. *Modifield after Kloprogge and Frost (1999).*

TABLE 6.2 Band Component Analysis of the Region Below 1200 cm^{-1} of Halloysite (10 Å) and Comparison with Dehydrated Halloysite

Hydrated halloysite Raman (Kloprogge and Frost, 1999)	Dehydrated halloysite Raman (Frost and Shurvell, 1997)	Ward's halloysite No. 12 IR (Calhoun and Carlisle, 1972)	Slovak halloysites IR (Madejová et al., 1997)	Assignment
–	–	1630 cm^{-1}	–	OH bend H_2O
–	1100 cm^{-1}	1100 cm^{-1}	1090 cm^{-1}	Si–O stretch
–	–	1020 cm^{-1}	1036 cm^{-1}	–
–	–	–	1013 cm^{-1}	–
944 cm^{-1}	–	–	–	–
910 cm^{-1}	–	910 cm^{-1}	912–914 cm^{-1}	Al–O libration
826 cm^{-1}	844 cm^{-1}	–	–	Si–O–Al deformation
794 cm^{-1}	779 cm^{-1}	–	795 cm^{-1}	OH translation
748 cm^{-1}	728 cm^{-1}	740 cm^{-1}	750 cm^{-1}	OH translation
710 cm^{-1}	693 cm^{-1}	700 cm^{-1}	690 cm^{-1}	OH translation
548 cm^{-1}	540 cm^{-1}	535 cm^{-1}	540 cm^{-1}	Si–O bend
503 cm^{-1}	510 cm^{-1}	–	–	Si–O bend
465 cm^{-1}	–	465 cm^{-1}	471 cm^{-1}	Si–O bend
430 cm^{-1}	442 cm^{-1}	435 cm^{-1}	434 cm^{-1}	Al–O stretch
–	396 cm^{-1}	–	–	Si–O deformation
359 cm^{-1}	–	–	–	H-bonded H_2O
332 cm^{-1}	–	–	–	H-bonded H_2O
275 cm^{-1}	–	–	–	O–H–O antisym. stretch
245 cm^{-1}	–	–	–	O–H–O sym. stretch

of the layers. This halloysite has been shown to have a tubular morphology. The libration mode of the inner Al–OH groups has been observed before as a single band at 910 cm^{-1}. In this study, however, a second band is observed at 944 cm^{-1} that is ascribed to the inner-surface Al–OH libration mode.

6.4 OTHER SPECTROSCOPIC CHARACTERISATIONS OF HALLOYSITE

6.4.1 X-Ray Photoelectron Spectroscopy

Soma et al. (1992) used XPS to study iron-containing halloysites from New Zealand. Their study confirmed that Fe commonly replaces Al in the octahedral sheet of halloysite. This process, however, appears to be nonstoichiometric in that the loss of Al is disproportionately higher than the gain in Fe. As a result, the positive layer-charge deficiency, and hence the cation exchange capacity, of halloysites would be expected to increase with Fe substitution. The XPS data further suggest that the layers within a halloysite crystal are built up like 'onion skins' and that the chemical composition varies from layer to layer.

The XPS survey scans of the clay minerals of the kaolin group are characterised by the peaks of O 1s, Si 2p and Al 2p (Fig. 6.5). No impurities other than a small amount of carbon were observed. The Si atoms in halloysite are tetrahedrally bonded to O atoms, and form Si–O–Si linkages within the tetrahedral sheet and Al–O–Si linkages between the tetrahedral and octahedral sheet. Furthermore, weak linkages are present between two adjacent sheets. The Si 2p occurs at 103.2 eV with a full width at half maximum (FWHM) of 1.6 eV. The Si 2p is considerably shifted compared to quartz (α-SiO_2), and the second nearest-neighbour Al atoms do have a significant influence

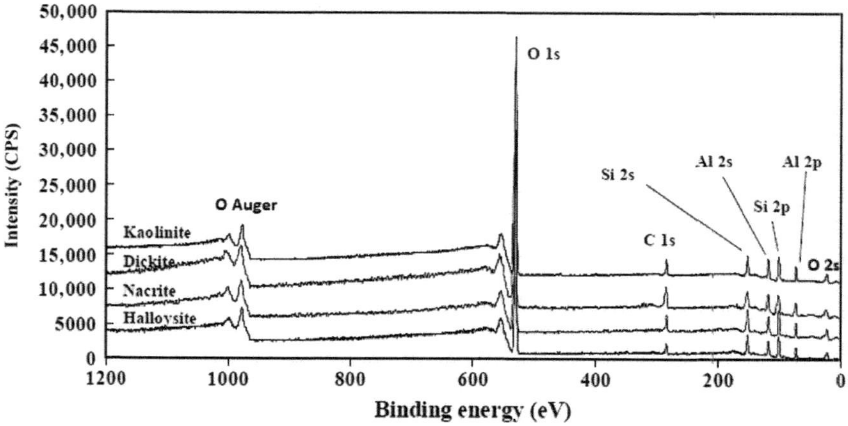

FIG. 6.5 XPS survey scans of the clay minerals of the kaolin group: kaolinite, dickite, nacrite and halloysite. *From Kloprogge and Wood (2015). Reproduced with kind permission of The Clay Minerals Society, publisher of Clays and Clay Minerals.*

on the electronic structure of the tetrahedrally bonded Si atoms. In halloysite, all aluminium atoms are present in the octahedral sheets in sixfold coordination by oxygen atoms and hydroxyl groups. An Al–O–Si linkage is formed between the SiO_4 tetrahedral sheet and the octahedral sheet. The Al 2p peak is found at 74.8 eV with a FWHM of 1.33 eV. A slight increase in the relative binding energy is observed from kaolinite to halloysite, reflecting a change in the stacking order due to the presence of interlayer water and the corresponding change in the van der Waals interactions between the siloxane sheet and the adjacent gibbsitelike sheet. The chemical bonding of the O atoms with Al and Si in the $Al_2Si_2O_5(OH)_4$ polymorphs is relatively simple. Three oxygen atoms of the tetrahedral sheet are shared by adjacent Si tetrahedral, while the fourth forms a link with the Al octahedra. The Al octahedra are coordinated by the remaining oxygen atoms and the hydroxyl groups. Because the charge density difference for each oxygen atom may not vary significantly, it is not expected to be able to resolve all these binding energy differences; however, the overall shapes of the O 1s peaks observed in the four polymorphs allow the identification of two peaks associated with the oxygen atoms and with the hydroxyl groups, respectively. It is not possible to distinguish between the inner and outer hydroxyl groups. In halloysite, a third, very weak peak was observed to be associated with interlayer water still present despite the ultrahigh vacuum in the instrument (Fig. 6.6).

6.4.2 Solid-State MAS-NMR

Newman et al. (1994) studied the ^{27}Al high-field MAS-NMR spectra of Te Akatea (New Zealand) halloysite. They showed that the major signal was found at $\delta = 0$ ppm assigned to sixfold coordinated Al, while a small signal at $\delta = 75$ (about 1% of total intensity) assigned to fourfold coordinated Al could either be due to a small amount of tetrahedral substitution in the halloysite or, alternatively, be due to a small amount of impurities. The Al^{IV} signal was not visible in the medium-field MAS-NMR spectra. Komarneni et al. (1985)

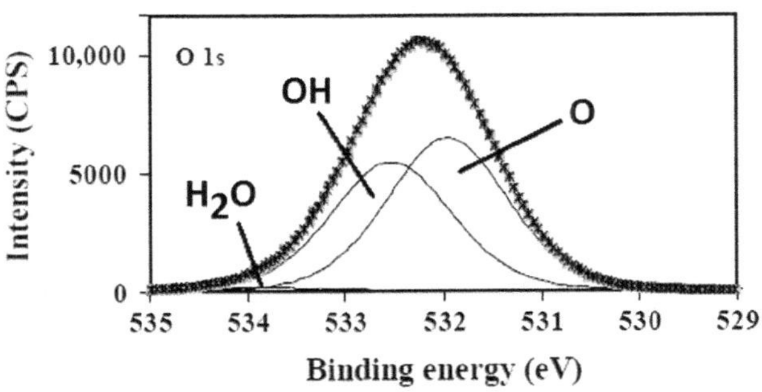

FIG. 6.6 Band component analysis of the O 1s high-resolution spectrum of halloysite. *From Kloprogge and Wood (2015). Reproduced with kind permission of The Clay Minerals Society, publisher of Clays and Clay Minerals.*

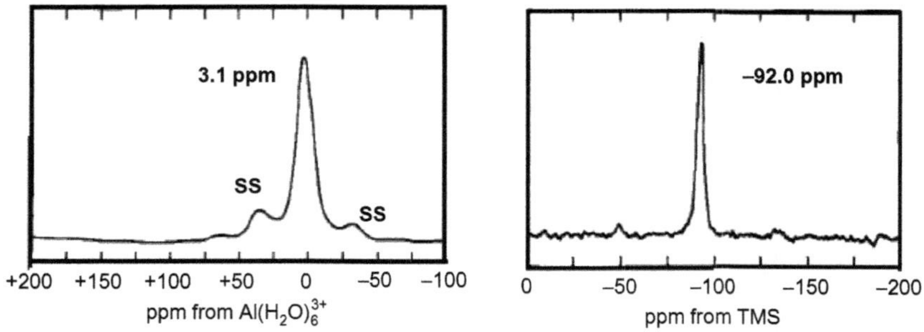

FIG. 6.7 ^{27}Al (*left*) and ^{29}Si (*right*) MAS-NMR spectra of halloysite (Utah), *SS* = spinning side band. *Modified after Komarneni et al. (1985).*

observed a single band at 3.1 ppm but found no evidence for fourfold coordinated Al in the halloysite sample from Utah (Fig. 6.7, left).

For halloysite from Choyo, Kumamoto, Japan, a single absorption with several poorly resolved peaks was observed around −90 ppm (Shimizu et al., 1988). This peak corresponds to the signal of Si nuclei in the Q^3 polymerization state; that is, with three Si atoms bonded to a SiO_4 tetrahedron. The peak width of this halloysite was broader than that of kaolinite, as reported by Barron et al. (1982) or Kinsey et al. (1985). Komarneni et al. (1985), however, reported that the ^{29}Si MAS-NMR spectrum of highly disordered halloysite showed a single sharp resonance at −92.0 ppm (Fig. 6.7, right).

A large gyromagnetic ratio and 100% natural abundance make ^{19}F one of the most sensitive nuclei for study by NMR. It is well known that F^- is structurally similar to OH^- in size and charge. Thus, structural hydroxyls in halloysite can be replaced by F. The natural fluorine content of clay minerals ranges from parts per million to a few percent. Levels of a few tens of ppm are sufficient for ^{19}F MAS-NMR spectroscopy because of the high ^{19}F sensitivity. The ^{19}F MAS-NMR spectrum had just one NMR line in each spectrum at around—133.7 ppm instead of separate resonances for the inner and interlayer fluorine sites. Additional experimental work on synthetic kaolinite provided strong evidence that the inner and interlayer fluoride ions have similar chemical shifts (Labouriau et al., 1995).

Early work by Kitagawa (1972) using wide-line NMR showed that the energy levels of water were clearly distinguished between adsorbed water and structural hydroxyl groups in clay minerals such as kaolinite, halloysite and montmorillonite. The water in the interlayer space of halloysite behaved less like liquid water than water present in the interlayer space of montmorillonite. Cruz et al. (1978) showed that in the electrically neutral interlayer space of halloysite, the water molecule does not exhibit a preferential orientation. The influence of the neutral surface was to provoke anisotropy of the molecular motions, which could be separated into a rapid 180-degree flipping motion about the C_2 axis and a slow tumbling about an axis perpendicular to C_2. The proton and deuteron spin–spin and spin–lattice relaxation times could be explained by this simple model.

6.4.3 ESR, Mössbauer and Cathodeluminescence Techniques

Several studies have been devoted to the use of ESR of halloysites (Komusinski et al., 1981; Chaikum and Carr, 1987; Nagasawa and Noro, 1987). Chaikum and Carr (1987) showed ESR spectra of three samples of halloysite (7 Å) (from New Zealand) covering the magnetic field range 0–4000 G (Fig. 6.8). In general, there are two principal resonances at $g=2.0$ and $g=4.2$. The lineshape and intensity of both resonances vary. The high-field resonance ($g=2.0$) showed more variations in lineshape and relative intensity. For Te Puke halloysite, in addition to a strong absorption at $g=2.0$, a weak six-line resonance centred at $g=2.0$ appeared. The corresponding signal in Matauri Bay halloysite showed a line-shape characteristic of a paramagnetic centre in an axially symmetric environment and an intensity of the central line much

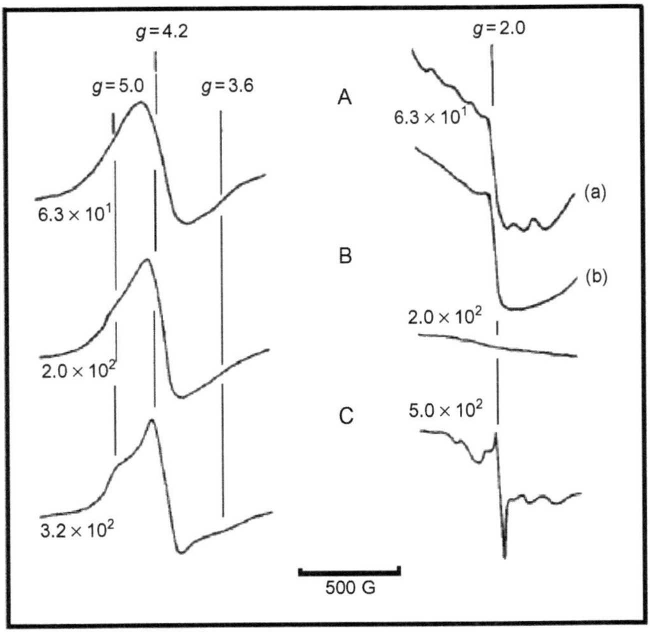

FIG. 6.8 ESR spectra of halloysite (7 Å): (A) Te Puke halloysite: untreated (a) and Na^+-exchanged (b); (B) untreated Kauri halloysite; (C) untreated Matauri Bay halloysite. *Modified after Chaikum and Carr (1987).*

smaller than that of the resonance in Te Puke halloysite. Similarities between ESR spectra of halloysite and kaolinite suggest that absorption lines in the $g \approx 4$ region originate from similar paramagnetic centres and assignment of the signals can be made in the same manner. Thus, the resonance at $g=5$ is due to Fe^{3+} ions in an octahedral site, which are influenced in some way by interactions between hydroxyl groups of the octahedral sheet and adjacent oxygen sheet or intercalated molecules. Another Fe^{3+} structural site is associated with an extremely distorted octahedral symmetry and corresponds to the ESR absorption line at $g=4.2$. A similar resonance was observed around $g=4.3$ by Komusinski et al. (1981) for three Polish and Slovak halloysites. Goetze et al. (2002) showed that after γ-irradiation, the intensity decreased due to the partial reduction of Fe^{3+} to Fe^{2+}. Resonances in the $g \approx 2$ region are not common to all halloysites. No general definite assignments was made for the strong absorption line at $g=2.0$, although the ESR signal for Matauri Bay halloysite is possibly due to a trapped positive hole or a trapped O_2^- ion (Coyne et al., 1989). Nagasawa and Noro (1987) attributed a broad signal around $g=2$ to iron impurities in their halloysite samples. The six-line resonance produced by Te Puke halloysite and removable by Na^+ ion saturation was attributed to Mn^{2+} ions at exchange sites on the clay mineral surfaces. Low-temperature lines exhibited by all samples are due to the presence of V^{4+} or VO^{2+} ions replacing Al^{3+} in the octahedral sheet (Chaikum and Carr, 1987).

Komusinski et al. (1981) used Mössbauer spectroscopy, in addition to ESR, to show that iron present in the crystal structure of halloysite was Fe^{3+} and in octahedral coordination (Fig. 6.9A). The line widths of the halloysites were also very narrow: $\Gamma=0.39$ mm/s for the Dunino (Poland) sample, and $\Gamma=0.48$ mm/s for the Michalovce (Slovakia) sample. The layers are stacked almost at random in these clay minerals, causing the symmetry of the unit cell to be pseudomonoclinic. This may create a crystal field with a symmetry similar to the nacrite or dickite, which is reflected in narrow line widths.

In contrast, De Roy et al. (1981) fitted their spectra of two New Zealand halloysites based on three different positions (Fig. 6.9B). The first Fe^{III} centre (about 75% of the total signal) was in octahedral coordination of oxygen atoms. The width of the bands was interpreted as being due to slight distortion of the sites. The second Fe^{III} centre (about 20% of total signal) was characterised by a large width and high quadrupole splitting, which was ascribed to exchangeable Fe^{3+}. The final centre was thought to be Fe^{II} (about 5% of the total signal) in a regular octahedral coordination. This centre was assigned to Fe^{2+} substituting for Al in the octahedral sheet.

Goetze et al. (2002) studied a series of clay minerals, including halloysite, with CL microscopy and spectroscopy. Halloysite CL was observed to be relatively weak and in the greenish-blue region, as opposed to deep blue for the three other clay minerals of the kaolin group. The spectrum showed a

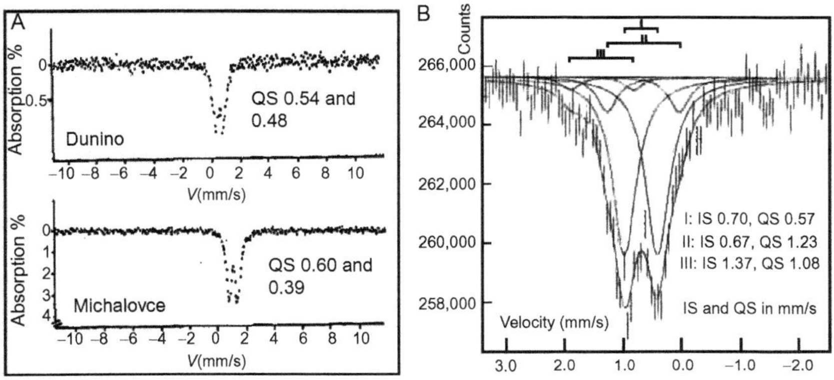

FIG. 6.9 (A) Room-temperature Mössbauer spectra Dunino halloysite and Michalovce halloysite. (B) Mössbauer spectrum and band component analysis of Te Puke halloysite (*IS*, isomeric shift; *QS*, quadrupole splitting). *Panel (A): From Komusinski et al. (1981). Reproduced with kind permission of The Clay Minerals Society, publisher of Clays and Clay Minerals. Panel (B): Modified after De Roy et al. (1981). Copyright Wiley-VCH Verlag GmbH & Co. KGaA. Reproduced with permission.*

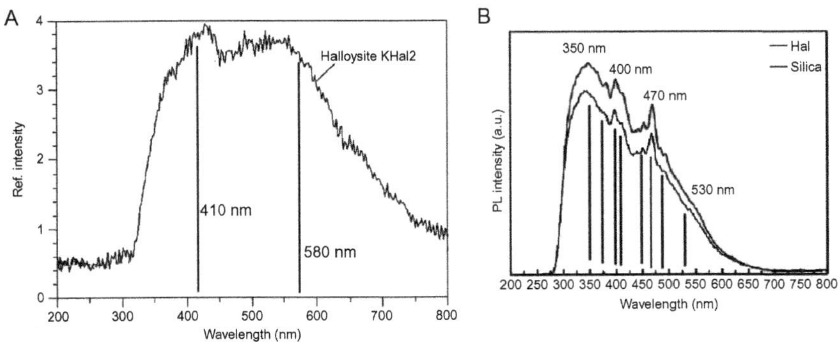

FIG. 6.10 (A) CL spectrum of halloysite taken on a pressed tablet of the powdered sample. The spectrum shows the characteristic blue emission at the UV–vis boundary (weak doublet) and a broad emission band around 580 nm. (B) PL spectra of tubular halloysite and silica samples under excitation at 286 nm. *Panel (A): Modified after Goetze et al. (2002). Panel (B): Modified after Zhang et al. (2012).*

weak doublet at the UV–vis boundary at 410 nm that decreased in intensity during electron radiation. In addition, a broad emission band was observed around 580 nm (Fig. 6.10A). It was thought that the 410-nm emission band might be due to radiation-induced paramagnetic defect centres in the halloysite structure, particularly that associated with Si–O centres. The origin of the characteristic 580-nm emission observed for halloysite and certain kaolinite samples remains unknown. Zhang et al. (2012) recently reported on the transformation of tubular halloysite into tubular silica through acid treatment and the resulting photoluminescence (PL) spectra of both materials (Fig. 6.10B). A main, intense PL band was observed at 350 nm with a shoulder at 375 nm and three weak bands at 450, 486 and 530 nm. Two weak emission bands were clearly distinguishable at 400 and 470 nm. Luminescence bands around 350 nm and a shoulder of 375 nm were attributed to some types of intrinsic defects induced by surplus oxygen in the coating, not to any kind of impurity. The violet light emission at 400 nm was ascribed to the intrinsic diamagnetic defect centre: a twofold coordinated silicon lone-pair centre (O·Si·O) clearly due to the high oxygen deficiency of the samples. The blue light emission at 470 nm was attributed to the neutral oxygen vacancy (Si·Si), while the strong peaks at 486 and 530 nm were probably due to the hydroxyl radicals (Si·OH) covered on the outer or inner surface of the tubular crystals.

6.5 CONCLUDING REMARKS

Halloysites have been studied with a large variety of spectroscopic techniques for many years. In particular, the use of IR and Raman spectroscopy have proved to be very useful in gaining a better understanding of the behaviour of the different hydroxyl groups and the interlayer water in halloysite in comparison to the other kaolin minerals, kaolinite, dickite and nacrite. IES has provided better insight into the thermal behaviour of halloysite to supplement more commonly used methods such as thermal gravimetric analysis, differential thermal analysis and solid-state NMR spectroscopy. Other spectroscopic techniques, such as ESR spectroscopy and Mössbauer spectroscopy, have led to a better understanding of, for example, paramagnetic iron impurities in the halloysite structure. Although the spectroscopic behaviour of halloysite is reasonably well understood, there are still outstanding questions, and it can be expected that in the future, the use of other spectroscopic techniques, such as those based on synchrotron radiation, will add to an even better understanding of this fascinating clay mineral.

REFERENCES

Barron, P.F., Wilson, M.A., Campbell, A.S., Frost, R.L., 1982. Detection of imogolite in soils using solid state silicon-29 NMR. Nature 299, 616–618.

Calhoun, F.G., Carlisle, V.W., 1972. Infrared spectra of selected colombian Andosols. Proc. Soil Crop Sci. Soc. Florida 31, 157–161.

Chaikum, N., Carr, R.M., 1987. Electron spin resonance studies of halloysites. Clay Miner. 22, 287–296.

Cheng, H., Frost, R.L., Yang, J., Liu, Q., He, J., 2010a. Infrared and infrared emission spectroscopic study of typical Chinese kaolinite and halloysite. Spectrochim. Acta A Mol. Biomol. Spectrosc. 77A, 1014–1020.

Cheng, H., Yang, J., Liu, Q., Zhang, J., Frost Ray, L., 2010b. A spectroscopic comparison of selected Chinese kaolinite, coal bearing kaolinite and halloysite—a mid-infrared and near-infrared study. Spectrochim. Acta A Mol. Biomol. Spectrosc. 77, 856–861.

Coyne, L.M., Costanzo, P.M., Theng, B.K.G., 1989. Luminescence and ESR studies of relationships between oxygen(1-)-centers and structural iron in natural and synthetically hydrated kaolinites. Clay Miner. 24, 671–693.

Craig, P., Dash, J., McGuire, A., Nagle, D., Reiswig, R., 1959. Nuclear resonance absorption of gamma rays in Ir^{191}. Phys. Rev. Lett. 3, 221.

Crowley, J.K., Vergo, N., 1988. Near-infrared reflectance spectra of mixtures of kaolin-group minerals: use in clay mineral studies. Clay Clay Miner. 36, 310–316.

Cruz, M.I., Letellier, M., Fripiat, J.J., 1978. NMR study of adsorbed water. II. Molecular motions in the monolayer hydrate of halloysite. J. Chem. Phys. 69, 2018–2027.

De Roy, G., Verhaert, I., Vansant, E.F., 1981. Allophane and halloysite minerals. Structural considerations. Recueil des Travaux Chimiques des Pays-Bas 100, 102–106.

Farmer, V.C., 1974. The Infrared Spectra of Minerals. Mineralogical Society, London.

Frost, R.L., 1995. Fourier transform Raman spectroscopy of kaolinite, dickite and halloysite. Clay Clay Miner. 43, 191–195.

Frost, R.L., 1997. The structure of the kaolinite minerals—a FT-Raman study. Clay Miner. 32, 65–77.

Frost, R.L., 1998. Hydroxyl deformation in kaolins. Clay Clay Miner. 46, 280–289.

Frost, R.L., Shurvell, H.F., 1997. Raman microprobe spectroscopy of halloysite. Clay Clay Miner. 45, 68–72.

Frost, R.L., Vassallo, A.M., 1996. The dehydroxylation of the kaolinite clay minerals using infrared emission spectroscopy. Clay Clay Miner. 44, 635–651.

Frost, R.L., Fredericks, P.M., Bartlett, J.R., 1993. Fourier transform Raman spectroscopy of kandite clays. SpEectrochIm ACTA A. 49A, 667–674.

Goetze, J., Ploetze, M., Goette, T., Neuser, R.D., Richter, D.K., 2002. Cathodoluminescence (CL) and electron paramagnetic resonance (EPR) studies of clay minerals. Mineral. Petrol. 76, 195–212.

Johnston, C.T., Sposito, G., Birge, R.R., 1985. Raman spectroscopic study of kaolinite in aqueous suspension. Clays Clay Miner. 33, 483–489.

Kinsey, R.A., Kirkpatrick, R.J., Hower, J., Smith, K.A., Oldfield, E., 1985. High resolution aluminum-27 and silicon-29 nuclear magnetic resonance spectroscopic study of layer silicates, including clay minerals. Am. Mineral. 70, 537–548.

Kitagawa, Y., 1972. An aspect of the water in clay minerals: an application of nuclear magnetic resonance spectrometry to clay mineralogy. Am. Mineral. 57, 751–764.

Kloprogge, J.T., 2005. Short introduction to infrared and Raman spectroscopy. In: Kloprogge, J.T. (Ed.), The Application of Vibrational Spectroscopy to Clay Minerals and Layered Double Hydroxides. The Clay Minerals Society, Aurora, CO, pp. 1–7.

Kloprogge, J.T., 2015. Comment on "A spectroscopic comparison of selected Chinese kaolinite, coal bearing kaolinite and halloysite—a mid-infrared and near-infrared study" and "Infrared and infrared emission spectroscopic study of typical Chinese kaolinite and halloysite" by Hongfei Cheng et al. (2010). Spectrochim. Acta A Mol. Biomol. Spectrosc. 136, 175–177.

Kloprogge, J.T., Frost, R.L., 1999. Raman microprobe spectroscopy of hydrated halloysite from a Neogene cryptokarst from southern Belgium. J. Raman Spectrosc. 30, 1079–1085.

Kloprogge, J.T., Frost, R.L., 2000. Infrared emission spectroscopy study of the dehydroxylation of 10 Å halloysite from a Neogene cryptokarst of southern Belgium. Geol. Belg. 2, 213–220.

Kloprogge, J.T., Wood, B.J., 2015. Chemical bonding and electronic structures of the $A_{12}SiO_5(OH)_4$ polymorphs kaolinite, dickite, nacrite and halloysite by X-photoelectron spectroscopy. Clay Science 19, 39–45.

Komarneni, S., Fyfe, C.A., Kennedy, G.J., 1985. Order-disorder in 1:1 type clay minerals by solid-state ^{27}Al and ^{29}Si magic-angle-spinning NMR spectroscopy. Clay Miner. 20, 327–334.

Komusinski, J., Stoch, L., Dubiel, S.M., 1981. Application of electron paramagnetic resonance and Mössbauer spectroscopy in the investigation of kaolinite-group minerals. Clays Clay Miner. 29, 23–30.

Labouriau, A., Kim, Y.-W., Chipera, S., Bish, D.L., Earl, W.L., 1995. A ^{19}F nuclear magnetic resonance study of natural clays. Clay Clay Miner. 43, 697–704.

Libowitzky, E., 1999. Correlation of O-H stretching frequencies and O-H O hydrogen bond lengths in minerals. In: Schuster, P., Mikenda, W. (Eds.), Hydrogen Bond Research. Springer, Vienna, pp. 103–115.

Loh, E., 1973. Optical vibrations in sheet silicates. J. Phys. C: Solid State Phys. 6, 1091–1104.

Lyon, R.J.P., Tuddenham, W.M., 1960. Infrared determination of the kaolin-group minerals. Nature 185, 835–836.

Madejová, J., Kraus, I., Tunega, D., Samajova, E., 1997. Fourier transform infrared spectroscopic characterization of kaolin-group minerals from the main Slovak deposits. Geol. Carpath. Clays 6, 3–10.

Michaelian, K.H., 1986. The Raman spectrum of kaolinite #9 at 21°C. Can. J. Chem. 64, 285–289.

Mössbauer, R.L., 1958. Kernresonanzfluoreszenz von Gammastrahlung in Ir^{191}. Z. Phys. A 151, 124–143.

Nagasawa, K., Noro, H., 1987. An electron spin resonance study of halloysites. Clay Sci. 6, 261–268.

Newman, R.H., Childs, C.W., Churchman, G.J., 1994. Aluminum coordination and structural disorder in halloysite and kaolinite by ^{27}Al NMR spectroscopy. Clay Miner. 29, 305–312.

Pampuch, R., 1966. Infrared study of the thermal transformation of kaolinite and the structure of metakaolin. Pr. Mineral. 6, 53–70.

Petit, S., Madejova, J., Decarreau, A., Martin, F., 1999. Characterization of octahedral substitutions in kaolinites using near infrared spectroscopy. Clay Clay Miner. 47, 103–108.

Petit, S., Decarreau, A., Martin, F., Bucher, R., 2004. Refined relationship between the position of the fundamental OH stretching and the first overtones for clays. Phys. Chem. Miner. 31, 585–592.

Shimizu, H., Watanabe, T., Henmi, T., Masuda, A., Saito, H., 1988. Studies on allophane and imogolite by high-resolution solid-state ^{29}Si- and ^{27}Al-NMR and ESR, Geochem. J. 22, 23–31.

Soma, M., Churchman, G.J., Theng, B.K.G., 1992. X-ray photoelectron spectroscopic analysis of halloysites with different composition and particle morphology. Clay Miner. 27, 413–421.

Stubičan, V., Roy, R., 1961. Proton retention in heated 1:1 clays studied by infrared spectroscopy, weight loss and deuterium uptake. J. Phys. Chem. 65, 1348–1351.

Wiewiora, A., Wieckowski, T., Sokolowska, A., 1979. The Raman spectra of kaolinite sub-group minerals and of pyrophyllite. Arch. Mineral. 35, 5–12.

Zhang, Y., Fu, L., Yang, H., 2012. Insights into the physicochemical aspects from natural halloysite to silica nanotubes. Colloids Surf. A 414, 115–119.

Chapter 7

Thermal-Treatment-Induced Deformations and Modifications of Halloysite

P. Yuan*
CAS Key Laboratory of Mineralogy and Metallogeny / Guangdong Provincial Key Laboratory of Mineral Physics and Materials, Guangzhou Institute of Geochemistry, Chinese Academy of Sciences (CAS), Guangzhou, China
Corresponding author: e-mail: yuanpeng@gig.ac.cn

7.1 INTRODUCTION

The mastery of the art and technology of firing clay for manufacturing pottery dates back as far as the Neolithic period. A recent archaeological study on the dating of early pottery from Xianrendong Cave, Wannian County, Jiangxi Province, China, has demonstrated that the technology of making pottery vessels has existed for at least 19,000–20,000 years (Wu et al., 2012). Even earlier, around 29,000–25,000 BC, simple ceramic objects were being fabricated in Gravettian culture, including the clay figurine Venus of Dolní Věstonice; such objects have been discovered in the Czech Republic (White, 1988; Vandiver et al., 1989). A very high degree of proficiency in making fine porcelain products by firing a mixture of the clay mineral, kaolinite and other minerals such as feldspar and quartz was achieved in ancient China as early as the Sui and Tang dynasties (AD 581–907) (Bushell, 1977); even the name 'china' originated from the transliteration of Changnan, the former name of the town famous for its porcelain industry, known today as Jingdezhen (Jingde Town) in Jiangxi Province, China.

However, understanding the science behind the firing of clay was not possible until the first half of the 20th century, when the development of microscopic and spectroscopic techniques enabled the detailed observation and characterization of the structural changes and phase transformations occurring. Since then, numerous studies were conducted not only on the mechanisms for producing a variety of ceramic products, but also on the thermal behaviours of clay minerals (Bradley and Grim, 1951). Considerable interest

Developments in Clay Science, Vol. 7. http://dx.doi.org/10.1016/B978-0-08-100293-3.00007-8

has been focused on clay minerals with broad industrial applications, such as montmorillonite (Brown et al., 1987; Sarikaya et al., 2000) and palygorskite (Hayashi et al., 1969; Cases et al., 1991), as well as on the heat-induced changes of composition, structure and physicochemical properties of these clay minerals (Noyan et al., 2006). In particular, the thermal behaviours of kaolinite have received the most attention among researchers because of its importance in ceramic manufacturing over the centuries. In fact, the mineral name 'kaolinite' is derived from the Chinese Kao-Ling (高岭/高嶺, pinyin Gāolǐng), the name of an old village in Fuliang County, located near the city of Jingdezhen (Finlay, 2010).

Extensive work on the calcination of kaolinite has been conducted since Brindley and Nakahira (1959) pioneered the study of the kaolinite-to-mullite reaction sequence, which is the most critical process in the production of ceramics using kaolinite and related minerals as raw materials; some intermediate processes in this reaction sequence, such as the kaolinite-metakaolinite conversion, have also attracted great interest from researchers working on pozzolanic cement or related materials (Kakalia et al., 2001). Typical thermally induced mineral phase transformation processes in kaolinite, such as dehydroxylation, have been thoroughly investigated (Rocha and Klinowski, 1990), and the intermediate and final equilibrium states of the kaolinite–mullite reaction have been well defined through various methods, including infrared spectroscopy (Percival et al., 1974; Vassallo et al., 1992; Frost and Vassallo, 1996), magic angle spinning nuclear magnetic resonance (MAS NMR) (Brown et al., 1985; Mackenzie et al., 1985; Rocha and Klinowski, 1990; Massiot et al., 1995), transmission electron microscopy (TEM) (Bergaya et al., 1996; Lee et al., 1999), electron paramagnetic resonance (Djemai et al., 2001), X-ray diffraction (XRD) and thermal analysis (Dion et al., 1998; Kristof et al., 1998, 2002). On the basis of these studies, the complete process for the calcination of kaolinite has been outlined. As summarized by Smith et al. (1993), this process initially involves dehydroxylation between 600°C and 850°C, at which most of the OH groups are removed, leading to reduced coordination of the originally octahedral aluminium. The metakaolinite formed in this step is X-ray amorphous; however, the *ab* order is preserved while some disorder is introduced along the *c*-axis (Bergaya et al., 1996). A large exothermic event at approximately 980–1000°C is believed to be triggered by the removal of the last hydroxyls and subsequent formation of a distinct alumina-rich phase. High-temperature calcination (at approximately 1400°C) yields the final equilibrium products: a 3:2 mullite ($3Al_2O_3{\cdot}2SiO_2$) and cristobalite.

Much less attention has been devoted to the thermal-treatment-induced deformations and transformations of halloysite, even though it has long been known to be a hydrated polymorph of kaolinite. The thermal behaviours of halloysite have been reasonably postulated to be generally analogous to those of kaolinite because halloysite is structurally and chemically similar to kaolinite; this assumption has been confirmed by a few studies (eg, Smith et al., 1993; Yuan et al., 2012). However, the behaviour of halloysite under heating is not identical to that of kaolinite because the differences in composition, texture and morphology, leading to substantial differences in their thermal behaviours. These differences in composition, morphology and texture include the following:

1. *The presence of interlayer water*. Halloysite (10 Å) undergoes dehydration of interlayer water upon heating, whereas kaolinite does not, because kaolinite does not contain interlayer water. Thus, the response of interlayer water molecules in halloysite to certain intercalation tests has been used to differentiate halloysite from other kaolin minerals (Churchman and Theng, 1984; Theng et al., 1984; Churchman, 1990; Hillier and Ryan, 2002). The presence of strongly held water molecules in halloysite might also lead to important differences in its physical and chemical properties from those of other minerals in the kaolin family (Joussein et al., 2005).
2. *Morphology*. Halloysite possesses a tubular morphology with one-dimensional mesoporosity, in contrast to the platelike nature of kaolinite, which is actually nonporous. This morphological difference might lead to significant differences in textural properties, such as specific surface area (SSA) and porosity, upon the calcination of halloysite.
3. *Particle size*. The particle size of kaolinite is normally tens or even hundreds of times larger than that of tubular halloysite. This difference might result in some visible differences in thermal behaviours between kaolinite and halloysite during dehydroxylation and phase transformations. Historically, whether the alumina-rich phase formed in the exothermic event at approximately 980–1000°C during the calcination of kaolinite is a spinel-like γ-Al_2O_3 or an alumina-rich mullite has been a topic of debate (Okada et al., 1986; Brindley and Lemaitre, 1987; Sonuparlak et al., 1987). This question is also of interest with halloysite.

In this chapter, the results from previous studies concerning the aforementioned points are overviewed, and the thermal-treatment-induced deformations and modifications of halloysite and its potential applications are discussed.

7.2 DEHYDRATION OF HALLOYSITE UNDER THERMAL TREATMENT—EFFECTS OF TEMPERATURE

The existence of interlayer water in halloysite (10 Å) results in an approximately 3 Å difference in the d_{001} value between halloysite (10 Å) and halloysite (7 Å); this difference represents the thickness of a monolayer of water molecules. Therefore, the dehydration of halloysite (10 Å)—ie, the conversion from halloysite (10 Å) to halloysite (7 Å) by the loss of interlayer water—can be clearly identified by monitoring the d_{001} changes in the XRD patterns. Thermal analysis provides another straightforward method for observing the dehydration process of halloysite (10 Å) by detecting the changes resulting from the endothermic event related to dehydration. As shown in the differential thermal analysis (DTA) curves of three New Zealand halloysite reference samples (Opotiki, Te Puke and Matauri Bay) (Fig. 7.1), obvious endothermic peaks centred at approximately 52°C were resolved; these peaks correspond to the loss of the physically adsorbed water and the interlayer water. This result is typical of those reported in the literature; the dehydration of halloysite (10 Å) has been reported to normally occur at temperatures ranging from 50°C to 150°C, although, in some cases, relatively mild heating (eg, at 35°C) was observed to trigger the dehydration process (Joussein et al., 2006). Nevertheless, the complete removal of water molecules in halloysite requires high temperatures at approximately 450°C, as observed by Joussein et al. (2006) on both Opotiki and Te Puke halloysite samples because the mass losses between 50°C and 450–500°C of these samples were continuous. Brindley and Goodyear (1948) also observed that heating to 400°C was required for the complete removal of interlayer water from Indiana halloysite.

Joussein et al. (2006) comprehensively compared the difference in dehydration behaviour between two Ca^{2+}-saturated halloysites: Te Puke-Ca, predominantly composed of tubular halloysite; and Opotiki-Ca, composed of halloysite particles almost uniformly spheroidal in shape. Te Puke halloysite derived from rhyolite and andesite and was formed by weathering and hydrothermal alteration (Hughes, 1966). Opotiki halloysite derived from rhyolite and was formed by weathering (Kirkman, 1977). The two halloysite samples

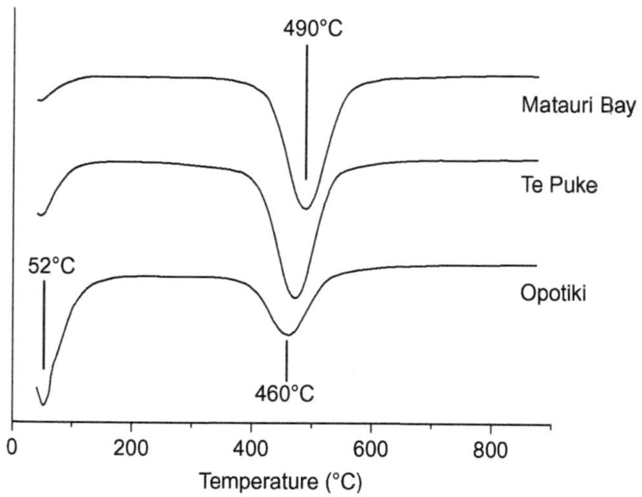

FIG. 7.1 Differential thermal analysis (DTA) curves of three New Zealand reference halloysite samples. *Reprinted with permission from Joussein et al. (2005). Copyright 2005 The Mineralogical Society.*

do not exhibit the same hydration state: Te Puke is partially hydrated, as indicated by the presence of a broad (001) reflection ranging from 7 to 10 Å, whereas Opotiki is fully hydrated, with a d_{001} value of about 10 Å (Fig. 7.2). For both halloysite samples, the increase in temperature from 25°C to 120°C eventually induced dehydration, indicated by the progressive replacement of the (001) reflection of 10 Å by that of about 7 Å (Fig. 7.2). However, the intermediate dehydration processes of the two halloysite specimens differed. The dehydration of the tubular Te Puke halloysite occurred through an intermediate (001) reflection with a d_{001} value between 7 and 10 Å, whereas the fully hydrated spheroidal Opotiki halloysite directly transitioned to the dehydrated phase without forming an intermediary phase. For Te Puke halloysite, the transition temperature for dehydration was 40°C (T40), where the 7 Å reflection became more intense than the 10 Å reflection; for Opotiki halloysite, this transition occurred at 55°C (T55). Full dehydration occurred when the 10 Å reflection completely disappeared, between 45°C and 55°C and between 70°C and 90°C for the Te Puke and Opotiki halloysites, respectively.

The aforementioned differences in dehydration behaviour of the Te Puke and Opotiki halloysite specimens have been attributed to the different states of the water molecules (Joussein et al., 2006). Costanzo et al. (1982, 1984),

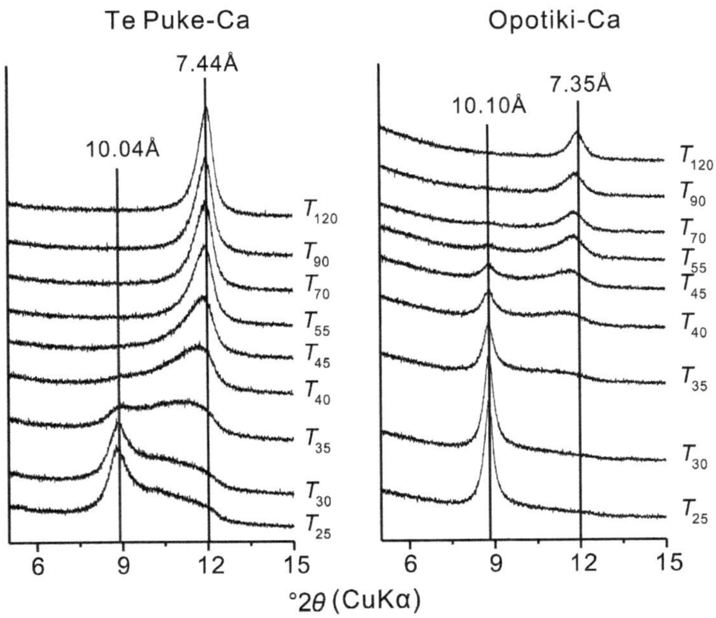

FIG. 7.2 XRD patterns of Ca-saturated Te Puke and Opotiki halloysite samples at various temperatures increasing from 25°C (T25) to 120°C (T120). *Reprinted with permission from Joussein et al. (2006). Copyright 2006 The Clay Minerals Society.*

Costanzo and Giese (1985) and Lipsicas et al. (1985) proposed a brief classification of the types of water molecules in halloysite or kaolinite hydrates. The interlayer water in halloysite was observed to occur in two different structural environments: (i) 'hole' water and (ii) 'associated' water. Hole water is located in ditrigonal cavities in the basal oxygen plane of the tetrahedral halloysite sheet, forming hydrogen bonds with the basal oxygen, while associated water is located at a different level in the interlayer space and has a high degree of mobility at ambient temperatures (Costanzo et al., 1982; Lipsicas et al., 1985). Through a computational analysis of the XRD profiles of the two halloysite samples (Fig. 7.3), Joussein et al. (2006) observed that the dehydration of the Te Puke halloysite proceeded via one intermediate hydration

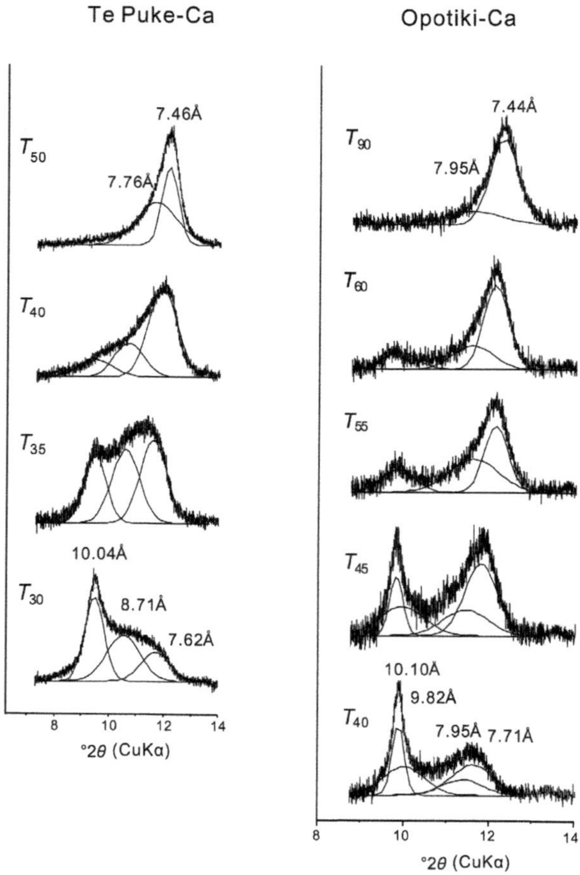

FIG. 7.3 Computational fitting of the XRD profiles obtained for Ca-saturated Te Puke samples heat-treated at 30°C, 35°C, 40°C and 50°C, and for Ca-saturated Opotiki samples heat-treated at 40°C, 45°C, 55°C, 60°C and 90°C. *Reprinted with permission from Joussein et al. (2006). Copyright 2006 The Clay Minerals Society.*

state, leading to a clear reflection at about 8.7 Å; the presence of this stable intermediate component resulted from the loss of associated water, which is in agreement with the aforementioned results. The presence of the 7.76 Å reflection at 50°C in the XRD pattern of Te Puke halloysite arises from the interstratification of the 8.7 and 7 Å phases. However, in Opotiki halloysite, the dehydration of the fully hydrated phase resulted in a dehydrated phase without the formation of the 8.7 Å intermediate. Instead, a reflection at 7.95 Å was detected through computational fitting (Fig. 7.3) and was interpreted as the interstratification between the 10 Å and 7 Å phases, or, more likely, between the 8.7 Å and 7 Å phases (Joussein et al., 2006). The 7.95 Å phase of Opotiki halloysite was still present at 90°C, whereas it disappeared at lower temperatures in Te Puke halloysite. This result demonstrates that the types of water molecules dominant in spheroidal Opotiki halloysite interact more strongly with the layers than the types in tubular Te Puke halloysite. Joussein et al. (2006) proposed two possible reasons for the strong interaction between the water molecules and Opotiki halloysite layers: (i) the greater number of structural defects in the spheroidal (Opotiki) than in the tubular (Te Puke) morphology, and (ii) the position of the vacant octahedral sites opposite the ditrigonal cavities in the tetrahedral sheet because of layer translations. These results demonstrated that the presence of different types of water molecules—associated and hole water—is an essential factor controlling the dehydration behaviour of halloysite samples.

The dehydration of halloysite (10 Å) could also result in structural and morphological changes in addition to the changes in the states of interlayer water molecules. Kohyama et al. (1978) investigated the dehydration-induced structural and morphological changes of halloysite sourced from Kusatsu, Gunma Prefecture, Japan, by using a conventional electron microscope equipped with an environmental cell. Their selected-area electron diffraction (SAED) observations showed that halloysite (10 Å) has two-layer periodicity in a monoclinic structure, which is almost the same as that of halloysite (7 Å). This observation indicated that the dehydration of the interlayer water did not greatly change or did not affect the structure of halloysite (10 Å); however, during dehydration, clear stripes approximately 50–100 Å in width, appeared along the halloysite tube axis, and the diameters of the tubular particles increased by approximately 10%. These authors proposed an idealized model for the morphological changes in halloysite (10 Å) during dehydration on the basis of TEM observations and related calculations. In Fig. 7.4, the *left* and *right* images show the hydrated and dehydrated forms, respectively; the *upper* images present cross-sectional views and the *lower* shows simulated TEM images of the tubes. Each layer represents a 'unit' equivalent to or slightly larger than the 'crystallite' thickness, and the width of stripes observed in the images corresponds to the 'crystallite' thickness. The unit consists of a small number of unit cells of 20.7 or 14.9 Å thick. The number of unit cells

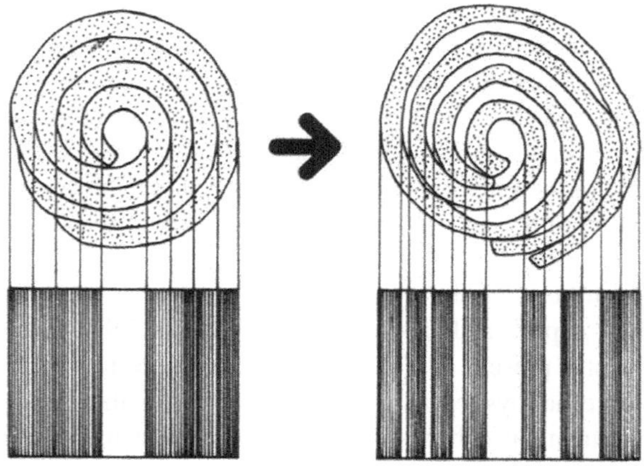

FIG. 7.4 Idealized morphological change of halloysite (10 Å) particle by dehydration. *Upper*, cross-sectional views; *lower*, transmitted views of the tubes. *Reprinted with permission from Kohyama et al. (1978). Copyright 1978 The Clay Minerals Society.*

ranges from 2 to 10, as determined from the width of stripes in the actual TEM images and from the XRD patterns. In the case of fully hydrated halloysite (10 Å), the assembly of layers along the *c*-axis direction with thicknesses of 50–100 Å are tightly connected to each other; they separate during dehydration, leading to visible stripes along the tube axis (Fig. 7.4). This separation is attributed to shrinkage of the unit cell from 20.7 to 14.9 Å, and the thickness of each particle also shrinks by approximately 30%. However, the diameter of the halloysite (7 Å) expands by approximately 10–20%. This contrast was explained as follows: the rolled configuration of halloysite gradually loosens because the attractive forces between the layers are weakened by the formation of crevices during layer separation; as a result, the diameter of the halloysite (10 Å) tube tends to expand (Kohyama et al., 1978). However, the thickness along the *c*-axis contracts under dehydration when the rolled layer loosens. Moreover, the dimensions of the *a*- and *b*-axes remain unchanged. Accordingly, an inherent strain was proposed to favour expansion when the rolled configuration of tubular halloysite is dehydrated. This process is analogous to a book rolled by one hand that expands upon relaxation of the grip.

In addition to heating temperature, factors such as environmental humidity also significantly affect the dehydration of halloysite. Kohyama et al. (1978) showed that halloysite (10 Å) dehydrated to halloysite (7 Å) at approximately 40°C at 40% relative humidity (RH) and at 65°C at 100% RH. Moreover, the dehydration of halloysite (10 Å) is considered an irreversible process; ie, normally, the interlayer water cannot be restored by treatment with water after it has been lost. This irreversibility implies that halloysite samples should be treated with great care and stored under controlled humidity (Slansky, 1983) to avoid dehydration. The storage of halloysite (10 Å) in a sealed container in contact with water or in a water-saturated atmosphere has been suggested as being necessary to maintain the fully hydrated state (Giese, 1988). More details are reported in Chapter 2 on the dehydration/rehydration behaviour of halloysite.

7.3 STRUCTURAL CHANGES AND PHASE TRANSFORMATIONS OF HALLOYSITE UNDER CALCINATION

The most dramatic structural changes during the calcination of halloysite start at approximately 450°C, the temperature at which dehydration finishes and dehydroxylation begins, as confirmed by the results of various studies (Smith et al., 1993; Yuan et al., 2012; Ouyang et al., 2014). In a XRD study aimed to monitor the changes under heating of a highly pure halloysite sample sourced from Shanxi Province, China, Yuan et al. (2012) observed that the XRD pattern of Hal-400 (halloysite heated at 400°C; this nomenclature rule is applied hereafter in this chapter) was very similar to that of unheated halloysite. However, in the case of Hal-500, the intensities of the main reflections decreased significantly. The XRD patterns of Hal-600 to Hal-900 displayed broad diffraction maxima because of dehydroxylation and subsequent formation of metahalloysite, an XRD-amorphous product. In addition, a broad reflection centred at approximately 21.1° 2θ ($d=0.421$ nm), which was attributed to the dissociative amorphous SiO_2 that segregated from the metahalloysite, was observed in the patterns of these samples. In the XRD patterns of Hal-1000 and Hal-1100, three broad reflections (with d values of approximately 0.240, 0.198 and 0.140 nm) assigned to nanocrystalline γ-Al_2O_3 were observed (Yuan et al., 2012). Overall, the XRD results reported in previous studies are consistent with the abovedescribed process, although in some cases, impurity phases such as cristobalite in raw halloysite resulted in the disturbance of resolution on the signals of intermediate amorphous SiO_2 and γ-Al_2O_3 (Smith et al., 1993).

The structural evolution of halloysite during calcination has been clearly captured through thermogravimetric analysis and differential scanning calorimetry, which have shown that the dehydroxylation of halloysite occurs at approximately 500–550°C, which is normally below the dehydroxylation temperature of kaolinite (Frost and Vassallo, 1996). Thermal analysis has also revealed a significant exothermic event at approximately 980–1000°C during the calcination of halloysite. This exotherm is believed to be due to the formation of a distinct alumina-rich phase; however, in the case of kaolinite, the issue of whether the alumina-rich phase formed in this event is a spinel-like γ-Al_2O_3 or an alumina-rich mullite has been debated (Okada et al., 1986; Brindley and Lemaitre, 1987; Sonuparlak et al., 1987). Unfortunately, the resolution of this problem lies beyond the limited capabilities of thermal analysis or XRD because the two mentioned phases have similar thermal behaviours or produce similar diffraction patterns. To uncover the process leading to the exothermic event at approximately 1000°C, Smith et al. (1993) conducted a comprehensive ^{29}Si and ^{27}Al MAS NMR study. They observed that the chemical shift and intensity distribution of the alumina phase formed at 1000°C is entirely consistent with transitional alumina and, most likely, γ-Al_2O_3; they also observed that the intermediate five-coordinated alumina (Al^V, with a chemical shift at approximately 26 ppm; see Fig. 7.5) disappears during the crystallization of γ-Al_2O_3. These authors proposed that the phase separation of γ-Al_2O_3 and amorphous silica occurred at such a small scale that the diffraction signals from such small domains, even those of crystalline γ-Al_2O_3, would be very weak. This proposal is in good agreement with the actual XRD observations of high-purity halloysite (Yuan et al., 2012; Ouyang et al., 2014), where the identified reflections from γ-Al_2O_3 were very broad, implying a nanocrystalline nature.

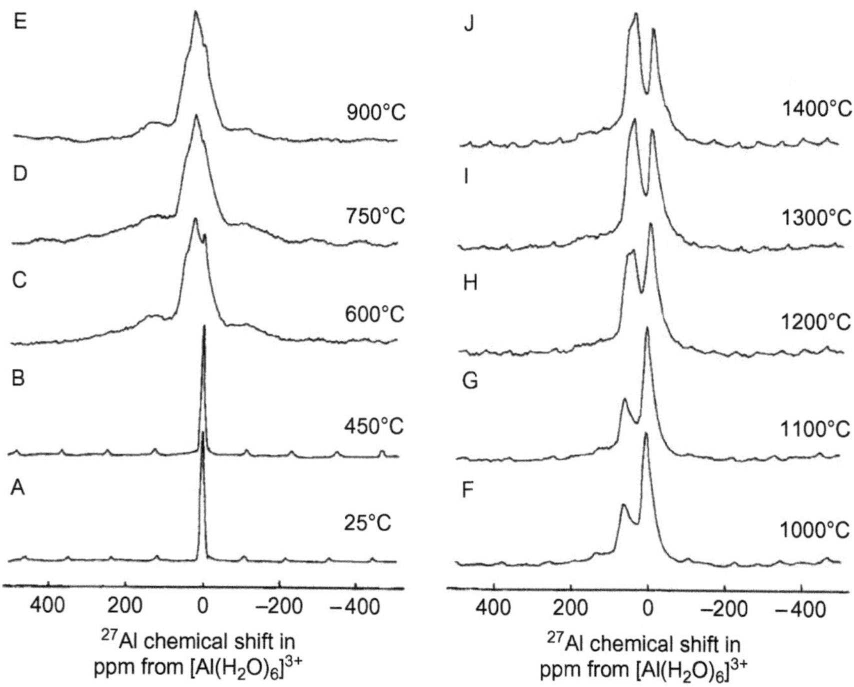

FIG. 7.5 ^{27}Al MAS NMR spectra of halloysite after heat treatment for 2 h at various temperatures. *Reprinted with permission from Smith et al. (1993). Copyright 1993 Springer-Verlag.*

The existence of the γ-Al_2O_3 phase was observed by Yuan et al. (2012) in a TEM study on Hal-1000 and Hal-1100. In the TEM observation on Hal-1100 (Fig. 7.6), two broad diffraction rings in the SAED pattern (*inset* in Fig. 7.6A) were observed and corresponded to the (400) and (440) reflections of γ-Al_2O_3 according to JCPDS reference data (No. 00-010-0425). This result confirms the formation of γ-Al_2O_3 nanoparticles during calcination at approximately 1000–1100°C. A single halloysite nanotube in Hal-1100 is shown in Fig. 7.6B, and the nanocrystalline γ-Al_2O_3 domains of this nanotube were directly observed in high-resolution imaging (see Fig. 7.6C, where the domain is indicated by an *arrow*). On the basis of measurements from noise-filtered high-resolution TEM images resulting from FFT processing (*inset* of Fig. 7.6C), the (440) planes of nanocrystalline γ-Al_2O_3 were identified, with an interplanar distance (the distance between planes of atoms of the same type) of 0.135 nm. The width and the length of the observed nanocrystalline γ-Al_2O_3 domains were approximately 5 nm and 5–15 nm, respectively, as observed in different TEM images. The small size of the nanocrystalline

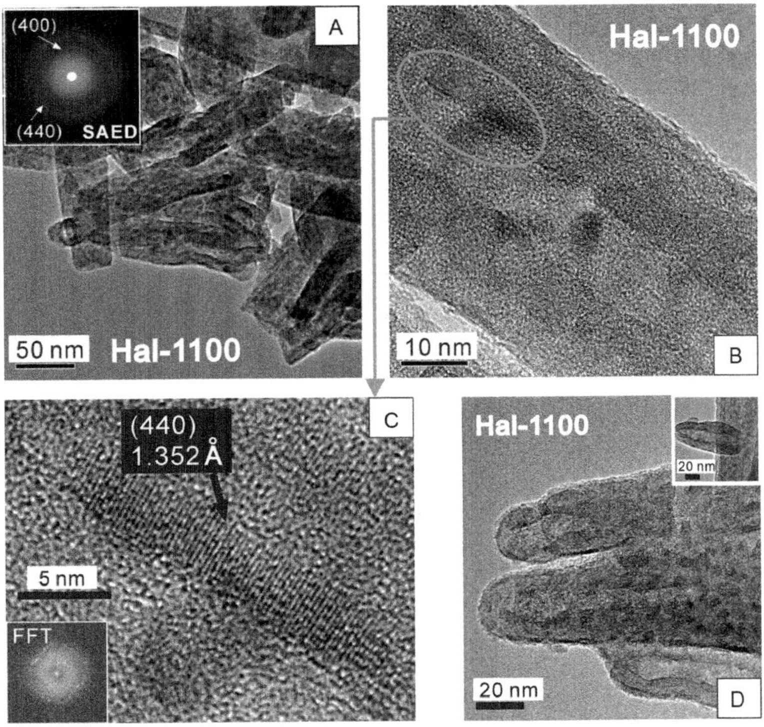

FIG. 7.6 TEM images of calcined halloysite sourced from Shanxi, China. *Reproduced with permission from Yuan et al. (2012). Copyright 2012 The Clay Mineral Society.*

γ-Al_2O_3 particles was consistent with the broad XRD diffractions, and their presence strongly supports the attribution of the exotherm at approximately 1000°C to the formation of γ-Al_2O_3.

Informative details of the structural changes of halloysite under calcination were also obtained by detecting the signals in the skeletal vibration region of Fourier transform infrared (FTIR) spectroscopy. The peak assignments in the FTIR spectra to vibrational modes in the heating products of halloysite are summarized in Table 7.1; these values were extracted from several publications on halloysite, alumina and mullite (Frost and Vassallo, 1996; Morterra and Magnacca, 1996; Koretsky et al., 1997; Madejová and Komadel, 2001; Voll et al., 2001; Joussein et al., 2005). As an example, in the case of the Shanxi halloysite sample (Yuan et al., 2012), Hal-120 and Hal-400 produced similar spectra, as expected (Fig. 7.7A and B); however, Hal-500 exhibited some obvious differences because of dehydroxylation. The intensity of the Al–O–Si deformation band at 546 cm^{-1} gradually weakened in the spectra of Hal-500 (Fig. 7.7C) and Hal-600 (Fig. 7.7D), and was completely absent in the spectrum of Hal-700 (Fig. 7.7E). These results indicate that dehydroxylation was accompanied by the destruction of Al–O–Si bonds in halloysite, which led to the phase separation of amorphous SiO_2 and Al_2O_3 in metahalloysite. The Si–O–Si bending band (at 470 cm^{-1}) broadened and gradually shifted towards lower wavenumbers as the calcination temperature was increased from 500°C to 900°C (Fig. 7.7C–G). This result indicates progressive disordering in metahalloysite and an increase in the mean bond angle in the Si–O–Si network. A similar mechanism for calcined kaolinite was proposed by MacKenzie et al. (1985) on the basis of ^{29}Si MAS NMR results. However, the Si–O–Si stretching band appeared again at 470 cm^{-1} in the spectra of Hal-1000 and Hal-1100 (Fig. 7.7H and I), but, in this case, it was assigned to amorphous silica (Zhuravlev, 1993). The narrowing of the band at 470 cm^{-1} was observed in the spectrum of Hal-1000, which indicated some ordering of the Si–O–Si network. This reordering was attributed to the possible role of amorphous silica in the formation of nanocrystalline mullite or cristobalite at 1000–1100°C.

The Si–O absorption bands (690, 756 and 1102 cm^{-1}) in the spectra of Shanxi halloysite followed nearly identical trends as functions of increasing calcination temperature. Their intensities in the spectrum of metahalloysite (Hal-500 and above) were substantially weaker than those in the spectrum of Hal-120 and were accompanied by an obvious wavenumber shift (Fig. 7.7), possibly indicating the breakdown of the ordered Si_2O_5 network. Moreover, the originally separated band at 1102 cm^{-1} could not be distinguished from the in-plane Si–O stretching band (initially at 1028 cm^{-1}) in the spectrum of metahalloysite (Fig. 7.7) because the latter band shifted to progressively higher wavenumbers as the calcination temperature was increased (to approximately

TABLE 7.1 Positions and Assignments of the Infrared Vibration Bands of Halloysite

Position (cm^{-1})	Assignments
3745	O–H stretching of silanols in amorphous silica
3700	O–H stretching of inner-surface hydroxyl groups[a]
3673	O–H stretching of inner-surface hydroxyl groups[b]
3663	O–H stretching of the aluminols in alumina
3626	O–H stretching of inner hydroxyl groups
3250	O–H stretching of water
2933	CH_2 stretching
1102	Perpendicular Si–O stretching
1028–1096	In-plane Si–O–Si stretching
912	Al–O–H deformation of inner hydroxyl groups
906	Al–O vibration associated with four coordinated Al
816	Al–O vibration associated with six coordinated Al
756	Perpendicular Si–O stretching
738	Al–O vibration associated with four coordinated Al
690	Perpendicular Si–O stretching
672	Al–O vibration associated with five coordinated Al
556/560	Al–O stretching vibration associated with six coordinated Al
546	Deformation of Al–O–Si
470	Deformation of Si–O–Si

[a]*Vibration with a transition moment nearly perpendicular to the (001) plane.*
[b]*Vibration with a transition moment lying in the (001) plane (Farmer, 1998).*

1094 cm^{-1} for calcination at 900°C; see Fig. 7.7G). This phenomenon was attributed to the progressive disordering of the calcined halloysite at higher temperatures. For Hal-1100, the Si–O stretching band at 1096 cm^{-1} began to form a shoulder at higher wavenumbers (approximately 1210 cm^{-1}), which was ascribed to the presence of amorphous silica (Shoval et al., 2001), consistent with the expulsion of silica from metahalloysite at this temperature.

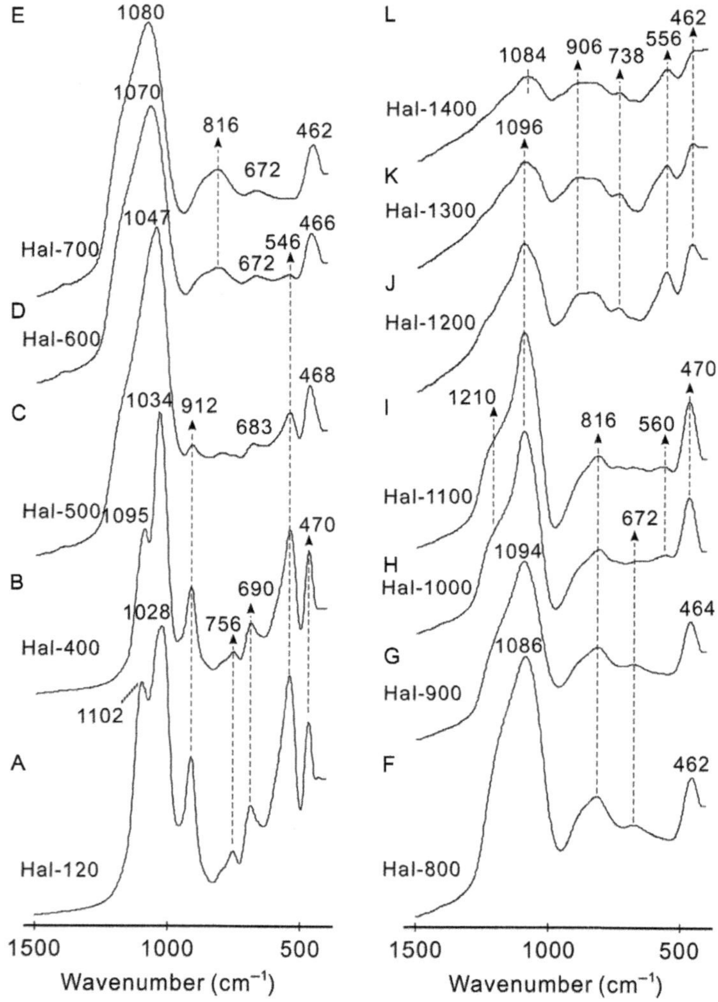

FIG. 7.7 FTIR absorbance spectra of the heating products of Shanxi halloysite. *Reproduced with permission from Yuan et al. (2012). Copyright 2012 The Clay Mineral Society.*

In the spectra of halloysite calcined at 600°C and above (Fig. 7.7), the broad band at approximately 816 cm^{-1} was assigned primarily to Al–O stretching associated with octahedrally coordinated Al (Al^{VI}); this band became more pronounced with increasing temperature up to 1100°C. Tetrahedrally coordinated Al (Al^{IV}) in γ-Al_2O_3 may also contribute to this broad band at high temperatures (1000°C and 1100°C) according to Morterra and Magnacca (1996). As the calcination temperature was increased to 1100°C, two new bands appeared at 906 and 738 cm^{-1} and were assigned to Al^{IV}–O stretching in mullite (Voll et al., 2002). Simultaneously, the intensity of the original band at 816 cm^{-1} slightly decreased with increasing treatment temperature (Fig. 7.7J–L).

The very weak band at 560 cm^{-1} in the spectra of Hal-1000 and Hal-1100 (Fig. 7.7H and I) was assigned to the Al-O stretching vibration associated with Al^{VI} in γ-Al_2O_3 (Voll et al., 2001; Yuan et al., 2012). The occurrence of this band reflected the formation of nanocrystalline γ-Al_2O_3. The absorption band at approximately 672 cm^{-1}, resolved in the spectra of halloysite calcined at 600–900°C (Fig. 7.7D–G), was assigned to the Al–O vibration of Al^V in metahalloysite. This assignment was based on the hypothesis that Al^V existed in disordered metahalloysite formed at 600–900°C, as demonstrated by the ^{27}Al MAS NMR study by Smith et al. (1993). Several high-field NMR studies on various aluminosilicates (Bunker et al., 1991), SiO_2–Al_2O_3 glasses (Sato et al., 1991) and calcined kaolinite (Massiot et al., 1995; He et al., 2004) have shown that the occurrence of Al^V is much more common in disordered systems than would otherwise be assumed given its rarity in crystalline compounds. This Al coordination results from the loss of hydroxyls, which reduces the coordination of aluminium during the transformation from $AlO_2(OH)_4$ to AlO_4. In the spectra of Hal-1000 and Hal-1100 (Fig. 7.7H and I), the 672 cm^{-1} band was no longer clearly evident, indicating the disappearance of the intermediate Al^V. This result is in good agreement with ^{27}Al MAS NMR studies of halloysite and kaolinite calcined at high temperatures (Smith et al., 1993; Massiot et al., 1995), in which only two Al signals, attributed to Al^{IV} and to newly formed Al^{VI}, were observed.

For the Shanxi halloysite products calcined at temperatures higher than 1200°C, some intense structural changes and phase transformations were observed. In the XRD patterns of Hal-1200, the reflections from γ-Al_2O_3 disappeared, and those of mullite were resolved instead (Yuan et al., 2012; Ouyang et al., 2014). Heating at higher temperatures led to more prominent mullite diffractions, and the (101) reflection of cristobalite (at $d = 0.408$ nm), in particular, was clearly visible after heating at 1400°C (Yuan et al., 2012). The intensity of the Si–O–Si stretching band at 462 cm^{-1} was significantly reduced (Fig. 7.7J–L) in the spectra of Hal-1200, Hal-1300 and Hal-1400; similarly, Al^{VI}–O stretching (556 cm^{-1}) of mullite was clearly observed, signifying the depletion of the amorphous silica during the formation of mullite. In the spectrum of Hal-1400, a broad and asymmetric band centred at approximately 1084 cm^{-1} was detected and attributed to Si–O vibrations in both mullite (Voll et al., 2001) and cristobalite (Vassallo et al., 1992; Shoval et al., 1997, 2001).

The intense structural changes and phase transformations were clearly observed in TEM micrographs (Fig. 7.8). The mullite nanoparticles in Hal-1200 appeared to have coalesced but exhibited rodlike morphology, and they appeared to be associated with the amorphous SiO_2 substrate, although the

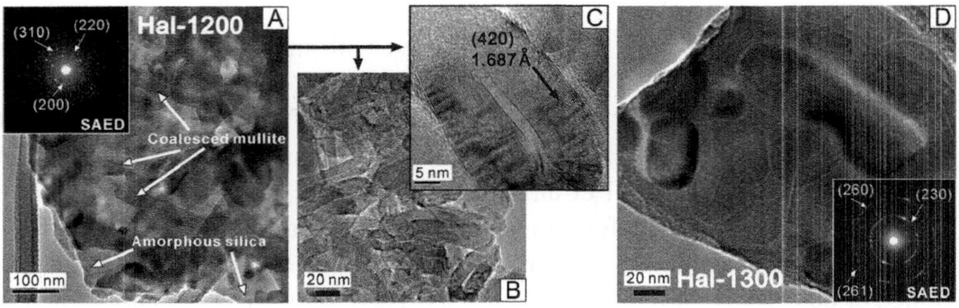

FIG. 7.8 TEM images of calcined halloysite sourced from Shanxi, China. *Reproduced with permission from Yuan et al. (2012). Copyright 2012 The Clay Mineral Society.*

boundaries between the two phases were difficult to identify (Fig. 7.8). The association of mullite with silica was verified by SAED patterns, which showed a diffuse halo (corresponding to the amorphous SiO_2) together with the diffraction rings corresponding to the (200), (220) and (310) reflections of mullite (*inset* in Fig. 7.8A). Under high-resolution imaging (Fig. 7.8B), mullite nanoparticles were observed to be elliptical in shape and some were observed to possess hollow structures (Fig. 7.8B and C). An interplanar distance of 0.169 nm, corresponding to the (420) planes of mullite, was observed (Fig. 7.8C). The nuclei of the mullite phase were preferentially oriented relative to the parent nanocrystalline γ-Al_2O_3 to minimize the interfacial energy; these nuclei thus grew via the interdiffusion of silica and alumina. Yuan et al. (2012) suggested that the elliptical form of the mullite particles might have been inherited from a tubular structure, such as that of Hal-1100, in which ordered γ-Al_2O_3 could react with the surrounding free amorphous silica at their interface, indicated by the scale of Fig. 7.8C; thus, the small hollow structure (about $2 \times$ about 20 nm) was ascribed to the collapsed lumen of the nanotube cross section. The large-scale formation of mullite nanoparticles accompanying the extensive structural transformation indicated that the nearly complete destruction of the tubular halloysite products occurred at approximately 1200°C.

In the TEM images of Hal-1300 (Fig. 7.8D), the transitional hollow structures of the mullite nanoparticles observed in Hal-1200 were no longer evident. Most mullite particles appeared to be nonporous and were approximately 30–50 nm long. They were confirmed as mullite by SAED (*inset* in Fig. 7.8D). The mullite nanoparticles in Hal-1400 also exhibited rodlike morphologies, and their average particle size was approximately 60 nm (Yuan et al., 2012). Energy dispersive X-ray spectroscopy analysis indicated that the mullite was a final 3:2 mullite resulting from the reaction between the dissociative silica and the previously formed alumina-rich mullite (Smith et al., 1993; Yuan et al., 2012). TEM characterization of a single euhedral mullite nanoparticle clearly identified the (200) and (020) planes of mullite and showed that the angle between them was 88.5° (Yuan et al., 2012), demonstrating the high crystallinity of mullite at this temperature.

7.4 DEFORMATIONS IN TEXTURE AND MORPHOLOGY OF HALLOYSITE UNDER CALCINATION AND RELATED MODIFICATIONS IN SURFACE REACTIVITIES

7.4.1 Deformations in Texture and Morphology

In contrast to the similarities between kaolinite and halloysite with respect to their dehydroxylation and phase transformation behaviour under heating, the morphological changes of kaolinite and halloysite are not comparable because halloysite particles possess a tubular morphology, whereas kaolinite particles exhibit a platelike morphology. Furthermore, kaolinite is actually nonporous, not taking into account the pores between particles in aggregates, whereas hollow halloysite nanotubes exhibit a one-dimensional porous structure. At least two types of pores are present in halloysite: (i) the inherent mesopores or macropores contributed by lumens, the inner cavities of the halloysite nanotubes (Churchman et al., 1995); and (ii) the mesopores formed during dehydration as a result of the separation of the originally tightly connected rolled layers in fully hydrated halloysite (this leads to the creation of slit-shaped longitudinal pores with a width of several nanometers, as previously mentioned) (Kohyama et al., 1978). In addition, voids are also formed among the aggregated halloysite particles (Liu et al., 2007; Forsgren et al., 2010). In this sense, the changes in morphology and in the textural characteristics (eg, porosity) of the calcined halloysite are highly worthy of investigation.

Moreover, the particle thickness of kaolinite is tens to hundreds of times larger than the wall thickness of halloysite nanotubes. As a result, the SSA of and the number of surface groups in a unit mass of kaolinite are much less than those in a unit mass of halloysite. For example, halloysite possesses a relatively high SSA, as well as a total pore volume (V_{Pore}) that is much higher than that of platy kaolinite, but usually lower than that of montmorillonite. These characteristics imply that significant changes in porosity may occur during calcination; changes in physicochemical properties, such as the surface reactivity, would therefore be expected. However, the literature contains very few studies on the porosity and surface reactivity changes of halloysite under heating. Yuan et al. (2012) investigated the morphological, textural and surface properties of calcined Shanxi halloysite, which has also been used for aforementioned investigation on structural changes and phase transformations. They showed that heating at $T \leq 400°C$ did not induce any evident morphological changes in the halloysite particles. Surface mottling, however, was observed in Hal-600, Hal-700 and Hal-800. The mottled appearance was likely caused by the structural disordering associated with dehydroxylation and the phase separation of amorphous SiO_2 and Al_2O_3. The tubular morphology remained largely intact in the particles of Hal-900, demonstrating the thermal resistance of the sample at temperatures up to 900°C. However, at 900°C, the phenomenon of surface mottling of halloysite particles became universal, reflecting a high degree of structural disordering. Hal-1000 and Hal-1100 (Fig. 7.6) exhibited substantial changes in morphology, including significant distortions in their tubular outlines. Some of the nanotubes in Hal-1100 even closed off, normally at one end and occasionally at both ends (*inset* in Fig. 7.6D). These phenomena signified substantial collapse of the tubular structure.

In addition, nitrogen adsorption–desorption isotherms were used to measure the porosity and SSA of Shanxi halloysite because they provide a more complete and representative picture of porosity than microscopic methods such as TEM. According to the classifications of the IUPAC (International Union of Pure and Applied Chemistry) all isotherms (Fig. 7.9A) of the original and heated halloysite products were of type IV with H3 hysteresis loops (Gregg and Sing, 1982; Sing and Williams, 2004). The hysteresis loops were associated with the filling and emptying of the mesopores by capillary condensation. Three primary pore populations centred at approximately 3.5, 11 and 22 nm are shown in the pore size distribution (PSD) curve of Hal-120 (Fig. 7.9B). Two possible origins for the 3.5-nm mesopores were proposed (Yuan et al., 2012): the first was the formation of artificial pores resulting from the tensile strength effect, as interpreted in detail by Groen et al. (2003); the second was the formation of longitudinal mesopores, with widths of several nanometers, during the dehydration of tubular halloysite (Kohyama et al., 1978). The 11 and 22 nm mesopores were identified as the lumens of halloysite. Notably, nanotubes with lumen sizes of approximately 20 nm were observed by TEM, whereas those of 11 nm were not, probably because of the sampling method used in TEM characterization. Other mesopores, with diameters ranging from 20 to 40 nm, were attributed to the interparticle porosity in the bundles.

Hal-400 through Hal-900 exhibited nitrogen adsorption–desorption isotherms and PSD curves similar to those of Hal-120, consistent with the TEM observations of intact tubular morphologies for temperatures up to 900°C. However, the SSA (78.1 m^2/g) and V_{Pore} (0.33 cm^3/g) values of Hal-900 were approximately 10% larger than those of Hal-120 (72.5 m^2/g and 0.30 cm^3/g). This change may be due to the formation of micropores during surface breakage resulting from structural rearrangement or transformation. A similar SSA increase, from approximately 35.3–37.8 m^2/g, in kaolinite by thermal treatment has been reported (Melo et al., 2010). However, calcination of halloysite at 1000°C and 1100°C resulted in significant decrease of SSA and porosity values (*inset* in Fig. 7.9A) because of the collapse of the tubular structures at these temperatures, consistent with the aforementioned TEM

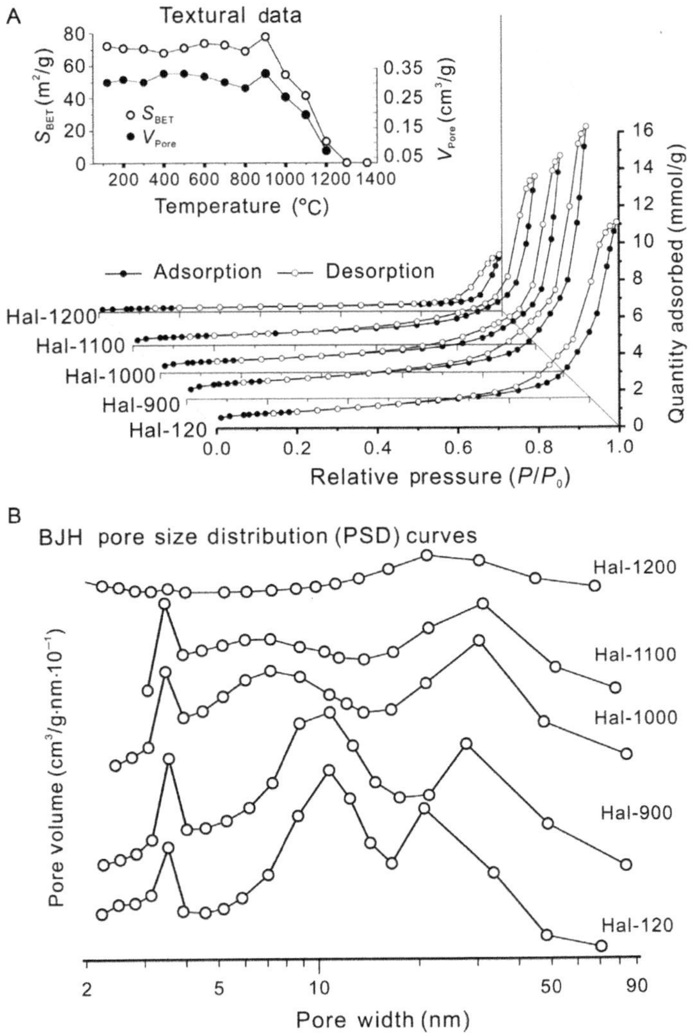

FIG. 7.9 (A) N_2 adsorption–desorption isotherms and textural data (the *inset*) of heating products of Shanxi halloysite. (B) PSD curves of heating products of Shanxi halloysite. *Reprinted with permission from Yuan et al. (2012). Copyright 2012 The Clay Mineral Society.*

observations. Hal-1200 exhibited small values for these textural features (13.2 m^2/g and 0.07 cm^3/g, respectively), consistent with the nearly complete collapse of the tubular structure at 1200°C. The SSA values of Hal-1300 and Hal-1400 were very small because large mullite particles were formed from sintering.

The high thermal persistence of the tubular structure of halloysite is important for the design of novel halloysite-based nanocomposites or ceramic materials. For example, high thermal stability can be expected for nanosized products of halloysite-metal oxide reactions, although they have rarely been investigated (Antill, 2003). Moreover, the curving and closing of the ends of the tubes may pave the way for applications in the encapsulation of materials within guest-loaded halloysite-derived nanotubes for controlled-release purposes. This encapsulation technique has been attempted using several methods (Lvov et al., 2008; Abdullayev and Lvov, 2010). It is noteworthy that small variations in the critical thermal stability temperature (ie, the highest temperature point at which the porosity of halloysite remains almost intact) exist among natural halloysite specimens from different sources. For example, Ouyang et al. (2014) observed that the natural halloysite sample from Hunan, China, retained its tubular morphology at 1100°C and that the product heated to 1100°C exhibited a larger SSA (65.7 m^2/g) and a larger V_{Pore} (0.28 cm^3/g) than unheated halloysite (56.8 m^2/g and 0.22 cm^3/g). Notably, the XRD patterns of the heated products of this sample were similar to those of the halloysite sample used by Yuan et al. (2012), which suggests similar phase transformation processes. The difference between the temperatures where the two halloysite samples retained their tubular morphologies may be related to the diversity in morphological parameters, such as length, internal/external diameter and tube-wall thickness of the halloysite nanotubes, and to the presence of a small amount of impurity minerals affecting the behaviour of halloysite upon heating. In addition, measurements of the mechanical properties, such as the Young modulus of a halloysite nanotube, has been achieved using methods based on TEM (Lu et al., 2010) and atomic force microscopy (Lecouvet et al., 2013). It could be anticipated that elucidating more accurately the relationship between the mechanical properties and structural changes of halloysite during calcination could be realized in the future.

7.4.2 Changes in Surface Reactivities

Yuan et al. (2012) used FTIR and diffuse reflectance infrared Fourier-transform (DRIFT) spectroscopy to demonstrate that, in addition to morphological and textural changes, significant changes in surface groups occurred during the calcination of halloysite. DRIFT has been shown to be more powerful than FTIR for the detection of the hydroxyl-stretching region in silicate minerals; furthermore, it works with small interference through sample preparation and is sensitive for the detection of surface groups (Frost and Johansson, 1998; Yuan et al., 2004). Similar to the other characterization methods discussed here, the DRIFT spectra of Shanxi halloysite heating

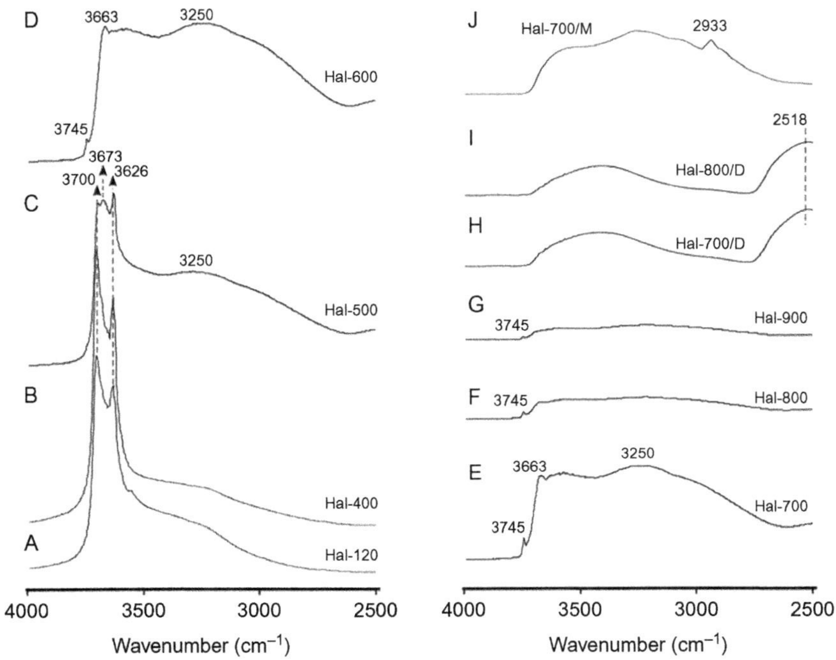

FIG. 7.10 DRIFT spectra of the heating products (A, B, C, D, E, F and G); deuterated products (H and I) and an APTES-modified product (J) of Shanxi halloysite. *Reprinted with permission from Yuan et al. (2012). Copyright 2012 The Clay Mineral Society.*

products did not show any evident changes for temperatures up to 400°C (Fig. 7.10A and B); however, after the halloysite was heated at 500°C, the bands due to the vibrations of both inner-surface hydroxyl groups (at 3700 cm^{-1}) and inner hydroxyl groups (3626 cm^{-1}) were much less intense (Fig. 7.10C). Moreover, a band at 3673 cm^{-1} was resolved. This latter band was attributed to vibrations of the inner-surface hydroxyl groups (Farmer, 1998; Frost and Johansson, 1998) and was hidden in the lower-temperature spectra by the band at 3700 cm^{-1}, which was broad and asymmetrically skewed towards lower wavenumbers. The 912 cm^{-1} band in the FTIR spectrum of Hal-500, resulting from Al–O–H bending involving the inner hydroxyl groups, was also substantially reduced in intensity (Fig. 7.7C). The hydroxyl-group bands were further weakened, and a new band at 3745 cm^{-1} was resolved (Fig. 7.10D) in the DRIFT spectrum of Hal-600. This band was assigned to the isolated surface silanols of the amorphous silica (Morrow and Mcfarlan, 1992; Yuan et al., 2004), which was expelled from the halloysite. For Hal-700, the band at 3745 cm^{-1} was more intense (Fig. 7.10E), indicating the continued formation of amorphous silica. The 3663 cm^{-1} band resolved in the spectrum of Hal-700 was attributed to the remaining aluminols of the amorphous alumina in metahalloysite (Morterra and Magnacca, 1996; Koretsky et al., 1997). These hydroxyl vibration bands were significantly weakened (Fig. 7.10F) at 800°C because of the loss of hydroxyls by condensation between the OH groups. Absorption at 3745 cm^{-1} by the silanols was still present in the spectrum of Hal-900, but the intensity of the band sharply decreased (Fig. 7.9G). No band was observed at 3745 cm^{-1} in the spectrum of Hal-1100, signifying complete dehydroxylation.

The surface reactivities of the calcined Shanxi halloysite were evaluated via deuteration and organosilane modification experiments (Yuan et al., 2012). In the spectra of the deuterated halloysite heating products (denoted as Hal-700/D and Hal-800/D, corresponding to Hal-700 and Hal-800, respectively), both the 3745 and 3673 cm^{-1} bands were absent, as expected (Fig. 7.10H and I). Instead, a broad band of OD vibrations from several environments was centred at approximately 2510 cm^{-1} (Drago, 1992), indicating the replacement of the OH groups by OD groups. In the organosilane modification experiment, γ-aminopropyltriethoxysilane (APTES) was used, and in the DRIFT spectrum of the modified halloysite sample, Hal-700/M (APTES-modified Hal-700), a new CH_2-stretching band appeared at 2933 cm^{-1} in the hydroxyl region (Fig. 7.10J). The O–H stretching band of the silanol groups at 3745 cm^{-1} was not clearly resolved, implying that the modification was accompanied by the consumption of surface silanols. Moreover, the intensity of the aluminol band at 3663 cm^{-1} was substantially diminished, suggesting that the aluminols formed during calcination were also available for grafting. These results demonstrated that grafting occurred between the surface hydroxyl groups of the calcined halloysite and the hydrolysed APTES. The SSA and V_{Pore} values of Hal-700/M were 60.5 m^2/g and 0.22 cm^3/g, respectively, both of which were less than those of Hal-700 (73.0 m^2/g and 0.29 cm^3/g). This was because the grafted APTES on the halloysite surface decreased the porosity. Similar phenomena were also observed in the cases of Hal-800/M and Hal-900/M. All of these results confirmed the successful amine modification of the surfaces of calcined halloysites.

The aforementioned results indicated that the original siloxane groups on the external surface were partially or fully replaced by hydroxyl groups after the halloysite underwent calcination. This finding provides some novel insights into the organomodification of halloysite. In fact, up to now, little attention has been paid to modification of the external surface of halloysite nanotubes, although some attempts to modify the interlayer space by intercalation have been reported (Tonle et al., 2007), because the siloxane surface is generally nonreactive. Yuan et al. (2008) used APTES to modify the surface of unheated halloysite. The hydrolysed APTES was observed to be grafted to the aluminol groups of the internal tube surface, but also were attached to the external siloxane surface primarily by hydrogen bonds (for more information, see Chapter 8). However, the grafting of organosilane onto the external surface was very minor because the total number of external surface defect sites, where aluminol and silanol groups were located and available for grafting, was very small (Yuan et al., 2008). The appearance of hydroxyl groups on the external surface of calcined halloysite signifies new possibilities for organosilane modification, and some related applications can be postulated. First, because of their availability for ion-exchange and silylation reactions, the hydroxyl groups could act as new active sites for attracting and fixing functional guests. Second, the organomodification of the external tube surface of calcined halloysite could be used as a particularly versatile yet site-specific approach to improve the affinity between the calcined halloysite and the polymer substrate in halloysite polymer nanocomposites, although this approach has not yet been reported in the research literature. Therefore, these potential external surface-based applications might extend the utility of current lumen-based applications of halloysite.

7.5 SOME APPLICATIONS OF HEAT-TREATED HALLOYSITE

Numerous applications or potential uses of halloysite are directly tied to its thermally induced behaviour. Historically, research into halloysite has focused on its ceramic properties (ie, its performance as a source material for the manufacture of ceramics or porcelain) because early in the 1930s, halloysite was recognized as chemically and structurally similar to kaolinite (Ross and Kerr, 1934), and kaolinite was undoubtedly the main mineral source in the ceramics industry. Hample and Cutler (1953) investigated the differences between the ceramic properties of halloysite and kaolinite, in particular discovering the reasons for the large shrinkage during the calcination of halloysite sourced from several deposits such as the Dragon Iron Mine. Moreover, the large total shrinkage of halloysite resulting from high temperatures was attributed to the packing characteristics of halloysite particles and to the collapse of their tubular structure.

In addition to efforts aimed at using halloysite as a substitute for kaolinite, some efforts have been made to evaluate the possibility of using halloysite as a main source for the manufacture of some clay polymer nanocomposite (CPN) materials. In one study on halloysite polymer nanocomposites (Alhuthali and Low, 2013), a clear relationship was observed between the addition of halloysite and the enhanced thermal stability and flammability in CPN; the good thermal stability and flame retardation properties of the CPN were attributed to the inherent properties and structure of halloysite. Other research has focused on the use of halloysite as a material for the preparation of high-performance ceramics, such as silicon aluminium oxynitride (sialon) ceramics, which exhibit excellent engineering properties, including high hardness and strength, good thermal shock resistance and low thermal expansion (Lee and Cutler, 1979; Neal et al., 1994; Zhang et al., 2000). For example, Neal et al. (1994) prepared β-sialon ($Si_{6-z}Al_zO_zN_{8-z}$; with $0<z<4.2$) by the carbothermal reduction and nitridation of halloysite; they described the overall reaction for the formation of β-sialon ($z=3$) from halloysite as

$$3\left[Al_2Si_2O_5(OH)_4 \cdot 2H_2O\right] + 15C + 5N_2 \rightarrow 2Si_3Al_3O_3N_5 + 15CO + 12H_2O \tag{7.1}$$

This study showed that the carbothermal reduction and nitridation of halloysite to form β-sialon is a series of complex reactions intimately associated with the structural and phase evolution of halloysite, among other factors. The carbothermal reduction and nitridation reaction proceeded via the structural breakdown of halloysite to form mullite and silica. Afterwards, SiC was formed through the reaction of silica with CO that diffused into the halloysite particles. Silicon in the mullite then reacted to form more SiC and Si_3N_4, leaving a mullite framework devoid of silicon. The β-sialon then formed by the reaction of alumina in the mullitelike framework with SiC, Si_3N_4 and nitrogen.

Recently, several studies focused on applying heated halloysite for adsorption purposes have been reported. Kadi et al. (2012) used unheated and heated Algerian halloysites as adsorbents for Pb(II). The thermal evolution behaviour of Algerian halloysite under heating treatment was analogous to those of the Shanxi halloysite reported by Yuan et al. (2012), and the decrease of SSA started at approximately 800°C (from 63 m^2/g at 200°C to 54 m^2/g at 800°C). The Pb(II) adsorption capacity was observed to be directly related to the SSA of the heated halloysite. In some other cases, heat treatments are more frequently combined with additional activation methods such as acid treatments (see Chapter 4 for more details concerning the acid-treatment of halloysite) to achieve improved textural characteristics. Belkassa et al. (2013) used HCl to treat halloysite that had been heated at 600°C; they subsequently measured a higher SSA, which increased from approximately 61 m^2/g to a maximum of 503 m^2/g. This value far exceeded that of the halloysite product subjected to acid treatment alone because the heat pretreatment resulted in dehydroxylation and amorphization, which was beneficial to the extraction of Al and to the formation of greater microporosity in the product. The resulting materials performed well in the adsorption of crystal violet, where rapid adsorption was observed and the adsorption capacity reached 70 mg/g. On the basis of a similar treatment method that consisted of combining heat and acid treatments, halloysite was also modified for the adsorption and release of ofloxacin, an orally administered generic antibacterial drug (Wang et al., 2014). More recently, Shu et al. (2015) combined heat treatment with acid or alkali treatment of halloysite and used the product as an adsorbent for methylene blue; the acid- and alkali-modified halloysite products exhibited excellent adsorption capacities of 426.7 and 249.5 mg/g, respectively. Additional applications of modified halloysites for adsorption uses are described in Chapter 21.

7.6 CONCLUDING REMARKS

Heating of nanotubular halloysite results in a series of substantial changes in its crystalline structure and phases. Dehydration of halloysite readily occurs under heating at low temperature from approximately 35°C; however, dehydroxylation of halloysite occurs from approximately 500°C to approximately 900°C, leading to a loss of long-range order and increasing disconnection of the silica and alumina originally in the tetrahedral and octahedral sheets, respectively. Formation of nanocrystalline γ-Al_2O_3 occurs under calcination at approximately 1000–1100°C, and the eventual formation of mullite and cristobalite occur at temperatures greater than 1200°C.

Textural and morphological changes closely accompany the structural changes during the heating process. The dehydration of halloysite leads to the separation of originally tightly connected rolled layers in fully hydrated halloysite and creates slit-shaped longitudinal mesopores in the dehydrated product. However, the tubular structure of halloysite overall demonstrates high thermal persistence, and the morphology and the porosity almost did not change until 900°C or even at higher temperatures. Higher calcination temperatures result in the breakdown of the tubular morphology and significantly decrease the porosity and SSA as the halloysite nanotubes become distorted, accompanying with destroy of structure.

Changes in the surface properties of halloysite are also apparent after heating. Dehydration results in an almost irreversible loss of interlayer water in halloysite (10 Å), leading to the formation of halloysite (7 Å) and significantly different reactivities, such as performance in intercalation reactions, compared with that of halloysite (10 Å). Another important change during calcination is the formation of hydroxyl groups at the outer surface of the halloysite nanotubes at approximately 600–900°C because of the disconnection and further disordering of silica and alumina originally in the tetrahedral and octahedral sheets. These hydroxyl groups become available for organosilane grafting, enabling potential uses involving attracting or loading functional guests, developing halloysite-based functional materials and increasing the affinity between calcined halloysite and polymer substrates in CPN.

In summary, thermal treatment results in extensive changes in the structure, texture, porosity, surface reactivity and other physicochemical properties of nanotubular halloysite. The details of these changes are significant not only for understanding the characteristics of halloysite, but also for ensuring the highly efficient application of precious halloysite resources, as well as for facilitating the design of novel halloysite-based functional materials in the near future.

REFERENCES

Abdullayev, E., Lvov, Y., 2010. Clay nanotubes for corrosion inhibitor encapsulation: release control with end stoppers. J. Mater. Chem. 20, 6681–6687.

Alhuthali, A., Low, I.M., 2013. Water absorption, mechanical, and thermal properties of halloysite nanotube reinforced vinyl-ester nanocomposites. J. Mater. Sci. 48, 4260–4273.

Antill, S.J., 2003. Halloysite: a low-cost alternative nanotube. Aust. J. Chem. 56, 723.

Belkassa, K., Bessaha, F., Marouf-Khelifa, K., Batonneau-Gener, I., Comparot, J.D., Khelifa, A., 2013. Physicochemical and adsorptive properties of a heat-treated and acid-leached Algerian halloysite. Colloids Surf. A 421, 26–33.

Bergaya, F., Dion, P., Alcover, J.F., Clinard, C., Tchoubar, D., 1996. TEM study of kaolinite thermal decomposition by controlled-rate thermal analysis. J. Mater. Sci. 31, 5069–5075.

Bradley, W., Grim, R., 1951. High temperature thermal effects of clay and related materials. Am. Mineral. 36, 182–201.

Brindley, G.W., Goodyear, J., 1948. X-ray studies of halloysite and metahalloysite. Part II. The transition of halloysite to metahalloysite in relation to relative humidity. Mineral. Mag. 28, 407–422.

Brindley, G.W., Lemaitre, J., 1987. Thermal, oxidation and reduction reactions of clay minerals. In: Newman, A.C.D. (Ed.), Chemistry of Clays and Clay Minerals. Mineralogical Society, London, pp. 319–370.

Brindley, G.W., Nakahira, M., 1959. The kaolinite-mullite reaction series: I, A survey of outstanding problems. J. Am. Ceram. Soc. 42, 311–314.

Brown, I.W.M., Mackenzie, K.J.D., Bowden, M.E., Meinhold, R.H., 1985. Outstanding problems in the kaolinite-mullite reaction sequence investigated by Si-29 and Al-27 solid-state nuclear magnetic resonance: II, High-temperature transformations of metakaolinite. J. Am. Ceram. Soc. 68, 298–301.

Brown, I.W.M., Mackenzie, K.J.D., Meinhold, R.H., 1987. The thermal-reactions of montmorillonite studied by high-resolution solid-state Si-29 and Al-27 NMR. J. Mater. Sci. 22, 3265–3275.

Bunker, B.C., Kirkpatrick, R.J., Brow, R.K., Turner, G.L., Nelson, C., 1991. Local structure of alkaline earth boroaluminate crystals and glasses: II, B-11 and Al-27 MAS NMR spectroscopy of alkaline-earth boroaluminate glasses. J. Am. Ceram. Soc. 74, 1430–1438.

Bushell, S.W., 1977. Chinese Pottery and Porcelain. Oxford University Press, Kuala Lumpur.

Cases, J.M., Grillet, Y., Francois, M., Michot, L., Villieras, F., Yvon, J., 1991. Evolution of the porous structure and surface area of palygorskite under vacuum thermal treatment. Clay Clay Miner. 39, 191–201.

Churchman, G.J., 1990. Relevance of different intercalation tests for distinguishing halloysite from kaolinite in soils. Clay Clay Miner. 38, 591–599.

Churchman, G.J., Theng, B.K.G., 1984. Interactions of halloysites with amides: mineralogical factors affecting complex formation. Clay Miner. 19, 161–175.

Churchman, G.J., Davy, T.J., Aylmore, L.A.G., Gilkes, R.J., Self, P.G., 1995. Characteristics of fine pores in some halloysites. Clay Miner. 30, 89–98.

Costanzo, P.M., Giese, R.F., 1985. Dehydration of synthetic hydrated kaolinites: a model for the dehydration of halloysite(10 Å). Clay Clay Miner. 33, 415–423.

Costanzo, P.M., Giese, R.F., Lipsicas, M., Straley, C., 1982. Synthesis of a quasi-stable hydrated kaolinite and heat capacity of interlayer water. Nature 296, 549–551.

Costanzo, P.M., Giese, R.F., Lipsicas, M., 1984. Static and dynamic structure of water in hydrated kaolinites: I, The static structure. Clay Clay Miner. 32, 419–428.

Dion, P., Alcover, J.F., Bergaya, F., Ortega, A., Llewellyn, P.L., Rouquerol, F., 1998. Kinetic study by controlled-transformation rate thermal analysis of the dehydroxylation of kaolinite. Clay Miner. 33, 269–276.

Djemai, A., Balan, E., Morin, G., Hernandez, G., Labbe, J.C., Muller, J.P., 2001. Behavior of paramagnetic iron during the thermal transformations of kaolinite. J. Am. Ceram. Soc. 84, 1017–1024.

Drago, R.S., 1992. Physical Methods for Chemists, second ed. Saunders College Publishing, Mexico.

Farmer, V.C., 1998. Differing effects of particle size and shape in the infrared and Raman spectra of kaolinite. Clay Miner. 33, 601–604.

Finlay, R., 2010. The Pilgrim Art: Cultures of Porcelain in World History. University of California Press, California.

Forsgren, J., Jamstorp, E., Bredenberg, S., Engqvist, H., Stromme, M., 2010. A ceramic drug delivery vehicle for oral administration of highly potent opioids. J. Pharm. Sci. 99, 219–226.

Frost, R.L., Johansson, U., 1998. Combination bands in the infrared spectroscopy of kaolins—a DRIFT spectroscopic study. Clay Clay Miner. 46, 466–477.

Frost, R.L., Vassallo, A.M., 1996. The dehydroxylation of the kaolinite clay minerals using infrared emission spectroscopy. Clay Clay Miner. 44, 635–651.

Giese, R.F., 1988. Kaolin minerals: structures and stabilities. In: Bailey, S.W. (Ed.), Hydrous Phyllosilicates (Exclusive of Micas). Reviews in Mineralogy, vol. 19. Mineralogical Society of America, Washington, DC, pp. 29–66.

Gregg, S.J., Sing, K.S.W., 1982. Adsorption, Surface Area and Porosity, second ed. Academic Press, New York.

Groen, J.C., Peffer, L.A.A., Perez-Ramirez, J., 2003. Pore size determination in modified micro- and mesoporous materials. Pitfalls and limitations in gas adsorption data analysis. Microporous Mesoporous Mater. 60, 1–17.

Hampel, B.F., Cutler, I.B., 1953. Some ceramic properties of halloysite. J. Am. Ceram. Soc. 36, 30–34.

Hayashi, H., Otsuka, R., Imai, N., 1969. Infrared study of sepiolite and palygorskite on heating. Am. Mineral. 53, 1613–1624.

He, H.P., Guo, J.G., Zhu, J.X., Yuan, P., Hu, C., 2004. Si-29 and Al-27 MAS NMR spectra of mullites from different kaolinites. Spectrochim. Acta A Mol. Biomol. Spectrosc. 60, 1061–1064.

Hillier, S., Ryan, P.C., 2002. Identification of halloysite (7 Å) by ethylene glycol solvation: the 'MacEwan effect'. Clay Miner. 37, 487–496.

Hughes, I.R., 1966. Mineral changes of halloysite on drying. New Zeal. J. Sci. 9, 103–113.

Joussein, E., Petit, S., Churchman, J., Theng, B., Righi, D., Delvaux, B., 2005. Halloysite clay minerals—a review. Clay Miner. 40, 383–426.

Joussein, E., Petit, S., Fialips, C.I., Vieillard, P., Righi, D., 2006. Differences in the dehydration-rehydration behavior of halloysites: new evidence and interpretations. Clay Clay Miner. 54, 473–484.

Kadi, S., Lellou, S., Marouf-Khelifa, K., Schott, J., Gener-Batonneau, I., Khelifa, A., 2012. Preparation, characterisation and application of thermally treated Algerian halloysite. Microporous Mesoporous Mater. 158, 47–54.

Kakalia, G., Perrakib, T., Tsivilis, S., Badogiannisa, E., 2001. Thermal treatment of kaolin: the effect of mineralogy on the pozzolanic activity. Appl. Clay Sci. 20, 73–80.

Kirkman, J.H., 1977. Possible structure of halloysite disks and cylinders observed in some New Zealand rhyolitic tephras. Clay Miner. 12, 199–215.

Kohyama, N., Fukushima, K., Fukami, A., 1978. Observation of the hydrated form of tubular halloysite by an electron microscope equipped with an environmental cell. Clay Clay Miner. 26, 25–40.

Koretsky, C.M., Sverjensky, D.A., Salisbury, J.W., DAria, D.M., 1997. Detection of surface hydroxyl species on quartz, γ-alumina, and feldspars using diffuse reflectance infrared spectroscopy. Geochim. Cosmochim. Acta 61, 2193–2210.

Kristof, J., Frost, R.L., Horvath, E., Kocsis, L., Inczedy, J., 1998. Thermoanalytical investigations on intercalated kaolinites. J. Therm. Anal. Calorim. 53, 467–475.

Kristof, J., Frost, R.L., Kloprogge, J.T., Horvath, E., Mako, E., 2002. Detection of four different OH-groups in ground kaolinite with controlled-rate thermal analysis. J. Therm. Anal. Calorim. 69, 77–83.

Lecouvet, B., Horion, J., D'Haese, C., Bailly, C., Nysten, B., 2013. Elastic modulus of halloysite nanotubes. Nanotechnology 24, 8.

Lee, G., Cutler, I.B., 1979. Sinterable sialon powder by reaction of clay with carbon and nitrogen. Am. Ceram. Soc. Bull. 58, 869–871.

Lee, S., Kim, Y.J., Moon, H.S., 1999. Phase transformation sequence from kaolinite to mullite investigated by an energy-filtering transmission electron microscope. J. Am. Ceram. Soc. 82, 2841–2848.

Lipsicas, M., Straley, C., Costanzo, P.M., Giese, R.F., 1985. Static and dynamic structure of water in hydrated kaolinites: II, The dynamic structure. J. Colloid Interface Sci. 107, 221–230.

Liu, M.X., Guo, B.C., Du, M.L., Cai, X.J., Jia, D.M., 2007. Properties of halloysite nanotube-epoxy resin hybrids and the interfacial reactions in the systems. Nanotechnology 18, 455–703.

Lu, D., Chen, H.B., Wu, J.S., Chan, C.M., 2010. Direct measurements of the Young's Modulus of a single halloysite nanotube using a transmission electron microscope with a bending stage. J. Nanosci. Nanotechnol. 11, 7789–7793.

Lvov, Y.M., Shchukin, D.G., Mohwald, H., Price, R.R., 2008. Halloysite clay nanotubes for controlled release of protective agents. ACS Nano 2, 814–820.

Mackenzie, K.J.D., Brown, I.W.M., Meinhold, R.H., Bowden, M.E., 1985. Outstanding problems in the kaolinite-mullite reaction sequence investigated by Si-29 and Al-27 solid-state nuclear magnetic-resonance: I, Metakaolinite. J. Am. Ceram. Soc. 68, 293–297.

Madejová, J., Komadel, P., 2001. Baseline studies of the clay minerals society source clays: infrared methods. Clay Clay Miner. 49, 410–432.

Massiot, D., Dion, P., Alcover, J.F., Bergaya, F., 1995. Al-27 and Si-29 MAS NMR study of kaolinite thermal decomposition by controlled rate thermal analysis. J. Am. Ceram. Soc. 78, 2940–2944.

Melo, J.D.D., Costa, T.C.D., de Medeiros, A.M., Paskocimas, C.A., 2010. Effects of thermal and chemical treatments on physical properties of kaolinite. Ceram. Int. 36, 33–38.

Morrow, B.A., Mcfarlan, A.J., 1992. Surface vibrational modes of silanol groups on silica. J. Phys. Chem. 96, 1395–1400.

Morterra, C., Magnacca, G., 1996. A case study: surface chemistry and surface structure of catalytic aluminas, as studied by vibrational spectroscopy of adsorbed species. Catal. Today 27, 497–532.

Neal, G.S., Smith, M.E., Trigg, M.B., Drennan, J., 1994. Solid-state NMR examination of the formation of β-sialon by carbothermal reduction and nitridation of halloysite clay. J. Mater. Chem. 4, 245–251.

Noyan, H., Oenal, M., Sarikaya, Y., 2006. The effect of heating on the surface area, porosity and surface acidity of a bentonite. Clay Clay Miner. 54, 375–381.

Okada, K., Otsuka, N., Ossaka, J., 1986. Characterization of spinel phase formed in the kaolin-mullite thermal sequence. J. Am. Ceram. Soc. 69, C251–C253.

Ouyang, J., Zhou, Z., Zhang, Y., Yang, H., 2014. High morphological stability and structural transition of halloysite (Hunan, China) in heat treatment. Appl. Clay Sci. 101, 16–22.

Percival, H., Duncan, J., Foster, P., 1974. Interpretation of the kaolinite-mullite reaction sequence from infrared absorption spectra. J. Am. Ceram. Soc. 57, 57–61.

Rocha, J., Klinowski, J., 1990. Si-29 and Al-27 magic-angle-spinning NMR studies of the thermal transformation of kaolinite. Phys. Chem. Miner. 17, 179–186.

Ross, C.S., Kerr, P.F., 1934. Halloysite and allophane: US Geol. Survey Prof. Paper 185-G., pp. 135–148.

Sarikaya, Y., Onal, M., Baran, B., Alemdaroglu, T., 2000. The effect of thermal treatment on some of the physicochemical properties of a bentonite. Clay Clay Miner. 48, 557–562.

Sato, R.K., Mcmillan, P.F., Dennison, P., Dupree, R., 1991. High-resolution Al-27 and Si-29 MAS NMR investigation of SiO_2-Al_2O_3 glasses. J. Phys. Chem. 95, 4483–4489.

Shoval, S., Champagnon, B., Panczer, G., 1997. The quartz-cristobalite transformation in heated chert rock composed of micro and crypto-quartz by micro-Raman and FT-IR spectroscopy methods. J. Therm. Anal. 50, 203–213.

Shoval, S., Boudeulle, M., Yariv, S., Lapides, I., Panczer, G., 2001. Micro-Raman and FT-IR spectroscopy study of the thermal transformations of St. Claire dickite. Opt. Mater. 16, 319–327.

Shu, Z., Chen, Y., Zhou, J., Li, T., Yu, D., Wang, Y., 2015. Nanoporous-walled silica and alumina nanotubes derived from halloysite: controllable preparation and their dye adsorption applications. Appl. Clay Sci. 112–113, 17–24.

Sing, K.S.W., Williams, R.T., 2004. Review: the use of molecular probes for the characterization of nanoporous adsorbents. Part. Part. Syst. Charact. 21, 71–79.

Slansky, E., 1983. Halloysite in a weathered profile at Port Macquarie, New South Wales. Trans. R. Soc. South Aust. 107, 177–185.

Smith, M.E., Neal, G., Trigg, M.B., Drennan, J., 1993. Structural characterization of the thermal transformation of halloysite by solid-state NMR. Appl. Magn. Reson. 4, 157–170.

Sonuparlak, B., Sarikaya, M., Aksay, I.A., 1987. Spinel phase formation during the 980°C exothermic reaction in the kaolinite-to-mullite reaction series. J. Am. Ceram. Soc. 70, 837–842.

Theng, B.K.G., Churchman, G.J., Whitton, J.S., Claridge, G.G.C., 1984. Comparison of intercalation methods for differentiating halloysite from kaolinite. Clay Clay Miner. 32, 249–258.

Tonle, I.K., Diaco, T., Ngameni, E., Detellier, C., 2007. Nanohybrid kaolinite-based materials obtained from the interlayer grafting of 3-aminopropyltriethoxysilane and their potential use as electrochemical sensors. Chem. Mater. 19, 6629–6636.

Vandiver, P.B., Soffer, O., Klima, B., Svoboda, J., 1989. The origins of ceramic technology at Dolní Věstonice, Czechoslovakia. Science 246, 1002–1008.

Vassallo, A.M., Coleclarke, P.A., Pang, L.S.K., Palmisano, A.J., 1992. Infrared emission spectroscopy of coal minerals and their thermal transformations. Appl. Spectrosc. 46, 73–78.

Voll, D., Lengauer, C., Beran, A., Schneider, H., 2001. Infrared band assignment and structural refinement of Al-Si, Al-Ge, and Ga-Ge mullites. Eur. J. Mineral. 13, 591–604.

Voll, D., Angerer, P., Beran, A., Schneider, H., 2002. A new assignment of IR vibrational modes in mullite. Vib. Spectrosc. 30, 237–243.

Wang, Q., Zhang, J.P., Zheng, Y., Wang, A.Q., 2014. Adsorption and release of ofloxacin from acid- and heat-treated halloysite. Colloids Surf. B 113, 51–58.

White, R., 1988. The upper paleolithic: a human revolution. In: Calhoun, D., Britannica, E. (Eds.), 1989 Yearbook of Science and the Future. Encyclopaedia Britannica, Inc., Chicago, pp. 30–49.

Wu, X.H., Zhang, C., Goldberg, P., Cohen, D., Pan, Y., Arpin, T., Bar-Yosef, O., 2012. Early pottery at 20,000 years ago in Xianrendong Cave, China. Science 336, 1696–1700.

Yuan, P., Wu, D.Q., He, H.P., Lin, Z.Y., 2004. The hydroxyl species and acid sites on diatomite surface: a combined IR and Raman study. Appl. Surf. Sci. 227, 30–39.

Yuan, P., Southon, P.D., Liu, Z., Green, M.E.R., Hook, J.M., Antill, S.J., Kepert, C.J., 2008. Functionalization of halloysite clay nanotubes by grafting with γ-aminopropyltriethoxysilane. J. Phys. Chem. C 112, 15742–15751.

Yuan, P., Tan, D., Annabi-Bergaya, F., Yan, W., Fan, M., Liu, D., He, H., 2012. Changes in structure, morphology, porosity, and surface activity of mesoporous halloysite nanotubes under heating. Clay Clay Miner. 60, 561–573.

Zhang, C., Komeya, K., Tatami, J., Meguro, T., Cheng, Y.B., 2000. Synthesis of Mg-α SiAlON powders from talc and halloysite clay minerals. J. Eur. Ceram. Soc. 20, 1809–1814.

Zhuravlev, L.T., 1993. Surface characterization of amorphous silica—a review of work from the former USSR. Colloids Surf. A 74, 71–90.

Chapter 8

Surface Modifications of Halloysite

D. Tan[a,b], P. Yuan[b,*], D. Liu[b] and P. Du[b]
[a]*Key Laboratory of Solid Waste Treatment and Resource Recycle, Ministry of Education, Southwest University of Science and Technology, Mianyang, China*
[b]*CAS Key Laboratory of Mineralogy and Metallogeny / Guangdong Provincial Key Laboratory of Mineral Physics and Materials, Guangzhou Institute of Geochemistry, Chinese Academy of Sciences (CAS), Guangzhou, China*
[*]*Corresponding author: e-mail: yuanpeng@gig.ac.cn*

8.1 INTRODUCTION

Halloysite is a naturally occurring dioctahedral 1:1 clay mineral that belongs to the kaolin group. The unit layer of kaolin group minerals is composed of one SiO_4 tetrahedral sheet and one $AlO_2(OH)_4$ octahedral sheet. In the SiO_4 tetrahedral sheet, the SiO tetrahedron is linked to three adjacent ones by sharing three basal-oxygen atoms to form an infinite two-dimensional sheet with hexagonal-like holes embedded into the sheet. In the $AlO_2(OH)_4$ octahedral sheet, AlO octahedrons are connected by sharing edges, and these edge-shared octahedrons form sheets with hexagonal or pseudohexagonal symmetry. Because the unit layer is linked by sharing the apical-oxygen of the SiO tetrahedron with the AlO octahedron, one surface of the unit layer consists entirely of the basal-oxygen atoms of the tetrahedral sheet, and the other surface is composed of the OH groups of the octahedral sheet. The unit layers stack in the *c*-direction to form kaolin group minerals (ie, kaolinite, dickite and nacrite). However, for halloysite, the adjacent unit layers are separated by one monolayer of water molecules. Halloysite is dominantly tubular in morphology, which results from the wrapping of the halloysite layers under favourable geological conditions. This topic is discussed in detail in Chapters 2 and 16.

In recent decades, tubular halloysite has attracted significant scientific interest because of its excellent physicochemical properties (detailed in Chapter 4). For instance, halloysite has been widely used as a filler in polymers, as a carrier for the loading and controlled release of guest molecules and as an adsorbent for pollution remediation. All these applications are detailed in other chapters of this book. However, the practical applications of halloysite are severely restrained by its own limitations, which lead to

Developments in Clay Science, Vol. 7. http://dx.doi.org/10.1016/B978-0-08-100293-3.00008-X

low loading content of the guest molecules on halloysite and a fast release of the guest molecules from halloysite. For example, when halloysite is used as a carrier, only a weak affinity (ion exchange, hydrogen bonding and van der Waals forces) is observed between halloysite and guest molecules. Moreover, hydrophilic halloysite, with the water contact angle being as low as 10 degrees ±3 degrees (Abdullayev et al., 2009) may exhibit low compatibility with certain organic and nonpolar polymer matrixes when used as filler for polymers. In addition, because of the low extent of isomorphous substitution by Al for Si in the tetrahedral sheet, halloysite has a low cationic exchange capacity, which leads to low adsorption capacity for heavy metals and organic contaminants when it is used as an adsorbent for pollution remediation.

To improve the performance of halloysite in the abovementioned situations, the surface properties of halloysite must be modified. In general, surface modification refers to the introduction of functional groups onto the surface of a host material, which can be conducted via physical modification (wrapping the modifier onto the host by van der Waals forces, hydrogen bonding and electrostatic attraction) or chemical modification (by covalently attaching the modifier to the host). After surface modification, the physical properties (hydrophilicity/hydrophobicity, dispersion, rheology, etc.) and chemical properties (reactivity, biotoxicity, electrochemistry, etc.) of the host can be carefully tailored and, consequently, promote the performance of the host when used for specific applications.

Based on this description, tubular halloysite has four types of surface (Fig. 8.1): (i) an internal lumen surface, composed of hydrophilic aluminol (Al–OH) groups; (ii) an external surface, composed of hydrophobic siloxane

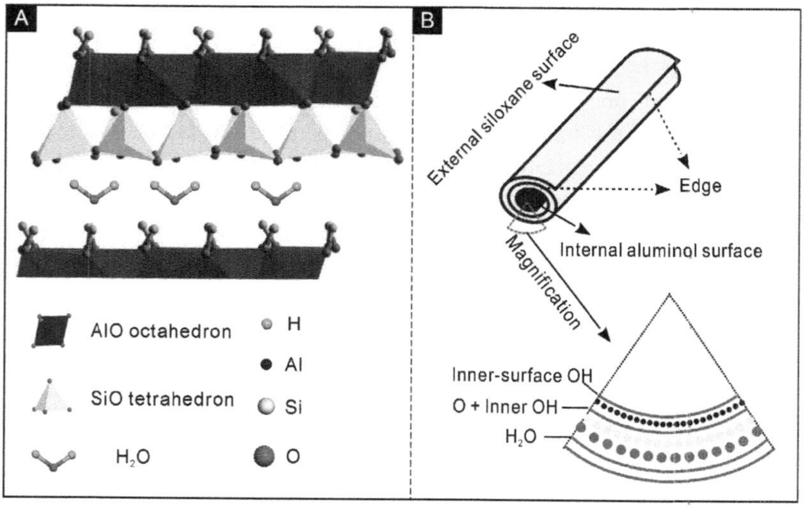

FIG. 8.1 Schematic diagram of (A) crystalline structure of halloysite, and (B) structure of a halloysite particle. *Adapted with permission from Yuan et al. (2015).*

(Si–O–Si) groups; (iii) an interlayer surface that is chemically the same as the internal lumen surface; and (iv) an edge surface, terminated by Al–OH and Si–OH groups. Regarding the internal lumen surface, the exposed Al–OH groups have high chemical activity and are available for covalent grafting with many organic compounds (eg, organosilane, organophosphonic acid). This method is preferentially recommended because covalent grafting enables durable immobilization of the reactive organic groups on the lumen surface of halloysite with high thermal stability and high resistance to hydrolysis and leaching.

For the external surface, the siloxane surface is theoretically recognized as chemically inert and is unavailable for chemical grafting by organic compounds. Isomorphous substitution occurs in naturally occurring halloysite samples, which results in a permanent negative charge on the siloxane surface of halloysite. In addition, some Al–OH and Si–OH groups are located on surface defects of halloysite. These hydroxyl groups allow halloysite to carry a pH-dependent surface charge. Therefore, the external surface of halloysite exhibits a negative zeta potential over a pH range of 2.5–8.5 (Veerabadran et al., 2007). This allows to modify the external surface properties of halloysite by physically adsorbing some specific cations (eg, cation surfactants, cationic polymers and biopolymers). Because this physical modification method usually results in secondary forces such as van der Waals, hydrogen and electrostatic force between the halloysite particles and the modifier, the physical adsorption modifications result in halloysite with a lower thermal stability and lower resistance to washing with solvents compared with the covalent-grafting modification method. However, physical adsorption modifications could be easily accomplished using a simple mixture under mild conditions (eg, room temperature, aqueous solution and stirring) and are advantageous for choosing the multitudinous modifier; therefore, physical adsorption modifications have been widely investigated.

Although the interlayer inner surface Al–OH groups are chemically active, they are normally inaccessible for guest molecules because of blocking by interlayer hydrogen bonding and limited interlayer space. However, some small organic molecules and some monovalent cation salts can be directly intercalated into the interlayer space of halloysite to modify the interlayer surface. Based on this direct intercalation, some large organic molecules can be intercalated into the interlayer space of halloysite by replacing the preintercalated molecules. In addition, some organic compounds could be covalently grafted onto these Al–OH groups by using preintercalated halloysite as a precursor under harsh conditions. Although this surface modification is subject to certain limitations (eg, severe conditions, tedious fabrication, time-consuming), it is still worth investigating for potentially promising halloysite applications in controlled loading and release and pollution remediation.

Quantitative investigation of the edge surface of halloysite has been rarely reported. However, it is reasonable to assume that halloysite has the same or less of an edge surface than that in kaolinite because halloysite can be recognized to be formed by the rolling of the kaolinite layers. The area of edge surface of kaolinite accounts for approximately 10% of the total kaolinite surface area, calculated from the pH-dependent surface charge, geometric consideration and scanning force microscopy analysis (Wieland and Stumm, 1992; Zhou and Gunter, 1992; Brady et al., 1996). Thus, the edge surface of halloysite is rather minor in comparison with the other three types of surface. This chapter will not discuss the modification of the edge surface of halloysite because of its low proportion; rather, it will mainly focus on the surface modifications of halloysite on the basis of the aforementioned three types of surface. Moreover, the application of surface-modified halloysite will be briefly reviewed.

8.2 CHEMICAL MODIFICATION OF THE INTERNAL LUMEN SURFACE

8.2.1 Grafting of Organosilane onto the Internal Aluminol Groups

Silane coupling agents commonly used to modify the surface properties of inorganic materials (silica, metallic oxide and silicate) have the general formula of X_3SiR, where R is an organofunctional group (eg, $-(CH_2)_3NH_2$) and X is a hydrolysable group (eg, –Cl, $-OCH_3$ or $-OC_2H_5$). The silane coupling agent is prone to hydrolysis, which results in the generation of Si–OH groups. These Si–OH groups can condense with the surface hydroxyl groups of inorganic materials by eliminating water molecules under proper conditions. Therefore, the organofunctional group (R) is covalently grafted onto the surface of the inorganic material and consequently modifies the surface properties. In the case of halloysite, many types of silane coupling agents, eg, 3-aminopropyltriethoxysilane (APTES) (Yuan et al., 2008; Wan et al., 2009; Joo et al., 2012; Hemmatpour et al., 2015; Jana et al., 2015; Peixoto et al., 2016), 3-methacryloxypropyltrimethoxysilane (Liu et al., 2011), 3-(2-aminoethyl)aminopropyltrimethoxysilane (AEAPS) (Barrientos-Ramirez et al., 2011; Zhu et al., 2012), γ-glycidoxypropyltrimethoxysilane (Liu et al., 2008), *n*-octadecyltrimethoxysilane (Wu et al., 2013) and 3-azidopropyltrimethoxysilane (Riela et al., 2014), have been used to adjust the surface properties of halloysite. A general procedure for the organosilane modification of halloysite is as follows. First, the organosilane is dissolved in organic solvent (toluene, ethanol, cyclohexane, etc.). Then, halloysite powder is added into the solution and the obtained dispersion is refluxed under constant stirring. The solid phase in the resultant mixture is filtered and extensively washed with fresh solvent to remove the excess organosilane. After drying, the resulting solid is organosilane-modified halloysite.

Yuan et al. (2008) systematically investigated the mechanisms of grafting APTES onto the lumen surface of halloysite and the factors that influenced

APTES grafting. The general mechanisms of APTES grafting on halloysite can be described as follows: (i) the hydroxyl groups of the hydrolysed APTES and halloysite condensate and form covalent Al–O–C bonds, which result in the direct grafting of APTES; and (ii) some of the hydrolysed APTES condenses with each other to form oligomers, which form a cross-linked network in the lumen through grafting by covalent and hydrogen bonds with the already grafted APTES. Another grafting mechanism was recently published by Peixoto et al. (2016) who showed that the APTES can form lateral polymerization via NH_2-silicon oligomerization. This grafting mechanism was directly verified by the presence of a new Si environment (Si–N) in the Si 2p curve of X-ray photoelectron spectroscopy (XPS) spectra, indicated by the downward shift of the binding energy of Si $2p_{3/2}$ and Si $2p_{1/2}$ from 103.1 and 103.7 eV to 102.6 and 103.2 eV, respectively.

Because the lumen surface is covered by Al–OH groups, APTES is mainly grafted onto the lumen surface in which a layer of aminopropyl groups is introduced. The attached aminopropyl groups obviously decrease the polarity and hydrophilic properties of halloysite and enhance the compatibility and affinity of the halloysite with organic compounds. However, the formation of oligomers and cross-linked network in the lumen during APTES modification could substantially sacrifice the porosity of halloysite; the specific surface area (SSA) and pore volume (V_{Pore}) of halloysite decreased from 49.5 m^2/g and 0.25 cm^3/g to 19.7 m^2/g and 0.11 cm^3/g, respectively. Due to blocking by interlayer hydrogen bonding and the limited interlayer space of halloysite, APTES is not grafted onto the interlayer surface Al–OH groups, which is shown by the unchanged d_{001} value of the halloysite after APTES modification. The inert Si–O–Si groups on the external surface of halloysite are unavailable for APTES grafting. However, some defects (surface breakage, crystallographic defects) were directly observed on the halloysite surface through TEM and AFM observation (Yuan et al., 2008) (Fig. 8.2), which indicated that some of the Al–OH

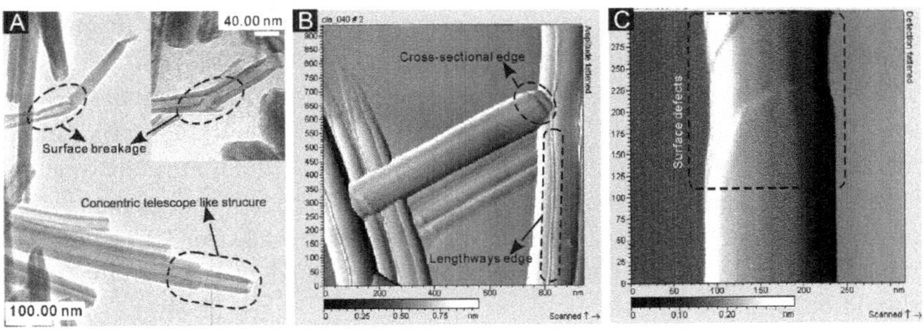

FIG. 8.2 (A) TEM image of the surface breakage of halloysite; (B and C) AFM images of the edges and surface defects of halloysite. *Adapted with permission from Yuan et al. (2008). Copyright (2008) American Chemical Society.*

and Si–OH groups are exposed on the external surface of halloysite and that these hydroxyl groups are extra reaction sites for APTES modification.

The extent of APTES grafting on halloysite is mainly determined by the extent of the APTES hydrolysis and the amount of surface hydroxyl groups on the halloysite. Thus, pretreatments (heating, evacuation, NaOH etching, H_2O_2 treatment, etc.) that alter the water content and the density of the hydroxyl groups of halloysite could significantly influence the grafting of APTES on halloysite. For instance, a thermal pretreatment at 400°C was shown to completely remove the physically adsorbed water and the interlayer water from halloysite, leading to a restriction of APTES oligomerization and the subsequent formation of a cross-linked network in the lumen. Consequently, the lumen space of halloysite was preserved and the SSA and V_{Pore} of the APTES-grafted heated halloysite were 37.0 m^2/g and 0.21 cm^3/g, respectively (Yuan et al., 2008). Evacuation pretreatment can remove the air from the lumen and is beneficial for enhancing APTES loading into the lumen, and for increasing the extent of grafting or oligomerization. Moreover, NaOH etching pretreatment can etch the external siloxane surface of halloysite and enhances the quantity of the hydroxyl groups on the surface of halloysite by exposing more Al–OH and Si–OH and consequently increasing the amount of grafted APTES on halloysite (Vahedi and Pasbakhsh, 2014). In addition, Ng et al. (2011) observed that the amount of Si–OH on the surface of halloysite could be significantly enhanced by H_2O_2 pretreatment, which was demonstrated by time-of-flight secondary ion mass spectrometry. In addition to the pretreatment conditions, the origin of the halloysite exerted a remarkable influence on APTES grafting. Halloysite with a high SSA had a higher grafted APTES content than halloysite with a small SSA because the halloysite with a larger SSA had a greater number of hydroxyl groups on the surface, which allowed more APTES grafting (Yuan et al., 2008).

Based on the abovementioned studies, it is possible to control and fine-tune the surface properties of halloysite for a specific application by selecting an organosilane with a particularly desirable organofunctional group. Moreover, the organosilane modification of halloysite could be adjusted to an optimized extent by carefully tailoring the pretreatment conditions to meet the requirements for a specific application. For example, in the field of loading and controlled release of guest species, the halloysite lumen should be preserved as much as possible, meaning the restriction of the formation of oligomers and a cross-linked network in the lumen. Thus, thermal pretreatment is highly recommended. In the field of organic pollutant adsorption, the grafting of greater amounts of organosilane on halloysite results in better pollution remediation efficiency. Thus, evacuation and NaOH etching pretreatments should be strongly advised.

Because organosilane modification introduces a layer of functional groups on the halloysite surface, the modified halloysite exhibits potential chemical activity for the organic compounds that are unable to directly react with halloysite. For example, the grafting of aminosilanes on the halloysite surface develop amino groups, which have a high chemical activity and are good candidates for serving as active sites that react with other active molecules to further modify the halloysite surface properties. Joo et al. (2012) synthesized a carboxylic acid-functionalized halloysite (Hal-COOH) from an APTES-modified halloysite (amino-functionalized halloysite, Hal-NH_2) by reacting succinic anhydride and Hal-NH_2 in dimethyl formamide. This modification significantly promoted halloysite dispersion in aqueous solution according to direct observations of the solution state in multiple angle polarized dynamic light scattering experiments. In a neutral solution (deionized water, pH 7), the flexible side chains of Hal-COOH generated three types of hydrogen bond (HB) networks through interactions among the amide and carboxyl groups of the side chains and the halloysite surface oxygen: HB-1, chain–chain and chain–surface interactions; HB-2, chain–surface interactions; and HB-3, chain–chain interactions (Fig. 8.3A). These interactions led to the aggregation of Hal-COOH. In acidic solution (HCl solution, pH 1), the extra protons potentially mediated the interactions between Hal-COOH via HB-4; however, the protons in solution also tended to form positively charged oxonium complexes with the oxygen atoms connected to two Si atoms on the surface of the Hal-COOH, which resulted in an electrostatic repulsive force (RF-1) and reduced HB-1 and HB-3 (Fig. 8.3B). Thus, the Hal-COOH was dispersed better in the acidic solution than in the neutral solution. In basic solution (NaOH solution, pH 12), the carboxyl groups of Hal-COOH were deprotonated; therefore, the negatively charged side chains electrostatically repulsed each other (RF-1) and the halloysite surface oxygen (RF-2) (Fig. 8.3C). These two repulsive forces resulted in the monodispersion of Hal-COOH in basic solution.

Other researchers have also used APTES-modified halloysite as a precursor to react with α-bromoisobutyryl bromide (BiBB) under severe conditions (chloroform solution, N_2 atmosphere, low temperature, etc.). This modification was achieved through an acylation reaction between the amino group of Hal-NH_2 and the –Br group of BiBB. The –Br group on the surface of halloysite could be used as an initiator to trigger the polymerization of monomer and graft polymer chains on the surface of halloysite. The introduction of polymer chains, such as polyacrylonitrile (Li et al., 2008), polystyrene (Li et al., 2009) and poly(*N*,*N*-dimethylaminoethyl methacrylate) (PDMAEMA) (Hemmatpour et al., 2015), could significantly change the surface properties of the halloysite. For example, PDMAEMA chains are a weak electrolyte ($pK_a = 7.3–7.5$) because of their tertiary amine groups. Thus, Hal-PDMAEMA exhibits a pH-dependent dispersion stability in aqueous solutions. At low pH, the tertiary amine groups are protonated and become positively charged polyelectrolytes, which results in repulsion between the polymer chains and therefore more water is adsorbed among the chains, which promotes the extent of swelling. Thus, the aqueous dispersion of Hal-PDMAEMA is stable in acidic solutions (pH 1) even after a long period of sonication. However, at pH 10, the polymer chains

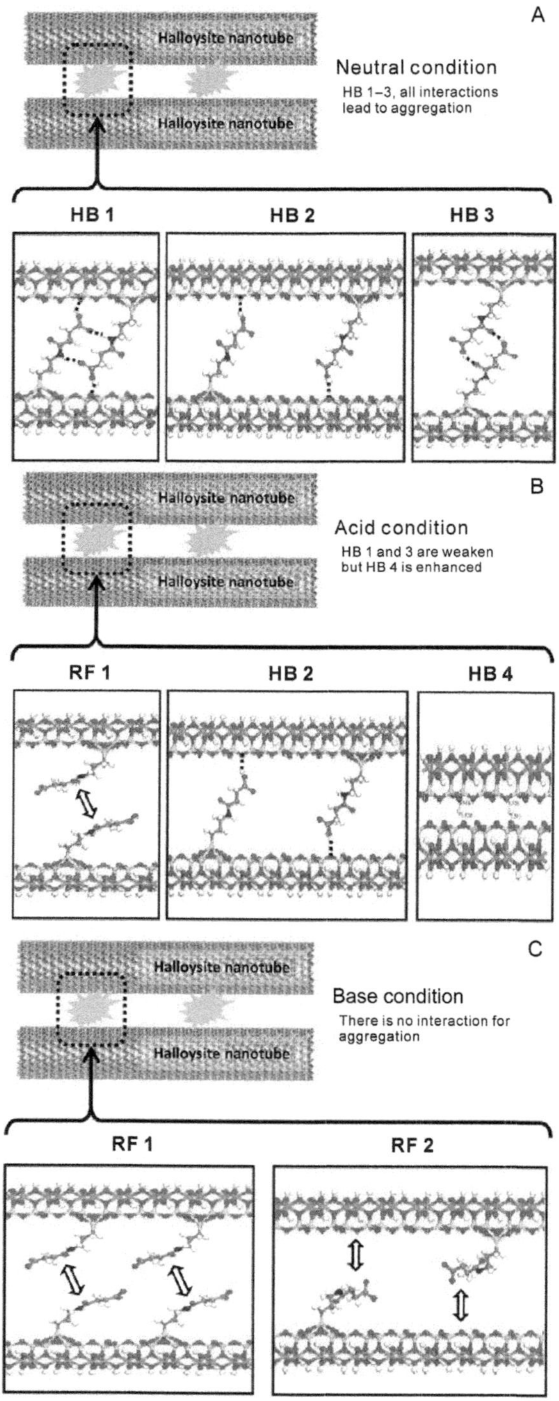

FIG. 8.3 See legend on next page.

are deprotonated and consequently collapse, resulting in sedimentation of Hal-PDMAEMA in a basic aqueous medium.

By combining the abovementioned surface modifications of halloysite, excellent halloysite dispersion can be achieved in both acidic (for carboxylic acid-functionalized halloysite) and basic (for PDMAEMA-grafted halloysite) solution, in which the promising application of halloysite in clay polymer nanocomposites (CPN), catalysts, drug delivery systems and cosmetics could be predictable.

8.2.2 Grafting of Other Organic Compounds onto Internal Aluminol Groups

In addition to organosilane, organophosphonic acid can react with hydroxyl groups and be grafted on the surfaces of metal oxides (Al_2O_3, TiO_2, etc.) (Luschtinetz et al., 2008, 2009). Luschtinetz et al. (2008) investigated the chemical adsorption of phosphonic acid (PA) and ethylphosphonic acid (EPA) on the surface of corundum (α-Al_2O_3), bayerite (β-$Al(OH)_3$) and boehmite (γ-AlOOH) using a density-functional-based tight-binding method. PA and EPA could react with a surface hydroxyl group via an acid–base condensation mechanism to form P–O–Al bonds and eliminate one water molecule. Consequently, the PA and EPA were chemically adsorbed onto the surface as monodentate, bidentate and tridentate coordination complexes (Fig. 8.4).

The lumen surface of halloysite similar to the surface of α-Al_2O_3 can also be modified by organophosphonic acid. Yah et al. (2012a) used octadecylphosphonic acid (ODP) to modify the lumen surface of halloysite. The ODP is dissolved in an ethanol–water mixture (v/v = 4:1, pH 4), and the halloysite powder is added into the solution. The dispersion is then transferred to a vacuum jar and evacuated within three cycles for the maximum loading of ODP into the halloysite lumen. The grafting reaction lasts for approximately 1 week at room temperature under constant stirring. After solid–liquid separation and drying, the resultant solid is the ODP-modified halloysite.

The ODP was not grafted onto the external siloxane surface of halloysite because no Si–O–P bond (a binding energy of 103 eV) is present in the XPS spectra of the ODP-modified halloysite, which indicated that no Si–O–P bond formed. Similar to the grafting of organosilane on halloysite, the interlayer surface Al–OH groups were unavailable for ODP grafting, as shown by the unchanged d_{001} value of the ODP-modified halloysite. The ODP was grafted

FIG. 8.3 Computer simulation of interactive and repulsive properties of Hal-COOH in (A) neutral, (B) acidic and (C) basic solution. Coloured (different shades of grey) spheres stand for the following: *white* for hydrogen, *red* (grey in the print version) for oxygen, *purple* (grey in the print version) for aluminium, *yellow* (light grey in the print version) for silicon, *blue* (dark grey in the print version) for nitrogen and *black* for carbon atoms. *Modified from permission from Joo et al. (2012). Copyright (2012) American Chemical Society.*

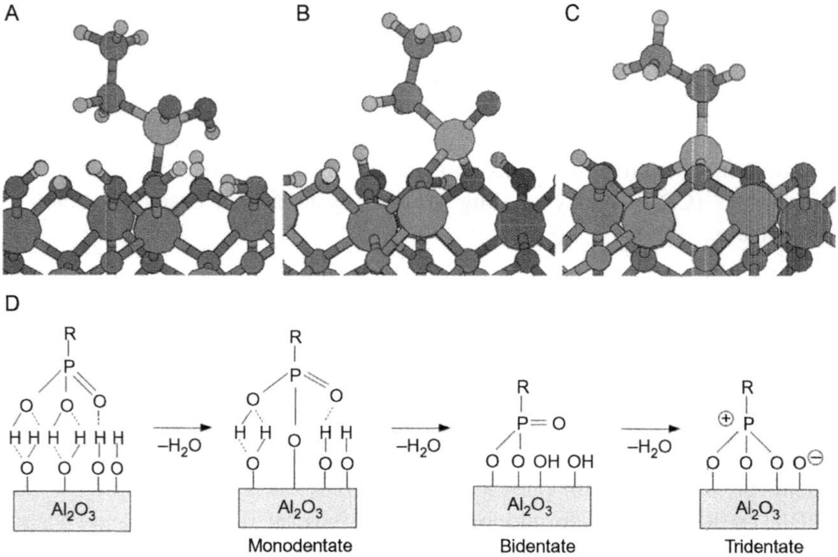

FIG. 8.4 (A) Monodentate, (B) bidentate and (C) tridentate binding coordination of ethylphosphonic acid on the surface of α-Al_2O_3; (D) stepwise condensation reaction between the phosphonic acid headgroup and the aluminium oxide surface. *Reprinted with permission from Luschtinetz et al. (2008).*

onto the lumen surface of halloysite, the TEM observations demonstrated that the lumen was covered with a less dense layer and the texture analysis results demonstrated that the SSA and V_{Pore} of the ODP-modified halloysite decreased from 43.0 m^2/g and 0.53 cm^3/g to 27.6 m^2/g and 0.38 cm^3/g, respectively. Similar to the reactions of organophosphonic acids between metal oxides, ODP could act as a monodentate, bidentate or tridentate ligand on the lumen surface of halloysite. However, the ^{31}P CP MAS NMR spectrum of the ODP-modified halloysite showed chemical upfield shifts of 26.8 and 22.7 ppm for ODP-modified halloysites from 33.0 ppm for pure ODP, which suggested that the primary ODP attachment modes are bidentate and tridentate. The lack of detected monodentate Al–O–P linkages in the ^{31}P CP MAS NMR spectrum resulted from removing the weakly bound monodentate compounds by sonication and from rinsing during the modification process.

This ODP modification endowed halloysite with a hydrophobic lumen and a hydrophilic external surface. Compared with unmodified halloysite, the modified halloysite had a higher adsorption capacity for hydrophobic ferrocene because the hydrophobic tubular core acts like a sponge, embedding ferrocene into its alkyl chains. However, the modified halloysite had a lower adsorption capacity for hydrophilic ferrocenecarboxylic acid because of favourable hydrogen bonding in the unmodified halloysite.

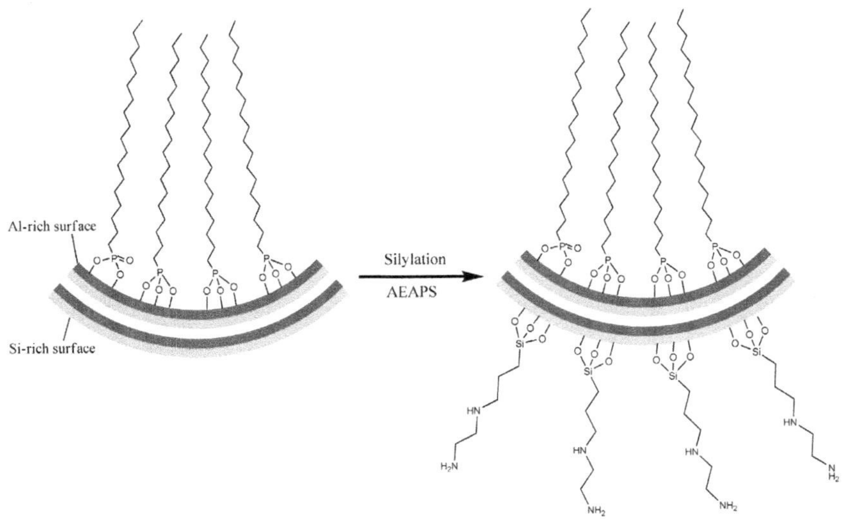

FIG. 8.5 Schematic illustration of the bifunctionalization of the lumen surface and the external surface of halloysite by ODP and subsequent AEAPS silylation. *Reprinted with permission from Yah et al. (2012a). Copyright (2012) American Chemical Society.*

In addition, the halloysite already modified by ODP on the internal lumen surface could be further grafted by AEAPS on the external surface because some Al–OH and Si–OH groups exist at the edges or surface defects, resulting in bifunctionalization of the halloysite (Fig. 8.5). This bifunctionalized halloysite could enable different potential applications, from water purification to drug immobilization and controlled release.

Likewise, Yah et al. (2012b) modified halloysite by immobilizing a type of catecholic anchor, 2-bromo-*N*-[2-(3,4-dihydroxyphenyl)ethyl]-isobutyryl amide (Dopa, a dopamine derivative), onto the internal lumen surface. Similar to the grafting of organosilane and ODP on halloysite, the XPS and ^{13}C solid-state NMR spectra showed that Dopa is covalently grafted onto the lumen surface but is not intercalated into the interlayer space or grafted onto the external surface of halloysite. However, Dopa theoretically binds to both the silicalike external surface and the aluminalike internal lumen surface of halloysite through silica–catechol bonds and alumina–catechol bonds, respectively. Because the noncovalent silica–catechol bonds are much weaker, with a binding energy of 14 kcal/mol, the silica–catechol bonds are readily removed by sonication and tetrahydrofuran–water solution rinses, leading to a lack of immobilization of Dopa on the external surface of halloysite. The Dopa-modified halloysite bearing –Br group is further used as an atom-transfer radical polymerization initiator for the selective polymerization of

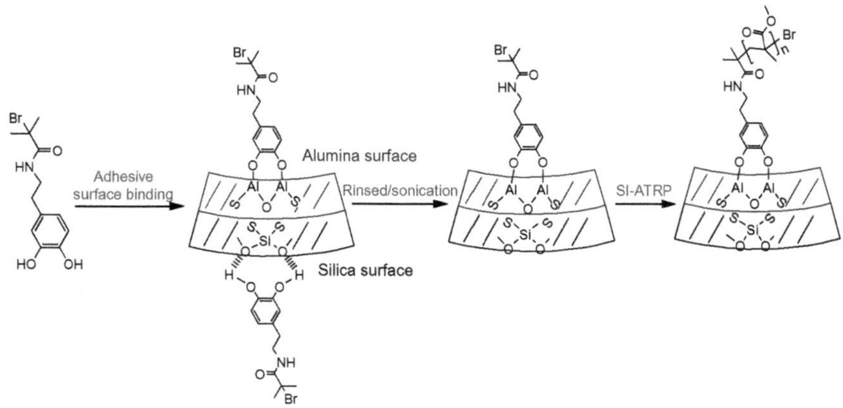

FIG. 8.6 Selective modification of halloysite lumen with a Dopa biomimetic initiator, and subsequent polymerization. *Reprinted with permission from Yah et al. (2012b). Copyright (2012) American Chemical Society.*

polymethyl methacrylate in the lumen of halloysite because of the selective immobilization of Dopa in the lumen (Fig. 8.6).

8.3 MODIFICATION OF THE EXTERNAL SURFACE

8.3.1 Surfactant, Polymer and Biopolymer Coatings

Surfactants contain one or more polar groups and a long aliphatic chain and are usually applied as a modifier to synthesize organoclay. For halloysite, the principle of surfactant treatment is the preferential adsorption of the polar groups of the surfactant onto the high-energy surface of the halloysite by electrostatic interaction. Lee and Kim (2002) coated the hexadecyl trimethylammonium bromide (HDTMA) surfactant on the surface of halloysite. The HDTMA formed micellelike clusters on the halloysite surface. The introduction of a long alkyl chain of HDTMA significantly increased the hydrophobicity of the halloysite, resulting in a good adsorption capacity for the organic pollutant, naphthalene. The adsorption of naphthalene is driven by the distribution of naphthalene in the surfactant tails, which results in nearly linear naphthalene adsorption isotherms. In addition, the HDTMA-modified halloysite is used as an adsorbent for the removal of hydrophilic anionic contaminants, eg, Cr(VI) (dichromate) (Wang et al., 2010) and nitrate (Xi et al., 2010). The HDTMA-modified halloysite exhibited a rapid adsorption rate for Cr(VI), approaching 90% of the maximum adsorption capacity within 5 min, and this adsorption followed the Langmuir model.

Similar to surfactants, polymeric dispersants have macromolecular chains with functional groups such as –OH, $-NH_2$, –COOH and $-SO_3H$, which can help to anchor the particle surface through hydrogen bonds, electrostatic forces or both. Lvov et al. (2002) adsorbed poly(ethylenimine) (PEI) on the surface of halloysite to form a monolayer with a thickness of 54 nm and then alternately adsorbed halloysite and PEI to generate thin films with a thickness of 720 nm (approximately 14 sets of halloysite–PEI monolayers). Due to the loosely packed halloysite in the monolayer, the thin film consists of a network with 50% empty space, which could be used as a loading space for the subsequent controlled release of guest molecules. After coating the polycations, the negatively charged external surface of halloysite becomes positively charged and is available for further coating by polyanions (Fig. 8.7A). Using the layer-by-layer (LBL) assembly method, Veerabadran et al. (2009) adsorbed polycations [poly(allylamine) hydrochloride, PAH; PEI; chitosan] and polyanions [sodium poly(styrene sulphonate), PSS; poly(acrylic acid), PAA; gelatin] on the external surface of halloysite. The sequential polycation/polyanion assembly altered the surface ξ-potential of halloysite from a positive value to a negative value, and then to a positive value (ranging from −30 to +60 mV) (Fig. 8.7C). This LBL assembly could result in the formation of organized shells on the external surface of halloysite. When monitoring with the quartz crystal microbalance, the bilayer thickness of hydrated PEI/PAA and PAH/PSS was approximately 2.5 ± 0.5 nm. Therefore, the LBL coating of the organic shells on the halloysite surface could enlarge the diameter of the halloysite with high controllability by tuning the numbers of the coating bilayer. More important, this LBL coating did not change the cytotoxicity of the halloysite. The polymer-coated halloysite and the raw halloysite

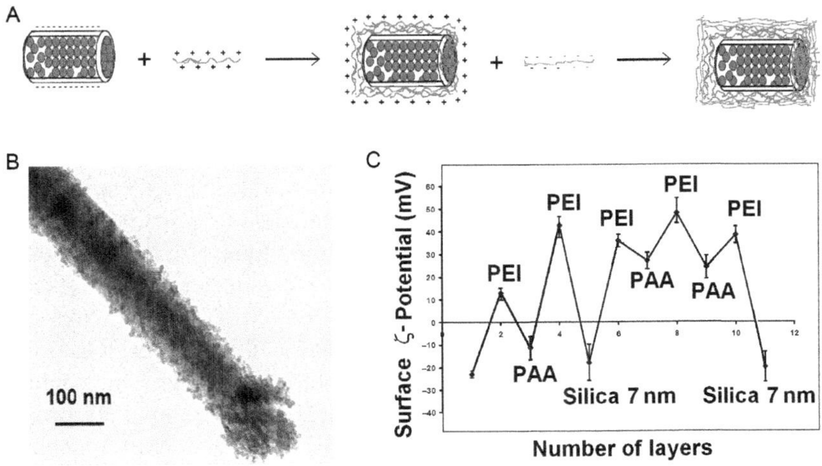

FIG. 8.7 (A) Schematic of the LBL deposition of the polyelectrolytes; (B) TEM image of the encapsulated tube with an LBL shell of PEI/PSS/PEI/silica; and (C) alteration of the halloysite surface potential during the LBL process. *Modified with permission from Abdullayev and Lvov (2011).*

both yielded approximately 98% live cells (fibroblast cells and human breast cancer cells), enabling the safe use of polymer-coated halloysite in pharmaceutics. Because the polymer coating blocks the end of the halloysite nanotube, the diffusion of the preloaded drug from the lumen of halloysite will decrease significantly, and consequently will result in a prolonged release.

In addition to surfactants and polymers, some other biomacromolecules and naturally occurring biopolymers are used as modifiers to adjust the surface properties of halloysite. Shamsi and Geckeler (2008) wrapped DNA on the surface of the halloysite nanotube using a mechanochemical reaction in the solid state. The DNA-wrapped halloysite was obtained by simply milling DNA and halloysite. After milling, the halloysite nanotubes were cut into shorter lengths, but the DNA remained intact and completely covered the surface of the halloysite. This modification afforded halloysite with high aqueous solubility and revealed suitability for some potential biomedical applications. In addition, Chang et al. (2011) wrapped amylose on the surface of halloysite to improve its biocompatibility. The obtained halloysite–amylose nanocomposite exhibited good dispersion in a DMSO/H_2O solution. Moreover, the coating of chitosan on halloysite was studied (Zhai et al., 2013; Fu et al., 2015). Chitosan is a linear polycationic biopolymer of glucosamine with abundant amino and hydroxyl groups. Chitosan could be effectively assembled onto the negatively charged external surface of halloysite under acidic conditions. The halloysite chitosan hybrid material perfectly combines the following advantages of the inorganic and organic materials: the three-dimensional CPN with hierarchically porous structures are potential carriers for guest loading and controlled release; and the functional groups on chitosan chains, which serve as electrostatic interaction and coordination sites, allow the hybrid material to serve as an adsorbent in pollution remediation and as a cross-linker and reinforcing filler in CPN.

8.3.2 Organosilane Modification of Calcined Halloysite

Yuan et al. (2008) discussed the presence of Al–OH and Si–OH on the edges and surface defects of the halloysite and demonstrated that these hydroxyl groups are available for organosilane modification. In comparison with the Al–OH on the lumen surface of halloysite, the amount of these hydroxyl groups is considered negligible, despite the increase in these hydroxyl group numbers by NaOH etching (Vahedi and Pasbakhsh, 2014) and H_2O_2 treatment (Ng et al., 2011). Recently, Yuan et al. (2012a) found that the Al–OH groups on the lumen surface and interlayer inner surface of halloysite could be completely removed under thermal treatment at 600–900°C; however, original siloxane groups on the external surface could be partially or entirely replaced by hydroxyl groups, which was indicated by the presence of the Si–OH stretching vibration at 3745 cm^{-1} in the DRIFT spectra of the calcined halloysite (see Chapter 7 for more details). The formation of these hydroxyl groups occurred before the breakdown of the tubular structure and was accompanied by the disconnection and further disordering of the silica and alumina originally present in the tetrahedral and octahedral sheets (Yuan et al., 2012a). These newly formed hydroxyl groups at the external surface of calcined halloysite had high chemical reactivity, similar to the Al–OH groups on the lumen surface of halloysite, and achieved covalent grafting of organosilanes on the external surface of the calcined halloysite. This organosilane modification on the external surface of calcined halloysite is believed to substantially enhance the dispersion of halloysite in the polymer substrate and increase the interaction between halloysite and the polymer, which would promote the performance of halloysite in CPN.

8.4 MODIFICATION OF THE INTERLAYER SURFACE

8.4.1 Intercalation of Guest Molecules into the Interlayer Space

The interlayer space of swelling clays can accommodate guest molecules (eg, metal ions, surfactants) through ion exchange. Although exchangeable cations for kaolin group minerals are lacking, the kaolinite and halloysite can intercalate guest molecules, ranging from organic chemicals to inorganic salts. The intercalated guest molecules can be divided into the following three groups according to the point of clay interaction (Frost and Kristof, 1997):

1. The first group consists of compounds that can either accept or donate protons to form strong hydrogen bonds with the inner surface Al–OH groups in the octahedral sheet and the basal-oxygen atoms in the tetrahedral sheets: eg, formamide (FA) (Joussein et al., 2007), urea (Nicolini et al., 2009), hydrazine (Horvath et al., 2003) and acetamide (Carr and Chih, 1971). For example, the intercalation of FA increased the d_{001} value of halloysite from 7.2–7.4 to 10.2 Å. The FA intercalation is rapidly completed in 1 h for halloysite but begins to intercalate in kaolinite after 4 h of reaction (Churchman et al., 1984). This phenomenon is caused by the differences in the crystal structures of halloysite and kaolinite. Depending on the extent of the FA intercalation in the halloysite and kaolinite, the dehydrated halloysite can be effectively differentiated and quantified from kaolinite. The $-NH_2$ group and C=O group of FA can act as a proton donor and proton acceptor, respectively. Therefore, intercalated FA can form hydrogen bonds with both the basal-oxygen atoms and the inner surface Al–OH groups. The formation of hydrogen bonds was demonstrated by infrared (IR) and Raman spectroscopy (Frost et al., 1999), and by density function theory computations (Song et al., 2013). Using the molecular simulation, the conformation of the intercalated FA is as follows: the molecular plane of most of the FA molecules is basically parallel to the clay surface and the $\delta_{CH \rightarrow O}$ bond vectors generally point towards the inner surface Al–OH groups of the adjacent octahedral sheet (Fig. 8.8) (Rutkai and Kristóf, 2008).

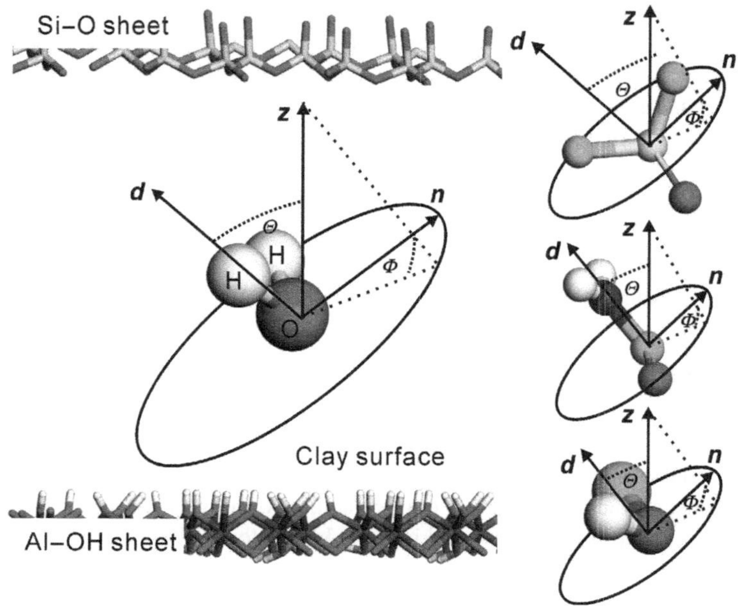

FIG. 8.8 Definition of the orientation of molecules. From *top* to *bottom*: The graphical representations of the molecules correspond to DMSO, formamide and methanol, respectively. *Modified with permission from Rutkai and Kristóf (2008).*

2. The second group consists of molecules with strong dipole moment (eg, dimethyl sulphoxide (DMSO)). The high dipolarity of DMSO (4.3 D) enables it to readily intercalate into the interlayer space of halloysite and kaolinite. This intercalation is much faster in halloysite than in kaolinite because of the lower structure order and the smaller numbers of interlayer hydrogen bonding in the halloysite (Olejnik et al., 1968; Costanzo and Giesse, 1986). Due to its lack of labile hydrogen atoms, DMSO acts only as a proton acceptor, not as a proton donator. Therefore, hydrogen bonds are formed between the oxygen atoms of DMSO and the interlayer inner surface Al–OH groups, which result in a higher mobility of DMSO in the interlayer space than FA, urea and other substances that bond to both the Al–OH and basal-oxygen surface of halloysite. This mobility causes DMSO to arrange itself in the interlayer space to achieve the least amount of expansion and is consistent with minimum interactions between adjacent DMSO molecules. The conformation of DMSO in the interlayer space is as follows: the oxygen atom forms hydrogen bonds with Al–OH groups, the sulphur atom points towards the tetrahedral sheets and one methyl group is partially placed into hexagonal-like holes (Fig. 8.8). This conformation has been demonstrated by XRD (Olejnik et al., 1968), NMR (Hayashi, 1997), Raman and DRIFT spectra (Frost et al., 1998) and by molecular simulation (Fang et al., 2005; Rutkai and Kristóf, 2008).

3. The third group is the alkali salts of short-chain fatty acids (in particular, acetic and propionic acids). Potassium acetate (KAc), for example, is readily intercalated into the interlayer space of halloysite and kaolinite using an immersion or milling method and often approaches 100% intercalation. This intercalation is strongly related to the crystallinity of halloysite and kaolinite, the concentration of the intercalating agent, the ageing time and the pH (Cheng et al., 2010). The KAc and halloysite interact through hydrogen bonds that are formed between the inner surface Al–OH groups and the acetate ions. This interaction was directly demonstrated by the presence of the vibration at 3306 cm^{-1} in the Raman spectra (Frost et al., 2000). According to neutron pair distribution function analysis, White et al. (2011) revealed that the hydrated KAc was present as a single layer in the interlayer space of kaolinite and proposed three possible configurations for water molecules and KAc molecules (Fig. 8.9). The weak interactions and high water solubility of KAc render the intercalated KAc prone to deintercalation by water rinse.

The intercalation of guest molecules into the interlayer space of halloysite can enlarge its d_{001} value (see references in Carr et al., 1978; Joussein et al., 2005). Generally, the intercalation process has no influence on the morphology of halloysite except for the intercalation of phenylphosphonic acid (PPA). Tang et al. (2011) found that the intercalation of PPA could cause the halloysite nanotubes to unfold and become nanoplatelets. The extent of this transformation is related to the duration of PPA intercalation. After a PPA treatment of 50 h, some of the original tubular halloysite particles were transformed into flat halloysite platelets without destroying the crystal structure (Fig. 8.10B and D). However, after a PPA treatment of 100 h, most of

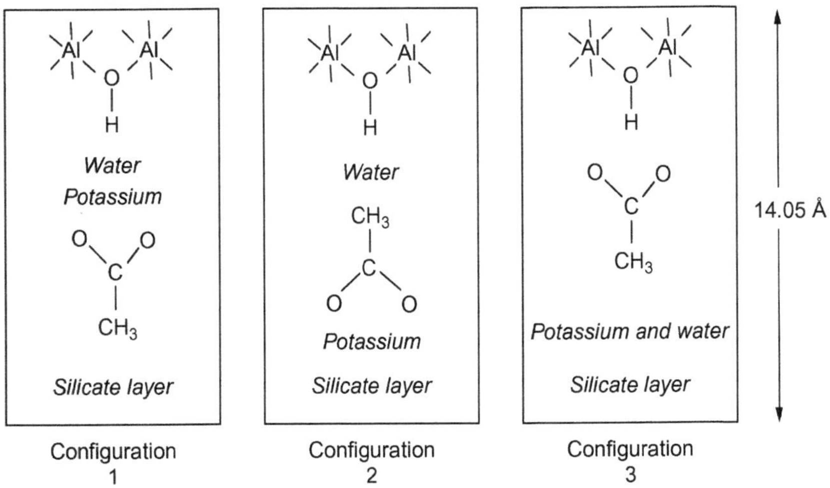

FIG. 8.9 Possible configurations for hydrated potassium acetate within the expanded kaolinite structure. *Reprinted with permission from White et al. (2011). Copyright (2011) American Chemical Society.*

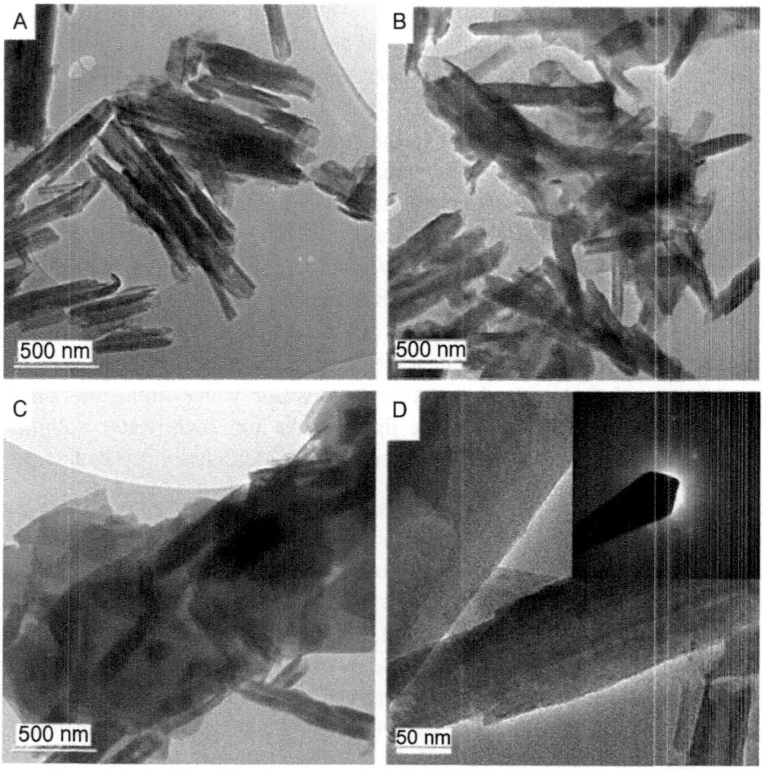

FIG. 8.10 TEM images of halloysite particles treated by PPA for different durations: (A) as received; (B and C) after PPA treatment for 50 and 100 h, respectively; (D) electron diffraction pattern after PPA treatment of 50 h, showing the crystal structure of halloysites. *Reprinted with permission from Tang et al. (2011).*

the halloysite particles became platelets, with some even becoming dispersed flakes (caused by the exfoliation of aluminosilicate layers) (Fig. 8.10C). For kaolinite, the intercalation of PPA expanded the basal reflection of kaolinite to 15.4 Å, the same as that in halloysite (Breen et al., 2002). Unlike the morphology transformation in the case of halloysite, this intercalation had no influence on the platy morphology of the kaolinite. However, rolling of the platy kaolinite was previously observed by some researchers to occur through the intercalation/deintercalation of long-chain alkylamine. This transformation of platy kaolinite to halloysite-like nanotubes was caused by weakening the interlayer hydrogen bonding due to sufficient delamination of the kaolinite layers (see Chapter 17 for more detail on this topic). However, the mechanism for halloysite unfolding due to PPA intercalation remains unclear. The mutual transformation of the platy and tubular morphologies of aluminosilicate layers suggests substantial possibilities of scientific interest.

The abovementioned three groups of guests can be directly intercalated into the interlayer space of halloysite and kaolinite. Due to their weak interactions (hydrogen bonding), the intercalated guests are unstable and can readily be rinsed out by water or organic solvent. Using this process, some large molecules that cannot achieve direct intercalation can be intercalated into the interlayer space using the displacement method. For kaolinite, the intercalation of large molecules via the one-step displacement or multistep displacement methods has been widely reported. DMSO preintercalated kaolinite and KAc preintercalated kaolinite are commonly used as precursors, and various guests have been intercalated into the interlayer spaces of kaolinite, such as polyacrylamide (Sugahara et al., 1990), ε-caprolactam (Komori et al., 1999a), poly(vinylpyrrolidone) (Komori et al., 1999b), alkylamines (Komori et al., 1999c), *p*-nitroaniline (*p*NA) (Kuroda et al., 1999; Takenawa et al., 2001), benzamide (Gardolinski et al., 2000), nylon 6 (Matsumura et al., 2001), cyclic imides (Elbokl and Detellier, 2008), some quaternary ammonium salts (Kuroda et al., 2011; Yuan et al., 2013). These new organic–kaolinite complexes have some special characteristics and broaden the possible applications of kaolinite. For instance, using *N*-methylformamide (NMF) intercalated kaolinite as a precursor, Kuroda et al. (1999) intercalated *p*NA into the interlayer space of kaolinite by using a two-step guest displacement method (Fig. 8.11). The intercalation of *p*NA increased the basal reflection of kaolinite from 0.72 to 1.50 nm. The intercalated *p*NA formed a monomolecular arrangement, with the molecular axis inclined to the kaolinite layers by about 60 degrees. The noncentrosymmetric orientation of the *p*NA molecules in the interlayer space of kaolinite endowed the kaolinite–*p*NA intercalation compounds with the properties of the second active harmonic generation. Moreover, Komori et al. (1999c) intercalated alkylamines into the interlayer space of kaolinite using the displacement method. The basal reflection reached up to 5.75 nm when octadecylamine was used. The intercalated alkylamines formed a bilayer arrangement with their alkyl chains nearly perpendicular to the layers of kaolinite. By substituting alkylamines with quaternary ammonium salts, Kuroda et al. (2011) found that the basal reflection of kaolinite increased to 3.95 nm, and kaolinite layers spontaneously rolled to form halloysite-like nanotubes.

For halloysite, it is rational to regard the direct intercalation of small molecules as similar to that of kaolinite. However, this analogy is not suitable for the indirect intercalation of large molecules because the tubular structure of the halloysite may obstruct the displacement process. For this reason, the intercalation of large molecules into halloysite is rarely reported.

8.4.2 Grafting of Organics in the Interlayer Space

Although the surface hydroxyl groups of the internal lumen surfaces, edges and external surfaces of halloysite could be grafted by organic compounds (Yuan et al., 2008; Yah et al., 2012a,b), the interlayer inner surface Al–OH groups were unavailable for direct grafting because they were blocked by

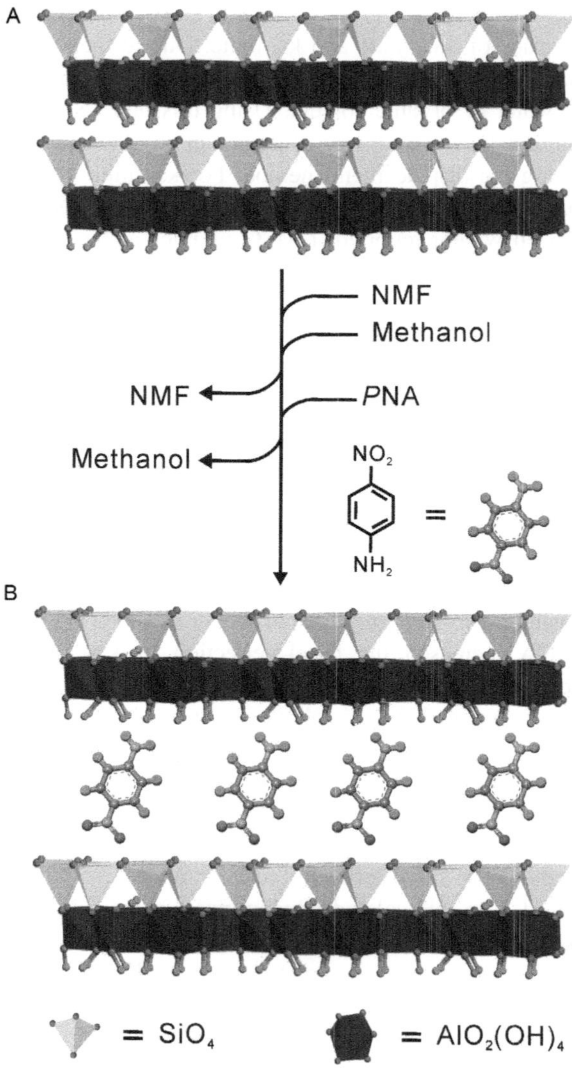

FIG. 8.11 (A) Structure of kaolinite and (B) ideal structure of kaolinite–*p*NA intercalation compound. *Modified from Kuroda et al. (1999). With the permission of the Royal Society of Chemistry.*

strong hydrogen bonding between the adjacent layers. Fortunately, this interlayer hydrogen bonding could be weakened by the direct intercalation of some small organic molecules such as DMSO, FA and KAc, resulting in the grafting of organic compounds onto the interlayer inner surface Al–OH groups. The modification of the interlayer surface of kaolinite has been widely reported. The modifiers include methanol (Tunney and Detellier,

1996; Komori et al., 2000), diols (ethylene glycol, propanediol, butanediol) (Tunney and Detellier, 1993, 1994; Gardolinski and Lagaly, 2005a,b), polyols (D-sorbitol, adonitol) (Brandt et al., 2003), ethanolamine (Tunney and Detellier, 1997; Letaief and Detellier, 2007), diethanolamine (DEA), triethanolamine (TEA) (Matusik and Bajda, 2013; Matusik and Wścisło, 2014), tris(hydroxymethyl)aminomethane (de Faria et al., 2010), pyridine-carboxylic acids (de Faria et al., 2009), APTES (Tonlé et al., 2007; Yang et al., 2012), 1-(2-hydroxyethyl)-3-methylimidazolium (Tonlé et al., 2009) and terbium pyridine-picolinate (de Faria et al., 2011). The modification of the interlayer surface of halloysite is assumed to be analogous to that of kaolinite because halloysite and kaolinite have similar crystal structures.

Tunney and Detellier (1996) first investigated the grafting of various alcohols in the interlayer space of kaolinite. These authors grafted methanol into the interlayer inner surface Al–OH groups of kaolinite by reacting methanol with DMSO or NMF-intercalated kaolinite at temperatures between 200°C and 270°C. Komori et al. (2000) further improved this modification under mild conditions (room temperature, 7 days stirring) using a displacement reaction with NMF-intercalated kaolinite as an intermediate. The mechanism for grafting methoxy groups on the interlayer inner surface Al–OH groups of kaolinite is that methanol molecules are intercalated into the interlayer space of kaolinite by replacing the preintercalated DMSO or NMF, which enlarges the d_{001} value of kaolinite (in a wet state) to 1.11 nm. During the stirring and drying processes, methoxy modification occurs through condensation between every third interlayer inner surface Al–OH group and one methanol molecule, forming one Al–O–C bond and one H_2O molecule (Fig. 8.12). The methyl groups point away from the surface, and portions of the methyl

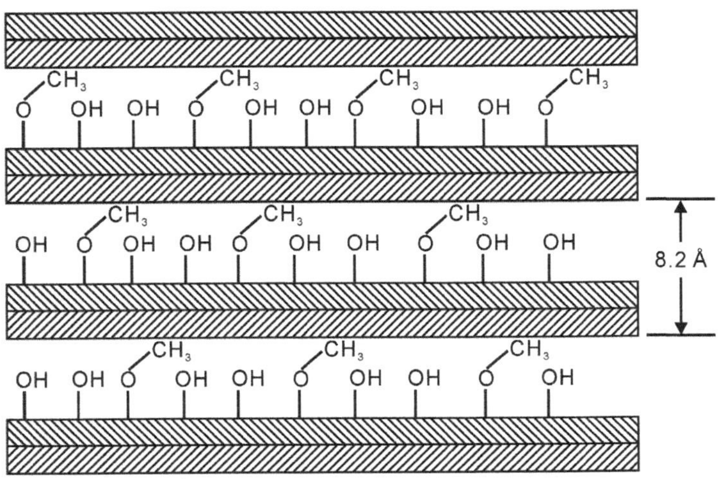

FIG. 8.12 Schematic representation of methoxy-modified kaolinite. *Reprinted from Tunney and Detellier (1996). With the permission of the Royal Society of Chemistry.*

groups are keyed into the $(SiO)_6$ macro-ring of the adjacent silicate sheet. Because of the introduction of the methoxy groups into the interlayer space, the basal reflection of methoxy-modified kaolinite (in a dry state) is 0.82–0.86 nm. This d_{001} value depends on the number of interlayer water molecules generated by the interaction between the methanol molecules and Al–OH groups. For example, a d_{001} value of 0.82 nm is observed for the methoxy-modified kaolinite reacted under high temperature (200–270°C) because of the absence of water molecules in the interlayer space (Tunney and Detellier, 1996); however, a d_{001} value of 0.86 nm is observed for methoxy-modified kaolinite reacting at room temperature because of the presence of 0.58 mol of water in each kaolinite unit, corresponding to the chemical formula $Al_2(OH)_{3.64}Si_2O_5{\cdot}0.58H_2O$ and calculated by CHN and TG analysis (Komori et al., 2000). The covalently grafted methoxy groups are highly stable, resistant to thermal decomposition at temperatures of up to 350°C in both air and N_2 atmosphere, and resistant to water hydrolysis.

In addition to alcohols, a new type of organic compound, aminoalcohols, was developed as a modifier to be grafted onto Al–OH groups on the interlayer inner surface of halloysite and kaolinite. Letaief and Detellier (2007) investigated the grafting of ethanolamine, DEA and TEA on the interlayer space of kaolinite preintercalated with DMSO at high temperatures (170–180°C) and in an N_2 atmosphere. The results showed that ethanolamine is intercalated, but not grafted, on the interlayer surface because the intercalated ethanolamine is easily rinsed out by water. In addition, the ^{13}C CP/MAS NMR, ^{27}Al MAS NMR and IR results demonstrated that DEA and TEA were covalently grafted onto the interlayer surface of kaolinite and were highly resistant to thermal decomposition at temperatures of up to 300°C and to extensive hydrolysis during washing. The grafting of TEA is particularly interesting because TEA is grafted to one Al–OH group of each unit structure of halloysite, kaolinite or both, bridging two adjacent unit structures and allowing the nongrafted third hydroxyl group to interact with the siloxane macro-ring of the opposing layer [corresponding to the chemical formula $Al_2(OH)_3Si_2O_5[H\text{-}(OCH_2CH_2)_3N]_{0.51}$]. The nongrafted Al–OH groups and the introduced amino group of TEA could be available for further modification by iodomethane quaternization (Matusik and Bajda, 2013) and serve as a chelating site for metal ions (Matusik and Wścisło, 2014).

Although most organosilanes are much larger than the short chains of alcohols and aminoalcohols, organosilane can be grafted onto the interlayer inner surface Al–OH groups of kaolinite under rigorous reaction conditions. Tonlé et al. (2007) first studied the grafting of APTES onto the interlayer surface of kaolinite and found that this modification occurred only when DMSO-intercalated kaolinite was used as a precursor in an N_2 atmosphere and at a high temperature (ie, 195°C). The unmodified kaolinite, urea-intercalated kaolinite or air atmosphere resulted in no APTES grafting onto the interlayer surface

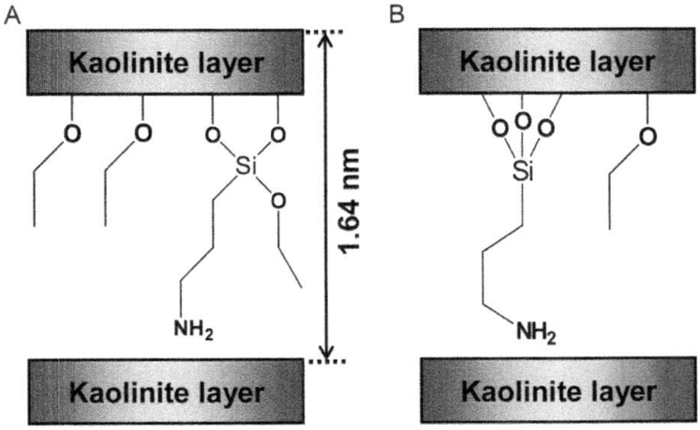

FIG. 8.13 Schematic representation showing the structural possibilities of the grafting of organosilyl and ethoxy groups on the kaolinite interlayer space. (A) Bidentate and (B) tridentate grafting of APTES onto the interlayer surface of kaolinite. *Reprinted with permission from Tonlé et al. (2007). Copyright (2007) American Chemical Society.*

of kaolinite. The grafting of APTES increased the interlayer space of kaolinite from 1.12 nm (for DMSO-intercalated kaolinite) to 1.64 nm. ^{13}C MAS NMR and ^{29}Si CP/MAS NMR demonstrated that the following types of grafting mechanism occurred: (i) ethoxylation, occurring between the ethanol (generated by the hydrolysis of the APTES) and the interlayer surface Al–OH groups at high temperatures; and (ii) silylation, occurring between the APTES and interlayer surface Al–OH groups. In the case of silylation, the grafted APTES mainly presented bidentate grafting (Fig. 8.13A) with the concurrent grafting of the two ethoxy groups; and also tridentate grafting (Fig. 8.13B), with the concurrent grafting of one ethoxy group and the loss of two ethanol molecules. The silylation was roughly equally distributed between the bidentate and tridentate fixations. Following 24 h of washing with water, the Al–O–Si bond was partially hydrolysed, which resulted in the coexistence of basal reflections at 1.6 and 1.0 nm.

Yang et al. (2012) investigated the influence of reaction temperature on the grafting of APTES onto the interlayer surface Al–OH groups of kaolinite. Under low reaction temperatures (175°C, 185°C and 195°C), the APTES replaced the preintercalated DMSO and was intercalated into the interlayer space of kaolinite. The intercalated APTES mainly formed a pseudobilayer arrangement in the interlayer space that corresponded to a basal reflection of 1.73–1.89 nm and formed hydrogen bonds between the amino-head of APTES and the Al–OH groups of kaolinite. In addition, a portion of the APTES was covalently grafted onto the interlayer surface Al–OH groups and acted as a cross-linked monolayer arrangement, corresponding to a basal reflection of 1.00 nm (Fig. 8.14). High reaction temperatures are beneficial for grafting APTES, which is indicated by the higher intensity of the reflection at 1.00 nm

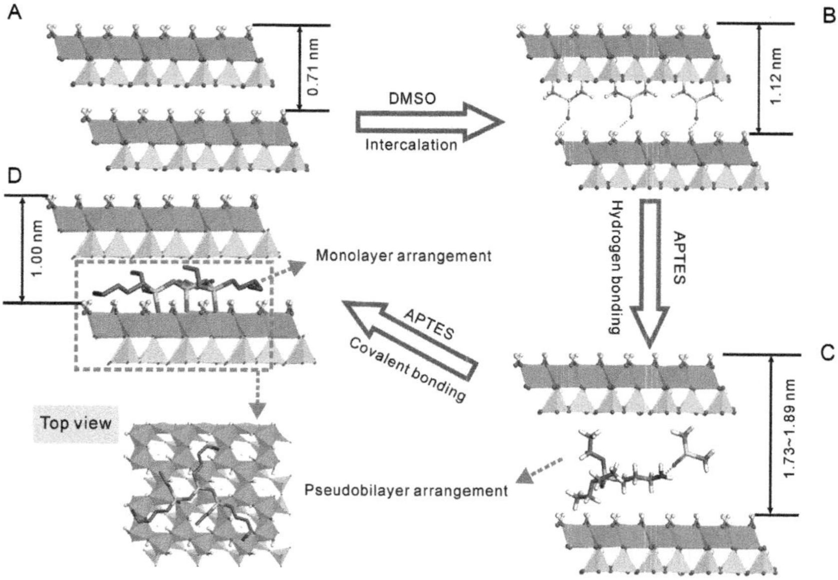

FIG. 8.14 Schematic representation of the interlayer grafting of APTES on kaolinite. (A) raw kaolinite, (B) DMSO-intercalated kaolinite, (C) APTES-grafted kaolinite with a pseudobilayer arrangement and (D) APTES-grafted kaolinite with a monolayer arrangement. *Reprinted with permission from Yang et al. (2012).*

from APTES-grafted kaolinite generated at higher reaction temperatures. When a high reaction temperature was used (220°C), the hydrogen-bound APTES was condensed with the Al–OH groups and covalently grafted onto the interlayer surface of kaolinite. Because the amine groups of APTES did not participate in the grafting process and the alkyl groups were pointed away from the surface, the aminosilane molecules were turned over from their initial amine-down position in the hydrogen-bound state to the amine-up position.

Unlike the alcohols and the aminoalcohols, which were grafted on the interlayer surfaces of both halloysite and kaolinite, interlayer surface modification by APTES was not observed for tubular halloysite under the same conditions, potentially due to the steric hindrance of the APTES molecule and the microstructure of the tubular halloysite.

Grafting organics in the interlayer spaces of halloysite and kaolinite could significantly enhance the ability of these interlayer spaces to accommodate other molecules. For instance, methoxy modification introduced a plane of methoxy groups into the interlayer space of halloysite and kaolinite weakening the original hydrogen bonds between adjacent layers. This interlayer space became available for further intercalation. For instance, the interlayer space of the methoxy-modified halloysite and kaolinite is available for loading guest molecules, such as 5-fluorouracil (Tan et al., 2014a) and amitrole

(Tan et al., 2015a,b). The intercalated guests exhibited high levels of loading and controlled release because of the development of this additional loading space and the significant diffusional restrictions of the lamellar structure.

8.5 APPLICATIONS OF SURFACE-MODIFIED HALLOYSITE

8.5.1 Halloysite Polymer Nanocomposite

Halloysite nanotubes are expected to be a promising filler in CPN due to their high SSA, high aspect ratio, good dispersion and excellent mechanical properties. The incorporation of halloysite into polymers had three main effects on the properties of the resulting CPN, which are summarized as follows: (i) halloysite acted as a reinforcing agent in the CPN and significantly improved its mechanical properties (tensile strength, impact strength, flexural property, etc.); (ii) halloysite acted as a flame-retardant agent and significantly increased the thermal stability and flame retardancy of the CPN; and (iii) halloysite affected the crystallization behaviour of semicrystalline polymers, such as polypropylene and polyamide-6. Two crucial factors determine the performance of halloysite–polymer nanocomposites: good dispersion of the halloysite in the polymer matrix and a desirable interfacial affinity between the halloysite and the polymer (Du et al., 2010; Liu et al., 2014). However, the hydrophilic halloysite is not very compatible with and does not interact strongly with the hydrophobic polymer matrix, leading to the formation of micron-sized aggregates of halloysite nanotubes in a polymer matrix. The surface modification of halloysite can promote the performance of halloysite in CPN through enhancing the affinity between the halloysite and polymer matrix and improving the dispersion of halloysite nanotubes within the polymer matrix.

The intercalation of organic molecules into the interlayer space of halloysite is an effective method for increasing halloysite dispersion in the polymer matrix (Deng et al., 2009; Tang et al., 2011). Tang et al. (2011) reported that the intercalation of PPA into the interlayer space of halloysite could promote the dispersion of halloysite in epoxy and the formation of multiple microcracks and plastic deformations in the interfaces, resulting in an increase in the fracture toughness of CPN by 78.3% without sacrificing other basic properties (such as strength, modulus, glass transition temperature and thermal stability).

The organosilane modification of halloysite is also a promising method for promoting the mechanical properties of CPN because the introduced functional groups (eg, an aminopropyl group) on the surface of halloysite can increase its hydrophobicity and enhance its compatibility and affinity with the polymer matrix (Deng et al., 2009; Pasbakhsh et al., 2010; Barrientos-Ramirez et al., 2011; Raman et al., 2013; Vahedi and Pasbakhsh, 2014). The APTES-modified halloysite could notably increase the impact strength of epoxy by 207% compared with neat epoxy (Vahedi and Pasbakhsh, 2014). This improvement is attributed to better dispersion of halloysite in the epoxy matrix and to the stronger interactions between the APTES-modified halloysite and epoxy generated between the amine groups and the epoxy network.

In addition, modifying the surface of halloysite by chitosan coating is an easy and brilliant method for promoting the performance of halloysite in CPN because the chitosan coating could implant abundant amino and hydroxyl groups onto the external surface of the halloysite, which would increase the dispersion and the interactions of halloysite with the polymer matrix. Fu et al. (2015) simultaneously enhanced the tensile strength and elongation at break of waterborne polyurethane (WPU) by using chitosan-coated halloysite as reinforcing and cross-linking filler. This increase was attributed to the uniform dispersion of the modified halloysite in the WPU matrix and to the strong interfacial interactions and hydrogen bonding interactions between chitosan-modified halloysite and WPU. The chitosan-modified halloysite could be used as a potential filler of CPN for mass-scale industrial applications because of the inexpensive raw material (chitosan and halloysite) and the ready ability to modify halloysite by chitosan coating with commercial feasibility.

8.5.2 Controlled Loading and Release of Guest Molecules

Halloysite nanotubes can be used as hosts for the loading and controlled release of various guests, including inorganic salts and organic substances, ranging from simple organic molecules to high-molar-mass biochemical molecules because of their mesoscopic or macroscopic lumen, high biocompatibility and low cytotoxicity. Two drawbacks for using raw halloysite as a carrier were identified: low loading capacity and fast release of guest molecules. The low loading capacity is caused by (i) the weak interactions (ion exchange, hydrogen bonding and van der Waals forces) between the guest molecule and the halloysite; and (ii) the limited loading space, with which guest molecules are mainly encapsulated into the lumen but cannot be loaded into the interlayer space of halloysite, except for some small organic molecules. The fast release is caused by the weak interactions and by insufficient restriction of the large lumen of halloysite to the outward diffusion of the lumen-loaded guest. To overcome these two drawbacks, and to broaden the use of halloysite as a carrier in various application fields, it is important to modify the surface properties of halloysite in order to increase the loading capacity of halloysite by enhancing the affinity between the guest molecules and the halloysite or by creating additional loading space, and also to prolong the release of guests by reducing their diffusion from halloysite.

The affinity between the guest molecules and halloysite can be enhanced by introducing specific functional groups on the surface of halloysite through organosilane modification. For example, when halloysite was used as a carrier for ibuprofen, ibuprofen participated only in hydrogen bonding and van der Waals interactions with unmodified halloysite, which resulted in low loading content and fast release of ibuprofen in unmodified halloysite. For APTES-modified halloysite, electrostatic attractions were generated between the carboxylate ($–COO^-$) of ibuprofen and the protonated aminopropyl group ($-NH_3^+$) of the grafted APTES, which was formed by proton transfer from the carboxyl groups of ibuprofen to the aminopropyl groups of the grafted APTES. This strong interaction increased the loading of ibuprofen from 11.7 to 14.8 mass% and significantly prolonged the release of ibuprofen (Fig. 8.15), which was achieved by introducing more encapsulated ibuprofen into the lumen and by retarding the dissolution and diffusion of ibuprofen from the lumen of the APTES-modified halloysite (Tan et al., 2014a). This increase in loading capacity and the decreased release of guest by APTES modification was also achieved for orange II (Yuan et al., 2012b).

The interlayer space of halloysite can be developed as an additional loading space for guest molecules by weakening the hydrogen bonding between the adjacent layers through methoxy modification of the interlayer space of halloysite. Tan et al. (2014b, 2015a,b) demonstrated that methoxy modification could develop the interlayer space of halloysite and kaolinite as additional loading spaces for the anticancer drug 5-fluorouracil and the herbicide amitrole. The guest replaced the methanol molecules and intercalated them into the interlayer space of halloysite in a monolayer arrangement. Methoxy modification significantly promoted the loading capacity of halloysite and kaolinite; for example, the loading of amitrole onto methoxy-modified halloysite was as high as 30.5 mass% and was 74.3% greater than that on unmodified halloysite. In addition, the loading of 5-fluorouracil on methoxy-modified kaolinite reached up to 55.4 mass%, which was 147.3% greater than that on unmodified kaolinite. The lamellar structures of halloysite

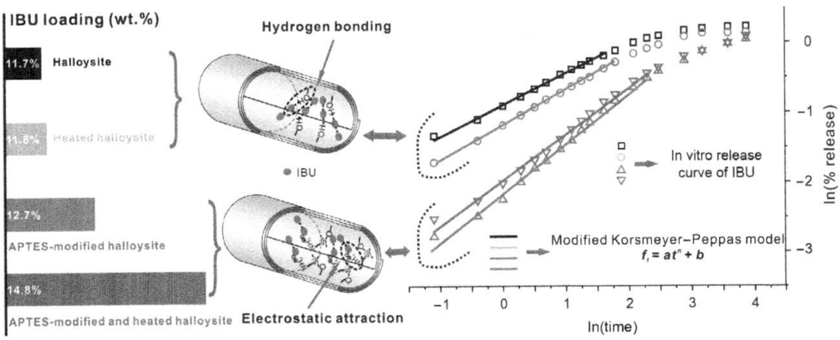

FIG. 8.15 Schematic representation of the effect of APTES modification on the loading and release of ibuprofen on halloysite. *Reprinted with permission from Tan et al. (2014b).*

and kaolinite had restricted the diffusion of the intercalated guest molecules, which resulted in a prolonged release.

Because the guests were mainly loaded into the lumen of halloysite, the formation of organic shells and coatings on the external surface of halloysite obviously slowed the outward diffusion of the lumen-loaded guest molecule and achieved prolonged release. Veerabadran et al. (2009) first loaded dexamethasone into the lumen of halloysite and then coated organic shells on the negatively charged external surface of halloysite via LBL assembly. The organized shell acted as a diffusion barrier and significantly decreased the release of dexamethasone from halloysite. In addition, this decrease in diffusion was significantly related to (i) the type of polyelectrolytes, with strong polyelectrolytes with a high degree of ionization (PEI and PSS) offering remarkable retardation of diffusion compared with weak polyelectrolytes (PAH and PAA); and (ii) the molar mass of the coated polyelectrolytes, with higher molar mass polyelectrolytes having a weaker permeability and leading a slower diffusion of dexamethasone.

8.5.3 Pollution Remediation

The use of halloysite as an adsorbent for pollution remediation has obtained considerable attention because it is economically viable and environmentally friendly and has a high SSA and abundant surface hydroxyl groups. The pollutants adsorbed by halloysite can be categorized into two types: (i) ionic pollutant, including heavy metal ions and cationic dyes; and (ii) organic contaminants.

In principle, halloysite always exhibited low adsorption capacity for metal ions and hydrophobic organic contaminants because of its low cationic exchange capacity and hydrophilic nature. Because surface modification can introduce special functional groups on halloysite, which can serve as new adsorption sites for pollutants and increase the hydrophobicity of halloysite, the performance of modified halloysite as an adsorbent has been significantly promoted. For instance, Matusik and Wścisło (2014) modified the interlayer space of halloysite by grafting DEA and TEA and found that the adsorption capacity for Cd(II) had been increased by approximately 10 times. This improvement in metal removal was attributed to two steps: gradual diffusion of the ions into the interlayer space and the subsequent attraction by the amine nitrogen. In addition, Liu et al. (2012) incorporated halloysite into alginate (biopolymer, with abundant carboxylic groups) to prepare porous beads. This hybrid material was an effective adsorbent for removing methylene blue. The maximum adsorption capacity was approximately 250 mg/g, which was much greater than that of the unmodified halloysite (84.3 mg/g) (Zhao and Liu, 2008). Moreover, the hybrid material exhibited high reusability; thus, the removal efficiency of methylene blue was maintained above 90% after 10 successive adsorption–desorption cycles. The excellent performance of the hybrid material was attributed to the perfect combination of the advantages of halloysite and alginate.

8.6 CONCLUDING REMARKS

Tubular halloysite has three types of surfaces: the internal lumen surface, covered by Al–OH groups; the external siloxane surface, with a negative surface charge; and the interlayer surface, blocked by hydrogen bonding. These surfaces have specific reactivities and chemical environments that allow site-specific surface modifications. The surface modification of halloysite can be classified into the following three categories:

1. The chemical modification of the lumen surface by covalently grafting organic compounds (namely, organosilane and organophosphonic acid) onto Al–OH groups. This modification can selectively tune the lumen surface of halloysite by introducing specific functional groups onto the lumen surface.
2. The physical modification of the external surface by coating cationic substances (namely, surfactants, polymers and biopolymers) onto the negatively charged surface. This modification can effectively enhance the dispersibility, hydrophobicity and biocompatibility of halloysite.
3. The modification of the interlayer surface by direct and indirect intercalation of organic compounds or by grafting alcohols or aminoalcohols onto the interlayer inner surface Al–OH groups. This modification can weaken interlayer hydrogen bonding and further develop the interlayer space as an additional space for adsorption and loading.

The surface modification of halloysite can regulate the physicochemical properties of halloysite (surface potential, hydrophilicity/hydrophobicity, cationic exchange capacity, chemical reactivity, etc.) and significantly promote the performance of halloysite in practical applications. For the halloysite–polymer nanocomposite, surface modification of halloysite (by such actions as grafting, coating and intercalation) can increase the mechanical properties of the CPN by enhancing the dispersion and affinity of the modified halloysite in the polymer matrix. For controlled loading and release, the surface modification of halloysite can increase the loading and slow the release of guest molecules by enhancing the affinities of these molecules and by developing additional loading space. For pollution remediation, surface modification can improve the adsorption capacity for a pollutant by increasing the number of adsorption sites and altering the hydrophobicity and hydrophilicity of the halloysite.

ABBREVIATIONS

AEAPS	3-(2-aminoethyl)aminopropyltrimethoxysilane
APTES	3-aminopropyltriethoxysilane
BiBB	α-bromoisobutyryl bromide
CPN	clay polymer nanocomposite
DEA	diethanolamine

DMSO	dimethyl sulphoxide
Dopa	2-bromo-*N*-[2-(3,4-dihydroxyphenyl)ethyl]-isobutyryl amide
EPA	ethylphosphonic acid
FA	formamide
HB	hydrogen bond
HDTMA	hexadecyl trimethylammonium bromide
KAc	potassium acetate
LBL	layer-by-layer
NMF	*N*-methylformamide
ODP	octadecylphosphonic acid
PA	phosphonic acid
PAA	poly(acrylic acid)
PAH	poly(allylamine) hydrochloride
PDMAEMA	poly(*N*,*N*-dimethylaminoethyl methacrylate)
PEI	poly(ethylenimine)
***p*NA**	*p*-nitroaniline
PPA	phenylphosphonic acid
PSS	sodium poly(styrene sulphonate)
RF	repulsive force
SSA	specific surface area
TEA	triethanolamine
V_{Pore}	pore volume
WPU	waterborne polyurethane

REFERENCES

Abdullayev, E., Lvov, Y., 2011. Halloysite clay nanotubes for controlled release of protective agents. J. Nanosci. Nanotechnol. 11 (11), 10007–10026.

Abdullayev, E., Price, R., Shchukin, D., Lvov, Y., 2009. Halloysite tubes as nanocontainers for anticorrosion coating with benzotriazole. ACS Appl. Mater. Interfaces 1 (7), 1437–1443.

Barrientos-Ramirez, S., de Oca-Ramirez, G.M., Ramos-Fernandez, E.V., Sepulveda-Escribano, A., Pastor-Blas, M.M., Gonzalez-Montiel, A., 2011. Surface modification of natural halloysite clay nanotubes with aminosilanes. Application as catalyst supports in the atom transfer radical polymerization of methyl methacrylate. Appl. Catal. A Gen. 406 (1–2), 22–33.

Brady, P.V., Cygan, R.T., Nagy, K.L., 1996. Molecular controls on kaolinite surface charge. J. Colloid Interface Sci. 183 (2), 356–364.

Brandt, K.B., Elbokl, T.A., Detellier, C., 2003. Intercalation and interlamellar grafting of polyols in layered aluminosilicates. D-Sorbitol and adonitol derivatives of kaolinite. J. Mater. Chem. 13 (10), 2566–2572.

Breen, C., D'Mello, N., Yarwood, J., 2002. The thermal stability of mixed phenylphosphonic acid/water intercalates of kaolin and halloysite. A TG–EGA and VT-DRIFTS study. J. Mater. Chem. 12 (2), 273–278.

Carr, R.M., Chih, H.W.A., 1971. Complexes of halloysite with organic compounds. Clay Miner. 9, 153–166.

Carr, R., Chaikum, N., Patterson, N., 1978. Intercalation of salts in halloysite. Clay Clay Miner. 26 (2), 144–152.

Chang, P.R., Xie, Y.F., Wu, D.L., Ma, X.F., 2011. Amylose wrapped halloysite nanotubes. Carbohydr. Polym. 84 (4), 1426–1429.

Cheng, H., Liu, Q., Yang, J., Du, X., Frost, R.L., 2010. Influencing factors on kaolinite–potassium acetate intercalation complexes. Appl. Clay Sci. 50 (4), 476–480.

Churchman, G.J., Whitton, J.S., Claridge, G.G.C., Theng, B.K.G., 1984. Intercalation method using formamide for differentiating halloysite from kaolinite. Clay Clay Miner. 32 (4), 241–248.

Costanzo, P., Giesse, R., 1986. Ordered halloysite; dimethylsulfoxide intercalate. Clay Clay Miner. 34 (1), 105–107.

de Faria, E.H., Lima, O.J., Ciuffi, K.J., Nassar, E.J., Vicente, M.A., Trujillano, R., Calefi, P.S., 2009. Hybrid materials prepared by interlayer functionalization of kaolinite with pyridine-carboxylic acids. J. Colloid Interface Sci. 335 (2), 210–215.

de Faria, E.H., Ciuffi, K.J., Nassar, E.J., Vicente, M.A., Trujillano, R., Calefi, P.S., 2010. Novel reactive amino-compound: Tris(hydroxymethyl)aminomethane covalently grafted on kaolinite. Appl. Clay Sci. 48 (3), 516–521.

de Faria, E.H., Nassar, E.J., Ciuffi, K.J., Vicente, M.A., Trujillano, R., Rives, V., Calefi, P.S., 2011. New highly luminescent hybrid materials: terbium pyridine–picolinate covalently grafted on kaolinite. ACS Appl. Mater. Interfaces 3 (4), 1311–1318.

Deng, S., Zhang, J., Ye, L., 2009. Halloysite–epoxy nanocomposites with improved particle dispersion through ball mill homogenisation and chemical treatments. Compos. Sci. Technol. 69 (14), 2497–2505.

Du, M.L., Guo, B.C., Jia, D.M., 2010. Newly emerging applications of halloysite nanotubes: a review. Polym. Int. 59 (5), 574–582.

Elbokl, T.A., Detellier, C., 2008. Intercalation of cyclic imides in kaolinite. J. Colloid Interface Sci. 323 (2), 338–348.

Fang, Q., Huang, S., Wang, W., 2005. Intercalation of dimethyl sulphoxide in kaolinite: molecular dynamics simulation study. Chem. Phys. Lett. 411 (1–3), 233–237.

Frost, R.L., Kristof, J., 1997. Intercalation of halloysite: a Raman spectroscopic study. Clay Clay Miner. 45 (4), 551–563.

Frost, R.L., Kristof, J., Paroz, G.N., Kloprogge, J.T., 1998. Molecular structure of dimethyl sulfoxide intercalated kaolinites. J. Phys. Chem. B. 102 (43), 8519–8532.

Frost, R.L., Lack, D.A., Paroz, G.N., Tran, T.H.T., 1999. New techniques for studying the intercalation of kaolinites from Georgia with formamide. Clay Clay Miner. 47 (3), 297–303.

Frost, R.L., Kristof, J., Mako, E., Kloprogge, J.T., 2000. Modification of the hydroxyl surface in potassium-acetate-intercalated kaolinite between 25 and 300 °C. Langmuir 16 (19), 7421–7428.

Fu, H., Wang, Y., Chen, W., Xiao, J., 2015. Reinforcement of waterborne polyurethane with chitosan-modified halloysite nanotubes. Appl. Surf. Sci. 346, 372–378.

Gardolinski, J.E.F.C., Lagaly, G., 2005a. Grafted organic derivatives of kaolinite: I. Synthesis, chemical and rheological characterization. Clay Miner. 40 (4), 537–546.

Gardolinski, J.E.F.C., Lagaly, G., 2005b. Grafted organic derivatives of kaolinite: II. Intercalation of primary n-alkylamines and delamination. Clay Miner. 40 (4), 547–556.

Gardolinski, J.E., Ramos, L.P., de Souza, G.P., Wypych, F., 2000. Intercalation of benzamide into kaolinite. J. Colloid Interface Sci. 221 (2), 284–290.

Hayashi, S., 1997. NMR study of dynamics and evolution of guest molecules in kaolinite/dimethyl sulfoxide intercalation compound. Clay Clay Miner. 45 (5), 724–732.

Hemmatpour, H., Haddadi-Asl, V., Roghani-Mamaqani, H., 2015. Synthesis of pH-sensitive poly (N,N-dimethylaminoethyl methacrylate)-grafted halloysite nanotubes for adsorption and controlled release of DPH and DS drugs. Polymer 65, 143–153.

Horvath, E., Kristof, J., Frost, R., Redey, A., Vágvölgyi, V., Cseh, T., 2003. Hydrazine-hydrate intercalated halloysite under controlled-rate thermal analysis conditions. J. Therm. Anal. Calorim. 71 (3), 707–714.

Jana, S., Das, S., Ghosh, C., Maity, A., Pradhan, M., 2015. Halloysite nanotubes capturing isotope selective atmospheric CO_2. Sci. Rep. 5, 8711.

Joo, Y., Jeon, Y., Lee, S.U., Sim, J.H., Ryu, J., Lee, S., Lee, H., Sohn, D., 2012. Aggregation and stabilization of carboxylic acid functionalized halloysite nanotubes (HNT-COOH). J. Phys. Chem. C 116 (34), 18230–18235.

Joussein, E., Petit, S., Churchman, J., Theng, B., Righi, D., Delvaux, B., 2005. Halloysite clay minerals: a review. Clay Miner. 40 (4), 383–426.

Joussein, E., Petit, S., Delvaux, B., 2007. Behavior of halloysite clay under formamide treatment. Appl. Clay Sci. 35 (1), 17–24.

Komori, Y., Matsumura, A., Itagaki, T., Sugahara, Y., Kuroda, K., 1999a. Preparation of a kaolinite-ε-caprolactam intercalation compound. Clay Sci. 11 (1), 47–55.

Komori, Y., Sugahara, Y., Kuroda, K., 1999b. Direct intercalation of poly(vinylpyrrolidone) into kaolinite by a refined guest displacement method. Chem. Mater. 11 (1), 3–6.

Komori, Y., Sugahara, Y., Kuroda, K., 1999c. Intercalation of alkylamines and water into kaolinite with methanol kaolinite as an intermediate. Appl. Clay Sci. 15 (1–2), 241–252.

Komori, Y., Enoto, H., Takenawa, R., Hayashi, S., Sugahara, Y., Kuroda, K., 2000. Modification of the interlayer surface of kaolinite with methoxy groups. Langmuir 16 (12), 5506–5508.

Kuroda, K., Hiraguri, K., Komori, Y., Sugahara, Y., Mouri, H., Uesu, Y., 1999. An acentric arrangement of p-nitroaniline molecules between the layers of kaolinite. Chem. Commun. (22), 2253–2254.

Kuroda, Y., Ito, K., Itabashi, K., Kuroda, K., 2011. One-step exfoliation of kaolinites and their transformation into nanoscrolls. Langmuir 27 (5), 2028–2035.

Lee, S.Y., Kim, S.J., 2002. Adsorption of naphthalene by HDTMA modified kaolinite and halloysite. Appl. Clay Sci. 22 (1–2), 55–63.

Letaief, S., Detellier, C., 2007. Functionalized nanohybrid materials obtained from the interlayer grafting of aminoalcohols on kaolinite. Chem. Commun. (25), 2613–2615.

Li, C.P., Liu, J.G., Qu, X.Z., Guo, B.C., Yang, Z.Z., 2008. Polymer-modified halloysite composite nanotubes. J. Appl. Polym. Sci. 110 (6), 3638–3646.

Li, C., Liu, J., Qu, X., Yang, Z., 2009. A general synthesis approach toward halloysite-based composite nanotube. J. Appl. Polym. Sci. 112 (5), 2647–2655.

Liu, M., Guo, B., Du, M., Lei, Y., Jia, D., 2008. Natural inorganic nanotubes reinforced epoxy resin nanocomposites. J. Polym. Res. 15 (3), 205–212.

Liu, C., Luo, Y.F., Jia, Z.X., Zhong, B.C., Li, S.Q., Guo, B.C., Jia, D.M., 2011. Enhancement of mechanical properties of poly(vinyl chloride) with polymethyl methacrylate-grafted halloysite nanotube. Express Polym. Lett. 5 (7), 591–603.

Liu, L., Wan, Y., Xie, Y., Zhai, R., Zhang, B., Liu, J., 2012. The removal of dye from aqueous solution using alginate-halloysite nanotube beads. Chem. Eng. J. 187, 210–216.

Liu, M., Jia, Z., Jia, D., Zhou, C., 2014. Recent advance in research on halloysite nanotubes-polymer nanocomposite. Prog. Polym. Sci. 39 (8), 1498–1525.

Luschtinetz, R., Oliveira, A.F., Frenzel, J., Joswig, J.O., Seifert, G., Duarte, H.A., 2008. Adsorption of phosphonic and ethylphosphonic acid on aluminum oxide surfaces. Surf. Sci. 602 (7), 1347–1359.

Luschtinetz, R., Frenzel, J., Milek, T., Seifert, G., 2009. Adsorption of phosphonic acid at the TiO2 anatase (101) and rutile (110) surfaces. J. Phys. Chem. C 113 (14), 5730–5740.

Lvov, Y., Price, R., Gaber, B., Ichinose, I., 2002. Thin film nanofabrication via layer-by-layer adsorption of tubule halloysite, spherical silica, proteins and polycations. Colloids Surf. A 198, 375–382.

Matsumura, A., Komori, Y., Itagaki, T., Sugahara, Y., Kuroda, K., 2001. Preparation of a kaolinite–nylon 6 intercalation compound. Bull. Chem. Soc. Jpn. 74 (6), 1153–1158.

Matusik, J., Bajda, T., 2013. Immobilization and reduction of hexavalent chromium in the interlayer space of positively charged kaolinites. J. Colloid Interface Sci. 398, 74–81.

Matusik, J., Wścisło, A., 2014. Enhanced heavy metal adsorption on functionalized nanotubular halloysite interlayer grafted with aminoalcohols. Appl. Clay Sci. 100, 50–59.

Ng, K.M., Lau, Y.T.R., Chan, C.M., Weng, L.T., Wu, J., 2011. Surface studies of halloysite nanotubes by XPS and ToF-SIMS. Surf. Interface Anal. 43 (4), 795–802.

Nicolini, K.P., Fukamachi, C.R.B., Wypych, F., Mangrich, A.S., 2009. Dehydrated halloysite intercalated mechanochemically with urea: thermal behavior and structural aspects. J. Colloid Interface Sci. 338 (2), 474–479.

Olejnik, S., Aylmore, L.A.G., Posner, A.M., Quirk, J.P., 1968. Infrared spectra of kaolin mineral-dimethyl sulfoxide complexes. J. Phys. Chem. 72 (1), 241–249.

Pasbakhsh, P., Ismail, H., Fauzi, M.N.A., Abu Bakar, A., 2010. EPDM/modified halloysite nanocomposites. Appl. Clay Sci. 48 (3), 405–413.

Peixoto, A.F., Fernandes, A.C., Pereira, C., Pires, J., Freire, C., 2016. Physicochemical characterization of organosilylated halloysite clay nanotubes. Microporous Mesoporous Mater. 219, 145–154.

Raman, V.S., Rooj, S., Das, A., Stockelhuber, K.W., Simon, F., Nando, G.B., Heinrich, G., 2013. Reinforcement of solution styrene butadiene rubber by silane functionalized halloysite nanotubes. J. Macromol. Sci. A 50 (11), 1091–1106.

Riela, S., Massaro, M., Colletti, C.G., Bommarito, A., Giordano, C., Milioto, S., Noto, R., Poma, P., Lazzara, G., 2014. Development and characterization of co-loaded curcumin/triazole-halloysite systems and evaluation of their potential anticancer activity. Int. J. Pharm. 475 (1), 613–623.

Rutkai, G., Kristóf, T., 2008. Molecular simulation study of intercalation of small molecules in kaolinite. Chem. Phys. Lett. 462 (4–6), 269–274.

Shamsi, M.H., Geckeler, K.E., 2008. The first biopolymer-wrapped non-carbon nanotubes. Nanotechnology 19 (7), 075604.

Song, K.-h., Wang, X., Qian, P., Zhang, C., Zhang, Q., 2013. Theoretical study of interaction of formamide with kaolinite. Comput. Theor. Chem. 1020, 72–80.

Sugahara, Y., Satokawa, S., Kuroda, K., Kato, C., 1990. Preparation of a kaolinite-polyacrylamide intercalation compound. Clay Clay Miner. 38 (2), 137–143.

Takenawa, R., Komori, Y., Hayashi, S., Kawamata, J., Kuroda, K., 2001. Intercalation of nitroanilines into kaolinite and second harmonic generation. Chem. Mater. 13 (10), 3741–3746.

Tan, D., Yuan, P., Annabi-Bergaya, F., Liu, D., Wang, L., Liu, H., He, H., 2014a. Loading and in vitro release of ibuprofen in tubular halloysite. Appl. Clay Sci. 96, 50–55.

Tan, D., Yuan, P., Annabi-Bergaya, F., Liu, D., He, H., 2014b. High-capacity loading of 5-fluorouracil on the methoxy-modified kaolinite. Appl. Clay Sci. 100, 60–65.

Tan, D., Yuan, P., Annabi-Bergaya, F., Dong, F., Liu, D., He, H., 2015a. A comparative study of tubular halloysite and platy kaolinite as carriers for the loading and release of the herbicide amitrole. Appl. Clay Sci. 114, 190–196.

Tan, D., Yuan, P., Annabi-Bergaya, F., Liu, D., He, H., 2015b. Methoxy-modified kaolinite as a novel carrier for high-capacity loading and controlled-release of the herbicide amitrole. Sci. Rep. 5, 8870.

Tang, Y.H., Deng, S.Q., Ye, L., Yang, C., Yuan, Q.A., Zhang, J.N., Zhao, C.B., 2011. Effects of unfolded and intercalated halloysites on mechanical properties of halloysite-epoxy nanocomposites. Compos. A Appl. Sci. Manuf. 42 (4), 345–354.

Tonlé, I.K., Diaco, T., Ngameni, E., Detellier, C., 2007. Nanohybrid kaolinite-based materials obtained from the interlayer grafting of 3-aminopropyltriethoxysilane and their potential use as electrochemical sensors. Chem. Mater. 19 (26), 6629–6636.

Tonlé, I.K., Letaief, S., Ngameni, E., Detellier, C., 2009. Nanohybrid materials from the grafting of imidazolium cations on the interlayer surfaces of kaolinite: application as electrode modifier. J. Mater. Chem. 19 (33), 5996–6003.

Tunney, J.J., Detellier, C., 1993. Interlamellar covalent grafting of organic units on kaolinite. Chem. Mater. 5 (6), 747–748.

Tunney, J.J., Detellier, C., 1994. Preparation and characterization of two distinct ethylene glycol derivatives of kaolinite. Clay Clay Miner. 42 (5), 552–560.

Tunney, J.J., Detellier, C., 1996. Chemically modified kaolinite. Grafting of methoxy groups on the interlamellar aluminol surface of kaolinite. J. Mater. Chem. 6 (10), 1679–1685.

Tunney, J.J., Detellier, C., 1997. Interlamellar amino functionalization of kaolinite. Can. J. Chem. 75 (11), 1766–1772.

Vahedi, V., Pasbakhsh, P., 2014. Instrumented impact properties and fracture behaviour of epoxy/modified halloysite nanocomposites. Polym. Test. 39, 101–114.

Veerabadran, N.G., Price, R.R., Lvov, Y.M., 2007. Clay nanotubes for encapsulation and sustained release of drugs. Nano 2 (2), 115–120.

Veerabadran, N.G., Mongayt, D., Torchilin, V., Price, R.R., Lvov, Y.M., 2009. Organized shells on clay nanotubes for controlled release of macromolecules. Macromol. Rapid Commun. 30 (2), 99–103.

Wan, C., Li, M., Bai, X., Zhang, Y., 2009. Synthesis and characterization of photoluminescent Eu(III) coordination halloysite nanotube-based nanohybrids. J. Phys. Chem. C 113 (36), 16238–16246.

Wang, J.H., Zhang, X.A., Zhang, B., Zhao, Y.F., Zhai, R., Liu, J.D., Chen, R.F., 2010. Rapid adsorption of Cr (VI) on modified halloysite nanotubes. Desalination 259 (1–3), 22–28.

White, C.E., Provis, J.L., Gordon, L.E., Riley, D.P., Proffen, T., van Deventer, J.S., 2011. Effect of temperature on the local structure of kaolinite intercalated with potassium acetate. Chem. Mater. 23 (2), 188–199.

Wieland, E., Stumm, W., 1992. Dissolution kinetics of kaolinite in acidic aqueous solutions at 25°C. Geochim. Cosmochim. Acta 56 (9), 3339–3355.

Wu, H., Watanabe, H., Ma, W., Fujimoto, A., Higuchi, T., Uesugi, K., Takeuchi, A., Suzuki, Y., Jinnai, H., Takahara, A., 2013. Robust liquid marbles stabilized with surface-modified halloysite nanotubes. Langmuir 29 (48), 14971–14975.

Xi, Y., Mallavarapu, M., Naidu, R., 2010. Preparation, characterization of surfactants modified clay minerals and nitrate adsorption. Appl. Clay Sci. 48 (1), 92–96.

Yah, W.O., Takahara, A., Lvov, Y.M., 2012a. Selective modification of halloysite lumen with octadecylphosphonic acid: new inorganic tubular micelle. J. Am. Chem. Soc. 134 (3), 1853–1859.

Yah, W.O., Xu, H., Soejima, H., Ma, W., Lvov, Y., Takahara, A., 2012b. Biomimetic dopamine derivative for selective polymer modification of halloysite nanotube lumen. J. Am. Chem. Soc. 134 (29), 12134–12137.

Yang, S., Yuan, P., He, H., Qin, Z., Zhou, Q., Zhu, J., Liu, D., 2012. Effect of reaction temperature on grafting of γ-aminopropyl triethoxysilane (APTES) onto kaolinite. Appl. Clay Sci. 62–63, 8–14.

Yuan, P., Southon, P.D., Liu, Z., Green, M.E.R., Hook, J.M., Antill, S.J., Kepert, C.J., 2008. Functionalization of halloysite clay nanotubes by grafting with γ-aminopropyltriethoxysilane. J. Phys. Chem. C 112 (40), 15742–15751.

Yuan, P., Tan, D.Y., Annabi-Bergaya, F., Yan, W.C., Fan, M.D., Liu, D., He, H.P., 2012a. Changes in structure, morphology, porosity, and surface activity of mesoporous halloysite nanotubes under heating. Clay Clay Miner. 60 (6), 561–573.

Yuan, P., Southon, P.D., Liu, Z.W., Kepert, C.J., 2012b. Organosilane functionalization of halloysite nanotubes for enhanced loading and controlled release. Nanotechnology 23 (37), 375705.

Yuan, P., Tan, D., Annabi-Bergaya, F., Yan, W., Liu, D., Liu, Z., 2013. From platy kaolinite to aluminosilicate nanoroll via one-step delamination of kaolinite: effect of the temperature of intercalation. Appl. Clay Sci. 83–84, 68–76.

Yuan, P., Tan, D., Annabi-Bergaya, F., 2015. Properties and applications of halloysite nanotubes: recent research advances and future prospects. Appl. Clay Sci. 112–113, 75–93.

Zhai, R., Zhang, B., Wan, Y., Li, C., Wang, J., Liu, J., 2013. Chitosan–halloysite hybrid-nanotubes: horseradish peroxidase immobilization and applications in phenol removal. Chem. Eng. J. 214, 304–309.

Zhao, M.F., Liu, P., 2008. Adsorption behavior of methylene blue on halloysite nanotubes. Microporous Mesoporous Mater. 112 (1–3), 419–424.

Zhou, Z., Gunter, W.D., 1992. The nature of the surface charge of kaolinite. Clay Clay Miner. 40 (3), 365–368.

Zhu, H., Du, M.L., Zou, M.L., Xu, C.S., Fu, Y.Q., 2012. Green synthesis of Au nanoparticles immobilized on halloysite nanotubes for surface-enhanced Raman scattering substrates. Dalton Trans. 41 (34), 10465–10471.

Chapter 9

Physicochemical Properties of Imogolite

A. Fernandez-Martinez[a,b,*] and L.J. Michot[c]
[a]*ISTerre, CNRS, Grenoble, France*
[b]*ISTerre, Univ. Grenoble Alpes, Grenoble, France*
[c]*Sorbonne Universités, UPMC Univ. Paris 06, CNRS, Laboratoire PHENIX, Case 51, 4 place Jussieu, Paris, France*
**Corresponding author: e-mail: Alex.Fernandez-Martinez@ujf-grenoble.fr*

9.1 INTRODUCTION

The surface structure of imogolite, as initially reported by Cradwick et al. (1972) from electron diffraction data, and corroborated by ^{27}Al and ^{29}Si nuclear magnetic resonance (NMR) studies (Goodman et al., 1985; Ildefonse et al., 1994; Brown et al., 1999; Kang et al., 2010; Yucelen et al., 2011), can be described as an external curved gibbsite layer populated with orthosilicate anions on the inner surface. The presence of aluminol and silanol amphoteric groups gives imogolite interesting acid–base properties that, combined with its nanotubular and porous structure, make it attractive for a range of industrial applications from gas storage or catalysis to drug release or as adsorbent for pollutants from aqueous solutions. The small diameter of the inner pore (~1 nm) and its hydrophilic character (Bonelli et al., 2013b) make imogolite an interesting model material for the study of 1D confinement (Belorizky et al., 2010), as opposed to other nanotubular materials such as carbon or boron nitride nanotubes, which are of a hydrophobic character.

The first reports on the physicochemical properties of imogolite come from the soil science community. Imogolite is an important component in typically weathered horizons of volcanic soil such as andosols, in strong association with organic matter, leading to strong organo-clay mineral complexes (Parfitt, 2009). Imogolite aggregates, typically organized pseudohexagonal bundles, are able to adsorb large quantities of water thanks to imogolite surface hydrophilicity. Water contents of up to 50% in mass have been reported by thermogravimetric studies. Little is known about any possible swelling effects linked to the mesostructure of these aggregates

Developments in Clay Science, Vol. 7. http://dx.doi.org/10.1016/B978-0-08-100293-3.00009-1

and to the impact of the high hydration on the stability of volcanic soil (Zang et al., 2011). Its formation as a weathering product of volcanic glass makes imogolite a potential constituent of 'jaboncillo', a clayey material with up to 500% water content in mass, which is present in the Mexico DF basin and considered to be responsible for the shrinkage of the soil in the region (Gutiérrez-Castorena et al., 2005). Besides its strong interaction with organic moieties, imogolite forms strong complexes with ions of environmental interest such as uranyl, Co^{2+}, Cu^{2+}, Ni^{2+} or PO_4^{3-} in soil and natural water. Its OH-populated surface, as well as the fact that it can bear a positive and a negative charge simultaneously under certain pH solution conditions, make it is one of the few soil components that is able to adsorb anions and cations at the same time.

Recent advances in imogolite and imogolite-like materials have been performed by research groups in the fields of physical chemistry of materials for adsorption and catalysis. The increasing interest in imogolite and imogolite-like materials is exemplified by the growing number of published papers in the field dealing with gas adsorption and surface modifications. In spite of recent advances, this review highlights aspects that clearly deserve more research, such as the origin of the structural charge, the control of point defects and the relative hydrophobicity of the external surface. Advances in these areas are important for the future development of functional materials based on imogolite and imogolite-like materials.

9.2 SURFACE PROPERTIES OF IMOGOLITE

9.2.1 Surface Charge

Adsorption at mineral surfaces is a crucial phenomenon in agricultural and environmental science, as it largely controls the transfer and speciation of dissolved and complexed species in soil. It also plays a major role in numerous industrial applications such as catalysis, food science and cosmetics. Understanding adsorption requires a thorough analysis of the structural, textural and surface properties of the involved minerals. As far as this latter parameter is concerned, one key quantity to be measured for oxide surfaces is the surface charge of the particle. For oxide surfaces, such a quantity can have two main origins: it can be structural or linked to the presence at the surface of amphoteric surface groups. When looking at the structure of imogolite (discussed further in Chapter 10) and considering spectroscopic information, the existence of structural charge should a priori be ruled out, but (as will be shown in the following section) such a point remains debated. Information about surface charge can be obtained from three experiments (potentiometric titration, ion adsorption and electrophoretic mobility) which probe various components of the electrical double layer. The tubular structure of imogolite and the different surface chemistry of its inner and outer surfaces make these measurements difficult to interpret. However, studies using combinations of these techniques can help unravel the specific contribution of each surface to charge and electrokinetic phenomena.

Generally, two structural features (presence of amphoteric groups and isomorphic substitutions) are at the origin of a surface charge of clay minerals. The pH-dependent net proton charge, σ_H, originated from the difference between the moles of protons and hydroxide ions complexed by surface functional groups (Sposito, 2008) is originated by the amphoteric aluminol and silanol groups located on the outer and inner surfaces of imogolite, respectively. The presence of these functional groups gives imogolite the character of a variable charge clay mineral. On the other hand, permanent charge, σ_0, originates from isomorphic substitutions of ions of different charges. However, only in synthetic imogolite, isomorphic Fe^{3+} and Ge^{4+} substitutions have been reported in the Al^{3+} and Si^{4+} sites, respectively, which do not result in any net charge (Ookawa et al., 2006; Levard et al., 2009b; Avellan et al., 2014; Baker et al., 2014). Fe^{3+} substitutions have been reported by different authors in both Si- and Ge-imogolite (McBride, 1984; Ookawa et al., 2006; Avellan et al., 2014; Baker et al., 2014). Si imogolites are able to incorporate up to 9% at Fe^{3+} in their layer before iron oxides are precipitated, a concentration that is lower for Ge-imogolite ($\sim$1%) (Avellan et al., 2014).

Even though these isomorphic substitutions bear no net structural charge, they could modify adsorption properties of imogolite through changes in their surface reactivity. Until now, no studies have addressed these potential effects. The only study reporting different physical properties of Fe-doped imogolite is one by Alvarez-Ramirez (2009) that uses density functional theory (DFT)–based electronic structure calculations. These authors reported a decrease in the band gap from 4.6 to 2.6 eV for the silicon and from 4.2 to 1.0 eV for germanium, which puts these Fe-doped imogolites down to the semiconductor band gap range. This fact, as well as recent advances in the synthesis methods (Levard et al., 2008, 2009b), make imogolite a potential candidate for applications in electronics, a field that remains to be further explored.

Natural imogolite does not bear any structural charge originating from isomorphic substitutions ($\sigma_0=0$) according to the structure published by Cradwick et al. (1972) from electron diffraction experiments, which are corroborated by using NMR (Goodman et al., 1985; Ildefonse et al., 1994). However, in spite of the lack of isomorphic substitutions, Gustafsson (2001) proposed that a weak positive surface charge is present in the outer surface, a hypothesis later tested and confirmed by Guimaraes et al. (2007) using electronic structure calculations based on the DFT. This charge is caused by steric effects related to the curved structure that induce a nonsymmetric repartition of the electronic density around Si^{4+} and Al^{3+} cations, with longer Al–O bond lengths at the external surface of the gibbsitelike layer. Quantification of this weak permanent charge by DFT methods yields values for the calculated Mulliken charges that are only 6% higher for the inner H atoms than for the external ones. Experimental confirmation of the existence of this permanent charge, however, remains a challenge. One possible method is evaluation of the net adsorbed ion charge, Δq, determined from adsorption experiments, at a value of the net adsorbed proton charge, σ_H, equal to zero. The electroneutrality condition yields:

$$\Delta q = -\sigma_H - \sigma_0 \tag{9.1}$$

The application of this method for the case of imogolite is presented in Fig. 9.1. Potentiometric titration data from Karube et al. (1998) is combined with a net adsorbed ion charge measured by Tsuchida et al. (2005). The intercept at $\sigma_H = 0$ results in a value of $\sigma_0 = -16 \pm 10$ mmol$_c$ kg^{-1}.

This value is negligible if compared with typical values of permanent charge of other clay minerals such as smectites or vermiculites, with values ranging from about 800 to 2000 mmol$_c$ kg^{-1} (Sposito et al., 1998). Gustafsson (2001) links the existence of this permanent charge with the ability to adsorb anions; the presence of defects in the structure or the ill-coordinated atoms at the tube ends are other mechanisms usually used to explain the adsorption of anions. Another structural feature that should not be ignored is point defects. The existence of different types of vacancies in both the octahedral and tetrahedral layers has been described by Yucelen et al. (2012a) for synthetic imogolite samples (Fig. 9.2). These vacancies, as well as the consequent bond undersaturation and proton complexation, are potentially at the origin of a permanent charge. However, the low values for

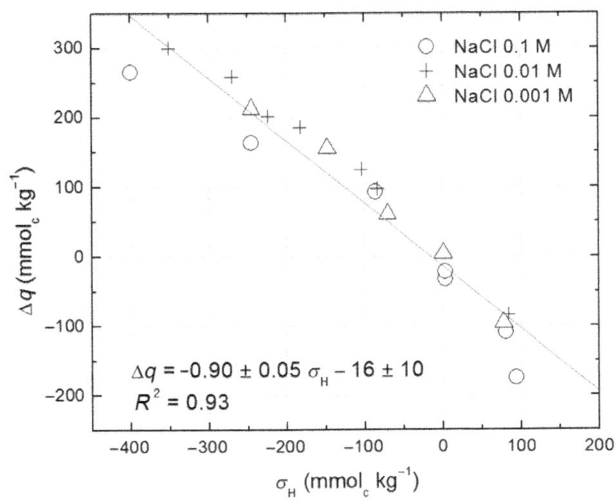

FIG. 9.1 Plot of the net adsorbed ion charge against the net adsorbed proton charge for natural imogolite, constructed with data from Tsuchida et al. (2005) and Karube et al. (1998). A specific surface area of 818 m^2 g^{-1} estimated from a hydrated (15% mass/mass) atomistic model of imogolite has been considered.

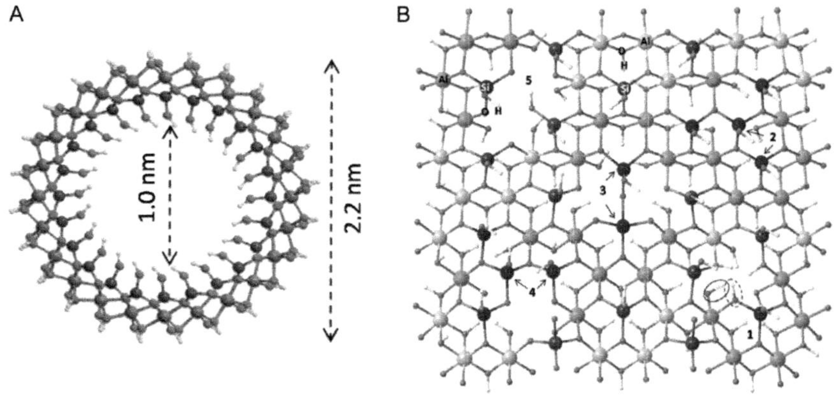

FIG. 9.2 (A) Section view of the imogolite structure. Hydrogen (*white*), oxygen (*red; grey* in the print version), aluminium (*green* (*light grey* in the print version)) and silicon (*blue* (*dark grey* in the print version)). (B) Wall structure of the imogolite nanotube (unrolled), showing various types of defects (coloured areas (*light* and *dark grey* in the print version)) identified from NMR experiments. *Adapted with permission from Yucelen et al. (2012a). Copyright 2011 American Chemical Society.*

σ_0 obtained from the plot in Fig. 9.1 indicate that these vacant sites occur at very low concentrations in natural imogolite.

In order to ascertain the net proton charge of individual surfaces, and thus the different reactivity of the internal and external surfaces, some authors have established parallelisms between charging phenomena at the external layers of imogolite and gibbsite (Arai et al., 2006). However, there is some controversy about the potentiometric titration results from gibbsite. While Hiemstra and Van Riemsdijk (1999) proposed that $>AlOH^0$ groups are mainly uncharged over a wide range of pH, Bickmore et al. (2004) reported acidity constants for the protonation of different $>AlOH^0$ sites ranging from pK_a −5.1 to 10.8. In addition, the reactivity of $>AlOH^0$ sites forming the imogolite outer layer is expected to differ from that of gibbsite due to the curvature of the structure and the slightly different Al–O bond lengths and different water structures. A useful model accounting for surface structure in acid site reactivity is the MUSIC model (Hiemstra et al., 1989). This model uses the undersaturation or oversaturation of surface oxygen atoms from the bond-valence model to calculate acidity constants of surface sites. Application of this method to the defect-free octahedral layer of imogolite, using molecular modelling, yields a different reactivity than for gibbsite, which is one pH-unit more acidic (Fernandez-Martinez, 2009). This difference is attributed to differences in the hydrogen bond network around surface oxygen atoms. The parallelism between gibbsite (planar) and imogolite external layer (curved) is therefore not applicable, and more structural information about imogolite–water interfaces is needed to better constrain acidity and complexation models.

Potentiometric titration is the most common method used to determine the net adsorbed proton charge at interfaces of mineral colloids in aqueous dispersions. Titration of imogolite dispersions results in a development of a positive net proton charge at pH values lower than 7—the so-called point of zero net proton charge (PZNPC). These results fall between values obtained for natural imogolite dispersions, which range from PZNPC = 6 (Tsuchida et al., 2005) to PZNPC = 8.4 (Su et al., 1992). The origin of these different values could be due to differences in the purification methods for natural imogolite, which is usually difficult to separate from other minerals derived from soil aggregates and weathered horizons.

9.2.2 Electrokinetic Phenomena

Surface charge is also at the origin of electrokinetic phenomena observed in imogolite dispersions. Early measurements of surface charge characteristics by electrophoretic measurements were carried out by Horikawa (1975a,b) on natural samples, with the goal of separating allophane from imogolite in natural mixtures, showing that natural imogolite displayed an isoelectrical point (IEP; ie, the pH value at which the electrophoretic mobility is zero) of around 10. This value was confirmed later for both natural and synthetic samples (Harsh et al., 1992; Karube et al., 1992; Su and Harsh, 1993; Tsuchida et al., 2005) (Fig. 9.3). Such a feature appears truly typical of imogolite and was recently proposed as an easy way to assess the quality of synthesis

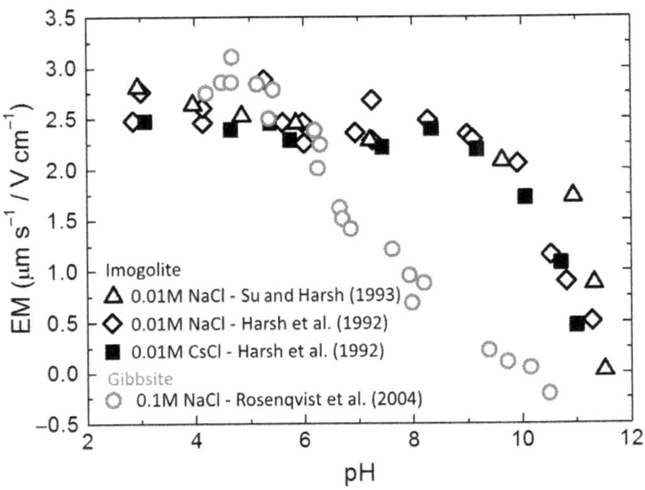

FIG. 9.3 Electrophoretic mobility (EM) of imogolite and gibbsite as a function of pH (Harsh et al., 1992; Su and Harsh, 1993; Rosenqvist et al., 2002).

procedures (Arancibia-Miranda et al., 2011). Potentiometric titration experiments (Harsh et al., 1992) revealed values of point of zero salt effects (PZSE; ie, the pH value at which potentiometric titration curves performed at different ionic strengths intersect) that were less than the IEP by at least 3 pH units, a feature that was less marked for allophanes of similar Al/Si ratios (allophane has a different atomic structure than imogolite; different electrokinetic and charging phenomena are expected). This suggests strongly that the tubular nature of imogolite plays a significant role in determining its surface charge characteristics. As far as the difference between IEP and PZSE is concerned, the most likely explanation is that electrophoretic mobility measurements that probe the electrical charge at the hydrodynamical slip plane are not sensitive to the silanol groups present in the internal pore, whereas potentiometric titrations that probe the surface charge are sensitive to their presence (Tsuchida et al., 2005). The dissociation pK_a of silanol groups is much less than that of aluminol groups (Tossell and Sahai, 2000), which leads to an upward shift of the IEP. Electrokinetic and surface charge properties are linked to the presence of the structural charge discussed in the previous section. Since its presence is not entirely understood, questions remain open regarding the origin of such phenomena.

A very informative comparison between the electrophoresis mobility of gibbsite and imogolite as a function of pH is shown in Fig. 9.3. The positive mobility of imogolite in a wide pH range is similar to that of gibbsite. The difference in behaviour, with a plateau observed at high pH values for imogolite, could be due to the ability to form neutral ion pairs at the >AlOH sites of the gibbsite basal plane reported by Rosenqvist et al. (2002), which could be different for imogolite. The mobility of gibbsite is actually explained by these authors as resulting from a charge present at the particle edges. From this perspective, and considering that edge sites are only a very small fraction of the total number of imogolite surface sites due to the high aspect ratio, it is possible that $>AlOH^0$ sites on the imogolite surface have a greater ability to form charged complexes than the same sites at the gibbsite surface. More work remains to be done in order to ascertain charging phenomena for both surfaces of imogolite.

9.3 CHEMISORPTION AND PHYSISORPTION

In view of its peculiar structure, imogolite is clearly a very specific adsorbent that may display unusual properties. The thermal properties of imogolite that control the way under which adsorption experiments are performed will be briefly described here, with a focus on the adsorption of nonspecific probes that provide information about the textural properties of imogolite in terms of surface area and pore volume. The adsorption of reactive molecules that probe the acid–base properties of the materials will then be analysed before focusing on the specific case of H-bonded adsorbates, particularly H_2O, and on the adsorption of metals and metalloids.

9.3.1 Gas Adsorption Properties

9.3.1.1 Determination of Textural Properties Through Adsorption of Nonreactive and Non-H-Bonded Molecules

In view of the particular structure of imogolite, its gas adsorption properties were studied soon after its discovery, with the goal of better assessing the potential of this material in various applications such as molecular sieves, catalysis and remediation. One of the main problems associated with gas adsorption studies of imogolite materials is linked to the fact that under atmospheric conditions, the inner part (lumen) of the tubes is filled with structured water molecules, the removal of which is required to properly assess the accessible surface area. This can be achieved by pretreating the material at elevated temperatures under vacuum. However, the temperature used for outgassing must remain lower than that related to the collapse of the structure. Studying the thermal properties of imogolite, then, has been recognized as an important issue almost since the discovery of this clay mineral. The use of thermogravimetric curves was even proposed to determine the imogolite content in andosols (Wada and Tokashiki, 1972; Horikawa et al., 2002). Dehydration and dehydroxylation mechanisms of imogolite nanotubes have been examined in numerous studies combining thermogravimetric analysis and differential thermal analysis, ^{29}Si and ^{27}Al NMR and IR spectroscopy (Van der Gaast et al., 1985; Wilson et al., 1988; MacKenzie et al., 1991; Kang et al., 2010). Beyond 250°C, dehydroxylation reversibly transforms Q_3 silicons into Q_4 and irreversibly transforms octahedral aluminium into pentacoordinated and tetrahedral aluminium. Dehydroxylation at temperatures higher than 400°C results in partial pore collapse. Heat treatment under vacuum at 250°C–300°C, then, appears as an optimal pretreatment condition for maximizing the accessible pore volume while maintaining the integrity of the nanotubular structure. Further difficulties in assessing textural properties are associated with the fact that at least three types of pores can be envisioned in imogolite powder, the surface of which will be probed by gas adsorption measurements: (i) internal pores; (ii) intertubular pores, because the tubes are almost systematically associated as bundles; and (iii) interbundle pores. Early measurements on a natural imogolite sample by Egashira and Aomine (1974) revealed that the maximal BET surface area derived from nitrogen adsorption measurements at 77 K ($\approx$400 m^2 g^{-1}) was obtained after outgassing at 300°C under vacuum and that the adsorption isotherms were of type I according to the IUPAC classification (Gregg and Sing, 1982), showing that imogolite is predominantly microporous (ie, with pore diameters less than 2 nm). It must then be emphasized that the surface area derived from the BET equation must be considered as an equivalent surface area for microporous samples such as imogolite, as the assumptions underlying the BET equation are not adapted to such materials. The first measurements on synthetic imogolite were performed by Adams (1980) using both volumetric and chromatographic measurements.

A similar nitrogen surface area was obtained as that previously discussed by Egashira and Aomine (1974), and a clear effect of gaseous probe size on adsorption was observed, a bulky molecule such as perfluorotributylamine with a kinetic diameter of 1.02 nm (Ackerman et al., 1993) being nearly nonadsorbed on synthetic imogolite. Nitrogen adsorption measurements were also carried out by Werpy et al. (1989, 1990) on the so-called tubular silicate layer silicate complex (TSLS; Johnson et al., 1988); ie, a sample where synthetic imogolite was intercalated in the interlayer of a swelling clay mineral. In such conditions, interbundle pores should not be present, and gas adsorption measurements revealed two types of pores: internal and intertubular pores that were both in the micropore range. A detailed study of nitrogen adsorption on synthetic Al–Si and Al–Si–Ge synthetic imogolite was carried out by Ackerman et al. (1993). The importance of outgassing temperature was also revealed in this study, where synthetic samples displayed significantly higher adsorption capacities than their natural counterparts (ie, after outgassing at 275°C, the specific surface areas (SSA) were around 400 and 300 $m^2\ g^{-1}$ for synthetic and natural samples, respectively), as displayed in Fig. 9.4. This difference was assigned to the better ordering of synthetic samples (also revealed by their lower mesoporosity) that consequently possess more intertubular pores. In addition, the germanium sample displayed higher pore volume in agreement with its larger pore opening. Still, in the absence of a fully consistent model to describe pore size distribution in microporous samples, no exact pore size distribution could be deduced from these experiments, even if pore sizes around 4 and 9 Å could be deduced. Methane adsorption experiments were also performed in this work and revealed tendencies similar to those deduced from nitrogen adsorption. Since these early works, the SSA of synthetic samples have been measured in various publications, mainly using nitrogen adsorption at 77 K after outgassing at ≈300°C, and have revealed SSA most often around 400 $m^2\ g^{-1}$, but values as high as 1400 $m^2\ g^{-1}$ have also been reported (Parfitt and Henmi, 1980).

9.3.1.2 Determination of Acid–Base Properties Through Adsorption of Reactive Molecules

To assess the catalytic potential of various materials, it is extremely relevant to test the adsorption of different reactive molecules that probe the acidic and basic sites of the materials, and in particular the type and abundance of acidic Lewis and Brønsted species. From a catalytic point of view, the most interesting features of imogolite-type materials are their nanoporous structure and the inner surface rich in SiOH groups (Imamura et al., 1993, 1996). Indeed, external, octahedrally coordinated Al atoms should not exhibit any significant acidity, in contrast with inner SiOH, which is expected to be more reactive given the low-deprotonation pK_a and the high electrostatic potential in the inner pore. Such studies have been carried out since 2009, mainly by the Italian

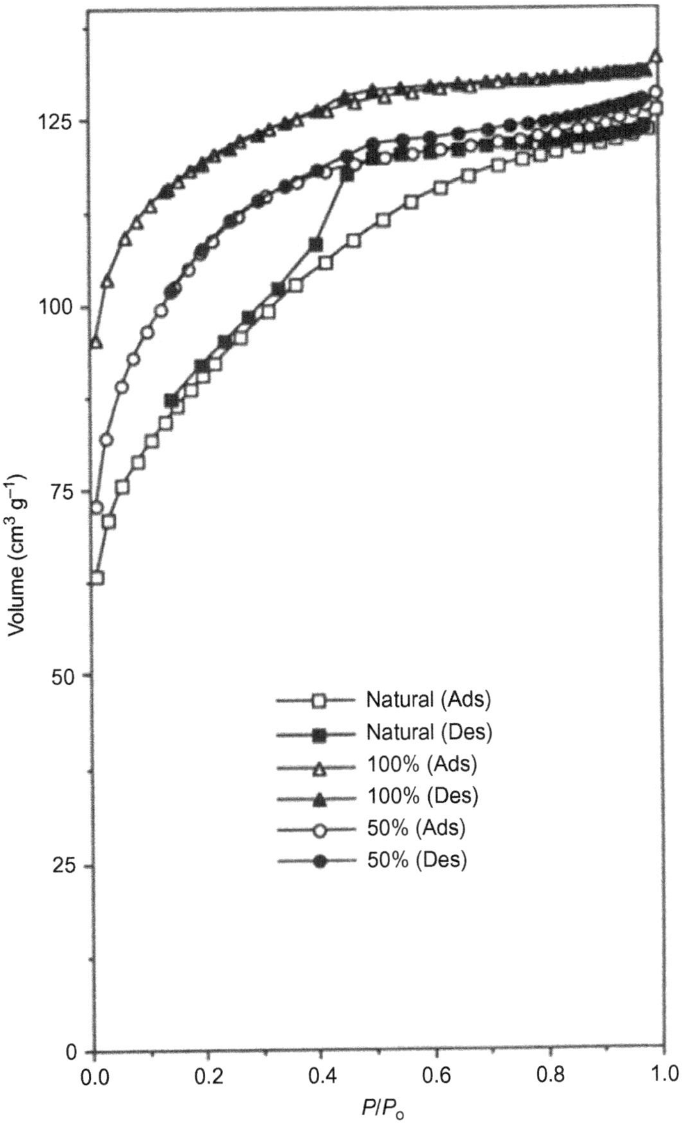

FIG. 9.4 Nitrogen adsorption/desorption isotherms at 77 K for natural and synthetic imogolites (100% Si and 50% Si). *Reprinted with permission from Ackerman et al. (1993). Copyright 1993 American Chemical Society.*

team of Politecnico di Torino using infrared studies of various probes including NH_3, CO and CO_2 (Bonelli et al., 2009, 2013a,b; Bottero et al., 2011; Zanzottera et al., 2012a,b). Different infrared spectra are obtained for imogolite outgassed at 300°C after the adsorption of NH_3 (Fig. 9.5A) and for CO

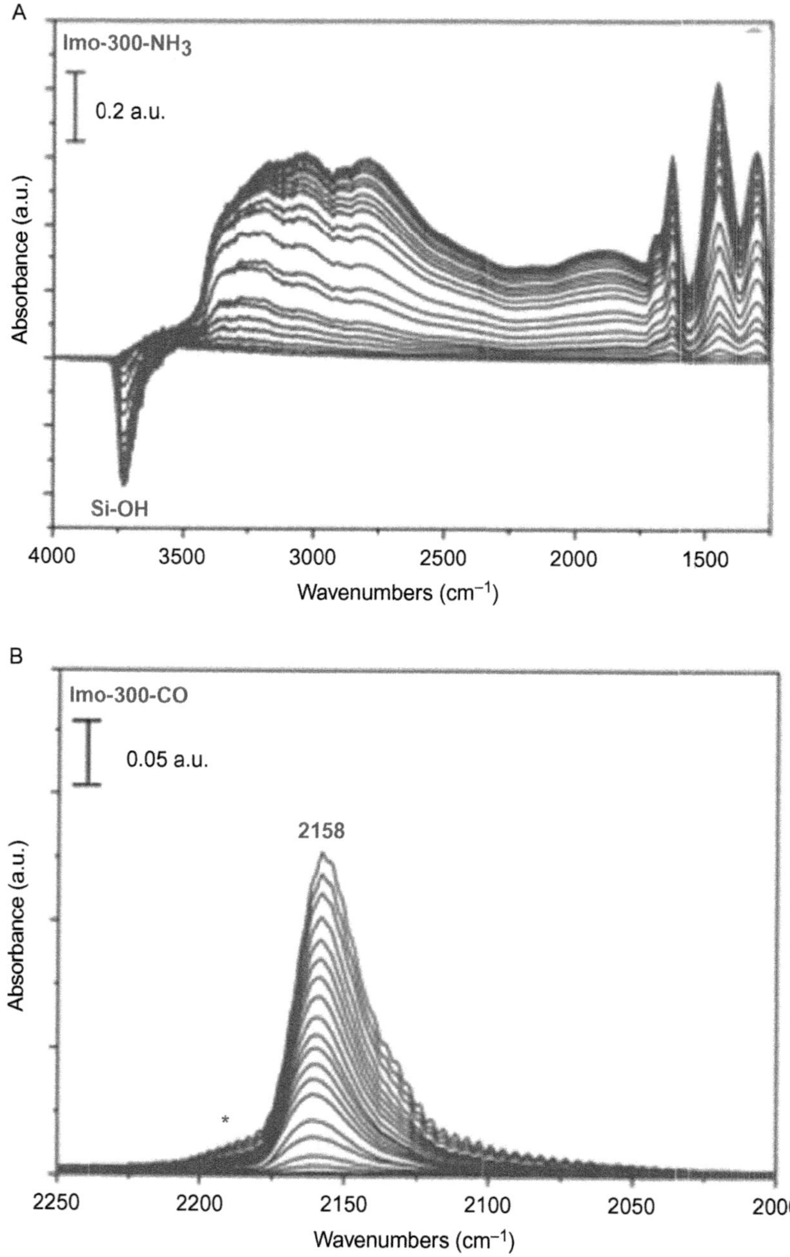

FIG. 9.5 Fourier transform infrared spectra of imogolite obtained after dosing (A) NH_3 and (B) CO, in the 0–20 mbar pressure range. *Reprinted with permission under the Creative Commons Attribution License from Zanzottera (2012).*

(Fig. 9.5B). The interaction of gaseous ammonia at high equilibrium pressure in the OH stretching region reveals the interaction of NH_3 molecules with silanol groups, as the peak at 3745 cm^{-1} decreases due to the formation of NH_4^+ groups. A new broad band located at 3200–3000 cm^{-1} appears and indicates extensive H-bonding. An additional band at 2750 cm^{-1} also appears, which can be assigned to the stretching mode of the NH group of the ammonium species under its monosolvated form $N_2H_7{}^+$ and is observed on protonic zeolites (Zanzottera, 2012). This is further confirmed by the shift of the 1450 cm^{-1} band towards 1460 cm^{-1} at higher ammonia pressure, a feature generally assigned to the formation of ammonium species on Brønsted acidic sites. Finally, the bands at 1625 and 1300 cm^{-1} are assigned to the asymmetric and symmetric bending mode of gaseous NH_3 molecules coordinated with external Al^{3+} Lewis sites. This confirms that significant acidic sites associated to silanol groups are present in imogolite. As far as basic sites are concerned, CO is a relevant probe. Only rather weak interactions between CO and imogolite can be deduced from the data displayed in Fig. 9.5B, which confirms that in terms of catalytic activity, imogolite could likely be used as an acidic catalyst. Still, catalytic tests using either methanol or phenol revealed a non-negligible conversion of methanol, but a negligible degree of phenol conversion, likely due to the difficulty of phenol molecules to access the whole pore space of imogolite. Still, the potential of imogolite as a catalyst clearly deserves further investigation, and the role of confinement of its acid–base properties remains to be studied in detail.

9.3.1.3 Water and H-Bonded Liquid Adsorption on Imogolite

In view of its structure, the way in which hydrogen-bonded liquids (and particularly water) adsorb on imogolite is particularly interesting. Of particular interest from the fundamental point of view is the study of the structure and dynamics of water adsorbed in the internal pore of imogolite. This has been described from proton NMR relaxometry data as a case of almost-ideal 1D diffusion (Belorizky et al., 2010). Furthermore, recent developments in synthetic procedures using either silicon or germanium provide variable tube diameters and then a wide range of 1D confinement (Levard et al., 2008, 2009b). On the other hand, the external surface of imogolite provides a perfect model to study the influence of curvature on water adsorption and water structure at oxide surfaces.

Early measurements of water adsorption by natural imogolite were carried out by Wada and Yoshinaga (1969) and Wada and Henmi (1972). They revealed very significant water uptake with strong adsorption at low relative pressure and adsorbed amounts of more than 30 g of water per 100 g of imogolite for relative water pressures around 0.85. Surprisingly, since then, very few studies have provided experimental measurements of water uptake in these materials. A recent study of Konduri et al. (2008) combined experiments and grand canonical Monte Carlo simulations to better understand water uptake in single-walled imogolite nanotubes. Their computed and experimental adsorption isotherms (Zang et al., 2010) exhibited strong upward inflections at low partial pressures, and most adsorbed molecules appeared to be located in the pores of the nanotube, displaying remarkable hydrophilicity. This strong hydrophilicity of the tube lumen was also evident in the molecular dynamics (MD) simulations of Creton et al. (2008a,b) in which water molecules inside the nanotube were shown to interact strongly with the silanol surface, building a first layer with very low mobility For water concentrations corresponding to relative humidities up to 40%, no significant diffusion along the tube axis was calculated. In addition, more recent simulations (Zang et al., 2010) showed that the flexibility of the inner silanol groups played a significant role in the way in which hydrogen-bonded molecules adsorb in this region. These simulations predicted a very strong selectivity of water vs methanol in terms of adsorption that could be of some use for potential applications such as alcohol dehydration. Diffusivities for water, methanol and ethanol molecules inside the tubes were also computed by MD simulations (Zang et al., 2009) in a rigid network of either Si-imogolite or Ge-imogolite. Diffusion in the latter was higher due to the larger pore diameter. In filled tubes, diffusion coefficients for water are significantly reduced in comparison to their bulk value, whereas mixtures of water and methanol displayed a segregation behaviour in which water is preferentially located closer to tube wall. In that regard, experiments in quasi-elastic neutron-scattering could be of interest to better constrain the diffusion coefficients deduced from simulation studies, especially considering their expected high anisotropy.

In contrast, water adsorption on the external surfaces of imogolite tubes display different behaviour. This is first suggested by the shape of the experimental adsorption isotherms (Wada and Henmi, 1972; Konduri et al., 2008; Kang et al., 2011). Indeed, the isotherms strongly flatten after the initial step related to water adsorption inside the tube. The external surface of imogolite then appears as rather hydrophobic. This was confirmed by molecular simulations carried out by Creton et al. (2008a) who concluded that 'water interacts weakly with the external aluminium hydroxide surface, which leads to water clustering in the external void. The molecules are highly mobile and their structural and dynamical characteristics strongly resemble those in the bulk liquid water. The behaviour of intertubular water reveals no dependence on water content (relative humidity). All these features permit to qualify the external imogolite surface as hydrophobic'. The long-time dynamics of this water was also analysed by NMR relaxometry (Levitz et al., 2008), which revealed that the motions of water along the tube could be analysed by intermittent Brownian dynamics with relatively short ($\approx$4 ns) residence time of water molecules at the imogolite surface. This relative hydrophobicity of the external surface of imogolite may explain some of the geochemical features of this clay mineral, such as strong interaction with natural organic matter and the ability to form bundles (Fernandez-Martinez, 2009). Furthermore, the hydrophilic or hydrophobic character of imogolite internal and external surfaces, respectively, could provide advantages for potential applications in contaminant remediation or drug delivery systems. Such a point clearly deserves further investigations both in terms of experiments and simulations.

9.3.2 Metal/Metalloid Adsorption in the Liquid Phase

The high surface area and amphoteric character of imogolite surfaces make it an important component of the soil where it is present. There is general agreement that imogolite is precipitated in situ in B horizons of andosols, where there is a low content of Al-complexing organic compounds and where the low-leaching rate favours the polymerization of Al and its coprecipitation with monomeric Si (Ugolini and Dahlgren, 1991). A biotic route has also been proposed for the formation of imogolite directly from plagioclase (Tazaki et al., 2006), from observations of imogolite-bearing biofilms at the surface of incubated pumice grains. Imogolite is thus formed in wet environments of volcanic regions, where it interacts with toxic ions and nutrients, affecting their biogeochemical cycling. One example of imogolite interaction with phosphate, an important nutrient, has been described by Theng et al. (1982) and Parfitt et al. (1974). An adsorption capacity of 120–250 μmol/g was reported. This high adsorption capacity, which is equivalent to half of the surface sites of the imogolite surface populated by phosphate molecules, contrasts with values for adsorption capacity of iron oxides, which are four orders of magnitude lower (Geelhoed et al., 1997; Antelo et al., 2010). The studies from Theng et al. (1982) and Parfitt et al. (1974) suggest that ligand-exchange reactions with surface aluminol groups are responsible for most of the adsorption. However, the high adsorption levels reached suggest that adsorption at the diffuse ion swarm may also be present.

A similar type of ligand-exchange mechanism has been reported by Arai et al. (2006) for the adsorption of uranyl at the imogolite–water interface. These authors used X-ray absorption spectroscopy and wet chemistry methods to elucidate the structure of single uranyl and ternary uranyl-carbonate surface complexes (Fig. 9.6). The short U–Al interatomic distances found (~3.3 Å) suggest the formation of bidentate mononuclear complexes, with different coordination environments for the carbonate molecules. An outer-sphere complex, where the uranyl is adsorbed via electrostatic interactions through carbonate bridges, was also identified. These multiple structures highlight the potential of the imogolite surface to display multiple binding environments due to its high surface area and high surface density on surface sites (~18 surface OH/nm^2, according to the structure reported by Cradwick et al., 1972).

A different adsorption mechanism was identified by Levard et al. (2009a) for Ni in natural imogolite samples and in synthetic Ge-imogolite samples, used as analogues. The goal of the study was to elucidate the role of poorly

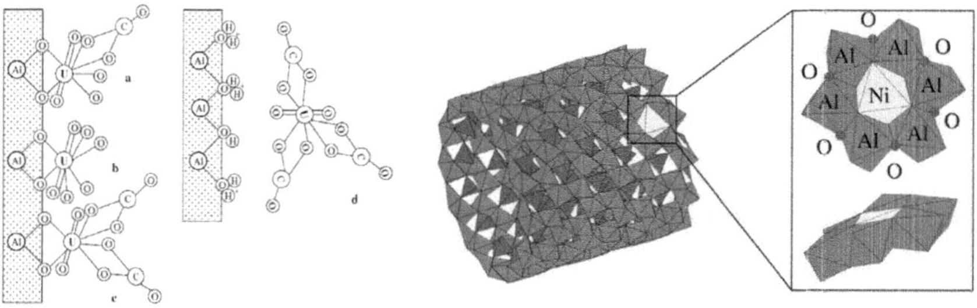

FIG. 9.6 *Left*: schematic representation of the uranyl-carbonate complexes identified by Arai et al. (2006) at the imogolite–water interface from X-ray absorption spectroscopy data. (A) U(VI)-monocarbonato ternary complex via bidentate mononuclear U(V)–O–Al linkage. (B) Bidentate mononuclear binary U(VI) surface complex. (C) bis-Carbonato U(VI) ternary complex via bidentate mononuclear U(VI)–O–Al linkage. (D) Planar tris-Carbonato U(VI) outer-sphere surface complex (with three carbonate groups facing the surface of the imogolite outer wall) (Arai et al., 2006). *Right*: polyhedral representation of the imogolite structure, highlighting the incorporation mechanism of a Ni^{2+} ion into an Al^{3+} vacancy site. *Reprinted from Levard et al. (2009a). Copyright 2009, with permission from Elsevier.*

crystalline aluminosilicate phases present in andosols from Reunion Island as Ni scavengers. Reunion Island soil displays high Ni concentrations when compared with other soil, with average levels of 206 mg kg^{-1}, as opposed to an average level of 22 mg kg^{-1} in soil found elsewhere in the world. Therefore, an assessment of the mobility of Ni in this soil, including the adsorption mechanisms onto mineral phases, was necessary. By comparing X-ray absorption spectroscopy data from natural samples with data from the Ge-analogues, Levard et al. (2009a) revealed an incorporation mechanism of Ni into the gibbsitelike imogolite layer in the vacancies of Al^{3+} (Fig. 9.6). This mechanism is similar to that observed in studies of Ni incorporation onto other mineral phases, where the formation of mixed Ni–Al hydroxide phases was identified (Scheidegger et al., 1997). It shows the potential of the imogolite structure to accommodate other ions in its structure via a mechanism initiated by adsorption process.

The adsorption of different bivalent metals by imogolite precursors was also studied by Denaix et al. (1999) using potentiometric titrations. The competition between protons and metallic cations can be measured using this method by quantifying the release of protons to the solution in titrations made in the presence of metallic cations. The results show that while Cd^{2+} does not show any specific effect, Cu^{2+} and Pb^{2+} have a strong affinity for imogolite adsorption sites. The adsorption mechanism for Cu^{2+} was studied by Clark and McBride (1984a,b). A specific adsorption at binuclear AlOH sites was described by electron spin resonance. Based on similar studies about the incorporation of metallic ions in vacancies of manganese oxide nanoparticles that show a size dependence of the adsorption/incorporation mechanism

(Peña et al., 2010), it could be hypothesized that the incorporation of metallic cations with ionic sizes similar to Ni^{2+} could happen through a similar incorporation mechanism as that shown by Levard et al. (2009a).

Monovalent ion adsorption was studied by Clark and McBride (1984a,b), who reported the retention of Cl^- and ClO_4^- by imogolite at solution pH values where a positive charge is not observed. This phenomenon was not observed for a large organic spin probe, 3-carboxy-2,2,5,5-tetramethyl-1-pyrrolidinyloxyl –COOH, which suggests that concomitant intercalation of Na^+ and Cl^- in the tubes could explain the observed enhanced Cl^- uptake by a charge compensation effect of Na^+ in the inner pores. This effect was thoroughly investigated by Su et al. (1992), who observed the same effect in the absence of Na^+ adsorption. These authors concluded that a specific adsorption mechanism for Cl^- could be at play, with the formation of strong and immobile inner-sphere complexes. Guerra et al. (2011) studied the retention of Rb^+ by imogolite in the framework of a study on radionuclide retention, revealing a high adsorption capacity. These results show the potential of imogolite for its use in desalination technologies, and as potential scavenger of monovalent radionuclides. The widespread occurrence of allophane and imogolite in Japanese soil and the results from the abovementioned studies make imogolite a potential cheap alternative for radionuclide trapping that deserves more attention.

9.4 CONCLUSIVE REMARKS

Imogolite and imogolite-like materials have been the subject of renewed interest in the last years due to their potential industrial applications in many fields (such as catalysis, water decontamination and gas storage). Recent advances in synthesis methods (Levard et al., 2008, 2009b) and better understanding of the formation mechanisms (Yucelen et al., 2011, 2012b) are promising milestones that could help with further development. However, much remains to be done regarding fine synthesis details, such as the control of vacancy concentration or texture. Results from ion adsorption in the liquid phase have described different adsorption mechanisms, including the incorporation of Ni^{2+} into Al^{3+} vacant sites (Levard et al., 2009a). While the precise charge compensation mechanism remains to be described, this observed substitution opens the door to the fabrication of mixed-metal, imogolite-like materials with different electronic properties, as suggested by theoretical studies (Alvarez-Ramirez, 2009). The control of imogolite hydrophilicity is another milestone that was reached recently by a chemical modification of the inner SiOH groups by hydrophobic moieties (Bonelli et al., 2013b; Amara et al., 2015). This discovery opens the door for applications in the trapping of organic pollutants in the liquid and gas phases. All these advances, as well as the increasing interest in imogolite in the materials science literature, are very promising for the use of imogolite-like materials for various technological purposes.

REFERENCES

Ackerman, W.C., Smith, D.M., Huling, J.C., Kim, Y.W., Bailey, J.K., Brinker, C.J., 1993. Gas vapor adsorption in imogolite—a microporous tubular aluminosilicate. Langmuir 9, 1051–1057.

Adams, M.J., 1980. Gas chromatographic adsorption studies on synthetic imogolite. J. Chromatogr. A 188, 97–106.

Alvarez-Ramirez, F., 2009. Theoretical study of (OH)(3)N2O3MOH, M=C, Si, Ge, Sn and N=Al, Ga, In, with imogolite-like structure. J. Comput. Theor. Nanosci. 6, 1120–1124.

Amara, M.S., Paineau, E., Rouzière, S., Guiose, B., Krapf, M.-E.M., Taché, O., Launois, P., Thill, A., 2015. Hybrid, tunable-diameter, metal oxide nanotubes for trapping of organic molecules. Chem. Mater. 27, 1488–1494.

Antelo, J., Fiol, S., Pérez, C., Mariño, S., Arce, F., Gondar, D., López, R., 2010. Analysis of phosphate adsorption onto ferrihydrite using the CD-MUSIC model. J. Colloid Interface Sci. 347, 112–119.

Arai, Y., McBeath, M., Bargar, J.R., Joye, J., Davis, J.A., 2006. Uranyl adsorption and surface speciation at the imogolite-water interface: self-consistent spectroscopic and surface complexation models. Geochim. Cosmochim. Acta 70, 2492–2509.

Arancibia-Miranda, N., Escudey, M., Molina, M., Teresa Garcia-Gonzalez, M., 2011. Use of isoelectric point and pH to evaluate the synthesis of a nanotubular aluminosilicate. J. Non Cryst. Solids 357, 1750–1756.

Avellan, A., Levard, C., Kumar, N., Rose, J., Olivi, L., Thill, A., Chaurand, P., Borschneck, D., Masion, A., 2014. Structural incorporation of iron into Ge-imogolite nanotubes: a promising step for innovative nanomaterials. RSC Adv. 4, 49827–49830.

Baker, L.L., Nickerson, R.D., Strawn, D.G., 2014. XAFS study of Fe-substituted allophane and imogolite. Clays Clay Miner. 62, 20–34.

Belorizky, E., Fries, P.H., Guillermo, A., Poncelet, O., 2010. Almost ideal 1D water diffusion in imogolite nanotubes evidenced by NMR relaxometry. Chemphyschem 11, 2021–2026.

Bickmore, B.R., Tadanier, C.J., Rosso, K.M., Monn, W.D., Eggett, D.L., 2004. Bond-valence methods for pKa prediction: critical reanalysis and a new approach. Geochim. Cosmochim. Acta 68, 2025–2042.

Bonelli, B., Bottero, I., Ballarini, N., Passeri, S., Cavani, F., Garrone, E., 2009. IR spectroscopic and catalytic characterization of the acidity of imogolite-based systems. J. Catal. 264, 15–30.

Bonelli, B., Armandi, M., Garrone, E., 2013a. Surface properties of alumino-silicate single-walled nanotubes of the imogolite type. Phys. Chem. Chem. Phys. 15, 13381–13390.

Bonelli, B., Zanzottera, C., Armandi, M., Esposito, S., Garrone, E., 2013b. IR spectroscopic study of the acidic properties of alumino-silicate single-walled nanotubes of the imogolite type. Catal. Today 218, 3–9.

Bottero, I., Bonelli, B., Ashbrook, S.E., Wright, P.A., Zhou, W., Tagliabue, M., Armandi, M., Garrone, E., 2011. Synthesis and characterization of hybrid organic/inorganic nanotubes of the imogolite type and their behaviour towards methane adsorption. Phys. Chem. Chem. Phys. 13, 744–750.

Brown, G.E., Henrich, V.E., Casey, W.H., Clark, D.L., Eggleston, C., Felmy, A., Goodman, D.W., Grätzel, M., Maciel, G., McCarthy, M.I., Nealson, K.H., Sverjensky, D.A., Toney, M.F., Zachara, J.M., 1999. Metal oxide surfaces and their interactions with aqueous solutions and microbial organisms. Chem. Rev. 99, 77–174.

Clark, C.J., McBride, M.B., 1984a. Chemisorption of Cu(II) and Co(II) on allophane and imogolite. Clays Clay Miner. 32, 300–310.

Clark, C.J., McBride, M.B., 1984b. Cation and anion retention by natural and synthetic allophane and imogolite. Clays Clay Miner. 32, 291–299.

Cradwick, P.D., Wada, K., Russell, J.D., Yoshinag, N., Masson, C.R., Farmer, V.C., 1972. Imogolite, a hydrated aluminum silicate of tubular structure. Nat. Phys. Sci. 240, 187–189.

Creton, B., Bougeard, D., Smirnov, K.S., Guilment, J., Poncelet, O., 2008a. Molecular dynamics study of hydrated imogolite. 1. Vibrational dynamics of the nanotube. J. Phys. Chem. C 112, 10013–10020.

Creton, B., Bougeard, D., Smirnov, K.S., Guilment, J., Poncelet, O., 2008b. Molecular dynamics study of hydrated imogolite—2. Structure and dynamics of confined water. Phys. Chem. Chem. Phys. 10, 4879–4888.

Denaix, L., Lamy, I., Bottero, J.Y., 1999. Structure and affinity towards Cd2+, Cu2+, Pb2+ of synthetic colloidal amorphous aluminosilicates and their precursors. Colloids Surf. A Physicochem. Eng. Asp. 158, 315–325.

Egashira, K., Aomine, S., 1974. Effects of drying and heating on the surface area of allophane and imogolite. Clay Sci. 4, 231–242.

Fernandez-Martinez, A., 2009. Physics of Natural Nanoparticles–Water Interfaces: Chemical Reactivity and Environmental Implications. Ph.D. Thesis, Université Joseph-Fourier, Grenoble (France).

Geelhoed, J.S., Hiemstra, T., Van Riemsdijk, W.H., 1997. Phosphate and sulfate adsorption on goethite: single anion and competitive adsorption. Geochim. Cosmochim. Acta 61, 2389–2396.

Goodman, B.A., Russell, J.D., Montez, B., Oldfield, E., Kirkpatrick, R.J., 1985. Structural studies of imogolite and allophanes by aluminum-27 and silicon-29 nuclear magnetic resonance spectroscopy. Phys. Chem. Miner. 12, 342–346.

Gregg, S.J., Sing, K.S.W., 1982. Adsorption, Surface Area & Porosity, second ed. Academic Press, London. http://dx.doi.org/10.1002/bbpc.19820861019/abstract. http://onlinelibrary.wiley.com/.

Guerra, D.L., Batista, A.C., Viana, R.R., Airoldi, C., 2011. Adsorption of rubidium on raw and MTZ- and MBI-imogolite hybrid surfaces: an evidence of the chelate effect. Desalination 275, 107–117.

Guimaraes, L., Enyashin, A.N., Frenzel, J., Heine, T., Duarte, H.A., Seifert, G., 2007. Imogolite nanotubes: Stability, electronic, and mechanical properties. ACS Nano 1, 362–368.

Gustafsson, J.P., 2001. The surface chemistry of imogolite. Clays Clay Miner. 49, 73–80.

Gutiérrez-Castorena, M.D.C., Stoops, G., Ortiz Solorio, C.A., López Avila, G., 2005. Amorphous silica materials in soils and sediments of the Ex-Lago de Texcoco, Mexico: an explanation for its subsidence. Catena 60, 205–226.

Harsh, J.B., Traina, S.J., Boyle, J., Yang, Y., 1992. Adsorption of cations on imogolite and their effect on surface-charge characteristics. Clays Clay Miner. 40, 700–706.

Hiemstra, T., Van Riemsdijk, W.H., 1999. Surface structural ion adsorption modeling of competitive binding of oxyanions by metal (hydr)oxides. J. Colloid Interface Sci. 210, 182–193.

Hiemstra, T., Dewit, J.C.M., Vanriemsdijk, W.H., 1989. Multisite proton adsorption modeling at the solid-solution interface of (hydr)oxides—a new approach. 2. Application to various important (hydr)oxides. J. Colloid Interface Sci. 133, 105–117.

Horikawa, Y., 1975a. Electrokinetic phenomena of aqueous suspensions of allophane and imogolite. Clay Sci. 4, 255–263.

Horikawa, Y., 1975b. Electrokinetic behavior in allophane-imogolite mixtures in alkaline media. Clay Sci. 4, 265–269.

Horikawa, Y., Kitamura, T., Honna, T., 2002. Semiquantitative determination of imogolite and Al-rich allophane (Si/Al = 1:2) in some volcanic ash soils in San' in region by a combination of thermogravimetry-differential thermal analysis and trimethylsilylation analysis of soil clays. Soil Sci. Plant Nutr. 48, 779–786.

Ildefonse, P., Kirkpatrick, R.J., Montez, B., Calas, G., Flank, A.M., Lagarde, P., 1994. ^{27}Al MAS NMR and aluminum X-ray absorption near edge structure study of imogolite and allophanes. Clays Clay Miner. 42, 276–287.

Imamura, S., Hayashi, Y., Kajiwara, K., Hoshino, H., Kaito, C., 1993. Imogolite: a possible new type of shape-selective catalyst. Ind. Eng. Chem. Res. 32, 600–603.

Imamura, S., Kokubu, T., Yamashita, T., Okamoto, Y., Kajiwara, K., Kanai, H., 1996. Shape-selective copper-loaded imogolite catalyst. J. Catal. 160, 137–139.

Johnson, I.D., Werpy, T.A., Pinnavaia, T.J., 1988. Tubular silicate-layered silicate intercalation compounds: a new family of pillared clays. J. Am. Chem. Soc. 110, 8545–8547.

Kang, D.-Y., Zang, J., Wright, E.R., McCanna, A.L., Jones, C.W., Nair, S., 2010. Dehydration, dehydroxylation, and rehydroxylation of single-walled aluminosilicate nanotubes. ACS Nano 4, 4897–4907.

Kang, D.-Y., Zang, J., Jones, C.W., Nair, S., 2011. Single-walled aluminosilicate nanotubes with organic-modified interiors. J. Phys. Chem. C 115, 7676–7685.

Karube, J., Nakaishi, K., Sugimoto, H., Fujihira, M., 1992. Electrophoretic behavior of imogolite under alkaline conditions. Clays Clay Miner. 40, 625–628.

Karube, J., Sugimoto, H., Fujihira, M., Nakaisbi, K., 1998. Stability and charge characteristics of allophane and imogolite. Trans. Jpn. Soc. Irrig. Drain. Reclam. Eng. 196, 673.

Konduri, S., Tong, H.M., Chempath, S., Nair, S., 2008. Water in single-walled aluminosilicate nanotubes: diffusion and adsorption properties. J. Phys. Chem. C 112, 15367–15374.

Levard, C., Rose, J., Masion, A., Doelsch, E., Borschneck, D., Olivi, L., Dominici, C., Grauby, O., Woicik, J.C., Bottero, J.Y., 2008. Synthesis of large quantities of single-walled aluminogermanate nanotube. J. Am. Chem. Soc. 130, 5862–5863.

Levard, C., Doelsch, E., Rose, J., Masion, A., Basile-Doelsch, I., Proux, O., Hazemann, J.L., Borschneck, D., Bottero, J.Y., 2009a. Role of natural nanoparticles on the speciation of Ni in andosols of la Reunion. Geochim. Cosmochim. Acta 73, 4750–4760.

Levard, C., Masion, A., Rose, J., Doelsch, E., Borschneck, D., Dominici, C., Ziarelli, F., Bottero, J.Y., 2009b. Synthesis of imogolite fibers from decimolar concentration at low temperature and ambient pressure: a promising route for inexpensive nanotubes, J. Am. Chem. Soc. 131, 17080–17081.

Levitz, P., Zinsmeister, M., Davidson, P., Constantin, D., Poncelet, O., 2008. Intermittent Brownian dynamics over a rigid strand: heavily tailed relocation statistics in a simple geometry. Phys. Rev. E 78, 030102.

MacKenzie, K.J.D., Bowden, M.E., Meinhold, R.H., 1991. The structure and thermal transformations of allophanes studied by 29 Si and 27 Al high resolution solid-state NMR. Clays Clay Miner. 39, 337–346.

McBride, M.B., 1984. Iron substitution in aluminosilicate sols synthesized at low pH. Clay Miner. 19, 1–8.

Ookawa, M., Inoue, Y., Watanabe, M., Suzuki, M., Yamaguchi, T., 2006. Synthesis and characterization of Fe containing Imogolite. Clay Sci. 12 (Supplement 2), 280–284. http://ci.nii.ac.jp/vol_issue/nels/AA00607148/ISS0000390342_en.html.

Parfitt, R.L., 2009. Allophane and imogolite: role in soil biogeochemical processes. Clay Miner. 44, 135–155.

Parfitt, R.L., Henmi, T., 1980. Structure of some allophanes from New Zealand. Clays Clay Miner. 28, 285–294.

Parfitt, R.L., Thomas, A.D., Atkinson, R.J., Smart, R.S.C., 1974. Adsorption of phosphate on imogolite. Clays Clay Miner. 22, 455–456.

Peña, J., Kwon, K.D., Refson, K., Bargar, J.R., Sposito, G., 2010. Mechanisms of nickel sorption by a bacteriogenic birnessite. Geochim. Cosmochim. Acta 74, 3076–3089.

Rosenqvist, J., Persson, P., Sjoberg, S., 2002. Protonation and charging of nanosized gibbsite (alpha-Al(OH)(3)) particles in aqueous suspension. Langmuir 18, 4598–4604.

Scheidegger, A., Lamble, G., Sparks, D., 1997. Spectroscopic evidence for the formation of mixed-cation hydroxide phases upon metal sorption on clays and aluminum oxides. J. Colloid Interface Sci. 186, 118–128.

Sposito, G., 2008. The Chemistry of Soils, second ed. Oxford University Press, New York.

Sposito, G., Skipper, N.T., Sutton, R., Park, S.H., Soper, A.K., Greathouse, J.A., 1998. Surface geochemistry of the clay minerals. Proc. Natl. Acad. Sci. U.S.A 96, 3358–3364.

Su, C.M., Harsh, J.B., 1993. The electrophoretic mobility of imogolite and allophane in the presence of inorganic anions and citrate. Clays Clay Miner. 41, 461–471.

Su, C., Harsh, J.B., Bertsch, P.M., 1992. Sodium and chloride sorption by imogolite and allophanes. Clays Clay Miner. 40, 280–286.

Tazaki, K., Morikawa, T., Watanabe, H., Asada, R., Okuno, M., 2006. Microbial formation of imogolite. Clay Sci. 12, 245–254.

Theng, B.K.G., Rusell, M., Churchman, G.J., Parfitt, R.L., 1982. Surface properties of allophane, halloysite, and imogolite. Clays Clay Miner. 30, 143–149.

Tossell, J.A., Sahai, N., 2000. Calculating the acidity of silanols and related oxyacids in aqueous solution. Geochim. Cosmochim. Acta 64, 4097–4113.

Tsuchida, H., Ooi, S., Nakaishi, K., Adachi, Y., 2005. Effects of pH and ionic strength on electrokinetic properties of imogolite. Colloids Surf. A Physicochem. Eng. Asp. 265, 131–134.

Ugolini, F.C., Dahlgren, R.A., 1991. Weathering environments and occurrence of imogolite and allophane in selected andisols and spodosols. Soil Sci. Soc. Am. J. 55, 1166–1171.

Van der Gaast, S.J., Wada, K., Wada, S.I., Kakuto, Y., 1985. Small-angle X-ray powder diffraction, morphology, and structure of allophane and imogolite. Clays Clay Miner. 33, 237–243.

Wada, K., Henmi, T., 1972. Characterization of micropores of imogolite by measuring retention of quaternary ammonium chlorides and water. Clay Sci. 4, 127–136.

Wada, K., Tokashiki, Y., 1972. Selective dissolution and difference infrared spectroscopy in quantitative mineralogical analysis of volcanicash soil clays. Geoderma 7, 199–213.

Wada, K., Yoshinaga, N., 1969. Structure of imogolite. Am. Mineral. 54, 50–71.

Werpy, T.A., Michot, L.J., Pinnavaia, T.J., 1989. Adsorption properties of a tubular silicate—layered silicate intercalation complex formed from imogolite and montmorillonite. Clay Res. 8, 47–52.

Werpy, T.A., Michot, L.J., Pinnavaia, T.J., 1990. New tubular silicate—layered silicate nanocomposite catalyst: microporosity and acidity. In: Baker, R.T.K., Murrell, L.L. (Eds.), Novel Materials in Heterogeneous Catalysis. American Chemical Society, Washington DC, pp. 119–128.

Wilson, M.A., Wada, K., Wada, S.I., Kakuto, Y., 1988. Thermal transformations of synthetic allophane and imogolite as revealed by nuclear magnetic resonance. Clay Miner. 23, 175–190.

Yucelen, G.I., Choudhury, R.P., Vyalikh, A., Scheler, U., Beckham, H.W., Nair, S., 2011. Formation of single-walled aluminosilicate nanotubes from molecular precursors and curved nanoscale intermediates. J. Am. Chem. Soc. 133, 5397–5412.

Yucelen, G.I., Choudhury, R.P., Leisen, J., Nair, S., Beckham, H.W., 2012a. Defect structures in aluminosilicate single-walled nanotubes: a solid-state nuclear magnetic resonance investigation. J. Phys. Chem. C 116, 17149–17157.

Yucelen, G.I., Kang, D.-Y., Guerrero-Ferreira, R.C., Wright, E.R., Beckham, H.W., Nair, S., 2012b. Shaping single-walled metal oxide nanotubes from precursors of controlled curvature. Nano Lett. 12, 827–832.

Zang, J., Konduri, S., Nair, S., Sholl, D.S., 2009. Self-diffusion of water and simple alcohols in single-walled aluminosilicate nanotubes. ACS Nano 3, 1548–1556.

Zang, J., Chempath, S., Konduri, S., Nair, S., Sholl, D.S., 2010. Flexibility of ordered surface hydroxyls influences the adsorption of molecules in single-walled aluminosilicate nanotubes. J. Phys. Chem. Lett. 1, 1235–1240.

Zang, J., Nair, S., Sholl, D.S., 2011. Osmotic ensemble methods for predicting adsorption-induced structural transitions in nanoporous materials using molecular simulations. J. Chem. Phys. 134, 184103.

Zanzottera, C., 2012. Hybrid Organic/Inorganic Nanotubes of Imogolite Type. Ph.D. Thesis, Politecnico di Torino, Torino, Italy.

Zanzottera, C., Armandi, M., Esposito, S., Garrone, E., Bonelli, B., 2012a. CO2 adsorption on aluminosilicate single-walled nanotubes of imogolite type. J. Phys. Chem. C 116, 20417–20425.

Zanzottera, C., Vicente, A., Celasco, E., Fernandez, C., Garrone, E., Bonelli, B., 2012b. Physico-chemical properties of imogolite nanotubes functionalized on both external and internal surfaces. J. Phys. Chem. C 116, 7499–7506.

Chapter 10

Characterisation of Imogolite by Microscopic and Spectroscopic Methods

A. Thill*
Laboratoire Interdisciplinaire sur l'Organisation Nanométrique et Supramoléculaire, NIMBE, CEA, CNRS, University Paris-Saclay, Paris, France
Corresponding author: e-mail: antoine.thill@cea.fr

10.1 INTRODUCTION

In early studies, a gel-like substance has been noticed around weathered pumice grains in some volcanic soils in Japan (Miyauchi and Aomine, 1966). This substance was characterised as threadlike particles, and the term 'imogolite' was introduced by Yoshinaga and Aomine (1962). In this first seminal article, most of the methods used to characterise imogolite were already applied: X-ray diffraction (XRD), transmission electronic microscopy (TEM), infrared (IR) spectroscopy and thermogravimetric (TG) analysis. The authors also pointed out some distinctive visual aspects of the imogolite mineral compared to the already known allophanes (which was the term used to describe amorphous aluminosilicates). Indeed, it forms a gel-like structure upon precipitation and centrifugation, whereas allophane forms a solid deposit.

It has been quickly possible to produce synthetic imogolite and also imogolite-like and hybrid nanotubes (Farmer et al., 1977; Wada and Wada, 1982). These nanotubes, now available by lab synthesis as very pure dispersion in water, are perfectly transparent and display a characteristic birefringence when observed between cross polar. Upon drying, they tend to form birefringent films. If special care is taken (for example, by using freeze-drying), a cottonlike solid product is obtained (Fig. 10.1).

It is already clear from Fig. 10.1 that imogolites are nanoparticles that do not scatter or absorb visible light. A simple visual inspection of an imogolite sample allows for assessing the quality of the synthetic nanotubes. Indeed, the sample should be water-clear and viscous upon concentration. Inspection between crossed polarisers allows visualising the presence of the nanotubes

Developments in Clay Science, Vol. 7. http://dx.doi.org/10.1016/B978-0-08-100293-3.00010-8

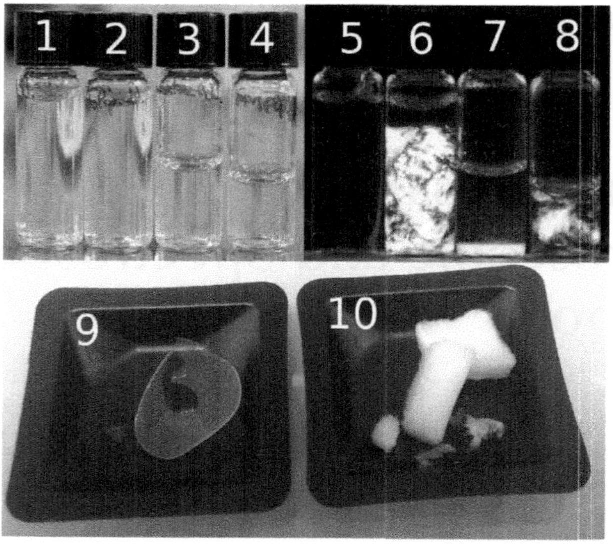

FIG. 10.1 Pictures of synthetic imogolites; from *left* to *right*: 1, Al–Ge long, double-walled (DW) imogolite obtained using Amara et al. (2013) recipes before dialysis; 2, the same sample after dialysis; 3, Al–Si imogolite after concentration by tangential ultrafiltration; 4, the same after dialysis; 5–8, the samples observed between crossed polarisers; 9, self-supported Al–Ge long DW imogolite film obtained by ambient temperature drying; and 10, cottonlike solid obtained by freeze-drying of Al–Ge long, DW imogolite.

through their birefringence. In this chapter, the main imogolite characterisation methods will be discussed in light of the most recent discoveries.

10.2 MICROSCOPIC METHODS

10.2.1 Atomic Force Microscopy and Scanning Tunneling Microscopy

The first observation of an imogolite nanotube by atomic force microscopy (AFM) was performed by Yamamoto et al. (2001). The imogolite nanotubes are adsorbed by electrostatic attraction on a silicon wafer from a 0.05 mass% dispersion at pH 3. On a line profile for an isolated imogolite nanotube, a height of slightly less than 2 nm is observed. However, these authors also claimed to observe a height of 2.6 nm without giving any details on what might have caused this value. Exley et al. (2002) studied the formation of hydroxyaluminosilicates at room temperature, with some of them having a Si/Al ratio of 0.5 (which is similar to imogolite). Using AFM, they observed an amorphous precipitate having a rectangular form with a length of about 40 nm and a height of 1–2 nm. Tani et al. (2004) were the first to propose a dedicated article on AFM observation of imogolite. They deposited a diluted drop (20 mg L^{-1}) of a synthetic imogolite on a glass surface. The deposited drops were dried at ambient temperature for 1 day or 1 month. These authors, by using contact mode scanning, observed that the imogolite deposit is very sensitive to moisture. Therefore, they used tapping mode observation, which led to more stable images. The imogolite dispersion was prepared by sonication of a freeze-dried powder. Imogolite was always present on the surface as bundles with a diameter of about 70 nm. The diameter of the bundles was not very sensitive to the ultrasonic treatment applied to prepare the imogolite dispersion. On the contrary, the average length of the bundles was significantly reduced when the sonication time increased. The drying conditions of the sample also appeared to be very important, as the bundle size was reduced by 21% after 1 month of drying time and further reduced by 24 h of oven drying. Ohrai et al. (2005) did a similar study on imogolite adsorbed on a gold 111 surface. They showed that the structure of the deposit was different from that observed on glass and also that it depended on the sonication duration (ie, on the length of the imogolite bundles). In some conditions, aligned structures interpreted as single nanotubes were analysed. These authors showed that the average horizontal distance between tubes was significantly larger than their expected diameter. However, they also observed a vertical height of 2–3 nm by tapping mode. Yamamoto et al. (2005) used AFM to observe imogolite polyvinyl alcohol (PVA) hybrid materials, which kept a fibre/bundle aspect up to a mass ratio imogolite/PVA of 1:20, the structure of the bundles being strongly affected by the PVA content. Jiravanichanun et al. (2008) prepared imogolite hybrid material with poly[disodium 2,5-bis(3-sulfonatopropoxy)-1,4-phenylene-alt-1,4-phenylene]. They prepared films using layer-by-layer or spin-coating techniques. The film thickness was estimated by AFM. These authors also demonstrated the particular alignment of the imogolite bundles obtained when the films were prepared by spin coating.

In the great majority of these AFM experimental studies, imogolite appeared as bundles of several tens of nanometers. Using alternative sample preparation techniques, several authors managed to have AFM images that unambiguously showed isolated imogolite nanotubes. Ookawa et al. (2006) showed AFM images of perfectly isolated imogolite nanotubes containing Fe atoms. The samples were prepared by drying a more diluted dispersion droplet on a mica surface. The droplet was then dried overnight in a desiccator. The diameter of the nanotube is determined to be 2.2–2.4 nm using height perpendicular section of the nanotubes in tapping mode images.

Maillet et al. (2011) and Liu et al. (2012) studied the length distribution of imogolite nanotubes using AFM. The sample was prepared using adsorption from diluted dispersion on a freshly cleaved mica surface. Using a new synthesis method, Amara et al. (2013) demonstrated by the same AFM technique that very long Al–Ge DW imogolite can be obtained. Diagrams of length vs height obtained from tapping mode images show that the nanotubes are distributed on vertical lines (Maillet et al., 2011). As expected, they all had the same diameter, but AFM seems to give a smaller height compared to the other techniques, as well as a nonnegligible standard deviation around the mean value. The AFM pictures of individual adsorbed nanotubes clearly demonstrated a very strong polydispersity in length with values spanning from 20 to about 100 nm for Al–Ge-based imogolite. Length histograms displayed a nonsymmetric shape, with many small and fewer long nanotubes. The same length distribution was obtained by Liu et al. (2012) in a study on the length effect on the toxicity of Al–Ge imogolite nanotubes.

The used of freshly cleaved mica sheets as a sample support seem to prevent the formation of bundles. In this case, the nanotubes interact strongly through electrostatic forces with the mica surface and sustain the drying step without surface aggregation due to lateral capillary forces (Thill and Spalla, 2002).

Scanning tunneling microscopy (STM) was also used to image imogolite nanotubes. Indeed, Park et al. (2007) observed perfectly aligned hybrid imogolite grafted by octadecyl phosphonic acid (OPA). They showed that the hybrid imogolite self-assemble at the air/water interface in very regular regions of aligned nanotubes. They determined the average distance between grafted nanotubes with very good precision and demonstrated that OPA chains interpenetrate, giving a tube distance of 5.1 nm smaller than the expected distance without chain interpenetration (7.2 nm).

In conclusion, AFM is a great tool to observe the morphology of imogolite on a solid surface. The morphology is, however, very sensitive to the nature of the substrate, the drying method, the relative humidity (RH) and the method used for sample preparation. Due to convolution with the AFM probe size, horizontal distances are not easily related to nanotube diameter except on very well-organised domains. The height scans are unbiased, but the obtained size is systematically smaller than the expected value. It is probable that significant nanotube deformation occurs after adsorption and dehydration of the nanotubes. AFM is, however, one of the rare methods that allows for measuring the imogolite length distribution with good precision.

To our knowledge, no in situ AFM observations of imogolite in water have been reported up to now. This would be an interesting challenge to detect the possible effect of drying on the deformation of the observed nanotubes. AFM has also not been used yet to probe the mechanical characteristic of imogolite. That will be a future challenge of this technique and other more adapted methods, such as surface force apparatus.

10.2.2 Transmission Electron Microscopy

The first observation of imogolite using TEM was performed by Yoshinaga and Aomine (1962) in the 0.2-μm clay fraction of three Japanese soils. They described imogolite as threadlike particles with a diameter of 10–20 nm. Miyauchi and Aomine (1966) showed that imogolite forms birefringent aggregates observed between crossed polarisers. After moderate sonication, the aggregates gave threadlike particles. The finest observed fibres were 10 nm in diameter. Wada et al. (1970), using electron diffraction (ED) with TEM, were the first to notice a characteristic distance of about 2 nm below the threadlike particle size. At that time, the interpretation of XRD pattern was not clear and they were talking about a 'chain' structure. Russell et al. (1969) confirmed the proposed dimension by ED study of randomly oriented and aligned threadlike imogolite particles. Thanks to ED, these authors identified a 8.4-Å repeat distance parallel to the fibre axis and a 23-Å distance perpendicular to it. They also proposed a chain structure for the atomistic description of imogolite. Later, Cradwick et al. (1972) published an article elucidating the structure of imogolite thanks to the full interpretation of ED patterns of aligned imogolite fibres. Wada and Wada (1982) used high-resolution TEM to study the structural modifications induced by substitution of germanium for silicon in synthetic imogolite. They demonstrated that the substitution was possible at all ratios and that an imogolite-like structure was obtained. From the TEM pictures, a diameter of 2.8 and 3.3 nm is obtained for Ge/(Si+Ge) ratio of 0 and 1, respectively. These authors also noticed that the presence of Ge significantly reduced the length of the obtained nanotubes. Ackerman et al. (1993) managed to produce perfectly aligned imogolite films. The ED diffraction of such films revealed a 2.5-nm diameter. Bursill et al. (2000) made several observations on the sample preparation steps of microscopic grids. A careful control of the concentration and drying conditions allows for producing very different textures. Using ED, they largely confirmed the structure proposed by Cradwick et al. (1972). These authors tried to increase the resolution of the TEM images, but this is extremely difficult to do because the isolated nanotubes happen to be very fragile. They managed to obtain such images by embedding isolated nanotubes in an amorphous aluminosilicate phase. In TEM images of very high resolution, they noticed that the nanotube sections tend to be faceted. Koenderink et al. (1999) observed by TEM that when as-prepared synthetic imogolite is centrifuged at 25,000 rpm, the sample contains two different components in the sediment and in the supernatant, respectively. The sediment contains threadlike particles with diameters between 5 and 25 nm and average lengths of about 0.5 μm, composed of imogolite associated in bundles that resemble closely the purified natural imogolite. The supernatant contains nanoparticles attributed to proto-imogolite (small nanoparticles at the expense of which the imogolite nanotubes occur). Wilson et al. (2001) were the first to publish a kinetic study in which TEM images were acquired from 0 to 119 h of synthesis reaction. They showed that proto-imogolite sheets appeared after 7 h and that imogolite fibres are observed after 22.5 h. Imogolite seems to grow at the expense of sheets. After 53 h, the imogolite nanotubes composed the majority of the sample. Later, Mukherjee et al. (2005) performed the same type of experiment for both Al–Si and Al–Ge synthetic imogolites. They observed the formation of Al–Si nanotubes after 10 h and confirmed that the formation of nanotubes goes along with the decrease of an early formed phase.

This observation is very similar in the case of Al–Ge nanotubes except that the length of the tubes remains very small (at about 10 nm). Arancibia-Miranda et al. (2011) compared different synthesis methods for Al–Si imogolite. They observed a necklace structure as an intermediate stage between proto-imogolite and fully formed nanotubes.

Yang and Su (2007) showed that using standard procedure to prepare TEM grids may result in heterogeneous deposits. For example, after 3 h reaction time, the central region of a dried droplet does not contain imogolite nanotubes; but the border of the same droplet clearly shows nanotubes and their characteristic ED patterns. After a longer reaction time, the authors showed that an entangled network of nanotubes could be observed everywhere on the droplets. They argued that it was impossible to give any information on the nanotube length from such images. These authors proposed an alternative method to prepare the TEM grids. They let the diluted droplets dry in an atmosphere saturated with ethanol. With this method, evaporation was evenly distributed on the droplet surface, and they also observed that it prevented the nanotubes from forming bundles. Using this method, Yang and Su (2007) allowed the observation of individual nanotubes and thus the measurements of length distribution. Despite its effectiveness, this method was rarely used by other authors for imogolite studies (Shikinaka et al., 2015). Barona et al. (2013) showed that TEM grid prepared from water and hexane dispersion gave completely different morphologies. The sample prepared from water showed the classically observed bundles, whereas individual nanotubes were obtained from the hexane dispersion.

Yamamoto et al. (2001) were the first to explore the morphology of functionalised imogolite external surface with TEM. They showed that their threadlike structure and dispersability were conserved after grafting with poly(methyl methacrylate). Bottero et al. (2011) observed for the first time TEM images of imogolite with a modified internal surface. Here again, the overall structure of imogolite was observed to have threadlike nanotube bundles.

TEM is a very effective method to explore the morphology and purity of imogolite samples. However, care must be taken during the sample preparation stage, as it influences directly the morphology of the deposited imogolite. In particular, except when special drying methods are used (Yang et al., 2008), the classical preparation procedure leads to the formation of threadlike nanotube bundles, hindering the observation of individual imogolite. It is also very difficult to obtain high-resolution images due to the electron beam damage on the imogolite.

10.2.3 Cryo-TEM

A very popular method in biology was introduced later for the observation of imogolite. For such observations, a thin amorphous ice layer containing the imogolite dispersion was prepared. For this purpose, a carbon grid containing holes was used. A few microlitres of the sample were deposited on such grids, which were often treated by glow discharge to improve wettability. After removing the excess water, each grid was quickly dipped in liquid ethane. From this stage, the grid was kept at the liquid N_2 temperature. Special care must be taken to prevent crystalline ice deposition on the sample. Special observation modes (low dose) are generally available on cryo-microscopes and can be used to limit the electron dose to the sample. All preliminary steps before acquisition were performed in a side zone. The beam was placed on the sample only for final exposure. Examples of cryo-TEM images (Fig. 10.2) were obtained from different kind of imogolite nanotubes.

Cryo-TEM was first applied by Maillet et al. (2010) for the observation of Al–Ge imogolite. Thanks to this method, very precise images of isolated nanotubes were recorded. These authors discovered that Al–Ge imogolites exist in two forms: single-walled (SW) and double-walled (DW). These first observations were extended by Thill et al. (2012a,b) to explain the formation mechanism of such morphologies. Yucelen et al. (2011) used cryo-TEM to study the influence of ligands on the precursors and nanotube curvature. Thanks to a high-sensitivity camera, they managed to work very close to focus in order to measure unbiased distances and observed that the nanotube diameter can be significantly changed depending on added acids from 2.2 to 2.8 nm. Kang et al. (2014) observed imogolite with a modified internal cavity by aminomethyltriethoxysilane by both low-resolution TEM and high-resolution cryo-TEM.

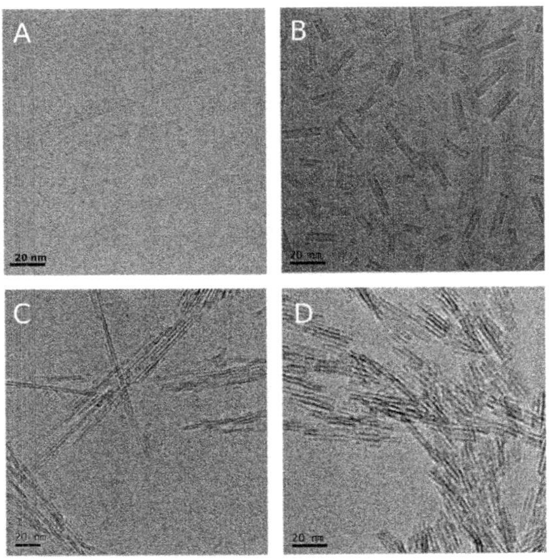

FIG. 10.2 Cryo-TEM images obtained for (A) imogolite, (B) double-walled Ge imogolite, (C) hybrid imogolite with Si–CH_3 groups inside their cavities and (D) hybrid Al–Ge imogolite with Ge–CH_3 groups inside their cavities.

Although cryo-TEM is a difficult technique, due to its delicate sample preparation and handling stages, it is particularly well suited for imogolite observation. It allows obtaining high-resolution images of nanotubes without bundle formation due to drying. To our knowledge, only Thill et al. (2012a,b) have published images of proto-imogolite in the case of Al–Ge imogolite, so there is room for mechanistic and kinetic studies of imogolite formation using cryo-TEM in the future.

10.3 SPECTROSCOPIC METHODS

10.3.1 Fourier Transformed Infrared Spectroscopy

Fourier transformed infrared (FTIR) spectroscopy is one of the most popular methods for the characterisation of imogolite. Yoshinaga and Aomine (1962) showed that imogolite has a spectroscopic signature very similar to allophane between 600 and 4000 cm^{-1} with a wide absorption band around 1000 cm^{-1}. Russell et al. (1969) showed that this band in fact contained a doublet at 995 and 932 cm^{-1} when imogolite is hydrated. The wavenumber of these bands shifts to 1010 and 920 cm^{-1} and a new band appears at 833 cm^{-1} upon dehydration under vacuum. They showed after deuteration and dehydration of imogolite samples that bands at 3730, 3640, 3583 and 833 cm^{-1} of imogolite are shifted to 2760, 2690, 2640 and 690 cm^{-1}, respectively. These bands are attributed to OH groups. The new band appearing upon dehydration is attributed to a low-frequency shift due to a decrease in hydrogen bonding experienced by OH groups. Creton et al. (2008) attributed the stretching vibrations of internal OH groups to the range 3200–3400 cm^{-1} and the outer OH groups to the 3700–3800 cm^{-1} region. Recently, Bonelli et al. (2013) and Amara et al. (2015) showed that upon replacing OH internal groups by CH_3 groups, the OH-stretching region did indeed shift toward higher wavenumbers. This tends to confirm that the internal OH groups are those contributing in the 3200–3400 cm^{-1} region.

Farmer et al. (1977) showed that the presence of imogolite structure was recognised in soil through the measurement of a 348-cm^{-1} absorption band. It is indeed a signature of the 'imogolite structure' rather than imogolite mineral, as this band is also found in proto-imogolite and in some allophanes (Bishop et al., 2013). The distinction between imogolite and allophane is the shape of the absorption band near 1000 cm^{-1}, which is attributed to Si–O stretching vibrations. Indeed, allophane and proto-imogolite, which does not have the fibre structure, generally display a single absorption band at about 940 cm^{-1}. Wilson et al. (2001) published a kinetic study where FTIR spectra were recorded from the initial mixing of precursors to the full formation of imogolite. They confirmed that the initial spectra had a single band in the 1000 cm^{-1} region. The appearance of a doublet after 22.5 h of reaction corresponded to the occurrence of fibres. Arancibia-Miranda et al. (2011)

did the same experiments with two different synthesis protocols. In both cases, the same results were obtained, but at different times after mixing. In the first recipe, the pH was adjusted to pH 4.5 after NaOH addition with a mixture of HCl and CH_3COOH. In this case, the doublet at about 1000 cm^{-1} was observed after 24 h. In the second recipe, the doublet was observed after only 4 days (Table 10.1).

Bishop et al. (2013) attributed the bands below 700 cm^{-1} to Si–O–Al_δ and Si–O_δ. However, Wada and Wada (1982) showed that upon substitution of Si

TABLE 10.1 Assignment of Main IR Vibrations

Wavenumber (cm^{-1})	Wavelength (μm)	Feature	Comments
3450–3730	2.68–2.90	OH_v and H_2O_v	
1600–1645	6.08–6.25	H_2O_δ	Not observed after evacuation at 20°C under vacuum (20 mmHg) or after heating at 150°C
1430–1485	6.73–6.99	OH_δ from O_3SiOH	Single band in natural imogolite, doublet in synthetic imogolite
1380	7.24	n.a.	Sharp band in transmission spectra
990–1010	9.9–1.01	Si–O–Al_δ	
910–943	1.06–1.1	Si–O–Al_δ	These two bands are characteristic of the imogolite tubular shape
900–920	1.09–1.11	Ge–O–Al_δ	
790–820	1.22–1.27	Ge–O–Al_δ	
683–695	1.44–1.46	Various vibrations, particularly Al–O	Not affected by deuteration or Ge/Si substitution
550–590	1.69–1.82	Various vibrations, particularly Al–O	Not affected by deuteration or Ge/Si substitution
470–490	2.04–2.13	Various vibrations, particularly Al–O	Not affected by deuteration or Ge/Si substitution
420–428	2.34–2.38	Various vibrations, particularly Al–O	Not affected by deuteration or Ge/Si substitution
335–348	2.87–2.99	Various vibrations, particularly Al–O	Not affected by deuteration or Ge/Si substitution

with Ge atoms, none of these bands were shifted. Several authors attempted to compute the spectroscopic characteristic of imogolite using molecular dynamic models (Tamura and Kawamura, 2002; Alvarez-Ramírez, 2007; Creton et al., 2008). Tamura and Kawamura (2002) and Creton et al. (2008) confirmed the attribution of the bands near 1000 cm^{-1} to Si–O stretching vibrations. Creton et al. (2008) stated that the silicon atom is not in a symmetric T_d site, but explained the structure of the vibration bands by a C_{3v} symmetry. Indeed, the three Si–O–Al bonds do not have the same behaviour as the Si–OH bond. Attribution of the bands below 700 cm^{-1} is more difficult. Indeed, in this region, almost all atoms display vibrations. Comparison of the experimental and calculated spectra allows for concluding that the AlO6 octahedron does not behave as a symmetric isolated entity. Creton et al. (2008) also demonstrated that considering the superposition of two C_{3v} sites (upper and lower part of the octahedra) is not enough to explain the experimental spectra. The contribution of AlO6 octahedron is rather complex because the local environment and bonding induce a descent in symmetry and also because the octahedron cannot be considered as isolated. In particular, some modes of the oxygen atoms in Si–O–Al above 800 cm^{-1} indicate coupling between the Al–O and Si–O bond stretching. Finally, FTIR is an efficient fingerprint to assess the formation of imogolite. The spectra of imogolite and allophane are very similar; however, the occurrence of a doublet in the 1000 cm^{-1} region seems to be characteristic of imogolite.

Several other IR characteristics are often associated with imogolite. For example, a very sharp band is often observed near 1380 cm^{-1} (Kuroda and Kuroda, 2010; Bishop et al., 2013). Kuroda and Kuroda (2010) attributed this band to the stretching vibration of NO_3^- which is the counterion of aluminium used in their synthesis. Bishop et al. (2013) observed this band for natural samples. Similarly, when perchloric acid or aluminium perchlorate was used during the synthesis, several IR bands near 1100 cm^{-1} were observed (Koenderink et al., 1999) due to residual perchlorate ions.

In some syntheses, a relatively sharp band at 1070 cm^{-1} has been observed (Denaix et al., 1999; Hu et al., 2004; Tani et al., 2004). Tani et al. (2004) attributed this band to pseudoboehmite.

10.3.2 Nuclear Magnetic Resonance

Nuclear magnetic resonance (NMR) is a technique that can probe the local environment of atoms. It is often used for silicon and aluminium in clay minerals. Using Si^{29} NMR, it is possible to assess the condensation of silicate. Indeed, single tetrahedra (Q^0) display chemical shifts in the range of −66 to −72 ppm. When the condensation degree increases, the chemical shift is also reduced: −78 to −82 ppm for double tetrahedra (Q^1), −86 to −88 ppm for the tetrahedra chain (Q^2) and −91 to −95 ppm for cyclic layered structures (Q^3). In clay minerals, Si atoms are linked to other Si atoms, but also to Al atoms through oxygen bridges. Following the same notation, a Si tetrahedra coordinated with Al is noted Q^1(Al). In the imogolite structure, the Si tetrahedra is linked to three Al atoms Q^3(3Al). Barron et al. (1982) were the first to use NMR for studying imogolite. They showed that for a purified imogolite, the Si^{29} NMR spectrum gave a single sharp line at −78 ppm. As there is no equivalent Q^3(3Al) local environment described in other aluminosilicates, they tried to extrapolate the effect of Al in Q^4 environments. Lippmaa et al. (1980) published the spectra of many Q^4 aluminosilicates. A Q^4(3Al) site gives a chemical shift of −88 ppm compared to Q^4 at −107 ppm. Substitution of three Si by Al thus gives a difference in chemical shifts of 19 ppm. Barron et al. (1982) extrapolated the same for Q^3 sites and thus predicted a chemical shift for Q^3(3Al) of −72 to −76 ppm, which is in good agreement with the experimental value observed at −78 ppm.

Goodman et al. (1985) did Si^{29} NMR and Al^{27} NMR measurements on synthetic imogolite and proto-imogolite. They found the same sharp peak at -78 ± 1 ppm corresponding to the Q^3(3Al) local environment introduced by Cradwick et al. (1972) for both imogolite and proto-imogolite. It demonstrates that silicon in proto-imogolite has already the same local structure as the final imogolite nanotubes. These authors also showed that the Si^{29} NMR spectra of natural imogolite can contain contribution at higher shifts. Such contributions are similar to the Si^{29} NMR spectra obtained from gels with an Al/Si ratio of 1. The Al^{27} NMR spectrum of imogolite contains a sharp, 5.8-ppm peak corresponding to Al in sixfold coordination. A smaller contribution at 63 ppm due to Al in fourfold coordination is also observed. As for Si^{29} NMR, the spectra from imogolite and proto-imogolite are very similar. The main peak is similar in the gibbsite spectrum, which gives a sharp peak at 7.8 ppm.

MacKenzie et al. (1989) studied the Si^{29} and Al^{27} NMR spectra of imogolite upon heating. They showed that the imogolite structure starts to be modified through dehydroxylation at temperatures higher than 300°C. Due to the large distance between adjacent Si tetrahedra, the creation of an Si–O–Si bond between them is very unlikely. MacKenzie et al. (1989) proposed a dehydroxylation mechanism involving the breakage of the nanotubes. The most likely mechanism involved a unique tube that would collapse after the wall breakage through the creation of Q^4(3Al) sites in good agreement with the −93-ppm chemical shift observed after dehydroxylation and before mullite crystallisation.

Ackerman et al. (1993) used Si^{29} NMR to characterise synthetic imogolite samples. They observed the same −78 ppm sharp peak that was characteristic of the Q^3(3Al) Si atom environment in imogolite. The purity of the obtained product was compared to purified natural imogolite, which contains a small peak at −86 ppm due to the presence of amorphous silica gel. They also showed that the local structure was not changed by the careful dehydration step used before gas adsorption experiments. Jaymes and Douy (1995), while studying the synthesis of mullite precursors, observed by Si^{29} and Al^{27} NMR that at a hydrolysis ratio of OH/Al = 2.4, the observed spectra contained the characteristic −78 ppm peak of Q^3(3Al) sites. They concluded that the obtained particles were more related to allophane than imogolite due to their very small size.

Ildefonse et al. (1994) used Al^{27} NMR and X-ray adsorption spectroscopy to study imogolite and allophane. They showed that synthetic imogolite at an Al/Si ratio of 2 contained a single peak at 6.7 ppm due to Al^{VI} and no peak due to Al^{IV}. In comparison, the natural imogolite had a wider Al^{VI} peak and showed an additional contribution at 63 ppm due to Al^{IV}. Denaix et al. (1999) compared two syntheses with decimolar or millimolar initial aluminium concentrations before and after heating. They showed that proto-imogolite contains mainly a sharp peak at −78 ppm and a small contribution around −100 ppm due to polymerised silica. This contribution disappeared after the heat treatment. Wilson et al. (2001) did a kinetic study of the formation of imogolite. The Si^{29} NMR spectra were recorded from 0 to 120 h of reaction time. These authors also showed that a −83 ppm peak corresponding to amorphous Al/Si gel slowly disappeared at the expense of the −78-ppm sharp imogolite peak. For Al^{27} NMR, the dominant peak was always the Al^{VI} peak, at about 3.4 ppm. However, a peak at about 60 ppm due to Al^{IV} was also present at up to 22.5 h of reaction. They also noticed that the main peak was initially skewed to low ppm, becoming more symmetric from 7 h onward. They suggested that at early reaction times, more than one type of octahedral aluminium site existed. The structure became more homogeneous as the reaction proceeded. The same observation was made by Levard et al. (2011) as a function of the hydrolysis ratio for the synthesis of Al–Ge imogolite. They showed that the skewness of the peak decreased with the hydrolysis ratio up to OH/Al = 2.5. In the same study, they also showed that at low hydrolysis ratios, Al^{IV} was observed and accounted for up to 6.7% of the Al atoms. From OH/Al = 1.75 and onward, the contribution of Al^{IV} disappeared.

Kang et al. (2010) used H^1, Si^{29} and Al^{27} NMR to assess the transformation of imogolite upon heating from 25°C to 400°C. By proton NMR, they differentiated between two types of protons: protons associated with isolated surface hydroxyls, giving a peak 1.8 ppm; and protons associated with physisorbed water, giving a peak at about 6 ppm. At about 100°C, the two peaks had approximately the same height. Below 100°C, the signal was clearly dominated by water, while from 200°C onward, OH protons dominated. Using Si^{29} NMR, the researchers clearly demonstrated that dehydroxylation occurred mainly between 300°C and 400°C. At 400°C, the −78 ppm peak decreased, and a large −90 ppm peak dominated. As explained in McKenzie et al. (1989), this is attributed to the Q^4(3Al) site, in keeping with a tube collapse with dehydroxylation between internal silanol groups. Kang et al. (2010) also showed that this process is largely reversible. Upon hydration, the −90 ppm peak significantly decreased and the −78 ppm peak dominated again. This could be explained by a hydrolysis of the Si–O–Si bond created between internal silanol during dehydroxylation. On the other hand, Al^{27} NMR showed that between 300°C and 400°C, new peaks associated with Al^{V} and Al^{IV} coordination appeared. Upon rehydration, these peaks remained, meaning that the dehydroxylation affecting the Al local environment was irreversible. This could explain why the gas adsorption capacity of the imogolite heated at 400°C was not fully recovered. The formation of five and four coordinated aluminium species was further confirmed by Hatakeyama et al. (2011) using MQMASS NMR.

NMR spectroscopy appears very useful for assessing the quality of the synthetic (or natural) products (Bonelli et al., 2009; Levard et al., 2011). Indeed, the relative intensity of the −78 and −90 ppm peaks reveals the presence of other products or internal defects. The same conclusion can be made from the presence of 60 ppm peak in Al^{27} NMR spectra. Chemni et al. (2013) showed that using HF in the synthesis of imogolite contributes to reduce these side peaks significantly. Yucelen et al. (2012) published a detailed work using advanced NMR techniques to explore the defect structure in imogolite. They proposed that some of the contributions in the −90-ppm peak of Si^{29} NMR could arise from internal nanotube defects rather than from side products, as has often been speculated.

NMR has also been used to explore the grafting of different molecules onto imogolite (Bottero et al., 2011; Zanzottera et al., 2012b; Kang et al., 2014). Bottero et al. (2011) showed that when TEOS is replaced by methyltriethoxysilane, an imogolite-like nanotube is formed in which the internal silanol, which gives the sharp −78-ppm peak, is replaced by Si–CH_3 groups with a sharp −42-ppm peak. Zanzottera et al. (2012b) showed that 3-aminopropyltriethoxysilane (3-APS) was adsorbed on the external surface of hybrid imogolite. This grafting gave additional peaks in the Si^{29} NMR spectra that corresponded to 3-APS molecules grafted with one, two and three Si–O–Al bonds (T1, T2 and T3) at −50, −55 and −69 ppm, respectively. An additional contribution at −64 ppm was attributed to polymerised 3-APS molecules.

10.3.3 X-Ray Absorption Including X-Ray Absorption Near Edge Structure and Extended X-Ray Absorption Fine Structure

X-ray absorption (XAS) is also a technique able to probe local structure. Ildefonse et al. (1994) were the first to explore the X-ray absorption near edge structure (XANES) at the Al K edge in the range 1550–1600 eV. Comparing XANES and Al^{27} NMR, they showed that XANES is sensitive to the $Al^{IV}/Al^{tot.}$ ratio, NMR being more precise, especially for small ratios. These authors also showed that aluminium occurs in a single octahedral site. As XANES requires access to a synchrotron X-ray source and does not give significantly different information compared to NMR, it has no longer been used for studying aluminium local structure in imogolite.

More recently, Levard et al. (2008) used XANES and extended X-ray absorption fine structure (EXAFS) spectroscopy at the Ge K edge to probe the local structure of imogolite synthesised using germanium instead of silicon. Indeed, Ge NMR is not very practical and requires very intense magnetic fields. Using Ge K edge EXAFS, these authors showed (i) that the nanotube obtained with germanium have the same local structure as silicon based nanotubes and (ii) that each Ge is surrounded by an average of 5.7 Al atoms. In a further publication, Levard et al. (2010) made a similar study as a function of reaction time and also compared the nondialysed and dialysed samples. They showed that the average number of Al atoms in the first coordination sphere of Ge atoms increased from 4 in the proto-imogolite to 5.9 in the fully formed nanotubes. The average distance of the Ge–O bond being 1.75 Å and the average Al–Ge distance being 3.26 Å. A series of experiments was also performed to assess the effect of the hydrolysis ratio $R = OH/Al$. These authors showed that the local environment of Ge is not affected by the hydrolysis ratio. It remains in an unaffected tetrahedral environment. On the contrary, the Al–Ge distance is strongly affected by the hydrolysis ratio: (i) for $R = 0$, the average distance is 3.28 Å, with 2.9 Al neighbours and (ii) for $R = 2$, the average Al–Ge distance is 3.26 Å, with 5.9 Al neighbours very close to the six expected Al neighbours for an infinitely long imogolite nanotube.

10.3.4 X-Ray Photoelectron Spectroscopy

X-ray photoelectron spectroscopy (XPS) is a surface-sensitive quantitative technique which enables to probe the chemical composition and chemical state of a material. It has been initially used for allophanes (Kawano et al., 1993; Kawano and Tomita, 1994; He et al., 1995; Childs et al., 1997). It was first used for imogolite by Mukherjee et al. (2005). The XPS spectra allowed to estimate the composition of the sample using the Al2p (71.8 eV) and Si2p (99.4 eV) bands. They found ratios of $Al/Si = 1.7$ and $Al/Ge = 2.67$ for imogolite and Al–Ge imogolite, respectively. All the other XPS studies on imogolite concerns their functionalisation by various molecules (Jiravanichanun et al., 2008; Qi et al., 2008; Ma et al., 2012; Zanzottera et al., 2012b). Qi et al. (2008) confirmed the presence of osmium and assessed its degree of oxidation used for catalysis. Jiravanichanun et al. (2008) confirmed the presence of imogolite in a hybrid thin film through the presence of Si2p (103.6 eV) and Al2p (75.8 eV). Ma et al. (2012) showed that the peak corresponding to Al2p binding energy of pristine imogolite (74.3 eV) becomes wider and can be decomposed into two contributions (74.3 and 76.3 eV) upon binding with dodecylphosphoric acid. The second contribution is attributed to an increase of the positive charge of Al atoms due to the formation of Al–O–P bond. Zanzottera et al. (2012b) showed that the N1s line is only present on samples where 3-APS was grafted. The presence of a Cl2p line is also noticed on both methyl imogolite and 3-APS grafted methyl imogolite (Zanzottera, 2012).

10.4 SCATTERING METHODS

10.4.1 Dynamic Light Scattering

Dynamic light scattering (DLS) allows for measuring the diffusion coefficient of nanoparticles. Estimation of their size and shape can then be indirectly assessed. Particularly for anisotropic particles, the diffusion can be described with a translational (D) and a rotational (T) diffusion coefficient. T and D are linked by $L^2T/D–K$; $K=2.04–2.1$ according to Kajiwara et al. (1988) and $K=9$ according to Mukherjee et al. (2005). T decreases quickly with L (L^{-3} dependence). If the time required to circumscribe a sphere $\tau=2\pi^2/T$ is less than the delay time of the autocorrelator (typically ~1 μs), the diffusion can be considered as the one of a sphere whose diameter is the length of the nanotube. Kajiwara et al. (1988) studied the DLS signal of imogolite dispersion at various concentrations. From DLS, the obtained average length of the imogolite nanotubes of 118–154 nm at 12.7 and 21.5 g L^{-1}, respectively.

Mukherjee et al. (2005) used DLS to follow the growth kinetics of imogolite and Al–Ge imogolite. They showed that Al–Ge imogolite nanotubes are very short (below 20 nm) and that the nanotubes can be seen as equivalent spheres of diameter L. For imogolite, a full fit of the autocorrelation function has to be performed. The length distribution and the flexibility of the nanotubes have to be taken into account. The evolution of the average length as a function of synthesis time is proposed from these measurements. Mukherjee et al. (2007) used the same method to explore the growth kinetic of Al–Ge imogolite at different temperatures (65°C, 75°C, 85°C and 95°C). Yucelen et al. (2011) used DLS as a function of reaction time during the ageing and heating stages. They showed that the scattering signal contains two contributions. A fast dynamic due to small nanotubes and a slow dynamic attributed to ~200-nm particles. The composition of these large particles is not clear, and their number concentration is too low to explain a significant part of the nanotube formation. Yucelen et al. (2011) attribute the slow dynamic contribution to the formation, immediately after heating, of a loosely connected gel-like network of partially condensed aluminosilicate precursors.

DLS is a very powerful technique when carefully employed. Much useful information on the structure and dynamic of imogolite dispersions could be obtained from such methods in the future. However, to be fully exploited, DLS of anisotropic nanoparticle takes great advantage of multiangle measurements and special data treatment, which is not easy to obtain on many commercial instruments.

10.4.2 X-Ray-Based Analysis

10.4.2.1 XRD by Imogolite and Imogolite Bundles

From the earliest discovery of imogolite, it has been recognised that these particles are thread like and tend to form bundles (Yoshinaga and Aomine, 1962).

To understand the XRD diagrams, it is interesting to discuss first the ED patterns. Indeed, ED obtained from aligned bundles of imogolites clearly allowed attribution of diffraction contributions (Russell et al., 1969; Cradwick et al., 1972; Wada et al., 1979; Barrett et al., 1990; Bursill et al., 2000). Russell et al. (1969) interpreted the sharp reflections in line parallel to the fibre axis as the second, fourth and sixth orders of a 8.4-Å repeat unit. They also noticed that the diffraction pattern contain reflections perpendicular to the thread axis. They attributed these peaks to the second, third and fourth order of a reflection arising from a repeat distance of 23 Å. Cradwick et al. (1972) were the first to propose a tubular structure for imogolite. Their results confirmed the first interpretation of Russell et al. (1969). In brief, the diffraction pattern is characterised by three sharp reflections corresponding to the second, fourth and sixth orders of a 8.4-Å repeat distance at $2\theta=21.1$, 43.1 and 66.8 degrees (Cu K_α). It contains also two diffuse intense reflections at d-value of 3.3 and 2.3 Å [$2\theta=27$, 39 degrees (Cu K_α)] attributed by Cradwick et al. (1972) to planes (071) and (063), respectively. Barrett et al. (1990) published the diffraction pattern of aligned imogolite and showed an excellent agreement between the experimental reflection and a two-dimensional (2D) hexagonal packing with a centre-to-centre distance of 2.78 nm. The hexagonal packing of the imogolite was also observed by Yang and Su (2007) with exactly the same centre-to-centre separation of 2.78 nm.

In XRD experiments, the sharp reflection corresponding to the 8.4 Å repeat distance are almost never very clearly observed. It is interesting to notice that all these reflections at $2\theta>20$ degrees (Cu K_α) seem to be rather stable between natural imogolite and different synthetic imogolites. The reflections at d-values 3.3 and 2.3 Å are even observed in proto-imogolite and allophane (Arai et al., 2006; Iyoda et al., 2012; Levard et al., 2011). These two reflections can be associated to the imogolite local structure. The 002, 004 and 006 reflections being only present in the tubular form (not in allophane neither proto-imogolite) and are often too weak to be observed.

The stability of the XRD signal among different imogolite samples (natural and synthetic) is no longer true for the low-angle part of the XRD diagram ($2\theta<20$ degrees). Indeed, in this part, the diffraction is sensitive to the size and mutual arrangment of the imogolite nanotubes and to the RH. This has been noticed very early by Farmer et al. (1983) who compared 'wet' imogolite films and films dried at 100°C (Fig. 10.3) and studied in details by Van der Gaast et al. (1985). In this region of the diagram, the signal combines information from the individual nanotube shape and from their mutual organisation. In the 'wet' film, the diffraction pattern is well described by the reflection from hollow cylinders with external radius 1.32 nm and internal radius 0.72 nm plotted in dotted lines. In the 100°C dried films additional contributions appeared. The theoretical positions of reflections of a 2D hexagonal phase with a repeat distance of 26.5 Å are noted. In this particular case, the physical size of the nanotubes and the repeat distance of the hexagonal phase

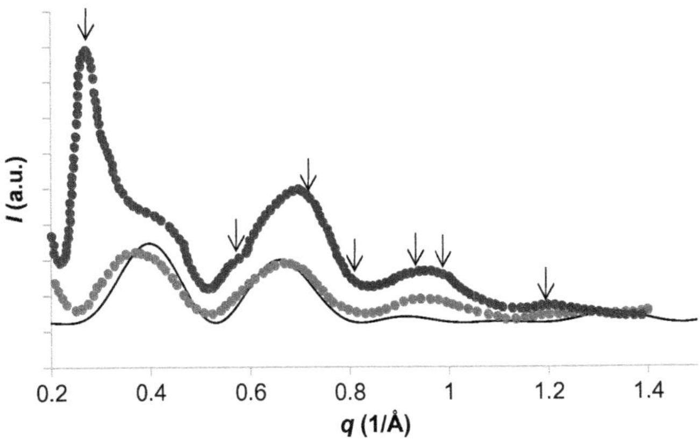

FIG. 10.3 Diffraction pattern for 'wet' film of synthetic imogolite (*red* (*grey* in the print version) *curve*) and a 100°C dried film (*blue* (*dark grey* in the print version) *curve*). The *arrows* indicate the theoretical positions of the reflections of a 2D hexagonal phase with a repeat distance of 2.65 nm. The continuous lines show the reflection computed for a hollow cylinder model ($r_i = 0.72$ nm, $r_o = 1.32$ nm). *Adapted from Farmer et al. (1983).*

seem to match rather correctly. This is not always the case, though, due to the presence of residual water/ions or because of nanotube deformation (Cradwick et al., 1972; Amara et al., 2014).

Mukherjee et al. (2005) discussed XRD patterns of imogolite and Al–Ge imogolite. For the first time, these authors attributed the scattering pattern at a low angle to a monoclinic packing with an angle of 78 degrees. They also observed a reflection at the *d*-value of 8.51 Å, which they attributed to the (001) plane. Following this published result, several authors used this assignment (Hongo et al., 2013). This peaks indexing is surprising, as in the Cradwick et al. (1972) structure, only even orders of the 8.5-Å repeat distance give reflections. The second oscillation due to the form factor of synthetic nanotubes is often misinterpreted because its maximum falls at a *d*-value of about 8.5 Å. Wada and Wada (1982) published diffraction patterns from natural, synthetic imogolite and Al–Ge imogolite. They showed that both synthetic imogolite and Al–Ge imogolite have a reflection at the *d*-value at 8.7 and 8.5 Å, respectively. In this study, the same reflection from natural imogolite appears at 7.8 Å due to its smaller diameter. Concerning the Al–Ge imogolite, it has been discovered by Maillet et al. (2010) that these nanotubes are obtained as DW structures when synthesised at low concentrations. The diffraction pattern from DW Al–Ge imogolite has a characteristic shape, with three oscillations at about *d*-value 12.5, 8.9 and 7.4 Å. Wada and Wada (1982), Mukherjee et al. (2005) and Levard et al. (2011) observed these reflections for Al–Ge imogolite, which should not be mistaken for the 8.4-Å repeat distance of the *c*-axis, which only gives diffraction signal at even orders.

Alvarez-Ramírez (2007) explored in detail the diffraction patterns by bundles of imogolite nanotubes in both monoclinic and hexagonal arrangements. He showed that diffraction is very sensitive to the nanotube radius and intertube distance, and he concluded that the best agreement between the calculation and experimental data is obtained for imogolite with 12 units $a=b=24.8$ Å and $\gamma=64$ degrees. Kang et al. (2010) published a detailed calculation of the XRD pattern as a function of the hydration state of the nanotubes. They gave several possibilities to explain the diffraction data. The hexagonal packing afforded a nice explanation for the shape of the diffraction data between $2\theta=8$ and 10 degrees. This was not the case for the monoclinic packing with $\gamma=74$ degrees.

Recently, Amara et al. (2014) analysed by small-angle X-ray scattering (SAXS) the diffraction obtained by powders of Al–Ge SW nanotubes, and they showed that these nanotubes have an external diameter of 20.3 Å. When dried in the form of a well-organised powder containing large bundles, the nanotubes arrange on a 2D hexagonal lattice having a lattice parameter of 39.7 Å. The hexagonal arrangement was also directly observed on TEM cross section of the powder. In this case, the lattice parameter is less than the physical diameter of the nanotubes. These authors explained that this is due to a deformation of the nanotubes, which became hexagonal when arranged in large bundles. Deformation of imogolite has also been anticipated based on numerical studies (Tamura and Kawamura, 2002; Creton et al., 2008) for imogolite.

The (10) reflection of the 2D hexagonal arrangement of imogolite nanotubes associated in bundles is often used to obtain the nanotube diameter using $a=\frac{4\pi}{q_{10}\sqrt{(3)}}=\frac{2d_{10}}{\sqrt{(3)}}$ (Farmer et al., 1983; Zanzottera et al., 2012a). It has been observed very early on that the position of this reflection is very different from one sample to another and also very sensitive to temperature increases for the same sample (Yoshinaga and Aomine, 1962; Wada and Wada, 1977; Van der Gaast et al., 1985). The surface charge of imogolite could lead to anion trapping in the imogolite powder, which could also affect the structure of the powder (Gustafsson, 2001; Amara et al., 2015). Indeed, in synthetic imogolite obtained from aluminium perchlorate salt, characteristic peaks at about 1100 cm^{-1} show the presence of ClO_4^- anions even after thorough dialysis. Finally, it has been recently discovered that the imogolite nanotubes deformed when assembled in bundles (Amara et al., 2014). The influence of all these parameters makes it difficult to rely on the $d_{(10)}$ reflection to have a precise measurement of the diameter. For example, errors of 10% in the estimation of the diameter could be made when nanotubes are packed in very small bundles, as reported in Chapter 11.

10.4.2.2 Small-Angle X-Ray Scattering

In SAXS, the intensity is collected in transmission geometry, and the direct X-ray beam crosses the sample. It is generally blocked by a beam-stop to protect the detector. The scattering signal is collected on a fixed 2D detector, and the SAXS intensity is obtained through radial averaging. Compared to classical XRD experiments, there is no variation of the scattering volume as a function of the scattering angle.

Kajiwara et al. (1988) were the first to publish SAXS data from a 4 weight % dispersion of imogolite nanotube at pH 3 in acetic acid. They showed, using the Guinier regime of the small-angle part of the scattering curve, that the nanotubes have a diameter of 2.52 nm, which is significantly higher than what has been determined by XRD on dry powder (2.3 nm). Hoshino et al. (1996) used SAXS to study the organisation of imogolite nanotubes in concentrated dispersion. Using Guinier and Porod extrapolation of the measured scattering at small and large angles, respectively, they computed the distance distribution function $P(r)$ and compared the experimental $P(r)$ to different packing models. They concluded that experimental $P(r)$ are qualitatively well described by hexagonal packing of imogolite nanotubes to form a raftlike imogolite sheet. Levard et al. (2010) used SAXS to follow the formation of Al–Ge imogolite nanotubes in situ. Indeed, the possible synthesis of Al–Ge nanotubes at concentrations larger than 0.1 mol L^{-1} makes it possible to directly record the SAXS intensity of the dispersed nanotubes in water. Such experiments are impossible for imogolite, which are generally synthesised in mmol L^{-1} concentration range. They showed that the volume fraction of scattering objects is constant from the beginning of the synthesis. They also proposed a model for the proto-imogolite structure. Indeed, the scattering curve is well explained by a curved sheet corresponding to a nanotube piece containing an average of 200 Al atoms. Such a structure for proto-imogolite has been anticipated very early (Farmer et al., 1979). According to Levard et al. (2010), proto-imogolite contains Ge lacuna and does not have the same curvature as the final nanotube. Maillet et al. (2010) discovered using SAXS and cryo-TEM that Al–Ge imogolite can form both SW and DW nanotubes (Fig. 10.4). The scattering intensity in the small-angle part is very well described by homogeneous models consisting of hollow cylinders. Looking back at the XRD data since this discovery tends to show that most of the Al–Ge nanotubes obtained by previous research were most probably DW. Maillet et al. (2011) used SAXS to follow the growth kinetic of Al–Ge imogolite. They showed that the proto-imogolite coexists with the nanotubes and that their concentration decreases with the nanotube growth. The overall volume fraction is constant during the transformation of proto-imogolite into nanotubes. The diameter of the nanotubes is the same from the beginning to the end of the process. Thill et al. (2012a,b) used SAXS to explain the mechanism controlling the formation of SW or DW Al–Ge nanotubes and proposed methods to obtain both structures.

Amara et al. (2014) showed that SAXS and XRD give different diameters for Al–Ge SW nanotubes, as already observed for imogolite by Kajiwara et al. (1988). Using a detailed analysis of the XRD data, they demonstrated that the nanotube deform upon drying. They adopted a hexagonal shape in a 2D

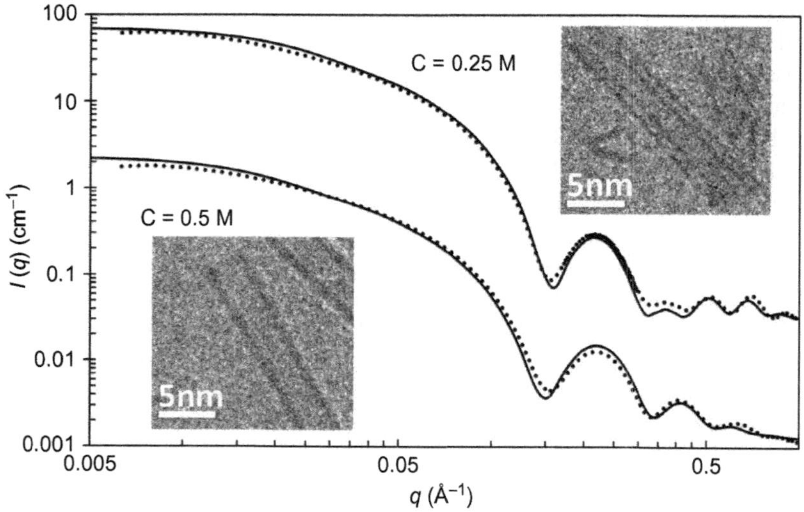

FIG. 10.4 SAXS curves from Al–Ge imogolite synthesised at initial aluminium concentrations of $C=0.25$ and 0.5 M. Insets show close-up cryo-TEM pictures of individual nanotubes. At $C=0.25$ M, DW nanotubes are mainly formed, while SW nanotubes dominate at $C=0.5$ M. *Reprinted with permission from Maillet et al. (2010). Copyright 2010 American Chemical Society.*

hexagonal packing. Considering a constant perimeter, the contraction occurring from a circle to a hexagon would be given by a factor of $\frac{\pi}{2\sqrt{(3)}}=0.906$. In the case of Kajiwara et al. (1988), the ratio is $2.3/2.52=0.91$ in very good agreement with this prediction. For Amara et al. (2014) the ratio is $3.97/4.06=0.977$. This ratio is larger than expected, meaning that the packing parameter of the hexagonal nanotubes is too high. The authors explained this 0.1-nm difference by the presence of ClO_4^- anions between nanotube walls. The presence of these anions is confirmed by IR spectra.

SAXS has been used by Boyer et al. (2014) to characterise the structure of hybrid imogolite with an hydrophobic internal cavity in which Si–OH groups are replaced by Si–CH_3 groups (methyl imogolite). They showed that these nanotubes are larger than standard synthetic imogolite and that their cavities do not contain bulk water. Indeed, the electronic density of the internal cavity is determined to be about $0.1\,e^-\,Å^{-3}$, whereas the electronic density of bulk water is $0.33\,e^-\,Å^{-3}$. SAXS data also showed that the hybrid nanotubes are not as perfectly dispersed as for synthetic imogolite. Amara et al. (2015) did the same characterisation of both methyl imogolite and Al–Ge methyl imogolite. They confirmed that the diameter of methyl imogolite is larger than the synthetic imogolite. But surprisingly, Al–Ge methyl imogolite have a smaller diameter than SW Al–Ge imogolite. The electronic density inside Al–Ge methyl imogolite is also reduced compared to bulk water.

Amara et al. (2015) showed also that SAXS is able to demonstrate the trapping of hydrophobic molecules inside the hybrid nanotubes. They used electron-rich molecules (bromopropanol) to follow the modification of the internal electronic density. Upon the addition of bromopropanol, the internal electronic density increased from 0.1 to 0.33 e^{-} Å^{-3}. It is not really possible to say if this increase is only due to bromopropanol entering the nanotubes, or if bromopropanol modified the internal cavity wetting properties, allowing water to enter the cavity. In this study, Amara et al. (2015) measured the SAXS intensity of a concentrated dispersion of synthetic imogolite. Using a fit of the whole scattered intensity with a homogeneous hollow cylinder model, they determined an external diameter of 2.68 nm that was slightly larger than the diameter determined from natural imogolite by Kajiwara et al. (1988).

SAXS is not very well suited to measure the length and length distribution of imogolite. But it is probably the best way to characterise their diameter. It does not require drying, which is known to cause deformation. The information is obtained from a scattering volume containing more than 10^{10} nanotubes. Therefore, the data are true average morphologies.

10.5 CHEMICAL AND MASS ANALYSIS

10.5.1 Chemical Composition

The chemical composition of allophane and imogolite has been measured mainly by atomic absorption spectrometer and agrees to a large extent with the average theoretical formula $(OH)_3Al_2O_3Si(OH)$ with Si/Al = 0.5 proposed by Cradwick et al. (1972). The different measured Si/Al values obtained from the research literature are presented in Table 10.2.

According to Levard et al. (2010) and Yucelen et al. (2011), before the nanotube formation, the Si/Al ratio tends to be less than the theoretical value of 0.5. This is because Si atoms adopt a Q^3(3Al) coordination, which is only possible in the lacuna of a dioctahedral sheet. The theoretical ratio for a nanotube as a function of its length would be given by $\frac{\text{Si}}{\text{Al}} = \frac{1}{2+(\delta/L)}$, where L is the nanotube length and δ is approximately 8.5 Å, as the repeat distance along the c-axis of imogolite. This ratio is always less than 0.5 and increases with the nanotube length. This is in agreement with the fact that very short Al–Ge imogolite nanotubes have Ge/Al ratio of 0.48 (Wada and Wada, 1982), 0.37 (Mukherjee et al., 2005) and 0.472 (Levard et al., 2011). For imogolite, however, the ratio is very often larger than 0.5. This is probably explained by its coexistence with other aluminosilicate and amorphous SiO_2 species.

10.5.2 Gas Adsorption

Gas adsorption is a very powerful and popular technique for the characterisation of porous materials. It has been applied by many research groups to study

TABLE 10.2 Measured Si-Al Ratios for Natural and Synthetic Imogolites

References	Si/Al (mol ratio)	Comment
Yoshinaga and Aomine (1962)	0.53	Natural imogolite
Russel et al. (1969)	0.75	Natural imogolite
Wada and Yoshinaga (1969)	0.53–0.58	Natural imogolite
Jenne (1972)	0.54–0.81	Natural imogolite and allophane
Cradwick et al. (1972)	0.5	Proposed theoretical formula
Kitagawa (1974)	0.69	Natural imogolite
Farmer et al. (1977)	0.58	Synthetic imogolite
Parfitt et al. (1980)	0.5	
Theng et al. (1982)	0.505	
Wada and Wada (1982)	0.515 and 0.485	Imogolite and Al–Ge imogolite, respectively
Clark and McBride (1984)	0.525	
Inoue and Huang (1985)	0.405–0.735	
Van der Gaast et al. (1985)	0.56–0.84	
Goodman et al. (1985)	0.5 and 0.588	Synthetic and natural imogolite, respectively
MacKenzie et al. (1989)	0.555	
Johnson et al. (1988)	0.53	
Garcia-Hernandez and Rodriguez Rodriguez (1990)	0.465	
Su and Harsh (1993)	0.495	Synthetic imogolite
Denaix et al. (1999)	0.526	
Bursill et al. (2000)	0.58	
Mukherjee et al. (2005)	0.58 and 0.37	Imogolite and Al–Ge imogolite, respectively
Nakanishi et al. (2008)	0.532–0.559	Effect of dialysis time
Levard et al. (2010)	0.467, 0.513	Al–Ge imogolite $t=0$ and 5 days
Yucelen et al. (2011)	0.356–0.387	Av. Si/Al from ESI ageing and heating, respectively
Levard et al. (2011)	0.781–0.472	Al–Ge imogolite as a function of OH/Al

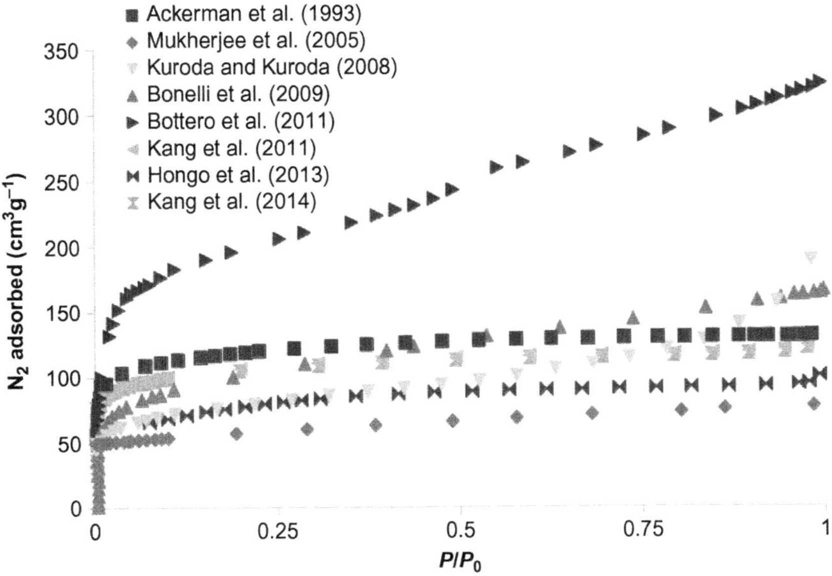

FIG. 10.5 Nitrogen adsorption isotherms for imogolites and hybrid imogolites.

the porosity of imogolite. The N_2 adsorption isotherms for several imogolite samples are reported in Fig. 10.5.

In a very detailed study, Ackerman et al. (1993) tested N_2 adsorption on two synthetic imogolites with 100% Si and with 50% Si/50% Ge. They defined three types of pores in imogolite powder: (i) the C-type, corresponding to mesopores between imogolite bundles; (ii) the B-type, corresponding to micropores between nanotubes in a bundle; and (iii) the A-type, corresponding to the micropore inside the nanotube. These authors showed that high-temperature outgassing is required to empty the smallest pores (the B-type). Indeed, they observed an increase of low-pressure gas adsorption between 225°C and 275°C outgassing temperature. For synthetic imogolite, they found a specific surface area (SSA) of 398 m^2g^{-1} and a mean pore radius of 1.01 nm. This SSA value is mainly due to the internal surface of the imogolite. The very well-organised films used by Ackerman et al. (1993) contained no mesoporosity, and B-type pores in well-organised nanotubes correspond to a very small surface. Ackerman et al. (1993) also tested the adsorption of CO_2 and CH_4.

Ohashi et al. (2004) measured the CH_4 adsorption capacity of natural and synthetic imogolite and showed that the synthetic imogolite could adsorb 50.6 mg ml^{-1} at 4.05 MPa, which is higher than natural compressed methane (28 mg ml^{-1} at 4 MPa). They also measured the adsorption isotherm of water, showing that synthetic imogolite adsorbs 40 mass% of water at $P/P_0=0.9$, with a small hysteresis between $P/P_0=0.1$ and 0.9. The adsorption isotherm shows a sharp increase between $P/P_0=0.9$ and 1, with a maximum adsorption of 80 mass%.

Mukherjee et al. (2005) measured the N_2 adsorption isotherm on synthetic imogolite and Al–Ge imogolite. The samples were outgassed at 250°C, and the measurements were between P/P_0 10^{-3} and 1. Despite the largest radius measured by XRD for Al–Ge imogolite, the maximum adsorbed volume is less than for imogolite on almost the entire P/P_0 range. Bottero et al. (2011) measured the N_2 adsorption of hybrid imogolite, whose internal surface is modified with CH_3 groups. They showed that the radius increased, as in the case of Al–Ge nanotubes, but the adsorption capacity was greatly enhanced. Recently, using modification of the internal surface functions, Kang et al. (2014) showed that it is possible to greatly enhance the adsorption selectivity of hybrid imogolite between CO_2 and CH_4.

10.5.3 Water Adsorption and Thermogravimetric Analysis

Yoshinaga and Aomine (1962) reported a free water content of 19.7% at 50% RH for imogolite. The total water content, including structural water, being $H_2O/Al_2O_3=4.96$ (26.9 mass%). After dehydration at 150°C, a value of H_2O/Al_2O_3 ratio between 2.0 and 2.8 was given by Wada and Yoshinaga (1969) and Parfitt et al. (1980). The water content depends on the sample preparation environment. For example, the degree of nanotube alignment significantly changes the water content (Yoshinaga et al., 1968; Farmer et al., 1983). Thermal analyses (TA) of imogolite showed that a strong endothermic peak at 170°C–190°C occurs due to the large amount of water associated with imogolite, and that a second and smaller endothermic peak at 425°C–435°C is attributed to the dehydroxylation of the imogolite structure (Yoshinaga and Aomine, 1962). MacKenzie et al. (1989) confirmed the previous results (ie, 25.5% mass loss below 200°C and 13% endothermic mass loss with a maximum rate at 400°C). The second endothermic peak corresponds to the loss of the low angles that are characteristic features of imogolite on XRD measurements. Yoshinaga and Aomine (1962) claimed that this peak is characteristic of imogolite. Kitagawa (1974), in a TA detailed study of allophane, showed that an exothermic peak due to dehydroxylation was also observed, but at a smaller temperature of 320°C. Yoshinaga et al. (1973) measured differential thermal analysis (DTA) curves for natural samples. They observed several endothermic peaks at 271°C, 402°C and 462°C, and they attribute the 271°C peak to the presence of gibbsite and the 462°C peak to the presence of layered aluminosilicate. Henmi and Yoshinaga (1981) examined the effect of dry grinding on the imogolite structure and showed that upon grinding for only 2 min, the endothermic peaks corresponding to dehydroxylation shifted significantly from 405°C to 393°C, while the first dehydration peak, at about 120°C, was not affected. Abidin et al. (2008) showed that the second endothermic peak of DTA curves could be used to assess the quality of synthetic imogolite.

The mass loss of imogolite as a function of temperature has also been used to characterise hybrid imogolite. The grafting of OPA on imogolite has been investigated by TA (Yamamoto et al., 2001; Bac et al., 2009). Bottero et al. (2011) compared the TA analyses of imogolite and hybrid imogolite with methyl on the inner surface. They showed that the loss of water occurred at lower temperatures. An endothermic mass loss at higher temperatures, at about 325°C, was also observed. Kang et al. (2011) measured DTA curves for several hybrid nanotubes and showed that an additional peak between 450°C and 600°C can be observed for the hybrid nanotubes.

At very high temperatures, between 935°C and 955°C depending on the samples, a sharp exothermic peak is observed. This peak is caused by the crystallisation of mullite.

To understand more precisely the interaction between water and imogolite, a few groups published water adsorption isotherms (Fig. 10.6). Zhang et al. (2011) used osmotic ensemble computation methods to predict the water adsorption curves on different imogolite bundle configurations. They showed that in a bundle of densely packed imogolite nanotubes, water is adsorbed at very low partial pressure between the nanotubes. At a pressure threshold of 0.1 kPa ($P/P_0 \sim 3 \cdot 10^{-2}$), the nanotubes are filled with water. After this threshold, the model predicts a saturation of the water content. To explain the continuous increase of the water content observed experimentally, Zhang et al. (2011) considered a swelling of the imogolite nanotube bundles from 24 to 30 Å.

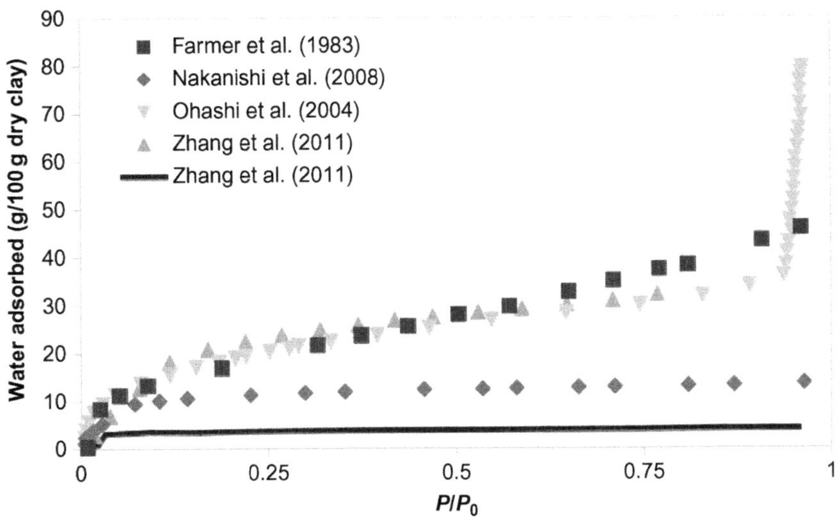

FIG. 10.6 Water adsorbed by imogolite as a function of partial water pressure. The *solid line* marks the result of osmotic ensemble computation by Zhang et al. (2011).

10.6 CONCLUDING REMARKS

Since its discovery in 1962, the characterisation of imogolite has progressed tremendously. IR spectroscopy allows a clear identification of the imogolite structure. It can also give clues about the purity of the sample, along with simple visual inspection. XRD is the most often used technique to characterise imogolite. However, its interpretation is rather complex, as the diffraction signal is modified by the shape of the nanotubes, by their wetting and by the nature of their mutual arrangement, which can span from disordered 'mikado-like' powders to well-aligned bundles. Due to this complexity, some inconsistencies exist in the research literature, especially for the determination of the imogolite diameter. In addition, it has been pointed out very recently that imogolite nanotubes may deform upon drying, which adds another parameter to the complex interpretation of XRD diffraction patterns. Therefore, even if XRD is the most common technique used to quantify the diameter of imogolite, it may not be the most adapted method. Recently, SAXS proved to be a very powerful technique to measure the structure of imogolite nanotubes. Indeed, when used in liquid dispersion, it prevents all deformation and drying artefacts. Concerning the microscopic techniques, many progresses have also been performed. The introduction of cryo-TEM to prevent most of the sample preparation modifications has proved to be very useful to obtain high-resolution images of isolated nanotubes. It is probable that this technique will be used fruitfully to explore the formation mechanism of imogolite in the near future. A remaining analytical challenge for imogolite is to develop a method allowing high-throughput determination of imogolite length distribution. Today, only tedious microscopic-based complex techniques (eg, AFM and TEM) allow for accessing this information. Other simpler techniques (such as DLS and viscosity/mass concentration correlations) only give access to an average length. Satisfying this analytical need could significantly heighten our understanding of the kinetics of imogolite growth and could help validate the recently developed models.

Nowadays, a powerful set of tools allows us to make progress on the synthesis, modification and use of imogolite, an exceptional clay mineral nanotube with great scientific and technological potential.

REFERENCES

Abidin, Z., Matsue, N., Henmi, T., 2008. A new method for nano tube imogolite synthesis. Jpn. J. Appl. Phys. 47, 5079–5082.

Ackerman, W.C., Smith, D.M., Huling, J.C., Kim, Y.-W., Bailey, J.K., Brinker, C.J., 1993. Gas/vapor adsorption in imogolite: a microporous tubular aluminosilicate. Langmuir 9, 1051–1057.

Alvarez-Ramírez, F., 2007. Ab initio simulation of the structural and electronic properties of aluminosilicate and aluminogermanate nanotubes with imogolite-like structure. Phys. Rev. B 76, 125421.

Amara, M.-S., Paineau, E., Bacia-Verloop, M., Krapf, M.-E., Davidson, P., Belloni, L., Levard, C., Rose, J., Launois, P., Thill, A., 2013. Single-step formation of micron long $(OH)_3Al_2O_3GeOH$ imogolite-like nanotubes. Chem. Commun. 49, 11284–11286.

Amara, M.S., Rouzière, S., Paineau, E., Bacia-Verloop, M., Thill, A., Launois, P., 2014. Hexagonalization of aluminogermanate imogolite nanotubes organized into closed packed bundles. J. Phys. Chem. C 118, 9299–9306.

Amara, M.-S., Paineau, E., Rouzière, S., Guiose, B., Krapf, M.E.M., Taché, O., Launois, P., Thill, A., 2015. Hybrid tunable-diameter metal-oxide nanotubes for organic molecules trapping. Chem. Mater. 27, 1488–1494.

Arai, Y., McBeath, M., Bargar, J.R., Joye, J., Davis, J.A., 2006. Uranyl adsorption and surface speciation at the imogolite–water interface: self-consistent spectroscopic and surface complexation models. Geochim. Cosmochim. Acta 70, 2492–2509.

Arancibia-Miranda, N., Escudey, M., Molina, M., Garcia-Gonzales, M.T., 2011. Use of isoelectric point and pH to evaluate the synthesis of a nanotubular aluminosilicate. J. Non-Cryst. Solids 357, 1750–1756.

Bac, B.H., Song, Y., Kim, M.H., Lee, Y.B., Kang, I.M., 2009. Surface-modified aluminogermanate nanotube by OPA: synthesis and characterization. Inorg. Chem. Commun. 12, 1045–1048.

Barona, G.N.B., Lim, J., Choi, M., Jung, B., 2013. Interfacial polymerization of polyamide-aluminosilicate SWNT nanocomposite membranes for reverse osmosis. Desalination 325, 138–147.

Barrett, S.M., Budd, P.M., Price, C., 1990. The synthesis and characterization of imogolite. Eur. Polym. J. 27, 609–612.

Barron, P.F., Wilson, M.A., Campbell, A.S., Frost, R.L., 1982. Detection of imogolite in soils using solid state ^{29}Si NMR. Nature 299, 616–618.

Bishop, J.L., Rampe, E.B., Bish, D.L., Abidin, Z., Baker, L.L., Matsue, E., Henmi, T., 2013. Spectral and hydration properties of allophane and imogolite. Clays Clay Miner. 61, 57–74.

Bonelli, B., Bottero, I., Ballarini, N., Passeri, S., Cavani, F., 2009. IR spectroscopic and catalytic characterization of the acidity of imogolite-based systems. J. Catal. 264, 15–30.

Bonelli, B., Armandi, A., Garrone, E., 2013. Surface properties of alumino-silcate single-walled nanotubes of the imogolite type. Phys. Chem. Chem. Phys. 15, 13381–13390.

Bottero, I., Bonelli, B., Ashbrook, S.E., Wright, P.A., Zhou, W., Tagliabue, M., Armandi, M., Garrone, E., 2011. Synthesis and characterization of hybrid organic/inorganic nanotubes of the imogolite type and their behaviour towards methane adsorption. Phys. Chem. Chem. Phys. 13, 744–750.

Boyer, M., Paineau, E., Bacia-Verloop, M., Thill, A., 2014. Aquous dispersion state of amphiphilic hybrid aliminosilicate nanotubes. Appl. Clay Sci. 96, 45–49.

Bursill, L.A., Peng, J.L., Bourgeois, L.N., 2000. Imogolite: an aluminosilicate nanotube material. Philos. Mag. A 80, 105–117.

Chemni, A., Brendlé, J., Marichal, C., Lebeau, B., 2013. A novel fluoride route for the synthesis of aluminosilicate nanotubes. Nanomaterials 3, 117–125.

Childs, C.W., Inoue, K., Seyama, H., Soma, M., Theng, B.K.G., Yuan, G., 1997. X-ray photoelectron spectroscopy characterization of silica springs allophane. Clay Miner. 32, 565–572.

Clark, C.J., McBride, M.B., 1984. Cation and anion retention by natural and synthetic allophane and imogolite. Clays Clay Miner. 32, 291–299.

Cradwick, P.D.G., Farmer, V.C., Russell, J.D., Masson, C.R., Wada, K., Yoshinaga, N., 1972. Imogolite, a hydrated aluminium silicate of tubular structure. Nat. Phys. Sci. 240, 187–189.

Creton, B., Bougeard, D., Smirnov, K.S., Guilment, J., Poncelet, O., 2008. Molecular dynamics study of hydrated imogolite 2. Structure and dynamics of confined water. Phys. Chem. Chem. Phys. 10, 4879–4888.

Denaix, L., Lamy, I., Bottero, J.Y., 1999. Structure and affinity towards Cd^{2+}, Cu^{2+}, Pb^{2+} of synthetic colloidal amorphous aluminosilicates and their precursors. Colloids Surf. A 158, 315–325.

Exley, C., Schneider, C., Doucet, F.J., 2002. The reaction of aluminium with silicic acidic solution: an important mechanism in controlling the biological availability of aluminium. Coord. Chem. Rev. 228, 127–135.

Farmer, V.C., Fraser, A.R., Tait, J.M., 1977. Synthesis of imogolite: a tubular aluminium silicate polymer. J. Chem. Soc. Chem. Commun. 462–463.

Farmer, V.C., Fraser, A.R., Tait, J.M., 1979. Characterization of the chemical structure of natural and synthetic aluminosilicate gels and sols by infrared spectroscopy. Geochim. Cosmochim. Acta 43, 1417–1420.

Farmer, V.C., Adams, M.J., Fraser, A.R., Palmieri, F., 1983. Synthetic imogolite: properties, synthesis, and possible applications. Clay Miner. 18, 459–472.

Garcia-Hernandez, J.E., Rodriguez Rodriguez, R.A., 1990. Altération expérimentale des cendres basaltiques dans l'oxalate d'ammonium 10^{-4} M. Science du sol 28, 193–201.

Goodman, B.A., Russell, J.D., Montez, B., Oldfield, E., Kirkpatrick, R.J., 1985. Structural studies of imogolite and allophane by aluminium-27 and silicon-29 nuclear magnetic spectroscopy. Phys. Chem. Miner. 12, 342–346.

Gustafsson, J.P., 2001. The surface chemistry of imogolite. Clays Clay Miner. 49, 73–80.

Hatakeyama, M., Hara, T., Ichikuni, N., Shimazu, S., 2011. Characterization of heat-treated synthetic imogolite by ^{27}Al MAS and ^{27}Al MQMAS solid-state NMR. Bull. Chem. Soc. Jpn. 84, 656–659.

He, H., Barr, T.L., Klinowski, J., 1995. ESCA and solid-state NMR studies of allophane. Clay Miner. 30, 201–209.

Henmi, T., Yoshinaga, N., 1981. Alteration of imogolite by dry grinding. Clay Miner. 16, 139–149.

Hongo, T., Sugiyama, J., Yamazaki, A., Yamasaki, A., 2013. Synthesis of imogolite from rice hush ash and evaluation of its acetaldehyde adsorption ability. Ind. Eng. Chem. Res. 52, 2111–2115.

Hoshino, H., Urakawa, H., Donkai, N., Kajiwara, K., 1996. Simulation of mesophase formation of rodlike molecule, imogolite. Polym. Bull. 36, 257–264.

Hu, J., Kannangara, K., Wilson, M.A., Reddy, N., 2004. The fused silicate route to protoimogolite and imogolite. J. Non-Cryst. Solids 347, 224–230.

Ildefonse, P., Kirkpatrick, R.J., Montez, B., Calas, G., Flank, A.M., Lagarde, P., 1994. ^{27}Al MAS NMR and aluminum X-ray adsorption near edge structure study of imogolite and allophanes. Clays Clay Miner. 42, 276–287.

Inoue, K., Huang, P.M., 1985. Influence of citric acid on the formation of short range ordered aluminosilicates. Clays Clay Miner. 33, 312–322.

Iyoda, F., Hayashi, S., Arakawa, S., John, B., Okamoto, M., Hayashi, H., Yuan, G., 2012. Synthesis and adsorption characteristics of hollow spherical allophane nano-particles. Appl. Clay Sci. 56, 77–83.

Jaymes, I., Douy, A., 1995. Homogeneous precipitation of mullite precurssors. J. Sol-Gel Sci. Technol. 4, 7–13.

Jenne, E.A., 1972. Surface charge density dependency on Al_2O_3 content in imogolite. Clays Clay Miner. 20, 101–103.

Jiravanichanun, N., Yamamoto, K., Yonemura, H., Yamada, S., Otsuka, H., Takahara, A., 2008. Fabrication of conjugated polymer hybrid films with radially oriented aluminosilicate nanofibers by spin-assembly. Bull. Chem. Soc. Jpn. 81 (12), 1663–1668.

Johnson, I.D., Werpy, T.A., Pinnavaia, T.J., 1988. Tubular silicate-layered silicate intercalation compounds: a new family of pillared clays. J. Am. Chem. Soc. 110, 8545–8547.

Kajiwara, K., Urakawa, H., Donkai, N., 1988. Lyotropic mesophase formation of imogolite. Bull. Inst. Chem. Res., Kyoto Univ. 66, 165–175.

Kang, D.Y., Zang, J., Wright, I.R., McCanna, A.L., Jones, C.W., Nair, S., 2010. Dehydratation, dehydroxilation, and rehydroxylation of single-walled aluminosilicate nanotubes. ACS Nano 4, 4897–4907.

Kang, D.-Y., Zang, J., Jones, C.W., Nair, S., 2011. Single-walled aluminosilicate nanotubes with organic-modified interiors. J. Phys. Chem. C 115, 7676–7685.

Kang, D.-Y., Brunelli, N.A., Yucelen, G.I., Venkatasubramanian, A., Zang, J., Leisen, J., Hesketh, P.J., Jones, C.W., Nair, S., 2014. Direct synthesis of single-walled aminoaluminosilicate nanotubes with enhanced molecular adsorption selectivity. Nat. Commun. 5, 3342.

Kawano, M., Tomita, K., 1994. Growth of smectite from leached layer during experimental alteration of albite. Clays Clay Miner. 42, 7–17.

Kawano, M., Tomita, K., Kamino, Y., 1993. Formation of clay-minerals during low temperature experimental alteration of obsidian. Clays Clay Miner. 41, 431–441.

Kitagawa, Y., 1974. Dehydratation of allophane and its structural formula. Am. Miner. 59, 1094–1098.

Koenderink, G.H., Kluijtmans, S.G.J.M., Philipse, A.P., 1999. On the synthesis of colloidal imogolite fibers. J. Colloid Interf. Sci. 216, 429–431.

Kuroda, Y., Kuroda, K., 2008. Layer-by-layer assembly of imogolite nanotubes and polyelectrolytes into core-shell particles and their conversion to hierarchically porous spheres. Sci. Technol. Adv. Mater. 9, 025018.

Kuroda, Y., Kuroda, K., 2010. Expansion of intertubular mesopores of imogolite nanotubes by thermal decomposition of an imogolite–poly(sodium 4-styrenesulfonate) composite. Chem. Lett. 40, 46–48.

Levard, C., Rose, J., Masion, A., Doelsch, E., Borschneck, D., Olivi, L., Dominici, C., Grauby, O., Woicik, J.C., Bottero, J.Y., 2008. Synthesis of large quantity of single-walled aluminogermanate nanotube. J. Am. Chem. Soc. 130, 5862–5863.

Levard, C., Rose, J., Thill, A., Masion, A., Doelsch, E., Maillet, P., Spalla, O., Olivi, L., Cognigni, A., Ziarelli, F., Bottero, J.Y., 2010. Formation and growth mechanism of imogolite-like aluminogermanate nanotubes. Chem. Mater. 22, 2466–2473.

Levard, C., Masion, A., Rose, J., Doelsch, E., Borschneck, D., Olivi, L., Chaurand, P., Dominici, C., Ziarelli, F., Thill, A., Maillet, P., Bottero, J.Y., 2011. Synthesis of Ge-imogolite: influence of the hydrolysis ratio on the structure of the nanotubes. Phys. Chem. Chem. Phys. 13, 14516–14522.

Lippmaa, E., Magi, M., Samoson, A., Engelhardt, G., Grimer, A.R., 1980. Structural studies of silicates by solid-state ^{29}Si NMR. J. Am. Chem. Soc. 102, 4889–4893.

Liu, W., Chaurand, P., Di Georgio, C., Di Méo, M., Thill, A., Auffan, M., Masion, A., Borschneck, D., Chaspoul, F., Gallice, P., Botta, A., Bottero, J.Y., Rose, J., 2012. Influence of the length of imogolite-like nanotubes on their cytotoxicity and genotoxicity towards human dermal cells. Chem. Res. Toxicol. 25 (11), 2513–2522.

Ma, W., Yah, W.O., Otsuka, H., Takahara, A., 2012. Surface functionalization of aluminosilicate nanotubes with organic molecules. Beilstein J. Natotechnol. 3, 82–100.

MacKenzie, K.J., Bowden, M.E., Brown, J.W.M., Meinhold, R.H., 1989. Structural and thermal transformation of imogolite studied by ^{29}Si and ^{27}Al high-resolution solid-stated magnetic nuclear resonance. Clays Clay Miner. 37, 317–324.

Maillet, P., Levard, C., Larquet, E., Mariet, C., Spalla, O., Menguy, N., Masion, A., Doelsch, E., Rose, J., Thill, A., 2010. Evidence of double-walled Al-Ge imogolite-like nanotubes. A cryo-TEM and SAXS investigation. J. Am. Chem. Soc. 132, 1208–1209.

Maillet, P., Levard, C., Spalla, O., Masion, A., Rose, J., Thill, A., 2011. Growth kinetic of single and double-walled aluminogermanate imogolite-like nanotubes: an experimental and modelling approach. Phys. Chem. Chem. Phys. 13, 2682–2689.

Miyauchi, N., Aomine, S., 1966. Mineralogy of gel-like substance in the pumice bed in Kanuma and Kitakami districts. Soil Sci. Plant Nutr. 12 (5), 19–22.

Mukherjee, S., Bartlow, V.M., Nair, S., 2005. Phenomenology of the growth of single-walled aluminosilicate and aluminogermanate nanotubes of precise dimensions. Chem. Mater. 17, 4900–4909.

Mukherjee, S., Keesuk, K., Nair, S., 2007. Short, highly ordered, single-walled mixed oxide nanotubes assemble from amorphous nanoparticles. J. Am. Chem. Soc. 129, 6820–6826.

Nakanishi, R., Suzuki, M., Inukai, K., Maeda, M., Wada, S.I., 2008. Effects of dialysis on the formation of imogolite from highly concentrated hydroxyaluminum-silicate solution. Trans. Mater. Res. Soc. Jpn. 33, 493–496.

Ohashi, F., Tomura, S., Akaku, K., Hayashi, S., Wada, S.I., 2004. Characterization of synthetic imogolite nanotubes as gas storage. J. Mater. Sci. 39, 1799–1801.

Ohrai, Y., Gozu, T., Yoshida, S., Takeushi, O., Iijima, S., Shigekawa, H., 2005. Atomic force microscopy on imogolite, aluminosilicate nanotube, adsorbed on Au(111) surface. Jpn. J. Appl. Phys. 44, 5397–5399.

Ookawa, M., Inoue, Y., Watanabe, M., Suzuki, M., Yamaguchi, T., 2006. Synthesis and characterization of Fe containing imogolite. Clay Sci. 12, 280–284.

Parfitt, R.L., Furkert, R.J., Henmi, T., 1980. Identification and structure of two types of allophane from volcanic ash soils and tephra. Clays Clay Miner. 28, 328–334.

Park, S., Lee, Y., Kim, B., Lee, J., Jeong, Y., Noh, J., Takahara, A., Sohn, D., 2007. Two-dimensional alignment of imogolite on a solid surface. Chem. Commun. 28, 2917–2919.

Qi, X., Yoon, H., Lee, S.-H., Yoon, J., Kim, S.-J., 2008. Surface-modified imogolite with 3-APS-OsO_4 complex: synthesis, characterization and its application in the dihydroxylation of olefins. J. Ind. Eng. Chem. 14, 136–141.

Russell, J.D., McHardy, W.J., Fraser, A.R., 1969. Imogolite: a uniquealuminosilicate. Clay Miner. 8, 87–99.

Shikinaka, K., Koizumi, Y., Shigehara, K., 2015. Mechanical/optical behaviors of imogolite hydrogels depending on their compositions and oriented structures. J. Appl. Polym. Sci. 132 (12), 41691.

Su, C., Harsh, J.B., 1993. The electrophoretic mobility of imogolite and allophane in the presence of inorganic anions and citrate. Clays Clay Miner. 41, 461–471.

Tamura, K., Kawamura, K., 2002. Molecular dynamic modelingof tubular aluminum silicate: imogolite. J. Phys. Chem. B 106, 271–278.

Tani, M., Liu, C., Huang, P.M., 2004. Atomic force microscopy of synthetic imogolite. Geoderma 118, 209–220.

Theng, B.K.J., Russell, M., Churchman, G.J., Parfitt, R.L., 1982. Surface properties of allophane, halloysite and imogolite. Clays Clay Miner. 30, 143–149.

Thill, A., Spalla, O., 2002. Capillary against adhesion forces during drying of particle submonolayers on a solid substrate. Langmuir 18, 4783–4789.

Thill, A., Maillet, P., Guiose, B., Spalla, O., Belloni, L., Chaurand, P., Auffan, M., Olivi, L., Rose, J., 2012a. Physico-chemical control over the single-or double-wall structure of aluminogermanate imogolite-like nanotubes. J. Am. Chem. Soc. 134, 3780–3786.

Thill, A., Guiose, B., Bacia-Verloop, M., Geertsen, V., Belloni, L., 2012b. How the diameter and structure of $(OH)_3Al_2O_3Si_xGe_{1-x}(OH)$ imogolite nanotubes are controlled by an adhesion versus curvature competition. J. Phys. Chem. C 116, 26841–26845.

Van der Gaast, S.J., Wada, K., Wada, S.I., Kakuto, Y., 1985. Small-angle X-ray powder diffraction, morphology, and structure of allophane and imogolite. Clay Miner. 33, 237–243.

Wada, S.I., Wada, K., 1977. Density and structure of allophane. Clay Miner. 12, 289–298.

Wada, S.I., Wada, K., 1982. Effects of substitution of germanium for silicon in imogolite. Clays Clay Miner. 30, 123–128.

Wada, K., Yoshinaga, N., 1969. The structure of imogolite. Am. Miner. 54, 50–71.

Wada, K., Yoshinaga, N., Yotsumoto, H., Ibe, K., Aida, S., 1970. High resolution electron micrographs of imogolite. Clay Miner. 8, 487–489.

Wada, S.I., Eto, A., Wada, K., 1979. Synthetic allophane and imogolite. J. Soil Sci. 30, 347–355.

Wilson, M.A., Lee, G.S.H., Taylor, R.C., 2001. Tetrahedral rehydratation during imogolite formation. J. Non-Cryst. Solid. 296, 172–181.

Yamamoto, K., Otsuka, H., Wada, S.I., Takahara, A., 2001. Surface modification of aluminosilicate nanofiber "imogolite" Chem. Lett. 30, 1162–1163.

Yamamoto, K., Otsuka, H., Wada, S.I., Sohn, D., Takahara, A., 2005. Transparent polymer nanohybrid prepared by in situ synthesis of aluminosilicate nanofibers in poly(vinyl alcohol) solution. Soft Matter 1, 372–377.

Yang, H., Su, Z., 2007. Individual dispersion of synthetic imogolite nanotubes via droplet evaporation. Chin. Sci. Bull. 52, 2301–2303.

Yang, H., Wang, C., Su, Z., 2008. Growth mechanism of synthetic imogolite nanotubes. Chem. Mater. 20, 4484–4488.

Yoshinaga, N., Aomine, S., 1962. Imogolite in some andosoils. Soil Sci. Plant Nutr. 8 (3), 22–29.

Yoshinaga, N., Yotsumoto, H., Ibe, K., 1968. An electron microscopic study of soil allophane with an ordered structure. Am. Miner. 53, 319–323.

Yoshinaga, N., Tait, J.M., Soong, R., 1973. Occurrence of imogolite in some volcanic ash soils of New Zealand. Clay Miner. 10, 127–130.

Yucelen, G.I., Choudhury, R.P., Vyalikh, A., Scheler, U., Beckham, H.W., Nair, S., 2011. Formation of single-walled aluminosilicate nanotubes from molecular precursors and curved nanoscale intermediates. J. Am. Chem. Soc. 133, 5397–5412.

Yucelen, G.I., Choudhury, R.P., Leisen, J., Nair, S., Beckham, H.W., 2012. Defects structures in aluminosilicates single-walled nanotubes: a solid-state nuclear magnetic resonance investigation. J. Phys. Chem. C 116, 17149–17157.

Zanzottera, C., 2012. Hybrid Organic/Inorganic Nanotubes of Imogolite Type. PhD Thesis, Politechnico di Torino, Italy.

Zanzottera, C., Armandi, M., Esposito, S., Garrone, E., Bonelli, B., 2012a. CO_2 adsorption on aluminosilicate single-walled nanotubes of imogolite type. J. Phys. Chem. C 116, 20417–20425.

Zanzottera, C., Vicente, A., Celasco, E., Fernandez, C., Garrone, E., Bonelli, B., 2012b. Physico-chemical properties of imogolite nanotubes functionalized on both external and internal surfaces. J. Phys. Chem. C 116, 7499–7506.

Zhang, J., Nair, S., Sholl, D.S., 2011. Osmotic ensemble methods for predicting adsorption-induced structural transitions in nanoporous materials using molecular simulations. J. Chem. Phys. 134, 184103.

Chapter 11

Deformations and Thermal Modifications of Imogolite

S. Rouzière*, M.S. Amara, E. Paineau and P. Launois
Laboratoire de Physique des Solides, CNRS, Univ. Paris-Sud, Université Paris-Saclay, Orsay Cedex, France
Corresponding author: e-mail: stephan.rouziere@u-psud.fr

11.1 INTRODUCTION

An imogolite is a natural aluminosilicate nanotube with an outer diameter of about 2.3 nm, discovered in Andosols (Yoshinaga and Aomine, 1962). Cradwick et al. (1972), in their seminal article, stated that aluminosilicate nanotubes have the chemical formula $(OH)_3Al_2O_3Si(OH)$ and that they are cylindrical single-walled (SW) nanotubes, formed of a rolled gibbsite layer with isolated orthosilicate groups replacing its inner hydroxyl surface (Fig. 11.1A). Imogolite can also be synthesised in solution (Farmer et al., 1977); in this case, they have a slightly larger diameter—about 2.8 nm (Wada et al., 1979). Imogolite-like SW aluminogermanate nanotubes, where silicon is replaced by germanium, were synthesised by Wada and Wada (1982); their outer diameter is about 4 nm. Since aluminosilicate and aluminogermanate nanotubes have similar structures, the term 'imogolite nanotube' will refer to both of them thereafter in this chapter. A continuous increase in the diameter of SW nanotubes was modelled by tuning the substitution ratio $x = [Si]/([Si] + [Ge])$ (Konduri et al., 2007). More recently, double-walled (DW) aluminogermanate nanotubes (Maillet et al., 2010) were observed and a synthesis procedure to obtain DW nanotubes with micrometric length was proposed (Amara et al., 2013). It should be underscored that imogolite nanotubes are synthesised with a high monodispersity in diameter. Molecular dynamics simulations and ab initio calculations confirmed the stability of the tubular structure and its monodispersity in diameter, which is explained in terms of a strain energy minimum (Tamura and Kawamura, 2002; Konduri et al., 2006; Guimaraes et al., 2007; Lourenço et al., 2014). Due to these unique properties of tuning and monodispersity of diameter, imogolite nanotubes are promising building blocks for nanotechnologies where scalable

Developments in Clay Science, Vol. 7. http://dx.doi.org/10.1016/B978-0-08-100293-3.00011-X

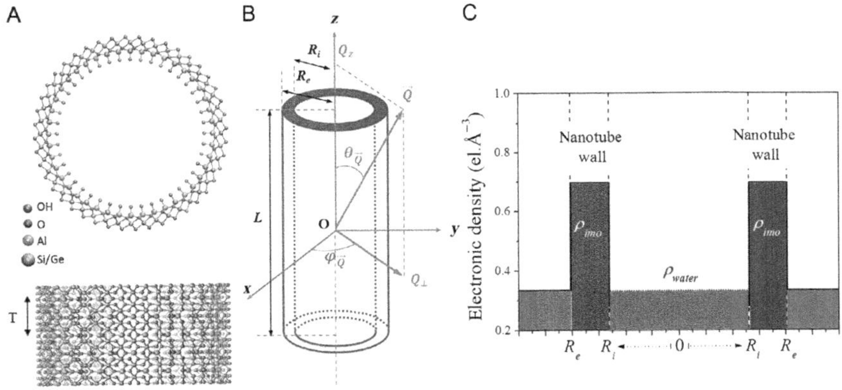

FIG. 11.1 (A) Top and side views of an imogolite nanotube. (B) Schematic representation of an imogolite nanotube with inner and outer radii R_i and R_e and length L. The notations used in the calculations of the scattered intensity are indicated. (C) Radial profile of the electronic density of the imogolite nanotube in dispersion.

and monodisperse objects are required for potential applications in molecular separation or catalysis (Bonelli et al., 2009; Zang et al., 2009, 2010).

In powders, thanks to their monodispersity, these nanotubes typically assemble in bundles (Cradwick et al., 1972; Farmer et al., 1983). While non-interacting imogolite nanotubes are cylindrical, deformations or further structural modifications of these nanotubes may occur due to intertube interactions in powders or under the action of temperature. A few works, based on numerical simulations (Tamura and Kawamura, 2002; Creton et al., 2008), point towards an ovalisation of aluminosilicate nanotube cross sections when they aggregate in bundles of a few nanotubes. In this chapter, the authors will present experimental results carried out on SW aluminogermanate nanotubes. Depending on the degree of organisation of the nanotubes into bundles, two phenomena (namely, ovalisation and hexagonalisation) are evident. These studies are essentially based on the X-ray scattering (XRS) technique and the development of specific methods to analyse XRS measurements. The XRS technique is a well-adapted tool that takes advantage of the high monodispersity in diameter of imogolite nanotubes in order to probe their shape modifications. Basic concepts of XRS formalism used specifically for these studies will be presented here, prior to focusing on the ovalisation and the hexagonalisation phenomena.

Besides interaction-induced deformations, high-temperature treatments of imogolite above 250°C lead to successive modifications of the nanotube structure. The dehydroxylation phenomenon of the nanotube walls occurs in the range 300–600°C, leading eventually to the formation of a lamellar structure that corresponds to the collapse of the tubular structure. Furthermore, higher-temperature treatment of aluminosilicate nanotubes above 900°C leads to a structural transformation towards a crystalline mullite-silica phase. Finally, these thermally induced structural transformations, from the dehydroxylation process to the lamellar and high-temperature mullite phase transformations, are reviewed based on the literature.

11.2 X-RAY SCATTERING FORMALISM

First, it is worth to recall that XRS is a nondestructive technique that yields statistical information over a macroscopic area of the sample under investigation. Besides the fact that it is a global technique, it is particularly interesting in the case of imogolite, compared to electron microscopy or diffraction techniques, as these objects are highly sensitive to electron irradiation. XRS experiments presented in the following discussion, for relatively small scattering wave-vectors, allow determining the cross-section shape and dimensions of the nanotubes, the type of organisation and the number of nanotubes in bundles. The basic concepts of XRS formalism used to obtain this information are detailed in this section.

11.2.1 Individual Nanotubes

Let us first consider a single nanotube, in vacuum, illuminated by a monochromatic X-ray beam. The corresponding XRS amplitude is equal to the Fourier transform of the electronic density $\rho\left(\vec{r}\right)$ (Guinier, 1963):

$$A_1\left(\vec{Q}\right) = \iiint \rho\left(\vec{r}\right) e^{i\vec{Q}\cdot\vec{r}} \mathrm{d}\vec{r} \tag{11.1}$$

where $\vec{Q}$ is the scattering wave-vector, defined as the difference between the wave-vectors of scattered and incident X-rays (the index '1' indicates that one is considering a single nanotube).

Expressing the global electronic density as the convolution product of electronic density around a nucleus and of atomic nuclei positions, the scattering amplitude writes

$$A_1\left(\vec{Q}\right) = \sum\nolimits_{k=\mathrm{O,\,H,\,Al,Si/Ge}} f_k(Q) \iiint \rho_k\left(\vec{r}\right) e^{i\vec{Q}\cdot\vec{r}} \mathrm{d}\vec{r} \tag{11.2}$$

where $f_k(Q)$ is the Fourier transform of atom k electronic density (namely, its X-ray form factor), and $\rho_k\left(\vec{r}\right)$ gives atom k positions. The scattered intensity is proportional to the squared modulus of the scattering amplitude:

$$I_1\left(\vec{Q}\right) \propto \left|A_1\left(\vec{Q}\right)\right|^2 \tag{11.3}$$

For small wave-vectors (ie, for wave-vector moduli smaller than 1 Å^{-1}), XRS is not sensitive to the detailed atomic structure of the nanotube (Rols et al., 1999; Amara, 2014). The nanotube can be approximated by an homogeneous cylinder characterised by its internal and external radii, R_i and R_e, by its length L and by the number of Si or Ge atoms per half its period T (Fig. 11.1A and B). A slightly more elaborate description of the nanotube, the core-shell cylinder model presented in Maillet et al. (2010), leads to a similar XRS diagram that confirms the validity of the simplest homogeneous cylinder approximation. There are $N_{Si/Ge}\,(OH)_3Al_2O_3Si_xGe_{1-x}(OH)$ entities over half the period. Let n_k be the number of atoms k within an entity ($n_O=7$, $n_H=4$, $n_{Al}=2$, $n_{Si/Ge}=1$). Eq. (11.2) becomes

$$A_1\left(\vec{Q}\right)=\widetilde{\rho}_{imo}(Q)\int_{cylinder} e^{i\vec{Q}\cdot\vec{r}}\,d\vec{r} \tag{11.4}$$

with $\widetilde{\rho}_{imo}(Q)=\dfrac{2N_{Si/Ge}\cdot\Sigma_k n_k f_k(Q)}{\pi T\left(R_e^2-R_i^2\right)}\left(f_{Si/Ge}(Q)=xf_{Si}(Q)+(1-x)f_{Ge}(Q)\right)$. Note that $\widetilde{\rho}_{imo}(Q=0)$ gives the average electronic density ρ_{imo} within the nanotube wall: $\widetilde{\rho}_{imo}(Q=0)=\rho_{imo}=\dfrac{2N_{Si/Ge}\cdot\sum_k n_k Z_k}{\pi T\left(R_e^2-R_i^2\right)}$, with Z_k as the atomic number of the atom k $\left(Z_{Si/Ge}=xZ_{Si}+(1-x)Z_{Ge}\right)$. For aluminogermanate nanotubes, one finds $\rho_{imo}=0.797$ electrons Å^{-3} (Amara et al., 2014).

Integration in Eq. (11.4) gives

$$A_1\left(\vec{Q}\right)=L\widetilde{\rho}_{imo}(Q)\frac{\sin\left(Q_z\frac{L}{2}\right)}{Q_z\frac{L}{2}}\left(F_{R_e}(Q_\perp)-F_{R_i}(Q_\perp)\right) \tag{11.5}$$

where $F_R(Q_\perp)=2\pi\dfrac{RJ_1(Q_\perp R)}{Q_\perp}$ is the form factor of a disc of radius R, J_1 is the cylindrical Bessel function of the first order and $Q_\perp$ and Q_z are the projections of the scattering vector $\vec{Q}$ perpendicular to the tube axis and along its axis, respectively (Fig. 11.1B).

The case of an imogolite nanotube in aqueous dispersion is sketched in Fig. 11.1C. The same calculations as the ones developed here lead to

$$A_1\left(\vec{Q}\right)=L(\widetilde{\rho}_{imo}(Q)-\widetilde{\rho}_{water}(Q))\frac{\sin\left(Q_z\frac{L}{2}\right)}{Q_z\frac{L}{2}}\left(F_{R_e}(Q_\perp)-F_{R_i}(Q_\perp)\right) \tag{11.6}$$

with $\widetilde{\rho}_{water}(Q)=\rho_{water}\dfrac{f_O(Q)+2f_H(Q)}{10}$, and $\rho_{water}=0.334$ electrons Å^{-3} being the electronic density of water.

Let us now consider a dispersion of noninteracting nanotubes with random orientations. The total XRS intensity is calculated as the average of the squared modulus of the scattering amplitude over all nanotube orientations:

$$I(Q) \propto \iint \left|A_1\left(\vec{Q}\right)\right|^2 \sin\theta_{\vec{Q}}\, \mathrm{d}\varphi_{\vec{Q}}\, \mathrm{d}\theta_{\vec{Q}}$$
$$\propto L^2 \left(\widetilde{\rho}_{\mathrm{imo}}(Q) - \widetilde{\rho}_{\mathrm{water}}(Q)\right)^2 \int \left(\frac{\sin\left(Q\frac{L}{2}\cos\theta_{\vec{Q}}\right)}{Q\frac{L}{2}\cos\theta_{\vec{Q}}}\right)^2$$
$$\left(F_{R_e}\left(Q\sin\theta_{\vec{Q}}\right) - F_{R_i}\left(Q\sin\theta_{\vec{Q}}\right)\right)^2 \sin\theta_{\vec{Q}}\mathrm{d}\theta_{\vec{Q}} \tag{11.7}$$

Integration is performed numerically for nanotubes of finite length and solved analytically for nanotubes of infinite length. In this case, it can be demonstrated, using $\lim_{L\to\infty} L\left(\frac{\sin\left(Q\frac{L}{2}\cos\theta_{\vec{Q}}\right)}{Q\frac{L}{2}\cos\theta_{\vec{Q}}}\right)^2 = 2\pi\delta\left(Q\cos\theta_{\vec{Q}}\right) = \frac{2\pi}{Q}\delta\left(\cos\theta_{\vec{Q}}\right)$, that Eq. (11.7) becomes

$$\widetilde{I}(Q) \propto \frac{1}{Q}\left(\widetilde{\rho}_{\mathrm{imo}}(Q) - \widetilde{\rho}_{\mathrm{water}}(Q)\right)^2 \int \delta\left(\cos\theta_{\vec{Q}}\right)\left(F_{R_e}\left(Q\sin\theta_{\vec{Q}}\right) - F_{R_i}\left(Q\sin\theta_{\vec{Q}}\right)\right)^2 \mathrm{d}\cos\theta_{\vec{Q}}$$

It follows that the intensity normalised to nanotube length $\widetilde{I}(Q) = \frac{I(Q)}{L}$ is written as

$$\widetilde{I}(Q) \propto \frac{1}{Q}\left(\widetilde{\rho}_{\mathrm{imo}}(Q) - \widetilde{\rho}_{\mathrm{water}}(Q)\right)^2 \left(F_{R_e}(Q) - F_{R_i}(Q)\right)^2 \tag{11.8}$$

In Fig. 11.2A, calculated XRS intensities for different nanotube lengths are shown. Scattered intensities present oscillations whose depths depend on the nanotube length. They are deeper for longer nanotubes. Moreover, to be compared to experimental diagrams, calculated curves have to be convoluted with the experimental resolution, whose full width at half maximum (FWHM) is typically 0.013 Å^{-1} in laboratory XRS experiments (Amara et al., 2014). The calculated intensity curve for infinite nanotubes, convoluted to the resolution, and the calculated curve for nanotubes with 450 Å length are found to be very similar [*red* (*grey* in the print version) line and black circles in Fig. 11.2A]. It follows that within experimental resolution, nanotubes longer than 450 Å can be approximated as infinite. The positions of the minima of the XRS diagrams allow one to determine internal and external nanotube radii. In Eq. (11.8), the minima of the oscillations correspond to Q-values that satisfy $F_{R_e}(Q) = F_{R_i}(Q)$; that is, $R_e J_1(QR_e) = R_i J_1(QR_i)$. Minima positions are thus determined by radii values, as illustrated in Fig. 11.2B for two sets of radii values of nanotubes of the same wall thickness.

A comparison between experimental and calculated XRS diagrams is shown in Fig. 11.3 for a dispersion of SW aluminogermanate nanotubes (Amara et al., 2014). Comparison of experimental and calculated intensities

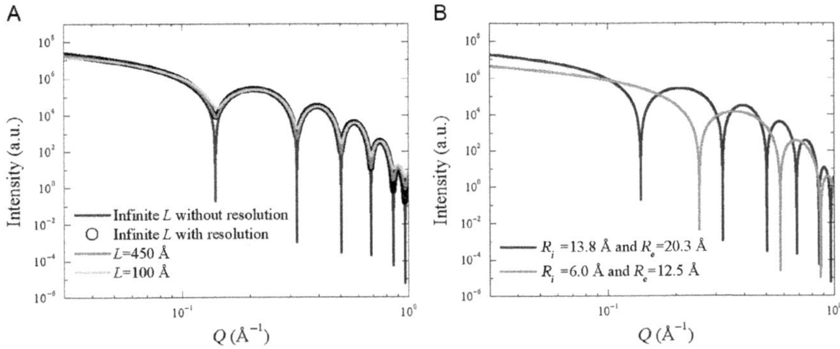

FIG. 11.2 (A) Calculated XRS intensities for different nanotube lengths $L = 100$ Å (*green* (*light grey* in the print version) *line*), 450 Å (*red* (*grey* in the print version) *line*) and infinite (*blue* (*dark grey* in the print version) *line*). The curve of the calculated intensity for infinite tubes convoluted by experimental resolution (FWHM$=0.013$ Å^{-1})—*black circles*—superimposes with the *red* (*grey* in the print version) *line*, meaning that XRS experiments do not give access to lengths greater than 450 Å. Nanotube radii are $R_i = 13.8$Å and $R_e = 20.3$Å. (B) Calculated XRS intensities, without resolution, of nanotubes of the same wall thickness with two different sets of inner and outer radii.

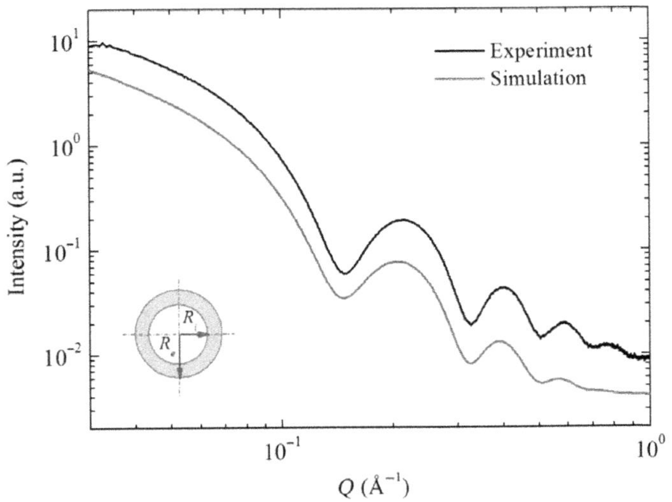

FIG. 11.3 XRS diagram of a dispersion of single-walled aluminogermanate nanotubes. *Black line*: experimental data; *red* (*grey* in the print version) *line*: calculated intensity for infinite nanotubes with internal radius $R_i = 13.8$Å and external radius $R_e = 20.3$Å, convoluted to the experimental resolution. It is translated vertically for the sake of clarity. The inset represents a sketch of an SW nanotube. *From Amara et al. (2014).*

allowed us to determine the nanotube radii to be equal to $R_i = 13.8\text{Å}$ and $R_e = 20.3\text{Å}$. The experimental curve is well reproduced using Eq. (11.8) after convolution of the calculated intensity with the resolution, meaning that most of the nanotubes in the dispersion are longer than 450 Å. In the following section, therefore, XRS formalism will only be discussed for infinite nanotubes.

11.2.2 Nanotubes Organised in Bundles

Imogolites in powders can assemble in bundles (Cradwick et al., 1972; Farmer et al., 1983; Amara et al., 2014). Nanotube packing in small bundles has been proposed to be either in a monoclinic or hexagonal arrangement (Mukherjee et al., 2005; Kang et al., 2010). For large bundles, and as is evident for the hexagonalisation of imogolites discussed in Section 11.4, one considers here that dried nanotubes are organised on a two-dimensional (2D) hexagonal lattice in the plane perpendicular to their long axis (Fig. 11.4A).

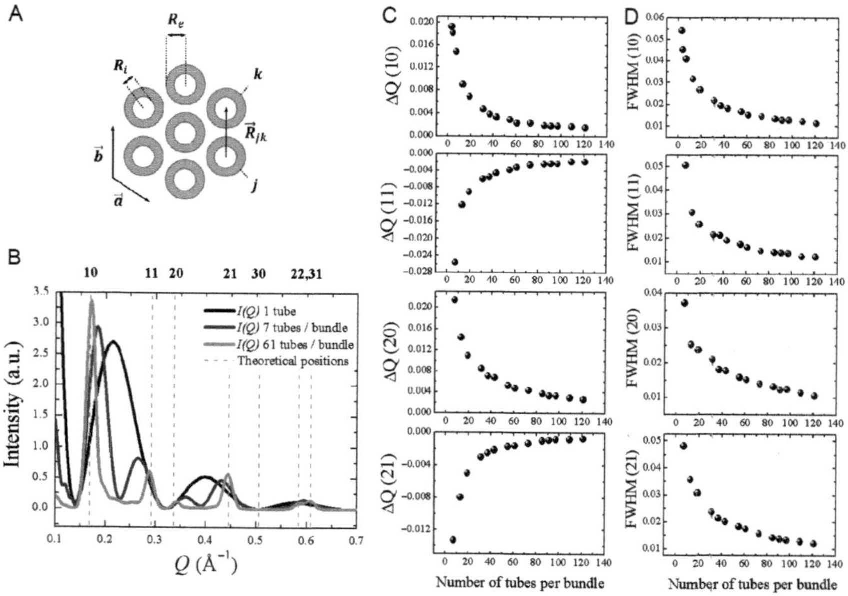

FIG. 11.4 (A) Nanotubes arranged on a 2D hexagonal lattice. (B) Calculations of XRS intensities for powders of the individual tubes (*black line*) and for bundles formed with 7 tubes (*blue* (*dark grey* in the print version) *line*) or 61 tubes (*red* (*grey* in the print version) *line*); nanotubes have inner and outer radii $R_i = 13.8\text{Å}$ and $R_e = 20.3\text{Å}$, and the hexagonal lattice parameter is $a = 43\text{Å}$; hk integers ($hk = 10$, 11, etc.) index the reciprocal lattice nodes of an infinite hexagonal 2D lattice. (C) Dependence of the difference $\Delta Q(hk) = Q_{hk}^{\text{calc}} - Q_{hk}^{\text{theo}}$ between calculated positions (derived from a Gaussian fit of the corresponding peak hk on the XRS diagram) and theoretical values (for an infinite lattice) as a function of the number of tubes in the bundle. (D) Dependence of hk reflection FWHM as a function of the number of tubes in the bundle.

It follows from Eqs (11.3) and (11.5) that the intensity scattered by an infinite cylindrical nanotube, renormalised to the tube length, is given by

$$\widetilde{I}_1\left(\vec{Q}\right) \propto \widetilde{\rho}_{\text{imo}}(Q)^2 (F_{R_e}(Q_\perp) - F_{R_i}(Q_\perp))^2 \delta(Q_z) \tag{11.9}$$

For cylindrical nanotubes, the organisation in bundles leads to a supplementary term $S\left(\vec{Q}\right) = \sum_{j,k} e^{i\vec{Q}\cdot\vec{R}_{jk}}$, called the structure factor in the expression of the scattered intensity:

$$\widetilde{I}_{\text{bundle}}\left(\vec{Q}\right) \propto \widetilde{I}_1\left(\vec{Q}\right) \sum_{j,k} e^{i\vec{Q}\cdot\vec{R}_{jk}} \tag{11.10}$$

where the sum runs over all nanotubes within the bundle, and $\vec{R}_{jk}$ is the vector between the two axes of the tubes j and k in the bundle, perpendicular to these axes. The intensity scattered by a powder formed by such bundles is obtained by integrating the intensity in Eq. (11.10) over all equiprobable bundle orientations. One easily obtains

$$\widetilde{I}(Q) \propto \frac{1}{Q} \widetilde{\rho}_{\text{imo}}(Q)^2 \left(F_{R_e}(Q) - F_{R_i}(Q)\right)^2 \sum_{j,k} J_0(QR_{jk}) \tag{11.11}$$

with J_0 being the cylindrical Bessel function of order zero $\left(J_0(x) = \frac{1}{2\pi}\int_0^{2\pi} e^{ix\cos(\varphi)}\,\mathrm{d}\varphi\right)$.

Calculations of XRS diagrams of powders of individual cylindrical nanotubes with radii $R_i = 13.8$ Å, $R_e = 20.3$ Å, and of small and large bundles (formed of 7 and 61 tubes, respectively, and with intertube distance a chosen as equal to 43 Å) are presented in Fig. 11.4B. The broad modulations of the curve corresponding to the powder of individual nanotubes give way to narrower reflections corresponding to the organisation within bundles. For a nanotube arrangement on an infinite planar hexagonal lattice with parameter a in the bundle, the reflections are located at the reciprocal positions $\vec{Q}_{hk} = h\,\vec{a}^* + k\,\vec{b}^*$, where $\vec{a}^*$ and $\vec{b}^*$ are the basis vectors in reciprocal space, with h and k being integers. The modulus of the scattering wave-vector is

$$Q_{hk} = \sqrt{h^2 + k^2 + hk}\,\frac{4\pi}{\sqrt{3}a} \tag{11.12}$$

Reflection positions depend on the combined effect of the structure factor and of the squared form factor of the nanotubes (that is, of the nanotube radii and of the number of nanotubes within the bundles). As has already been demonstrated in previous studies in the case of carbon nanotubes (Thess et al., 1996; Rols et al., 1999), the resulting reflection positions differ from the theoretical values given by Eq. (11.12), which is strictly valid only for an infinite number of tubes. The difference $\Delta Q(hk) = Q_{hk}^{\text{calc}} - Q_{hk}^{\text{theo}}$ between calculated positions and theoretical values is shown in Fig. 11.4C. For example, reflection positions are shifted towards higher Q-values for $hk=10$, 20, and towards smaller Q-values for $hk=11$, 21, depending of the sign of the local slope of the squared nanotube form factor (black line in Fig. 11.4B). The 30 reflection intensity is almost zero, as it falls very close to a minimum (zero-value) of the squared nanotube form factor. Moreover, the FWHM of the reflections decrease when the number of nanotubes in the bundles increases (Fig. 11.4D). It must be underscored that for relatively small bundles, the intertube distance cannot be simply deduced from the position of the 10 reflection. For instance, assuming that $a=\frac{\sqrt{3}}{4\pi Q_{10}}$ for bundles formed of seven nanotubes (Fig. 11.4B) leads to an underestimation of the intertube distance by more than 10%.

Therefore, the study of imogolite when nanotubes are assembled in bundles requires a careful analysis of XRS diagrams, including modelling. Such careful analysis has been applied in particular in the study of the hexagonalisation of aluminogermanate nanotubes presented in Section 11.4.

11.2.3 Typical Experimental Setup

XRS experiments described in Sections 11.3 and 11.4, later in this chapter, are performed using setups similar to the typical one drawn in Fig. 11.5, which includes an X-ray rotating anode generator coupled with multilayer optics providing a monochromatic and high-flux beam of 1×1 mm^2 at the sample position. The wavelength of the incident beam is $\lambda=0.1.542$ Å (0.709 Å, resp.) for a copper (molybdenum, resp.) anode. Imogolite nanotubes dispersed in water are put in a Kapton capillary with a 2.6-mm diameter. A few milligrammes of powder of dried imogolites are wrapped in a container made of 10-μm-thick aluminium foil. XRS images are recorded on a 2D detector, such as a mar345 image plate detector. Dedicated software programmes allow angular grouping of measured intensities in order to obtain XRS diagrams $I(Q)$. The accessible Q range is dependent on the experimental parameters, the wavelength λ, the distance sample to detector D, and the dimensions H of the detector,

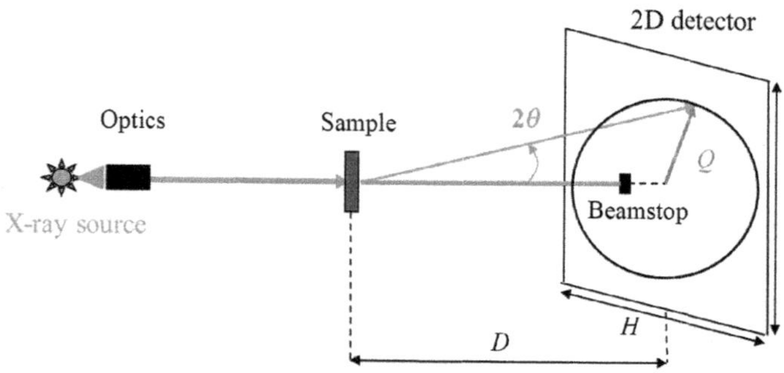

FIG. 11.5 Scheme of XRS experimental setup. D is the sample-to-detector distance, H the dimensions of the detector, 2θ the scattering angle and Q the corresponding scattering wavevector ($Q=\frac{4\pi\sin(\theta)}{\lambda}$, where λ is the wavelength of the incident X-ray beam).

through the relation $Q = \frac{4\pi}{\lambda}\sin\left(\frac{1}{2}\tan^{-1}\left(\frac{H}{2D}\right)\right)$. For the low Q-values corresponding to a large D-value, a vacuum chamber placed between the sample and the detector allows the minimisation of the small angle-scattering signal from the air, allowing one to obtain a high signal over the background ratio.

11.3 OVALISATION OF THE IMOGOLITE

Molecular dynamic calculations performed on imogolites, either hydrated or dried, suggest that they are highly deformable (Tamura and Kawamura, 2002; Creton et al., 2008). They show that the basis of interacting nanotubes can deform into an oval, as is illustrated in Fig. 11.6 for dried nanotubes in the work by Tamura and Kawamura (2002). Furthermore, calculations of the frequency values of radial breathing modes (Guimaraes et al., 2007; Konduri et al., 2007) suggest that imogolites are rather soft; the calculated Young's modulus falls in the 200–400 GPa range for both aluminosilicate and aluminogermanate nanotubes (Guimaraes et al., 2007; Lourenço et al., 2014), which is smaller than for other nanotubes, such as carbon or boron nitride nanotubes. Defects such as vacancies within the imogolite walls (Yucelen et al., 2012) could also favour a mechanism of deformation.

In the following discussion, the first experimental evidence of the ovalisation of imogolites was based on a detailed XRS analysis. The study was carried out on aluminogermanate nanotubes.

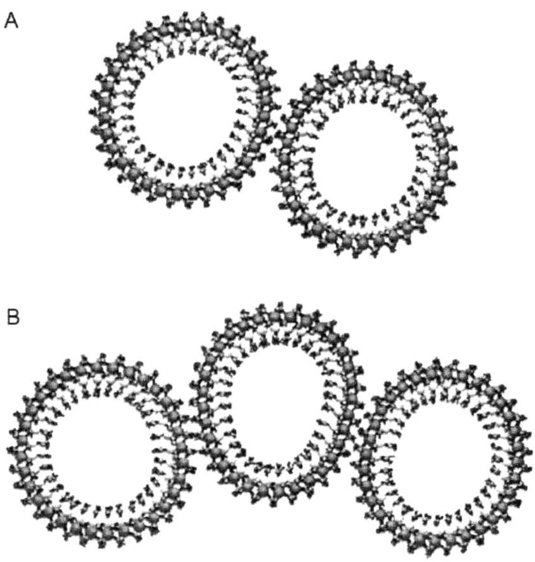

FIG. 11.6 Imogolite structures calculated from molecular dynamics simulations, viewed along the direction of the tube axis: (A) for two interacting nanotubes and (B) for three interacting nanotubes. *From Tamura and Kawamura (2002).*

Fig. 11.7 shows XRS diagrams of dispersion of aluminogermanate nanotubes and of powder obtained by drying the same dispersion and then heating it at 250°C to evaporate the residual water inside and around these hydrophilic nanotubes. Oscillations are as broad for the powder as for the dispersion, which shows that nanotubes remain individual and do not assemble in bundles in the powder. The oscillations, however, appear strongly attenuated for the powder. Note also that the minima positions are unchanged. In principle, this effect could be attributed to a reduction in length of the nanotubes (Fig. 11.2A), but that can be ruled out since drying cannot induce nanotube cutting. In fact, it points towards a deformation in the shape of the nanotubes, as demonstrated in the following discussion.

The state of individual nanotubes in the powder obtained after drying can be schematically described as shown in Fig. 11.8A, considering that nanotubes are randomly disordered in orientation and positions, but are locally in contact with each other. The XRS intensity of such dehydrated imogolite

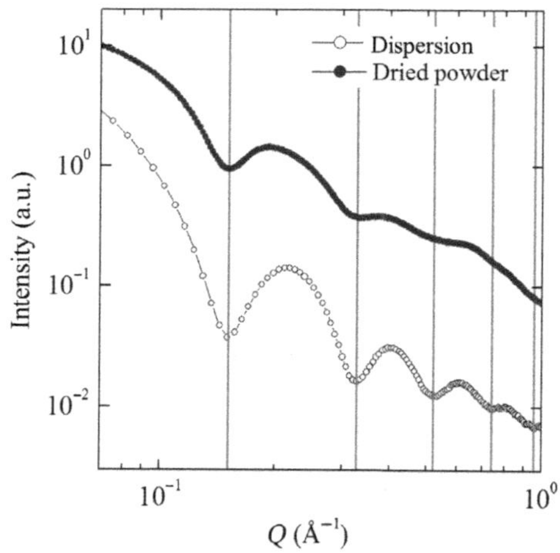

FIG. 11.7 XRS diagrams of a dispersion and a powder of individual aluminogermanate nanotubes. Vertical *red* (*grey* in the print version) *lines* mark the positions of the minima of oscillations for both curves. Curves are translated vertically for the sake of clarity.

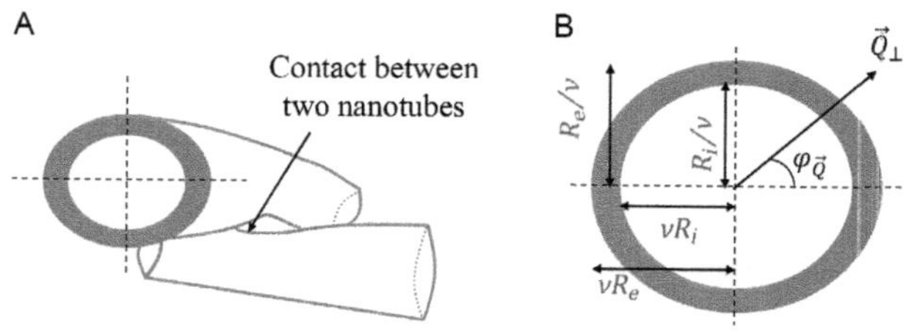

FIG. 11.8 (A) Representation of two nanotubes in contact, resulting in their local deformation. (B) Schematic view of the elliptic section of the deformed nanotube. The components of the scattering wave-vector ($Q_\perp, \varphi_{\vec{Q}}$) in the elliptic plane are indicated.

powder is directly proportional to the squared form factor of the nanotubes averaged over all orientations.

Nanotubes can deform around their contact points due to intertube interactions, resulting in a more or less oval cross-section shape of the nanotube, as predicted and illustrated in Fig. 11.6 (Tamura and Kawamura, 2002). Let us first consider the case of infinite-length nanotubes with an elliptical cross section instead of a circular one. Starting from a cylindrical nanotube taken as a reference with internal and external radii R_i and R_e (the adjustment of the dispersion diagram (Fig. 11.3) gives $R_i = 13.8\,\text{Å}$ and $R_e = 20.3\,\text{Å}$). In the first approximation, the surface perpendicular to the tube axis is assumed to be constant. The dimensions of the elliptical base section of the nanotube are then given by an elliptic deformation factor ν, nanotube dimensions being $(\nu R_i, \nu R_e)$ along the direction of one given elliptic axis and $\left(\frac{R_i}{\nu}, \frac{R_e}{\nu}\right)$ along the second elliptic axis (Fig. 11.8B).

Following the same steps as described here for cylindrical nanotubes, one easily shows that the X-ray intensity scattered by a powder of individual infinite nanotubes with an elliptic section basis is written as (Amara, 2014):

$$\widetilde{I}_\nu(Q) \propto \frac{\widetilde{\rho}_{\text{imo}}(Q)^2}{Q} \int_0^{2\pi} d\varphi_{\vec{Q}} \left(\frac{R_e J_1(Q'_\perp R_e) - R_i J_1(Q'_\perp R_i)}{Q'_\perp} \right)^2 \tag{11.13}$$

with $Q'_\perp = Q \cdot \sqrt{\nu^2 \cos^2\varphi_{\vec{Q}} + \frac{1}{\nu^2}\sin^2\varphi_{\vec{Q}}}$. Note that this expression contains the modulus of the wave-vector $\vec{Q}$ and not that of $\vec{Q}_\perp$; this is because for infinite-length tubes, the scattering is limited to the equatorial plane (Eq. 11.9).

Calculated XRS intensities from Eq. 11.13 are plotted in Fig. 11.9 for several values of the elliptic ν factor, with the case of the cylindrical nanotube corresponding to $\nu = 1$. It shows that increasing ovalisation is responsible for the progressive attenuation of the oscillations of the intensity.

In the hypothesis that the nanotube base section is elliptic along the entire length of the nanotube, one may now allow the elliptic deformation factor to vary from one tube to the other. A Gaussian distribution $G(\nu)$, centred around the value $\nu = 1$ corresponding to the case of a undistorted nanotube, is introduced:

$$G(\nu) = \frac{2}{\Delta}\sqrt{\frac{\ln 2}{\pi}} \exp\left\{ \frac{-4\ln 2}{\Delta^2}(\nu - 1)^2 \right\} \tag{11.14}$$

The variance Δ can be considered as an average degree of the deformation of the nanotubes.

Taking into account the distribution of the elliptic deformation, the scattered X-ray intensity by a powder of individual and uncorrelated nanotubes then becomes

$$\widetilde{I}(Q) \sim \int_{-\infty}^{+\infty} d\nu \widetilde{I}_\nu(Q) G(\nu) \tag{11.15}$$

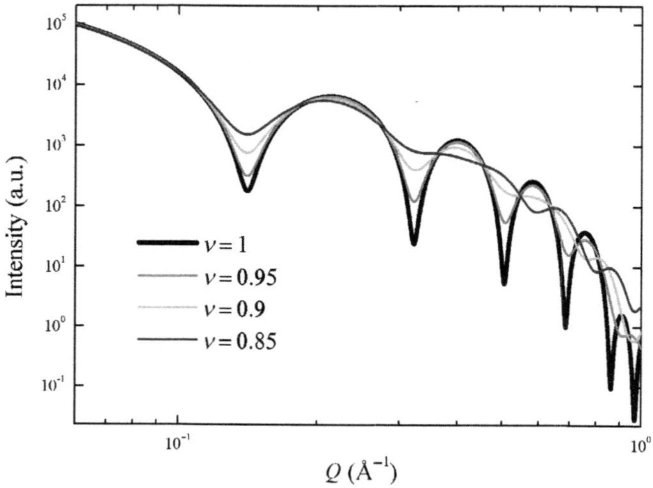

FIG. 11.9 Calculated XRS intensities, convoluted by the experimental resolution (FWHM=0.013 Å^{-1}), for different values of the elliptic factor ν, $\nu = 1$ (thick black line) corresponding to a cylindrical nanotube.

Calculations of the XRS intensity for different values of the parameter Δ are plotted in Fig. 11.10A and compared to the one from undistorted cylindrical nanotubes. The amplitude of minima of the intensity is steadily decreasing with increasing Δ value, while their positions remain unchanged since the distribution is centred around $\nu = 1$. The value $\Delta = 0.6$ gives a curve that compares well with the experimental one (Fig. 11.10B). This value should be considered as an upper limit. Indeed, it is certainly overestimated because the deformation is actually restricted to the localised regions of contact between nanotubes, whereas the calculation of XRS intensity was performed in the approximation of a uniform deformation along the whole length of the nanotube. The small lengths over which the deformation occurs also contribute to the attenuation of the oscillations of the curve, as discussed earlier (see Fig. 11.2).

As it is, the simple approach developed in this chapter shows that dried aluminogermanate imogolite nanotubes do ovalise. It has been predicted, based on molecular dynamics simulations, for aluminosilicate nanotubes and was not confirmed experimentally. Further simulations for aluminogermanate nanotubes and experimental studies on aluminosilicate nanotubes would be very interesting.

11.4 HEXAGONALISATION OF THE IMOGOLITE

A stronger deformation can be encountered when nanotubes are assembled in bundles, in which interactions between tubes are increased. The deformation of aluminogermanate nanotubes when self-assembled in large bundles

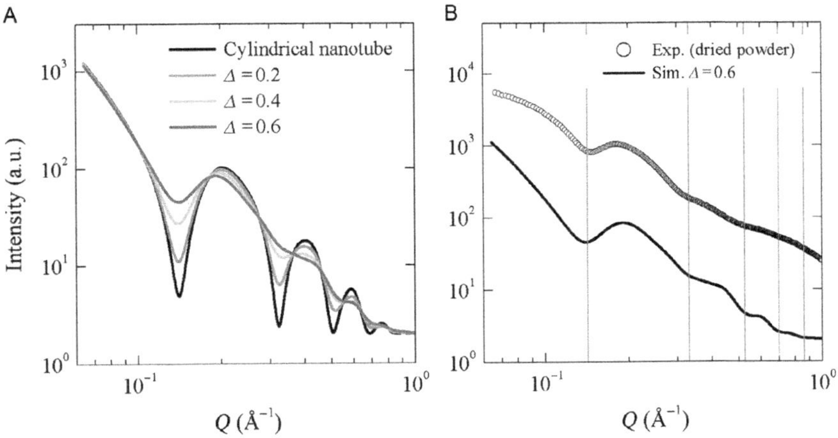

FIG. 11.10 (A) Calculations of XRS diagrams for different values of the degree of deformation Δ and of the cylindrical nanotube ($R_i = 13.8\,\text{Å}$, $R_e = 20.3\,\text{Å}$). (B) Measured XRS intensity for the dried powder of deformed individual imogolites and the calculated one with $\Delta = 0.6$, translated vertically for the sake of clarity. Vertical *red* (*grey* in the print version) *lines* mark the positions of the minima of the oscillations.

(Fig. 11.11A) has been studied by Amara et al. (2014). These authors gave the first experimental evidence of the hexagonalisation of the imogolite by using a systematic method developed to analyse XRS diagrams as a function of the nanotube shape. Unlike with ovalisation, hexagonalisation of the nanotube basis had not been predicted before this experimental finding and thus should motivate further theoretical studies.

After synthesis in dispersion, the sample was dried at room temperature and then heated at 250°C to remove all water remaining in the powder and to obtain a completely dehydrated powder, and yet this occurs before the dehydroxylation of the nanotubes (see the thermogravimetric (TG) analysis shown in Fig. 11.13). Removal of all the water from the powder allows one to avoid interference effects coming from water molecules on the XRS diagram, which otherwise would complicate the analysis (Kang et al., 2010).

XRS analysis performed on aluminogermanate nanotubes (Fig. 11.3) allows for determining that the dispersion contains cylindrical SW nanotubes with inner radius $R_i = 13.8\,\text{Å}$ and outer radius $R_e = 20.3\,\text{Å}$. For the experimental curve of the dehydrated powder shown in Fig. 11.11B, XRS analysis allows for assigning the reflections *hk* to a 2D hexagonal lattice according to the formalism detailed earlier in Section 11.2.2. Large bundles of about 60 nanotubes and a lattice constant $a = 39.7\,\text{Å}$ are found for the best agreement between the experimental diagram and the calculated one (Fig. 11.11B). This value of the lattice parameter, however, is not compatible with a cylindrical shape of the nanotubes. Indeed, the distance between two adjacent nanotubes in the bundle should be larger than twice

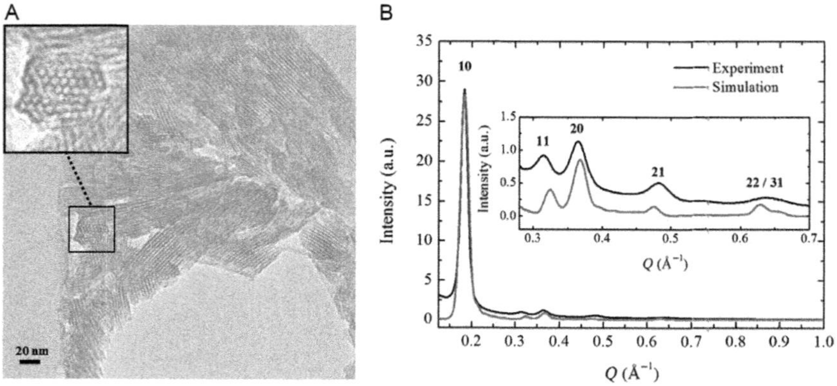

FIG. 11.11 (A) TEM image and (B) experimental XRS diagram of dehydrated powder of aluminogermanate nanotubes assembled in bundles formed by 60 tubes. The *hk* indices refer to the positions of the 2D hexagonal reciprocal lattice peaks. *From Amara et al. (2014).*

their outer radius ($2R_e = 40.6$Å). Therefore, Amara et al. (2014) proposed that dehydrated aluminogermanate nanotubes are deformed when assembled in bundles.

Deformation of the nanotubes assembled in large bundles is based on a model for the nanotube cross section consisting of a rounded-hexagonal surface that should respect the hexagonal symmetry of the 2D lattice and allows the lattice constant to be smaller than with cylindrical tubes. Rounded parts have an angular aperture of $2\varphi(\varphi \in [0, 30°])$ and the flat parts in between are characterised by inner and outer apothems d_0 and d'_0, as sketched in Fig. 11.12A. Therefore, the nanotube basis is circular ($R_i = d_0, R_i = d'_0$) for $\varphi = 30°$, whereas it becomes fully hexagonal for $\varphi = 0°$. Other deformations besides the rounded-hexagonal shape are excluded because they would imply a discrepancy with the hexagonal symmetry of the lattice determined experimentally or the introduction of an orientational disorder of deformed imogolites (ovalised ones, for instance), which would lead to a larger lattice constant.

The calculated intensity for a powder of bundles of nanotubes per nanotube unit length writes (Amara et al., 2014)

$$\widetilde{I}(Q) \propto \frac{\widetilde{\rho}_{\text{imo}}(Q)^2}{Q} \int_0^{2\pi} \mathrm{d}\varphi_{\vec{Q}} \left[\left| F_{\varphi, d_0, d'_0}\left(\vec{Q}\right) \right|^2 \sum\nolimits_{j,k \in \text{bundle}} \cos\left(\vec{Q} \cdot \vec{R}_{jk}\right) \right] \quad (11.16)$$

where $\varphi_{\vec{Q}}$ is the angle between the scattering wave-vector $\vec{Q}$, localised in the equatorial plane for infinite-length nanotubes; and the x-axis in Fig. 11.12A. The 2D form factor $F_{\varphi, d_0, d'_0}\left(\vec{Q}\right)$ is the Fourier transform of the rounded-hexagonal nanotube surface (Amara et al., 2014).

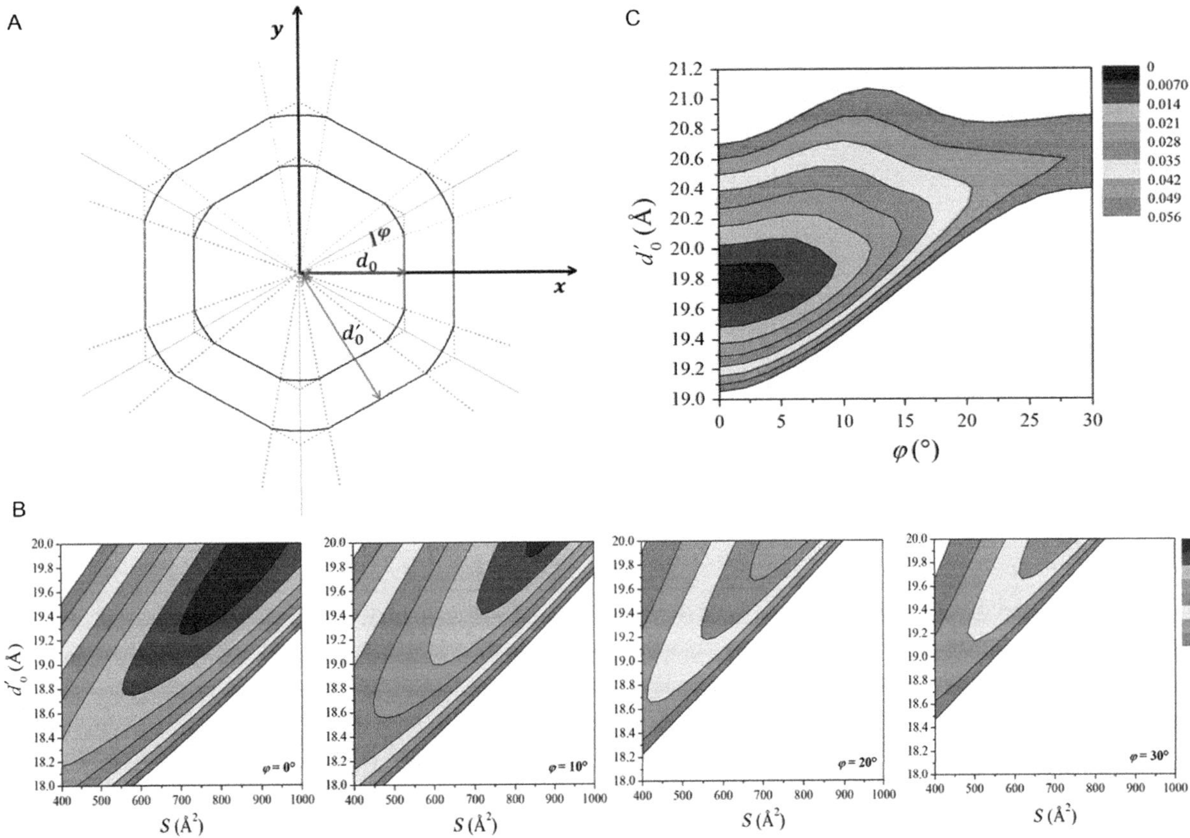

FIG. 11.12 (A) Rounded hexagonal model for a deformed single-walled imogolite nanotube. (B) Evolution of the standard deviation Δ as a function of d_0' and the surface S for values of $\varphi = 0$, 10, 20 and 30 degrees. (C) Evolution of the standard deviation Δ as a function of d_0' and φ for a constant surface $S = 830^{-2}$.

Calculated XRS intensities depend on three parameters φ, d_0 and d_0', (or, equivalently, φ and d_0') and S, which is the nanotube surface:

$$S = 6\left(d_0'^2 - d_0^2\right)\left[\frac{\varphi}{\cos^2\left(\frac{\pi}{6} - \varphi\right)} + \tan\left(\frac{\pi}{6} - \varphi\right)\right] \tag{11.17}$$

A systematic method of analysis is developed to determine the nanotube shape through the best adjustment of experimental data. The measured XRS reflections in Fig. 11.11 are fitted with Gaussian line-shaped functions. Maximum intensities $I_{\exp}(Q)$ of reflections 10, 11, 20, 21 and 22/31 (not separated) are labelled by integers ranging from 0 to 4. One defines for $i = 1$–4 (corresponding to $hk = 11$, 20, 21 and 22/31) the experimental ratio $r_{\exp} = \frac{I_{\exp}[i]}{I_{\exp}[0]}$, as well as the calculated one, $r_{\text{calc}} = \frac{I_{\text{calc}}[i]}{I_{\text{calc}}[0]}$, I_{calc}, being deduced from Eq. (11.16). Intensities are renormalised to the intensity of the most intense reflection in order to minimise the background contribution. The adjustment between the calculated intensities with the experimental one is evaluated by the standard deviation:

$$\Delta\left(\varphi, d_0', S\right) = \sqrt{\sum_{i=1}^{4}\left(r_{\exp}[i] - r_{\text{calc}}[i]\right)^2} \tag{11.18}$$

The variation of Δ as a function of d_0' and S is plotted for different φ values, from $\varphi = 0°$ (hexagonal shape) to $\varphi = 30°$ (cylindrical shape) in Fig. 11.12B. The absolute minimum value Δ is obtained for $\varphi = 0°$, which corresponds to the perfect hexagonal shape, and with a nanotube base surface $S \approx 800\,\text{Å}^2$. It is larger than the surface of an isolated cylindrical nanotube in solution ($S \approx 700\,\text{Å}^2$), probably due to the effect of perchlorate anions at the surface of the nanotubes in the dried powder (Gustafsson, 2001) which leads to a larger effective surface of the nanotube. The variation of Δ in the function of d_0' and φ for a constant surface $S = 830\,\text{Å}^2$ (Fig. 11.12C) indicates that its minimum value is equal to $d_0' = 19.8\,\text{Å}$, which is consistent with the measured lattice constant ($a = 2d_0'$).

The specific method for XRS analysis applied in this study clearly concludes that SW aluminogermanate nanotubes, when assembled in large bundles on a 2D hexagonal lattice, acquire a perfect hexagonal cross section instead of a circular one. The newly formed, honeycomblike material is expected to exhibit new mechanical and physicochemical properties that are different from the properties of the cylindrical nanotubes.

11.5 DEHYDROXYLATION AND HIGH-TEMPERATURE STRUCTURAL TRANSFORMATIONS

Starting from a powder of natural or synthetic nanotubes at room temperature, the successive processes taking place with increasing temperatures are

(i) dehydration of the hydrophilic imogolites, (ii) modification of the imogolite structure through dehydroxylation and the subsequent nanotube structure collapse phenomenon at temperatures above T at about 300°C and (iii) structural transformation towards a mullite crystalline phase at around 1000°C. Temperatures measured by TG and reported in the literature for dehydration, dehydroxylation and the structural transformation towards mullite crystalline phases are presented in Table 11.1.

The dehydroxylation phenomenon has been observed in aluminosilicate, as well as in aluminogermanate nanotubes within a broad temperature range (Table 11.1) by TG analyses (Yoshinaga and Aomine, 1962; MacKenzie et al., 1989; Donkai et al., 1992; Kang et al., 2010; Ma et al., 2012; Zanzottera et al., 2012). Discrepancies in temperatures may be explained by different experimental conditions, with measurements performed either under air, in an inert gas flux, or with different heating rates. An additional endothermic mass loss is also observed at 447°C by Zanzottera et al. (2012), who ascribed this second peak to the release of oxygen molecules caused by the thermal decomposition of the remaining perchlorate species ClO_4^- used in the synthesis. TG analyses were also conducted for a powder of aluminogermanate imogolite by Amara (2014), showing a similar behaviour for TG curves (Fig. 11.13A): a dehydration regime takes place up to 250°C, and then two endothermic mass loss peaks at 350°C and 415°C mark the dehydroxylation and the liberation of dioxygen molecules from the decomposition of perchlorate species, respectively. Let us point out that the whole dehydroxylation process is progressive over a large temperature range. In the case of aluminogermanate imogolite (Fig. 11.13A), this phenomenon related to the broad mass loss peak starts at around 300°C, as hinted by the inflexion point of the curves at this temperature, and ends around 550°C. Kang et al. (2010) studied the dehydration, dehydroxylation and rehydroxylation of SW aluminosilicate nanotubes with the combination of Fourier-transform infrared, nuclear magnetic resonance (NMR), TG and mass spectroscopy, nitrogen (N_2) physisorption and XRS techniques. In their study, complete dehydration is achieved at 250°C under vacuum, as a prerequisite to the partial dehydroxylation process, which was investigated up to 450°C, at which a proportion (about 30%) of the hydroxyl groups are still present and the tubular structure is preserved (Kang et al., 2010). The authors also reported by ^{29}Si NMR and N_2 physisorption measurements that the rehydroxylation mechanism may partially occur upon reexposure to water vapour. Recent numerical calculations (Da Silva Chagas et al., 2015) indicate that dehydroxylation of the inner surface silanols of the aluminosilicate nanotube should induce deformations of the nanotube before the recovery of the cylindrical structure with the subsequent rehydroxylation of the outer surface. These authors concluded that the control of the degree of dehydroxylation by heat treatment may allow one to adjust the electronic and mechanical properties of imogolite nanotubes.

TABLE 11.1 Comparison of the Thermal Stability for SW Aluminosilicates (AlSi) and Aluminogermanates (AlGe) Nanotubes

Imogolite nanotube					
Nature	*Type*	T_{H_2O}	T_{OH}	T_{exo}	**References**
AlSi	Natural	120–180	385–435	940–995	Yoshinaga and Aomine (1962), Yoshinaga (1968), Miyauchi and Aomine (1966), Horikawa (1975), MacKenzie et al. (1989), Donkai et al. (1992) and Abidin et al. (2008)
		About 120	410	940–950	Violante and Tait (1979)
	Synthetic	120	360–400	990	Wada et al. (1979), Wada and Wada (1982)
		100	320–360	990	Yang et al. (2007) and Abidin et al. (2008)
		120	About 400	850	Johnson et al. (1988)
		120	350–400	nd	Kang et al. (2010, 2011) and Zanzottera et al. (2012)
		40	330	nd	Bottero et al. (2011)
AlGe		120	400	nd	Wada and Wada (1982)
		65, 125	450	no	Amara (2014)

T_{H_2O}, T_{OH} and T_{exo} (°C units) refer to the temperatures of dehydration, dehydroxylation and transformation into a crystalline mullite phase deduced from TG analysis. *nd*, not determined; *no*, not observed.

The dehydroxylation process actually preludes the strongest modifications of the aluminosilicate nanotube structure. Further heating treatment was shown to lead either to the occurrence of an X-ray amorphous phase (Farmer et al., 1983; Donkai et al., 1992) or to a lamellar phase (Zanzottera et al., 2012).

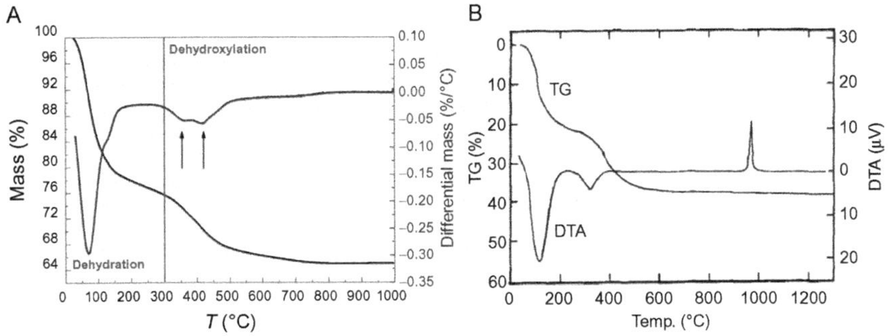

FIG. 11.13 (A) Thermogravimetric analysis of SW aluminogermanate imogolite (Amara, 2014). *Black arrows* point towards mass loss peaks assigned to dehydroxylation ($T=350°C$) and to the thermal decomposition of perchlorate anions ($T=415°C$). (B) TG analysis of SW aluminosilicate imogolite from Donkai et al. (1992).

The amorphous phase measured by XRS measurements on imogolite films at 350°C (Farmer et al., 1983) or around 500°C (Donkai et al., 1992) was assumed to correspond to the partial breakdown of the nanotube structure. At higher temperatures, TG analysis of SW aluminosilicate nanotubes exhibits an exothermic peak at around 960°C (Fig. 11.13B). It was related to the formation of γ-alumina (Yoshinaga and Aomine, 1962) or of the mullite ($3Al_2O_3 \cdot 2SiO_2$) crystalline phase in the temperature range from 980°C to 1600°C, as illustrated by the XRS measurements in Fig. 11.14 (Donkai et al., 1992). Note that the high-temperature exothermic peak reported at 960°C for aluminosilicate nanotubes is not observed for aluminogermanate nanotubes, at least until 1000°C (Fig. 11.12A).

In addition, Zanzottera et al. (2012) observed an irreversible structural transition towards a lamellar phase at around 400°C, explained by the collapse of SW aluminosilicate nanotubes. XRS diagram of the lamellar phase essentially shows a single reflection related to the distance between layers. From the value of the interlayer distance, the authors proposed the scheme of a layered structure containing microporous residual regions. The mechanism of nanotube collapse was discussed earlier by MacKenzie et al. (1989). In this study, two possible mechanisms for the tube collapse were proposed (Fig. 11.15). The first mechanism corresponds to a single-tube cleavage with condensation of two fragments from the breaking of the tube along its axis to form Si–O–Si bridges, causing the tubes to flatten. The second mechanism deals with a two-tube cleavage and condensation to form Si–O–Al bridges. The former mechanism (Fig. 11.15A) is accepted as the most likely, according

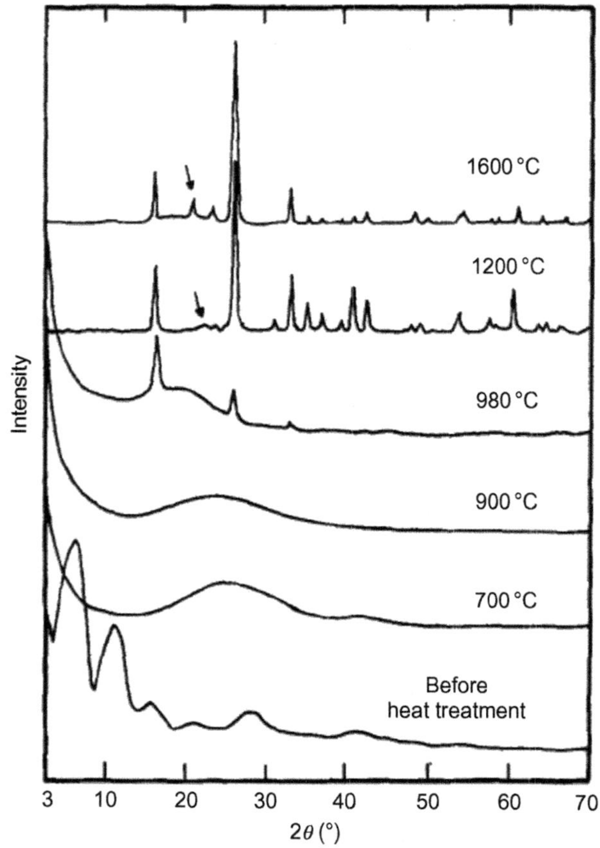

FIG. 11.14 X-ray diffraction diagrams of imogolite films heated at various temperatures up to 1600°C (CuK$_\alpha$ radiation). Diagrams of samples heat-treated at 980°C, 1200°C and 1600°C show narrow reflections that are assigned to the single mullite crystalline phase. *Arrows* point towards reflections attributed to a parasitic tridimite phase. *From Donkai et al. (1992).*

to the experimental NMR ^{29}Si shift (MacKenzie et al., 1989; Zanzottera et al., 2012).

The amorphous or the lamellar phases obtained from the thermal decomposition of imogolite were separately observed in the same temperature domain in different studies. Therefore, the mechanisms of formation of these phases and their possible relationship are still to be elucidated. Further in situ XRS experiments and their analysis based on structure modelling should bring invaluable information about the high-temperature structural transformations of imogolites.

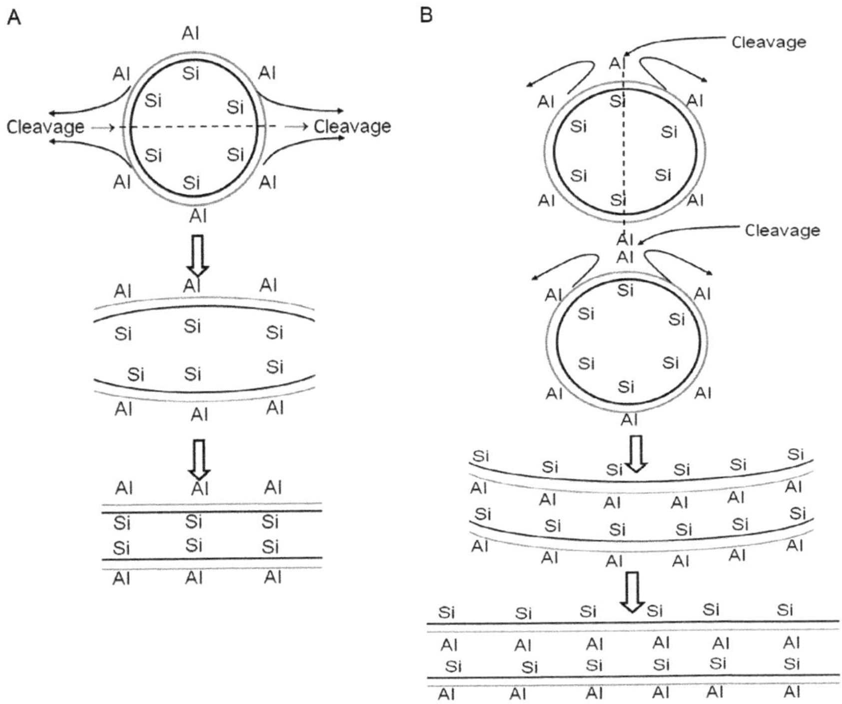

FIG. 11.15 The two collapsing mechanisms proposed by MacKenzie et al. (1989): (A) Single-tube cleavage with condensation of two fragments. (B) Two-tube cleavage and condensation. *Figure taken from Zanzottera et al. (2012).*

11.6 CONCLUDING REMARKS

Deformations of imogolite nanotubes can be accurately characterised via the XRS technique. XRS modelling is presented in some detail in this chapter to allow the nonspecialist to get a better understanding of the conclusions drawn from XRS data. Imogolite nanotubes are considered as cylindrical nanotubes in dispersion after synthesis. Ovalisation and hexagonalisation of dried aluminogermanate nanotubes have been evident and depend on the organisation of nanotubes. When nanotubes are assembled in large bundles, their cross sections are fully hexagonalised over the bundle height, whereas they are locally ovalised for nonbundled nanotubes with local contact points. Deformed imogolites most probably exhibit different electronic, mechanical and physicochemical properties from those of cylindrical nanotubes (for instance, in terms of adsorption or catalysis). These new forms of imogolite should thus motivate further theoretical and experimental studies, with the prospect of elaborating advanced materials with new potential applications. In addition, the XRS analysis should also be helpful for the study of functionalised imogolites (see Chapter 12) or DW imogolites.

High-temperature treatments of imogolite reveal the progressive dehydroxylation of the nanotube walls, leading to the occurrence of an amorphous phase or of a lamellar phase due to the collapse of the tube structure. Further higher-temperature treatment shows the structural transformation from amorphous phase into a mullite crystalline phase. However, a more complete description of these phases is needed to understand their formation mechanisms and their modifications. Control of thermal stability of imogolite and its different structural phases is a prerequisite for the elaboration of future materials and their applications.

REFERENCES

Abidin, Z., Matsue, N., Henmi, T., 2008. A new method for nanotube imogolite synthesis. Jpn. J. Appl. Phys. 47, 5079–5082.

Amara, M.S., 2014. Nanotubes d'imogolite et propriétés de l'eau confinée: organisation, structure et dynamique. Thesis, Université Paris XI, Orsay.

Amara, M.S., Paineau, E., Bacia-Verloop, M., Krapf, M.E., Davidson, P., Belloni, L., Levard, C., Rose, J., Launois, P., Thill, A., 2013. Single-step formation of micron long $(OH)_3Al_2O_3Ge(OH)$ imogolite-like nanotubes. Chem. Commun. 49, 11284–11286.

Amara, M.S., Rouzière, S., Paineau, E., Bacia-Verloop, M., Thill, A., Launois, P., 2014. Hexagonalisation of aluminogermanate imogolite nanotubes organized into closed-packed bundles. J. Phys. Chem. 118, 9299–9306.

Bonelli, B., Bottero, I., Ballarini, N., Passeri, S., Cavani, F., Garrone, E., 2009. IR spectroscopic and catalytic characterization of the acidity of imogolite-based systems. J. Catal. 264, 15–30.

Bottero, I., Bonelli, B., Ashbrook, S.E., Wright, P.A., Zhou, W., Tagliabue, M., Armandi, M., Garrone, E., 2011. Synthesis and characterization of hybrid organic/inorganic nanotubes of the imogolite type and their behaviour towards methane adsorption. Phys. Chem. Chem. Phys. 13, 744–750.

Cradwick, P., Farmer, V., Russel, J., Masson, C., Wada, K., Yoshinaga, N., 1972. Imogolite, a hydrated aluminum silicate of tubular structure. Nature Phys. Sci. 240, 187–189.

Creton, B., Bougeard, D., Smirnov, K., Guilment, J., Poncelet, O., 2008. Molecular dynamics study of hydrated imogolite. Phys. Chem. Chem. Phys. 10, 4879–4888.

Da Silva Chagas, M., Campos dos Santos, E., Lourenço, M.P., Gouvea, M.P., Duarte, H.A., 2015. Structural, electronic, and mechanical properties of inner surface modified imogolite nanotubes. Front. Mater. 2. art. 16, 1–10.

Donkai, N., Miyamoto, T., Kokubo, T., Tanei, H., 1992. Preparation of transparent mullite-silica film by heat-treatment of imogolite. J. Mater. Sci. 27, 6193–6196.

Farmer, V., Fraser, A., Tait, J., 1977. Synthesis of imogolite: a tubular aluminium silicate polymer. J. Chem. Soc. Chem. Commun. (13), 462–463.

Farmer, V., Adams, M., Fraser, A., Palmieri, F., 1983. Synthesis of imogolite: properties, synthesis, and possible applications. Clay Miner. 18, 459–472.

Guimaraes, L., Enyashin, A., Frenzel, J., Heine, T., Duarte, H.A., Seifert, G., 2007. Imogolite nanotubes: stability, electronic, and mechanical properties. ACS Nano 1, 362–368.

Guinier, A., 1963. X-Ray Diffraction in Crystals, Imperfect Crystals, and Amorphous Bodies. W.H. Freeman, San Francisco.

Gustafsson, J.P., 2001. The surface chemistry of imogolite. Clay Clay Miner. 49, 73–80.

Horikawa, Y., 1975. Electrokinetic phenomena of aqueous suspensions of allophane and imogolite. Clay Sci. 4, 255–263.

Johnson, I.D., Werpy, T.A., Pinnavaia, T.J., 1988. Tubular silicate-layered silicate intercalation compounds: a new family of pillared clays. J. Am. Chem. Soc. 110, 8545–8547.

Kang, D.Y., Zang, J., Wright, E.R., McCanna, A.L., Jones, C.W., Nair, A., 2010. Dehydration, dehydroxylation, and rehydroxylation of single-walled aluminosilicate nanotubes. ACS Nano 4, 4897–4907.

Kang, D.Y., Zang, J., Wright, E.R., Jones, C.W., Nair, A., 2011. Single-walled aluminosilicate nanotubes with organic-modified interiors. J. Phys. Chem. C 115, 7676–7685.

Konduri, S., Mukherjee, S., Nair, S., 2006. Strain energy minimum and vibrational properties of single-walled aluminosilicate nanotubes. Phys. Rev. B 74, 033401.

Konduri, S., Mukherjee, S., Nair, S., 2007. Controlling nanotube dimensions: correlations between composition, diameter, and internal energy of single-walled mixed oxide nanotubes. ACS Nano 1, 393–402.

Lourenço, M., Guimaraes, L., Da Silva, M., de Oliveira, C., Heine, H., Duarte, H.A., 2014. Nanotubes with well-defined structure: single and double-walled imogolite. J. Phys. Chem. C 118, 5945–5953.

Ma, W., Yah, W.O., Otsuka, H., Takahara, A., 2012. Surface functionalization of aluminosilicate nanotubes with organic molecules. Beilstein J. Nanotechnol. 3, 82–100.

MacKenzie, K.J., Bowden, M.E., Brown, J.W.M., Meinhold, R.H., 1989. Structure and thermal transformation of imogolite studied by ^{29}Si and ^{27}Al high-resolution solid-state nuclear magnetic resonance. Clay Clay Miner. 37, 317–324.

Maillet, P., Levard, C., Larquet, E., Mariet, C., Spalla, O., Menguy, N., Masion, A., Doelsch, E., Rose, J., Thill, A., 2010. Evidence of double-walled Al-Ge imogolite-like nanotubes. A Cryo-TEM and SAXS investigation. J. Am. Chem. Soc. 132, 1208–1209.

Miyauchi, N., Aomine, S., 1966. Mineralogy of gel-like substance in the pumice bed in Kanuma and Kitakami districts. Soil Sci. Plant Nutr. 12, 19–22.

Mukherjee, S., Bartlow, V.D., Nair, S., 2005. Phenomenology of the growth of single-walled aluminosilicate and aluminogermanate nanotubes of precise dimensions. Chem. Mater. 17, 4900–4909.

Rols, S., Almairac, R., Henrard, L., Anglaret, E., Sauvajol, J.-L., 1999. Diffraction by finite size crystalline bundle of single-walled nanotubes. Eur. Phys. J. B 10, 263–270.

Tamura, K., Kawamura, K., 2002. Molecular dynamics modeling of tubular aluminum silicate: imogolite. J. Phys. Chem. B 106, 271–278.

Thess, A., Lee, R., Nikolaev, P., Dai, H., Petit, P., Robert, J., et al., 1996. Crystalline ropes of metallic carbon nanotubes. Science 273, 483–487.

Violante, P., Tait, J.M., 1979. Identification of imogolite in some volcanic soils from Italy. Clay Miner. 14, 155–158.

Wada, S., Wada, K., 1982. Effects on substitution of germanium for silicon in imogolite. Clay Clay Miner. 30, 123–128.

Wada, S.I., Eto, A., Wada, K., 1979. Synthetic allophane and imogolite. J. Soil Sci. 30, 347–355.

Yang, H., Chen, Y., Su, Z., 2007. Microtubes via assembly of imogolite with polyelectrolyte. Chem. Mater. 19, 3087–3089.

Yoshinaga, N., 1968. Identification of imogolite in the filmy gel materials in the Imaichi and Shichihonzakura pumice beds. Soil Sci. Plant Nutr. 14, 238–246.

Yoshinaga, N., Aomine, S., 1962. Imogolite in some Ando soils. Soil Sci. Plant Nutr. 8, 22–29.

Yucelen, G.I., Choudhury, R.P., Leisen, J., Nair, S., Beckham, H.W., 2012. Defect structures in aluminosilicate single-walled nanotubes: a solid-state nuclear magnetic resonance investigation. J. Phys. Chem. C 116, 17149–17157.

Zang, J., Konduri, S., Nair, S., Sholl, D.S., 2009. Self-diffusion of water and simple alcohols in single-walled aluminosilicate nanotubes. ACS Nano 3, 1548–1556.

Zang, J., Chempath, S., Konduri, S., Nair, S., Sholl, D.S., 2010. Flexibility of ordered surface hydroxyls influences the adsorption of molecules in single-walled aluminosilicate nanotubes. J. Phys. Chem. Lett. 1, 1235–1240.

Zanzottera, C., Vicente, A., Armandi, M., Fernandez, C., Garrone, E., Bonelli, B., 2012. Thermal collapse of single-walled alumino-silicate nanotubes: transformation mechanisms and morphology of the resulting lamellar phases. J. Phys. Chem. C 116, 23577–23584.

Chapter 12

Surface Chemical Modifications of Imogolite

B. Bonelli*
INSTM Unit of Torino Politecnico, Politecnico di Torino, Turin, Italy
Corresponding author: e-mail: barbara.bonelli@polito.it

12.1 INTRODUCTION

One of the most important properties of imogolite is its unique nanoporous structure, which encompasses three families of pores and related surfaces (Ackerman et al., 1993). Imogolite nanopores may be classified into three categories: (i) pore A, with an inner diameter of ca. 1.0 nm, corresponding to the cavities of proper nanotubes; (ii) pore B, about 0.3–0.4-nm wide, corresponding to cavities among three nanotubes aligned in a bundle, not accessible to small molecules like water; and (iii) larger C pores, ie, disordered slit pores (mainly mesopores) among bundles (Fig. 12.1).

The formation of imogolite can be ideally described by considering a single sheet of gibbsite, $Al(OH)_3$, and substituting, on one side only, three OH groups with an orthosilicate unit O_3SiOH. As Si–O bonds are shorter than Al–O bonds, the gibbsitelike sheet curls up, eventually forming single-walled (SW) imogolite with chemical composition $Al_2SiO_3(OH)_4$ that can be written as $(OH)_3Al_2O_3SiOH$, going from the outer to the inner surface of nanotubes (Cradwick et al., 1972). The inner surface of nanotubes (namely, surface A) is lined by silanol (SiOH) groups; on the outer surface of nanotubes, both Al–O–Al and Al–OH–Al groups occur, which give it an amphoteric character (Bonelli et al., 2009). B pores are less accessible to small molecules like water due to their small dimension, at least in dried powder samples.

The length of imogolite varies widely (from 400 nm to several microns); the inner diameter of A pores is practically constant to 1.0 nm (as discussed later in this chapter); and the outer diameter varies between 2.0 and 2.7 nm for either natural or synthetic samples (Cradwick et al., 1972). The latter phenomenon is accounted for by considering the presence of different ions (ie, impurities), coming from either the soil or the synthesis batch, located

Developments in Clay Science, Vol. 7. http://dx.doi.org/10.1016/B978-0-08-100293-3.00012-1

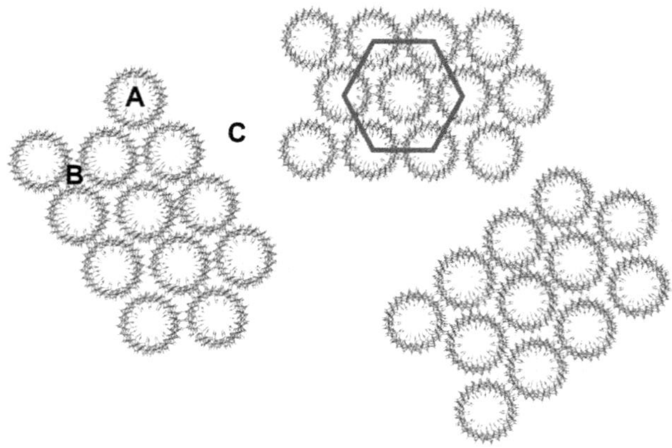

FIG. 12.1 Ideal representation of three bundles formed by imogolite and organized into a nearly hexagonal package. Pores A, B and C are pointed out only in one bundle, for the sake of clarity. *Adapted from Bonelli et al. (2013) with permission from the PCCP owner societies.*

within intertube spaces, and that, although they do not enter the formula of imogolite, they are responsible for variable intertube distances (Zanzottera et al., 2012b). Synthetic imogolite is usually produced at temperatures around 100°C, whereas synthesis at 25°C yields nanotubes with an external diameter comparable to that of natural imogolite (Wada, 1987). The inner diameter of A pores was first measured by electron diffraction experiments (Cradwick et al., 1972) and then confirmed by computer models (Pohl et al., 1996). Both studies showed that natural and synthetic nanotubes comprise 10 and 12 imogolite units in the cross section, respectively. More recently, Konduri et al. (2006) confirmed these structural findings by means of molecular dynamics simulations showing that the strain energy was at its minimum for nanotubes with 24 Al atoms in the cross section (indeed, corresponding to 12 imogolite units).

The packing of preformed nanotubes into bundles with nearly hexagonal symmetry gives rise to B pores, and the interconnection among bundles generates slit C mesopores (Fig. 12.1).

According to Ackerman et al. (1993), the network of natural imogolite is rather open and weblike, whereas by direct synthesis, more ordered structures form in which nanotubes are more aligned, and more densely packed and ordered over several nanometres. This may lead to lowered mesoporosity in synthetic samples, although a systematic study with a comparison of the porosity of natural and synthetic imogolite treated under the same experimental conditions is still lacking. In order to compare the features related to imogolite pore structure, the same experimental conditions should be adopted in terms of conditions of outgassing (under vacuum or an inert atmosphere), outgassing temperature, time and temperature ramp. This is because diffusion, especially of water molecules, is limited by the pore dimension. This is particularly important as far as pore A is concerned, being characterized by a high aspect ratio and high hydrophilicity. Both experimental (Bonelli et al., 2009) and molecular dynamics studies (Creton et al., 2008) showed that the inner surface of a nanotube was more hydrophilic than the outer one. To this end, the structure represented in Fig. 12.2 allows for making a few considerations over internal SiOH.

The calculated distance between two adjacent SiOH in the same circumference is 0.26 nm, and that between two SiOH of two adjacent circumferences is 0.40 nm (Bonelli et al., 2009). This leads to a calculated SiOH density of 9.1 OH nm^{-2}, ca. twice as much as the average SiOH density at the surface of hydrated amorphous silica, ca. 5 OH nm^{-2} (Iller, 1979). Taking into account the rigidity of the pseudocrystalline structure and the distances among them, internal SiOH should not show sizable interactions between them, in spite of the remarkable surface density. The hydrophilicity of A pores affects any possible application of imogolite in fields implying any diffusion of species within inner cavities, like catalysis, chromatography and adsorption, in that a preliminary dehydration step is required to remove molecular water. The same also holds for the functionalization of A pores in preformed imogolite (Kang et al., 2011).

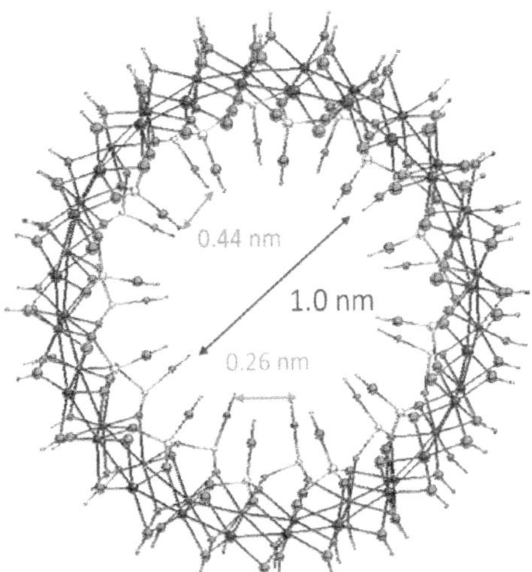

FIG. 12.2 Frontal view of an imogolite nanotube with two circumferences, both containing 12 imogolite units: the inner diameter of the A pore is 1.0 nm and distances are calculated of (i) 0.44 nm, between two SiOH belonging to two adjacent circumferences; and (ii) 0.26 nm, between two adjacent SiOH in the same circumference. *Adapted from Bonelli et al. (2009).*

An important issue to be considered during dehydration is the thermal stability of imogolite, in that the A surface becomes accessible to moieties of suitable size after prolonged evacuation at 270°C–300°C (Bonelli et al., 2009), but nanotubes collapse at slightly higher temperatures with the formation of a lamellar structure (MacKenzie et al., 1989). The latter, however, retains some residual porosity: Zanzottera et al. (2012b) have shown that after dehydration, dehydroxylation of both the inner and the outer surface of imogolite occurs above 300°C, giving rise to a buckled structure. This behaviour is in agreement with the nature of the material, which is a hydrated aluminium silicate with very rich hydroxyls population both inside and outside the nanotubes. Finally, a buckled structure with residual pores is obtained, as shown in Fig. 12.3.

The three kinds of pores of imogolite structure are characterized by different chemical composition and dimensions, so that three kinds of surface are given rise, summarized as follows:

(i) The A surface is lined by SiOH groups and is therefore very hydrophilic. It becomes accessible to gas molecules only after dehydration obtained by prolonged evacuation at 270°C–300°C (Bonelli et al., 2009);

(ii) The B surface is not accessible to even small molecules like water (Ackerman et al., 1993) in imogolite bundles, but it can be accessible in bundles of some chemically modified imogolite (Zanzottera et al., 2012a; Bonelli et al., 2013);

(iii) The C surface is related to slit mesopores among bundles. It has the same composition as the B surface and has an amphoteric character due to the presence of both Al–O–Al and Al–OH–Al groups, like an aluminium oxo/hydroxide.

The chemical composition of imogolite surfaces has an important effect, for instance, on its behaviour in water: at neutral pH values, the outer surface is positively charged, due to the protonation of Al–OH–Al groups, whereas

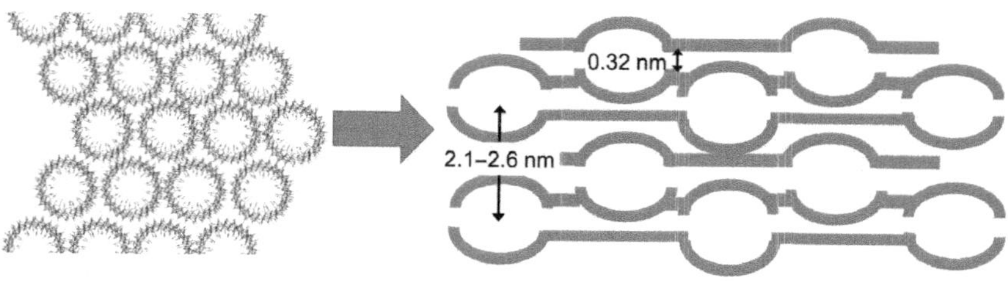

FIG. 12.3 Formation of a buckled structure by imogolite collapse at temperatures higher than 300°C. *Reprinted with permission from Zanzottera et al. (2012b). Copyright 2012 American Chemical Society.*

SiOH inside nanotubes are partially dissociated, according to the equilibria (Gustafsson, 2001):

$$Al(OH)Al + H^+ \leftrightharpoons Al(OH_2)^+ Al \quad (12.1)$$

$$\equiv SiO - H \leftrightharpoons \equiv SiO^- + H^+ \quad (12.2)$$

This feature accounts for the high dispersibility of imogolite in water and its precipitation in organic solvents. The behaviour of nanotubes in water is crucial: first, scientific interest for imogolite initially arose from its possible use for the removal of cations and anions from polluted water, in that the former should be able to interact with the inner surface and anions with the outer surface of nanotubes, respectively (Parfitt et al., 1974; Clark and McBride, 1984; Harsh et al., 1992; Denaix et al., 1999; Arai et al., 2006). For applications requiring highly dispersed nanotubes (eg, functionalization of the outer surface to get imogolite polymer nanocomposite materials), dispersion of imogolite in acidic solution is recommended (Yamamoto et al., 2001) such that positively charged nanotubes repel each other, finally preventing their aggregation.

12.2 MODIFICATION OF THE INNER PORES OF IMOGOLITE

As mentioned earlier in this chapter, A pores have a diameter of ca. 1.0 nm, the related surface being very hydrophilic. In order to render A pores accessible to probes, even small ones like CO, CO_2 and NH_3, imogolite powders must be dehydrated at 270°C–300°C. Once accessible to probes, inner SiOH displays a mild acidic character, like isolated SiOH at the surface of amorphous silica, as shown mainly by IR spectroscopy studies (Bonelli et al. 2009, 2013). The high aspect ratio of imogolite implies difficult diffusion of larger molecules, so that for practical applications, wider and less hydrophilic A pores could be useful. This goal may be achieved by either direct or post-synthesis modification of the inner A surface.

12.2.1 Direct Synthesis Methods

Direct synthesis methods are based on the replacement of the silicon precursor in the synthesis batch. In this way, nanotubes are obtained in which either Si is replaced by Ge or inner hydroxyls are replaced by organic functionalities that confer hydrophobic properties to a new A surface.

In the former case, the Si source (for instance, tetraethoxysilane, TEOS) is replaced by a Ge-precursor either like tetraethoxygermanium (TEOG) or germanium tetrachloride ($GeCl_4$), so that imogolite-like materials are created via the formula $(OH)_3Al_2Si_{1-x}Ge_xOH$. In such materials, hereafter referred to as 'Ge-imogolite', Si is either partially or totally replaced by Ge (Wada and Wada, 1982; Mukherjee et al., 2005, 2007; Konduri et al., 2007).

Another approach consists of modifying hydrophilicity of imogolite by replacing the Si precursor with an organosilane. When triethoxymethylsilane (TEMS $(CH_3CH_2O)_3Si–CH_3$) is used in place of TEOS, $(CH_3CH_2O)_4Si$, nanotubes are obtained with chemical composition $(OH)_3Al_2O_3SiCH_3$ (Bottero et al., 2011). The imogolite-like material obtained in this way, hereafter referred to as 'Me-imogolite', has a fully hydrophobic A surface, lined by methyl groups, and a (theoretically) unmodified hydrophilic outer surface. Besides the change of hydrophilic/hydrophobic properties of the material, this brings about a change in the inner diameter of the nanotubes, which ends up being >2.0 nm, so that Me-imogolite is actually a mesoporous material. When Me-imogolite form bundles, larger B pores are created due to the geometric arrangement shown in Fig. 12.4: the latter became accessible to probes like ammonia and CO_2 (Zanzottera et al., 2012a; Bonelli et al., 2013).

Me-imogolite synthesis is simple and has a higher yield with respect to that of proper imogolite. The procedure consists into modifying a traditional method for the synthesis of imogolite (Wada et al., 1979; Farmer et al., 1983): Al-*sec*-butoxide is used as the aluminium source in an acidic medium due to the presence of $HClO_4$, and TEOS, the silicon source in the original preparation, is replaced by TEMS. The Al:Si molar ratio used in this instance is 2:1·x, by which a slight excess of Si should prevent the formation of gibbsite, $Al(OH)_3$. A mixture is obtained that is dialyzed against deionized water for 4 days to remove salts formed during the reaction.

In principle, any silicon precursor containing the Si–C bond stable to hydrolysis would be suitable to obtain hybrid imogolite with an inner hydrophobic surface. However, to the best of our knowledge, any efforts to synthesize hybrid imogolite with other commercially available organic precursors have been unsuccessful. Recently, imogolite-like materials were obtained with up to 15% inner Si–OH groups replaced by $Si–CH_2–NH_2$ groups (hereafter referred to as *ANT*,

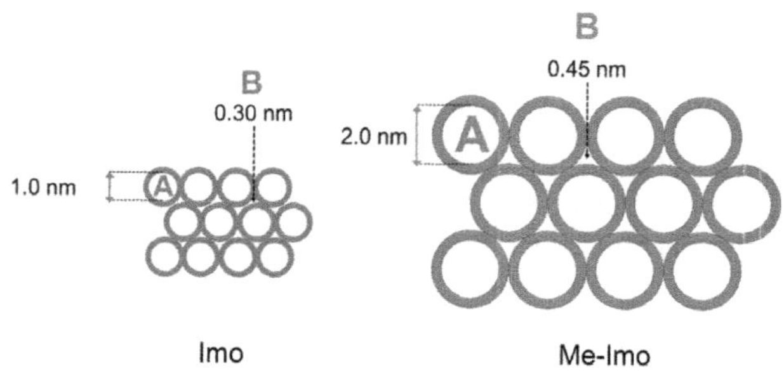

FIG. 12.4 Ideal representation of an imogolite and a Me-imogolite bundle, showing that the hexagonal packing gives rise to larger B pores with Me-imogolite (0.45 nm), having A pores with a diameter of 2.0 nm, with respect to imogolite. *Reprinted with permission from Zanzottera et al. (2012a). Copyright 2012 American Chemical Society.*

which stands for 'amine functionalized nanotubes') (Kang et al., 2014). In the latter case, a specifically synthesized precursor was used, namely $(CH_3CH_2O)_3Si–CH_2–NH_2$ (aminomethyltriethoxysilane, AMTES). Previous efforts made by using organosilanes with longer alkyl chains, like 3-aminopropyltrietoxysilane (3-APTES), were unsuccessful. The choice of the organic precursor is indeed crucial; one possible reason for this is the limited range of imogolite circumferences of stable size (Konduri et al., 2006). Nonetheless, only a fraction of inner SiOH groups was substituted in ANT, so residual ones need to be stabilized by intermolecular forces with surrounding functionalities.

It is possible to change further the chemical composition of A pores by producing hybrid Ge-containing nanotubes. For example, Amara et al. (2015) were able to obtain hybrid imogolite with composition $(OH)_3Al_2Si_{1-x}Ge_xCH_3$ by mixing TEMS and triethoxymethylgermane (TEMG) to a solution containing the Al precursor, acidified by $HClO_4$. By properly dosing the amount of TEMG, it is possible to fine-tune the diameter of the final material.

In conclusion, direct synthesis methods allow for modifying the inner surface of nanotubes in terms of the hydrophilic/hydrophobic properties, polarity and size of A pores. The obtained materials have well-defined characteristics, formation of by-products is limited and, at least with Me-imogolite, higher yields are obtained with respect to the synthesis of proper imogolite (Bottero et al., 2011).

12.2.2 Postsynthesis Methods

Postsynthesis modification may concern both the inner and the outer surfaces of nanotubes. In the former case, particular care has to be taken in removing molecular water, which is naturally present inside nanotubes, still preserving their structure and avoiding dehydroxylation. The dehydration step is crucial in that removal of molecular water starts at relatively low temperatures, but molecular water is fully removed only at about 270°C–300°C, presumably due to hindered diffusion along micron-long nanotubes (Bonelli et al., 2009; Zanzottera et al., 2012b). Nonetheless, the size of the organic linker is limited by the size of A pores. So far, only one paper has reported on the postsynthesis functionalization of inner pores of imogolite (Kang et al., 2011). These authors obtained stable imogolite, with a modified A surface, by grafting organic functionalities after careful dehydration of preformed imogolite. It resulted that the organic linkers substituted ca. 35% of inner SiOH groups. In order to get A pores free of molecular water, presynthesized imogolite was dehydrated under vacuum at 250°C for 24 h, and then transferred into a glove box. Using hexane as a solvent, the functionalizing reagent R (R = acetyl chloride, trimethylmethoxysilane or trichlorosilane) was added in the molar ratio R:imogolite hydroxyl groups = 2:1. The mixture was stirred in nitrogen atmosphere for 24 h; the flask was then connected to a vacuum line and treated at 180°C for 24 h to remove the solvent and unreacted reagent.

Several experimental techniques were adopted to study the physicochemical properties of the obtained materials, proving that the functionalization mainly concerned the inner surface of imogolite (ie, inner SiOH groups).

12.2.3 Properties of the Obtained Materials

Modification of the A surface of imogolite brings about changes in nanotube length, pore diameter and packing, specific surface area (SSA), hydrophilic/hydrophobic properties, thermal stability and accessibility to gases like CO_2 (Kang et al., 2014) and CH_4 (Bottero et al., 2011).

With respect to proper imogolite, Ge-imogolite are typically shorter (ca. 30 nm long), monodispersed in length and have a larger inner diameter of ca. 2.0 nm due to the larger atomic radius of Ge. Changing the Si/Ge ratio allows modulating both the inner diameter and length. Usually, shorter nanotubes are obtained with respect to proper imogolite; however, micron-long $(OH)_3Al_2Si_{1-x}Ge_xOH$ nanotubes have been obtained by Amara et al. (2013). Ge-imogolite may be either SW or double-walled (DW) (Maillet et al., 2010; Thill et al., 2012; Amara et al., 2013; Lourenço et al., 2014) at variance with proper imogolite, always forming SW nanotubes.

The presence of an inner hydrophobic surface, along with an outer hydrophilic one, made Me-imogolite a new example of an organic–inorganic hybrid. Bottero et al. (2011) showed that Me-imogolite has a higher degree of long-range order with respect to imogolite. High-resolution transmission electron microscopy (HRTEM) analysis indicated wider domains of aligned fibres and powder X-ray diffraction (XRD) patterns showed more intense reflections (Fig. 12.5) with respect to imogolite.

One possible explanation is that the equilibria described by Eqs (12.1) and (12.2) do not take place because of the absence of inner silanols. Fibres are, therefore, more loosely bound and freer to assume a close-packed configuration. The (100) reflection is at 3.37 degrees 2ϑ, the cell parameter a, corresponding to a centre-to-centre distance between two aligned nanotubes and assuming a hexagonal packing ($a=2^*d_{100}/\sqrt{3}$), is calculated to be 2.98 nm; with imogolite, the cell parameter is 2.68 nm (ie, slightly smaller). In the case of Ge-imogolite, the larger value of the distance between nanotubes was ascribed to the formation of nanotubes with a lower curvature, with the Ge–O bond being longer than the Si–O bond (Konduri et al., 2007). For Me-imogolite, two factors should be involved: a larger nanotube curvature could be caused by the presence of methyl groups, for either steric (the methyl group is larger than the hydroxyl group) or electronic effect. Sidorkin et al. (1999) reported that the calculated Si–O bond length in series of both $(HO)_3Si–X$ and $(CH_3O)_3Si–X$ compounds is affected by the electronegativity of the X substituent. The presence of methyl groups in place of hydroxyls could have the same effect, inducing a relaxation of the structure and a shift of energy minimum towards nanotubes with a higher number of units in the

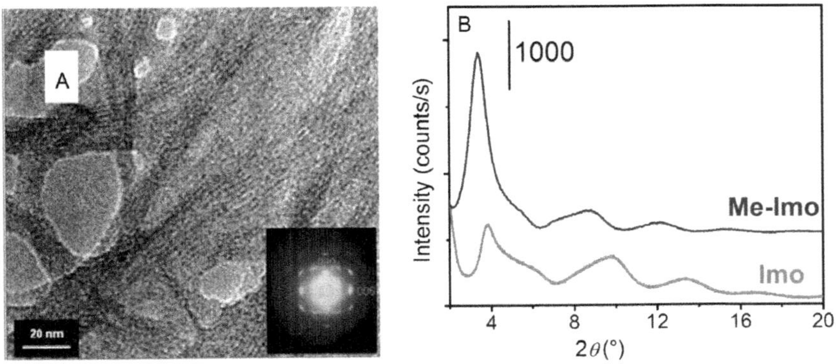

FIG. 12.5 (A) HRTEM micrograph of a Me-imogolite sample; *inset* shows selected area diffraction (SAED) patterns of a bundle formed by a few nanotubes, demonstrating the crystallinity within them. (B) XRD patterns of fresh Me-imogolite and imogolite samples. *Adapted from Bottero et al. (2011) with permission from the PCCP owner societies.*

circumference with respect to that calculated for imogolite (Konduri et al., 2006). Alternatively, the larger cell parameter could be ascribed to an increased distance between nanotubes, as mentioned previously. XRD patterns of the as prepared Me-imogolite and those of Me-imogolite outgassed at 150°C and 300°C showed that by increasing the temperature, the intensity of the reflections progressively decreases and a progressive loss of structural order occurs, starting at relatively low temperatures (Bottero et al., 2011). The reason for a facile loss of long-range ordering is probably the absence of strong interactions between nanotubes. Disordering of the system at low temperatures is not accompanied by any mass loss, as shown by thermogravimetric (TG) analysis (Bottero et al., 2011), which rules out dehydroxylation phenomena. Me-imogolite is more hydrophobic than imogolite; it releases molecular water much more easily, and milder thermal treatments (below 100°C) are required to make Me-imogolite nanovoids accessible to gas molecules. The SSA of Me-imogolite progressively increases with the outgassing temperature, reaching a maximum of 740 $m^2\ g^{-1}$ with Me-imogolite outgassed at 300°C (Bottero et al., 2011). The SSA and microporous volumes of Me-imogolite are twice those of imogolite treated at the same outgassing temperature, probably because of the different nanotube diameters. The pore size distributions (PSD) of Me-imogolite always show a family of nanopores with diameters in the 2.0–10-nm range, the most abundant being those with an inner diameter of ca. 2.0 nm. The intensity of the corresponding peak increases when the material is outgassed at increasing temperatures, whereas larger nanopores seem to be less affected by the thermal treatment (Bottero et al., 2011), indicating that Me-imogolite is a mesoporous material.

IR spectra of a self-supporting wafer of Me-imogolite outgassed at increasing temperatures, from room temperature (r.t.) to 300°C are reported in Fig. 12.6. Me-imogolite outgassed at r.t. shows a band at 1644 cm^{-1} assigned to the bending mode of water molecules mainly adsorbed at the outer surface of imogolite. This band disappears almost completely after a mild thermal treatment, in agreement with the results of both TG analysis (recording a mass loss at ca. 40°C) and N_2 adsorption/desorption isotherms showing a higher adsorbed volume by increasing the outgassing temperature (Bottero et al., 2011). The sharp band at 1275 cm^{-1} is due to the bending mode of methyl groups of inner Si–CH_3 (δ_{CH_3}); the bands at 2978 and 2919 cm^{-1} are assigned to the asymmetric (ν_{aCH_3}) and symmetric (ν_{CH_3}) stretching modes of the same groups, respectively. The broad and intense absorption in the OH stretch region (3800–3000 cm^{-1}) is assigned to outer Al(OH)Al groups.

Fig. 12.7 shows the comparison between the IR spectra of Me-imogolite and imogolite outgassed at 150°C, a temperature at which only the former is fully dehydrated. Accordingly, the band of molecular water is still present in more hydrophilic imogolite (shown by an arrow in the figure) and absent in more hydrophobic Me-imogolite. Comparison between the two curves in the O–H stretching region shows that absorption due to Al–OH–Al species extends ca. from about 3800 to 3000 cm^{-1}, whereas absorption due to SiOH engaged with by H-bonding adsorbed water is in the region between 3000 and 2800 cm^{-1}.

The presence of a hydrophobic inner surface, along with an outer hydrophilic one, could be exploited in adsorption processes aimed at gas separation (eg, hydrocarbon recovery from wet gaseous streams) or storage. For this

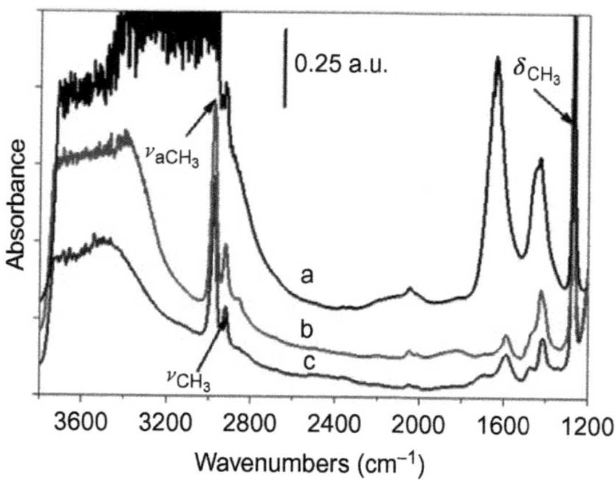

FIG. 12.6 IR spectra of Me-imogolite outgassed at: r.t. (a), 150°C (b) and 300°C (c). The *arrows* point out bands related to methyl groups of Si–CH_3 functionalities. *Adapted from Bottero et al. (2011) with permission from the PCCP owner societies.*

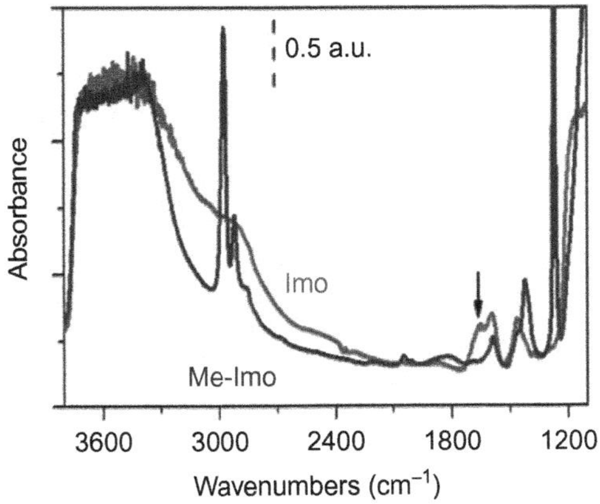

FIG. 12.7 IR spectra of Me-imogolite (*blue* (*dark grey* in print version) *curve*) and imogolite (*red* (*light grey* in print version) *curve*) outgassed at 150°C. The *arrow* points out the band due to residual molecular water on more hydrophilic imogolite. *Adapted from Bottero et al. (2011) with permission from the PCCP owner societies.*

reason, adsorption of methane was tested at 30°C on the two samples previously outgassed at 300°C, showing a larger amount of adsorbed methane in Me-imogolite (Bottero et al., 2011).

Another aspect related to the presence of A pores with a larger inner diameter is the consequent formation of larger B pores when Me-imogolite form bundles (Fig. 12.4); the resulting B pores of Me-imogolite are indeed larger with respect to imogolite (Zanzottera et al., 2012c). This allows probes like carbon dioxide and ammonia to enter the B pores, whereas these probes cannot access the B pores in imogolite (Bonelli et al., 2013).

Concerning ANT material (Kang et al., 2014), shorter nanotubes were observed by TEM analysis, with an average length of 50 nm. Both experimental and simulated XRD patterns of ANT with 15% aminomethyl substitution for inner SiOH groups showed that the presence of inner aminomethyl species causes a deviation from the corresponding patterns of imogolite, in that XRD peaks at higher angles are less prominent. Aminomethyl substitution brings also about a loss of micropore surface area, in agreement with the larger estimated volume of aminomethyl groups with respect to SiOH (38.1 Å^3 with respect to 16.9 Å^3, as estimated from the atomic van der Waals radii). The presence of the amino groups has an important consequence for the adsorption properties of ANT, which showed preferential adsorption towards carbon dioxide from both CO_2/CH_4 and CO_2/N_2 mixtures.

Modification of the A surface by postsynthesis (Kang et al., 2011) brings about a decrease of the inner pore volumes, and therefore of the inner surface area related to A pores. This phenomenon is likely due to the larger molecular size of the functionalizing reagent (R = acetyl chloride, trimethylmethoxysilane or trichlorosilane) with respect to SiOH. The postsynthesis functionalized imogolite showed well-defined XRD patterns, and no evidence emerged of either structure amorphization or alteration. Concerning thermal stability, dehydration and dehydroxylation occurred as with imogolite, but between 450°C and 600°C, an additional mass loss due to the degradation of the organic linker was observed. In agreement with the presence of organic functionalities within A pores, their hydrophilicity decreased, as shown by water adsorption isotherms (Kang et al., 2011).

12.3 MODIFICATION OF THE OUTER SURFACE OF IMOGOLITE

The reactivity of the outer surface of imogolite, unlike carbon nanotubes (CNT), allows a rather facile functionalization of preformed nanotubes by postsynthesis methods. Nanotube functionalization is of paramount importance for practical applications, especially for the production of clay polymer nanocomposites (CPN; Ma et al., 2012), films of aligned nanotubes (Park et al., 2007) or films with good transparency (Yamamoto et al., 2005b).

The following modifications of the outer surface of imogolite are reported in the literature: (i) grafting with 3-APTES (3-aminopropylsilane, NH_2–$(CH_2)_3$–$Si(OEt)_3$), the most used organosilane for the modification of inorganic materials and for the production of organic-inorganic hybrids; (ii) functionalization with alkylphosphonic acids; and (iii) isomorphic replacement of aluminium with Fe^{3+} ions. The latter process has been examined in only a few studies to date (Ookawa et al., 2006; Ookawa, 2012; Shafia et al., 2015), whereas many works have reported on the modification of the outer surface of imogolite with organic functionalities in order to produce hybrid (inorganic-organic) materials. Hydrophobization of the imogolite's outer surface is mandatory for the production of imogolite polymer nanocomposites, since it is necessary to improve imogolite compatibility with organic solvents (Ma et al., 2012). This would allow broader applications in the fabrication of nanocomposite materials.

12.3.1 Grafting of Organic Molecules

Johnson and Pinnavaia (1990) first reported on the functionalization with 3-APTES of the outer surface of imogolite in water at acidic pH values obtained by the presence of acetic acid. Acidic conditions allow a homogenous mixture of well-dispersed imogolites in water at acidic pH and ensure a fast and complete 3-APTES hydrolysis, in order to prevent 3-APTES polymerization and to guarantee the presence of mainly 3-APTES monomers and dimers in the reaction mixture. An excess of 3-APTES with respect to the nominal Al content in the synthesis of imogolite with respect to the monolayer content was used. A careful characterization of the obtained materials (mainly by IR spectroscopy and solid-state NMR) showed that reaction mainly occurred with outer Al–OH–Al groups rather than with inner SiOH groups. The latter groups should not be accessible to both 3-APTES monomers and dimers, due to the small diameter of A pores. However, the 3-APTES-grafted-imogolite was unstable in aqueous solution after several days of dialysis against deionized water, thus restricting its applications in water. The same authors showed that some functionalization likely also occurred at the inner surface of imogolite (Johnson and Pinnavaia, 1991). Two different kinetics of 3-APTES hydrolysis were indeed measured: a faster one assigned to hydrolysis of 3-APTES grafted at the outer surface of nanotubes, and a slower one, ascribed to hydrolysis of 3-APTES moieties grafted at inner SiOH groups. To this respect, two factors should be taken into account: (i) it is difficult to limit functionalization to the outer surface only because some SiOH groups (most likely at the mouth of the pores) may be accessible even to larger molecules; and (ii) the presence of water during grafting should be avoided in order to limit hydrolysis of the formed products. Two papers have reported on the grafting of 3-APTES on imogolite (Qi et al., 2008) and its methylated analogue Me-imogolite (Zanzottera et al., 2012c). In both cases, the reaction and the washing procedure that followed were carried out in toluene, an organic solvent that can prevent (undesired) hydrolysis phenomena. In the former case, imogolite reacted with 3-APTES, thus obtaining a material referred to as 'Imo-APTES'. Subsequently, osmium tetroxide (OsO_4) was immobilized by the amino groups at the outer surface of Imo-APTES. The resulting osmium-bound imogolite (Imo-APTES-OsO_4) was tested as a heterogeneous catalyst. Similar to the procedure of dehydration carried out before postsynthesis functionalization of the A surface (Kang et al., 2011), preformed imogolite was pretreated in vacuum, though at a lower temperature (ie, 100°C–110°C instead of 250°C). The obtained powder was dispersed in anhydrous toluene and treated with 3-APTES, refluxed at 100°C for 12 h under N_2 atmosphere, washed with toluene, filtered, dried and then functionalized with OsO_4 in *t*-BuOH (Qi et al., 2008). The presence of amino groups in Imo-APTES nanotubes was detected by means of UV–Vis spectroscopy following the reaction with ninhydrin (2,2-dihydroxyindane-1,3-dione). This procedure is known as the 'Kaiser test', in which ninhydrin is readily converted into a purple moiety by reaction with an alkyl amine. The Kaiser test allows for proving the functionalization of inorganic materials and is a rather simple check of the presence of amino groups in the functionalized material (Shemg et al., 1993). The presence of both Os(VI) and Os(IV) species due to the coordination by one and two amino groups, respectively, was confirmed by XPS analysis. The resulting catalyst was stable in air, a good catalytic activity was tested in the first run, and then a loss of activity occurred (Qi et al., 2008).

In order to avoid the reaction observed by Johnson and Pinnavaia (1991) between 3-APTES and inner SiOH groups Zanzottera et al. (2012c) carried out the functionalization with 3-APTES of Me-imogolite because it is expected that inner Si–CH_3 bonds should be stable to hydrolysis and do not react with 3-APTES. Preformed Me-imogolite was dried at 150°C for 4 h in vacuum and stirred in anhydrous toluene in the presence of 3-APTES in the molar ratio Me-imogolite:3-APTES = 1:0.3. The full functionalization of Me-imogolite should lead to a material with nominal composition NH_2–$(CH_2)_3$–$Si(O)_3Al_2O_3SiCH_3$. Zanzottera et al. (2012c) chose instead an amount of 3-APTES corresponding to one-third of the content of 3-APTES corresponding to the functionalization of all Al(OH)Al groups for three reasons: (i) the structure of imogolite shows that probably not all the external Al(OH)Al groups may react (eg, those within B pores); (ii) the bulky aminopropyl groups are likely to cover the surface even at lower loadings than the theoretical one; and (iii) a lower amount of 3-APTES should prevent the condensation of 3-APTES molecules with the formation of undesired phases. The successive synthesis steps were similar to those adopted by Qi et al. (2008), and the obtained material, hereafter referred to as 'Me-Imo-NH_2' was carefully characterized by several techniques, among which ^{29}Si MAS NMR spectroscopy clearly showed that functionalization led to the formation of the outer surface species reported in Fig. 12.8.

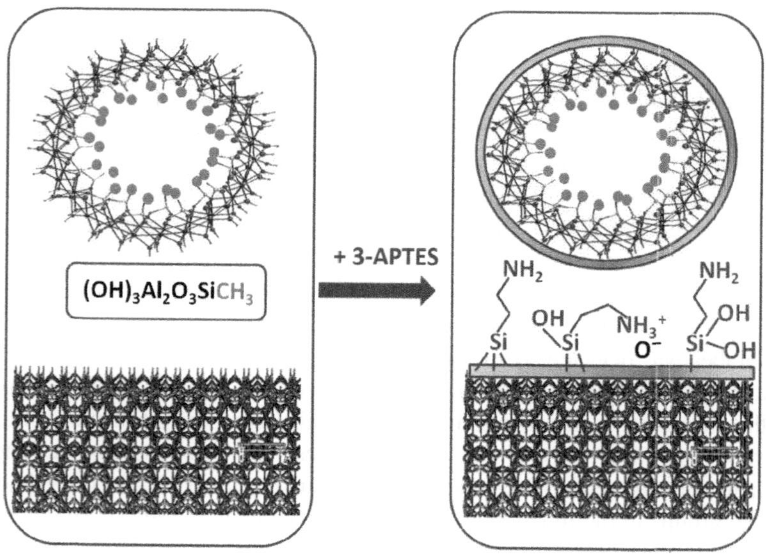

FIG. 12.8 Functionalization of Me-imogolite with 3-APTES in toluene leads to a material bearing three kinds of groups at the outer surface, as revealed by ^{29}Si MAS NMR spectroscopy. *Reprinted with permission from Zanzottera et al. (2012c). Copyright 2012 American Chemical Society.*

3-APTES reacted quantitatively with the outer surface by forming one- to three legged anchored species (Fig. 12.8); species protonated after the reaction with outer Al(OH)Al groups also formed: the reaction with diluted NaOH solution was able to restore the expected amine functionalities. As a whole, the functionalization of the outer surface brought about (i) a limited loss of surface area (from 530 to 450 $m^2\ g^{-1}$); (ii) the ability to interact with carbon dioxide by forming carbamates; and (iii) a moderate increase in thermal stability (Zanzottera et al., 2012c). The latter phenomenon was ascribed to a decrease in the number of outer Al–OH–Al groups, which may undergo dehydroxylation.

In addition to the reaction of an organosilane with outer Al–O–Al and Al–OH–Al groups, formation of organic-inorganic hybrids may be obtained by exploiting the high reactivity of outer Al–OH–Al groups with phosphoric acid (Yamamoto et al., 2005a). In order to get imogolite polymer nanocomposites with improved mechanical properties (namely, thermal stability and transparency), it is important to have modified imogolite well dispersed in hydrophobic polymer matrix. Usually, preformed nanotubes are dispersed in acidic solution in order to have a positively charged outer surface, due to the protonation of Al(OH)Al groups and repulsion between nanotubes, in order to improve their dispersion and allow the functionalization of the outer surface (Yamamoto et al., 2001). In this field, one of the most interesting results was obtained by Park et al. (2007), who were able to obtain a two-dimensional material by reactions between either octadecyl phosphonic acid (ODPA) or tetradecyl phosphonic acid (TDPA) and preformed imogolite. The latter were dispersed in acidic solution (pH 3.5–4.5), in order to have bundle disaggregation due to the protonation of Al–OH–Al groups and repulsion between $Al-(OH)_2{}^{+}-Al$. The solution was added to an aqueous solution of ODPA to obtain the imogolite ODPA nanocomposite, stirred for 2 days at r.t. and then washed with ethanol several times (a similar procedure was adopted to obtain the imogolite TDPA nanocomposite). The obtained imogolite polymer nanocomposites were well dispersed in organic solvent. Moreover, films of aligned nanotubes on a graphite support were obtained by the Langmuir–Blodgett technique. More ordered films were obtained with ODPA, likely due to the better intercrossing between longer alkyl chains with respect to TDPA. The surface-modified imogolite fibres ended up being well aligned with constant nanospacing, as visualized by scanning tunnelling microscopy (STM) (Park et al., 2007).

A further step is needed in order to obtain imogolite polymer nanocomposites and improve their affinity towards organic moieties. The outer surface of imogolite is functionalized with an organic group bearing a negatively charged head, which can react with the outer surface of imogolite, and a reactive end, which can undergo polymerization (Yamamoto et al., 2005a). By this way, imogolite poly(methyl methacrylate) hybrid materials were obtained after surface modification with an ester of phosphoric acid. The same authors tried to prepared hydrophobic imogolite polymer nanocomposites without prior modification of the outer surface, but the approaches were limited to hydrophilic polymers with an affinity for imogolite (Yamamoto et al., 2002). Besides alkylphosphonates, other organic anions can be exploited to interact with the outer surface of imogolite. Kuroda and Kuroda (2010) were able to increase the C surface among bundles by enlarging imogolite mesopores using poly(sodium 4-styrenesulfonate) as an expander. The organic anion interacts electrostatically with positively charged imogolite: an interwoven imogolite polymer nanocomposite is obtained, with expanded intertube mesopores that are expected to contribute to efficient diffusion, overcoming problems related to diffusion limitations.

12.3.2 Reactivity of Outer Surfaces

The reactivity of imogolite outer surface related to (i) its amphoteric properties, due to the presence of both Al–O–Al and Al–OH–Al groups; and (ii) the possibility of substituting external Al^{3+} ions with ions having suitable size and charge (eg, Fe^{3+}) will be considered.

The outer surface of imogolite has an amphoteric character. It can react with both acidic molecules [ie, with carbon dioxide (CO_2) by forming carbonatelike species], and basic molecules [ie, with ammonia (NH_3) by forming ammonium species]. With imogolite powder samples, such gaseous molecules will basically react with groups at the C surface since B pores in proper imogolite bundles are not accessible even to small probes. Instead, with Me-imogolite, B pores are accessible, since they are larger than with imogolite bundles (Fig. 12.9).

Concerning proper imogolite, the presence of carbonatelike species similar to those formed on common aluminium oxo/hydroxides (Morterra and Magnacca, 1996), deriving by the reaction with atmospheric carbon dioxide, was detected by IR spectra of as prepared imogolite, showing bands at 1595 and 1465 cm^{-1} (Bonelli et al., 2009). When CO_2 is dosed on imogolite outgassed at r.t. (Fig. 12.10), a sharp band forms at 2342 cm^{-1}, due to the asymmetric stretch of CO_2 molecules interacting with inner SiOH groups (Zanzottera et al., 2012a). At lower wavenumbers, less intense bands are seen at ca. 1650, 1555, 1460 and 1410 cm^{-1}, due to monodentate carbonates (1555 and 1410 cm^{-1}) and bicarbonates (1650 and 1460 cm^{-1}) formed at the outer surface of nanotubes. The latter bands are partially irreversible at r.t. (green (dotted grey in the printed version) dotted curve) and fully removed only by outgassing at 100°C (Fig. 12.10), indicating a rather strong interaction with the surface.

The same bands form after dosing CO_2 on Me-imogolite since the two materials have the same outer surface. With Me-imogolite, however, CO_2 was proved to interact with Al–OH–Al groups within B pores as well (Zanzottera et al., 2012a).

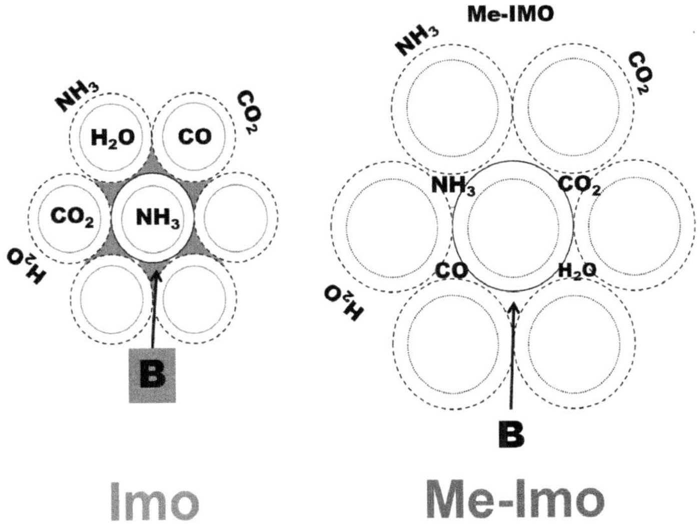

FIG. 12.9 B pores in imogolite are not accessible to small molecules, but in Me-imogolite, they can be probed by small gaseous species, like CO, NH_3, CO_2 and H_2O. *Adapted from Bonelli et al. (2013) with permission from the PCCP owner societies.*

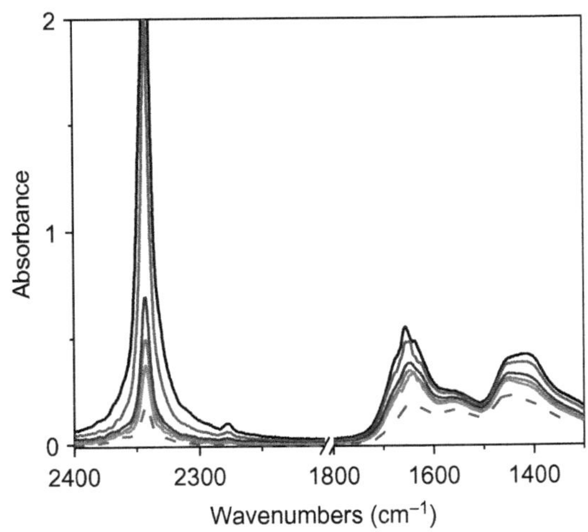

FIG. 12.10 IR spectra obtained after dosing CO_2 on imogolite outgassed at r.t. (equilibrium pressures: 0.00–30.0 mbar). *Reprinted with permission from Zanzottera et al. (2012a). Copyright 2012 American Chemical Society.*

The amphoteric nature of Al–OH–Al sites, therefore, may be appreciated when present in accessible B pores of Me-imogolite, in which they also behave as medium-strength acids, giving rise to the formation of ammonium species after dosing ammonia at r.t. (Bonelli et al., 2013). In this case, it is probable that the confinement effect in small nanocavities stabilizes the ammonium species, so that the acidic nature of such aluminol species is magnified (Fig. 12.11). Dosing ammonia on Me-imogolite gives rise, indeed, to the formation of bands due to the modes of NH_4^+ species (asterisks) at 1445 (bending), 3380 and 3280 (stretching) cm^{-1}. Ammonia also interacts with outer Al^{3+} ions, as shown by the formation of a band at 1625 cm^{-1} (arrow), due to the bending mode of ammonia molecules adsorbed on Lewis sites (Bonelli et al., 2013).

Octahedral Al^{3+} (Al(VI)) at the outer surface may be slightly distorted (at least in dehydrated samples) and so is able to adsorb not only ammonia, but weaker bases as well. When CO is dosed on dehydrated imogolite, a band is observed in the IR spectrum at 2190 cm^{-1}, due to CO adsorbed on weak Al^{3+} sites at the outer surface of imogolite (Bonelli et al., 2009). The observed position of the band indicates that such Al(VI) ions behave as weak Lewis sites since they are still highly coordinated.

Since small molecules like CO, NH_3 and CO_2 may also interact with inner SiOH, the outer surface of imogolite may be selectively studied by using bulkier probes, like phenol and 1,3,5-triethylbenzene (TEB) (Bonelli et al., 2013), and by following the corresponding adsorption process by IR spectroscopy.

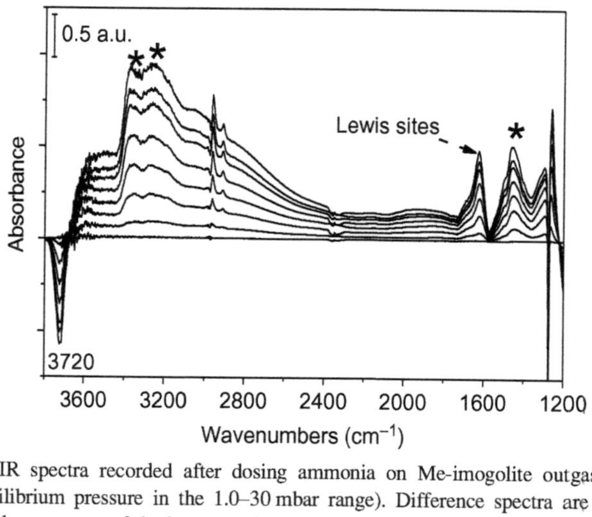

FIG. 12.11 IR spectra recorded after dosing ammonia on Me-imogolite outgassed at 300°C (ammonia equilibrium pressure in the 1.0–30 mbar range). Difference spectra are reported after subtraction of the spectrum of the bare sample. *Adapted from Bonelli et al. (2013) with permission from the PCCP owner societies.*

Phenol (Phe) is a dipolar molecule ($\mu = 1.54$ D) with a van der Waals diameter of 5.7 Å, which by virtue of its molecular dimensions could also test A pores; however, reactivity of Phe, followed by IR spectroscopy, mainly occurs with the outer surface, likely due to its hindered diffusion within A pores.

IR spectra of dehydrated imogolite after Phe adsorption are reported in Fig. 12.12. At low Phe partial pressure, the small negative peak seen in the hydroxyl range (Fig. 12.12A) at 3736 cm^{-1} was assigned to Al-related OH groups interacting with Phe molecules. Its small intensity suggests that the interaction with outer Al–OH–Al groups was limited. The shift measured at low coverage was of ca. 336 cm^{-1} (slightly lower with respect to that observed for phenol adsorbed on SiOH groups of amorphous silica, in agreement with the smaller acidity of those species). At higher coverage, a broad and intense absorption was observed, due to extended H-bonding caused by the formation of water, as observed in the lower wavenumber range (Fig. 12.12B). The IR spectrum of Phe in IR-transparent KBr pellets was reported as a comparison: in the free molecule, the band at 1595 and the shoulder at 1605 cm^{-1} are due to 8b and 8a modes of the aromatic ring, respectively, the bands at 1499 and 1472 cm^{-1} to the 19a and 19b modes, and finally the broad band below 1400 cm^{-1} to the in-plane C–O–H bending vibration. After adsorption of Phe on imogolite, a new band is seen at 1492 cm^{-1}, and is assigned to phenate ions formed by the abstraction of a proton by basic oxygen atoms located at the C outer surface (Al–OH–Al groups), so that reaction [Eq. (12.3)] occurs, involving also the formation of a water molecule adsorbed on aluminium:

$$C_6H_5-OH+Al-(OH)-Al \rightarrow C_6H_5-O-Al+Al(OH_2) \qquad (12.3)$$

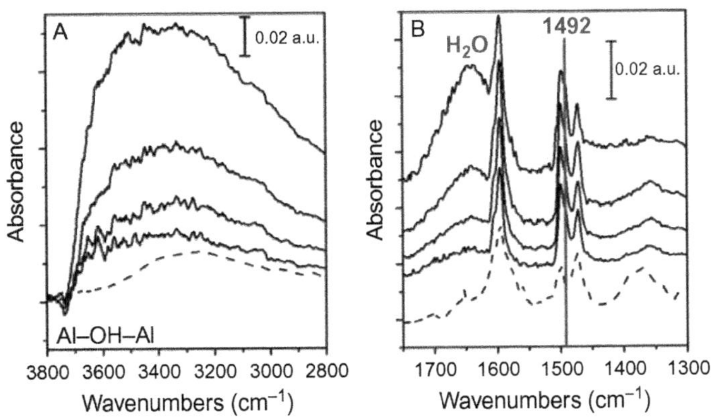

FIG. 12.12 IR spectra recorded after dosing Phe on imogolite outgassed at 300°C (equilibrium pressure in the 0.050–2.5 mbar range). Difference spectra are reported, after subtraction of the spectrum of the bare sample, in the OH stretching range (A) and in the 1750–1300 cm^{-1} range (B). *Adapted from Bonelli et al. (2013) with permission from the PCCP owner societies.*

Such Al–OH–Al groups are likely located at the C surface of imogolite and are probably also responsible of the reactivity with CO_2.

TEB has a van der Waals diameter of 0.92 nm and should mainly interact with the C surface. Upon adsorption of TEB on dehydrated imogolite, bands are seen at 1692, 1644, 1419 and 1369 cm^{-1} (Fig. 12.13). The latter two are perturbed with respect to the same bands in the free molecule (occurring at 1452 and 1374 cm^{-1}), whereas the former two, which are in the region of C=C bonds, indicate a noticeable perturbation suffered by the aromatic ring. Most likely, TEB molecules are adsorbed on (Lewis) Al^{3+} sites at the C surface. Concerning the OH region, a negative band is seen at 3736 cm^{-1}: the molecule size likely excludes an interaction with OH groups at the A surface and should be ascribed to Al–OH–Al species at the C surface, as observed for Phe adsorption.

Another kind of reactivity, only recently explored, is related to the possibility of partially replacing outer Al(VI) with Fe^{3+} ions (Ookawa et al., 2006; Ookawa, 2012; Shafia et al. 2015). Although the isomorphic substitution (IS) of iron for aluminium is a common process in all natural aluminosilicates, only a few papers deal with Fe-doped imogolite. When one Al(VI) is substituted by one Fe^{3+} at the outer surface of imogolite, in principle three Fe(OH)Al and three Fe–O–Al groups should form, as shown in Fig. 12.14. The presence of Fe^{3+} should impart the solid new chemical and solid-state properties, although the nature and the electronic effects generated by Al^{3+}/Fe^{3+} IS are not fully understood so far.

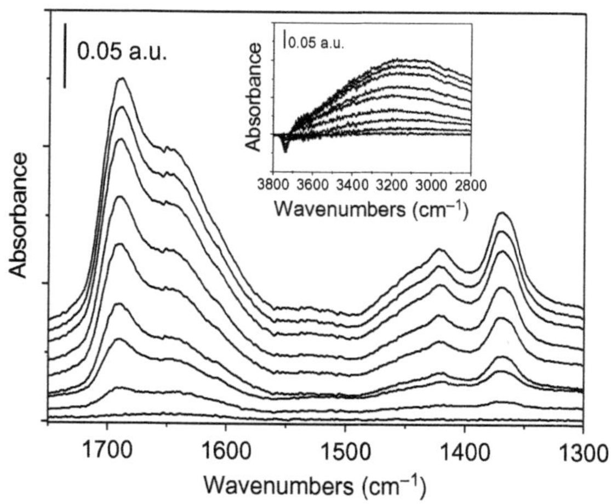

FIG. 12.13 IR spectra recorded after dosing TEB on imogolite outgassed at 300°C (equilibrium pressure in the 0.030–7.7 mbar range). Difference spectra are reported, after subtraction of the spectrum of the bare sample, in the 1750–1300 cm^{-1} range. *Inset*: same spectra, in the OH stretching range. *Adapted from Bonelli et al. (2013) with permission from the PCCP owner societies.*

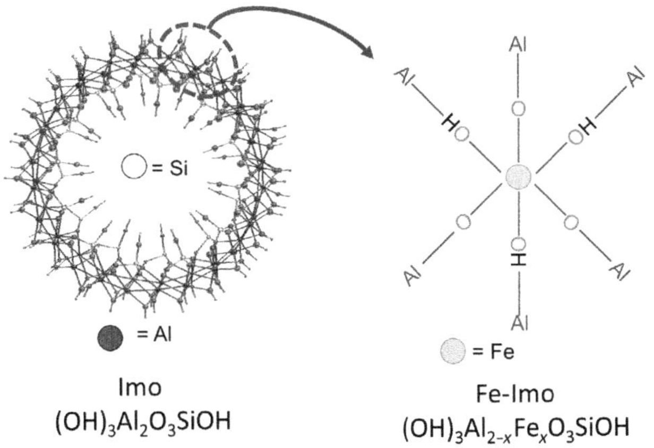

FIG. 12.14 Isomorphic substitution of one Al(VI) at the outer surface of imogolite by one Fe^{3+} ion gives rise to the formation of three Fe–OH–Al groups and three Fe–O–Al groups.

Pioneering theoretical calculations on Fe-doped imogolite showed that iron could either isomorphically substitute for aluminium or create defective sites both inside and outside imogolite (Alvarez-Ramírez, 2009). In the same study, it was shown that the presence of Fe could affect the Fermi level by reducing the band gap (E_g) of imogolite from 4.7 to 2.0–1.4 eV (ie, well into the visible range).

The first experimental studies on this subject (Ookawa et al., 2006; Ookawa, 2012) reported the synthesis of Fe-doped imogolite by directly adding the precursor ($FeCl_3$) to the synthesis mixture. The main result of this research was that formation of imogolite was preserved up to an overall 1.4 mass% Fe content. A recent report on Fe-doped aluminium-germanate nanotubes, isostructural with imogolite, showed that Al^{3+}/Fe^{3+} IS is indeed limited to 1.0 mass %, and formation of Fe oxo-hydroxides species unavoidably occurs at higher Fe content (Avellan et al., 2014). Recently, Shafia et al. (2015) studied the effect of Fe-doping on several properties of imogolite, including electronic conduction and magnetic properties. Samples were obtained by either direct synthesis or postsynthesis loading, with either 0.7 mass% or 1.4 mass% Fe. Textural properties were characterized by HRTEM, XRD and N_2 adsorption/desorption isotherms at −196°C. The presence of different iron species was studied by magnetic moment measurements and by three spectroscopies (Mössbauer, UV–vis and EPR). With 0.7 mass% Fe, Al^{3+}/Fe^{3+} IS is the main process, leading to a material with formula $(OH)_3Al_{1.975}Fe_{0.025}O_3SiOH$, featuring Fe(OH)Al groups at the outer surface. With 1.4 mass% Fe, some Fe(OH)Al groups probably act as nucleation seeds, due to the propensity of Fe to form Fe–O–Fe bridges, causing the formation of some Fe oxo-hydroxides nanoclusters also by direct synthesis, in agreement with Avellan et al. (2014). Comparison with the sample obtained by postsynthesis loading, in which only the formation of such iron oxo-hydroxides was expected, showed instead that whatever the synthesis procedure, the two samples at 1.4 mass% Fe did not differ much due to the occurrence of IS (which also occurred by postsynthesis loading). As a result, the two possible fates for Fe^{3+} cations in contact with imogolite systems—either IS to isolated octahedral Fe^{3+} species (Fe(VI)) or clustering to Fe oxides—depends more on the Fe amount rather than the preparation route. Fe^{3+}/Al^{3+} IS is the major process occurring in the sample at 0.7 mass% Fe content, produced by direct synthesis. When the amount of Fe increases, the content in isolated Fe^{3+} species increases to a limit of 0.9 mass%, as determined by magnetization curves at r.t., in fair agreement with what is observed for Fe-doped aluminium-germanates (Avellan et al., 2014). Higher Fe content leads to the formation of Fe_2O_3 clusters. Interestingly, when the postsynthesis loading procedure is adopted, a similar material is obtained, with the same maximum amount of Fe in the imogolite structure and Fe_2O_3 clusters.

The experimental work by Shafia et al. (2015) showed that imogolite has an E_g of 4.9 eV (Fig. 12.15), in fair agreement with the calculated value found by Alvarez-Ramírez (2009). In the presence of substitutional Fe^{3+} ions (curve 2, vide infra), E_g results in 2.8 eV: Fe-doped imogolite approaches the behaviour of a semiconductor, confirming, at least qualitatively, the theoretical calculations of Alvarez-Ramírez (2009). With curve 4, a further reduction of the band gap is observed ($E_g = 2.4$ eV); however, a precise determination of the band gap is hampered by the presence of Fe_2O_3 clusters, along with substitutional iron in the corresponding sample.

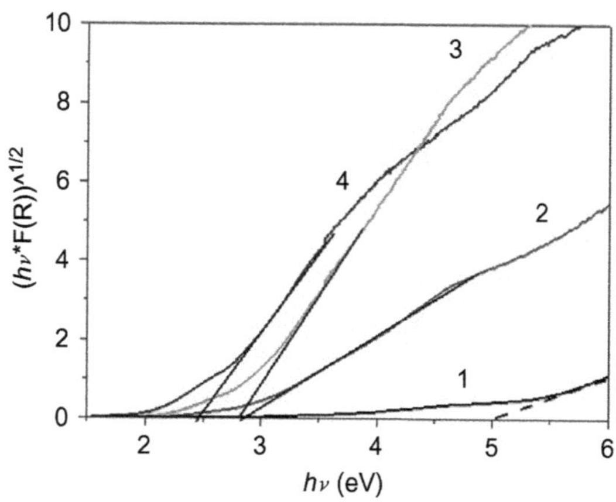

FIG. 12.15 Tauc's plot of imogolite (1), imogolite doped with 0.7 mass% Fe (2), imogolite doped with 1.4 mass% Fe (3) and imogolite loaded with 1.4 mass% by postsynthesis procedure (4). *Reprinted with permission from Shafia et al. (2015). Copyright 2015 Springer.*

The presence of isolated Fe(VI) sites in the loaded sample indicates that ionic exchange occurred between structural Al^{3+} in preformed imogolite and Fe^{3+} ions in water:

$$[Al(OH_2)^+Al] + Fe(H_2O)_6^{3+} = [Fe(OH_2)^+Al] + Al(H_2O)_6^{3+} \qquad (12.4)$$

According to Eq. (12.4), it is possible to change the composition of the imogolite outer surface by ionic exchanges with proper ions, avoiding more difficult procedures, like direct synthesis. Isolated Fe(VI) species coming from IS occur in high-spin states (as confirmed by both EPR and magnetic susceptibility measurements) and their full (octahedral) coordination hampers interaction with molecules like NO and is rather stable upon dehydration at 150°C. Mössbauer and EPR procedures ruled out the formation of Fe^{2+} species in both fresh and outgassed samples, respectively. Magnetic susceptibility measurements at r.t. indicated that imogolite samples have a diamagnetic response, as expected, whereas all Fe-doped samples display dominant paramagnetic behaviour.

The imogolite structure appears able to exchange dynamically with water solution Al and Fe ions. This fact indicates that ionic exchange is a viable method to introduce heteroatoms at the outer surface of imogolite, avoiding more complicated synthesis procedures. Besides Fe^{3+}, other metals with suitable charge and dimensions could be hosted at the outer surface of imogolite, including Cr^{3+} and Ti^{3+}. This result is of particular interest since Fe-doped imogolite could be useful for applications involving materials with both high porosity, high surface area and the presence of accessible transition metals, including adsorption and catalysis.

12.4 SURFACE PROPERTIES OF THE LAMELLAR PHASES DERIVING FROM IMOGOLITE THERMAL COLLAPSE

The structure of imogolite shown in Fig. 12.1 is unstable above 300°C and structural transformation occurs above this temperature, as described in Chapter 11 leading to a new porous material made from collapsed nanotubes (Fig. 12.15). The available surfaces still carry acid/base functionalities, as expected for a porous aluminium silicate. Such properties were studied by adsorption of gas molecules on collapsed imogolite, followed by IR spectroscopy (Bonelli et al., 2009). Adsorption of phenol led to the formation of phenolate species, as observed also with imogolite, indicating the presence of basic sites, whereas adsorption of ammonia showed the presence of both Lewis and Brønsted sites, indicating that an amphoteric surface was present and accessible to probes in the gas phase.

Adsorption of CO allowed figuring out the presence of different OH species, with a different nature with respect to Si–OH and Al–OH–Al groups of the parent imogolite. Fig. 12.16 displays different IR spectra recorded after dosing CO at nominal −196°C on collapsed imogolite. In the CO stretching

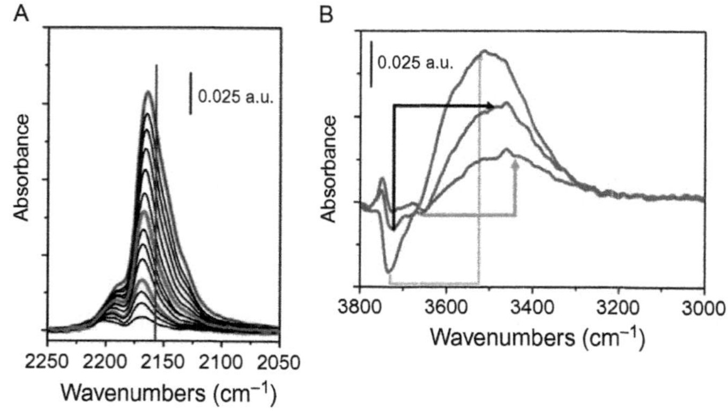

FIG. 12.16 (A) CO stretching range (2250–2050 cm^{-1}) of IR spectra obtained at nominal −196°C after dosing CO on collapsed imogolite (equilibrium pressure in the 0.50–15 mbar range); (B) the *red* (*dark grey* in print version) *curves* of section (A) are reported in the OH stretching range (3800–3000 cm^{-1}).

range (Fig. 12.16A), a minor band at 2203 cm^{-1} (shifting with coverage to 2188 cm^{-1}) and a main band at 2170 cm^{-1} (shifting to 2166 cm^{-1}) are observed. Both bands are reversible upon evacuation at low temperatures, indicating that interaction with rather weak sites: the 2203 cm^{-1} band was assigned to CO adsorbed onto coordinatively unsaturated Al^{3+}, as those occurring at the surface of transition aluminas (Morterra and Magnacca, 1996).

The 2170 cm^{-1} band was assigned to CO interacting with acidic hydroxyls. The latter were more acidic than Si–OH, since no band was observed at 2157 cm^{-1}, where CO adsorbing on silanols is expected (vertical bar in Fig. 12.16A). This finding was in agreement with the collapsing mechanism, which rules out the presence of accessible SiOH groups due to the formation of Al–O–(Si–O–Si)–O–Al (MacKenzie et al., 1989). Fig. 12.16B shows the OH stretching range of bold spectra of Fig. 12.16A: different spectra are seen, negative bands correspond to hydroxyls interacting with CO, and positive absorption to H-bond adducts. A broad negative band of hydroxyls interacting with CO was observed, indicating heterogeneity of sites. At least three components were observed at 3743, 3725 and 3660 cm^{-1}, and shifted downward by 233 [blue (light grey in print version) arrow], 245 (black arrow) and 210 cm^{-1} [green (dark grey in print version) arrow], respectively, in agreement with research reporting a shift ($\Delta\nu_{OH}$) of 200–240 cm^{-1} for a ν_{CO} of 2166–2170 cm^{-1} (Cairon et al., 1998).

The observed shifts (Fig. 12.16B) were too large to be due to Si–OH or Al–OH species, since SiOH suffered with CO a shift of about 100 cm^{-1} (Ghiotti et al., 1979), whereas hydroxyls at the surface of transition aluminas shifted downward by 70–95 cm^{-1} (Morterra and Magnacca, 1996; Cairon et al., 1998). Such bands were, therefore, ascribed to more acidic, Brønsted-like sites. With various zeolites in the course of dealumination, a band at 3670 cm^{-1} was indeed assigned to less acidic hydroxyls, shifted downward by about 200 cm^{-1} and still partially anchored to the zeolite framework (Cairon et al., 1998). In Al-rich micelle-templated silicates with similar composition, hydroxyls at 3660 cm^{-1} shifting with CO by ca. 220 cm^{-1} were ascribed to buried Si–O–Al(OH)–O–Si species (Bonelli et al., 2004). On amorphous aluminosilicates with Al/Si in the 1.0–2.0 range, 3748-cm^{-1} hydroxyls were observed to a shift of ca. 230 cm^{-1}, with the corresponding CO stretch mode at 2169 cm^{-1} (Daniell et al., 2000).

The lamellar structure proposed by MacKenzie et al. (1989) showed the presence of 5-coordinated Al and a few 4- and 6-coordinated sites and some residual hydroxyl groups, which therefore should be responsible for the acidic behaviour of this material and for bands observed in the OH stretch range. As a whole, a heterogeneous family of Brønsted sites of increased acidity forms with respect to those present at the surface of Imo: this feature may make collapsed imogolite an interesting acidic catalyst.

12.5 CONCLUDING REMARKS

The intrinsic properties of imogolite that limit practical applications are mainly related to the high aspect ratio of its nanotube, posing serious problems to gas diffusion and thermal stability. Chemical modification of imogolite allows for obtaining more interesting materials, at least from the point of view of gas adsorption and storage, catalysis and fabrication of CPN, among other applications.

Synthesis of modified imogolite materials leads to larger-diameter nanotubes, as with Ge-imogolite or Me-imogolite; controllable nanotube length (Ge-imogolite), tuneable hydrophilicity of the inner surface (Me-imogolite, postsynthesis grafted imogolite). By several means, hybrid imogolite with both organic and inorganic functionalities are obtained, which could be exploited in drug delivery or chromatography, for instance.

Reactivity of the outer surface is particularly useful for the production of imogolite polymer nanocomposites, which may have unique properties like high transparency, which is hardly attainable in CNT polymer nanocomposites. Moreover, at variance with CNT, imogolite is well dispersed in water, widening the possible modifications and simplifying the synthesis procedure.

Doping with transition metals is the most recent direction towards which researchers in the field are making big efforts. This could lead to the fabrication of semiconducting nanotubes with high SSA, which could have possible applications in photocatalysis.

Collapsed imogolite develops both medium-strength acidic and basic surface features, which result in an amphoteric behaviour.

ABBREVIATIONS

3-APTES	3-aminopropyltriethoxysilane
Al(VI)	octahedral Al
AMTES	aminomethyltriethoxysilane
ANT	amine functionalized nanotubes
BET	Brunauer–Emmett–Teller
DW	double-walled
EPR	electron paramagnetic resonance
Fe(VI)	octahedral Fe
HRTEM	high-resolution transmission electron microscopy
IS	isomorphic substitution
MAS NMR	magic-angle spinning nuclear magnetic resonance
ODPA	octadecyl phosphonic acid
Phe	phenol
PSD	pore size distribution
SAED	selected area diffraction
SSA	specific surface area
STM	scanning tunnelling microscopy
SW	single-walled
TDPA	tetradecyl phosphonic acid
TEB	1,3,5-triethylbenzene
TEMG	triethoxymethylgermane
TEMS	triethoxymethylsilane
TEOG	tetraethoxygermanium
TEOS	tetraethoxysilane
XRD	X-ray diffraction

REFERENCES

Ackerman, W.C., Smith, D.M., Huling, J.C., Kim, Y.-W., Bailey, J.K., Brinker, C.J., 1993. Gas/vapor adsorption in imogolite: a microporous tubular aluminosilicate. Langmuir 9, 1051–1057.

Alvarez-Ramírez, F., 2009. First principles studies of Fe-containing aluminosilicate and aluminogermanate nanotubes. J. Chem. Theory Comput. 5, 3224–3231.

Amara, M.-S., Paineau, E., Bacia-Verloop, M., Krapf, M.-E., Davidson, P., Belloni, L., Levard, C., Rose, J., Launois, P., Thill, A., 2013. Single-step formation of micron long $(OH)_3Al_2O_3GeOH$ imogolite-like nanotubes. Chem. Commun. 49, 11284–11286.

Amara, M.-S., Paineau, E., Rouzière, S., Guiose, B., Krapf, M.E.M., Taché, O., Launois, P., Thill, A., 2015. Hybrid tunable-diameter metal-oxide nanotubes for organic molecules trapping. Chem. Mater. 27, 1488–1494.

Arai, Y., McBeath, M., Bargar, J.R., Joye, J., Davis, J.A., 2006. Uranyl adsorption and surface speciation at the imogolite-water interface: self-consistent spectroscopic and surface complexation models. Geochim. Cosmochim. Acta 70, 2492–2509.

Avellan, A., Levard, C., Kumar, N., Rose, J., Olivi, L., Thill, A., Chaurand, P., Borschneck, D., Masion, A., 2014. Structural incorporation of iron into Ge–imogolite nanotubes: a promising step for innovative nanomaterials. RSC Adv. 4, 49827–49830.

Bonelli, B., Onida, B., Chen, J.D., Galarneau, A., Di Renzo, F., Fajula, F., Garrone, E., 2004. Spectroscopic characterisation of the strength and stability of the acidic sites of Al-rich microporous micelle-templated silicates. Microporous Mesoporous Mater. 67, 95–106.

Bonelli, B., Bottero, I., Ballarini, N., Passeri, S., Cavani, F., Garrone, E., 2009. IR spectroscopic and catalytic characterization of the acidity of imogolite-based systems. J. Catal. 264, 15–30.

Bonelli, B., Armandi, A., Garrone, E., 2013. Surface properties of alumino-silcate single-walled nanotubes of the imogolite type. Phys. Chem. Chem. Phys. 15, 13381–13390.

Bottero, I., Bonelli, B., Ashbrook, S.E., Wright, P.A., Zhou, W., Tagliabue, M., Armandi, M., Garrone, E., 2011. Synthesis and characterization of hybrid organic/inorganic nanotubes of the imogolite type and their behaviour towards methane adsorption. Phys. Chem. Chem. Phys. 13, 744–750.

Cairon, O., Chevreau, T., Lavalley, J.-C., 1998. Brønsted acidity of extraframework debris in steamed Y zeolites from the FTIR study of CO adsorption. J. Chem. Soc. Faraday Trans. 94, 3039–3047.

Clark, C.J., McBride, M.B., 1984. Cation and anion retention by natural and synthetic allophane and imogolite. Clay Clay Miner. 32, 291–299.

Cradwick, P.D.G., Farmer, V.C., Russell, J.D., Masson, C.R., Wada, K., Yoshinaga, N., 1972. Imogolite, a hydrated aluminium silicate of tubular structure. Nature Phys. Sci. 240, 187–189.

Creton, B., Bougeard, D., Smirnov, K.S., Guilment, J., Poncelet, O., 2008. Molecular dynamics study of hydrated imogolite. 2. Structure and dynamics of confined water. Phys. Chem. Chem. Phys. 10, 4879–4888.

Daniell, W., Schubert, L., Glöcker, R., Meyer, A., Noweck, K., Knözinger, H., 2000. Enhanced surface acidity in mixed alumina-silicas: a low-temperature FTIR study. Appl. Catal. A 196, 247–260.

Denaix, L., Lamy, I., Bottero, J.Y., 1999. Structure and affinity towards Cd^{2+}, Cu^{2+}, Pb^{2+} of synthetic colloidal amorphous aluminosilicates and their precursors. Coll. Surf. A 158, 315–325.

Farmer, V.C., Adams, M.J., Fraser, A.R., Palmieri, F., 1983. Synthetic imogolite: properties, synthesis, and possible applications. Clay Miner. 18, 459–472.

Ghiotti, G., Garrone, E., Morterra, C., Boccuzzi, F., 1979. Infrared study of low temperature adsorption. 1. Carbon monoxide on aerosil. An interpretation of the hydrated silica spectrum. J. Phys. Chem. 83, 2863–2869.

Gustafsson, J.P., 2001. The surface chemistry of imogolite. Clay Clay Miner. 49, 73–80.

Harsh, J.B., Traina, S.J., Boyle, J., Yang, Y., 1992. Adsorption of cations on imogolite and their effect on surface charge characteristics. Clay Clay Miner. 40, 700–706.

Iller, R.K., 1979. The Chemistry of Silica: Solubility, Polymerization, Colloid and Surface Properties and Biochemistry of Silica. John Wiley and Sons, New York.

Johnson, L.M., Pinnavaia, T.J., 1990. Silylation of a tubular aluminosilicate polymer (imogolite) by reaction with hydrolyzed (γ-aminopropyl)triethoxysilane. Langmuir 6, 307–311.

Johnson, L.M., Pinnavaia, T.J., 1991. Hydrolysis of (γ-aminopropyl)triethoxysilane-silylated imogolite and formation of a silylated tubular silicate-layered silicate nanocomposite. Langmuir 7, 2636–2641.

Kang, D.-Y., Zang, J., Jones, C.W., Nair, S., 2011. Single-walled aluminosilicate nanotubes with organic-modified interiors. J. Phys. Chem. C 115, 7676–7685.

Kang, D.-Y., Brunelli, N.A., Yucelen, G.I., Venkatasubramanian, A., Zang, J., Leisen, J., Hesketh, P.J., Jones, C.W., Nair, S., 2014. Direct synthesis of single-walled aminoaluminosilicate nanotubes with enhanced molecular adsorption selectivity. Nat. Commun. 5, 3342.

Konduri, S., Mukherjee, S., Nair, S., 2006. Strain energy minimum and vibrational properties of single-walled aluminosilicate nanotubes. Phys. Rev. B 74, 033401.

Konduri, S., Mukherjee, S., Nair, S., 2007. Controlling nanotube dimensions: correlation between composition, diameter, and internal energy of single-walled mixed oxide nanotubes. ACS Nano 1, 393–402.

Kuroda, Y., Kuroda, K., 2010. Expansion of intertubular mesopores of imogolite nanotubes by thermal decomposition of an imogolite–poly(sodium 4-styrenesulfonate) composite. Chem. Lett. 40, 46–48.

Lourenço, M.P., Guimarães, L., Da Silva, M.C., De Oliveira, C., Heine, T., Duarte, H.A., 2014. Nanotubes with well-defined structure: single-and double-walled imogolites. J. Phys. Chem. C 118, 5945–5953.

Ma, W., Yah, M.O., Otsuka, H., Takahara, A., 2012. Application of imogolite clay nanotubes in organic-inorganic nanohybrid materials. J. Mater. Chem. 22, 11887–11892.

MacKenzie, K.J., Bowden, M.E., Brown, J.W.M., Meinhold, R.H., 1989. Structural and thermal transformation of imogolite studied by ^{29}Si and ^{27}Al high-resolution solid-stated magnetic nuclear resonance. Clay Clay Miner. 37, 317–324.

Maillet, P., Levard, C., Larquet, E., Mariet, C., Spalla, O., Menguy, N., Masion, A., Doelsch, E., Rose, J., Thill, A., 2010. Evidence of double-walled Al-Ge imogolite-like nanotubes. A cryo-TEM and SAXS investigation. J. Am. Chem. Soc. 132, 1208–1209.

Morterra, C., Magnacca, G., 1996. A case study: surface chemistry and surface structure of catalytic aluminas, as studied by vibrational spectroscopy of adsorbed species. Catal. Today 27, 497–532.

Mukherjee, S., Bartlow, V.M., Nair, S., 2005. Phenomenology of the growth of single-walled aluminosilicate and aluminogermanate nanotubes of precise dimensions. Chem. Mater. 17, 4900–4909.

Mukherjee, S., Keesuk, K., Nair, S., 2007. Short, highly ordered, single-walled mixed-oxide nanotubes assemble from amorphous nanoparticles. J. Am. Chem. Soc. 129, 6820–6826.

Ookawa, M., 2012. Synthesis and characterization of Fe-imogolite as an oxidation catalyst. In: Valaškova, M., Martynkova, G.S. (Eds.), Clay Minerals in Nature—Their Characterization, Modification and Application. InTech, pp. 239–258.

Ookawa, M., Inoue, Y., Watanabe, M., Suzuki, M., Yamaguchi, T., 2006. Synthesis and characterization of Fe containing imogolite. Clay Sci. 12, 280–284.

Parfitt, R.L., Thomas, A.D., Atkinson, R.J., Smart, R.St.C., 1974. Adsorption of phosphate on imogolite. Clay Clay Miner. 22, 455–456.

Park, S., Lee, Y., Kim, B., Lee, J., Jeong, Y., Noh, J., Takahara, A., Sohn, D., 2007. Two-dimensional alignment of imogolite on a solid surface. Chem. Commun. 2917–2919.

Pohl, P.I., Faulon, J.-L., Smith, D.M., 1996. Pore structure of imogolite computer models. Langmuir 12, 4463–4468.

Qi, X., Yoon, H., Lee, S.-H., Yoon, J., Kim, S.-J., 2008. Surface-modified imogolite with 3-APS-OsO_4 complex: synthesis, characterization and its application in the dihydroxylation of olefins. J. Ind. Eng. Chem. 14, 136–141.

Shafia, E., Esposito, E., Manzoli, M., Chiesa, M., Tiberto, P., Barrera, G., Menard, G., Allia, P., Freyria, F.S., Garrone, E., Bonelli, B., 2015. Al/Fe isomorphic substitution versus Fe_2O_3 clusters formation in Fe-doped aluminosilicate nanotubes (imogolite). J. Nanoparticle. Res. 17, 336.

Shemg, S., Kraft, J.J., Schuster, S.M., 1993. A specific quantitative colorimetric assay for L-asparagine. Anal. Biochem. 211, 242–249.

Sidorkin, V.F., Shagun, V.A., Pestunovich, V.A., 1999. Stereoelectronic effects and the problem of the choice of model compounds for organic derivatives of a pentacoordinated silicon atom (taking silatranes as an example). Russ. Chem. Bull. 48, 1049–1053.

Thill, A., Maillet, P., Guiose, B., Spalla, O., Belloni, L., Chaurand, P., Auffan, M., Olivi, L., Rose, J., 2012. Physico-chemical control over the single-or double-wall structure of alumino-germanate imogolite-like nanotubes. J. Am. Chem. Soc. 134, 3780–3786.

Wada, S.I., 1987. Imogolite synthesis at 25°C. Clay Miner. 35, 379–384.

Wada, S.I., Wada, K., 1982. Effects of substitution of germanium for silicon in imogolite. Clay Clay Miner. 30, 123–128.

Wada, S.I., Eto, A., Wada, K., 1979. Synthetic allophane and imogolite. J. Soil Sci. 30, 347–355.

Yamamoto, K., Otsuka, H., Wada, S.I., Takahara, A., 2001. Surface modification of aluminosilicate nanofiber "imogolite". Chem. Lett. 30, 1162–1163.

Yamamoto, K., Otsuka, H., Wada, S.I., Takahara, A., 2002. Preparation of a novel (polymer/inorganic nanofiber) composite through surface modification of natural aluminosilicate nanofiber. J. Adhes. 78, 591–602.

Yamamoto, K., Osaka, H., Wada, S.I., Sohn, D., Takahara, A., 2005a. Preparation and properties of [poly(methyl methacrylate)/imogolite] hybrid via surface modification using phosphoric acid ester. Polymer 46, 12386–12392.

Yamamoto, K., Otsuka, H., Wada, S.I., Sohn, D., Takahara, A., 2005b. Transparent polymer nanohybrid prepared by in situ synthesis of aluminosilicate nanofibers in poly(vinyl alcohol) solution. Soft Matter 1, 372–377.

Zanzottera, C., Armandi, M., Esposito, S., Garrone, E., Bonelli, B., 2012a. CO_2 adsorption on aluminosilicate single-walled nanotubes of imogolite type. J. Phys. Chem. C 116, 20417–20425.

Zanzottera, C., Vicente, A., Armandi, M., Fernandez, C., Garrone, E., Bonelli, B., 2012b. Thermal collapse of single-walled alumino-silicate nanotubes: transformation mechanisms and morphology of the resulting lamellar phases. J. Phys. Chem. C 116, 23577–23584.

Zanzottera, C., Vicente, A., Celasco, E., Fernandez, C., Garrone, E., Bonelli, B., 2012c. Physico-chemical properties of imogolite nanotubes functionalized on both external and internal surfaces. J. Phys. Chem. C 116, 7499–7506.

Chapter 13

Liquid-Crystalline Phases of Imogolite and Halloysite Dispersions

P. Davidson* and I. Dozov
Laboratoire de Physique des Solides, UMR8502, CNRS, Univ. Paris-Sud, Université Paris-Saclay, Orsay Cedex, France
**Corresponding author: e-mail: patrick.davidson@u-psud.fr*

13.1 INTRODUCTION

Imogolite nanotubes are outstanding objects to work with because of their very high geometrical anisotropy and stiffness and their small polydispersity in diameter. Moreover, their surface chemical properties make them easily dispersible in polar solvents like water (see Chapters 3 and 9 for more information). Therefore, aqueous colloidal dispersions of imogolite nanotubes are readily obtained and make excellent examples of dispersions of very anisotropic, rodlike particles. On theoretical grounds, such systems are expected to display liquid-crystalline order, provided that sufficiently high concentrations can be reached. But the concentrations required are in fact quite small for such slender and stiff objects, as already demonstrated by dispersions of rodlike viruses (Dogic and Fraden, 2006; Grelet, 2014).

The notion of liquid-crystallinity is often associated with organic or biological matter because most liquid crystals are small organic molecules, surfactants or polymers. Nevertheless, liquid-crystalline dispersions of anisotropic rodlike (or disklike) mineral or metallic nanoparticles are now well documented and are sometimes called 'mineral liquid crystals' (Zocher, 1925; Gabriel and Davidson, 2000; Davidson and Gabriel, 2005; Lekkerkerker and Vroege, 2013). Most of the major liquid-crystalline phases displayed by organic compounds have already been observed as well with their mineral counterparts, sometimes as early as the beginning of the 20th century. Moreover, the use of mineral objects can give access to new physical properties, such as magnetism, that are quite difficult to obtain in organic systems. In this context, in retrospect, the observation of the liquid-crystalline properties of imogolite

Developments in Clay Science, Vol. 7. http://dx.doi.org/10.1016/B978-0-08-100293-3.00013-3

dispersions by Kajiwara et al. (1986a) is hardly surprising. Paineau et al. (2016) have shown that the variety of liquid-crystalline phases exhibited by these imogolite dispersions is richer than one might expect.

From another point of view, imogolite nanotubes, thanks to their huge geometrical anisotropy, are being increasingly used to prepare clay polymer nanocomposites (CPN) with superior mechanical properties. For this purpose, their alignment on a macroscopic scale is much sought after. Therefore, the incidence of liquid-crystalline order on the properties of imogolite-based nanocomposite materials also deserves to be explored.

Imogolite nanotubes are just one example of tubular clay minerals and other dispersions of rodlike clay minerals have also been investigated to determine their liquid-crystalline properties. Moreover, imogolite nanotubes are hydrophilic counterparts of carbon nanotubes, which are usually hydrophobic. The phase diagram of aqueous dispersions of imogolite nanotubes then gives us a glimpse of the liquid-crystalline phases that carbon nanotubes could reveal if only they could properly be functionalized to make them easily dispersible in water at sufficiently large concentrations.

13.2 STRUCTURES OF LIQUID CRYSTALS

Before looking at the different liquid-crystalline phases of the dispersions of imogolite nanotubes, an introduction to the field of liquid crystals could probably be useful to the general reader. However, those already acquainted with this topic may skip this section and directly proceed to the next one.

By definition, a liquid-crystalline phase is a state of condensed matter that is both anisotropic, like crystals and fluid, like liquids (De Gennes, 1974; Chaikin and Lubensky, 2000). By extension, a liquid crystal is a substance that shows a liquid-crystalline phase. In a phase diagram, such phases are usually placed between the usual crystalline and liquid phases. Hence, liquid-crystalline phases are also called 'mesophases' (ie, intermediate phases). One can argue that there are two ways to melt a crystal: either by increasing its temperature or by dissolving it in a solvent. In the first case, which is for the liquid crystals of small molecules used in electro-optic devices such as displays, the liquid-crystalline phases obtained are called 'thermotropic'. The second case, which is for surfactants that form micelles and dispersions of rodlike or disklike particles such as polymers and nanoparticles, is that of 'lyotropic' liquid crystals, which will be covered later in this chapter.

There are actually different kinds of liquid-crystalline phases that differ by their properties of symmetry, and we will describe the most common ones in this chapter. For this purpose, anisotropic objects that can either be rodlike or disklike will be considered.

Starting from the usual (isotropic) liquid phase (Fig. 13.1A), upon decreasing temperatures for thermotropic liquid crystals or upon increasing concentrations for lyotropic ones, the first mesophase that may appear is the

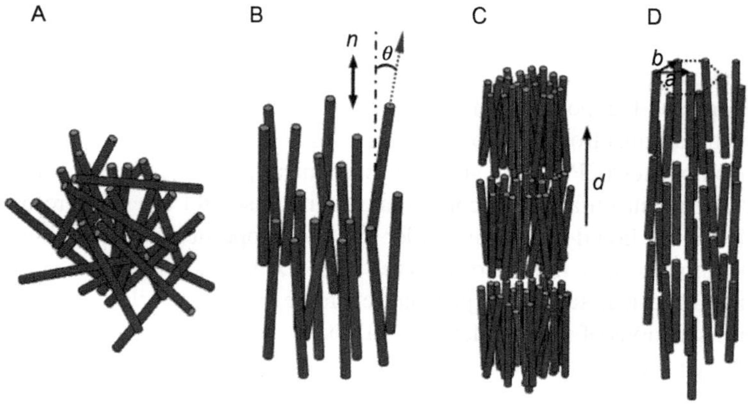

FIG. 13.1 Schematics of the structures of colloidal dispersions of rodlike particles in (A) the isotropic phase; (B) the nematic phase; (C) the smectic (or lamellar) phase; and (D) the columnar phase. Here, *n* is the nematic director, *θ* is the angle between a rod and the director (used to calculate the nematic-order parameter), *d* is the smectic period and *a* and *b* are the lattice vectors of the columnar phase.

'nematic' one. In this phase, all anisotropic objects are oriented along a common direction, called the 'director' *n* (Fig. 13.1B). Strictly speaking, the director is not a vector since it only represents a direction, with up-down symmetry. The nematic phase has long-range orientational order, which means that, in a monodomain of the nematic phase, there is a perfect correlation of the orientations of all the particles, whatever their positions. In contrast, the positions of the particles are only correlated over very short distances, as in a usual liquid. Thus, the nematic phase only has short-range positional order and the particles are free to diffuse in all directions, so that the nematic phase is fluid. Being both fluid and anisotropic, the nematic phase is liquid-crystalline by definition. The phase anisotropy gives rise to optical birefringence so that a nematic sample usually appears bright when observed between crossed polarizers. Nevertheless, thermal fluctuations affect the orientational order of the objects that are not perfectly aligned along their common average direction. To quantify this, the nematic-order parameter S is introduced. S is defined as: $S=\frac{1}{2}\langle 3\cos^2\theta - 1\rangle$, where θ is the angle between the axis of the object and the director and the angular brackets represent a statistical average over all objects and time. S takes values ranging from $S=1$ for a perfectly ordered nematic phase to $S=0$ for an isotropic liquid. The phase transition from the isotropic liquid to the nematic phase is generally of the first order (ie, with phase coexistence), with some jump of S at the transition.

Although in many cases, the different liquid-crystalline phases can be identified through the careful analysis of their optical textures in polarized-light microscopy, the most efficient technique for this purpose is X-ray scattering. Because the nematic phase only has short-range positional order, its scattering pattern displays diffuse halos. However, in contrast with the usual isotropic liquid phase, the scattering pattern of a nematic monodomain is anisotropic, as some of these diffuse halos are stronger in particular directions. This will be illustrated in the case of imogolite nanotube dispersions.

The Onsager model (Onsager, 1949; Vroege and Lekkerkerker, 1992), which is the main statistical physics model used to predict the occurrence of a lyotropic nematic phase in a dispersion of rodlike particles, will now be briefly described. The assumptions of this model are that (i) the particles are completely stiff and have very large aspect ratio, L/D, where L is the length of the particles and D their diameter; (ii) the transition occurs at low concentrations so that only two-body interactions are considered; (iii) the particles only interact through hard-core repulsions, meaning that the interaction potential is infinite when particles touch but vanishes otherwise. Therefore, there is no temperature scale in the problem, the system is called 'athermal', and temperature has no influence on the phase transition. Rather than the concentration, the volume fraction of particles, ϕ, (ie, the ratio of the volume of particles to the total volume of dispersion) is commonly used to build the phase diagrams. Under these assumptions, the Onsager model predicts a first-order isotropic/nematic phase transition with coexisting phases of volume fractions $\phi_N = 4.4D/L$ and $\phi_I = 3.3D/L$, as well as a nematic-order parameter jump from 0 to $S = 0.75$ at the transition. This immediately shows that, in order to observe liquid crystallinity in a dispersion of rodlike particles at low concentrations, the highest possible particle aspect ratio is desirable. Since its publication, the Onsager model inspired considerable work not only by analytical methods, but also by numerical simulations. These investigations confirmed the Onsager approach and made it more quantitative. In addition, the effect of particle flexibility was examined and shown to shift the phase transition to a higher volume fraction and to decrease the order parameter jump at the transition. Similar considerations also apply to disklike particles (Onsager, 1949). Moreover, Onsager also tackled the issue of electrostatic repulsions between rodlike particles in polar solvents. However, proper treatment of these interactions has actually proved to be quite involved and is still an active research field (Wensink, 2007; Jabbari-Farouji et al., 2013).

As the temperature is further decreased or as the concentration is further increased, another liquid-crystalline phase may appear. Like the nematic phase, the 'lamellar' or 'smectic' phase has long-range orientational order and is therefore birefringent, but it also has a one-dimensional, long-range positional order. In a monodomain of this mesophase, the anisotropic objects are all aligned on average in the same direction but, in addition, their centres are distributed in equidistant layers, with a lamellar period d (Fig. 13.1C). Within each layer, the centres of the anisotropic objects only show positional short-range order, and the phase is therefore fluid. In other words, the lamellar phase is a periodic stack of two-dimensional (2D) liquid layers that can freely slide past each other. Accordingly, the X-ray scattering pattern of the lamellar phase displays sharp equidistant diffraction lines (or reflections in the case of a lamellar monodomain) with spacing inversely proportional to the lamellar period, in addition to a diffuse scattering halo arising from the liquidlike character of the 2D layers.

Upon decreasing temperature or upon increasing concentration, another mesophase, called 'columnar', may also occur where the centres of the anisotropic objects are distributed in columns (Fig. 13.1D). Like the nematic and smectic ones, this phase has long-range orientational order, and it is therefore both anisotropic and birefringent. Moreover, the columns are liquid, and the positions of the objects in neighbouring columns are uncorrelated along the average direction of the columns. In other words, the columnar phase is a 2D assembly of liquid columns. Then, the X-ray scattering pattern of the columnar phase consists in a series of sharp diffraction lines arising from the 2D lattice and sometimes in a diffuse scattering halo due to the liquidlike order of the particles along the direction of the columns.

Note that the free energies of the lamellar and columnar phases are often very close, so predicting their relative stabilities is a theoretical and numerical challenge. Nevertheless, the phase diagrams of rodlike particles have already been predicted (Hentschke and Herzfeld, 1991; Bates and Frenkel, 1998; Wensink, 2007). However, the microscopic details of the phase constituents and of their interactions often determine the stabilities of these mesophases. Moreover, in the case of dispersions of nanoparticles, which is the focus of this chapter, the polydispersity distributions in diameter, length and thickness also play a crucial role. Note also that in the case of thermotropic molecular liquid crystals, many variations of the lamellar and columnar phases are known to occur when other degrees of freedom, such as a tilt angle, a polarity or chirality, also come into play.

13.3 THE NEMATIC PHASE OF IMOGOLITE NANOTUBES

Both natural and synthetic imogolite nanotubes are easily dispersed in water where they keep their outstanding one-dimensional morphology and light-scattering studies showed that they remain rather stiff in aqueous dispersion (Donkai et al., 1985). Imogolite nanotubes usually have very large aspect ratios, which typically range from 100 to 1000. Hence, aqueous dispersions of imogolite nanotubes are very good candidates to display liquid-crystalline phases. Indeed, the liquid-crystalline properties of imogolite dispersions were already recognized more than 30 years ago by a Japanese team of researchers (Kajiwara et al., 1986a,b). Samples of aqueous dispersions of imogolite nanotubes, in a given range of concentrations and held in test tubes left standing under the influence of gravity, spontaneously demix into two different phases within a few days. When observed between crossed polarizers, the bottom phase appears birefringent, whereas the top phase remains dark [ie, isotropic (Fig. 13.2)]. The proportion of the bottom phase increases with the overall concentration of imogolite nanotubes in the dispersion. When the test tubes are tilted, the interface between the two phases also tilts, showing that both phases are fluid rather than gel. The bottom phase being both fluid and birefringent is, by definition, liquid-crystalline. The same observations are made with samples held in flat glass capillaries suitable for polarized-light microscopy, which gives more insight in the nature and mechanism of the phase separation process. Starting from a freshly prepared sample that looks uniformly dark under the microscope, small birefringent droplets slowly form (Fig. 13.3). They are the first drops of the nematic phase and are sometimes called '*tactoids*', from the Greek word meaning 'to assemble' (Zocher, 1925). These nematic droplets are slightly denser than the isotropic liquid around them, and they slowly form a sediment at the bottom of the capillary, where they coalesce and make up the bottom birefringent phase. All these observations are completely consistent with a first-order isotropic/nematic phase transition occurring at thermodynamic equilibrium, as described by

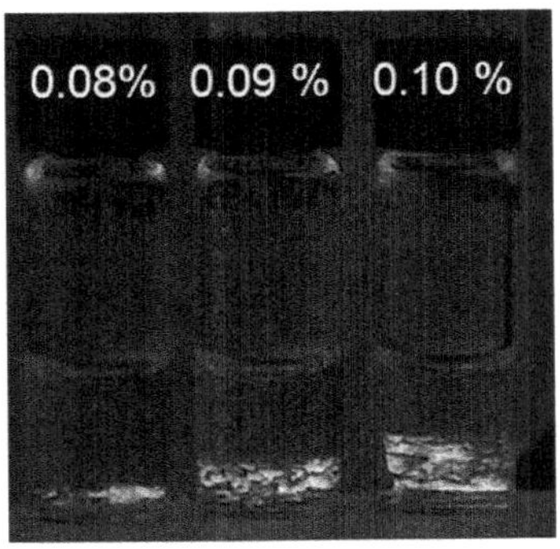

FIG. 13.2 Three test tubes filled with imogolite dispersions of volume fractions increasing from *left* to *right*, showing isotropic/nematic phase coexistence.

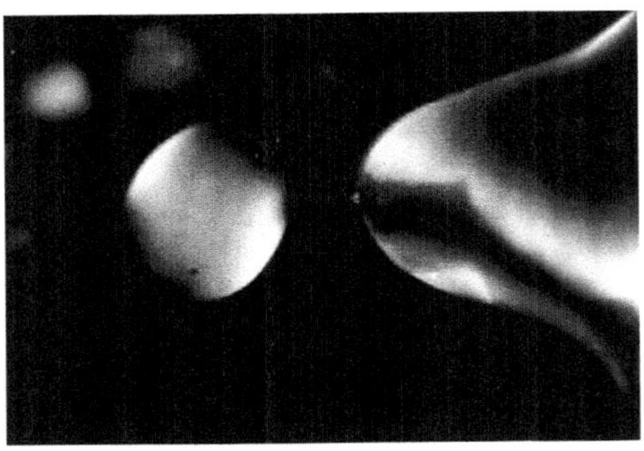

FIG. 13.3 Photograph in polarized-light microscopy of a nematic droplet in a sample of imogolite dispersion displaying isotropic/nematic phase coexistence.

the Onsager model of this phase transition (Onsager, 1949; Vroege and Lekkerkerker, 1992). Although still rather uncommon, examples of liquid-crystalline order in colloidal dispersions of mineral nanoparticles are now well documented (Gabriel and Davidson, 2000; Davidson and Gabriel, 2005; Lekkerkerker and Vroege, 2013). The further identification of this birefringent phase as being nematic relied only on the detailed examination of its optical texture in polarized-light microscopy, as no X-ray scattering experiment was performed on these samples at the time.

Aqueous dispersions of imogolite nanotubes were soon regarded as good model systems to test the validity of the Onsager theory. This requires comparing its predictions with the experimental volume fractions (ϕ_I and ϕ_N) at which the first droplets of the nematic phase (N) appear and the last droplets of the isotropic phase (I) disappear, upon increasing overall imogolite volume fractions. These two volume fractions correspond to the limits of the I/N biphasic region and are sometimes called 'A' and 'B' points. Interestingly, the nematic phase appears at very low volume fractions (of the order of 1% or even less, depending on the observed samples). For instance, in the earliest report of the nematic phase, it appeared at around 1% (Kajiwara et al., 1986b,c). This low value is a direct consequence of the large aspect ratio of imogolite nanotubes ($L/D \sim 200$, for this example). Other samples of longer imogolite nanotubes lead to the occurrence of the nematic phase at even lower volume fractions (Amara et al., 2013). The investigation of the dispersions of imogolite nanotubes led to the conclusion that the Onsager model of nematic-ordering agrees better with the experimental observations than the Flory model, which is another popular model of the N/I transition (Fig. 13.4). This confirms similar conclusions drawn from the study of other model systems like dispersions of viruses that are regarded as excellent model systems for such purposes due to their monodispersity (Dogic and Fraden, 2006; Grelet, 2014).

In stark contrast with viruses, natural samples of mineral nanoparticles often display a very large polydispersity that strongly affects the phase diagrams of their dispersions. Indeed, in general, even a moderately wide polydispersity distribution is known to widen the I/N biphasic gap of the dispersions (Vroege and Lekkerkerker, 1992). In practice, this leads to the onset of the nematic phase at even lower volume fractions (Kajiwara et al., 1986c). Upon increasing concentration, the first droplets of the nematic phase are enriched in longer particles, whereas the last droplets of the isotropic phase are richer in shorter particles. This 'fractionation' effect was confirmed by separating the two coexisting phases and analyzing their molecular weight distributions by sedimentation. In addition, a (I–N–N′) three-phase equilibrium, with two distinct nematic phases, was observed in these samples and tentatively explained by a possible nanotube bidispersity. This point remains a bit mysterious. The fractionation effect was used to reduce the polydispersity of an imogolite dispersion. The starting dispersion

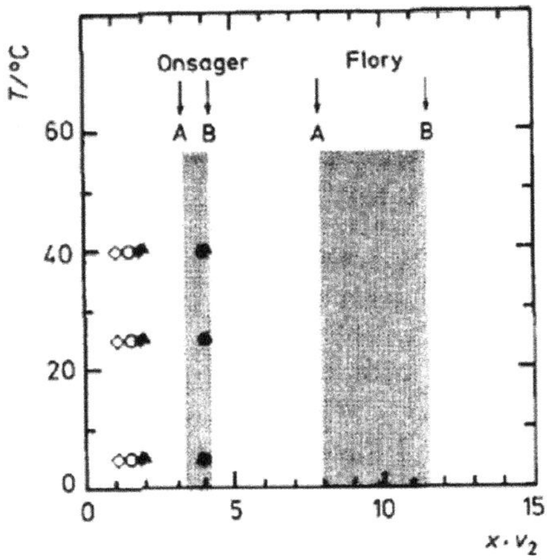

FIG. 13.4 Phase diagram of aqueous dispersions of imogolite. The *shaded areas* represent the predictions of the Onsager and Flory models. The *open* and *filled symbols* represent the onset and completion, respectively, of the isotropic/nematic transition (A and B points) for three different batches. Temperature has no influence on the phase transition within the range explored. *From Kajiwara et al. (1986c), with permission.*

was diluted so as to bring its volume fraction into the biphasic region, and phase separation occurred. The two coexisting phases were separately collected and the nematic (isotropic) phase contained longer (shorter) nanotubes and had narrower polydispersity distributions than the parent dispersion. By repeating this process a second time, the polydispersity was reduced from about 2.3 for the parent dispersion to about 1.6 (Donkai et al., 1993b).

As mentioned previously, the identification of the nematic phase of the dispersions of imogolite nanotubes was first achieved on the basis of their optical textures in polarized-light optical microscopy. However, the interpretation of these textures was not as straightforward as expected. Indeed, striated textures were quite often observed instead of the usual smooth schlieren or threaded textures of the nematic phases (Fig. 13.5) (Kajiwara et al., 1986b; Donkai et al., 1993a). Such striated textures are often the sign of a helicoidal organization in the liquid-crystalline phase, like that of the cholesteric phase, which is the chiral version of the nematic phase. The period of the striations then directly gives the pitch of the helicoidal organization. For example, dispersions of rodlike viruses or biopolymers, such as DNA, collagen, chitin and cellulose derivatives, often display these striated cholesteric textures. Moreover, mixtures in exactly equal proportions of left-handed and right-handed enantiomers do not show any striated textures, as the helicoidal pitch is inversely proportional to the net chirality of the sample. In this context, the observation of helicoidal organizations in dispersions of imogolite nanotubes of natural origin is highly surprising because these dispersions have no net chirality.

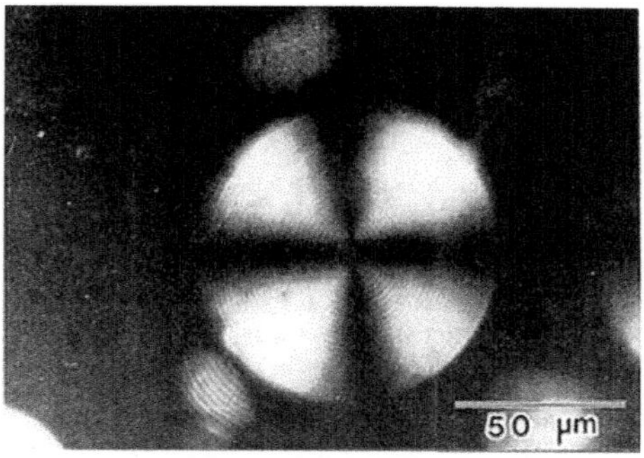

FIG. 13.5 Example of the striated texture in a liquid-crystalline domain of imogolite dispersion. *Adapted from Kajiwara et al. (1986b), with permission.*

Another possible explanation for the occurrence of the striated texture is that it would not be due to a helicoidal organization of the nanotubes (as in a cholesteric phase), but rather to the presence of periodic pleats. Such pleats were observed by electron microscopy, but this technique is badly invasive and it is not certain that the organization of the specimen really reflects that prevailing in solution (Donkai et al., 1993a; Hoshino et al., 1996). Moreover, the existence of this pleated texture implies that the nanotubes would first aggregate in rafts, which conflicts with more recent X-ray scattering studies of the dispersions (as discussed later in this chapter). In fact, nematic liquid crystals, as an example of 'soft matter', are prone to display all kinds of mechanical instabilities that give rise to striations in optical textures. For example, sheared solutions of polymer liquid crystals show undulation instabilities, after shearing is stopped, that are detected as banded textures in optical microscopy. There could then be other types of mechanisms that may explain the striated textures of imogolite nanotube dispersions. Nevertheless, these striated textures remain puzzling to this day. Note, however, that dispersions of synthetic imogolite nanotubes did not display the striated textures, but rather the usual schlieren nematic textures (Amara et al., 2013).

The most effective way of identifying a liquid-crystalline phase is by performing an X-ray scattering experiment (Warren, 1990; Guinier, 1994). Because the particles are far from each other in these dilute dispersions, the X-ray scattering signals are found in the region of low scattering angles, which corresponds to large scales. The specific technique of small-angle X-ray scattering (SAXS) must then be employed. In order to extract the most possible information from the SAXS patterns, the imogolite nanotubes should be aligned in dispersion in order to produce a nematic single domain. This can be achieved, for example, by shearing the dispersions in Couette cells used for rheology or by applying them in an a.c. electric field.

Very dilute dispersions of imogolite nanotubes, in the isotropic phase, give SAXS patterns where the scattered intensity regularly decreases with increasing scattering angle. When the dispersion is diluted enough, interparticle interferences can be safely neglected and the scattering arises only from intraparticle interference. This defines the so-called form factor, $P(q)$, of the particles [q is the scattering vector modulus $q=(4\pi/\lambda)\sin\theta$ where λ is the X-ray wavelength and 2θ the scattering angle].

In contrast, a typical SAXS pattern of a flow-aligned sample of nematic dispersion of imogolite nanotubes is shown in Fig. 13.6A. This pattern does not display any Bragg reflection or sharp diffraction line, but only diffuse scattering spots. This means that the nanotubes have no long-range positional order. Moreover, the pattern is anisotropic, which reflects the long-range orientational order of the nanotubes. Therefore, the liquid-crystalline phase must indeed be nematic.

A radial scan of the scattered intensity, starting from the centre of the pattern and going through the maximum of one of the diffuse spots, provides an $I(q)$ curve that displays a diffuse scattering peak. In a rough approximation, the position of the maximum of this peak gives a good estimate of the average distance between the nanotubes in the plane perpendicular to the nematic director (ie, the average direction of the nanotubes). A more elaborate (but still approximate for rodlike particles) treatment of the scattering assumes that it can be factorized as $I(q)=P(q)S(q)$, where $S(q)$ is the structure factor of the dispersion (Fig. 13.6C). Then, an inverse Fourier transform can be used to retrieve the pair correlation function, $g(r)$, of the nanotubes in the plane perpendicular to the director. The position of the first peak of $g(r)$ gives the average distance, d, between the nanotubes. Typically, d varies between 10 and 100 nm, depending on volume fraction. More precisely, in the limit of infinite one-dimensional objects, within the domain of stability of the nematic phase, d is expected to vary as $\phi^{-1/2}$, which is the signature of a 2D 'swelling law', as illustrated in Fig. 13.6D. The intercept of the swelling law provides an estimate of the outer radius of the nanotubes (Levitz et al., 2008).

An azimuthal scan of the scattered intensity along a circle centred on the direct beam and going through the maximum of the diffuse spot gives the dependence $I(\theta)$ of scattered intensity vs azimuthal angle θ (Fig. 13.6B). $I(\theta)$ shows peaks whose widths are directly related to the degree of orientational order of the nanotubes in the sample. If the alignment is good enough that the sample can be considered as a nematic single domain, then the nematic-order parameter can be directly inferred from the peak width and shape according to well-documented procedures (Davidson et al., 1995). In the case illustrated in Fig. 13.6, the nematic-order parameter measured in this way is $S=0.78\pm0.05$, which is a fairly high value but nevertheless agrees with the prediction of the Onsager model and the strongly first-order character of the I/N phase transition in dispersions of nanoparticles.

Applying an alternating electric field, as mentioned above, is an efficient way of aligning the nematic phase. This was done by using a homemade cell that allows for applying the electric field on samples held in capillary tubes, from outside; the electric field is directed along the capillary axis. Because

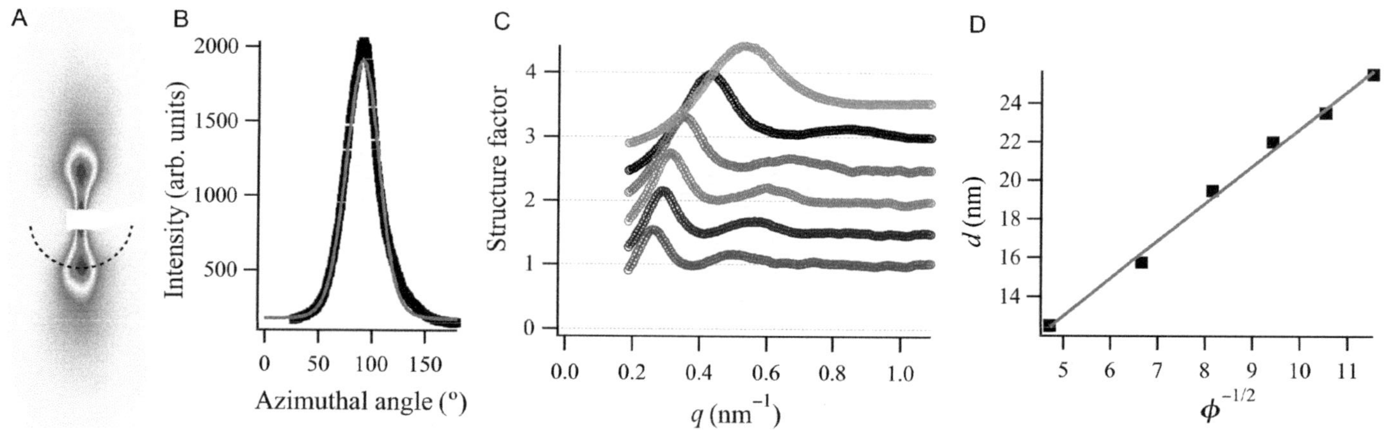

FIG. 13.6 (A) SAXS pattern of a flow-aligned nematic imogolite dispersion (the flow direction is horizontal); (B) *black data points*: scan of the scattered intensity vs azimuthal angle, $I(\theta)$, along the *dashed circle* in the SAXS pattern, *red* (*dark grey* in the print version) *curve*: fit used to obtain the nematic-order parameter; (C) structure factors $S(q)$, shifted for clarity, of imogolite dispersions of decreasing volume fraction from top to bottom; (D) 'swelling law' relating the first maximum of $g(r)$ to the volume fraction.

the dispersions usually have a significant ionic strength, ranging from 10^{-4} to 10^{-3} M, the ions tend to migrate and cancel the applied field. In order to avoid this phenomenon, a field with a large enough frequency (300–700 kHz) must be applied. The alignment of the nematic phase by the electric field is easily observed by polarized-light microscopy (Fig. 13.7). The sample initially shows a typical nematic threaded texture where the black and white regions are areas where the nematic director is respectively parallel or at 45 degrees with respect to the directions of the polarizer and analyzer. After applying the electric field for a few seconds, the texture becomes almost completely uniform, showing that the spatial variations in the orientation of the nanotubes have vanished and that a nematic single domain was produced. The birefringence, Δn, of this nematic domain could be measured ($\Delta n \sim 6 \times 10^{-5}$) and is proportional to the volume fraction, the nematic-order parameter S, and the specific birefringence, Δn_p, of the imogolite nanotube; therefore, one obtains $\Delta n_p \sim 0.02$.

The alignment of the nematic dispersions of imogolite nanotubes can also easily be studied by SAXS (Fig. 13.8). Starting from an unoriented nematic

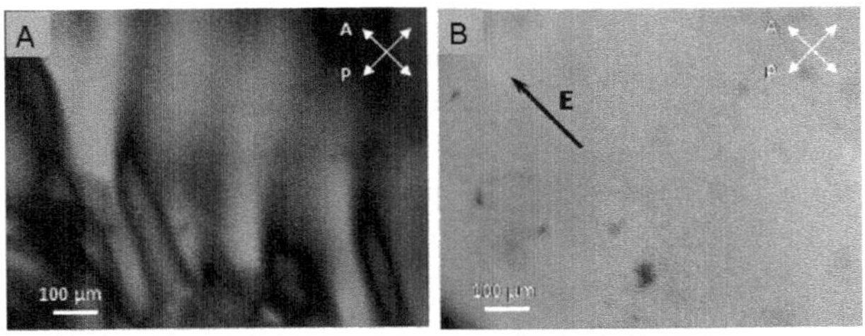

FIG. 13.7 Optical microscopy images illustrating the alignment of the nematic phase between crossed polarizers (*white arrows*), of imogolite in an electric field; (A) no field applied; (B) under an electric field (*black arrow*) $E = 2.10^4$ V/m at 500 kHz.

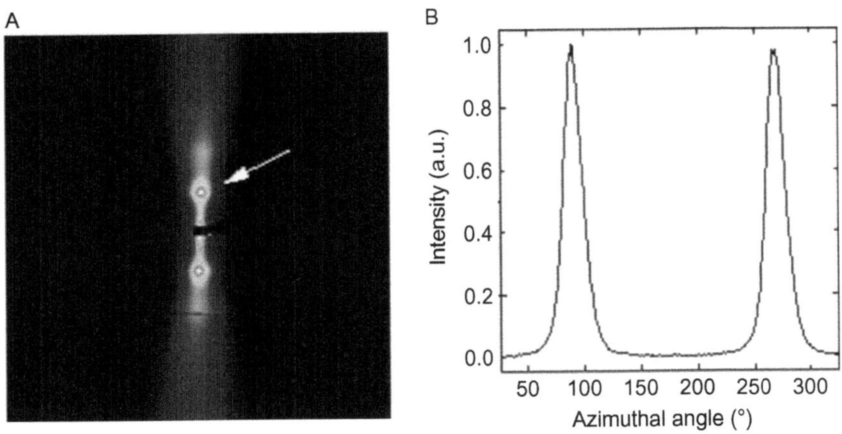

FIG. 13.8 (A) SAXS pattern of the nematic phase under an (horizontal) electric field $E = 100$ V/mm at 500 kHz; (B) azimuthal profile recorded along a circle going through the first diffuse scattering maximum (*white arrow*).

phase, the SAXS pattern becomes highly anisotropic when the electric field is switched on. The positions of the diffuse spots with respect to the direction of the electric field confirms that the imogolite nanotubes align parallel to the electric field, as expected from similar studies of dispersions of beidellite. Indeed, the predominant contribution to particle alignment by the electric field in an aqueous medium is that of the counterion cloud that has the maximum response along the particle's largest dimensions (ie, along the length in the case of nanotubes).

13.4 THE COLUMNAR PHASE OF IMOGOLITE NANOTUBES

Owing to the very small polydispersity in the diameter of imogolite nanotubes, a columnar liquid-crystalline phase can reasonably be expected to be found in the phase diagram of their dispersions. Indeed, both electron microscopy and X-ray scattering investigations of dried samples of imogolite nanotubes often reveal the presence of nanotube bundles where the nanotubes are well organized, at contact, on a 2D hexagonal lattice (Amara et al., 2014). However, it was not a priori quite clear whether the columnar phase could occur at volume fractions lower than that of the sol/gel transition beyond which spontaneous organization of the phase constituents is usually strongly hindered.

The study of optical textures by polarized-light microscopy is often the simplest method to explore the phase diagrams of lyotropic liquid crystals. Very recently, this technique has revealed the existence of a new optical texture (Fig. 13.9) in aqueous dispersions of both single-wall imogolite nanotubes and double-wall germanium-substituted imogolite nanotubes (Paineau et al., 2016). This new texture is quite distinct from the schlieren, threaded, or striated textures discussed previously, which are typical of the nematic or cholesteric phases, which lack any type of positional long-range order. However, even for experts in liquid crystals, the new optical texture is not clear enough to allow the unambiguous identification of the nature of the novel liquid-crystalline phase, as it could be of either a lamellar or columnar nature.

Proper identification of this new liquid-crystalline phase of dispersions of imogolite nanotubes was achieved by SAXS. In contrast with the nematic

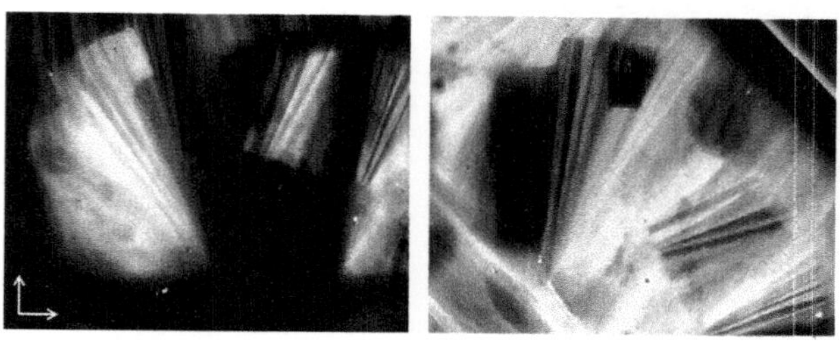

FIG. 13.9 Example of optical texture of the columnar phase. *Left*: capillary parallel to the polarizers (*white arrows*); *Right*: capillary at 45 degrees with respect to the polarizers.

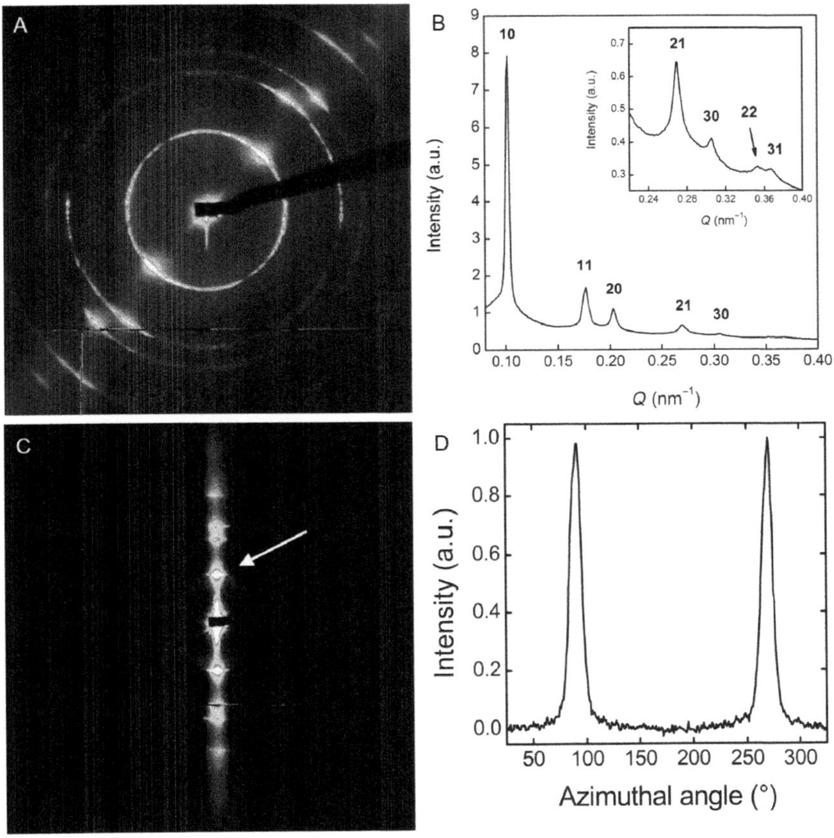

FIG. 13.10 SAXS patterns of the columnar phase: (A) in the absence of electric field; (B) radially averaged scattered intensity vs scattering vector modulus; (C) under an electric field $E = 100$ V/mm at 500 kHz; (D) azimuthal profile recorded along a circle going through the first hexagonal reflection (*white arrow*).

phase, the scattering pattern of this new phase (Fig. 13.10A) displays sharp diffraction lines that can all be indexed with a 2D hexagonal lattice. Therefore, the mesophase that displays the original optical texture in Fig. 13.9 is of the columnar hexagonal type. The SAXS patterns display as many as seven orders of reflection, which means that the positions of the imogolite nanotubes do not fluctuate much around their mean positions on the 2D lattice. In fact, modelling of the SAXS patterns has shown that the positional fluctuations are restricted to only a few percent of the lattice spacing.

Very interestingly, the volume fraction at which the columnar phase appears is typically of the order of 0.2%, which is very low. In fact, previous experimental investigations always reported the occurrence of columnar phases at very high volume fractions (~10–50%) in other dispersions of rodlike particles, either of organic or biological origin like DNA, polypeptides, polysaccharides, viruses or mineral nanoparticles like gibbsite or goethite (Bawden et al., 1936; Bernal and Fankuchen, 1941; Livolant and Leforestier, 1996; Vroege et al., 2006). Accordingly, all previous theoretical or numerical studies also predicted that the columnar phase is stable only at large volume fractions. Hence, the observation of the columnar phase at such high dilutions in imogolite nanotube dispersions makes them a quite outstanding system.

There are two main features that could explain why this colloidal system is so different from the previously reported ones. First, imogolite nanotubes have a very large aspect ratio because they are both very long and stiff. Indeed, not only is their overall length (also called the 'contour length') very great compared to their diameter, but their persistence length is as well, owing to their low flexibility. Other colloidal systems (such as dispersions of double-stranded DNA) can have very long contour length, but their persistence length is usually much less. Low values of the persistence length destabilize the liquid-crystalline phases and make them appear at higher concentrations. Second, most theoretical and simulation studies reported thus far have dealt with rodlike objects that interact through hard-core interactions, which means that if the particles are not in contact, they do not interact at all (Hentschke and Herzfeld, 1991; Bates and Frenkel, 1998). However, there are a few studies where the electrostatic repulsions between like-charged objects and their counterion clouds in polar solvents have been tentatively addressed (Wensink, 2007; Jabbari-Farouji et al., 2013). This is actually an extremely challenging problem because a suitable expression of the electrostatic potential between two charged nanorods was only proposed fairly recently (Trizac et al., 2002). Incorporating this potential within the theoretical and numerical investigation of charged assemblies is presently under way. Note that these two arguments can be extrapolated to other dispersions of slender, rodlike particles in polar solvents, like carbon nanotubes, if their surface could be modified to bear a large enough electrical charge. Finally, in principle, assemblies of rodlike particles can also display a lamellar liquid-crystalline phase. However, it was demonstrated that the polydispersity distributions in particle diameter and length directly control the respective stabilities of the columnar and lamellar phases (Vroege et al., 2006). If the length polydispersity is much less than the diameter polydispersity, the lamellar phase is favoured. In the opposite situation, the columnar phase prevails, which is precisely the case for the dispersion of imogolite nanotubes.

The columnar hexagonal phase of the dispersion of imogolite nanotubes could also be aligned, within seconds, in an ac electric field using the same setup already employed to align the nematic phase. Fig. 13.10C displays a very anisotropic SAXS pattern where all lattice reflections are concentrated along a direction perpendicular to that of the electric field. Again, such a pattern anisotropy proves that the imogolite nanotubes are oriented parallel to the direction of the electric field. An azimuthal scan of the scattered intensity, $I(\theta)$, can be fitted to a Gaussian distribution that describes the mosaïcity (ie, the distribution of columnar crystallites probed by the X-ray beam with respect to the field direction) of the sample. The full-width at half maximum of this distribution was only about 12 degrees, which is rather lower than other similar soft-matter systems and demonstrates the efficiency of the alignment by the electric field of this columnar mesophase of nanotubes.

As discussed next, imogolite nanotubes are often used to prepare CPN, and their orientation by various means is often pursued in order to improve mechanical properties. In this perspective, application of an ac electric field appears as an interesting and noninvasive technique to achieve this goal. Moreover, the use of the columnar liquid-crystalline phase allows for positioning the nanotubes precisely on the hexagonal lattice, which could pave the way to a cheap, 'bottom-up' process of nanostructuration.

13.5 ANISOTROPY OF CLAY POLYMER NANOCOMPOSITES BASED ON IMOGOLITE NANOTUBES

Thanks to their huge aspect ratio and good mechanical properties which are fairly comparable to those of carbon nanotubes, imogolite nanotubes have emerged as an interesting possibility for mineral particles to prepare CPN (Otsuka and Takahara, 2010; see also Chapter 23). Indeed, one of the present issues in this field is to reduce the amount of filler while preserving the desired properties of the composite materials. For this purpose, highly anisotropic particles are sought after in order to reduce their percolation threshold. Moreover, achieving macroscopic alignment of the particles is often useful to reduce the viscosity of the mixtures involved in processing the nanocomposites and to obtain anisotropic physical properties.

Early reports of imogolite-based nanocomposite materials were published by a Japanese group that investigated mixtures of imogolite nanotubes with either hydroxypropylcellulose (HPC) or poly(vinyl alcohol) (PVA) (Hoshino et al., 1992a,b). HPC is a stiff, rodlike polymer, and its aqueous dispersions form a liquid-crystalline phase on their own. As HPC is chiral, the mesophase is the helically twisted version of the nematic phase, called the 'cholesteric phase'. Moreover, HPC, like imogolite, has –OH surface groups that can be ionized in water, making HPC highly hydrosoluble. Indeed, mixtures of aqueous dispersions of HPC and imogolite do not segregate. Interestingly, the phase diagram of these mixtures does not show any linear behaviour of ϕ_I and ϕ_N, as adding only a few percent of imogolite to HPC dramatically changes the onset of nematic order (Fig. 13.11). Moreover, as soon as the relative proportion of imogolite reaches about 40%, the mixtures display the I/N transition at the same volume fraction as that of pure imogolite dispersions. Furthermore, no circular dichroism could be detected in mixtures with

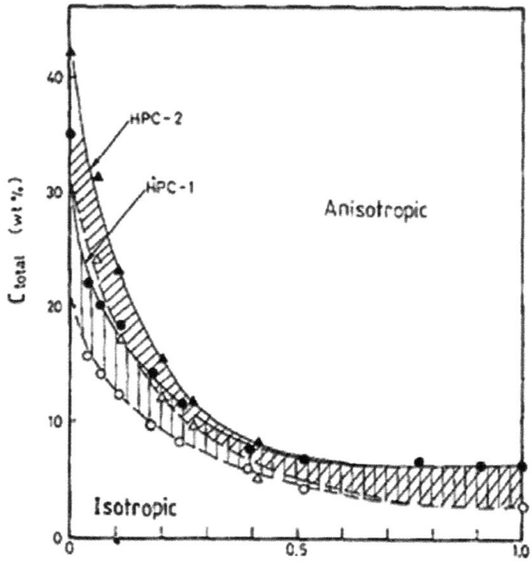

FIG. 13.11 Phase diagram of two different imogolite/HPC systems (*circles*: HPC-1; *triangles*: HPC-2). The *open* and *filled symbols* represent the onset and completion, respectively, of the isotropic/nematic transition ('A' and 'B' points). *Adapted from Hoshino et al. (1992b), with permission.*

imogolite content larger than 30%. Unfortunately, these puzzling observations were not discussed further. Nanocomposite films were cast from these mixtures, and their mechanical properties were examined. This study showed that, as expected, Young's modulus and the tensile strength clearly improved with the increasing imogolite proportion.

In contrast with HPC, PVA is a highly flexible polymer that adopts a random coil conformation in aqueous solvents. Therefore, PVA solutions do not display any liquid-crystalline phase, but only a usual isotropic liquid phase. However, interestingly, dispersions of both imogolite nanotubes and PVA in water did not segregate and remained stable. The nematic phase was observed as soon as the imogolite volume fraction reached a few percent in the mixtures. Oddly enough, the onset of nematic ordering was lower in the imogolite/PVA mixtures (down to 2.4%) than in the pure imogolite dispersion (6.8%). This effect, which was not fully explained in this study, might be due to the depletion effect of the PVA random coils on the colloidal stiff imogolite nanotubes. The stability of the mixed imogolite/PVA dispersions allowed producing transparent nanocomposite films by slow evaporation. The mechanical properties of these CPN were investigated only for small imogolite content. The results were slightly disappointing, as Young's modulus increased only slightly with increasing imogolite content in spite of imogolite nanotube alignment. However, a subsequent study of imogolite PVA nanocomposites claims improved thermal properties (Song et al., 2015). Note also that this kind of composites have been studied more recently for their porosity in water (Kang et al., 2012).

After the first two early Japanese studies (Hoshino et al., 1992a,b), a number of other publications described the elaboration of CPN based on imogolite nanotubes. These works often start with the surface modification of the nanotubes—a very important step indeed in this context. Homogeneous imogolite poly(methylmethacrylate) nanocomposites could thus be produced. Poly(acrylic acid) and polyacrylamide gels reinforced with imogolite nanotubes, functionalized in the former case, were also reported in other studies. Optical birefringence and X-ray scattering investigations showed that the large improvement in elastic properties of these CPN is due to the alignment of the imogolite nanotubes upon stretching (Fig. 13.12) (Lee et al., 2013; Shikinaka et al., 2013, 2015). Gelatin/imogolite composite gels were also recently developed for their mechanical properties (Teramoto et al., 2012) and polyamide/imogolite composite gels for desalination properties as well (Barona et al., 2013).

However, none of these studies explicitly addressed the issue of the liquid-crystalline nature of the CPN and its possible relevance to the improvement of their mechanical properties. Therefore, these questions remain to be explored.

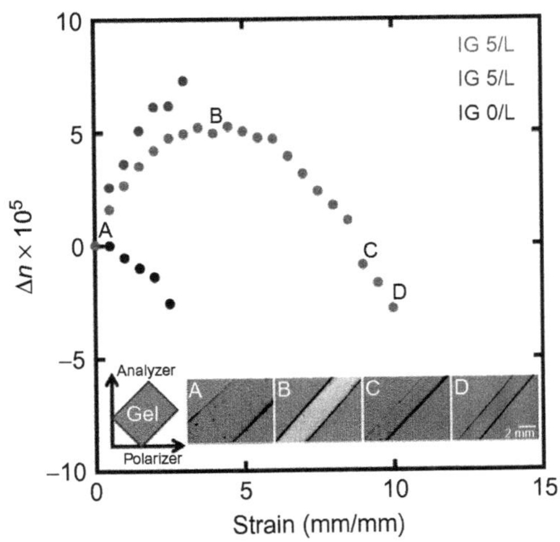

FIG. 13.12 Strain-induced birefringence of two different imogolite/polyacrylamide nanocomposite gels (labelled 'IG5/N' and 'IG5/L') and of an ordinary polyacrylamide gel (labelled 'IG0/L'). Insets are retardation colours (*different shades of grey* in the print version) of sample IG5/N at points A, B, C and D. *From Shikinaka et al. (2013), with permission.*

13.6 LIQUID-CRYSTALLINE PHASES OF HALLOYSITE, ANOTHER RODLIKE TUBULAR CLAY MINERAL

Like imogolite nanotubes, other rodlike clay mineral nanoparticles also lead to liquid-crystalline phases when dispersed in solvents. The case of sepiolite nanorods in particular has been studied by the group of van Duijneveldt (Zhang and van Duijneveldt, 2006; Yasarawan and van Duijneveldt, 2008, 2010; Woolston and van Duijneveldt, 2015a,b); it is, however, not discussed in this chapter because it falls out of the scope of the present volume, which only deals with nanotubes.

Halloysite is a natural clay mineral that occurs as nanotubes or nanoscrolls, but their diameter is usually a few times greater than that of imogolite nanotubes. Still, halloysite nanotubes have a rather large aspect ratio and can readily be dispersed in water, which makes their dispersions good candidates in the search for clay-based liquid crystals. Indeed, observations of samples in polarized light and X-ray scattering studies demonstrate the existence of a nematic phase, with a wide biphasic region spanning volume fractions from 1% to 25%, while the sol/gel transition occurs at about 35% (Luo et al., 2013). The unusual width of the biphasic region might be explained by the very large polydispersity of the particles. Zeta-potential measurements showed that the point of zero charge of these halloysite nanotubes lies around pH values of 2. The dispersions were prepared around neutral pH values, but the pH could vary widely while preserving the liquid-crystalline state. The same system was also used to prepare halloysite/cellulose nanocomposite fibres by a wet-spinning method. These highly aligned fibres show enhanced mechanical and moisture barrier properties compared to those of plain cellulose fibres (Luo et al., 2014). A liquid-crystalline phase was also detected, by optical and electron microscopies, in the drying rim of a droplet of an aqueous dispersion of halloysite nanotubes deposited on a substrate (Zhao et al., 2015). The concentration of the dispersion at the rim of the droplet, known as the 'coffee-ring' effect, provides a way of quickly scanning the phase diagram. However, drying must occur slowly enough to make sure that the dispersion always reaches an equilibrium state. The appearance of the liquid-crystalline phase was again discussed in the frame of the Onsager model.

13.7 CONCLUDING REMARKS

Rodlike clay minerals, thanks to their stiffness and huge aspect ratio, readily form liquid-crystalline phases when dispersed in water. Compared to organic or biological polymers, in spite of their inherent polydispersity in length, their availability in large amounts and good chemical stability make them interesting toy models to study the statistical physics of liquid-crystalline order. Conversely, exploiting the collective response of the nanoparticles to external stimuli, like flows or fields, in the liquid-crystalline state can open new ways of manipulating them for sophisticated physical studies, for nanostructuration purposes, or for elaborating anisotropic CPN. The articles referenced in this chapter still leave several questions open, such as the nature of the striated texture sometimes observed in the nematic phase of imogolite nanotubes, the origin of the columnar phase in highly diluted imogolite dispersions and the counterintuitive phase diagram of the imogolite/HPC dispersions. More general issues also naturally arise, such as the behaviour of clay mineral dispersions under confinement, the functionalization of clay mineral particles to study their self-assembly in apolar solvents, or the possibility to dope rodlike synthetic clay minerals with heavy chemical elements in order to impart the particles with interesting physical properties (ie, magnetism, luminescence) that could be combined with liquid-crystalline order.

ABBREVIATIONS

2D two-dimensional
Col columnar liquid-crystal phase
CPN clay polymer nanocomposite
HPC hydroxypropylcellulose
I isotropic liquid phase
N nematic liquid-crystal phase
PVA poly(vinyl alcohol)
SAXS small-angle X-ray scattering

REFERENCES

Amara, M.S., Paineau, E., Bacia-Verloop, M., Krapf, M.E.M., Davidson, P., Belloni, L., Levard, C., Rose, J., Launois, P., Thill, A., 2013. Single-step formation of micron long (OH)(3)Al2O3Ge(OH) imogolite-like nanotubes. Chem. Commun. 49, 11284–11286.

Amara, M.S., Rouziere, S., Paineau, E., Bacia-Verloop, M., Thill, A., Launois, P., 2014. Hexagonalization of aluminogermanate imogolite nanotubes organized into closed-packed bundles. J. Phys. Chem. C 118, 9299–9306.

Barona, G.N.B., Lim, J., Choi, M., Jung, B., 2013. Interfacial polymerization of polyamide-aluminosilicate SWNT nanocomposite membranes for reverse osmosis. Desalination 325, 138–147.

Bates, M.A., Frenkel, D., 1998. Influence of polydispersity on the phase behavior of colloidal liquid crystals, a Monte Carlo simulation study. J. Chem. Phys. 109, 6193–6199.

Bawden, F.C., Pirie, N.W., Bernal, J.D., Fankuchen, I., 1936. Liquid crystalline substances from virus infected plants. Nature 138, 1051–1052.

Bernal, J.D., Fankuchen, I., 1941. X-ray and crystallographic studies of plant virus preparations I. Introduction and preparation of specimens I. Modes of aggregation of the virus particles. J. Gen. Physiol. 25, 111–146.

Chaikin, P.M., Lubensky, T.C., 2000. Principles of Condensed Matter Physics. Cambridge University Press, England.

Davidson, P., Gabriel, J.C.P., 2005. Mineral liquid crystals. Curr. Opin. Colloid Interface Sci. 9, 377–383.

Davidson, P., Petermann, D., Levelut, A.M., 1995. The measurement of the nematic order-parameter by X-ray-scattering reconsidered. J. Phys. II 5, 113–131.

De Gennes, P.-G., 1974. The Physics of Liquid Crystals. The International Series of Monographs on Physics. Clarendon Press, Oxford England.

Dogic, Z., Fraden, S., 2006. Ordered phases of filamentous viruses. Curr. Opin. Colloid Interface Sci. 11, 47–55.

Donkai, N., Inagaki, H., Kajiwara, K., Urakawa, H., Schmidt, M., 1985. Dilute-solution properties of imogolite. Macromol. Chem. Phys. 186, 2623–2638.

Donkai, N., Hoshino, H., Kajiwara, K., Miyamoto, T., 1993a. Lyotropic mesophase of imogolite. 3. Observation of liquid-crystal structure by scanning electron nad novel polarized optical microscopy. Macromol. Chem. Phys. 194, 559–580.

Donkai, N., Kajiwara, K., Schmidt, M., Miyamoto, T., 1993b. Lyotropic mesophase of imogolite—molecular-weight fractionation and polydispersity effect. Makromol. Chem. Rapid Commun. 14, 611–617.

Gabriel, J.C.P., Davidson, P., 2000. New trends in colloidal liquid crystals based on mineral moieties. Adv. Mater. 12, 9–20.

Grelet, E., 2014. Hard-rod behavior in dense mesophases of semiflexible and rigid charged viruses. Phys. Rev. X 4, 021053.

Guinier, A., 1994. X-ray Diffraction in Crystals, imperfect Crystals, and Amorphous Bodies. Dover Publications, New York, USA.

Hentschke, R., Herzfeld, J., 1991. Isotropic, nematic, and columnar ordering in systems of persistent flexible hard-rods. Phys. Rev. A 44, 1148–1155.

Hoshino, H., Ito, T., Donkai, N., Urakawa, H., Kajiwara, K., 1992a. Lyotropic mesophase formation in PVA/imogolite mixture. Polym. Bull. 29, 453–460.

Hoshino, H., Yamana, M., Donkai, N., Sinigersky, V., Kajiwara, K., Miyamoto, T., Inagaki, H., 1992b. Lyotropic mesophase formation of HPC imogolite mixture. Polym. Bull. 28, 607–614.

Hoshino, H., Urakawa, H., Donkai, N., Kajiwara, K., 1996. Simulation of mesophase formation of rodlike molecule, imogolite. Polym. Bull. 36, 257–264.

Jabbari-Farouji, S., Weis, J.J., Davidson, P., Levitz, P., Trizac, E., 2013. On phase behavior and dynamical signatures of charged colloidal platelets. Sci. Rep. 3, 3559.

Kajiwara, K., Donkai, N., Fujiyoshi, Y., Hiraki, Y., Urakawa, H., Inagaki, H., 1986a. Some remarks on imogolite mesophase. Bull. Inst. Chem. Res. Kyoto Univ. 63, 320–331.

Kajiwara, K., Donkai, N., Fujiyoshi, Y., Inagaki, H., 1986b. Lyotropic mesophase of imogolite. 2. Microscopic observation of imogolite mesophase. Macromol. Chem. Phys. 187, 2895–2907.

Kajiwara, K., Donkai, N., Hiragi, Y., Inagaki, H., 1986c. Lyotropic mesophase of imogolite. 1. Effect of polydispersity on phase-diagram. Macromol. Chem. Phys. 187, 2883–2893.

Kang, D.Y., Tong, H.M., Zang, J., Choudhury, R.P., Sholl, D.S., Beckham, H.W., Jones, C.W., Nair, S., 2012. Single-walled aluminosilicate nanotube/poly(vinyl alcohol) nanocomposite membranes. ACS Appl. Mater. Interfaces 4, 965–976.

Lee, H., Ryu, J., Kim, D., Joo, Y., Lee, S.U., Sohn, D., 2013. Preparation of an imogolite/poly (acrylic acid) hybrid gel. J. Colloid Interface Sci. 406, 165–171.

Lekkerkerker, H.N.W., Vroege, G.J., 2013. Liquid crystal phase transitions in suspensions of mineral colloids: new life from old roots. Philos. Trans. R. Soc. A Math. Phys. Eng. Sci. 371, 20120263.

Levitz, P., Zinsmeister, M., Davidson, P., Constantin, D., Poncelet, O., 2008. Intermittent Brownian dynamics over a rigid strand: heavily tailed relocation statistics in a simple geometry. Phys. Rev. E 78, 030102.

Livolant, F., Leforestier, A., 1996. Condensed phases of DNA: structures and phase transitions. Prog. Polym. Sci. 21, 1115–1164.

Luo, Z., Song, H., Feng, X., Run, M., Cui, H., Wu, L., Gao, J., Wang, Z., 2013. Liquid crystalline phase behavior and sol–gel transition in aqueous halloysite nanotube dispersions. Langmuir 29, 12358–12366.

Luo, Z., Wang, A., Wang, C., Qin, W., Zhao, N., Song, H., Gao, J., 2014. Liquid crystalline phase behavior and fiber spinning of cellulose/ionic liquid/halloysite nanotubes dispersions. J. Mater. Chem. A 2, 7327–7336.

Onsager, L., 1949. The effects of shape on the interaction of colloidal particles. Ann. N. Y. Acad. Sci. 51, 627–659.

Otsuka, H., Takahara, A., 2010. Structure and properties of imogolite nanotubes and their application to polymer nanocomposites. Inorg. Met. Nanotubular Mater. Recent Technol. Appl. 117, 169–190.

Paineau, E., et al., 2016. A liquid-crystalline hexagonal columnar phase in highly-dilute suspensions of imogolite nanotubes. Nat. Commun. 7, 10271.

Shikinaka, K., Koizumi, Y., Kaneda, K., Osada, Y., Masunaga, H., Shigehara, K., 2013. Strain-induced reversible isotropic-anisotropic structural transition of imogolite hydrogels. Polymer 54, 2489–2492.

Shikinaka, K., Koizumi, Y., Shigehara, K., 2015. Mechanical/optical behaviors of imogolite hydrogels depending on their compositions and oriented structures. J. Appl. Polym. Sci. 132, 41691.

Song, H., Zhao, N., Qin, W., Duan, B., Ding, X., Wen, X., Qiu, P., Ba, X., 2015. High-performance ionic liquid-based nanocomposite polymer electrolytes with anisotropic ionic conductivity prepared by coupling liquid crystal self-templating with unidirectional freezing. J. Mater. Chem. A 3, 2128–2134.

Teramoto, N., Hayashi, A., Yamanaka, K., Sakiyama, A., Nakano, A., Shibata, M., 2012. Preparation and mechanical properties of photo-crosslinked fish gelatin/imogolite nanofiber composite hydrogel. Materials 5, 2573–2585.

Trizac, E., Bocquet, L., Aubouy, M., 2002. Simple approach for charge renormalization in highly charged macroions. Phys. Rev. Lett. 89, 248301.

Vroege, G.J., Lekkerkerker, H.N.W., 1992. Phase-transitions in lyotropic colloidal and polymer liquid-crystals. Rep. Prog. Phys. 55, 1241–1309.

Vroege, G.J., Thies-Weesie, D.M.E., Petukhov, A.V., Lemaire, B.J., Davidson, P., 2006. Smectic liquid-crystalline order in suspensions of highly polydisperse goethite nanorods. Adv. Mater. 18, 2565–2568.

Warren, B.E., 1990. X-Ray Diffraction. Dover Publications, New York, USA.

Wensink, H.H., 2007. Columnar versus smectic order in systems of charged colloidal rods. J. Chem. Phys. 126, 194901.

Woolston, P., van Duijneveldt, J.S., 2015a. Isotropic-nematic phase transition in aqueous sepiolite suspensions. J. Colloid Interface Sci. 437, 65–70.

Woolston, P., van Duijneveldt, J.S., 2015b. Isotropic-nematic phase transition of polydisperse clay rods. J. Chem. Phys. 142, 184901.

Yasarawan, N., van Duijneveldt, J.S., 2008. Dichroism in dye-doped colloidal liquid crystals. Langmuir 24, 7184–7192.

Yasarawan, N., van Duijneveldt, J.S., 2010. Arrested phase separation of colloidal rod-sphere mixtures. Soft Matter 6, 353–362.

Zhang, Z.X., van Duijneveldt, J.S., 2006. Isotropic-nematic phase transition of nonaqueous suspensions of natural clay rods. J. Chem. Phys. 124, 154910.

Zhao, Y., Cavallaro, G., Lvov, Y., 2015. Orientation of charged clay nanotubes in evaporating droplet meniscus. J. Colloid Interface Sci. 440, 68–77.

Zocher, H., 1925. Uber freiwillige Strukturbildung in Solen. (Eine neue Art anisotrop flüssiger Medien.). Z. Anorg. Allg. Chem. 147, 91–110.

Chapter 14

Molecular Simulation of Nanosized Tubular Clay Minerals

H.A. Duarte*

Grupo de Pesquisa em Química Inorgânica Teórica (GPQIT), Instituto de Ciências Exatas, Universidade Federal de Minas Gerais (UFMG), Belo Horizonte, MG, Brazil

Corresponding author: e-mail: duarteh@ufmg.br

14.1 INTRODUCTION

Nanostructured materials have been known to exist in nature for a long time. Pauling (1930), analysing the structure of some clay minerals, pointed out that hollow materials might exist in nature. The tubular shape of halloysite (Bates et al., 1950), chrysotile (Yada, 1971) and imogolite (Cradwick et al., 1972) have been determined. Naturally occurring nanotubular materials such as imogolite and halloysite are per se fascinating, but with the advent of nanotechnology, these natural nanotubes have become a target for developing advanced materials with enhanced properties. The aluminosilicate nanotubes are isolators, so applications in electronic devices are limited. However, these natural nanotubes are the perfect target for developing nanocatalysts, gas storage, controlled delivery systems, nanowires, clay polymer nanocomposite agents and many other relevant technological applications. Furthermore, the outer and inner surfaces of clay mineral nanotubes can be easily modified and different organic groups anchored, providing new chemical functionalities. Investigations based on sophisticated experimental techniques and computational chemistry have contributed to a complete understanding of the structure and chemical properties of clay mineral nanotubes at an atomistic level.

Computational chemistry techniques have been used to explore aspects such as the stability of clay mineral nanotubes and the role of hydrogen bonding at the surface, their electronic structure, their mechanical properties such as Young's moduli and the chemical properties of possibly synthesisable

Developments in Clay Science, Vol. 7. http://dx.doi.org/10.1016/B978-0-08-100293-3.00014-5

nanotubes. Such information has been shown to be essential for envisaging new strategies for developing different modified clay mineral nanotubes.

In this chapter, the contributions that computational chemistry have made over the last 10 years for understanding clay mineral nanotubes are presented. The different approaches used to model the clay mineral nanotubes are briefly reviewed. The most studied one, from the computational chemistry point of view, is imogolite, probably because it is a unique nanotube that is monodisperse and single-walled (SW) with a well-defined diameter. Computational chemistry investigations of other nanotubes have been reported, and their main achievements are presented here as well.

14.2 COMPUTATIONAL ASPECTS

The detailed fundamentals of the methodologies used to perform computer simulations of the nanostructured aluminosilicates are beyond the scope of this chapter. However, the main concepts behind the state-of-the-art computational chemistry techniques are important to highlight in order to present their uses and limitations. For details about the theory and methodology, see the cited references throughout the text.

The chemical model of aluminosilicate nanotubes is usually a cylinder with its respective unit cell, and the periodic boundary condition is applied along the cylinder axial. On the other axis, the size of the unit cell is large enough to avoid lateral interactions of the nanotubes. It is important to note that normally, the calculated nanotubes are considered to be in the gas phase without any effect of the solvent or lateral interactions due to the bundle formation, for example.

14.2.1 Force Field Simulations

The molecular and covalent bounded systems can be reasonably modelled as a set of atoms that are connected through force constants. The two-, three- and four-atom interactions due to bond lengths, angles and dihedrals are adequately modelled using suitable equations. For nonbonding interactions, such as van der Waals and hydrogen bonding, the Lennard–Jones potential is usually used. The topology of the systems is established by providing all the connections between the atoms and their types. Different parameters are used depending on the hybridisation involved and the type of the material. The set of parameters that describe the molecular structure is called a 'force field'. Experimental data such as geometries, heat of formation, reaction energies, vibrational frequencies and other properties are normally used to assess the quality of a force field. There are many different force fields, which have been developed for a class of systems such as proteins, materials and clay minerals

(Cygan et al., 2004; Zeitler et al., 2014). Normally, the total energy is adequately described for clay minerals by Eq. (14.1):

$$E = \underbrace{\frac{e^2}{4\pi\varepsilon_0}\sum_{i\neq j}\frac{q_i q_j}{r_{ij}}}_{\text{Coulomb}} + \underbrace{\sum_{i\neq j} D_{ij}^o\left[\left(\frac{R_{ij}^o}{r_{ij}}\right)^{12} - 2\left(\frac{R_{ij}^o}{r_{ij}}\right)^6\right]}_{\text{van der Waals}} + \underbrace{\sum_{i\neq j} k_{ij}^1\left(r_{ij}-r_0\right)^2}_{\text{bond stretch}} + \underbrace{\sum_{i\neq j\neq k} k_{ijk}^2\left(\theta_{ijk}-\theta_0\right)^2}_{\text{angle bend}}. \tag{14.1}$$

The partial charges q_i and q_j and the force constants k_{ij}^1 and k_{ijk}^2 are parameters normally derived from quantum-mechanical calculations. D_{ij}^o and R_{ij}^o are empirical parameters derived from structural and physical properties.

Due to the relative simplicity of the method, calculations of thousands of atoms can be performed. Force fields are often applied in molecular dynamics (MD), in which the system is evolved along the time at a given pressure and temperature. With the gradients of the force field and the velocities, the classical equations of motion can be established and integrated along the normal time for using periodic boundary conditions (Allen and Tildesley, 1987). The MD can be evolved up to nanoseconds using a time step in the range of femtosecond. MD can be used for investigating the dynamics of the solvent around the aluminosilicate nanotubes and the formation mechanism of nanotubes and nanoscrolls.

14.2.2 Density-Functional Theory

The density-functional theory (DFT) is probably the highest-level theory used for investigating large systems such as aluminosilicate nanotubes. Based on the published Hohenberg–Kohn theorems (Hohenberg and Kohn, 1964), the electron density is legitimized as the principal variable that permits calculating the total energy of an electronic system. Then Kohn and Sham (1965) published a method that permits the performance of DFT calculations in a computationally efficient manner. As explained by Levy (1982), the Kohn–Sham equations [Eqs (14.2)–(14.5)] use a noninteracting electron system as a reference for providing one-electron Kohn–Sham orbitals, ψ_i, that lead to the electron density of the real interacting electron system:

$$\left(-\frac{1}{2}\nabla^2 + v_{\text{ef}}(r)\right)\psi_i = \varepsilon_i\psi_i; \tag{14.2}$$

where

$$v_{\text{ef}}(r) = \sum_A^M -\frac{Z_A}{|r-R_A|} + \int\frac{\rho(r')}{|r-r'|}dr' + v_{\text{xc}}(r); \tag{14.3}$$

and

$$v_{\mathrm{XC}}(r)=\frac{\delta E_{\mathrm{XC}}[\rho]}{\delta\rho}; \tag{14.4}$$

$$\rho(r)=\sum_{i}^{N} n_i\psi_i^2(r). \tag{14.5}$$

In Eq. (14.2), the first term is the kinetic energy operator and the second term is the effective local potential felt by the electrons. The effective potential, Eq. (14.3), includes the external potential due to the nuclei charges, the Coulomb term due to the classical repulsion between the electrons and the exchange-correlation (XC) potential, which includes all nonclassical interactions between the electrons. Z_A, N and M are the atomic number of atom A, the number of electrons and the number of atoms, respectively. The XC potential is defined by Eq. (14.4) as the derivative of the XC energy functional, and the electron density, $\rho(r)$, is defined by Eq. (14.5), where n_i is the occupation of the KS orbitals.

The Schrödinger equation Eq. (14.2) is solved by expanding the one-electron KS orbital in a set of basis functions $\{\phi_\mu\}$:

$$\psi_i=\sum_{\mu} C_{\mu i}\phi_\mu. \tag{14.6}$$

The basis functions can be localized using Gaussian, Slater or numerical basis sets, depending on the strategy used for implementing the method in the different program packages.

The XC functional [Eq. (14.4)] is simultaneously the strength and weakness of the DFT method. The Hohenberg–Kohn theorems show that an XC functional of the electron density does exist, and consequently, so does the total electronic energy functional. However, the exact form of the XC functional is still unknown, although many properties of the exact XC functional have been reported (Parr and Yang, 1994). Different approximations for the XC functional define the different DFT methods. Most of the XC functionals are based on the homogeneous electron gas approach. The simplest XC functional named SVWN (for Slater, Vosko, Wilk and Nusair) is due to the work of Slater (1951) and Vosko et al. (1980), based on a local approach of the homogeneous electron gas approximation. The generalized gradient approximation (GGA) takes into account part of the inhomogeneity of the electron density of the molecular calculations incorporating the electron density gradients in the XC functional. The GGA XC functional proposed by PBE (named for Perdew, Burke and Ernzerhöf; Perdew et al., 1996) is nonempirical and largely used for solid-state calculations and molecular systems. The hybrid XC functional includes part of the exact exchange term of the Hartree–Fock method. The three-parameter hybrid method called B3LYP (named for Becke, Lee, Yang and Parr; Becke, 1988, 1993; Lee et al., 1988) is largely used mainly for organic molecular systems.

14.2.3 Self-consistent-Charge Density-Functional Tight-Binding Method

Frauenheim and collaborators developed the density-functional tight-binding (DFTB) method (Porezag et al., 1995; Seifert et al., 1996) that is an approximate DFT method (Oliveira et al., 2009). Basically, the tight-binding approximation is evoked, a minimal basis set is used for all atoms and the three-centre integrals present in the Hamiltonian (Eq. 14.2) are disregarded. Therefore, only one- and two centre integrals are present in the Hamiltonian and can be easily tabulated in a pairwise manner. The set of the tabulated integrals are called 'Slater-Koster files'. The DFTB method has been applied with success for investigating materials such as carbon and simpler inorganic nanotubes (Frauenheim et al., 1995; Seifert et al., 2001). However, for more heterogeneous systems where charge transfer may occur, the DFTB method fails. An extension of the DFTB method, called the 'self-consistent-charge density-functional tight-binding method (SCC-DFTB)' was developed (Elstner et al., 1998). The SCC-DFTB total energy is given by Eq. (14.7):

$$E_{\text{SCC}} = \sum_{i}^{N} n_i \langle \psi_i | \hat{H}^0 | \psi_i \rangle + \frac{1}{2} \sum_{\alpha,\beta}^{M} \gamma_{\alpha\beta} \Delta q_\alpha \Delta q_\beta + E_{\text{rep}}. \tag{14.7}$$

The SCC-DFTB method allows charge transfer to occur through a self-consistent procedure driven by $\gamma_{\alpha\beta} = \gamma_{\alpha\beta}\left(U_\alpha, U_\beta, |\vec{R}_\alpha - \vec{R}_\beta|\right)$, which is related to the Hubbard parameter, U_α, of each atom present in the structure. It is interesting to note that the Hubbard parameter is related to the hardness of the atoms. Here, Δq_α is estimated from Mulliken population analysis, and ψ_i is expanded in a minimal basis set according to Eq. (14.6). The Hamiltonian $\hat{H}^0$ is exactly the same as in the standard DFTB scheme, as shown here:

$$H^o_{\mu\nu} = \begin{cases} \varepsilon_\mu^{\text{free atom}}, & \mu = \nu \\ \left\langle \phi_\mu \left| -\frac{1}{2}\nabla^2 + \upsilon_{\text{ef}}\left[\rho_0^A + \rho_0^B\right] \right| \phi_\nu \right\rangle, & \mu \in \{A\}, \nu \in \{B\} \\ 0, & \text{otherwise} \end{cases} \tag{14.8}$$

Eq. (14.8) depends only on atoms A and B following the tight-binding approach. In this instance, ρ_0^A and $\varepsilon_\mu^{\text{free atom}}$ are the electron density of the atom A and one-particle energy for the free atom, respectively.

It is important to highlight that the lack of the three-centre integrals in the Hamiltonian leads to the necessity to incorporate a repulsive contribution, E_{rep}, to the total energy estimate. This repulsive contribution is incorporated in order to reproduce the DFT total energy of reference systems. E_{rep} is usually fitted to a polynomial function or to a series of splines. The parameterization of the repulsive contribution is crucial for the quality of the results to be obtained and to its transferability between different systems. Parameter sets for general and specific purposes are available (see Köskinen and Mäkinen,

2009). The SCC-DFTB method provides an efficient and reasonably accurate way to calculate the electronic energy of large systems.

14.3 IMOGOLITES

Carbon nanotubes and other inorganic nanotubes are normally polydisperse, with a large range of diameters. Many studies have shown that the strain energy necessary to roll a monolayer into a tube decreases smoothly with the increase in diameter, such as for carbon, MoS_2, TiS_2 and GaS nanotubes (Seifert et al., 2001; Kohler et al., 2004; Enyashin et al., 2007). The synthesis of the nanotubes normally leads to a kinetic polydisperse and multiwalled (MW) product. However, the imogolite nanotube is monodisperse, uncapped, and SW, with a well-defined diameter. Understanding this intriguing property of imogolite was probably the ultimate goal for these theoretical investigations.

14.3.1 Imogolite Model

Imogolite has the composition of $(HO)_3Al_2O_3SiOH$, which is the sequence of atoms going from the outer to the inner surface of the nanotube. Hereafter, this is referred as 'imogolite-Si', while 'imogolite-M' denotes other minerals sharing the same imogolite-like local structure, in which silicon was replaced by atom *M*. The hypothetical monolayer is built by taking a sheet of gibbsite structure and replacing the hydroxyl groups in one vacant site by orthosilicate anions (Fig. 14.1A). The convention for labelling carbon nanotubes is

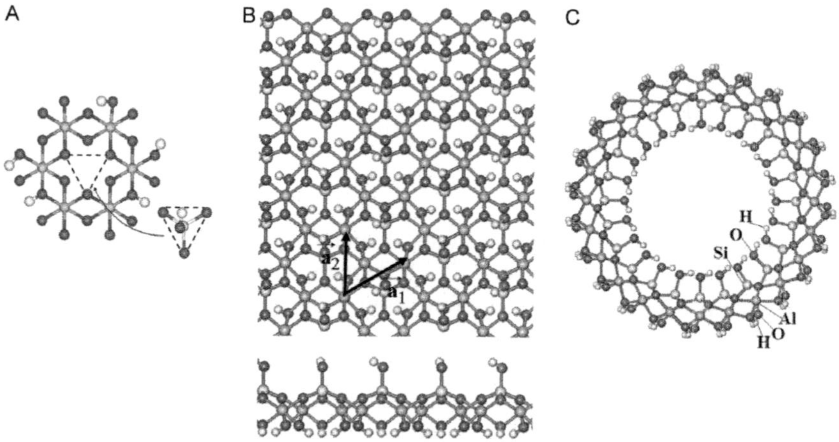

FIG. 14.1 (A) Gibbsite unit and its vacant site where the silanol group is placed; (B) hypothetical 2D imogolite layer with vectors $\mathbf{a}_1$ and $\mathbf{a}_2$; (C) zigzag (12,0) imogolite nanotube. *White atoms*, H; *red* (*dark grey* in the print version), O; *blue* (*grey* in the print version), Al; *yellow* (*light grey* in the print version), Si. *Adapted with permission from Guimarães et al. (2007). Copyright 2007 American Chemical Society.*

normally adopted for the aluminosilicate nanotubes. In Fig. 14.1, the lattice vectors $\mathbf{a}_1$ and $\mathbf{a}_2$ of the hexagonal lattice are used to define the rolling direction **B** in the two-dimensional (2D) lattice, where $\mathbf{B}=n\mathbf{a}_1+m\mathbf{a}_2$. Two classes of nanotubes have been considered in the computer calculations: armchair (n, n) and zigzag (n, 0).

The diameter of the nanotubes is easily estimated from theoretical simulations and experiments. However, it is important to note that the estimates can vary depending on the model or reference used. Normally in the computer simulations, the diameter is based on the outer hydrogen layer of imogolite, and experimental estimates are usually based on small-angle X-ray scattering (SAXS), gas adsorption and cryo-electron microscopy (cryo-EM) (Maillet et al., 2010; Thill et al., 2012b), which is only slightly affected by light atoms such as hydrogen. Therefore, the nanotubes are normally estimated to have smaller diameters compared to the computer-simulated structure.

14.3.2 Imogolite: Aluminosilicate Nanotubes

Tamura and Kawamura (2002) were among of the first researchers to perform calculations on the molecular models of imogolite using an interatomic potential model. For comparison, they also calculated the hypothetical gibbsite nanotube. Presumably, they have constructed models of zigzag (n, 0) symmetry. In their computer simulations, these authors showed that the total energy of the nanotubes have a minimum of $n=16$, with a diameter of 2.93 nm. Imogolite is more stable than the hypothetical gibbsite due to the tetrahedra (silanol group) bounded near the vacant octahedral sites of the gibbsite framework that stabilize the tubular structure. Konduri et al. (2006) used the clay force field (CLAYFF) (Cygan et al., 2004) to investigate in detail the strain energy, defined as the energy necessary to roll a planar monolayer to the nanotube. The contributions of the stretching potentials of the Al–O ($V_{\text{Al–O}}$) and Si–O ($V_{\text{Si–O}}$) bonds to the total energy behave in a different manner with respect to the diameter. The $V_{\text{Al–O}}$ decreases with the increase of the diameter, since it is approaching the ideal gibbsite structure. However, $V_{\text{Si–O}}$ increases with the diameter of the structure, since the ideal tetrahedral of the silanol is deformed. The sum of these two contributions leads to the minimum total energy, with a diameter of 2.26 nm.

Guimarães et al. (2007) used the SCC-DFTB method for investigating the electronic, mechanical and structural properties of imogolite. The quantum-mechanical approach used here permitted the exploration of the electronic properties and the stability of the nanotubes. They investigated the armchair (n, n) and zigzag (n, 0) chiralities. The structures have been fully optimized, showing that the cylindrical structure is a stable local minimum. Born–Oppenheimer MD simulations at 300 K was performed using the potential

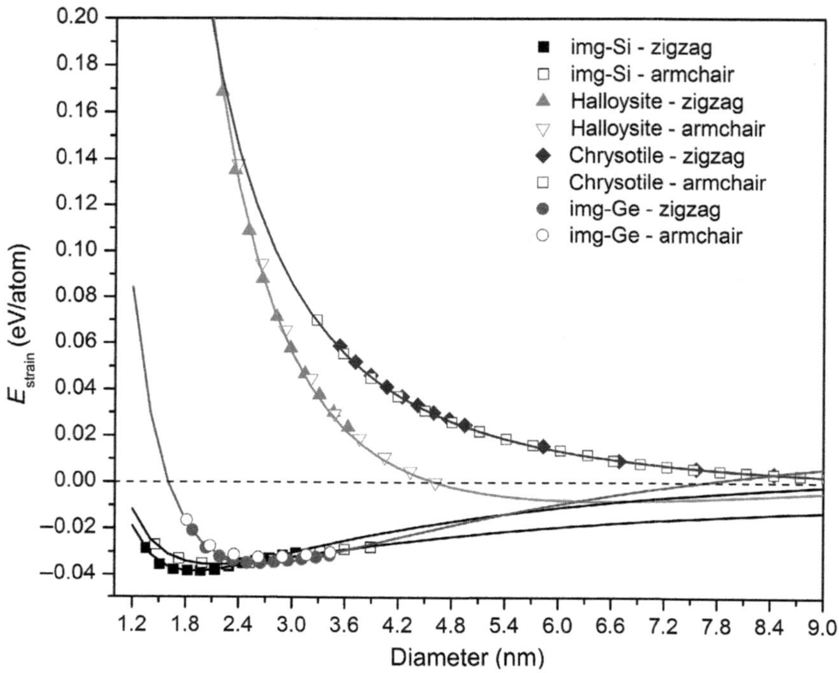

FIG. 14.2 Comparison of the different naturally occurring nanotubes. *Squares* mark imogolite-Si, *circles* mark imogolite-Ge, *triangles* mark halloysite and *diamonds* mark chrysotile; and solid objects are zigzag (*n*, 0) nanotubes and open objects are armchair (*n*, *n*) nanotubes. *Strain energy curves were obtained from Guimarães et al. (2007, 2010); Lourenço et al. (2012, 2014).*

energy surface (PES) obtained from the SCC-DFTB method. This is important to note that at the quantum-mechanical level, the bonds can be broken and distorted. The strain energy per atom was estimated by taking as a reference the ideal planar layer. A comparison of the strain energy curves of the different imogolite-like nanotubes, halloysite and chrysotile is reported in Fig. 14.2.

The strain energy curves of the different nanotube configurations as a function of the diameter have a minimum for the zigzag and armchair conformations. The zigzag configurations are the most stable for the $n < 16$. The (12,0) zigzag is the most stable, with a diameter of about 1.97 nm, taking as a reference the oxygen atoms on the outer surface. Recently, the imogolite-Si nanotubes were recalculated using a new set of Slater-Koster files, and the estimated value for (12,0) zigzag was 2.16 nm (Lourenço et al., 2014) in close agreement with the classical MD simulations (Tamura and Kawamura, 2002; Konduri et al., 2006). The imogolite band gap of about 10 eV confirms that it is an isolator with wide band gap material. The Young's modulus was estimated to be about 242 GPa, which is similar to other nanotubes such as chrysotile but less stiff than other nanotubes such as carbon (Hernandez et al., 1999). The radial breathing mode was estimated to be about 54 cm^{-1} for the zigzag (12,0) nanotube, but these results indicate that with more than 17 gibbsite units in the circumference, the nanotube could became instable since imaginary frequencies of the radial breathing mode were predicted. MD of the hydrated imogolite-Si were performed containing 12 water molecules inside of the nanotube and 82 water molecules in the intertubular void corresponding to mass percentage of about 14.1% similar to the experimental low relative humidity (Creton et al., 2008). The results pointed out that the breathing motions of the nanotube does not reduce to a single mode but rather a narrow band in the range of 30–50 cm^{-1} is observed containing modes involving all atoms of the nanotube.

These are important contributions from the computational chemistry investigations for understanding the properties of imogolites such as to be monodisperse, SW and with well-defined dimensions. The minimum in the strain energy curves and the mild conditions that the imogolite are synthesized in aqueous solution over 72 h of reflux lead invariably to the thermodynamic most stable structure with well-defined diameter. Cradwick et al. (1972) reported that natural imogolites are composed by 10 gibbsite units and the first imogolite nanotubes synthesized by Farmer et al. (1977) contained 12 gibbsite units around the circumference according to X-ray diffraction (XRD) analysis. Indeed, the experimental conditions (temperature, ionic strength, type of ions in the solution) can significantly change the diameter of the nanotube (Yucelen et al., 2012). This is a unique characteristic compared to the other nanotubes that may be considered as kinetic product of the syntheses since they are not the global minimum structure. The planar is normally the most stable for carbon and other inorganic nanotubes (Lourenço et al., 2012). It seems that for one-atom-thick nanotube wall, such as carbon nanotubes, it is not possible to obtain a structure with a minimum in the strain energy curve. The four-atom-thick imogolite nanotube wall with nonsymmetric bond strengths leads to a difference between the outer and inner surface tensions, and an optimal curvature is found for the nanotube.

Density-functional calculations have been performed for the imogolite. Alvarez-Ramírez (2007) and Li et al. (2008) investigated the electronic structure of the imogolite using localized numerical basis sets and the PW91 (named for Perdew and Wang; Perdew et al., 1992) and PBE XC functional, respectively. The calculations indicate a direct band gap with a value of 4.7 (PW91) and 3.67 (PBE) eV. However, this value might be taken with caution since it is known that the GGA XC functional underestimates the band gap. The projected density of states indicates that close to the Fermi level the valence band is mostly contributed by the Si–OH groups and the conduction band is mostly contributed by the Al–OH groups. Demichelis et al. (2010) used the roto-translational symmetry to perform DFT calculations using localized Gaussian basis sets with the three-parameter B3LYP hybrid XC functional. They found similar results compared to the SCC-DFTB calculations. The (10,0) zigzag nanotube was found to be the most stable with inner diameter of 1.45 nm and the (8,8) armchair is less stable of about 10.6 kJ mol^{-1} per formula unit. The B3LYP estimated value for the band gap is 7.2 eV. This is considered the best value since B3LYP is known to estimate accurate band gaps, and this value is in between the two extremes of PBE XC functional and the SCC-DFTB estimates. The experimental and calculated diameters of imogolites are reported in Table 14.1.

TABLE 14.1 Experimental and Calculated Diameters of SW and DW Imogolite-Si and SW and DW Imogolite-Ge Nanotubes

Nanotube	Method	Diameter (Å)	Band gap (eV)	Structure
SW: Imogolite	Exp.[a]	23.0		
	PBE/DFT[b]	17.50	5.7	(9,0)/(12,0)
	B3LYP/DFT[c]	14.52[d]	7.2	(10,0)
	PBE/DFT[e]	17.18		(8,0)
	PW91/DFT[f]	23.48	4.8	(12,0)
	SCC-DFTB[g]	19.72	10.0	(12,0)
	SCC-DFTB[h]	21.6	10.2	(12,0)
DW: Imogolite	SCC-DFTB[h]	18.50 (internal) 32.42 (external)	8.1	(10,0)@(19,0)
SW: Ge-imogolite	Exp.	33.0[i]	3.6	
		33.0[j]		
		30.04±0.10[k]		
		30[l]		
		35.0±1.6[m]		
		23.0±5.0[n]		
		38[o]		
	PW91/DFT[f]		4.8	
	SCC-DFTB	26.46[h]	9.6	(14,0)

Continued

TABLE 14.1 Experimental and Calculated Diameters of SW and DW Imogolite-Si and SW and DW Imogolite-Ge Nanotubes—Cont'd

Nanotube	Method	Diameter (Å)	Band gap (eV)	Structure
DW: Ge-imogolite	Exp.	24.0±1.0 (internal)[m] 40.0±1.0 (external)[m]		
		30.0±5.0 (external)[n]		
		26 (internal)[o] 43.0 (external)[o]		
	SCC-DFTB	23.42 (internal)[h] 37.46 (external)[h]	8.5	(12,0)@(21,0)

[a]*Mukherjee et al. (2005).*
[b]*Zhao et al. (2009).*
[c]*Demichelis et al. (2010).*
[d]*Internal diameter. The external diameter can be roughly estimated to be 20 Å.*
[e]*Lee et al. (2011).*
[f]*Alvarez-Ramírez (2007).*
[g]*Guimarães et al. (2007).*
[h]*Lourenço et al. (2014).*
[i]*Wada and Wada (1982).*
[j]*Mukherjee et al. (2005).*
[k]*Levard et al. (2008).*
[l]*Levard et al. (2010).*
[m]*Maillet et al. (2010), diameter estimated from SAXS.*
[n]*Maillet et al. (2011), diameter estimated from AFM.*
[o]*Thill et al. (2012b).*

One important aspect of the imogolite is the role of the hydrogen bonding between the silanol groups in the inner surface of the nanotube in the stabilization of the structure. The orientation of the hydrogen bonding is an important aspect that has been investigated by Demichelis et al. (2010) and Lee et al. (2011). These authors argue that the zigzag is favoured with respect to the armchair due to the hydrogen bonding in the inner surface. The different orientation of the hydrogen bonding can explain up to 2 kJ mol^{-1} per unit cell of difference in the strain energy curves. However, it seems that the favoured zigzag structure cannot be explained solely on the basis of the hydrogen bonding network in the inner surface of the imogolite, as discussed by Guimarães et al. (2013). Zhao et al. (2009) also performed PBE DFT calculations using numerical localized basis sets and estimated the band gap to be 5.7 eV. They found two minima in the strain energy curves for (9,0) and (12,0) zigzag nanotubes and tried to correlate with the natural and synthetic imogolites. However, analysing the results of the Demichelis et al. (2010)

shows that the hydrogen bonding configuration in the inner surface is very sensitive to the starting structures used for geometry optimization. In fact, with the increase of the diameter the orientation of the hydrogen bonding can be changed leading to an extra stabilization and the artefact of appearing a second minimum in the strain energy curve.

14.3.3 Aluminogermanate Nanotubes

For the first time, Wada and Wada (1982) synthesized imogolite-like nanotubes in which the silicon atoms were replaced by germanium atoms. The aluminogermanate nanotube (imogolite-Ge) has a larger external diameter of about 4.0 nm and it can be synthesized in large amounts as was shown by Levard et al. (2008, 2011). Alvarez-Ramírez (2007) performed a detailed analysis of the imogolite structure with different silicon-germanium content. The XRD pattern of imogolite-like structures was simulated, and they are in good agreement with the available experimental data for imogolite-Si and imogolite-Ge. The packing of the nanotubes in bundles has been discussed and its effect on the XRD pattern analysed. The band gap of the imogolite-like structure with different contents of germanium varies between 4.3 eV (for imogolite-Ge) up to 4.8 eV (imogolite-Si) at the PW91/plane wave level of theory. Konduri et al. (2007) also investigated imogolite with a different silicon-germanium content using a modified CLAYFF force field. The results indicated that a larger diameter for the imogolite-Ge is favoured in comparison to the imogolite-Si. They discussed the effect of increasing Ge content on the XRD pattern and on the outer diameter.

By controlling the pH, concentration and ionic strength, it is possible to synthesize the double-walled (DW) imogolite-Ge (Maillet et al., 2010; Thill et al., 2012b). Lourenço et al. (2014) used the SCC-DFTB calculations to investigate the electronic, mechanical and structural properties of the SW and DW imogolite-Ge. The zigzag (n, 0) SW imogolite-Ge nanotubes are favoured in comparison to the armchair (n, n) SW nanotubes. The (14,0) SW imogolite-Ge nanotube was predicted to be the most stable, with an external diameter of about 2.46 nm, while the (11,0) imogolite-Si nanotube is the most stable, with an external diameter of about 2.16 nm. It is important to note that computer simulations are performed on idealized structures calculated at the gas phase, while synthetic nanotubes can vary their diameters according to the electrolyte used and experimental conditions, as has been shown by Yucelen et al. (2012).

The (p, 0)@(q, 0) DW imogolite-Ge nanotubes have been calculated at the SCC-DFTB level of theory with the aim of understanding why the synthesis of DW imogolite-Ge was successful and DW imogolite-Si seems to not be achievable. Fig. 14.3 indicates clearly that the most favoured DW imogolite-Ge occurs for the combination (p, 0)@($p+9$, 0) nanotube, and the most stable is the (12,0)@(21,0) DW imogolite-Ge nanotube.

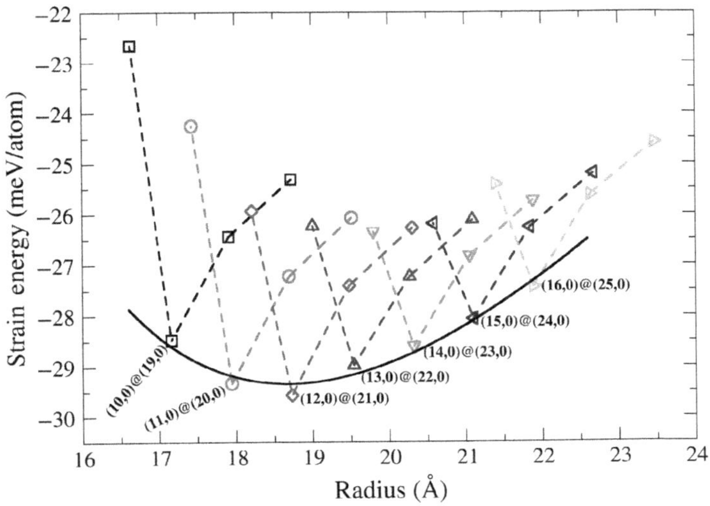

FIG. 14.3 Strain energy per atom as a function of the outer radius for double-walled (p, 0)@(q, 0) imogolite-Ge nanotubes, taking into account different sizes and wall interactions: $q-p=8$, 9, 10 and 11. *Dashed lines* are related to the same inner nanotube but different outer nanotube. The reference used to calculate the double-walled nanotubes strain energies was the bilayer. *Reproduced with permission from Lourenço et al. (2014). Copyright 2014 American Chemical Society.*

The external diameter is estimated to be about 3.75 nm. It is interesting to observe that it does not contain the (14,0) nanotube, which is the SW most stable nanotube. This is a characteristic of the imogolites that are synthesized in aqueous solution and in mild conditions leading to the thermodynamic product. Two contributions are involved in the DW nanotube formation: (i) the stabilizing contribution due to the hydrogen bonding interaction between the surfaces of the nanotubes; (ii) the destabilizing contribution due to the strain energy of the (12,0) and (21,0), which are not the most stable compared to the SW (14,0) nanotube. In fact, when one compares to the hypothetical DW imogolite-Si nanotubes (Fig. 14.4), it is clear that larger strain energy is paid to form the DW imogolite-Si and the stabilizing hydrogen bonding interaction is much smaller due to the larger distances between the two tube walls of about 0.207 nm. This value must be compared to the 0.143-nm measurement for the DW imogolite-Ge nanotube. A similar explanation was given by Thill et al. (2012a,b) based on qualitative analytical models of the imogolite energy as a function of radius. The band gap is predicted to be smaller with respect to the imogolite-Si, and the estimates are about 9.6 and 8.5 eV for the SW and DW imogolite-Ge nanotubes. This is an upper bound value since SCC-DFTB overestimates band gaps.

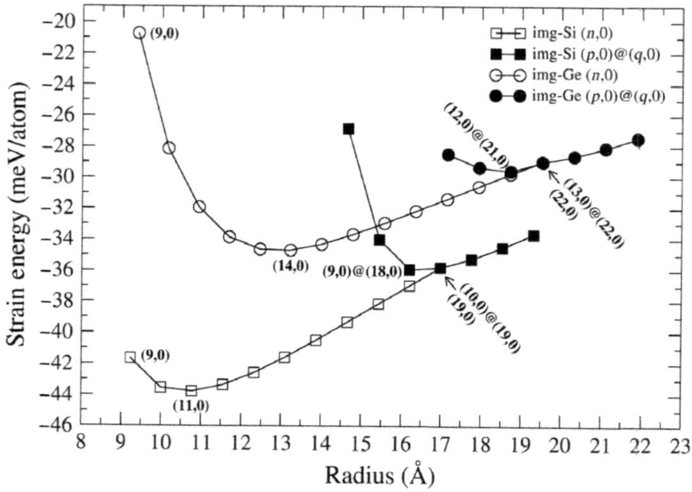

FIG. 14.4 Strain energy curves for the SW and DW nanotubes of the imogolite-Si and imogolite-Ge systems. *Reproduced with permission from Lourenço et al. (2014). Copyright 2014 American Chemical Society.*

14.3.4 Other Imogolite-like Nanotubes

It is disturbing that since the synthesis of the aluminogermanates nanotubes in 1982 by Wada and Wada, no other imogolite-like nanotubes have been successfully synthesized. The chemical intuition indicates that any replacement of the group III elements (Al, Ga, In) or group IV elements (C, Si, Ga, Sn) should be stable and synthesisable. Alvarez-Ramirez (2009) investigated the electronic and structural properties of such hypothetical structures showing that the band gap could vary from 4.6 eV for Al–Si (imogolite-Si) to 2.6 eV for In–Ge, as expected. Although all these possible imogolite-like structures are predicted to be stable with different external diameters and electronic properties, the right experimental conditions to achieve such a syntheses remains to be designed. Actually, it has been shown that the syntheses of the imogolites (imogolite-Si or imogolite-Ge) are very sensitive to pH, ionic strength and concentrations. The nanotubes are formed upon self-organizing mechanisms from the hydrolysis of the species, oligomerization and the formation of proto-imogolites (Levard et al., 2010; Yucelen et al., 2011). The understanding of the chemical speciation in aqueous solution seems very important to establish the right conditions for nanotube formation.

Guimarães et al. (2013) were interested for the replacement of the silanol by other fragments other than the $Ge(OH)_4$. Actually, any element that forms tetrahedron in aqueous solution is, in principle, a candidate to replace the silanol fragment. They have investigated the possible imogolite-like structures replacing the fragment $(O_3SiOH)^{3-}$ by $PO_4{}^{3-}$, $PO_3{}^{3-}$, $AsO_4{}^{3-}$ and $AsO_3{}^{3-}$. The basic idea is that on the inner side of the nanotube, there is the —P=O group instead of –Si–OH, which could be easily reduced to –P. The same stands for the —As=O, which could be reduced to –As. For all these structures, a minimum is observed for the strain energy curves with respect to the diameter. The band gap is much decreased to the reduced species, such as phosphite and arsenite (about 5.2 and 4.2 eV, respectively). For the imogolite-P-ate (phosphate), the band gap is predicted to be about 10 eV (similarly to imogolite-Si) and for the imogolite-As-ate, the band gaps are about 7.6 eV. It is interesting to note that the zigzag $(n, 0)$ structures are more stable than the armchair structures (n, n), similar to imogolite-Si and imogolite-Ge. This indicates that the hydrogen bonding present in the imogolite-Si and imogolite-Ge is not responsible for stabilizing the zigzag with respect to the armchair. The internal hydrogen bonding of imogolite-Si is responsible for the larger stabilization and more pronounced minimum in the strain energy surface.

Duarte et al. (2012) speculated that the pK_a of the precursor reactants could indicate the feasibility of a different imogolite-like structure. Table 14.2 lists the pK_a of the components of the possible species that could be used to

TABLE 14.2 Possible Elements to Form Imogolite-like Nanotubes

Element	Species	pK_a
Group III elements		
Al(III)	$[Al(H_2O)_6]^{3+}$	5.52
Ga(III)	$[Ga(H_2O)_6]^{3+}$	2.85
In(III)	$[In(H_2O)_6]^{3+}$	3.7
Group IV elements		
Si(IV)	$Si(OH)_4$	9.84
Ge(IV)	$Ge(OH)_4$	9.16
Sn(IV)	$Sn(OH)_4$[a]	–
Group V elements		
P(V)	H_3PO_4	2.12; 7.21; 12.67
P(III)	H_3PO_3	1.3; 6.7
As(V)	H_3AsO_4	2.19; 6.94; 11.5
As(III)	H_3AsO_3	9.2

The aqueous species are indicated with its respective pK_a.
[a]*Little information is available due to the very low solubility of $SnO_2(s)$ (Seby et al., 2001).*

form imogolite-like structures, and that has been investigated by computational chemistry, as have the species and the pK_a of possible candidates to replace silicon or aluminium in the imogolites.

Carbon and nitrogen species are not considered since they lead to the carbonate and nitrate species which are trigonal planar and not suitable for replacing the silicon. As it was pointed out by Duarte et al. (2012), the species $Si(OH)_4$ and $Ge(OH)_4$ have similar pK_a; therefore, the replacement of the Si by Ge is easily achieved using a very similar protocol of synthesis. Other elements, such as Ga(III) and In(III), are more acidic (small pK_a), and therefore, the synthesis must be done in a more acidic medium in order to obtain similar chemical speciation. For the replacement of the Si, only As(III) seems to be possible using a similar protocol, since the pK_a is very similar to the Si(IV). The understanding of the chemical speciation and the formation mechanism of imogolite at a molecular level seems to be a key point to manage the synthesis of other imogolite-like structures. Important insights have been provided about this mechanism using different experimental techniques such as ^{27}Al and ^{29}Si nuclear magnetic resonance (NMR), electrospray ionization mass spectrometry (ESI-MS), dynamic light scattering (DLS), extended X-ray absorption fine structure (EXAFS) and inductive coupled plasma atomic emission spectroscopy (ICP-AES) (Levard et al., 2010, 2011; Yucelen et al., 2011, 2012).

Recently, González et al. (2014) used classical MD to investigate the self-rolling of an aluminosilicate sheet model into an imogolite. They used flat models of the aluminosilicate sheets with $9<N<24$ (where N is the number of the hexagons to form the nanotube) and CLAYFF force field (Cygan et al., 2004) to perform the MD in different temperatures (10–368 K). These authors were able to reproduce the results showing a minimum in the strain energy curve. They found that the length of the unit cell of the tube varies with the temperature between 8.39 Å at 10 K up to 8.50 Å at 368 K. Furthermore, the starting flat models of different sizes were evolved for up to 200 ps at different temperatures. A small increase of temperature was sufficient to shift the minimum from the $n=10$ to $n=11$. At 368 K, the most stable nanotube is shifted to the $N=12$. The tendency to form nanoscrolls is observed ($n=15$) for higher temperatures. A precursor of the DW nanotubes has also been observed for higher temperatures and larger models ($N>15$). It is important to note that Thill et al. (2012a) showed that for the synthesis of $(OH)_3Al_2O_3Si_xGe_{1-x}OH$ nanotubes with $x=0.10$, the formation of nanoscrolls also occurs. In spite of the fact that González et al. (2014) investigated only the imogolite-Si system, to the best of our knowledge, this is the first attempt to understand the mechanism of imogolite formation considering the effect of the temperature. However, the mechanism of the imogolite formation involves the formation of proto-imogolites, which are self-organizing to form nanotubes, allophanes or amorphous aluminosilicates depending of the pH, ionic strength, temperature and concentration (Yucelen et al., 2011, 2012,

2013). The acidic constants of the hydrolysed species forming aluminates and silicates are key points for understanding the whole process. The computer modelling of such complex systems is still a challenge, and much effort has to be made towards understanding the imogolite formation at the atomistic level.

14.3.5 Modification of Imogolite

The internal and external surfaces of imogolite-Si can be easily modified using different strategies such as creating vacancies, decorating it with different organic groups, changing the reactants to form modified structures or through thermal treatments. The reaction with organophosphonic acids and organosilanes (Bac et al., 2009; Ma et al., 2011, 2012; Amara et al., 2015) modifies the external nanotube surfaces. New developments for modifying the interiors have been reported (Bottero et al., 2011; Kang et al., 2011, 2014; Zanzottera et al., 2012; Amara et al., 2015) enhancing the adsorption properties of the material. A different way of modifying the interior has been suggested by Kang et al. (2010) through dehydroxylation using controlled thermal treatment. These authors have shown that the dehydroxylation occurs predominantly in the interiors of the nanotube.

Teobaldi et al. (2009) investigated the hydroxyl vacancies in the outer- and inner-nanotube surfaces using PW91/plane wave calculations. The results indicated that the structure of the nanotube is only slightly modified upon hydroxyl vacancies but occupied and unoccupied states are created, reducing the band gap to 1.8 and 1.1 eV for imogolite-Si and imogolite-Ge, respectively.

The orientation and the distances between radial and axial inner hydroxyl surfaces are shown in Fig. 14.5. Kang et al. (2011) carried out reactions of the

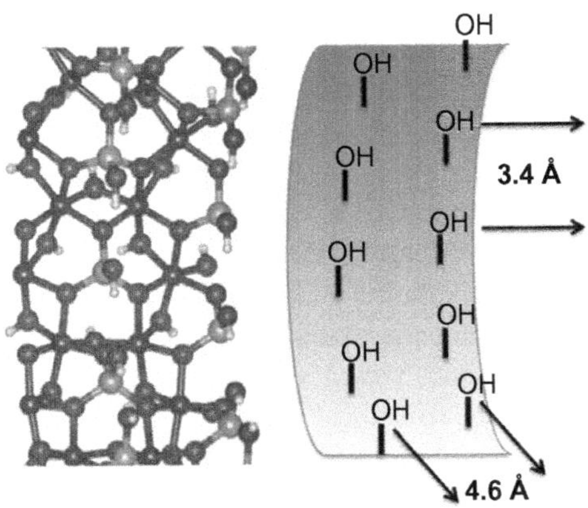

FIG. 14.5 Inner-surface hydroxyl orientation and the main distances between the radial and axial inner hydroxyl. *Adapted from da Silva et al. (2015).*

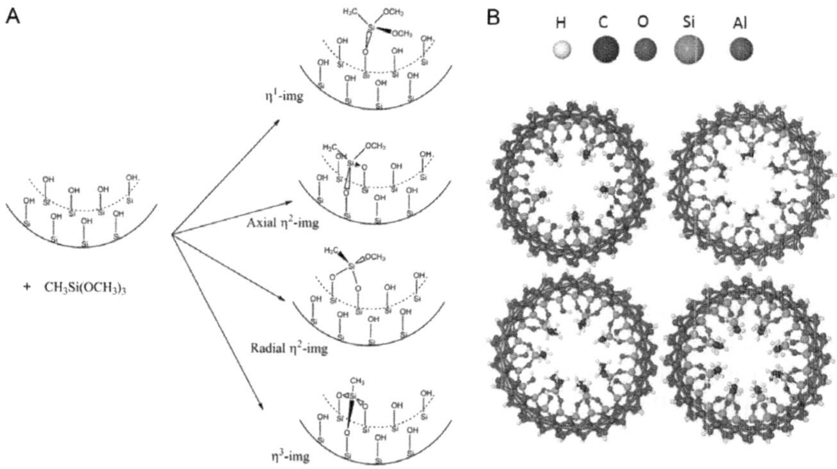

FIG. 14.6 (A) Inner-surface representations for the modified zigzag (12,0) imogolite nanotube. (B) The optimized structures of the substituted (12,0) imogolite. *Adapted from da Silva et al. (2015).*

methyltrimethoxysilane, $CH_3Si(OCH_3)_3$, with imogolite-Si as shown at Fig. 14.6, leading to the formation of methanol. From ^{29}Si NMR data, they have shown that about 24–38% of the silanols are substituted, leading mostly to η^2-imogolite. SCC-DFTB calculations showed that $CH_3Si(OCH_3)_3$ prefers radial η^2-imogolite, in which the two silanol groups in the same radial line are substituted (da Silva et al., 2015). The η^1-imogolite product leading to one substitution is about 9.1 kJ mol^{-1} (unit cell)$^{-1}$ larger in energy with respect to the most favoured η^2-imogolite product. Furthermore, the axial η^2-imogolite and η^3-imogolite products involving silanol groups in different radial lines are not favoured. The distance of 4.6 Å between silanol groups of different radial lines leads to bonding stress. Calculations were performed for different substitutions (up to 66% of the silanol groups). The calculations performed by da Silva et al. (2015) indicated that the reaction energy is not much affected by the number of substitutions, about 122 kJ mol^{-1} (unit cell)$^{-1}$ on average. Other properties, such as the band gap and Young's modulus, are slightly affected.

However, the dehydroxylation of the silanol groups, leading to Si–O–Si bonding in the inner surface due to the thermal treatment, affects substantially the geometry, band gap and Young's modulus of the material, as shown in Fig. 14.7. Kang et al. (2010) using ^{29}Si and ^{27}Al NMR experiments showed that up to 73% of the silanol groups were dehydroxylated. Recent calculations performed by da Silva et al. (2015) for dehydroxylation of up to 50% of the silanol groups indicated that the nanotube is much deformed. For 50% of dehydroxylation, the nanotube is hexagonalized and the reaction energy of

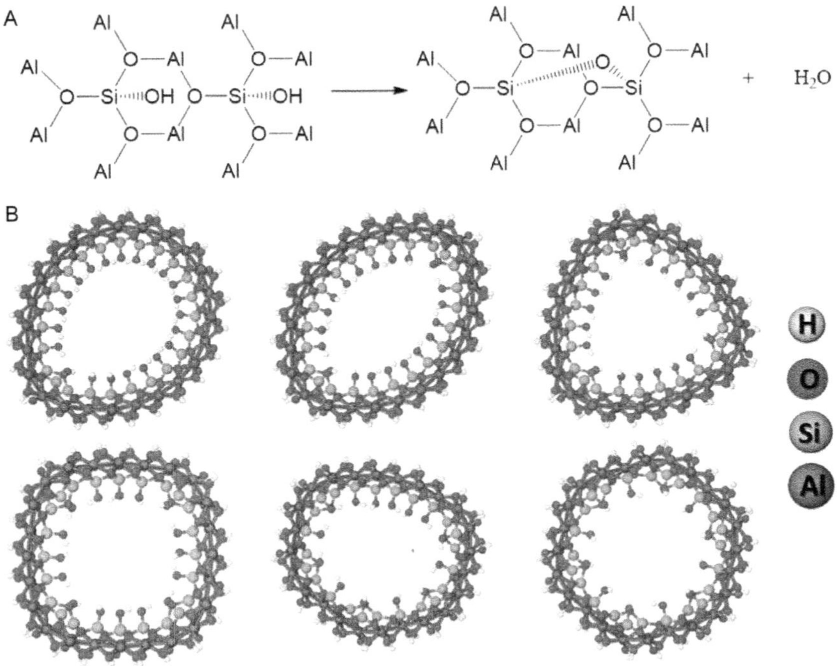

FIG. 14.7 (A) Dehydroxylation reaction scheme in the inner surface. (B) Optimized structures of the dehydroxylated zigzag (12,0) imogolite nanotube. *Adapted with permission from da Silva et al. (2015).*

dehydroxylation is increased with the number of dehydroxylations in the unit cell. The dehydroxylated nanotube, with 50% of the silanol groups modified, presents a bang gap of about 3.8 eV, which must be compared to 10 eV of the ideal structure. The bulk volume and the Young's modulus remain practically the same. These authors suggested that the band gap of the imogolite could be controlled by heat treatment upon dehydroxylation. If one takes into account that SCC-DFTB method overestimates the band gap, it is reasonable to conclude that the dehydroxylated imogolite is a semiconductor.

14.4 HALLOYSITE

Halloysite nanotubes are more abundant in nature than imogolites. They are mined from natural deposits, and therefore, their use in technological applications has been developed (Rawtani and Agrawal, 2012). Halloysite is a 1:1 layer with the formula $Al_2Si_2O_5(OH)_4$ and it is polydisperse, with lengths of 500–1000 nm and diameters varying between 15 and 100 nm (Veerabadran et al., 2007; Lvov et al., 2008). Halloysite is chemically similar to kaolinite, and it is usually MW, but it can also present as nanoscrolls or nanorolls

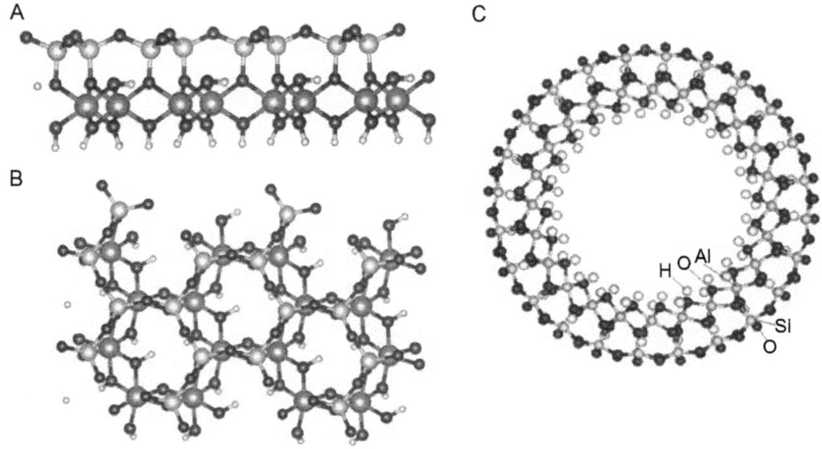

FIG. 14.8 (A) Side and (B) top view of the monolayer. (C) (12,0) Halloysite nanotube. *Adapted with permission from Guimarães et al. (2010). Copyright 2010 American Chemical Society.*

(Ece and Schroeder, 2007). Fig. 14.8 shows the cross-sectional views of halloysite. SW halloysite nanotube models have been calculated using SCC-DFTB (Guimarães et al., 2010). The strain energies have been calculated for up to (20,0) halloysites with external diameters of about 3.63 nm. The band gap was estimated to be about 8.4 eV, and the Young's modulus about 280 GPa. A comparison of the strain energy curves for different aluminosilicate nanotubes was shown in Fig. 14.2 earlier in this chapter. In the range of the diameters investigated, no minimum was found in the strain energy curves. However, extrapolating the curve to larger diameters, it is predicted that a very shallow minimum may exist. In fact, this has been used to explain the large range of diameters normally found for halloysite. Furthermore, according to the SCC-DFTB calculations, no preference is found for zigzag or armchair nanotubes.

The hydrated and anhydrous halloysite spiral nanotubes were also calculated using SCC-DFTB (Ferrante et al., 2015). Models consisting of an inner diameter of 5 nm and overlapping arms between one-half and one-third of a revolution were built. Water molecules in the hydrated halloysite nanospirals have been added following the $Al_2Si_2O_5(OH)_4{\cdot}2H_2O$ stoichiometry. The bond lengths are not affected by the spiralization of the kaolinite layer. The spiralization of the kaolinite favours the approach of the water to the surface, leading to some degree of disorder. In the overlapping of the arms, the water molecules are between an upper layer of aluminium oxide and a lower layer of silicon oxide. Water molecule pairs make bridges that connect the two surfaces of the arms. For the anhydrous halloysite, a larger number of AlO–H…$(O{-}Si)_2$ interactions compensates for the absence of water making hydrogen bonds. It has been argued that the disorder of the hydrogen bonding is related to the many minima in the PES with similar energies and very tiny barriers. Therefore, it is more a dynamic linkage between the two surfaces of the arms than a static one.

14.5 CHRYSOTILE AND NANO-FIBRIFORM SILICA

Chrysotile is easily found in nature, and it is one of the most common nanostructured silicates. Synthetic and natural MW nanotubes and scrolls have been reported on (Yada, 1967, 1971). The inner and outer diameters have been stated to be in the range of 1–10 and 10–50 nm, respectively, and lengths in the millimetre range (Falini et al., 2004). The formula of chrysotile is $Mg_3Si_2O_5(OH)_4$. The layer of Mg is normally associated with brucite ($Mg(OH)_2$) and the Si layer with tridymite (SiO_2) (Fig. 14.9). In addition, SW models have been calculated using SCC-DFTB for both possibilities with tridymite in the outer or inner side of the nanotube (Lourenço et al., 2012). The SCC-DFTB calculations indicated that the tridymite on the inner side is at least 30 meV/atom more stable than the other option. Fig. 14.2 also showed the strain energy curves for zigzag and armchair of SW chrysotile nanotubes calculated in the range of (17,0)–(45,0) and (9,9)–(29,29), corresponding to external diameters in the range between

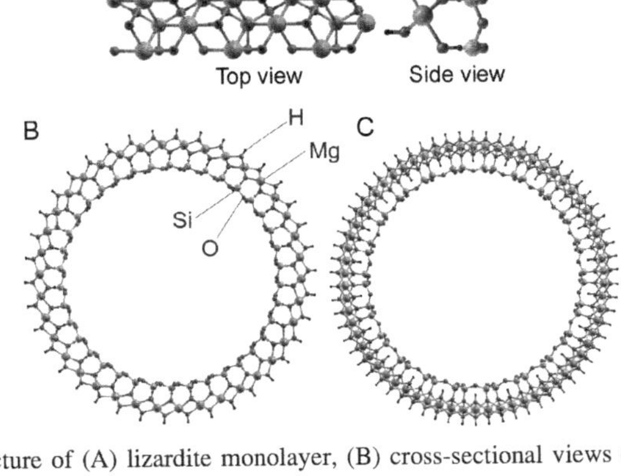

FIG. 14.9 Structure of (A) lizardite monolayer, (B) cross-sectional views of the zigzag (19,0) and (C) armchair (11,11) chrysotile nanotubes. *Reproduced with permission from Lourenço et al. (2012). Copyright 2012 American Chemical Society.*

3.2 and 9.4 nm. The strain energy curves do not present minima. Their stability is not affected by the chirality. It is estimated to be an insulator with a band gap of 10 eV and Young's modulus are estimated to be in the range of 261–323 GPa.

D'arco et al. (2009) performed B3LYP/DFT calculations on SW chrysotile models. Helical symmetry was used for calculating nanotubes in the range of (17,17)–(24,24) armchair. The band gap was estimated to be 6.4 eV, which is about 3.6 eV less than the SCC-DFTB estimated values.

Nano-fibriform silica nanotubes (SNT) can be synthesized by removing the brucite layer using acid leaching of chrysotile (Wang et al., 2006a,b). The external surface of the SNT can also be easily modified using organosilanes to produce a functionalized nanotubular silica (MSNT) material (Wang et al., 2009). da Silva et al. (2013) used the same SCC-DFTB approach to investigate the electronic, structural and mechanical properties of SNT and MSNT (Fig. 14.10A). Armchair and zigzag SNT with inner diameter in the

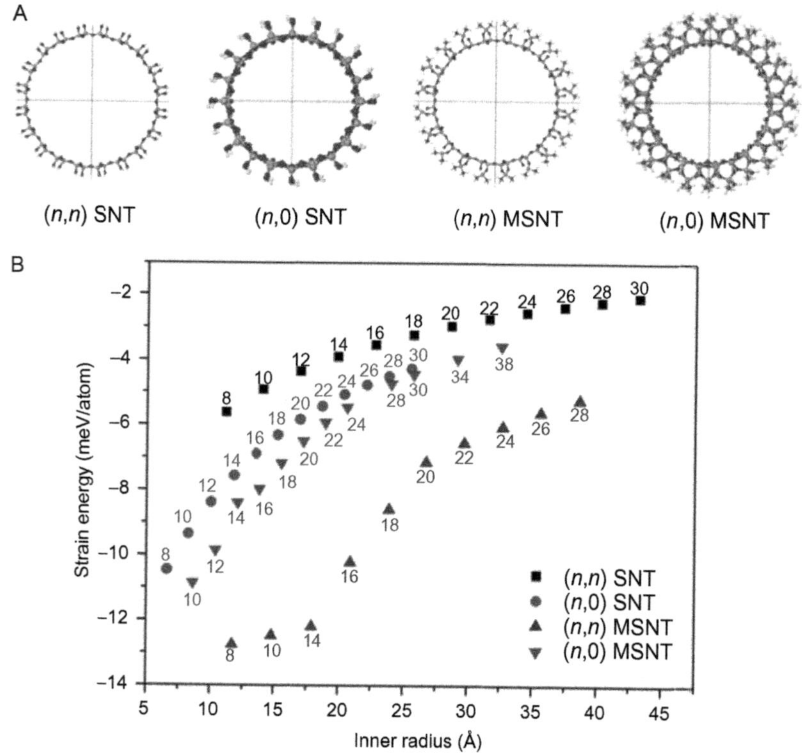

FIG. 14.10 (A) Side views of the armchair (n, n) and zigzag (n, 0) of SNT and MSNT; (B) below is the strain energy curves with respect to the inner radius for armchair and zigzag SNT and MSNT. *Adapted with permission from da Silva et al. (2013).*

range of 1.6–7.2 nm have been calculated. SNT and modified silica nanotubes (MSNT) are insulators with band gaps estimated to be about 8–10 eV. The strain energy curves showed at Fig. 14.10B present different behaviour than for the other inorganic nanotubes. First, the zigzag is predicted to be more stable for SNT, but upon functionalization with organosilanes, armchair MSNT are more stable than the respective zigzag MSNT. It is interesting to observe that these nanotubes are always more stable than the respective planar layer, as is the case for imogolites. But no minima are observed for SNT and MSNT. The strain energy decreases smoothly up to the smallest nanotube investigated (about 1.6 nm in diameter). According to da Silva et al. (2013), decreasing the diameter of the system will make it collapse to the silica structure. Young's modulus is predicted to be about 232–260 GPa for the most stable armchair SNT and 150–194 GPa for the zigzag SNT. For the MSNT, the Young's modulus is estimated to be in the range of 77–89 GPa for the most stable zigzag and 110–140 GPa for the armchair MSNT. These values must be compared to the chrysotile values of 261–323 GPa. The brucite layer outer of the chrysotile nanotubes increases the stiffness of the nanotube. However, it also increases the strain energy favouring the layered structure of the chrysotile called lizardite.

14.6 CONCLUDING REMARKS

Computational chemistry investigations of imogolites, halloysites and chrysotiles have been reported using different methodologies. Detailed analysis of the structure and energetics of the rolling of the ideal layer to form the nanotube has been provided. The comparison of the strain energy curves of the different nanotubes shown at Fig. 14.2 is very important for understanding the uniqueness of imogolite. The gibbsite is like a template for developing new nanotubes, in which, depending on the group placed on the octahedral site, replacing the silanol could lead to nanotubes with enhanced properties. From the theoretical point of view, the replacement of the silanol by germanate, phosphate, phosphite, arsenate and arsenite leads to stable structures. The aluminogermanate was already synthesized; however, the synthesis of new imogolite-like structures is still a challenge. The DW imogolite-Ge nanotube also presents a minimum strain energy curve for the (12,0)@(21,0) nanotube. The imogolite-Ge presents a more flat strain energy curve and a more efficient hydrogen bonding interaction between the two tubes, favouring its formation. The same is not predicted for imogolite-Si; since the strain energy curve presents a more pronounced minimum and higher energetic cost, the DW imogolite-Si nanotube formation is not favoured. The synthesis of nanocables based on carbon nanotube@imogolite has been proposed based on computer simulations (Kuc and Heine, 2009). The concept is interesting since the electron transport in the carbon nanotubes would not be affected by the environment due to the presence of the isolator imogolite shielding the nanowire. The synthesis of imogolite in the presence of an emulsion of carbon nanotubes could lead to the nanocable. It is expected that self-organization of the proto-imogolites around the carbon nanotube can be achieved by controlling the temperature and ionic strength of the system.

Halloysite and chrysotile have also been investigated by means of theoretical calculation. The lack of a minimum in the strain energy curves makes them behave like other inorganic NT that are polydisperse with a wide range of diameters. Insights about the electronic, structural and mechanical properties of SW halloysite and chrysotile nanotubes have been provided.

One crucial point is the chemical modification of the inner and outer surfaces of the nanotubes. The computational modelling can provide insights about the effect on the geometry, electronic structures and mechanical properties. Some theoretical calculations have been reported about the modified structures, indicating that the dehydroxylation of the imogolite could be used to fine-tune the band gap. On the other hand, functionalization of the inner silanol groups with organosilanes does not change significantly the structure and band gap of the imogolite.

Much effort has been made to understand the mechanism of the formation of imogolites from experimentation. However, the computational modelling of the formation mechanism of the imogolites (or halloysites) in aqueous solution is a challenge. The system is very complex, with an initial oligomer formation of aluminates and silicates, followed by the formation of larger aggregates called ‘proto-imogolites’, which is self-organized to form the nanotube. The structure and the acidic properties of these species are crucial for modelling the nanotube formation. Indeed, the type of acid used (HCl, $HClO_4$, CH_3COOH) in the synthesis can significantly change the diameter of the nanotube (Yucelen et al., 2012). However, the solvent effects, the role of the water and ions to the formation of such species are very difficult to be adequately modelled. In fact, the importance of the understanding of this mechanism at a molecular level is an important motivation for improving and developing new approaches for studying species in solution.

In spite of the many theoretical calculations reported in the present chapter, computational chemistry has not been fully explored as a way to investigate the aluminosilicate nanotubes. Theoretical calculations can predict spectroscopic data such as NMR, X-ray photoelectron emission, Fourier transform infrared spectroscopy (FTIR), RAMAN, ultraviolet visible (UV-vis) and many other techniques, and the results can be compared directly with the experiments. Actually, a combined experimental/theoretical approach based on sophisticated spectroscopic techniques could fulfil the lack of information about the acid/base behaviour, dynamics of ions and molecules inside of the nanotube, charge distribution between outer and inner-nanotube surfaces at aqueous solution and the mechanism of formation of nanotube and nanoscrolls.

ABBREVIATIONS

B3LYP	three-parameter hybrid method expressed for Becke, Lee, Yang and Parr
CLAYFF	Clay Force Field developed by Cygan, R. T.; Liang, J. J.; Kalinichev, A. G.
DFT	density-functional theory
DFTB	density-functional tight-binding
DW	double-walled
GGA	generalized gradient approximation
MD	molecular dynamics
MSNT	modified silica nanotubes
MW	multiwalled
NMR	nuclear magnetic resonance
NT	nanotubes
PBE	expressed for Perdew, Burke and Ernzerhöf
PES	potential energy surface
PW91	expressed for Perdew and Wang
SAXS	small-angle X-ray scattering
SCC-DFTB	self-consistent-charge density-functional tight-binding
SNT	silica nanotubes
SVWN	expressed for Slater, Vosko, Wilk and Nusair
SW	single-walled
XC	exchange-correlation
XRD	X-ray diffraction

REFERENCES

Allen, M.P., Tildesley, D.J., 1987. Computer Simulation of Liquids. Oxford Science Publications, London.

Alvarez-Ramírez, F., 2007. Ab initio simulation of the structural and electronic properties of aluminosilicate and aluminogermanate nanotubes with imogolite-like structure. Phys. Rev. B 76, 125421.

Alvarez-Ramirez, F., 2009. Theoretical study of (OH)(3)N_2O_3MOH, M = C, Si, Ge, Sn and N = Al, Ga, In, with imogolite-like structure. J. Comput. Theor. Nanosci. 6, 1120–1124.

Amara, M.S., Paineau, E., Rouziere, S., Guiose, B., Krapf, M.-E.M., Tache, O., Launois, P., Thill, A., 2015. Hybrid, tunable-diameter, metal oxide nanotubes for trapping of organic molecules. Chem. Mater. 27, 1488–1494.

Bac, B.H., Song, Y., Kim, M.H., Lee, Y.-B., Kang, I.M., 2009. Surface-modified aluminogermanate nanotube by OPA: synthesis and characterization. Inorg. Chem. Commun. 12, 1045–1048.

Bates, T.F., Hildebrand, F.A., Swineford, A., 1950. Morphology and structure of endellite and halloysite. Am. Mineral. 35, 463–484.

Becke, A.D., 1988. Density-functional exchange-energy approximation with correct asymptotic-behavior. Phys. Rev. A 38, 3098–3100.

Becke, A.D., 1993. Density-functional thermochemistry. III. The role of exact exchange. J. Chem. Phys. 98, 5648–5652.

Bottero, I., Bonelli, B., Ashbrook, S.E., Wright, P.A., Zhou, W., Tagliabue, M., Armandi, M., Garrone, E., 2011. Synthesis and characterization of hybrid organic/inorganic nanotubes of the imogolite type and their behaviour towards methane adsorption. Phys. Chem. Chem. Phys. 13, 744–750.

Cradwick, P.D., Wada, K., Russell, J.D., Yoshinag, N., Masson, C.R., Farmer, V.C., 1972. Imogolite, a hydrated aluminium silicate of tubular structure. Nat. Phys. Sci. 240, 187–189.

Creton, B., Bougeard, D., Smirnov, K.S., Guilment, J., Poncelet, O., 2008. Molecular dynamics study of hydrated imogolite. 1. Vibrational dynamics of the nanotube. J. Phys. Chem. C 112, 10013–10020.

Cygan, R.T., Liang, J.J., Kalinichev, A.G., 2004. Molecular models of hydroxide, oxyhydroxide, and clay phases and the development of a general force field. J. Phys. Chem. B 108, 1255–1266.

da Silva, M.C., dos Santos, E.C., Lourenço, M.P., Duarte, H.A., 2013. Structural, mechanical and electronic properties of nano-fibriform silica and its organic functionalization by dimethyl silane: a SCC-DFTB approach. J. Mol. Model. 19, 1995–2005.

da Silva, M.C., dos Santos, E.C., Lourenço, M.P., Gouvea, M.P., Duarte, H.A., 2015. Structural, electronic, and mechanical properties of inner surface modified imogolite nanotubes. Front. Mater. 2, 1–10.

D'arco, P., Noel, Y., Demichelis, R., Dovesi, R., 2009. Single-layered chrysotile nanotubes: a quantum mechanical ab initio simulation. J. Chem. Phys. 131, 204701.

Demichelis, R., Noel, Y., D'arco, P., Maschio, L., Orlando, R., Dovesi, R., 2010. Structure and energetics of imogolite: a quantum mechanical ab initio study with B3LYP hybrid functional. J. Mater. Chem. 20, 10417–10425.

Duarte, H.A., Lourenço, M.P., Heine, T., Guimarães, L., 2012. Clay mineral nanotubes: stability, structure and properties. In: Innocenti, A., Kamarulzaman, N. (Eds.), Stoichiometry and Materials Science—When Numbers Matter. InTech, Rijeka, pp. 1–24. Chapter 1.

Ece, O.I., Schroeder, P.A., 2007. Clay mineralogy and chemistry of halloysite and alunite deposits in the Turplu area, Balikesir, Turkey. Clays Clay Miner. 55, 18–35.

Elstner, M., Porezag, D., Jungnickel, G., Elsner, J., Haugk, M., Frauenheim, T., Suhai, S., Seifert, G., 1998. Self-consistent-charge density-functional tight-binding method for simulations of complex materials properties. Phys. Rev. B 58, 7260–7268.

Enyashin, A.N., Gemming, S., Seifert, G., 2007. Simulation of inorganic nanotubes. In: Gemming, S., Schreiber, M., Suck, J.B. (Eds.), Materials for Tomorrow: Theory, Experiments and Modelling. Springer Series in Materials Science, vol. 92. Springer-Verlag, Berlin, pp. 33–59. Chapter 2.

Falini, G., Foresti, E., Gazzano, M., Gualtieri, A.E., Leoni, M., Lesci, I.G., Roveri, N., 2004. Tubular-shaped stoichiometric chrysotile nanocrystals. Chemistry 10, 3043–3049.

Farmer, V.C., Fraser, A.R., Tait, J.M., 1977. Synthesis of imogolite: a tubular aluminium silicate polymer. J. Chem. Soc. Chem. Commun. 462–463.

Ferrante, F., Armata, N., Lazzara, G., 2015. Modeling of the halloysite spiral nanotube. J. Phys. Chem. C 119, 16700–16707.

Frauenheim, T., Jungnickel, G., Kohler, T., Stephan, U., 1995. Structure and electronic-properties of amorphous-carbon: from semimetallic to insulating behavior. J. Non-Cryst. Solids 182, 186–197.

González, R.I., Ramírez, R., Rogan, J., Valdivia, J.A., Munoz, F., Valencia, F., Ramírez, M., Kiwi, M., 2014. Model for self-rolling of an aluminosilicate sheet into a single-walled imogolite nanotube. J. Phys. Chem. C 118, 28227–28233.

Guimarães, L., Enyashin, A.N., Frenzel, J., Heine, T., Duarte, H.A., Seifert, G., 2007. Imogolite nanotubes: stability, electronic, and mechanical properties. ACS Nano 1, 362–368.

Guimarães, L., Enyashin, A.N., Seifert, G., Duarte, H.A., 2010. Structural, electronic, and mechanical properties of single-walled halloysite nanotube models. J. Phys. Chem. C 114, 11358–11363.

Guimarães, L., Pinto, Y.N., Lourenço, M.P., Duarte, H.A., 2013. Imogolite-like nanotubes: structure, stability, electronic and mechanical properties of the phosphorous and arsenic derivatives. Phys. Chem. Chem. Phys. 15, 4303–4309.

Hernandez, E., Goze, C., Bernier, P., Rubio, A., 1999. Elastic properties of single-wall nanotubes. Appl. Phys. A Mater. Sci. Process. 68, 287–292.

Hohenberg, P., Kohn, W., 1964. Inhomogeneous electron gas. Phys. Rev. B 136, B864.

Kang, D.Y., Zang, J., Wright, E.R., Mccanna, A.L., Jones, C.W., Nair, S., 2010. Dehydration, dehydroxylation, and rehydroxylation of single-walled aluminosilicate nanotubes. ACS Nano 4, 4897–4907.

Kang, D.Y., Zang, J., Jones, C.W., Nair, S., 2011. Single-walled aluminosilicate nanotubes with organic-modified interiors. J. Phys. Chem. C 115, 7676–7685.

Kang, D.-Y., Brunelli, N.A., Yucelen, G.I., Venkatasubramanian, A., Zang, J., Leisen, J., Hesketh, P.J., Jones, C.W., Nair, S., 2014. Direct synthesis of single-walled aminoaluminosilicate nanotubes with enhanced molecular adsorption selectivity. Nat. Commun. 5, 3342.

Kohler, T., Frauenheim, T., Hajnal, Z., Seifert, G., 2004. Tubular structures of GaS. Phys. Rev. B 69, 193403.

Kohn, W., Sham, L.J., 1965. Self-consistent equations including exchange and correlation effects. Phys. Rev. 140, 1133.

Konduri, S., Mukherjee, S., Nair, S., 2006. Strain energy minimum and vibrational properties of single-walled aluminosilicate nanotubes. Phys. Rev. B 74, 033401.

Konduri, S., Mukherjee, S., Nair, S., 2007. Controlling nanotube dimensions: correlation between composition, diameter, and internal energy of single-walled mixed oxide nanotubes. ACS Nano 1, 393–402.

Köskinen, P., Mäkinen, V., 2009. Density-functional tight-binding for beginners. Comput. Mat. Sci. 47, 237–253.

Kuc, A., Heine, T., 2009. Shielding nanowires and nanotubes with imogolite: a route to nanocables. Adv. Mater. 21, 4353–4356.

Lee, C.T., Yang, W.T., Parr, R.G., 1988. Development of the Colle-Salvetti correlation-energy formula into a functional of the electron-density. Phys. Rev. B 37, 785–789.

Lee, S.U., Choi, Y.C., Youm, S.G., Sohn, D., 2011. Origin of the strain energy minimum in imogolite nanotubes. J. Phys. Chem. C 115, 5226–5231.

Levard, C.M., Rose, J.R.M., Masion, A., Doelsch, E., Borschneck, D., Olivi, L., Dominici, C., Grauby, O., Woicik, J.C., Bottero, J.-Y., 2008. Synthesis of large quantities of single-walled aluminogermanate nanotube. J. Am. Chem. Soc. 130, 5862–5863.

Levard, C., Rose, J., Thill, A., Masion, A., Doelsch, E., Maillet, P., Spalla, O., Olivi, L., Cognigni, A., Ziarelli, F., Bottero, J.Y., 2010. Formation and growth mechanisms of imogolite-like aluminogermanate nanotubes. Chem. Mater. 22, 2466–2473.

Levard, C., Masion, A., Rose, J., Doelsch, E., Borschneck, D., Olivi, L., Chaurand, P., Dominici, C., Ziarelli, F., Thill, A., Maillet, P., Bottero, J.Y., 2011. Synthesis of Ge-imogolite: influence of the hydrolysis ratio on the structure of the nanotubes. Phys. Chem. Chem. Phys. 13, 14516–14522.

Levy, M., 1982. Electron densities in search of hamiltonians. Phys. Rev. A 26, 1200–1208.

Li, L., Xia, Y., Zhao, M., Song, C., Li, J., Liu, X., 2008. The electronic structure of a single-walled aluminosilicate nanotube. Nanotechnology 19, 175702.

Lourenço, M.P., De Oliveira, C., Oliveira, A.F., Guimarães, L., Duarte, H.A., 2012. Structural, electronic, and mechanical properties of single-walled chrysotile nanotube models. J. Phys. Chem. C 116, 9405–9411.

Lourenço, M.P., Guimarães, L., da Silva, M.C., De Oliveira, C., Heine, T., Duarte, H.A., 2014. Nanotubes with well-defined structure: single- and double-walled imogolites. J. Phys. Chem. C 118, 5945–5953.

Lvov, Y.M., Shchukin, D.G., Mohwald, H., Price, R.R., 2008. Halloysite clay nanotubes for controlled release of protective agents. ACS Nano 2, 814–820.

Ma, W., Kim, J., Otsuka, H., Takahara, A., 2011. Surface modification of individual imogolite nanotubes with alkyl phosphate from an aqueous solution. Chem. Lett. 40, 159–161.

Ma, W., Yah, W.O., Otsuka, H., Takahara, A., 2012. Surface functionalization of aluminosilicate nanotubes with organic molecules. Beilstein J. Nanotechnol. 3, 82–100.

Maillet, P., Levard, C.M., Larquet, E., Mariet, C., Spalla, O., Menguy, N., Masion, A., Doelsch, E., Rose, J.R.M., Thill, A., 2010. Evidence of double-walled Al-Ge imogolite-like nanotubes. A cryo-TEM and SAXS investigation. J. Am. Chem. Soc. 132, 1208–1209.

Maillet, P., Levard, C., Spalla, O., Masion, A., Rose, J., Thill, A., 2011. Growth kinetic of single and double-walled aluminogermanate imogolite-like nanotubes: an experimental and modeling approach. Phys. Chem. Chem. Phys. 13, 2682–2689.

Mukherjee, S., Bartlow, V.A., Nair, S., 2005. Phenomenology of the growth of single-walled aluminosilicate and aluminogermanate nanotubes of precise dimensions. Chem. Mater. 17, 4900–4909.

Oliveira, A.F., Seifert, G., Heine, T., Duarte, H.A., 2009. Density-functional based tight-binding: an approximate DFT method. J. Braz. Chem. Soc. 20, 1193–1205.

Parr, R., Yang, W., 1994. Density-Functional Theory of Atoms and Molecules. Oxford Science Publications, New York.

Pauling, L., 1930. The structure of the chlorites. Proc. Natl. Acad. Sci. U.S.A. 16, 578–582.

Perdew, J.P., Chevary, J.A., Vosko, S.H., Jackson, K.A., Pederson, M.R., Singh, D.J., Fiolhais, C., 1992. Atoms, molecules, solids, and surfaces: applications of the generalized gradient approximation for exchange and correlation. Phys. Rev. B 46, 6671–6687.

Perdew, J.P., Burke, K., Ernzerhof, M., 1996. Generalized gradient approximation made simple. Phys. Rev. Lett. 77, 3865–3868.

Porezag, D., Frauenheim, T., Kohler, T., Seifert, G., Kaschner, R., 1995. Construction of tight-binding-like potentials on the basis of density-functional theory: application to carbon. Phys. Rev. B 51, 12947–12957.

Rawtani, D., Agrawal, Y.K., 2012. Multifarious applications of halloysite nanotubes: a review. Rev. Adv. Mater. Sci. 30, 282–295.

Seby, F., Potin-Gautier, M., Giffaut, E., Donard, O.F.X., 2001. A critical review of thermodynamic data for inorganic tin species. Geochim. Cosmochim. Acta 65, 3041–3053.

Seifert, G., Porezag, D., Frauenheim, T., 1996. Calculations of molecules, clusters, and solids with a simplified LCAO-DFT-LDA scheme. Int. J. Quant. Chem. 58, 185–192.

Seifert, G., Kohler, T., Hajnal, Z., Frauenheim, T., 2001. Tubular structures of germanium. Solid State Commun. 119, 653–657.

Slater, J.C., 1951. A simplification of the Hartree–Fock method. Phys. Rev. 1, 385–390.

Tamura, K., Kawamura, K., 2002. Molecular dynamics modeling of tubular aluminum silicate: imogolite. J. Phys. Chem. B 106, 271–278.

Teobaldi, G., Beglitis, N.S., Fisher, A.J., Zerbetto, F., Hofer, A.A., 2009. Hydroxyl vacancies in single-walled aluminosilicate and aluminogermanate nanotubes. J. Phys. Condens. Matter 21, 195301.

Thill, A., Guiose, B., Bacia-Verloop, M., Geertsen, V., Belloni, L., 2012a. How the diameter and structure of (OH)(3)Al2O3SixGe1-xOH imogolite nanotubes are controlled by an adhesion versus curvature competition. J. Phys. Chem. C 116, 26841–26849.

Thill, A., Maillet, P., Guiose, B., Spalla, O., Belloni, L., Chaurand, P., Auffan, M., Olivi, L., Rose, J., 2012b. Physico-chemical control over the single- or double-wall structure of aluminogermanate imogolite-like nanotubes. J. Am. Chem. Soc. 134, 3780–3786.

Veerabadran, N.G., Price, R.R., Lvov, Y.M., 2007. Clay nanotubes for encapsulation and sustained release of drugs. Nano 2, 115–120.

Vosko, S.H., Wilk, L., Nusair, M., 1980. Accurate spin-dependent electron liquid correlation energies for local spin-density calculations—a critical analysis. Can. J. Phys. 58, 1200–1211.

Wada, S., Wada, K., 1982. Effects of substitution of germanium for silicon in imogolite. Clays Clay Miner. 30, 123–128.

Wang, L., Lu, A., Wang, C., Li, X., Zheng, X., Zhao, D., Liu, R., 2006a. Porous properties of nano-fibriform silica from natural chrysotile. Acta Geol. Sin. Engl. Ed. 80, 180–184.

Wang, L.J., Lu, A.H., Wang, C.Q., Zheng, X.S., Zhao, D.J., Liu, R., 2006b. Nano-fibriform production of silica from natural chrysotile. J. Colloid Interface Sci. 295, 436–439.

Wang, L., Lu, A., Xiao, Z., Ma, J., Li, Y., 2009. Modification of nano-fibriform silica by dimethyldichlorosilane. Appl. Surf. Sci. 255, 7542–7546.

Yada, K., 1967. Study of chrysotile asbestos by a high resolution electron microscope. Acta Crystallogr. 23, 704–707.

Yada, K., 1971. Study of microstructure of chrysotile asbestos by a high resolution electron microscope. Acta Crystallogr. A 27, 659–664.

Yucelen, G.I., Choudhury, R.P., Vyalikh, A., Scheler, U., Beckham, H.W., Nair, S., 2011. Formation of single-walled aluminosilicate nanotubes from molecular precursors and curved nanoscale intermediates. J. Am. Chem. Soc. 133, 5397–5412.

Yucelen, G.I., Kang, D.-Y., Guerrero-Ferreira, R.C., Wright, E.R., Beckham, H.W., Nair, S., 2012. Shaping single-walled metal oxide nanotubes from precursors of controlled curvature. Nano Lett. 12, 827–832.

Yucelen, G.I., Kang, D.-Y., Schmidt-Krey, I., Beckham, H.W., Nair, S., 2013. A generalized kinetic model for the formation and growth of single-walled metal oxide nanotubes. Chem. Eng. Sci. 90, 200–212.

Zanzottera, C., Vicente, A., Celasco, E., Fernandez, C., Garrone, E., Bonelli, B., 2012. Physicochemical properties of imogolite nanotubes functionalized on both external and internal surfaces. J. Phys. Chem. C 116, 7499–7506.

Zeitler, T.R., Greathouse, J.A., Gale, J.D., Cygan, R.T., 2014. Vibrational analysis of brucite surfaces and the development of an improved force field for molecular simulation of interfaces. J. Phys. Chem. C 118, 7946–7953.

Zhao, M., Xia, Y., Mei, L., 2009. Energetic minimum structures of imogolite nanotubes: a first-principles prediction. J. Phys. Chem. C 113, 14834–14837.

Part III

Synthesis of Nanosized Tubular Clay Minerals

Chapter 15

Why a 1:1 2D Structure Tends to Roll?

A Thermodynamic Perspective

L. Belloni[a,*] and A. Thill[b]
[a]*LIONS, NIMBE, CEA, CNRS, Université Paris-Saclay, CEA/Saclay, Gif-sur-Yvette, France*
[b]*Laboratoire Interdisciplinaire sur l'Organisation Nanometrique et Supramoleculaire, NIMBE, CEA, CNRS, University Paris-Saclay, Paris, France*
[*]*Corresponding author: e-mail: luc.belloni@cea.fr*

15.1 INTRODUCTION

This theoretical chapter tries to answer the following questions: What is the origin of the spontaneous curvature of the two-dimensional (2D) nanosized tubular structure? What are the relevant parameters that control the radius and length of the cylinders? What is the number of concentric walls in the most favourable geometry? What is the shape of the phase diagram concentration vs temperature? The privilege of the theoretical work is to invoke concepts and approaches that are quite general and can be applied to a wide spectrum of fields and applications, from natural to synthetic systems, from organic to inorganic materials, from surfactant micellar or lamellar objects up to the present nanosized tubular clay mineral superstructures. This chapter will focus on the equilibrium, structural and thermodynamical properties of imogolite nanotube dispersions. So, right now, the thermodynamic equilibrium is assumed to be reached and the (otherwise interesting) out-of-equilibrium and kinetic properties will be ignored. That statement is a strong one: while it is easy to understand that the equilibrium is obtained without difficulty for 'soft' surfactant systems, where the very mobile monomers can be exchanged very quickly between different nanometric objects, it is more difficult to accept the same assumption for 'hard', chemically reactive mineral organisations. The shape of the product observed at a given moment may depend on how the synthesis took place, the time of maturation, the response to modified external constraints (concentration, salinity, etc.) and the shift from single-walled (SW) to double-walled (DW) geometries (or vice versa)

Developments in Clay Science, Vol. 7. http://dx.doi.org/10.1016/B978-0-08-100293-3.00015-7

may require weeks or months. Again, these temporary regimes will be disregarded, and this chapter will deal *only* with the final equilibrium properties.

What kind of theory is being discussed here? The attentive reader has noticed that in this short introduction, the analogy with surfactant systems already has been invoked twice. This is not by chance: most of the theory presented in this chapter for mineral nanotubes borrows from the concepts introduced in the late 1970s for the self-assembly of surfactant architectures (Israelachvili et al., 1976), particularly cylindrical micelles. Rather than describing the clay mineral and the surrounding solvent molecules at the atomistic level with classical force fields and to perform long and heavy numerical simulations (see Chapter 14), the approach here will be performed at a more mesoscopic level of description, ignoring explicitly the details of the imogolite and of the solvent and replacing them implicitly with free energy expressions, bending modulus, surface tension asymmetry and adhesiveness. The price to pay with this coarse scale is that these different quantities are somewhat phenomenological parameters that must be input. The reward, however, is that the scale is able and sufficient to capture the whole thermodynamics of nanosized tubular colloids.

Section **15.2** deals with the energy (per unit of materials) of isolated, infinitely long nanotubes of different cross sections and determines which structure is the most favourable (ie, has the lowest energy). Section **15.3** investigates an imogolite dispersion and calculates its free energy, internal energy and entropy of mixing. The resulting complete phase diagram concentration–temperature is described in Section **15.4**, highlighting which nanosized geometry dominates for given applied conditions.

15.2 EQUILIBRIUM ENERGY OF A SINGLE NANOTUBE

This section details the theory presented by Thill et al. (2012b). The goal is to write the energy of an isolated imogolite nanoparticle of different possible shapes: 'SW', 'DW' or, more generally, 'Multi-walled' (MW), with $m=1$, 2, 3, ..., concentric walls (noted in the following mW) and scroll (SC). Each structure is described as a weakly flexible, homogeneous solid (corresponding to an imogolite layer) of thickness $2h$ ($h \approx 2$ Å) rolled into one, two or m concentric cylinders of a thick circular section or into a scroll of a thick spiral section. Following Guimares et al. (2007), each face is subjected to an interfacial tension that is much higher for one of the two faces, which will spontaneously become the internal face of the rolled layer. In the case of imogolite, the excess interfacial tension is due to the size mismatch between the silicon or germanium tetrahedra compared to the adsorption site on the gibbsite sheet. Similar excess interfacial tensions have been described for other minerals, such as titanate nanotubes (Zhang et al., 2003) and VO_x nanotubes (Krumeich et al., 1999). The spontaneous curvature is thus the result of the balance between the bending rigidity of the layer and the interfacial tension dissymmetry. Note that, in this section, the length L of the nanotubes is not a relevant parameter since, in order to avoid the penalty energy to form edges, infinitely long cylinders are chosen and one plays with the energy per unit of mass or per unit of total surface.

The following analytical expressions for the energy contain three terms, all implicitly of a free energy nature:

- *Elastic energy,* which favours flat layers. In *Materials Physics*, the energy per unit surface of sheet is $\frac{1}{2}YI/R^2$, where Y is the Young modulus; I the inertia moment, equal here to $2/3Lh^3$; and R is the local radius of curvature, measured at the centre of the layer, so $Yh^3/3R^2$. The total energy of the structure is then obtained by integrating along the curved trajectory s of the cylinder section and by multiplying by the nanotube length L. The Young modulus Y, in the range 200 GPa, accounts for the cohesiveness of the clay mineral and depends on its chemical composition.
- *Interfacial energy* is simply expressed as $\sigma_{ext} \times S_{ext}$ ($\sigma_{int} \times S_{int}$) for the external (internal) face of the layer (Fig. 15.1), where σ_i and S_i are the surface tension and the area of the water–solid interface, respectively. Obviously, $\sigma_{ext} < \sigma_{int}$! Once again, these somewhat phenomenological parameters depend on the surface chemical compositions and on the coupling with the surrounding solvent. That energy contribution will favour large curvatures.
- *Attraction between two facing layers* is the last contribution for MW or SC nanotubes. It imposes a fixed water thickness e between the adjacent external and internal faces and a negative energy $-U_{adh} \times S_{mean}$ ($U_{adh} > 0$), where S_{mean} represents the average area between the facing surfaces (Thill et al., 2012b). A good candidate for the physical origin of this attraction is the electrostatic interaction between two parallel surfaces carrying charges of opposite sign, especially in the asymmetrical regime, where the absolute surface charge densities differ, $Z_+ \neq |Z_-|$ (Paillusson et al., 2011). In planar or cylindrical geometry, the Poisson–Boltzmann (PB) theory exhibits a nonmonotonous pressure Π vs separation d law: at a large separation (much larger than the Debye length), the pressure is negative and proportional to the two (effective) charge densities, and follows a screened coulombic law. As d decreases, the excess salt ions are expelled from the water gap. At short distances, only the

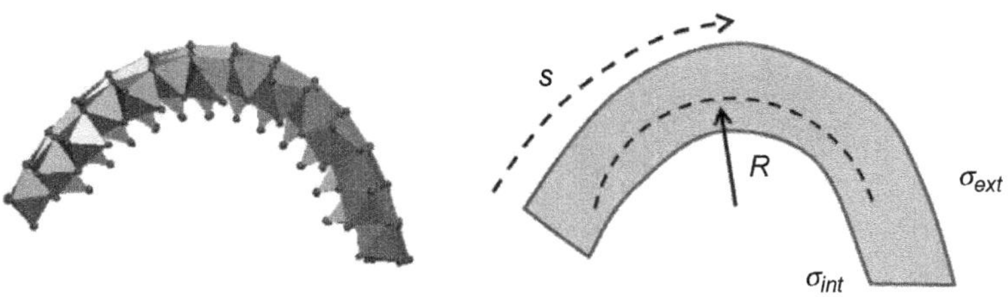

FIG. 15.1 Cross section of the layer. (*Left*) Atomic model. (*Right*) Continuous model.

counterions remain, which must be present in order to equilibrate the balance between the positive and the negative surfaces. The positive osmotic pressure of these ions dominates. Eventually, Π behaves as $|Z_+ + Z_-|/d$ when d becomes shorter than the Gouy–Chapman length. Therefore, the pressure and free energy curves present a negative minimum for a given gap that defines the parameters e and U_{adh}. Fig. 15.2 gives an example of PB calculation in the cylindrical geometry. In practice, $e \approx 2$ Å, $U_{adh} \approx 0.01$ J/m^2.

These three contributions are added and the resulting energy divided by the nanotube solid volume V to define the energy per unit volume, or by the total surface S of the layer ($V = S2h$) to obtain the energy by unit surface, hereafter simply designated as E. The minimisation of E with respect to the size of the

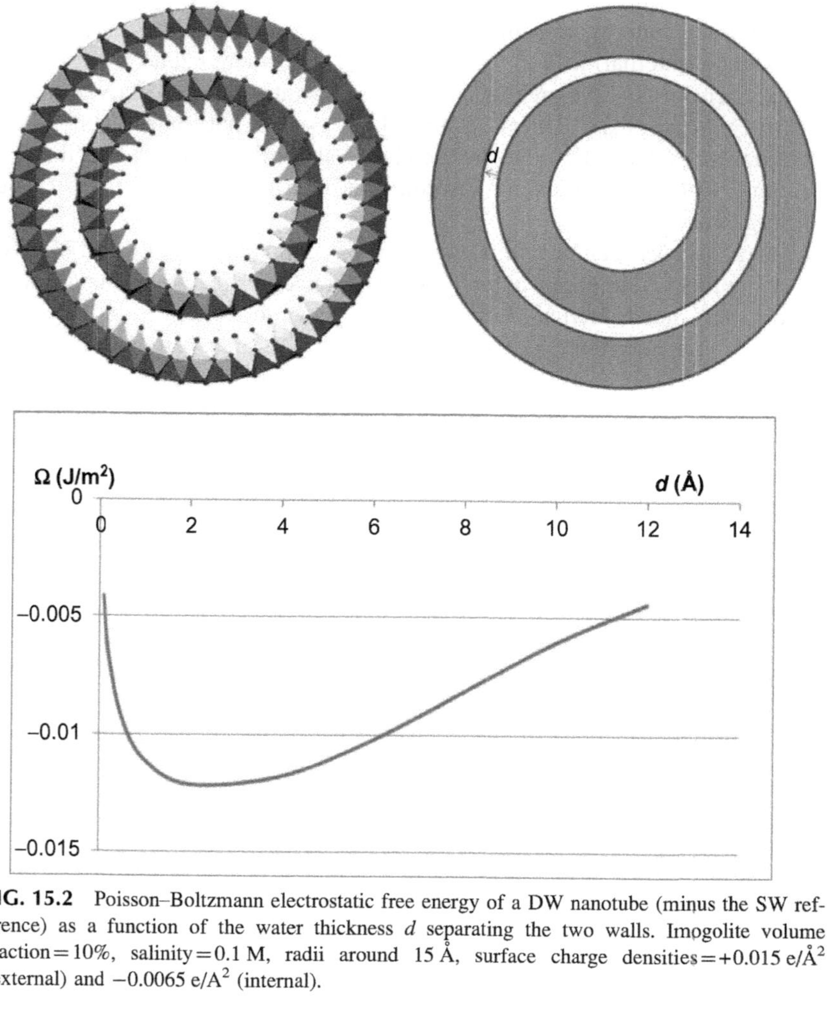

FIG. 15.2 Poisson–Boltzmann electrostatic free energy of a DW nanotube (minus the SW reference) as a function of the water thickness d separating the two walls. Imogolite volume fraction = 10%, salinity = 0.1 M, radii around 15 Å, surface charge densities = +0.015 e/Å^2 (external) and −0.0065 e/A^2 (internal).

cylinder section and to the number of walls m in the MW shape or the number of revolutions n in the SC will lead to the most favourable geometry (ie, lowest energy).

15.2.1 Single-Walled (SW) Case

For the SW, the situation is very simple: The radius of curvature of the section coincides everywhere with the radius R of the cylinder and the external/internal radii are $R \pm h$. Thus:

$$\text{Energy} = \frac{Yh^3}{3R^2} 2\pi RL + \sigma_{ext} 2\pi(R+h)L + \sigma_{int} 2\pi(R-h)L \tag{15.1}$$

Dividing by the total surface $S = 2\pi RL$ leads to the desired expression for E:

$$E = \frac{Yh^3}{3R^2} - \frac{\sigma h}{R} + \Sigma \tag{15.2}$$

where

$$\begin{aligned} \Sigma &= \sigma_{ext} + \sigma_{int} \\ \sigma &= \sigma_{int} - \sigma_{ext} > 0 \end{aligned} \tag{15.3}$$

Note that E is a harmonic function of the curvature $1/R$. The surface tension sum Σ brings a constant shift to the energy and is reproduced in all further expressions. This irrelevant factor thus has no impact on the competition between the different structures and will be ignored in the following discussion. On the other hand, the surface tension difference σ is the key factor that determines the spontaneous curvature. Eq. (15.2) can be expressed as

$$E = E_0 + \frac{1}{2} K_C \left(\frac{1}{R} - \frac{1}{R_0} \right)^2 \tag{15.4}$$

$$E \approx E_0 + \frac{1}{2} K (R - R_0)^2 \tag{15.5}$$

with the radius R_0, the equilibrium energy E_0 at the minimum and the binding modulus K are given by

$$\begin{aligned} R_0 &= \frac{2Yh^2}{3\sigma} \\ E_0 &= -\frac{3\sigma^2}{4Yh} \\ K &= \frac{K_C}{R_0{}^2} = \frac{27\sigma^4}{8Y^3h^5} \end{aligned} \tag{15.6}$$

This simple analytical expression given in Eq. (15.4) is able to quantitatively capture the energy curve $E(R)$ (Fig. 15.3) extracted from sophisticated atomistic numerical simulations for a variety of clay mineral systems

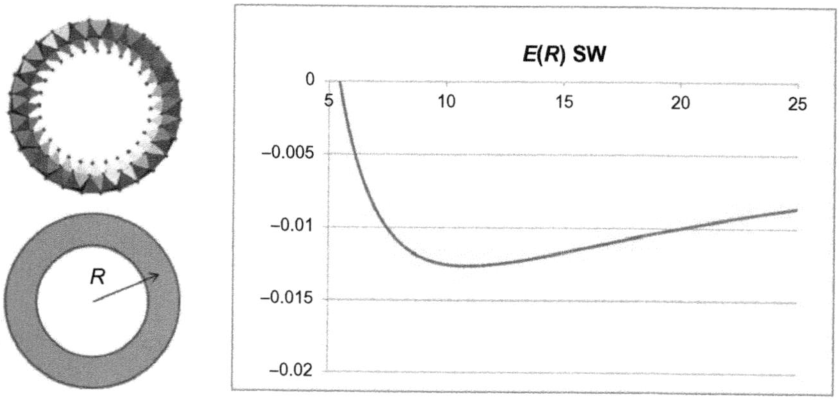

FIG. 15.3 Cross section of an SW imogolite and the corresponding continuous model. Energy curve $E(R)$ for the SW structure (R in Å).

(Guimares et al., 2007). The harmonic expression in R [Eq. (15.5)] is only valid close to the minimum. The experimental value $R_0 \approx 15$ Å corresponds to $\sigma \approx 4$ N/m and $K_C \approx 200\ kT$, where the thermal energy kT will 'measure' all the contributions, as discussed in Section **15.3** later in this chapter. It is interesting to note that the corresponding modulus K_C in surfactant systems lies in the range 1 kT.

15.2.2 Double-Walled (DW) Case

The section of the nanotube now contains two concentric thick circles of radii R_1 and R_2 ($R_1 < R_2$, measured at the centre of each layer). Since a water layer of thickness e is imposed between them, $R_2 - R_1 = e + 2h$, noted as $2a$ in the following, which defines the new parameter $a = e/2 + h$. The total energy and energy per unit area become

$$\begin{aligned}
&\text{Energy} = \frac{Yh^3}{3}\left(\frac{1}{R_1} + \frac{1}{R_2}\right)2\pi L + (\Sigma(R_1 + R_2) - 2\sigma h)2\pi L - U_{adh}\frac{R_1 + R_2}{2}2\pi L \\
&S = (R_1 + R_2)2\pi L \\
&E = \frac{Yh^3}{3R_1R_2} - \frac{2\sigma h}{R_1 + R_2} - \frac{U_{adh}}{2}
\end{aligned} \tag{15.7}$$

This time, the two radii cannot match simultaneously the previous equilibrium value R_0. The price to pay this extra binding energy must be counterbalanced by a strong enough surface–surface attraction. Note the coefficient ½ in front of this U_{adh} contribution: one attraction for two surfaces.

The minimisation of E [Eq. (15.7)] with respect to the mean radius $R = ½(R_1 + R_2)$ ($R_1 = R - a$ and $R_2 = R + a$) at constant a leads to the following equation for the DW equilibrium radius R_{min}:

$$\frac{(R_1R_2)^2}{R^3} = R_0 \tag{15.8}$$

This is a fourth-degree polynomial. Assuming that $a \ll R_0$ (a thin water layer) allows expanding the solution in powers of the first dimensionless parameter $\alpha = a/R_0$:

$$R_{min} \approx R_0\left[1 + 2\alpha^2 - 5\alpha^4 + 24\alpha^6 \ldots\right] \tag{15.9}$$

The energy at the minimum is

$$E_{min}(\mathrm{DW}) = E_0 + \frac{Ka^2}{2}\left[1 - 3\alpha^2 + 13\alpha^4 \ldots\right] - \frac{U_{adh}}{2} \tag{15.10}$$

15.2.3 Multi-Walled (MW) Case

The generalisation presented here is straightforward: the cross section of the nanotube is made of m concentric thick circles of increasing radii $R_1 < R_2 < \cdots < R_m$, with the constant gap $2a$ between two successive radii (Fig. 15.4). The adhesion energy between walls i and $i+1$ reads $-U_{adh}\frac{1}{2}(R_i + R_{i+1})2\pi L$. If R still represents the mean radius of the ensemble ($R = 1/m\Sigma_i R_i$), the radii are expressed as $R_i = R + (2i - m - 1)a$; $i = 1$, m (the extreme values being $R \pm (m-1)a$ in particular) and the energy E per total unit area becomes

$$E = \frac{Yh^3}{3}\frac{\frac{1}{m}\sum_{i=1}^{m}\frac{1}{R_i}}{R} - \frac{\sigma h}{R} - \frac{m-1}{m}U_{adh} \tag{15.11}$$

This time, there are $(m-1)$ attractions for m layers. One gains adhesion by increasing the number m of layers, but at the same time, one loses elastic energy because the different radii must deviate more and more from the

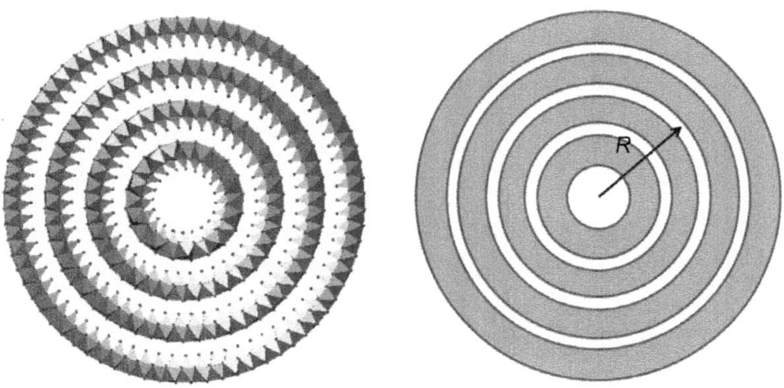

FIG. 15.4 Cross section of an MW imogolite and the corresponding MW continuous model.

minimum value R_0. The optimum geometry results from a balance between these two effects.

Minimisation with respect to R and m requires a numerical procedure in general. Meanwhile, a few useful and interesting analytical expressions can be derived in some limited cases.

15.2.3.1 ma≪R_0, *thin MW*

First, if the gap a is so small that all radii remain close to R_0 (namely, $m\alpha \ll 1$, a more severe assumption than for the DW case), the expression of E can be expanded as before and cut at the second order. The minimum is obtained at the radius R_{min} (fixed m):

$$R_{min} = R_0\left[1+\frac{2}{3}\left(m^2-1\right)\alpha^2\ldots\right] \tag{15.12}$$

and the mW equilibrium energy reads as

$$\begin{aligned} E_{min}(m\mathrm{W}) &= E_0+\frac{Ka^2}{2}\left[\frac{m^2-1}{3}\ldots\right]-\frac{m-1}{m}U_{adh} \\ E_{min}(m\mathrm{W}) &= E_0-U_{adh}+\frac{Ka^2}{2}\left[\frac{m^2-1}{3}\ldots+\frac{2{n_0}^3}{3m}\right] \end{aligned} \tag{15.13}$$

where a second dimensionless factor, n_0, has been advantageously introduced in the last line:

$$n_0 = \left(\frac{3U_{adh}}{Ka^2}\right)^{1/3} \tag{15.14}$$

which measures the adhesion in terms of intrinsic binding energy and quantifies the previously evoked balance. The preferred mW will be located in the α, n_0 phase diagram.

Finally, the minimum of the energy [Eq. (15.13)] with respect to m is obtained at $m=n_0$ (that was the motivation for that precise definition of n_0!). Since m must be an integer, contrary to the parameter n_0, one concludes that the most favourable mW corresponds to the nearest integer m_{min} of n_0. The energy of this optimum mW is

$$\mathrm{E}_{\mathrm{MW}}(m_{min}) \approx E_0-U_{adh}+\frac{Ka^2}{2}\left[{n_0}^2-\frac{1}{3}+\left(m_{min}-n_0\right)^2\right] \tag{15.15}$$

Note that the term in brackets always remains slightly below ${n_0}^2$, even when n_0 is a half-integer. Again, this limiting law is valid in the domain $n_0\alpha \ll 1$.

15.2.3.2 R_0≪a, *thick single layer*

This case, which is the opposite of the previous one, is much less realistic: the water thickness is much greater than the SW radius, $\alpha \gg 1$. The SW wins for

$\alpha^2 n_0{}^3 < 3/2$, and an infinity of circles (starting at $R_1 = \sqrt{\frac{1}{2} a R_0}$) wins for $\alpha^2 n_0{}^3 > 3/2$.

15.2.3.3 $R_0 \gg a$ and $R_0 \ll ma$ (so $m \gg 1$), thin single layer, but thick whole MW

That is a more realistic case. A single water layer is thin compared to the SW radius, but the juxtaposition of many concentric sheets motivated by a strong attraction leads to a total thickness $2ma$ that is much larger than R_0. The discrete sum of the curvatures, $\Sigma_i 1/R_i$, then can be replaced by a continuous integral, $\int di/R_i = \ln(R_m/R_1)/(2a)$. This special type of MW becomes strictly equivalent to a long nanoscroll, which is the subject of the next discussion.

15.2.4 Scroll (SC) Case

In the case of a nanoscroll SC (Thill et al., 2012a), the curve in the plane perpendicular to the scroll axis, which describes the centre of the layer, is a spiral expressed by the equation $r = r_0 + k\theta$ in polar coordinates r, θ (Fig. 15.5). The spiral constant k is determined by imposing again a fixed gap $2a$ between two successive revolutions, $k2\pi = 2a$ (the water thickness is defined less rigorously than in the MW case because the two normals to the facing surfaces do not coincide exactly). The spiral is characterised, for instance, by the extreme radii r_0 and $r_1 = r_0 + k\theta_1 = r_0 + 2na$, or by r_0 and the total number of

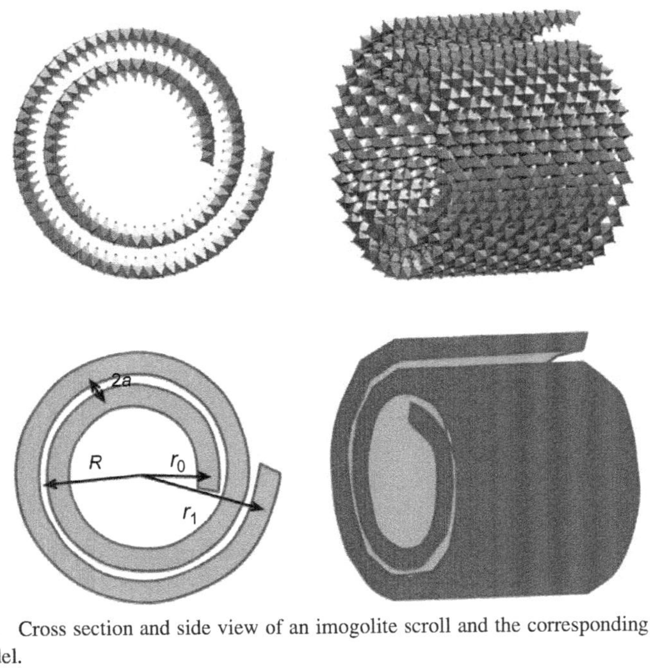

FIG. 15.5 Cross section and side view of an imogolite scroll and the corresponding continuous scroll model.

revolutions $n = \theta_1/2\pi$, which may not be an integer. In polar notation, the local radius of curvature of the spiral reads as $(r^2+k^2)^{3/2}/(r^2+2k^2)$, and the local length element $ds = (r^2+k^2)^{1/2}d\theta$. The integration on the total length of the scroll gives

$$E = \frac{\left\{ \begin{array}{l} \dfrac{Yh^3}{3}\dfrac{1}{k}\left[\sinh^{-1}\left(\dfrac{r}{k}\right) + \dfrac{8}{3}\dfrac{r}{(r^2+k^2)^{1/2}} + \dfrac{k^2}{3}\dfrac{r}{(r^2+k^2)^{3/2}}\right]_{r_0}^{r_1} \\ -\sigma h\dfrac{1}{k}\left[r + k\tan^{-1}\dfrac{r}{k}\right]_{r_0}^{r_1} - U_{adh}\dfrac{r_0+r_1}{2}(\theta_1 - 2\pi) \end{array} \right\}}{\dfrac{k}{2}\left[\sinh^{-1}\left(\dfrac{r}{k}\right) + \dfrac{r(r^2+k^2)^{1/2}}{k^2}\right]_{r_0}^{r_1}} \quad (15.16)$$

Fortunately, this complicated expression can be simplified very easily by assuming that the spiral grows slowly—namely, that $k \ll r$ everywhere. This approximation, equivalent to $\alpha \ll 1$, is valid in practice (Thill et al., 2012a). In that limit, polar radius and radius of curvature coincide and the energy E becomes

$$E \approx \frac{Yh^3}{3}\frac{\dfrac{1}{2na}\ln\dfrac{r_1}{r_0}}{R} - \frac{\sigma h}{R} - \frac{n-1}{n}U_{adh} \quad (15.17)$$

This is a function of the mean radius $R = (r_0 + r_1)/2$ and of n (or the total thickness of the scroll $2A$ with $A = na$). It is instructive to compare formally the energy of an mW [Eq. (15.11)] to that of an SC [Eq. (15.17)] of the same mean radius R and with a number of revolutions n identical to m. The only formal difference concerns the elastic contribution. For the spiral, the logarithm term is nothing but the integral of $1/r$ over the interval $[r_0, r_1] = [R - na, R + na]$. For mW, it is replaced by the sum over the discrete array $r_0 + a$, $r_0 + 3a$, …, $r_1 - 3a$, $r_1 - a$ of mesh size $2a$. This classical, 'open' formula, which involves the internal half-integer abscissa, is known as the 'extended midpoint rule' in the domain of numerical quadrature. For the convex $1/r$ function, it slightly underestimates the integral. As a consequence, we can immediately state that mW always will be slightly more favourable than an SC of the same section radius and length.

In general, the minimisation with respect to R and n requires a numerical procedure. Once again, a few interesting limiting cases will be explored more deeply next.

15.2.4.1 $na \ll R_0$

As for mW, the total thickness of the scroll $2A$ is smaller than R_0, $\beta \equiv n\alpha \ll 1$. Expansion in β gives, for the optimal size of the scroll:

$$R_{min} \approx R_0\left[1 + \frac{2}{3}\beta^2 - \frac{13}{45}\beta^4 \ldots\right] \quad (15.18)$$

In addition, the minimal energy (at fixed n) reads as

$$E_{min} = E_0 + \frac{KA^2}{2}\left[\frac{1}{3} - \frac{11}{45}\beta^2\ldots\right] - U_{adh}\left(1 - \frac{1}{n}\right) \tag{15.19}$$

Finally, minimisation of Eq. (15.19) with respect to A or β or n leads to the optimum length n_{min}:

$$n_{min} \approx n_0\left[1 + \frac{22}{45}\gamma^2\ldots\right] \tag{15.20}$$

where $\gamma \equiv n_0\ \alpha = (3U_{adh}a/\sigma h)^{1/3}$ is independent of Y. Once again, the characteristic number of loops n_0 appears naturally. The minimal energy of the winning nanoscroll is

$$\begin{aligned} E_{min}(SP) &= E_0 - U_{adh} + \frac{K{A_0}^2}{2}\left[1 - \frac{11}{45}\gamma^2\ldots\right] \\ &= E_0 - U_{adh} + \frac{Ka^2}{2}\left[{n_0}^2\ldots\right] \end{aligned} \tag{15.21}$$

It comes as no surprise, then, to learn that the energy of the optimal SC is very close to, but slightly above, that of the optimal mW [Eq. (15.15)], with $n_{min} \approx m_{min} \approx n_0$. Again, it is important to repeat that this expression is valid in the domain $n_0\alpha \ll 1$ only.

15.2.4.2 $R_0 \gg a$ *and* $R_0 \ll na$ *(so* $n \gg 1$*), namely,* $\alpha \ll 1$ *and* $\beta \gg 1$

Here, as said previously, mW and SC (with $n = m$) coincide. At fixed n, the optimal internal radius (r_0 or R_1) behaves as follows:

$$r_0 \approx \frac{R_0}{4}\left[1 + \frac{\ln(8\sqrt{e}\beta)}{4\beta}\right] \tag{15.22}$$

In addition, the minimum energy becomes

$$E_{min}(n) \approx -U_{adh} + \frac{\sigma h}{R_0}\left[\frac{\ln(8\beta)}{4\beta^2} - \frac{1}{\beta}\left(1 - \frac{1}{4\beta}\right) + \frac{\gamma^3}{3\beta}\right] \tag{15.23}$$

A final minimisation with respect to n or β leads to the following implicit equation for the optimum n:

$$\frac{\ln(8\sqrt{e}\beta)}{2\beta} = 1 - \frac{\gamma^3}{3} \tag{15.24}$$

The left side is supposed to be small and positive (since $\beta \equiv n\alpha \gg 1$). That is possible only if the product $\gamma \equiv n_0\alpha$ remains slightly below $\sqrt[3]{3} = 1.44$. Beyond that threshold, the optimal structure contains an infinite number of walls or revolutions, with the first radius starting at $R_0/4$.

A final remark about the scroll case is that the cross section of the Scroll structure is a thick, open curve, which means that its two free edges are in contact with the solvent. This is a fundamental difference between mW and SC which has been ignored up to now. An interface energy penalty of the form $2 \times \sigma_{edge} \times 2h \times L$ should have been added, which does not exist for the closed cross section of MW. Omitting this contribution is valid, provided that $\sigma_{edge}/\sigma \ll 1$ and, less intuitively, $\ll \alpha^2 n_0^3$. On the other hand, the edge surface tension will play a key factor in the entropy of mixing and thermodynamics of an imogolite solution, as described in Section **15.3**.

The complete α, n_0 phase diagram showing the structure of least energy resulting from full numerical minimisation is established in Fig. 15.6. This diagram shows the successive stability regions between SW, DW and so on as the value of n_0 increases at constant α. For a given competition between elasticity and differential interfacial tension, the increase of U is stabilising more and more walls in the nanotubes. As expected from the previous analysis, the structure with m walls is stable at $n_0 \sim m$ in the limit of low α and no domain of winning SC exists in between. As α increases, the different frontiers collapse on the same curve $n_0\alpha = \sqrt[3]{3}$, above which the structure is always losing energy when it adds more and more revolutions or walls through a rolling or stacking mechanism. Thill et al. (2012b) described in detail how such a simple, general phase diagram is able to quantitatively accommodate a variety of real nanotubes made of different types of materials.

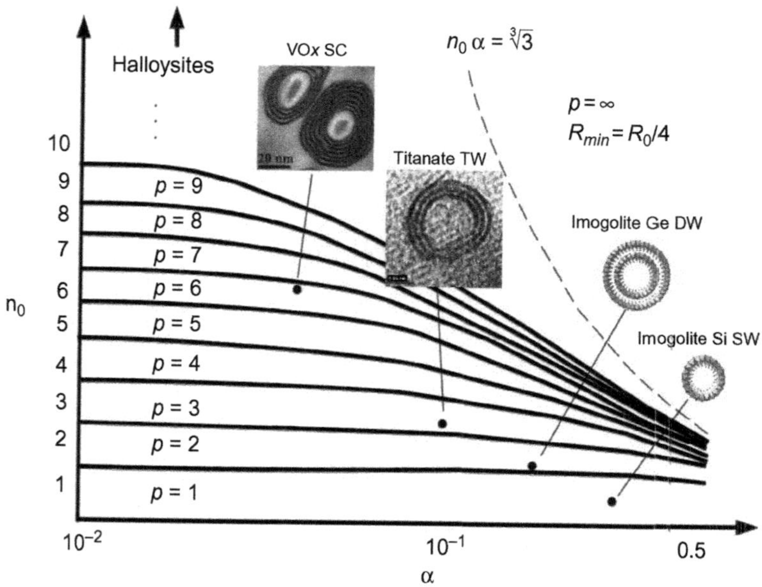

FIG. 15.6 Minimum energy diagram of MW mW nanotubes and SC scrolls as a function of the dimensionless parameters n_0 and α. *Reprinted with permission from Thill et al. (2012b). Copyright 2012 American Chemical Society, where the symbol* p *stands for* m.

15.3 ENTROPY OF THE MIXING OF PLATELETS AND NANOTUBES

Up to now, one was interested in the energy per unit area of a piece of infinitely long MW (or SC) cylinder, and one was seeking the structure and the size of minimal energy. That corresponded to the zero temperature limit, where energy U and free energy F of the solution coincide. Now, the temperature is finite and the entropy S will take place in $F=U-TS$. Thus, not only the structures of minimal energy will be investigated, but also the neighbouring shapes which, albeit less favourable, will count in the total balance thanks to the thermal disorder. One also is going to count the edge surface tension, which penalises the formation of short, finite tubes. In the same way, one will account for platelets or proto-imogolites, these small pieces of cylinders whose cross sections describe less than one entire loop each and thus are open (as a scroll with $n<1$). All these small objects are not favourable energetically (E is higher) but are favourable entropically because the smaller they are, the more numerous they are and the higher is the entropy. That is particularly true when kT is large and when the imogolite concentration is low (the entropy always wins at high dilution). For each temperature-concentration condition, the distribution of different structures/radii/curvatures inside the solution should be calculated, on the condition that all the pieces of imogolite are in thermodynamical equilibrium with each other. The theoretical approach reproduces what has been done, for instance, for the length distribution of surfactant cylindrical micelles (Israelachvili et al., 1976; May and Ben-Shaul, 2001). For simplicity, in the following discussion, the parameters involved in the energy balance (bending, surface tension, adhesion) will be assumed to be independent of the dispersion concentration. This approximation could break down for the adhesion of electrostatic origin when the counterions brought by the imogolites dominate the ionic strength.

15.3.1 Proto-Imogolites

One considers first m-proto-imogolites (called m-protos in the following text) seen as a stack of m pieces of cylinder of length L and of cross section curvilinear length s (averaged over the m layers) (Fig. 15.7). The total area of these layers is thus msL. For simplicity, the radius of curvature is assumed to be uniform along the nanotubes (it would be possible to investigate another source of disorder and entropy by saying that the curvature varies along the surface). The cross section of the object is thus constituted of m arcs of circle of mean radius R. This object is open and does not form a closed loop, so s and R are independent parameters and $s<2\pi R$. The surface energy is simply $msLE_m(R)$, where E_m is the energy per unit area considered in the previous section. As evoked earlier, one must now add the edge energy that comes

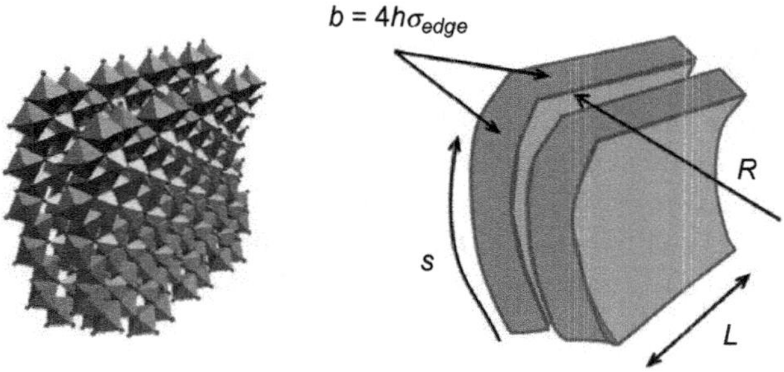

FIG. 15.7 Atomistic model of a DW proto-imogolite and the corresponding continuous model.

from the interface along the perimeter of the object between the layers of thickness $2h$ and the water solvent. With σ_{edge} designating the surface tension of this interface, the edge energy reads $\sigma_{edge}m(2s+2L)2h=mb(s+L)$, where $b=\sigma_{edge}4h$ is a line tension. Finally, the total energy of such m-proto becomes

$$U_m(L,s,R)=msLE_m(R)+mb(s+L) \tag{15.25}$$

One must now consider that the solution is made of a mixture of m-protos of different L, s, R values. One notes $N_m(L,s,R)dLdsdC$ as the concentration of m-protos whose length is L within dL, whose section length is s within ds and whose curvature is $C=1/R$ within dC $(=-dR/R^2)$. One must use dC instead of dR in order to avoid the possible divergences at $R\to\infty$ in the future integrals (which is expected—it is unreasonable to think in terms of concentration per unit of radius when the energy E reaches an asymptote at infinite radius). The free energy (per unit of volume) of this mixture (Israelachvili et al., 1976) now reads:

$$\begin{aligned}F_m(\text{proto})=&\iiint dLdsdCN_m(L,s,R)U_m(L,s,R)\\&+kT\iiint dLdsdCN_m(L,s,R)\left[\ln\frac{N_m(L,s,R)}{N_0}-1\right]\end{aligned} \tag{15.26}$$

The first term represents the internal energy U, and the second term is the entropy of mixing (times $-T$). As usual in such kind of density functional theory, one needs a reference, somewhat nebulous, quantity N_0 in this second term, which has the same unit as the N_m (namely, m^{-4}). The goal is to find the N_m distribution, which minimises the free energy expression [Eq. (15.26)], at a fixed total concentration of imogolite material. More precisely, one fixes the total volume fraction Φ or the total surface concentration of layer $S_{tot}=\Phi/2h$, which sums all surfaces msL:

$$S_{tot}=\iiint dLdsdCN_m(L,s,R)msL \tag{15.27}$$

Technically, one introduces a Lagrange multiplier μ and minimises with respect to the distribution N_m the new 'functional':

$$\Omega_m = F_m - \mu\iiint dLdsdCN_m(L, s, R)msL \tag{15.28}$$

The value of μ is then adjusted such that the total concentration verifies Eq. (15.27). More physically, μ represents simply the *chemical potential* of the imogolite materials. One recognises in Eq. (15.28) a Legendre transform from the canonical ensemble (where S_{tot} is fixed) to the grand-canonical ensemble (where μ is fixed). Minimising (in the sense of a functional) the grand potential (Eq. 15.28) leads to the desired equilibrium distribution:

$$N_m(L, s, R) = N_0 \exp[-\beta(U_m(L, s, R) - \mu msL)] \tag{15.29}$$

where $\beta = 1/kT$. The parameter μ is finally determined by the imposed total concentration (Eq. 15.27):

$$S_{tot} = \iiint dLdsdCN_0 \exp[-\beta m((E_m(R) - \mu)sL + b(s+L))]msL \tag{15.30}$$

Before going further, one can already insert Eq. (15.29) into Eq. (15.26) and obtain the following for the equilibrium free energy:

$$\begin{aligned} F_m &= -kT\iiint dLdsdCN_m(L, s, R) + \mu\iiint dLdsdCN_m(L, s, R)msL \\ F_m &= -kTC_{tot} + \mu S_{tot} \\ F_m &= -\Pi + \mu S_{tot} \end{aligned} \tag{15.31}$$

One recognises the classical expression where Π is the osmotic pressure of the solution containing the total concentration C_{tot} of m-protos. The grand potential (per unit of volume) Ω_m [Eq. (15.28)] equals $-\Pi$, as it should.

The main technical difficulty in the following discussion will be performing the triple integral [Eq. (15.30)] in order to determine μ as a function of the concentration S_{tot}. In short, the bounds are 0 and + infinity. Without losing generality, one could specify that the internal radius R_1 of the m-proto structure is automatically larger than h, so $C < 1/(h + ma)$ (beyond $E = +\infty$). In the same way, the section remains open, so $s < 2\pi R$. Finally, one could consider that the material layer is in fact constituted of elementary chemical patterns, the 'monomers', of size L_1 and s_1, and that the integral in L and s are replaced by discrete sums $L = iL_1$, $s = js_1$, i, j integers. In that case, $L_1 s_1 \mu$ represents *exactly* the chemical potential of these monomers. This discrete representation is interesting because it naturally introduces the notion of critical micellar concentration (*cmc*) but adds new parameters. So long as the true values of L_1 and s_1 (a few Å) remain 'small' (ie, smaller than the true L and R), one can ignore these subtleties and stay with the continuous representation (see the end of this proto-imogolite subsection for the dilute regime where the discrete monomeric description applies).

Two important points in Eq. (15.30) should be noted. First, μ *must* stay below all possible energies $E_m(R)$, so $\mu \leq E_{min,m}$ (if not, the coefficient in sL inside the exponent would be negative and the integral would diverge). The closer μ is to $E_{min,m}$ (from below), the higher the concentration (and vice versa). Second, nothing *here* (and here only) prohibits μ to reach exactly its upper bound $E_{min,m}$, the integrals in L and s staying convergent thanks to the edge penalty b (this is specific to the protos, where the two dimensions play a role simultaneously) (Israelachvili et al., 1976)! One can conclude now that the concentration of p-protos is *bound*. (Hopefully, that will be different for the MW nanotubes.)

For the integral in the curvature, one will assume that the energy per unit area E_m is harmonic in C, $E_m = E_{min,m} + ½K_{c,m}(C - C_{min,m})^2$, with $C_{min,m} = 1/R_{min,m}$ and $K_{c,m} = K_m R_{min,m}{}^4$. This is exact for the SW [see Eq. (15.4)], and this is reasonable for the mW, $m > 1$. The function in C is Gaussian, whose integral on 0–∞ gives [provided that $R_{min,m}$ is much larger than the radius root-mean-squared deviation $(\beta m K_m sL)^{-½}$, which is the case for the present 'hard' materials, except maybe in the marginal limit of very small objects]:

$$S_{tot} = \iint dL ds N_0 \sqrt{\frac{2\pi}{\beta m s L K_{c,m}}} \exp\left[-\beta m\left(\left(E_{\min,m} - \mu\right) sL + b(s+L)\right)\right] msL \tag{15.32}$$

The integral in L gives

$$\begin{aligned} S_{tot} &= \frac{N_0 \pi m}{(\beta m)^2 \sqrt{2K_{c,m}}} \int \frac{e^{-\beta mbs}\sqrt{s}}{[(E_{min,m} - \mu)s + b]^{3/2}} ds \\ S_{tot} &= \frac{N_0 \pi m}{(\beta m)^2 \sqrt{2K_{c,m}} (\beta mb^2)^{3/2}} G\left(x_m \equiv \frac{\beta mb^2}{E_{min,m} - \mu}\right) \end{aligned} \tag{15.33}$$

where G, an increasing function of x, accounts for the final integral in s:

$$G(x) = x^{3/2} \int_0^\infty \frac{e^{-xt}\sqrt{t}}{(t+1)^{3/2}} dt \tag{15.34}$$

The upper bound is ∞ (while $s < 2\pi R_{min,m}$), so it is implicitly assumed that the edge energy at the ends of the nanotubes is large compared to kT for a complete loop, $\beta mbR_{min,m} \gg 1$, which is valid in practice. The function $G(x)$ behaves as $x^{3/2}|\ln x|$ at small x and reaches the asymptote $\sqrt{\pi/2}$ at large x, which quantifies the maximum concentration of m-protos, $S^{max}_{proto,m}$, evoked previously. Note that this one varies as $1/m^{5/2}$. If the solution were constituted of m-protos only, it would be sufficient to invert the function G in Eq. (15.33) in order to express the chemical potential μ in the function of the concentration.

To conclude on the m-proto-imogolites, the mean size and surface of the objects, $\langle L \rangle = \langle s \rangle$ (number average) and $\langle Ls \rangle$, can be evaluated. At small x and large x, respectively:

$$\langle L\rangle_m = \langle s\rangle_m \approx \frac{1}{\beta mb|\ln x_m|};\quad \langle Ls\rangle_m \approx \frac{1}{2\beta m(E_{min,m}-\mu)} \ll \langle L\rangle_m \langle s\rangle_m;\quad x_m \ll 1$$
$$\langle L\rangle_m = \langle s\rangle_m \approx \frac{1}{2\beta mb};\qquad \langle Ls\rangle_m = \frac{1}{(2\beta mb)^2} = \langle L\rangle_m \langle s\rangle_m;\qquad x_m \gg 1 \tag{15.35}$$

Note that, contrary to an intuitive reasoning, the more costly the edges (large b) are, the smaller the mean sizes are. Moreover, at small x, $\langle Ls\rangle \ll \langle L\rangle\langle s\rangle$. This behaviour originates directly from the 2D structure (in L and s) of the energy [Eq. (15.25)] (Israelachvili et al., 1976).

Therefore, the answer to a high dilution (large negative values of μ) is the formation of numerous tiny layers. In that marginal limit, the discrete monomeric description evoked earlier could apply and the double integral [Eq. (15.32)] must be replaced by discrete sums $L = iL_1$, $s = js_1$. If $(E_{min,m} - \mu)s_1L_1$ is much larger than kT, only the first term $i = j = 1$ contributes, the solution is made of monomers *only*, and the real chemical potential $\mu_m = s_1L_1\mu$ is given by

$$\mu_m \approx s_1L_1E_{min,m} + (s_1 + L_1)b + kT\ln S_{tot} \tag{15.36}$$

The first two terms on the right side of Eq. (15.36) represent the energy of a monomer. Then, dimers and multimers ($ij > 1$) start to contribute to the sum [Eq. (15.32)] when $(E_{min,m} - \mu)s_1L_1$ becomes of the order of kT, so for concentrations above the *cmc*, where

$$cmc \approx \exp[-\beta(s_1 + L_1)b] \tag{15.37}$$

the more expensive the edge penalty, the lower the *cmc*.

15.3.2 MW Imogolites

The fundamental difference with the previous case of m-protos is that, now, the cross section of the mW nanotube is a closed contour, a complete circle (Fig. 15.8). So, once the length of the section s has been defined, the radius of curvature R is fixed, with $R = s/2\pi$. Two independent parameters are left (namely, L and s). Moreover, the edge penalty takes place only at the two ends of the tube, not along its length (after all, that is the motor for the formation of these nanotubes!). Compared to the m-proto equation [Eq. (15.25)], the energy of mW now reads:

$$U_{m\mathrm{W}}(L,s) = msLE_m(R) + mbs \tag{15.38}$$

As before, it will be considered that the solution is made of a mixture of mW of different L, s values. One notes that $N_{m\mathrm{W}}(L,s)dLds$ is the concentration of mW whose length is L within dL, and the section contour is s within ds. The free energy of this mixture is (per unit of volume of solution)

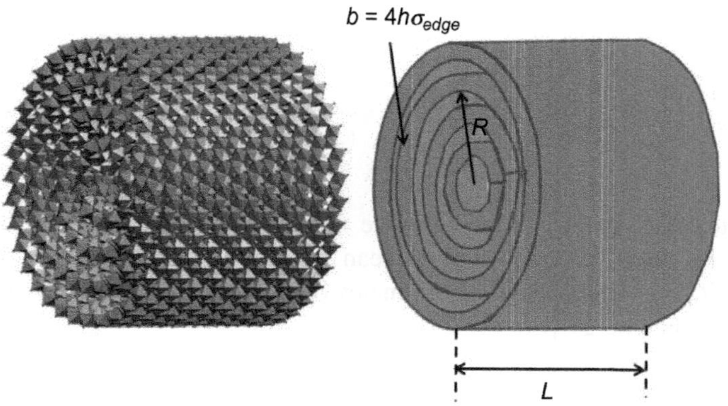

FIG. 15.8 Atomic model of a finite MW imogolite and the corresponding continuous model.

$$F_{mW} = \iiint dLdsN_{mW}(L,s)U_{mW}(L,s) + kT\iiint dLdsN_{mW}(L,s)\left[\ln\frac{N_{mW}(L,s)}{N_{0W}} - 1\right] \quad (15.39)$$

Once again, a reference parameter N_{0W} was introduced that has the same dimension as N_{mW}, m^{-5}, and is thus different from the previous N_0. For instance, $N_{0W} \approx N_0/(2\pi R_0)$.

One constructs the grand potential as Eq. (15.28) and minimises with respect to the function N_{mW}:

$$N_{mW}(L,s) = N_{0W}\exp\left[-\beta(U_{mW}(L,s) - \mu msL)\right] \quad (15.40)$$

Finally, the chemical potential μ is determined from the total concentration, as in Eq. (15.30):

$$S_{tot-mW} = \iint dLdsN_{0W}\exp\left[-\beta m\left(\left(E_m\left(R \equiv \frac{s}{2\pi}\right) - \mu\right)sL + bs\right)\right]msL \quad (15.41)$$

This time, the chemical potential *cannot* reach its upper bound exactly; otherwise, the integral in L would diverge (Israelachvili et al., 1976). Concentrations as high as desired can thus be explored by approaching the limit $E_{min,m}$ from below.

The integration in L is obvious:

$$S_{tot-mW} = \frac{N_{0W}m}{(\beta m)^2}\int_0^\infty \frac{e^{-\beta mbs}}{s(E_m(s/2\pi) - \mu)^2}ds \quad (15.42)$$

Here again, one could refine this by saying that the mean perimeter s must exceed $2\pi(h + ma)$. Then, it is assumed, as before, that the energy E is quadratic in $C = 2\pi/s$, so

$$S_{tot-m\mathrm{W}} = \frac{N_{\mathrm{OW}}m}{(\beta m)^2\left(\frac{1}{2}K_m R_{min,m}{}^2\right)^2} G'\left(B_m \equiv \beta m b 2\pi R_{min,m}, z_m \equiv \frac{\frac{1}{2}K_m R_{min,m}{}^2}{E_{min,m}-\mu}\right) \quad (15.43)$$

where the new $G'(B,z)$, increasing function of z, reads as follows after the change of variable $s = s_{min,m}y$:

$$G'(B,z) = \int_0^\infty \frac{e^{-By}}{y\left[\frac{1}{z^2} + \left(\frac{1}{y} - 1\right)^2\right]^2} dy \quad (15.44)$$

In this complicated integral, the neighbourhood of $y = 1$ corresponds to the optimum cylinders of minimal energy, $R = R_{min,m}$. One must pay the corresponding end energy $mbs_{min,m} \equiv kTB$, at the origin of the factor e^{-B}. The part $y \ll 1$ corresponds to the very thin cylinders, for which the curvature energy is much less favourable but the end energy is more negligible. The relative contribution of each species depends on the B- and z-values, namely on the concentration. Note that the variables x_m of the protos in Eq. (15.33) and z_m of the nanotubes in Eq. (15.43) are proportional, and both measure $(E_{min,m} - \mu)^{-1}$. More precisely, their ratio mixes line tension, bending and thermal energies:

$$\chi_m \equiv \frac{x_m}{z_m} = \frac{\beta m b^2}{\frac{1}{2}K_m R_{min,m}{}^2} \quad (15.45)$$

The important asymptotic behaviours for the function G' are

$$\begin{aligned} G'(B,z) &\approx z^2 \left| \ln\left(\sqrt{z}B\right)\right|; \quad z \ll 1 \\ G'(B,z) &\approx \frac{6}{B^4} + \frac{\pi}{2} e^{-B} z^{3/2}; \quad z \gg 1 \end{aligned} \quad (15.46)$$

Since in general $B \gg 1$ (an end energy costs more than kT, an assumption already made for the protos), the shape of the function G' is quite simple: it increases at small z (small concentrations) as z^2 up to the plateau $6/B^4$, this is the contribution of the thin cylinders. Then, at much larger concentrations of z, especially as B is large, begins the contribution of the cylinders of optimum size, which increases as $z^{3/2}$ and becomes the dominant term (at $z \gg e^{2/3B}/B^{8/3}$). In that important regime, the sizes of the nanotubes follow the laws $\langle R(m)\rangle \approx R_{min}(m)$ and $\langle L(m)\rangle \approx \exp(2/3mB)$. This time, the length of the nanotubes increases exponentially with the end energy. If the solution were constituted of mW only, for a special m value, it would be sufficient to inverse the function G' in Eq. (15.43) to extract μ from S_{tot}.

15.3.3 General Mixture of Protos and MW Imogolites

Finally, a general mixture of proto-imogolites and nanotubes with all possible values for m is considered. At thermodynamical equilibrium, all objects must correspond to the same chemical potential μ, which is determined by the total concentration. The general balance is obtained by gathering together Eqs (15.33) and (15.43):

$$S_{tot} = \sum_m \left\{ \begin{array}{l} \dfrac{N_0 \pi m}{(\beta m)^2 \sqrt{2K_{c,m}} (\beta m b^2)^{3/2}} G\left(x_m \equiv \dfrac{\beta m b^2}{E_{min,m} - \mu} \right) \\ + \dfrac{N_{0W} m}{(\beta m)^2 \left(\frac{1}{2} K_m R_{min,m}{}^2\right)^2} G' \left(B_m \equiv \beta m b 2\pi R_{min,m}, z_m \equiv \dfrac{\frac{1}{2} K_m R_{min,m}{}^2}{E_{min,m} - \mu} \right) \end{array} \right\} \quad (15.47)$$

With the simplistic hypothesis $N_{0W} \approx N_0/(2\pi R_0)$, which allows one to add protos and nanotubes, the equation that relates μ to S_{tot} becomes more compact:

$$S_{tot} = \frac{4N_0}{\beta^2} \sum_m \frac{1}{m K_m{}^2 R_{min,m}{}^5} \left\{ \frac{\pi}{2\chi_m{}^{3/2}} G(x_m) + G'(B_m, z_m) \right\} \quad (15.48)$$

Since the chemical potential must remain below all surface energies E, it must remain strictly below all minima $E_{min,m}$. If the minimum (with respect to m) of these minima is obtained for the special mW $m = m0$ [see Eq. (15.13)], that means that values of x_{m0} or z_{m0} that are as high as desired can be reached by approaching μ as close as possible to $E_{min,m0}$ from below, while all other x_m, z_m remain bounded, since $(E_{min,m} - \mu)^{-1} < (E_{min,m} - E_{min,m0})^{-1} \equiv 1/\Delta E_{min,m}$. As a consequence, at large enough concentrations, the structure of minimal energy will always win. On the other hand, by gradually diluting the solution, other structures may appear (perhaps of higher energy E), but are favoured by the end energy (smaller m) or by the entropy of mixing of small entities (protos).

15.4 DENSITY–TEMPERATURE PHASE DIAGRAM OF NANOSIZED CYLINDERS

In practice, Eq. (15.48) is solved graphically by plotting the different contributions of m-protos and mW as their total sum in function of, for instance, $z_{m0} \in [0,\infty]$, which measures the chemical potential (this is more clear in log–log scale since the different terms may differ by orders of magnitude). One visualises immediately the relative importance of the different structures in the function of z_{m0} (horizontal axis) or in the function of S_{tot} (vertical axis). Note that in practice, the identification of the experimental surface concentration S_{tot} (in m^2/m^3) with the theoretical one requires knowledge of the reference N_0. This is always the case in such a theoretical approach but is not really important; it is sufficient, for instance, to calibrate this parameter with an experimental measurement of proto-imogolites in a given concentration regime.

As reported in Section **15.2** earlier in this chapter, the phase diagram in pure surface energy E could be described by the two parameters α and n_0. In order to treat the entropy as well, it is sufficient to add two other parameters—for instance, the SW values $B \equiv B_1$ and $\chi \equiv \chi_1$:

$$B = \beta b 2\pi R_0$$
$$\chi = \frac{\beta b^2}{\frac{1}{2}K{R_0}^2} \equiv \frac{1}{2\pi^2}\frac{B^2}{\beta K_c} \tag{15.49}$$

where B represents the end energy of an optimal SW nanotube per kT and χ mixes that energy with the elastic energy of the layer $K_c = K{R_0}^4 = 2Yh^3/3$ in a complicated manner. A priori, $B \gg 1$ and $\beta K_c \ll 1$, but one can say nothing about χ. Moreover, the surface energies E and μ will be advantageously expressed in units of $½Ka^2$ (defining the dimensionless quantities E' and μ') since it was seen in Eq. (15.13) how $E'_{min,m}$ presents a simple expression that differs by a few units between neighbouring m values.

The systematic numerical results can now be explained with simple limiting laws.

In order to simplify the expressions, it is assumed that $\alpha = a/R_0 \ll 1$. In that regime, $R_{min,m}$ and K_m depend weakly on m, so $B_m \approx mB$ and $\chi_m \approx m\chi$. By identifying them with their SW counterparts (R_0 and K, respectively), Eq. (15.48) could be replaced by

$$S_{tot} = \frac{4N_0}{\beta^2}\frac{1}{K^2{R_0}^5}\sum_m \frac{1}{m}\left\{\frac{\pi}{2\chi^{3/2}m^{3/2}}G(x_m = m\chi z_m) + G'\left(mB, z_m = \frac{1}{\alpha^2\left(E'_{min,m} - \mu'\right)}\right)\right\} \tag{15.50}$$

At low 'chemical potential' z or x (high dilution), the protos (in $z^{3/2}$) always dominate the thin tubes (in z^2). The same trend subsists in the intermediate regime, where both types of structure reach their respective plateaus. Indeed, the ratio of their maximal concentration (thin tube/proto) is, according to Eqs (15.34) and (15.46), $(1/B^4)/(1/\chi^{3/2}) = 1/[B(\beta K_c)^{3/2}] \ll 1$. Therefore, the thin nanotubes can be ignored in the following discussion, and $G'(B,z)$ will be replaced by its pure mW contribution, $\pi/2\ e^{-B}z^{3/2}$ [see Eq. (15.46)]. It remains to treat the competition between m-protos and m optimum nanotubes, with the latter dominating at $z_m > z_m^{cut\text{-}off} = e^{2/3mB}/(m\chi)$ or $S_{tot} > S^{max}_{proto,m}$, and finally between different m values.

15.4.1 Most Favourable SW, $n_0 \approx 1$, So $m0 = 1$ (Energy and Entropy Favour SW)

On the left side of Fig. 15.9 are plotted the different contributions (m-protos and mW nanotubes) to the total concentration S_{tot} in Eq. (15.48) or Eq. (15.50) as a function of z_1 for $\alpha = 0.01$, $n_0 = 1$, $\chi = 0.05$ and $B = 10$. Note the log–log scale. This is the easiest case, as there are only SW protos and nanotubes, the latter winning above ${z_1}^{cut\text{-}off} \approx 10^4$ or $S^{max}_{proto,1} \approx 10^2$. The DW structures are penalised not only because x_2 and z_2 are bound (as already mentioned), but also because of the factor $1/m^{5/2}$ for the maximal concentration of protos and the factor e^{-mB} for the nanotubes. Finally, the composition of the

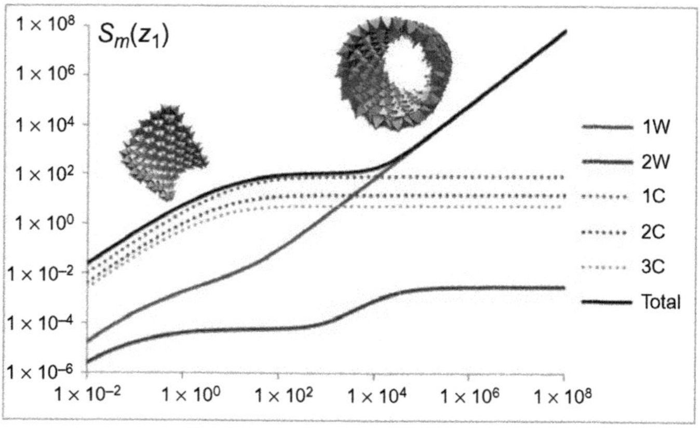

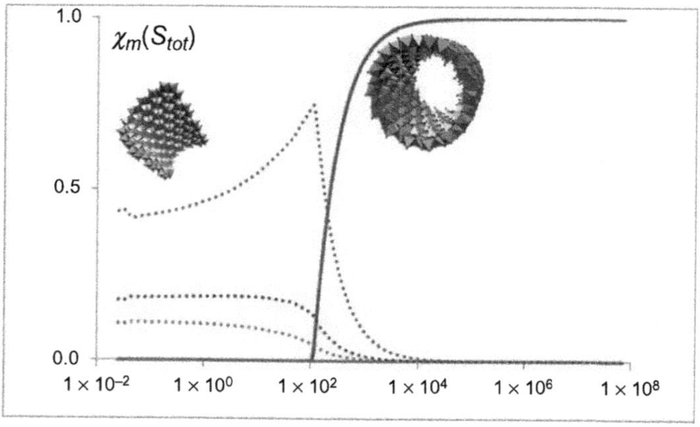

FIG. 15.9 Most favourable SW. (*Left*) Partial and total S_{tot} imogolite concentrations as a function of the chemical potential $z_1 \approx 1/(E_0 - \mu)$ (log–log plot). (*Right*) Relative composition of the solution in each structure as a function of total S_{tot} (semi-log plot). (*Dotted lines*) *m*-protos (1C, 2C, …); (*solid lines*) *m*W nanotubes (1W, 2W, …); total concentration in thick solid line; $\alpha = 0.01$, $n_0 = 1$, $\chi = 0.05$, $B = 10$.

imogolite solution is plotted (semi-log plot) in a function of total concentration S_{tot} on the right side of Fig. 15.9. The dilute regime contains proto-imogolites [mainly SW and a few *m*W ($m \geq 2$)], while the concentrated regime is dominated by SW nanotubes. In between, there is a narrow concentration regime around 10^2 where SW protos and nanotubes coexist.

15.4.2 Most Favourable DW, $n_0 \approx 2$, So $m0 = 2$ (Energy Favours DW but Entropy Favours SW)

This time, two opposed effects are competing. Of course, the SW protos always win at high dilution and the DW nanotubes always win at large z_2 or large concentrations but different scenarios may appear in between because the energetic advantage of the DW nanotubes [$(E_{min,2}-\mu)^{-1} > (E_{min,1}-\mu)^{-1}$ so z_1 is bound contrary to z_2] may be dominated by the entropic advantage (again, the factor $1/m^{5/2}$ for the protos and the factor e^{-B} for the nanotubes) in favour of the SW.

15.4.2.1 $(E_{min,1}-E_{min,2})/(\beta b^2) \ll e^{-2/3B}$ (small gap between SW and DW energies)

On the left side of Fig. 15.10 are plotted the different contributions (m-protos and mW nanotubes) to the total concentration S_{tot} as a function of z_2 for $n_0=2$, keeping the same values of α, χ and B as in Fig. 15.9. In that region of the phase space, the maximum x_1 parameter reachable by the 1-protos is high ($x_1^{max} \gg 1$), so the 1-protos dominate over the 2-protos even in the intermediate region, despite their less favourable E. For the nanotubes, the maximum value reachable by the SW parameter z_1 is much larger than its cutoff threshold, $z_1^{max} \gg z_1^{cut\text{-}off}$, so there is room for an intermediate concentration

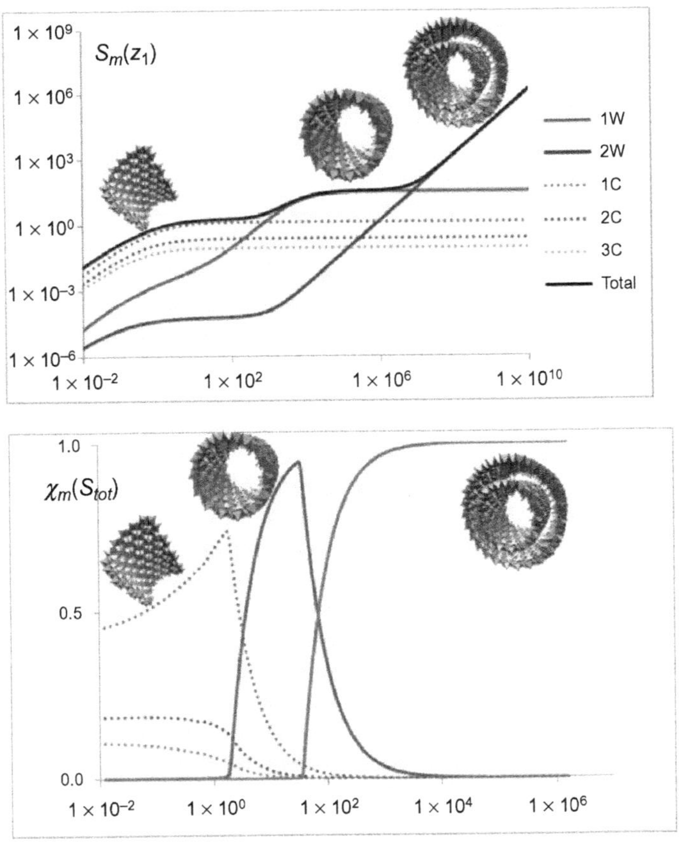

FIG. 15.10 Most favourable DW. Same legend as Fig. 15.9; $\alpha=0.01$, $n_0=2$, $\chi=0.05$, $B=10$.

regime where the SW nanotubes dominate. At high concentrations, the DW nanotubes eventually win, as expected. As a consequence, the total concentration S_{tot} curve presents two plateaus. Fig. 15.10 (right side) exhibits the relative composition of the solution with successive regions of dominating SW protos, SW nanotubes and DW nanotubes at increasing concentrations, with narrow two-phase regions.

15.4.2.2 $e^{-2/3B} \ll (E_{min,1} - E_{min,2})/(\beta b^2) \ll 1$ *(intermediate gap between SW and DW energies)*

The parameter α has been multiplied by 10 ($\alpha=0.1$), with the other parameters unchanged ($n_0=2$, $\chi=0.05$, $B=10$). The corresponding curves are presented in Fig. 15.11. The maximum x_1 parameter reachable by the 1-protos is still high, $x_1^{max} \gg 1$, so the 1-protos still dominate over the 2-protos even in the intermediate region, despite their less favourable E. On the other hand, the gap in energy E between 1W and 2W structures is too high (measured in terms of edge

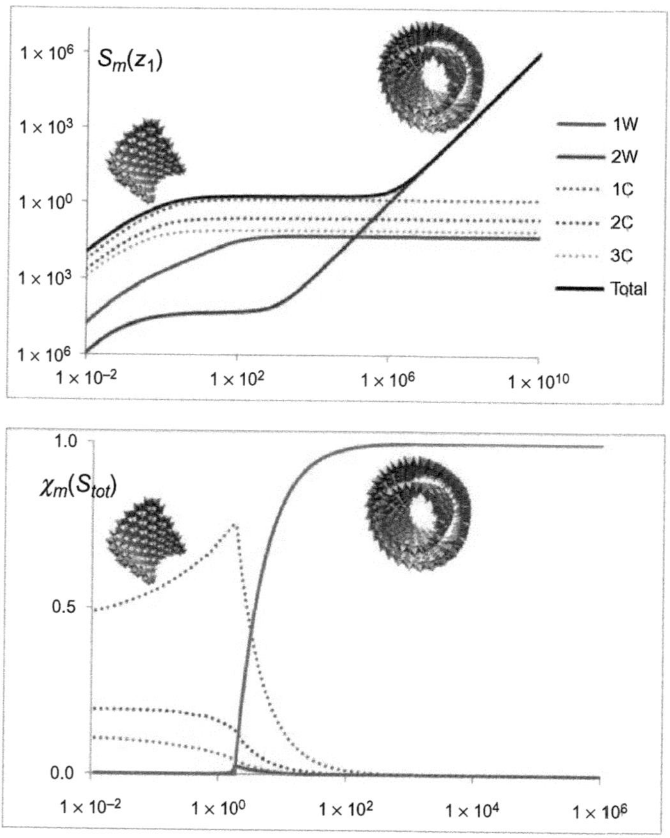

FIG. 15.11 Most favourable DW. Same legend as Fig. 15.9; $\alpha=0.1$, $n_0=2$, $\chi=0.05$, $B=10$.

penalty), so the z_1 upper bound is not large enough for the formation of 1W nanotubes ($z_1^{max} \ll z_1^{cut\text{-}off}$). As a consequence, the SW protos transform directly in DW nanotubes (a single plateau in the S_{tot} curve on the left side of Fig. 15.11, and a narrow concentration regime where 1W-protos and 2W nanotubes coexist on the right side of Fig. 15.11).

15.4.2.3 $1 \ll (E_{min,1} - E_{min,2})/(\beta b^2)$ (broad gap between SW and DB energies)

The parameter χ here equals 0.0005, with the other parameters keeping their previous values ($\alpha = 0.1$, $n_0 = 2$, $B = 10$). Now, the gap in energy E between 1W and 2W structures is so high that even the 1W-proto parameter x_1 cannot reach large values ($x_1^{max} \ll 1$), contrary to its 2W-proto competitor x_2; so the 1W-proto, winning at high dilution, is dominated by the 2W-proto in the intermediate region, before the apparition of the 2W nanotubes (see Fig. 15.12).

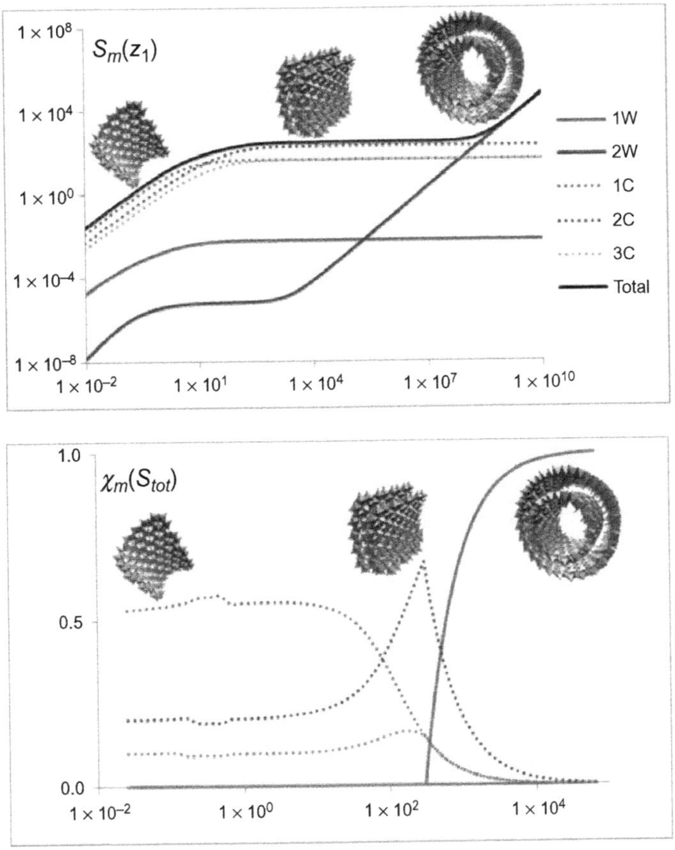

FIG. 15.12 Most favourable DW. Same legend as Fig. 15.9; $\alpha = 0.1$, $n_0 = 2$, $\chi = 0.0005$, $B = 10$.

The total concentration S_{tot} curve again presents two plateaus with successive regions of dominating 1W-protos, 2W-protos and 2W-tubes.

15.5 CONCLUDING REMARKS

This coarse-grain, mesoscopic thermodynamical description of the formation of MW clay mineral nanotubes has involved four types of energy: the elastic energy in terms of the Young modulus Y, the surface tension mismatch in σ, the surface–surface adhesion across a water layer $-U_{adh}$ and the edge energy σ_{edge} or line tension b. The infinitely long nanotube of lowest internal energy has been determined by comparing the first three types and introducing the dimensionless parameters α and n_0. Then, the free energy and the equilibrium composition of the whole solution have been obtained by adding the fourth energy, as well as the thermal reference kT, thus defining two new parameters: χ and B. It has been shown that, despite its simplicity and coarseness, this approach can account for a variety of nanoscopic structures and 'phase' transitions—some expected, other less obvious. This theory and the examples detailed in the last section of this chapter do not cover all the possible cases and instead must be viewed as a starting point for more refinements and richer phase diagrams.

REFERENCES

Guimares, L., Enyashin, A.N., Frenzel, J., Heine, T., Duarte, H.A., Seifert, G., 2007. Imogolite nanotubes: stability, electronic, and mechanical properties. ACS Nano 1, 362.

Israelachvili, J.N., Mitchell, D.J., Ninham, B.W., 1976. Theory of self-assembly of hydrocarbon amphiphiles into micelles and bilayers. J. Chem. Soc., Faraday Trans. 2 (72), 1525.

Krumeich, F., Muhr, H.J., Niederberger, M., Bieri, F., Schnyder, B., Nesper, R., 1999. Morphology and topochemical reactions of novel vanadium oxide nanotubes. J. Am. Chem. Soc. 121, 8324.

May, S., Ben-Shaul, A., 2001. Molecular theory of the sphere-to-rod transition and the second cmc in aqueous micellar solutions. J. Phys. Chem. B 105, 630.

Paillusson, F., Dahirel, V., Jardat, M., Victor, J.M., Barbi, M., 2011. Effective interaction between charged nanoparticles and DNA. Phys. Chem. Chem. Phys. 13, 12603.

Thill, A., Guiose, B., Bacia-Verloop, M., Geertsen, V., Belloni, L., 2012a. How the diameter and structure of (OH)3Al2O3SixGe1–xOH imogolite nanotubes are controlled by an adhesion versus curvature competition. J. Phys. Chem. C 116, 26841.

Thill, A., Maillet, P., Guiose, B., Spalla, O., Belloni, L., Chaurand, P., et al., 2012b. Physico-chemical control over the single- or double-wall structure of aluminogermanate imogolite-like nanotubes. J. Am. Chem. Soc. 134, 3780.

Zhang, S., Peng, L.M., Chen, Q., Du, G.H., Dawson, G., Zhou, W.Z., 2003. Formation mechanism of H2Ti3O7 nanotubes. Phys. Rev. Lett. 91, 256103.

Chapter 16

Formation Mechanisms of Tubular Structure of Halloysite

J. Niu*
School of Materials Science and Engineering, China University of Mining and Technology, Xuzhou, PR China
**Corresponding author: e-mail: jinan.niu@cumt.edu.cn*

16.1 INTRODUCTION

As a mineral from the kaolin group, most halloysites show a unique tubular morphology, similar to carbon nanotubes, which makes it important in many applications. For this reason, a great deal of attention was paid to the rolling mechanism leading to tubular halloysite. The proper explanation of the rolling mechanism will help in understanding the microstructure and geological origin of halloysite, further providing a basis for the controllable preparation and functional modification of halloysite. Moreover, the investigation on the rolling mechanism of kaolinite is helpful in revealing certain scientific propositions, such as the evolution of the environment and the origin of life (Cleaves et al., 2012; Zhou and Keeling, 2013).

So far, three rolling mechanisms of the kaolinite layer leading to tubular halloysite have been proposed: (i) the mismatch between the octahedral sheet and the tetrahedral sheet, by Pauling (1930); (ii) the attraction between interlayer hydroxyl groups in octahedrons, by Radoslovich (1963); and (iii) the surface tension of water, by Hope and Kittrick (1964). Besides tubular morphology (Honjo and Mihama, 1954), halloysite holds spherical (Birrell et al., 1955), fibre (de Souza Santos et al., 1965), crumpled lamellar (Wada and Mizota, 1982), cagelike (Jeong, 2000) and flat morphologies (Kunze and Bradley, 1964). The prevailing view is that the morphology of halloysite may be related to the content of impure elements, such as Fe and Ti (Singer et al., 2004), which causes a change in matching between tetrahedrons and octahedrons, affecting the degree of rolling in the layer (Churchman and Lowe, 2012). For most halloysites, the Al/Si atom ratio is greater than unity, which may suggest some instances of Al for Si substitution (Merino et al., 1989). The effects of octahedral and tetrahedral substitution on rolling in

Developments in Clay Science, Vol. 7. http://dx.doi.org/10.1016/B978-0-08-100293-3.00016-9

the unit layer will be described in this chapter, although the evidence for tetrahedral substitution is not unequivocal (Joussein et al., 2005).

Halloysite has versatile features, including large specific surface area (SSA), high porosity and tunable surface chemistry (Li et al., 2013), but its reserves are far less than those of another clay mineral with a similar layer structure—kaolinite (Costanzo et al., 1982). For this reason, efforts have been made over the last few decades to synthesize nanotubes from kaolinite (Singh and Mackinnon, 1996; Gardolinski and Lagaly, 2005; Matusik et al., 2009; Kuroda et al., 2011; Yuan et al., 2013). These experiments reproduce the rolling process, providing some direct evidence of the rolling mechanism of the kaolinite layer in halloysite through modern measurement techniques. Some attempts to directly observe the rolling process of the kaolinite layer in synthesis nanotubes are outlined in this chapter. Furthermore, because different natural kaolinite or halloysite samples may come from different geological periods or transformation stages, measurements of natural samples have attracted some attention, and that development will be described in this chapter as well.

16.2 FORMATION MECHANISMS OF THE TUBULAR STRUCTURE OF HALLOYSITE

16.2.1 Mismatch Between Octahedral and Tetrahedral Sheets

When Pauling (1930) proposed the first kaolinite structure, he pointed out that a kaolinite layer would have a tendency to curve due to a mismatch in two faces of the constituent layer. Here, the denoted two faces are a silicon–oxygen tetrahedral sheet (or simply a tetrahedral sheet) and an aluminium–oxygen octahedral sheet (or simply an octahedral sheet). The very close approximation in dimensions of the octahedral sheet in hydrargillite ($b=0.507$ nm and $a=0.865$ nm), as well as the complete tetrahedral sheet in β-tridymite (or β-cristobalite) ($a=0.503$ nm and $b=0.871$ nm) (Pauling, 1930), show that an unsymmetrical layer composed of these two would have a slight tendency to curve that could be overcome by forces occurring between layers.

According to Bates et al. (1950), when water molecules occur between the 1:1 layers, it weakens the interlayer force; therefore, the octahedral and tetrahedral sheets are free to approach their normal dimensions, which leads to a curvature of the kaolinite layer. These authors calculated the inner diameter of the resulting cylinder, which had the same order of magnitude as those of the smallest tubes that they observed.

Currently, the viewpoint regarding the mismatch between sheets has been widely recognized; however, some controversy remains focused on further structural details, such as how tetrahedral rotation and curving play a role in relieving stress, how to establish contact between the microstructure and macroscopic curling morphology and the role of water, all of which will be summed up in the following sections.

16.2.1.1 Tetrahedral Rotation and Tetrahedral Curving

After Newnham and Brindley (1956) found that the oversized silica layer was compressed by rotations of tetrahedrons when evaluating the structure of dickite, the structural disorder of kaolin minerals began to attract attention. Singh (1996) suggested that tubular halloysite formed by means of a tetrahedral curving mechanism rather than a tetrahedral rotation mechanism to correct the mismatch between the octahedral and tetrahedral sheets. Tetrahedral rotation reduces the lateral dimension of the tetrahedral sheet by shrinking the distances to an equal amount for basal oxygen, Si and apical oxygen in all the directions of the *a–b* plane; on the other hand, in the tetrahedral curving mechanism, the constriction occurs only along the rolled axis, and the apical oxygen could generate a greater constriction than the Si and basal oxygen, due to being relatively more inside the curvature (Fig. 16.1).

Both of the dimensions of tetrahedral and octahedral sheets correspond to the equilibrium between three different kinds of forces: cation–cation repulsion, anion–anion repulsion and cation–anion bonds. Cation–cation repulsion is the most influential force in causing individual departure from ideal structures because oxygen atoms only partially shield the cations from each other (Radoslovich, 1963; Bailey, 1980). Based on this finding, Singh (1996) thought that cation–cation repulsion played a major role in the determination of cell dimensions, stability and stacking of layer silicates. The Si–Si repulsion caused by tetrahedral curving and rotation in order to correct a given amount of mismatch was determined using Coulomb's law for the quantitative comparison of the two mechanisms:

$$F = \frac{q \cdot q'}{r^2} \tag{16.1}$$

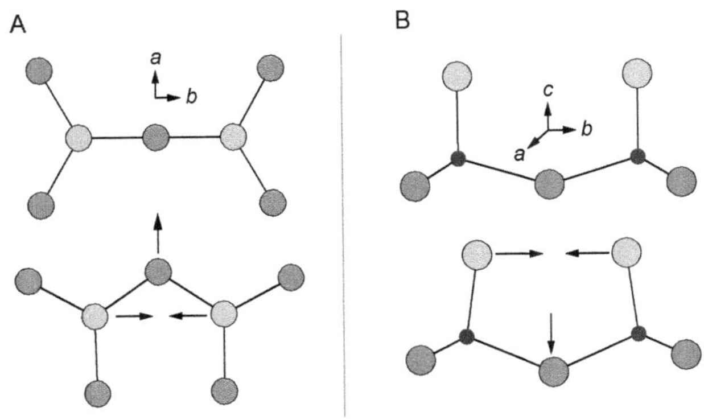

FIG. 16.1 Tetrahedral (A) rotation and (B) curving. *Reproduced with permission from Singh (1996). Copyright (1996) from The Clay Minerals Society, publisher of* Clays and Clay Minerals.

where F is the repulsion force, q and q' are the charges in electrostatic units and r is the distance between charges. The relative increase in repulsion can then be calculated from the relation

$$F_{\text{rel}} = \left(\frac{r_{\text{h}}}{r}\right)^2 \quad (16.2)$$

where F_{rel} is the relative repulsion, r_{h} is the Si–Si distance in a planar hexagonal tetrahedral sheet and r is the Si–Si distance in a curved or ditrigonal tetrahedral sheet. The value of r for the hexagonal planar and ditrigonal sheets can be calculated from the b dimension using the relation: $r=b/3$, where b is the cell dimension.

In order to calculate r for a curved tetrahedral sheet, the b dimension of a planar tetrahedral sheet (b_{tet}) can be determined from $b_{\text{tet}}=k\cdot(R+t)$, where R is the inner radius of the curvature of the 1:1 layer and t is the thickness of the 1:1 layer; and the b dimension of an octahedral sheet (b_{oct}) can be determined from $b_{\text{oct}}=k\cdot R$, where k is the angle of curvature in radians. Then R can be given by

$$R = \frac{b_{\text{oct}} \cdot t}{b_{\text{tet}} - b_{\text{oct}}} \quad (16.3)$$

The radius for the Si atom plane (R_{Si}) is

$$R_{\text{Si}} = R + t - h \quad (16.4)$$

where h is the distance between the Si atoms and the outer periphery of the tube with a value of 0.615 Å (Bates et al., 1950).

Applying this procedure determined the Si–Si distances for hexagonal, ditrigonal and curved tetrahedral sheets, as well as Coulomb repulsion before and after deformation. The results showed that the curving mechanism resulted in only 1% increase in the Coulomb repulsion between adjacent Si atoms, whereas tetrahedral rotation encountered 12 times greater repulsion in comparison to the curving mechanism to correct the same amount of mismatch.

Without interlayer water, the hydrogen bond between adjacent layers is strong, requiring the structure to correct the alignment between the basal oxygen and the outer hydroxyl (OH) groups, which leads to a shorter hydrogen bond (Bailey, 1988). This hydrogen bond provides an additional driving force for the tetrahedral rotation to correct the mismatch on both sides of the tetrahedral sheet, and therefore kaolinite cannot show a curved morphology. With interlayer water, Bailey (1990) pointed out that the water could minimize the interaction between adjacent layer surfaces, and tetrahedral rotation was blocked by the dynamic disorder of hydrogen bonds from H_2O molecules, so basal oxygen could not all rotate in the same direction and 'hole water' or exchangeable cations might be inserted into ring openings. In other words, interlayer water plays a dual role in the rolling of halloysite: (i) to block tetrahedral rotation and (ii) to relax interlayer hydrogen bonding. As a result, the hydrogen bond between adjacent layers will be greatly weakened, the basal oxygen side will contract at a lower amount than does the apical oxygen side and a rolling of the kaolinite layer will occur. It is noted that the mismatch also could be corrected by a combination of tetrahedral curving and rotation, and the ratio of two mechanisms could vary with the tube radius.

Tetrahedral curving and rotation mechanisms provide a generally reasonable explanation for the rolling/curving/curling/bending of kaolinite layers; however, the relationship between tetrahedral curving and macrorolling orientations of the layers is still unclear. This requires further study of the spatial distribution of tetrahedral curving in kaolinite layers, which will be discussed further in the next section.

16.2.1.2 How Water Molecules Enter the Interlayer Space

As stated earlier in this chapter, it is generally accepted that water molecules play an important role in halloysite rolling, but there is still some controversy over how water molecules enter the interlayer space. Bailey (1990) attributed the introduction of water and hydrated cations into the interlayer space to the substitution of Si by Al, producing a net negative charge and providing a driving force. However, using ^{27}Al nuclear magnetic resonance (NMR) spectroscopy, Newman et al. (1994) found the four coordinated Al (Al^{IV}) contents of two kaolinite and six halloysite samples were all $<1\%$ and provided no grounds for distinguishing between the two minerals. At the same time, the signals tentatively assigned to five coordinated Al (Al^{V}) were present in the spectra of kaolinites, but not in those of halloysites. Newman et al. (1994) suggested that this might be caused by the greater nonstoichiometric substitution of Fe for Al in the octahedral sheets with a higher permanent layer charge, which would attract extraneous cations to their water molecules of hydration. Differing from this point, Singh (1996) thought that the interlayer water in halloysite was driven into the interlayer space due to its activity, structural disorder, or both, and might not be specifically adsorbed on the tetrahedral side of the interlayer space.

Up to now, it has been difficult to determine how water molecules enter the interlayer space due to the lack of an effective means of observation; however, this remains a major issue to be examined with regard to the next step, which may be directly related to the interpretation of halloysite formation.

16.2.1.3 Rolling Orientations of the Layer

As mentioned previously, the tetrahedral curving mechanism establishes a relationship between the microstructure and the macroscopic curling morphology and has been widely recognized (Singh, 1996; Liu et al., 2012; Wei et al., 2012; White et al., 2012). However, the distribution of tetrahedral curving is not considered in the abovementioned studies, meaning that tetrahedral curving cannot be used to directly explain the different curling orientations of kaolinite layers.

Individual authors understand the definition of curling orientation differently, which can cause some confusion. For the purposes of clarification, the direction of the curling axis (or tube orientation) will be denoted in this chapter with the term 'curling orientation', which implies the distribution of the axial direction of curling with any deformation degree, and the direction of curling developing forward with the term 'curling direction', which represents the direction of the curling behaviour and is vertical to the curling orientation.

The common curling orientations of natural halloysite and synthetic halloysite-like nanotubes are [100] and [010] (Honjo and Mihama, 1954; Singh and Gilkes, 1992; Singh and Mackinnon, 1996; Kuroda et al., 2011). However, by means of selected area electron diffraction (SAED), Niu et al. (2014) found that there were four new curling orientations: [$1\bar{1}0$], [110], [310] and [$3\bar{1}0$]. In order to explain these orientations, Niu et al. (2014) proposed a new perspective, described as follows. In kaolinite, tetrahedral and octahedral sheets share a common plane of oxygen atoms. When silicon–oxygen tetrahedrons connect to aluminium–oxygen octahedrons, there are two matching types for the adjacent two tetrahedrons: (i) two tetrahedrons connect to only one octahedron (the two-one type) and (ii) two tetrahedrons connect to two octahedrons (the two-two type). For the two-one type, the distance between two tetrahedrons is undertaken by two oxygen atoms in an octahedron (as shown in Fig. 16.2A); for the two-two type, the distance between two tetrahedrons is undertaken by two oxygen atoms in two different octahedrons (Fig. 16.2B).

In gibbsite, which consists only of an octahedral sheet and can be referenced as the natural size of an octahedral sheet in kaolinite (Bates et al., 1950; Singh, 1996), the mean distances of two adjacent oxygen atoms in an octahedron and in two adjacent octahedrons are 0.280 and 0.295 nm, respectively (Balan et al., 2006). In kaolinite, the mean distance of two oxygen atoms in adjacent tetrahedrons (Bish, 1993) is 0.298 nm, which is larger than 0.280 and 0.295 nm. Because of the Si–Si electrostatic repulsion, it is preferable for the tetrahedral curving to occur in order to relax the size incompatibility in the hydrate (Singh, 1996). Therefore, the orientations of curling can be analysed according to the distributions of the two matching types.

There are four matchings for the two-one type (as ①, ②, ③ and ④, as shown in Fig. 16.2C) and two matchings for the two-two type (⑤ and ⑥, in Fig. 16.2C) in the kaolinite unit cell, which are all located on the edges of the six-membered ring. The six edges include three directions of tetrahedral curving: [010] (② and ④), [310] (① and ③) and [$3\bar{1}0$] (⑤ and ⑥); the corresponding orientations of curling are then determined to [100], [$1\bar{1}0$] and [110].

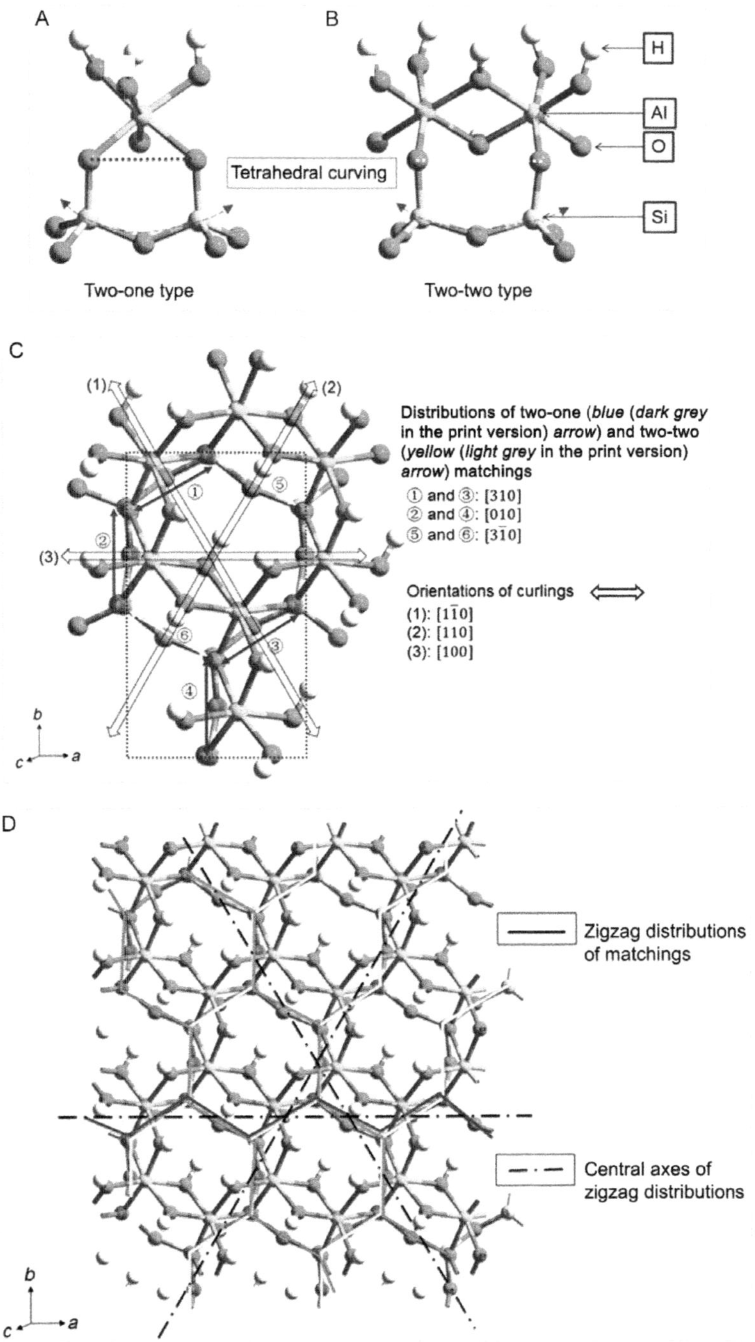

FIG. 16.2 Schematic diagram of curling orientations. (A) Two-one type and (B) two-two type matchings between octahedrons and tetrahedrons, (C) distributions of two matching types in a kaolinite cell, and (D) zigzag distributions of matchings in a kaolinite layer. *Reproduced with permission from Niu et al. (2014). Copyright (2014) from Elsevier, publisher of* Applied Clay Science.

Furthermore, through an analysis of a wider range of the kaolinite layer, the two-one and two-two types of matchings formed continuous zigzag distributions in the directions with [100], [$1\bar{1}0$] and [110] as the central axes, which may cause a strong tendency for the layer to curve in these directions, as shown in Fig. 16.2D. This produces curling with [010], [310] and [$3\bar{1}0$] orientations, which correspond exactly to the other three in the transmission electron microscopy (TEM) experiments (Niu et al., 2014). To summarize, kaolinite layers can curl in multiple directions for two structural reasons: (i) incompatible matching between tetrahedrons and octahedrons and (ii) hexagonal distributions and the zigzag distributions of two types of matchings. Although different curling orientations can be interpreted using these viewpoints, the ease of formation for the different types of curling are still difficult to compare, requiring further study to explain the probability of observing these curling orientations.

16.2.2 Attraction Among Interlayer Hydroxyl Groups in Octahedrons

Radoslovich (1963) proposed that the curvature of halloysite could not be explained solely in terms of a mismatch in dimensions between the tetrahedral and octahedral sheets. A tetrahedral sheet could readily reduce its dimensions by becoming more ditrigonal; consequently, there would be no mismatch with adjacent octahedral dimensions (or at least, the stresses caused by mismatching would be secondary). Radoslovich (1963) also pointed out the morphologies of different kaolin minerals were not directly related to subtle differences in the Si/Al ratio. An increase in the Al substitution to tetrahedral Si would result in slightly greater rotations in the already ditrigonal tetrahedral network; but this should not cause a more tubular morphology. However, a change in the Si/Al ratio might affect the OH content or interlayer bonding, and then the morphology. Radoslovich (1963) postulated that OH–OH bonds in the hydroxyl triads around vacant octahedral sites probably caused a contraction within the layer of surface hydroxyls. The polarizing power of Al was at a critically sensitive level, such that subtle structural changes might produce considerable variations in OH polarization, and therefore in interlayer OH–OH interaction. In halloysite, the presence of interlayer water disrupted the long OH–O bonds to the adjacent tetrahedral surface. Under these conditions, the immediate OH–O bonds were absent, forming a looser network of OH–OH and OH–H_2O. In other words, in halloysite, there is clearly the possibility of shortening some of the OH–OH distance around unoccupied sites, which may cause rolling in the layer (Fig. 16.3).

16.2.3 Surface Tension of Water

Hope and Kittrick (1964) proposed a unique view in accordance with their freeze-drying experiments. By comparing the morphologies of freeze-dried

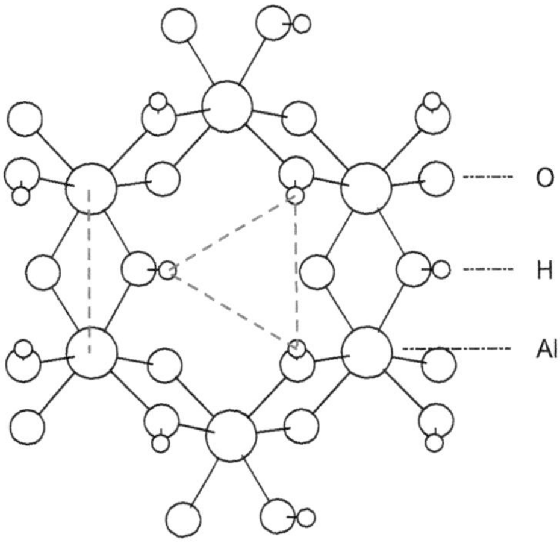

FIG. 16.3 Schematic representation of OH–OH attraction in halloysite. *Produced according to the description by Radoslovich (1963).*

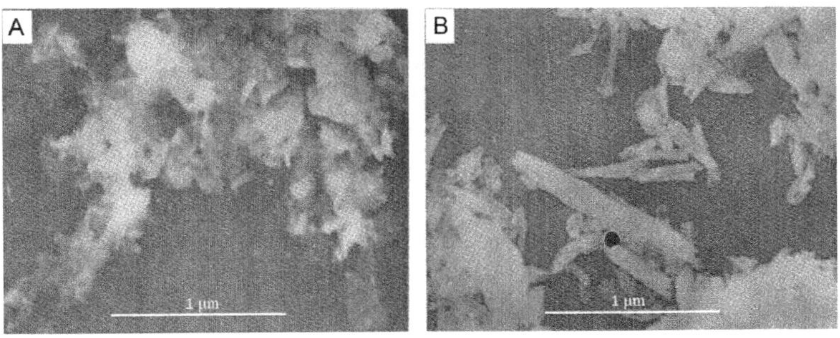

FIG. 16.4 Electron micrographs of freeze-dried (A) and non-freeze-dried (B) halloysite. *Reproduced with permission from Hope and Kittrick (1964). Copyright (1964) from Mineralogical Society of America, publisher of* American Mineralogist.

hydrated halloysite with halloysite dried at room temperature (Fig. 16.4), these authors found that halloysite occur as flat particles after freeze-drying, thus suggesting that (i) halloysite might be originally flat, and then curled due to surface tension that occurred during drying; or (ii) it might have been originally curled and then flattened as a result of being frozen. Hope and Kittrick (1964) were more willing to recognize the former based on the fact that surface tension can cause macroscopic objects to curl during drying. However, these authors also pointed out that surface tension alone did not appear to be sufficient to cause the formation of tubular halloysite, even though this force could be great. Montmorillonite with two tetrahedral sheets was taken as evidence for this proposition because it does not show similar tubular shapes as halloysite.

16.2.4 Comparison of Three Mechanisms

The three mechanisms proposed at different stages of the halloysite study may have greater or fewer limitations due to the lack of relevant technologies and knowledge. Each mechanism has its advisable role, which can be helpful in understanding why and how halloysite rolls. At present, the most widely accepted mechanism is the combination of tetrahedral curving and rotation associated with the situation of interlayer hydrogen bonding proposed by Singh (1996). Further development of this mechanism may be focused on the detailed relationship between tetrahedral curving and macro-rolling in the layer because even for the most usual [010] and [100] longest tube axes, the direction of tetrahedral curving is not entirely consistent with the direction of layer rolling—that is, the quantitative relationship between different orientations raised by Niu et al. (2014). For the OH–OH attraction mechanism, an opposite comment was made by Noro (1986): because the ionic radius of Fe^{3+} is larger than that of Al^{3+}, the size of the halloysite octahedra increases by about 6% when Fe^{3+} replaces Al^{3+} in octahedral positions. The resultant increase in the overall size of the octahedral sheet reduces the dimensional mismatch between the octahedral and tetrahedral sheets, allowing the layers to adopt a planar shape, such as in kaolinite (Bailey, 1990). This result directly conflicts with Radoslovich's point of view because interlayer OH–OH attraction does not lead to layer rolling. However, the effects of OH–OH attraction, such as the degree of influence on macrolayer rolling, the relationship between the direction of layer rolling and OH–OH attraction, and the comparison between the size increase of an octahedral sheet caused by Fe substitution and OH–OH attraction, may still require further evaluation. For the water surface tension mechanism, the hypothesis of the flat morphology of the original halloysite is obviously incorrect, and in comparing the morphology before and after freeze-drying, the volume expansion effect caused from transforming water into ice could be ignored. However, it is undeniable that the water surface tension can affect the curling of the kaolinite layer, and the next step in the research will be to investigate the role played by that surface tension.

In summary, future studies on the rolling mechanism of halloysite should take tetrahedral curving and rotation as its base, taking into account the less prominent role of interlayer OH–OH attraction and surface tension, as well as the influence of interlayer water. As Hope and Kittrick (1964) stated in their study on the surface tension of water, even though one type of force can be large, it alone may not be enough to cause the formation of tubular halloysite.

16.2.5 Other Factors Affecting Rolling

In addition to the three mechanisms described up to this point, there are some other factors that can affect rolling in halloysite, such as octahedral and tetrahedral substitutions. However, these factors still affect halloysite rolling by influencing these mechanisms. In the following subsections, they are described separately.

16.2.5.1 Octahedral Substitution

The isomorphous substitution of Fe for Al in the octahedral sheet has been reported in a great number of studies (Aomine and Wada, 1962; Wada and Mizota, 1982; Churchman and Theng, 1984; Quantin et al., 1988; Nagasawa and Noro, 1987; Muller and Calas, 1989; Hart et al., 2002). The prevailing view is that the radius of Fe is larger than that of Al, which causes an increase in the size of the halloysite octahedral, which reduces the dimensional mismatch between the octahedral and tetrahedral sheets, and thus the curvature of the kaolinite layer.

Moreover, this substitution may be nonstoichiometric. Watanabe et al. (1969) assumed some irregular replacements of Al by Fe in octahedral positions for a halloysite specimen from Japan. Using X-ray photoelectron spectroscopy (XPS), Soma et al. (1992) made a similar suggestion for a number of halloysites from New Zealand. Since the slope of the line correlating the Al/Si and Fe/Si ratios was less than unity, the octahedral substitution was apparently nonstoichiometric. As indicated by Soma et al. (1992), the resultant formation of octahedral vacancies might be expected to enhance the layer curvature.

In addition to Fe substitution for Al in the octahedral sheet, there are octahedral Ti and Cr substitutions in halloysite. Singer et al. (2004) found that the halloysites with a high content (up to 6.5%) of Ti in their aluminosilicate structure were platelike, and this phenomenon was due to that Al atoms in the octahedral sheets were partially replaced by Ti, which corrected the aforementioned dimensional mismatch. In addition, Maksimovic and White (1973) found some evidence for octahedral Cr substitution using infrared spectroscopy, cation exchange capacity (CEC) and X-ray diffraction (XRD) measurements. This substitution caused an increase in the *b* dimension of halloysite because Cr is larger than Al (Shannon, 1976); however, the further effect on halloysite rolling could not be clearly determined due to a lack of necessary morphology information.

Overall, there are numerous studies of substitutions of Fe for Al in the octahedral, while studies of substitutions of Ti or Cr for Al are rare, and the question of whether Ti or Cr occurs in the form of octahedral substitution remains uncertain. Due to the larger radius of these ions, their primary effect is to reduce the mismatch of the layer; however, it has been noted that possible nonstoichiometric substitutions may introduce some octahedral cation vacancies, which have the opposite effect on the correct mismatch of the layer and thus can promote sheet curling. As such, both effects need to be considered for octahedral substitution in future studies.

16.2.5.2 Tetrahedral Substitution

Tetrahedral substitution of Al for Si has also been found in many halloysite and kaolinite samples. To find evidence of four coordinations, Newman et al. (1994) investigated six halloysites and two kaolinites using ^{27}A1 NMR spectroscopy, finding Al^{IV} with content of <1% in both kaolinites and in five of the halloysites. Takahashi et al. (2001) found about 2% tetrahedral Al, and very likely some octahedral Al vacancies and Fe, in their halloysite samples using ^{27}A1 NMR spectroscopy; they thought that these defects resulted in high CEC and high SSA. Bailey (1990) suggested that interlayer water in the form of hydrated exchangeable cations can compensate for negative charges arising from tetrahedral Al for Si substitutions and block the tetrahedral rotation. Based on this proposal, Newman et al. (1994) concluded that the presence of interlayer water in halloysite could not be attributed to the substitution of Al for Si. For the effects of tetrahedral substitution on layer rolling, as stated by Merino et al. (1989), the greater the degree of Al-for-Si substitution in halloysite, the more bent (usually around the *b* axis) the 1:1 layers of this mineral becomes.

In summary, tetrahedral substitution of Al for Si occurs in most halloysites; however, the amount of the substitution is quite small and the same phenomenon also occurs in kaolinite. Therefore, further research is required as to whether and how it effects the rolling of halloysite.

16.3 EXPERIMENTAL OBSERVATION OF ROLLING OF KAOLINITE

According to the rolling mechanism of Singh (1996), the weakness of hydrogen bonds between adjacent layers can promote the rolling of the kaolinite layer. When this happens, the tetrahedral rotation mechanism transforms into a curving mechanism, the hydrogen bonds between opposite layers are broken, and other media enters into the interlayer space to form a new interface. Based on these occurrences, a number of methods have been proposed to prepare halloysite-like nanorolls by the rolling of kaolinite. In this chapter, the focus is on the theoretical explanation of the rolling phenomena, but additional discussion on the experimental details is reported in Chapter 17.

16.3.1 Rolling Mechanisms in Synthesis of Kaolinite Nanoroll

Early intercalation experiments (Hope and Kittrick, 1964; Wada, 1965; Sanchez and Gonzalez, 1970) can be considered as the first attempt in preparing kaolinite nanorolls. However, most intercalations (Ledoux and White, 1966; Olejnik et al., 1968, 1970) cannot achieve this because organic molecules with strong polarization attracts opposite layers, interlayer space is still not large enough, or both. Subsequently, kaolinite hydrate with a similar structure was synthesized. Costanzo et al. (1980, 1982) proposed to synthesize

1-nm kaolinite hydrate with reproducible crystallinity and even impurity, and expected it to have an ideal structure of halloysite (10 Å). After intercalating dimethylsulphoxide (DMSO) into Georgia kaolinite, these authors used fluoride ions to replace the interlayer hydroxyl; and then Costanzo et al. (1984) used water for another replacement to gain 1 nm of kaolinite hydrate. Costanzo et al. (1984) and Raythatha and Lipsicas (1985) found that the yield of kaolinite hydrate could increase significantly without fluoride ions when adding the new methanol-washing step after the DMSO intercalation. Tunney and Detellier (1994) used ethylene glycol for the second intercalation to kaolinite after it was intercalated by DMSO, followed by methanol washing and then water washing; finally, 0.84 nm of hydrate was gained. Amara (1997) intercalated potassium acetate into nacrite; after water washing, 0.84 nm of hydrate was gained. However, these hydrates have been not reported to display the similar tubular or rolling morphology of natural hydrated halloysite.

One exception is reported by Singh and Mackinnon (1996), who observed a change in the morphology of the kaolinite layer by repeating the potassium acetate intercalation and water washing several times. After hydration, kaolinite plates rolled along the major crystallographic directions to form tubes identical to proper tubular halloysite. Most tubes were elongated along the [010] crystallographic direction, while some were elongated along the [100] crystallographic direction, which indicates that the rolling occurred in response to a dynamic balance of intrinsic crystallographic or structural forces, rather than in response to a randomly generated external force. The characteristics of these experimentally formed tubes were consistent with those of natural halloysites. Therefore, Singh and Mackinnon (1996) proposed that planar kaolinite can be transformed to tubular halloysite.

In the model proposed by Bailey (1990), tetrahedral rotation in halloysites is blocked by water molecules and hydrated exchangeable cations seated in the hexagonal cavity of the basal oxygen plane. Therefore, halloysite layers roll in order to accommodate the lateral mismatch. This model suggests that tetrahedral rotation occurs prior to tetrahedral curving, which takes place only when the tetrahedral rotation is restricted. According to their experimental evidence, Singh and Mackinnon (1996) thought tubular halloysites might crystallize in their current rolled form and that a planar kaolin 1:1 layer could roll after crystallization. They also thought the rolling of the plates observed in the study could not be attributed to surface tension (Hope and Kittrick, 1964) because the thickness of rolled plates observed in the study varied between 10 and 50 nm.

Niu et al. (2014) found that significant changes occurred in the kaolinite layer relative to the original morphology, after repeating the DMSO–methanol–water treatment on kaolinite several times. There were not only single curling and two opposite parallel curling, as observed by Singh and Mackinnon (1996), but also two adjacent curling with an approximately

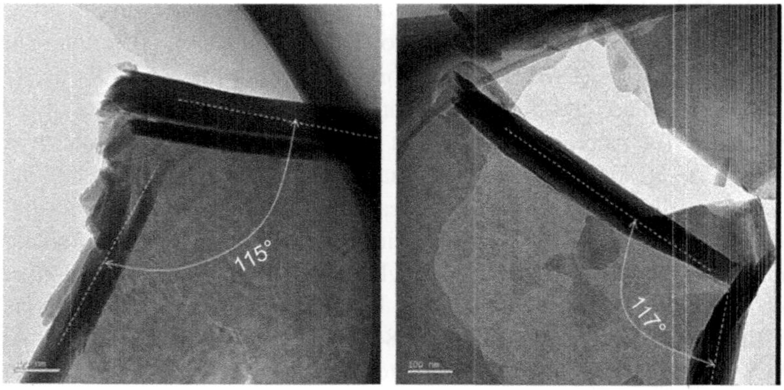

FIG. 16.5 Two adjacent curling orientations with about a 120-degree angle in one single layer. *Reprinted with permission from Niu et al. (2014). Copyright (2014) from Elsevier, publisher of* Applied Clay Science.

120-degree angle in one single layer (Fig. 16.5). Using the SAED examination, four new orientations of curling, $[1\bar{1}0]$, [110], [310] and $[3\bar{1}0]$ were observed in kaolinite hydrate. These findings renovated the previous viewpoint that kaolinite plates rolled only along the major crystallographic axes, [010] and [100], and thus need to be interpreted with a new mechanism.

As described in the previous section, the two matchings for the adjacent two tetrahedrons connecting to octahedrons had different distributions in the kaolinite layer, leading to various curling orientations. This model bridges the gap between macrorolling and microtetrahedral curving. It can be used in the future to analyse the probability of each curling based on the quantification to the difficulty of the occurrence of two matchings.

In addition to the method of repeated intercalation and water washing, some other efforts (first by modification and then intercalation) have been made over the last few decades to synthesize nanotubes from kaolinite such as (i) the multistep method of intercalating with n-alkylamines and deintercalating with toluene to modified kaolinite (Gardolinski and Lagaly, 2005; Matusik et al., 2009); or (ii) the one-step method of intercalating with quaternary ammonium salts to methoxy-modified kaolinite (Kuroda et al., 2011; Yuan et al., 2013; Xu et al., 2015). More details can be found in Chapter 17.

In order to explain the formation mechanism of kaolinite nanorolls, two models have been proposed by Gardolinski and Lagaly (2005) and Yuan et al. (2013). In the first model, the amine molecules in the grafted kaolinites were arranged such that the amino groups interacted with the interlayer hydroxyl groups. During intercalation, the grafted molecules changed their own arrangement from an almost parallel orientation, to the basal plane surfaces, to a perpendicular orientation. In this arrangement, the interaction between the alkyl chains of the grafted molecule and the amine was maximized, and the basal expansion needed to accommodate the amine bilayer was kept to a minimum. Because the large basal spacing of the intercalate impeded the formation of hydrogen bonds between the kaolinite layers, the obtained tubes were likely composed of rolled single kaolinite layers or contained only a few layers.

In Yuan et al. (2013) model, a monolayer of methanol molecules was intercalated between the undisturbed siloxane surface and the methoxy-modified inner surface in which the initial aluminol (AlOH) groups in the interlayer space were converted to aluminium oxymethyl (AlOMe) groups. After the cetyltrimethylammonium chloride (CTMACl) intercalation, the CTMACl molecules were arranged in the form of a tilted bilayer in the interlayer space of kaolinite, and the inclination angle of the alkyl chains of CTMACl was approximately 38.4 degrees, based on a stereochemical calculation. Hence, the kaolinite–CTMACl complex provided a sufficient expansion of the interlayer space, which promoted a transformation of the kaolinite layers into nanorolls.

In addition, for the one-step route, Kuroda et al. (2011) ascribed to the relatively weak interactions between the head groups of the guest species and the interlayer surface of methoxy-modified kaolinite to the successful synthesis of nanorolls, which is believed to allow the formation of a flexible array of interlayer guest species for swelling and is conducive for kaolinite plates transforming into nanorolls. The formation route to the nanoscrolls depends not only on the surface modification of kaolinite, but also on the structure of guest species; interaction between kaolinite layers, guest species and methanol should be balanced to achieve both sufficient intercalation and swelling. However, in the two-step route, only interaction between the intercalation compound and toluene should be adjusted to achieve sufficient swelling.

In summary, more effort was exerted to improve the yield of kaolinite nanorolls than to investigate the rolling mechanism based on application requirements. It seems that the rolling mechanism for both methods is due to the weakening of hydrogen bonds between adjacent layers as a result of the expansion of basal spacing. This enables tetrahedral curving to have an effect on layer rolling. There is no direct evidence for the roles of surface tension and OH–OH attraction.

16.3.2 Rolling in Natural Transformation from Kaolinite to Halloysite

Natural examples from different geological periods may involve information about rolling in kaolinite layers; therefore, some studies have been carried out to find corresponding evidence of kaolinite rolling, especially through intuitive electron microscope technology.

Singh and Gilkes (1992) observed that kaolinite plates within pseudomorphs of mica crystals had fractured at irregular intervals along the [100] crystallographic axis to produce laths elongated along the [010] axis. The laths near the edges of the pseudomorphs were less constrained by the pseudomorph and could roll to produce halloysite tubes (Fig. 16.6). The tubes showed two types of twinning: the first is that tubes and associated laths were joined side by side, and the other is that single tubes bifurcated into two individual tubes. Singh and Gilkes (1992) proposed that the first type of twinning occurred through the folding of adjacent laths that remained joined, while the second type occurred as a result of the exfoliation of a thick lath, followed by the folding of exfoliated lath fragments into tubes. These results support the idea that halloysite is a consequence of development within a kaolin pseudomorph after mica, and provide direct evidence for the rolling of halloysite layers when space became available as a result of partial dehydration, exfoliation or both.

Robertson and Eggleton (1991) used scanning electron microscopy (SEM) and TEM to investigate several mixed kaolinite and halloysite materials derived from the weathering of different feldspars. This showed that kaolinite plates were partly altered into parallel, polygonal rods of halloysite (7 Å), which had planar sides and a central void, and then partly fused with the surfaces of the kaolinite crystals (Fig. 16.7). In this proposed process, the transformation was initiated by a loss of structural rigidity at points along the kaolinite crystal, interpreted as hydration to halloysite. At these points, the mixed grain began to curl. During the early parts of this process, the remnants of kaolinite provided a localized rigidity, which restricted curvature and caused the halloysite to curl into a spiral with a polygonal cross section. This residual rigidity was lost through the gradual alternation of the relict kaolinite and the rods converted to those with circular or oval cross sections. The long axis of the halloysite spirals was parallel to the crystallographic axis of the kaolinite.

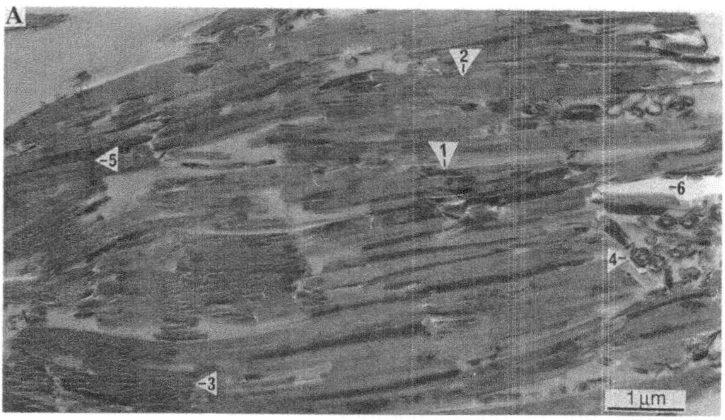

FIG. 16.6 TEM of a pseudomorph after mica. In the regions marked 4, the kaolinite plates fractured and partly or completely rolled or folded to form tubes. *Reproduced with permission from Singh and Gilkes (1992). Copyright (1992) from The Clay Minerals Society, publisher of* Clays and Clay Minerals.

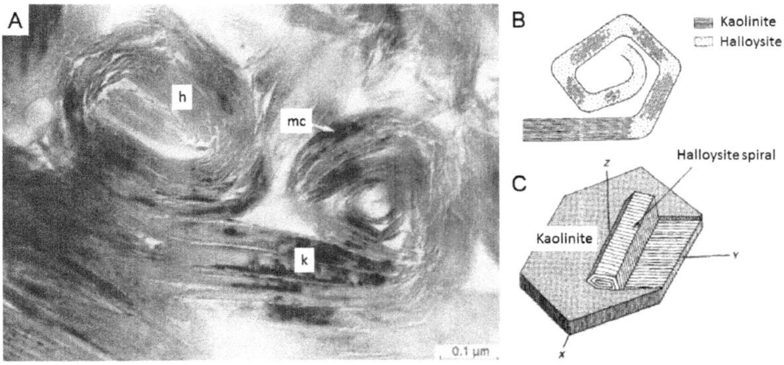

FIG. 16.7 (A) TEM of strips of kaolinite (k) continuous with polygonal and oval rolls of halloysite (h). Parts of the halloysite spirals contain straight sections, similar to kaolinite (mottled diffraction contrast, mc). (B) Model of halloysite spiral development. Hydration of kaolinite to halloysite has occurred at points along a kaolinite crystal. At these points, halloysite curls. (C) Halloysite spiral developed on and attached to a kaolinite plate by rolling up part of the plate. *Reproduced with permission from Robertson and Eggleton (1991). Copyright (1991) from The Clay Minerals Society, publisher of* Clays and Clay Minerals.

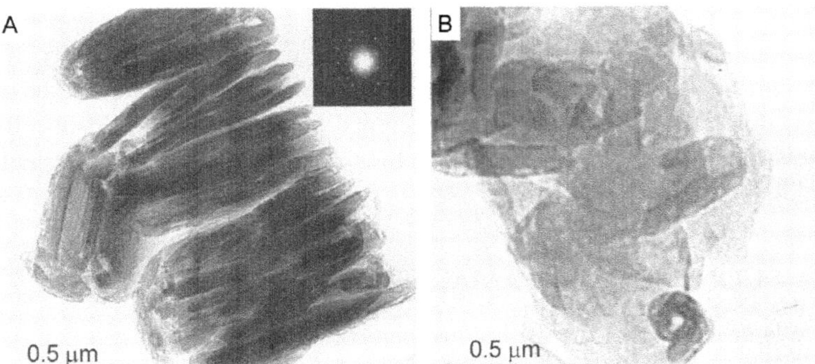

FIG. 16.8 (A) Aggregates of folded lath-shaped kaolinite, and the corresponding SAED pattern. (B) TEM image exhibiting rolled-up, lath-shaped kaolinite. *Reproduced with permission from Bobos et al. (2001). Copyright (2001) from The Clay Minerals Society, publisher of* Clays and Clay Minerals.

Bobos et al. (2001) identified the transformation of kaolinite to halloysite (7 Å) in a kaolinite deposit using XRD, Fourier transform infrared spectroscopy (FTIR) and TEM techniques. Both XRD patterns and FTIR spectra show the changes from kaolinite to halloysite (7 Å). The interlayer water content in the kaolinite structure increased during this transition. The lath-shaped kaolinite crystals were folded or rolled, and the halloysite exhibited an intermediate spherical shape formed from the rolled lath kaolinite (Fig. 16.8). These authors proposed that the transformation of kaolinite to halloysite (7 Å) is controlled by the surface reaction, in which exposing oxide surfaces to water caused the formation of surface OH groups that could be ionized. The pH of the medium influenced the degree of ionization. This reaction could contribute to the hydration of the kaolinite structure and subsequently the rolling mechanism described by Singh (1996).

The transformation between kaolinite and halloysite (especially from kaolinite to halloysite) has been the focus of attention in both artificial experiments and natural conversion studies. More and more evidence shows that tetrahedral curving plays a major role in the rolling of kaolinite layers. Due to the technical limitations of the measurements, the details of rolling cannot be monitored in real time, so they are inspected by investigations of the transformation's remains, which may continue into the future. Moreover, there are still two facts that need to be emphasized. First, different curling orientations have appeared in earlier experimental results prior to Niu et al. (2014); for example, different curling orientations were actually shown in an SEM image in a paper by Yuan et al. (2013), although the authors did not directly point out this phenomenon. This result suggests that the diversity of curling orientations is not an occasional case. If hydration status is not taken into consideration, the approaches involving no water (for example, the one-step method) may be more convenient when investigating the curling of kaolinite layers because of the greater curling amount, which can provide more opportunities to test and therefore serve as a basis for future research. Additionally, if multiple curling orientations are involved in natural samples cannot be determined due to a larger or smaller difference between kaolinite nanorolls and natural halloysite tubes, more research is required to reveal the orientation of halloysite tubes.

16.4 CONCLUDING REMARKS

There is no doubt that the rolling mechanism of the unit layer is a key issue in the formation of halloysite tubes or kaolinite nanorolls, for which a plausible theoretical framework has been established after years of development. Up to now, three possible mechanisms of rolling kaolinite into tubular halloysite have been proposed: (i) the mismatch between octahedral and tetrahedral sheets; (ii) the attraction between interlayer hydroxyl groups in octahedrons; and (iii) the surface tension of water. Of these, the first is the most widely accepted, but the effect of the other two mechanisms on rolling cannot be completely ruled out. For the first mechanism, several details are further explored: the tetrahedral rotation and curving effect, how water molecules enter into the interlayer spaces and the curling orientations. Some synthesis experiments producing halloysite-like structures, as well as observations regarding natural minerals containing information about the transformation from kaolinite to halloysite, also have been studied to collect more evidence for the rolling mechanism. These past studies have achieved fruitful results, but many issues remain, which are emphasized here:

(1) When the interlayer space of kaolinite or halloysite is small, interaction between layers is strong and tetrahedral rotation plays the dominant role, as in the case of kaolinite and halloysite (7 Å); and when the interlayer space is large, interaction between layers is weakened and tetrahedral curving plays a dominant role, as in the case of the kaolinite–CTMACl complex. But when the interlayer space is not particularly large (eg, 10 Å), the detailed role of water (or other molecules) is still unclear.
(2) It has been established that tetrahedral curving has certain distributions in the kaolinite layer that are not strictly in accordance with the crystallographic axes, and has two different matchings with octahedrons. However, a more detailed relevance between the macroscopic curling orientations of layers and tetrahedral curving has not been well understood.
(3) Many experiments have explored the effects of the substitution of Fe for Al in the octahedral sheet and Al for Si in the tetrahedral sheet; however, the detailed effect of these substitutions on the rolling of layers is still unclear. For example, the quantitative relationship between Fe substitution and the corresponding volume expansion of the octahedral sheet needs further research.

REFERENCES

Amara, A.B.H., 1997. X-ray diffraction, infrared and TGA/DTG analysis of hydrated nacrite. Clay Miner. 32, 463–470.

Aomine, S., Wada, K., 1962. Differential weathering of volcanic ash and pumice, resulting in formation of hydrated halloysite. Am. Mineral. 47, 1024–1048.

Bailey, S.W., 1980. Kaolin minerals. In: Brindley, G.W., Brown, G. (Eds.), Crystal structures of clay minerals and their X-ray identification. Mineralogical Society, London, pp. 1–123.

Bailey, S.W., 1988. Polytypism of 1:1 layer silicates. In: Bailey, S.W. (Ed.), Hydrous Phyllosilicates (Exclusive of Micas). Reviews in Mineralogy, vol. 19. Mineralogical Society of America, Washington DC, pp. 9–27.

Bailey, S.W., 1990. Halloysite—a critical assessment. Sci. Geol. Mem. 86, 89–98.

Balan, E., Lazzeri, M., Morin, G., Mauri, F., 2006. First-principles study of the OH-stretching modes of gibbsite locality: hypothetical structure calculated with DFT. Am. Mineral. 91, 115–119.

Bates, T.F., Hildebrand, F.A., Swineford, A., 1950. Morphology and structure of endellite and halloysite. Am. Mineral. 35 (7–8), 463–484.

Birrell, K.S., Fieldes, M., Williamson, K.I., 1955. Unusual forms of halloysite. Am. Mineral. 40, 122–124.

Bish, D.L., 1993. Rietveld refinement of the kaolinite structure at 1.5 K. Clay Clay Miner. 41, 738–744.

Bobos, I., Duplay, J., Rocha, J., Gomes, C., 2001. Kaolinite to halloysite-7 Å transformation in the kaolin deposit of São Vicente de Pereira, Portugal. Clay Clay Miner. 49 (6), 596–607.

Churchman, G.J., Lowe, D.J., 2012. Alteration, formation, and occurrence of minerals in soils. In: Huang, P.M., Li, Y., Sumner, M.E. (Eds.), Handbook of Soil Sciences, second ed. Properties and Processes, vol. 1. CRC Press (Taylor and Francis), Boca Raton, FL, pp. 20.1–20.72.

Churchman, G.J., Theng, B.K.G., 1984. Interactions of halloysites with amides: mineralogical factors affecting complex formation. Clay Miner. 19 (2), 161–175.

Cleaves II, H.J., Scott, A.M., Hill, F.C., Leszczynski, J., Sahai, N., Hazen, R., 2012. Mineral–organic interfacial processes: potential roles in the origins of life. Chem. Soc. Rev. 41 (16), 5502–5525.

Costanzo, P.M., Clemency, C.V., Giese, R.E., 1980. Low temperature synthesis of a 10-Å hydrate of kaolinite using dimethylsulfoxide and ammonium fluoride. Clay Clay Miner. 28, 155–156.

Costanzo, P.M., Giese, R.F., Lipsicas, M., Straley, C., 1982. Synthesis of a quasi-stable kaolinite and heat capacity of interlayer water. Nature 296, 549–551.

Costanzo, P.M., Giese, R.E., Clemency, C.V., 1984. Synthesis of a 10-Å hydrated kaolinite. Clay Clay Miner. 32, 29–35.

de Souza Santos, P., Brindley, G.W., de Souza Santos, H., 1965. Mineralogical studies of kaolinite-halloysite clays: part III. A fibrous kaolin mineral from Piedade, Sao Paulo. Brazil. Am. Mineral. 50, 619–628.

Gardolinski, J., Lagaly, G., 2005. Grafted organic derivatives of kaolinite: II. Intercalation of primary n-alkylamines and delamination. Clay Miner. 40 (4), 547–556.

Hart, R.D., Gilkes, R.J., Siradz, S., Singh, B., 2002. The nature of soil kaolins from Indonesia and Western Australia. Clay Clay Miner. 50 (2), 198–207.

Honjo, G., Mihama, K., 1954. A study of clay minerals by electron-diffraction diagrams due to individual crystallites. Acta Crystallogr. 7, 511–513.

Hope, E.W., Kittrick, J.A., 1964. Surface tension and the morphology of halloysite. Am. Mineral. 49, 859–866.

Jeong, G.Y., 2000. The dependence of localized crystallization of halloysite and kaolinite on primary minerals in the weathering profile of granite. Clay Clay Miner. 48, 196–203.

Joussein, E., Petit, S., Churchman, J., Theng, B., Righi, D., Delvaux, B., 2005. Halloysite clay minerals—a review. Clay Miner. 40 (4), 383–426.

Kunze, G.W., Bradley, W.F., 1964. Occurrence of a tabular halloysite in a Texas soil. In: Bradley, W.F. (Ed.), Proceedings of the 12th National Conference on Clays and Clay Minerals. Pergamon Press, Oxford, UK, pp. 523–528.

Kuroda, Y., Ito, K., Itabashi, K., Kuroda, K., 2011. One-step exfoliation of kaolinites and their transformation into nanoscrolls. Langmuir 27, 2028–2035.

Ledoux, R.L., White, J.L., 1966. Infrared studies of hydrogen bonding interaction between kaolinite surfaces and intercalated potassium acetate, hydrazine, formamide, and urea. J. Colloid Interface Sci. 21 (2), 127–152.

Li, C., Wang, J., Feng, S., Yang, Z., Ding, S., 2013. Low-temperature synthesis of heterogeneous crystalline TiO_2–halloysite nanotubes and their visible light photocatalytic activity. J. Mater. Chem. A 1 (27), 8045–8054.

Liu, H., Wang, C., Zou, S., Wei, Z., Tong, Z., 2012. Facile fabrication of polystyrene/halloysite nanotube microspheres with core–shell structure via pickering suspension polymerization. Polym. Bull. 69, 765–777.

Maksimovic, Z.J., White, J.L., 1973. Infrared study of chromium-bearing halloysites. In: Serratosa, J.M. (Ed.), Proceedings of the International Clay Conference. Division de Ciencias, C.S.I.C., Madrid, Spain, pp. 61–73

Matusik, J., Gaweł, A., Bielańska, E., Osuch, W., Bahranowski, K., 2009. The effect of structural order on nanotubes derived from kaolin-group minerals. Clay Clay Miner. 57 (4), 452–464.

Merino, E., Harvey, C., Murray, H.H., 1989. Aqueous-chemical control of the tetrahedral-aluminum content of quartz, halloysite, and other low-temperature silicates. Clay Clay Miner. 37 (2), 135–142.

Muller, J.P., Calas, G., 1989. Tracing kaolinites through their defect centers; kaolinite paragenesis in a laterite (Cameroon). Econ. Geol. 84 (3), 694–707.

Nagasawa, K., Noro, H., 1987. Mineralogical properties of halloysites of weathering origin. Chem. Geol. 60 (1), 145–149.

Newman, R.H., Childs, C.W., Churchman, G.J., 1994. Aluminium coordination and structural disorder in halloysite and kaolinite by ^{27}Al NMR spectroscopy. Clay Miner. 29, 305–312.

Newnham, R.E., Brindley, G.W., 1956. The crystal structure of dickite. Acta Crystallogr. 9 (9), 759–764.

Niu, J., Qiang, Y., Li, X., Liu, Z., Zhang, S., Feng, P., Ou, X., 2014. Morphology and orientation of curling of kaolinite layer in hydrate. Appl. Clay Sci. 101, 215–222.

Noro, H., 1986. Hexagonal platy halloysite in an altered tuff bed, Komaki city, Aichi prefecture, Central Japan. Clay Miner. 21, 401–415.

Olejnik, S., Aylmore, L.A.G., Posner, A.M., Quirk, J.P., 1968. Infrared spectra of kaolin mineral-dimethyl sulfoxide complexes. J. Phys. Chem. 72 (1), 241–249.

Olejnik, S., Posner, A.M., Quirk, J.P., 1970. The intercalation of polar organic compounds into kaolinite. Clay Miner. 8 (4), 421–434.

Pauling, L., 1930. The structure of the chlorites. Proc. Natl. Acad. Sci. U.S.A. 16, 578–582.

Quantin, P., Gautheyrou, J., Lorenzoni, P., 1988. Halloysite formation through in situ weathering of volcanic glass from trachytic pumices, Vico's volcano, Italy. Clay Miner. 23 (4), 423–437.

Radoslovich, E.W., 1963. The cell dimensions and symmetry of layer lattice silicates: VI. Serpentine and kaolin morphology. Am. Mineral. 48, 368–378.

Raythatha, R.H., Lipsicas, M., 1985. Mechanism of synthesis of 10-Å hydrated kaolinite. Clay Clay Miner. 33, 333–339.

Robertson, I.D.M., Eggleton, R.A., 1991. Weathering of granitic muscovite to kaolinite and halloysite and of plagioclase-derived kaolinite to halloysite. Clay Clay Miner. 39 (2), 113–126.

Sanchez, C.M., Gonzalez, G.S., 1970. Modification of the kaolinite crystal habit by dimethyl sulfoxide treatment. An. Edafol. Agrobiol. 29, 651–655.

Shannon, R.T., 1976. Revised effective ionic radii and systematic studies of interatomic distances in halides and chalcogenides. Acta Crystallogr. A 32 (5), 751–767.

Singer, A., Zarei, M., Lange, F.M., Stahr, K., 2004. Halloysite characteristics and formation in the northern Golan Heights. Geoderma 123, 279–295.

Singh, B., 1996. Why does halloysite roll?—a new model. Clay Clay Miner. 44 (2), 191–196.

Singh, B., Gilkes, R.J., 1992. An electron-optical investigation of the alteration of kaolinite to halloysite. Clay Clay Miner. 40, 212–229.

Singh, B., Mackinnon, I.D.R., 1996. Experimental transformation of kaolinite to halloysite. Clay Clay Miner. 44 (6), 825–834.

Soma, M., Churchman, G.J., Theng, B.K.G., 1992. X-ray photoelectron spectroscopic analysis of halloysites with different composition and particle morphology. Clay Miner. 27, 413–421.

Takahashi, T., Dahlgren, R.A., Theng, B.K.G., Whitton, J.S., Soma, M., 2001. Potassium-selective, halloysite-rich soils formed in volcanic materials from Northern California. Soil Sci. Soc. Am. J. 65 (2), 516–526.

Tunney, J.J., Detellier, C., 1994. Preparation and characterization of an 8.4 Å hydrate of kaolinite. Clay Clay Miner. 42, 473–476.

Wada, K., 1965. Intercalation of water in kaolin minerals. Am. Mineral. 50, 924–941.

Wada, S.I., Mizota, C., 1982. Iron-rich halloysite (10 Å) with crumpled lamellar morphology from Hokkaido, Japan. Clay Clay Miner. 30, 315–317.

Watanabe, Y., Kitagawa, Y., Sugo, S., 1969. Structural aspects of clay minerals occurring in some soils. In: Heller, L. (Ed.), Proceedings of the International Clay Conference. Israel University Press, Jerusalem, pp. 129–140.

Wei, Z., Wang, C., Liu, H., Zou, S., Tong, Z., 2012. Halloysite nanotubes as particulate emulsifier: preparation of biocompatible drug-carrying PLGA microspheres based on pickering emulsion. J. Appl. Polym. Sci. 125, 358–368.

White, R.D., Bavykin, D.V., Walsh, F.C., 2012. Spontaneous scrolling of kaolinite nanosheets into halloysite nanotubes in an aqueous suspension in the presence of GeO_2. J. Phys. Chem. C 116, 8824–8833.

Xu, H., Jin, X., Chen, P., Shao, G., Wang, H., Chen, D., Liu, H., Zhang, R., 2015. Preparation of kaolinite nanotubes by a solvothermal method. Ceram. Int. 41 (5), 6463–6469.

Yuan, P., Tan, D., Annabi-Bergaya, F., Yan, W., Liu, D., Liu, Z., 2013. From platy kaolinite to aluminosilicate nanoroll via one-step delamination of kaolinite: effect of the temperature of intercalation. Appl. Clay Sci. 83, 68–76.

Zhou, C., Keeling, J., 2013. Fundamental and applied research on clay minerals: from climate and environment to nanotechnology. Appl. Clay Sci. 74, 3–9.

Chapter 17

Halloysite-like Structure via Delamination of Kaolinite

J. Matusik*
Faculty of Geology, Geophysics and Environmental Protection, AGH University of Science and Technology, Krakow, Poland
**Corresponding author: e-mail: jakub_matusik@wp.pl*

17.1 INTRODUCTION

There is an increasing demand for mineral-based hybrid nanomaterials showing specific morphology which can be strictly controlled during their synthesis. Noncarbon-based fibrous or tubular structures of either natural or synthetic origin are versatile and relatively cheap and possess large specific surface areas (SSA) (Detellier and Schoonheydt, 2014). Their surface chemistry can be modified and sufficiently tuned for several applications by inducing additional active centres, immobilising appropriate molecules or both (Pan et al., 2011). Such materials may be employed for environmental and industrial purposes, which mainly include catalysis and adsorption. Moreover, the obtained nanoparticles may serve as reinforced fillers used in clay polymer nanocomposites (CPN) (Du et al., 2010; Matusik et al., 2011a).

The most known mesoporous materials (ie, with pore diameters in the 2–50 nm range) include derivatives of silica, such as MCM-41 (mobile crystalline material) and FSM-16 (folded sheet material) (Kresge et al., 1992; Inagaki et al., 1993; Lin and Mou, 1996). These materials do not possess a crystalline structure, but their porosity may be easily controlled by varying the synthesis conditions. Other nonmineral structures include lithium metasilicate (Li_2SiO_3) nanotubes, which could be prepared in hydrothermal conditions using aqueous alkaline solutions (Li and Yang, 2014). In the synthetic procedure, 1:1 layered nickel phyllosilicate nanotubes [$Ni_3Si_2O_5(OH)_4$] can be obtained at 250°C at a pressure of 10 MPa (McDonald et al., 2009). The nanostructures with lengths up to 200 nm were multiwalled, hollow and open-ended. A detailed description of the influence of synthesis conditions on the aluminogermanate nanotubes was given by Levard et al. (2008). Such structures are structural analogues of naturally occurring imogolite. It has

Developments in Clay Science, Vol. 7. http://dx.doi.org/10.1016/B978-0-08-100293-3.00017-0

been reported that the incorporation of Ge in the positions occupied by Si leads to the evolution of fibrous morphology. The nanotubular particles exhibited diameters equal to ~3.0 nm. Intercalation/deintercalation procedure performed at room temperature, which transforms the nanosheets into nanotubes, was described by Ma et al. (2004). The used inorganic nanosheets included titanium oxide (TiO_2), manganese oxide (MnO_2) and calcium niobate ($Ca_2Nb_3O_{10}$). An interesting inorganic lamellar material that shows curling in appropriate conditions includes tungsten disulfide (WS_2) and vanadium oxide (VO_x) (Chen et al., 2002; Li et al., 2002). Another protocol of creating nanotubes from kaolinite involved its use as a silica source (Dong et al., 2003). However, such treatment destroys the initial layered structure of the mineral.

Halloysite is a natural clay mineral exhibiting a 1:1 layered structure. Its layers are very often curled and show tubular morphology (Hendricks, 1938; Bates et al., 1950; Joussein et al., 2005). The mineral itself forms in environments with complex weathering processes (Kohyama et al., 1978; Robertson and Eggleton, 1991; Cho and Mermut, 1992; Singh and Gilkes, 1992). The tubular morphology of halloysite shows remarkable thermal resistance up to 900°C–1100°C (Yuan et al., 2012; Ouyang et al., 2014). The nanotube formation mechanism was connected to the lateral misfit between tetrahedral and octahedral sheets (Singh, 1996; Niu et al., 2014). The rolling of layers releases the strain connected with the misfit. This mechanism dominates when interlayer hydrogen bonds are weakened due to the presence of intercalated water molecules. In the case of kaolinite, where strong H-bonds hold the layers, the misfit correction takes place via the rotation of adjacent tetrahedras (Singh, 1996).

The natural deposits of halloysite in the world do not contain perfect and homogeneous natural nanotubes or very pure samples, with the presence of accessory minerals such as quartz or feldspars. The synthesis protocol can be an alternative, but the precipitation of halloysite in hydrothermal conditions from SiO_2 and Al_2O_3 containing solutions was shown to be very difficult. This is because the mineral is not thermodynamically stable in the natural environment (Joussein et al., 2005). In general, the conducted and reported syntheses led to preferential precipitation of kaolinite, which is energetically more favourable due to its reduced negative formation enthalpy. Many attempts were made to synthesise halloysite nanotubes from aluminosilicate gels and dilute solutions and through experimental alteration of geological materials (eg, volcanic glasses, feldspars and clays) in the laboratory (Van Oosterwyck-Gastuche and La Iglesia, 1978; Huertas et al., 1999; Fialips et al., 2000). All the attempts were unsuccessful, however, except the synthesis of an 'amorphous' aluminosilicate precipitate named 'protohalloysite', obtained by Hem and Lind (1974). In the 1970s, a patent has been proposed related to the possible synthesis of halloysite, but the reported synthetic material cannot be attributed to true halloysite (Robson, 1979). When kaolin minerals are obtained, kaolinite rather than halloysite is formed due to its relative stability. The main problems to drive the synthesis towards halloysite probably come from (i) the temperature of synthesis, (ii) kinetic aspect (ie, the time needed in controlled conditions) and (iii) thermodynamic stability. The appropriate synthesis conditions for halloysite should involve the use of solutions of very low concentrations (<2.4 mmol $Si(IV)_{aq}$/L and <1.7 $Al(III)_{aq}$/L) and low synthesis temperatures (<100°C) with regard to the halloysite hydration properties (La Iglesia and Martin-Vivaldi, 1975; La Iglesia and Van Oosterwyck-Gastuche, 1978; Van Oosterwyck-Gastuche and La Iglesia, 1978). Such conditions can help to avoid the crystallisation of kaolinite and formation of amorphous phases. Following the Arrhenius law, low temperature synthesis will take several hundred years to synthesise kaolin minerals from aluminosilicate gels. However, the temperature is not necessarily the only important factor. Indeed, many experiments clearly show the synthesis of kaolinite rather than halloysite regardless of the time and temperature involved. For example, Kittrick (1970) synthesised a small amount of kaolinite-like material at 25°C after 4 years starting from montmorillonite. Busenberg (1978) and Parham (1969) described halloysite-like minerals from alteration of feldspars in aqueous solutions at 25°C (50 days) and 78°C (140 days), and Bogatyrev et al. (1997) also obtained halloysite-like clay from gibbsite at 25°C in silicic solutions. In the case of the potential presence of halloysite, another issue is that the amount of synthesised clay mineral is very low and it is difficult to characterise samples accurately. This causes problems during the preparation of samples, observations and characterisation of particle morphology, which is not necessarily characteristic for halloysite or kaolinite. Thus, transmission electron microscopy (TEM) observations are only indicative. As reported by Joussein et al. (2005), when starting materials are of geological origin, especially volcanic glasses, one cannot exclude the initial presence of halloysite seeds, and the results should be examined with caution.

An interesting hydrothermal synthesis protocol was presented by White et al. (2012). In their procedure, in addition to Si(IV) and Al(III) as substrates, they used GeO_2 in different molar proportions in relation to Si(IV). The large ionic radius of Ge(IV), which could replace Si(IV), leads to the increase of misfit between the clay mineral sheets. As a result the observations showed formation of single-walled (SW) and multiwalled (MW) halloysite-like nanotubes. However, the amount of formed, pure halloysite nanotubes did not exceed 10% where the molar content of Ge(IV) was equal to 0.2. The content of Ge(IV) in the products obtained with the addition of Ge(IV) < 0.2 was below the detection limit of an energy-dispersive X-ray detector (EDX). This confirmed that the obtained nanotubular particles were pure aluminosilicates in chemical composition; however, their formation mechanism was ambiguous and was not solved in detail. The incorporation of higher amounts of Ge(IV) enhanced the formation of aluminogermanate nanotubes.

Due to problems with halloysite synthesis, other methods for aluminosilicate nanotube preparation were proposed using kaolinite as a substrate. These synthetic routes, reported during the last decade or so, were much more efficient and led to kaolinite-layer delamination and curling (Gardolinski and Lagaly, 2005b; Matusik et al., 2009; Kuroda et al., 2011). It is worth underlining that the described procedures do not destroy the initial kaolinite-layered structure. The morphology conversion was based on the intercalation/deintercalation processes, which mechanism was studied in detail in 2015 (Li et al., 2015). In general, several known chemical treatments leading to delamination and rolling of kaolinite layers may be divided into three categories: (i) one-step procedures, (ii) two-step procedures and (iii) polymer-induced treatment. The chapter summarises these synthetic routes, which allow for obtaining halloysite-like tubular structures from platy kaolinite particles, and provides information on their structural and textural features, as well as their thermal stability.

17.2 EARLY DELAMINATION STUDIES

One of the first historical descriptions showing delamination and rolling of 1:1 layers was by Hope and Kittrick (1964). These authors investigated the role of surface tension as the factor that affects halloysite morphology. Microphotographs of freeze-dried halloysite samples showed the dominating presence of platy particles. This was connected with the lack of surface tension due to water molecule sublimation. On the other hand, normal drying caused curling of layers, and numerous tubular particles were observed. Moreover, these authors reported that the obtained particles were similar in morphology to the particles synthesised in previous studies (eg, Gastuche et al., 1954) with a procedure involving kaolinite treatment with nitrobenzene in a reflux system for 305 h. These authors also reported a mechanism including a destroying of the interlayer hydrogen bonds due to the proton accepting properties of nitrobenzene. In conclusion, an observation was made that the nitrobenzene treatment and normal drying causes layers curling only when particles are sufficiently thin. When the particles are built from several layers, the curling takes place only at the edges to a small extent. This confirmed the important role of a delamination phenomenon defined as a separation of individual 1:1 layers. The influence of forced hydration on the rolling possibility of kaolinite particles was examined in detail based on a well-ordered Georgia kaolinite (KGa-1) from the Clay Minerals Society Repository (Singh and Mackinnon, 1996). The hypothesis reported and investigated theoretically in Singh (1996) was tested. The procedure involved repeated washing with potassium acetate CH_3COONH_4 and water, which caused the formation of kaolinite intercalate ($d_{001} = 14.1$ Å) in accordance with earlier studies (Wada, 1961). Also, interstratified particles giving reflection at $d = \sim 8.0$ Å were identified

(Churchman et al., 1972; Kohyama et al., 1978). The results show that the procedure leads to increase in stacking disorder and reduction of particle size. Most importantly the electron microphotographs showed the presence of tubular particles after 10 cycles of repeated washing. More extensive treatment (30 cycles) resulted in the formation of numerous curled particles with morphology identical to natural halloysite tubes (Fig. 17.1). Their thickness varied from 10 to 50 nm. In the discussion, the authors excluded the role of surface tension, as they observed a preferential curling direction of plates; ie, the tube axis was always parallel to major crystallographic directions. The study concluded that hydration is one of the most important factors leading to layer rolling. This was also shown earlier through TEM observations of kaolinite-to-halloysite transformation in geological samples (Robertson and Eggleton, 1991). Research on halloysite formation has indicated that the delamination and rolling of kaolinite layers in natural environments can be connected with near-surface, rapid mineral formation in solutions of high ionic strength. Thus, the formed disordered mineral, which lost its structural rigidity, can accommodate water molecules and subsequently form tubular morphology.

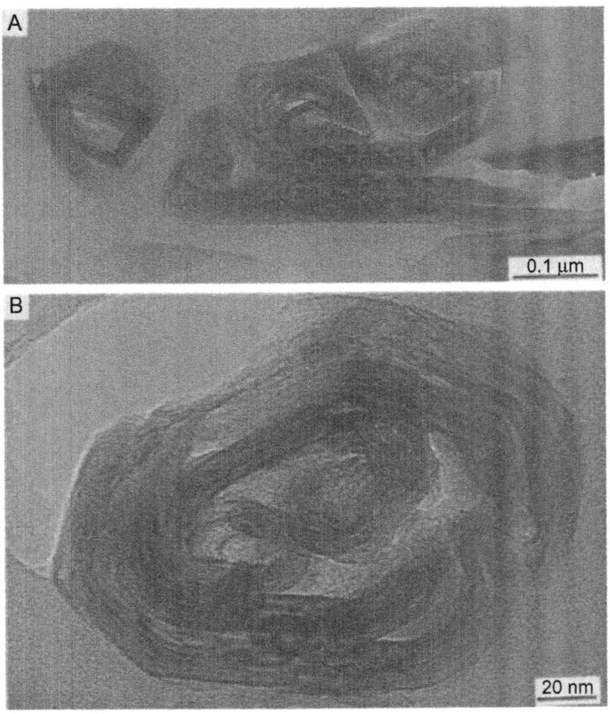

FIG. 17.1 Halloysite-like nanotubes prepared by repeated washing with potassium acetate and-water. More details can be found in the original paper. *Reprinted with permission from Singh and Mackinnon (1996). Copyright 2015 The Clay Minerals Society.*

It is worth underlining here that mechanochemical treatment coupled with intercalation may also lead to morphology changes in kaolinite (Sun et al., 2011). However, such treatment aimed at reducing the size of particles rather than controlling the formation of nanotubes.

The above mentioned experiments provide direct evidence of transformation from kaolinite layers to nanorolls and facilitate the understanding of the existence of natural tubular halloysite. However, kaolinite nanorolls with high yield are difficult to obtain through these methods, which greatly limits their large-scale applications in many areas.

17.3 TWO-STEP DELAMINATION AND ROLLING PROCEDURES

Two-step procedures involve the intercalation of alkylamines, followed by deintercalation of the organic guest by an appropriate solvent (Fig. 17.2). As the intercalating large molecules are usually hydrophobic, an appropriate preparation needs to be done prior to this process. This involves the synthesis of methoxy-kaolinite or diol-grafted derivatives from a kaolinite dimethyl sulphoxide (DMSO) preintercalate (Fig. 17.3). DMSO is one of the chemical compounds that can be directly intercalated and induces kaolinite swelling (Olejnik et al., 1968; Thompson and Cuff, 1985). This is due to its high dipole moment, which enables the formation of new H-bonds in the kaolinite layer between inner-surface OH groups of the mineral and oxygen atoms of DMSO. Although DMSO is nowadays most often used to obtain pre-intercalates of kaolin-group minerals, the first historical intercalation of kaolinite was done by Wada (1961) with the use of potassium acetate. Other molecules that have the ability to directly penetrate and subsequently intercalate kaolinite

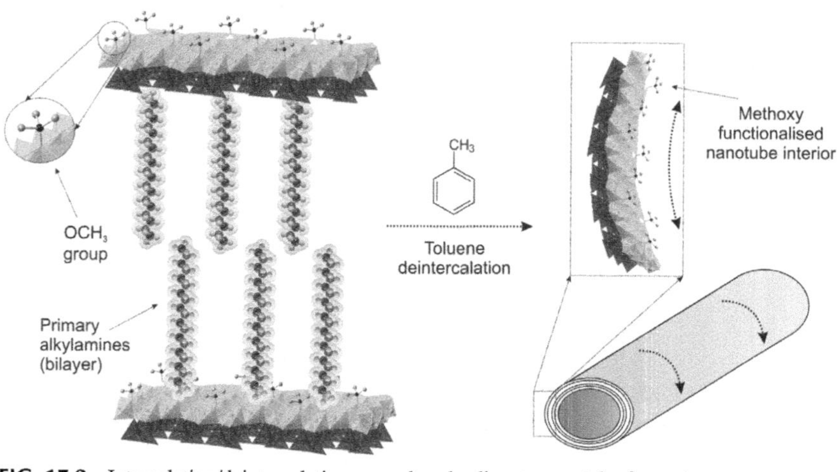

FIG. 17.2 Intercalation/deintercalation procedure leading to nanotube formation.

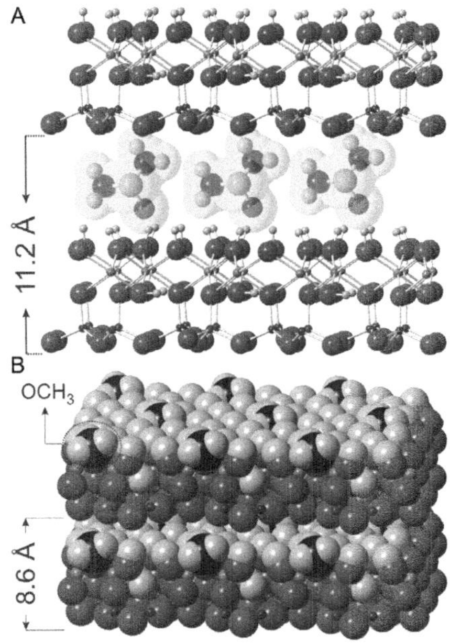

FIG. 17.3 Approximate structure of: (A) kaolinite DMSO intercalate and (B) methoxy-modified kaolinite.

interlayer space include urea (Weiss, 1963), formamide (Churchman et al., 1984) and hydrazine (Barrios et al., 1977). The obtained preintercalates of kaolinite can be used further for the intercalation of organic molecules, which could not be directly intercalated including alcohols (Tunney and Detellier, 1996b), diols (Gardolinski and Lagaly, 2005a), aminoalcohols (Letaief and Detellier, 2007), amines (Komori et al., 1999) or silylating agents (Avila et al., 2010). Moreover, it was shown that large polymer molecules, such as polyacrylonitrile, polyacrylamide, poly(ethylene glycol) and polystyrene, could be intercalated using DMSO-intercalated kaolinite as a starting material (Sugahara et al., 1988, 1990; Tunney and Detellier, 1996a; Elbokl and Detellier, 2006).

From the viewpoint of nanotubes formation the DMSO should be exchanged by organic compound which will increase hydrophobicity of the interlayer environment. Thus, a repeated washing of DMSO pre-intercalate with methanol could be done resulting in a formation of methoxy-kaolinite which in a dry state shows a reflection of ~8.6 Å (Tunney and Detellier, 1996b; Komori et al., 2000; Matusik et al., 2012b). The detailed investigations showed that part of inner-surface OH groups are exchanged by OCH_3 groups that form covalent Al–O–C bonds. Thus, the obtained grafted compound is an organic/inorganic mineral hybrid with a hydrophobic octahedral sheet. The material can be used for further intercalation of materials such as alkylamines or quaternary ammonium salts (Gardolinski and Lagaly, 2005b; Matusik and Kłapyta, 2013).

The first complete report on the intercalation/deintercalation synthetic routes was included in the PhD thesis by Gardolinski (2005). Afterwards, the most interesting results were published (Gardolinski and Lagaly, 2005a,b) and increased the body of knowledge about new interlayer grafted materials based on kaolinite, which subsequently served as a base for nanotubes synthesis. These results showed the susceptibility of inner-surface OH groups to react with selected organic molecules (ie, alkanols, diols and glycol mono-ethers). In most cases, the introduced molecules formed a monolayer in the interlayer space and were attached through a Al–O–C covalent bond to the octahedral sheet. The thermal stability of these intercalated materials was remarkably high (up to ~300°C), showing resistance to washing with organic solvents and water. The interlayer grafting was crucial for further intercalation of long-chain, primary amine molecules (Gardolinski and Lagaly, 2005b). The studies showed that the amine molecules were arranged in a bilayer, and therefore, the monitored d_{001} value increased significantly. The largest increase of ~64.2 Å was noticed for *n*-docosanamine molecule. The intercalated neutral amines were subjected to deintercalation, which was conducted using toluene, showing the highest removal efficiency. The conducted deintercalation led to the spontaneous delamination and rolling of kaolinite layers; the halloysite-like curled particles dominated in the treated samples. This was due to efficient breaking of the interlayer hydrogen bonds induced by a sevenfold to eightfold increase of the initial *d*-value of ~7.2 Å. The smallest external diameter of nanotubes was equal to ~25 nm, which closely corresponded to a single rolled kaolinite layer according to earlier calculations (Bates et al., 1950). As was evidenced, the solvent washing did not remove the grafted molecules. Thus, the nanotube interior, with exposed octahedral sheet, was tailored by grafted species, which controlled the chemical hydrophobic/hydrophilic character and subsequently the rheological behaviour of formed nanomaterials (Gardolinski and Lagaly, 2005a).

More recently, the investigations were carried out on the effect of the structural order of kaolin-group minerals on the efficiency of nanotube formation in a two-step procedure (Matusik et al., 2009, 2011b). In the experiments, three types of kaolin-group minerals from Polish deposits were chosen which differed in the degree of structural order (as expressed by the Hinckley index, given in parentheses): Maria III kaolinite (1.31), Jaroszów kaolinite (0.22) and halloysite from Dunino deposits (0.33). The results showed that the layer rolling depended very heavily on the efficiency of each intercalation and grafting step and also on the structural order of the starting mineral. The crucial step involved an appropriate selection of organic guest for the grafting reaction. When 1,3-butanediol was used, the rolling efficiency was higher for minerals with low structural order (Matusik et al., 2009). In all cases, the observed nanotubes had mean diameters close to ~30 nm in agreement with earlier studies. Their length was in the 150–500-nm range, and observations showed that this dimension depended on the initial size of the kaolinite particles. Thus, exceptionally, particles with length of ~1 μm were observed. As reported, the chemical treatment led to a decrease of structural order and particle size due to extensive delamination. These effects were much more pronounced for the poorly ordered Jaroszów kaolinite. The sample of halloysite from Dunino deposits apart from platy kaolinite particles contained natural, micrometre-sized nanotubes. The chemical treatment of this sample showed that the morphology of the natural tubes was not altered. In turn, the associated kaolinite grains were delaminated and formed stacks of layers that were subsequently rolled (Fig. 17.4). The use of methoxy-modified kaolinite for the intercalation/deintercalation procedure significantly increased the number of rolled layers in the case of well-ordered Maria III kaolinite (Fig. 17.4) (Matusik et al., 2011b). In turn, the rolling efficiency was very low for the kaolinite and halloysite of low structural order. The observations were in

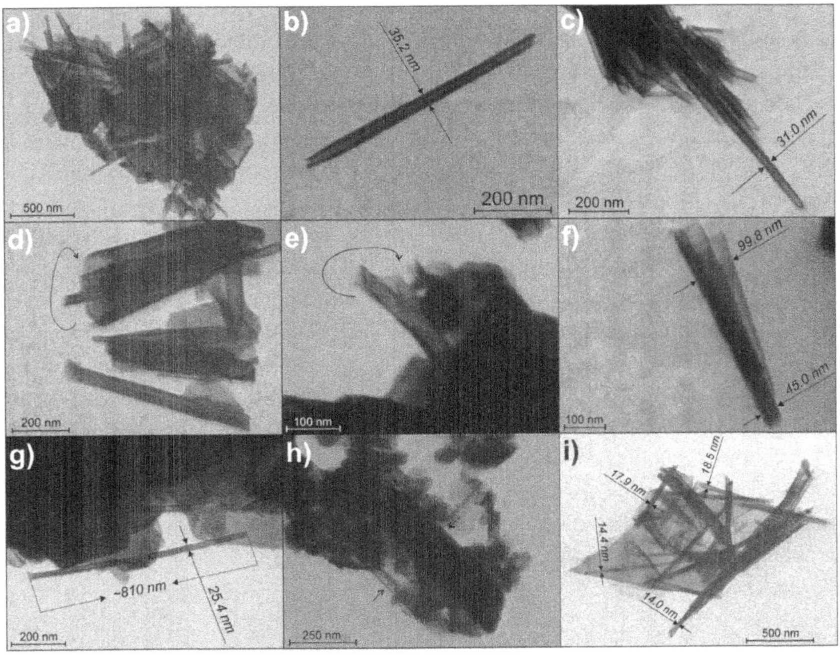

FIG. 17.4 TEM microphotographs of: (A–C) halloysite-like nanotubes, (D–F) partially rolled kaolinite layers, (G) an exceptionally long nanotube, (H) synthetic and natural halloysite nanotubes and (I) dickite-based nanotubes. *Reprinted with permission from Matusik et al. (2009, 2011b). Copyright 2015 The Clay Minerals Society. Panel (I): Reprinted with permission from Matusik et al. (2012a). Copyright 2015 Elsevier.*

contrast to these noticed when 1,3-butanediol was used (Matusik et al., 2009). Moreover, the amount of formed nanotubes increased when an amine with longer hydrocarbon chain was used (ie, octadecylamine instead of hexylamine). On the basis of N_2 adsorption/desorption measurements, Matusik et al. (2011b) provided information on the textural parameters of formed nanotubular particles that revealed their mesoporous character with insignificant microporosity. For the Maria III kaolinite, an increase of SSA (S_{BET}) and total pore volume (V_{tot}) was observed. The calculations showed an agreement between the mean diameter of the newly formed pores and the observed diameters of the nanotubes. This confirmed that the texture evolution was exclusively due to delamination and rolling phenomena. Such connection was not observed for the poorly ordered kaolin-group minerals, where mainly macro- and large mesopores were observed as interparticle and interaggregate spaces. Further studies showed that the intercalation/deintercalation can also lead to the formation of nanotubes using dickite as a starting mineral (Matusik et al., 2012a). The relatively thick dickite particles were delaminated through methoxy-modification and hexylamine/toluene treatment (Fig. 17.4).

A similar two-step approach leading to nanotube formation was reported in 2014 (Liu et al., 2014). These authors, however, used methanol washing to deintercalate alkylamine molecules. This step, in connection with ultrasonic treatment, caused delamination and 1:1 layer rolling. In accordance with earlier studies, these authors reported an increase in efficiency of nanotubes formation, along with an increase of the hydrocarbon chain lengths of intercalated alkylamines. The folding of kaolinite was also observed after intercalation of metallophthalocyanines, which are synthetic compounds similar to porphyrins (da Silva et al., 2014). As a result, the lumens of formed kaolinite nanotubes were occupied by introduced Co(II) tetracarboxyphthalocyanine, showing catalytic activity. The use of kaolinite nanotubes obtained by intercalation/deintercalation for catalysis was also reported by Nakagaki et al. (2006). For the deintercalation procedure, toluene was used, along with the addition of iron porphyrin. These authors reported high catalytic activity of the hybrid nanomaterials in oxidation reactions of cyclohexane and *n*-heptane. The kaolinite nanotubes were classified as a very useful nanoporous material for catalytic purposes, as well as other nanotubular and nanofibrous structures, such as natural halloysite, chrysotile, imogolite, fibrous silica, hydroxides, niobates, hydroxide salts and layered double hydroxides (Nakagaki and Wypych, 2007).

17.4 ONE-STEP DELAMINATION AND ROLLING PROCEDURES

One-step transformation of kaolinite morphology to halloysite-like particles for material purposes was reported by Kuroda et al. (2011). This procedure was designed in order to simplify the laboratory experiments leading to nanotube formation. The key to morphology change was the intercalation of quaternary ammonium salts into kaolinite interlayer space which, similar to forced hydration, weakens the interlayer bonding. The authors claimed that the intercalation directly caused rolling of the layers without the need of subsequent deintercalation of the guest species. The used organic compounds involved mainly alkyltrimethylammonium bromides and chlorides. Although the rolling was induced in one step through salt intercalation, the process itself cannot be performed directly using raw kaolinite (Matusik and Kłapyta, 2013; Matusik et al., 2013). The preintercalation needs to be carried out according to the methodology described in the previous section. The change in surface hydrophilic character of octahedral sheet to hydrophobic enables the intercalation of salts. Kuroda et al. (2011) showed that the efficiency of nanotube formation depends greatly on the type of guest molecule. In particular, the importance of the following factors was underlined: the ammonia nitrogen local environment (ie, the type of ammonium salt-head group), the hydrocarbon chain length and the charge balancing anion type (bromine, chlorine). The most extensive delamination and layers rolling were observed for the following intercalating compounds: hexadecyl trimethylammonium chloride, dodecyl trimethylammonium bromide and hexadecyl methylimidazolium chloride.

In other cases, partial rolling of kaolinite particles was observed or the nanotubes were not formed. The weak interaction between the organic cation and methoxy-modified octahedral sheet is a key factor for the rolling of kaolinite layers. When the interaction was too strong, which depended on the salt structure, an intercalate was formed with a retained platy morphology. Structural studies involving XRD measurements allowed for proposing a bilayer tilted arrangement of the alkyl ammonium salts in the interlayer space (Fig. 17.5). The mean inner and outer diameters of the obtained nanotubes were equal to 17 and 29 nm, respectively, while their length was in the 100–850 nm range (Fig. 17.6). The delamination of individual particles was also evidenced by a significant increase of S_{BET}. For the tested, well-crystallised KGa-1b kaolinite, the initial S_{BET} value was equal to 17 m^2/g, which is low and typical for kaolinite of high structural order. In turn, for the intercalate with hexadecyl trimethylammonium chloride, a value of 71 m^2/g was noted. Moreover, the mean diameter of nanotubes (~13 nm), visible in the pore size distribution diagrams, was in accordance with diameters observed by TEM. It is worth underlining that the morphology showed resistance to thermal treatment up to 600°C. This treatment, however, caused dehydroxylation and removal of organic guest species and subsequent formation of an amorphous aluminosilicate framework. For the hexadecyltrimethylammonium chloride (CTMACl) intercalated material, this was accompanied by a further increase of S_{BET} to 95 m^2/g.

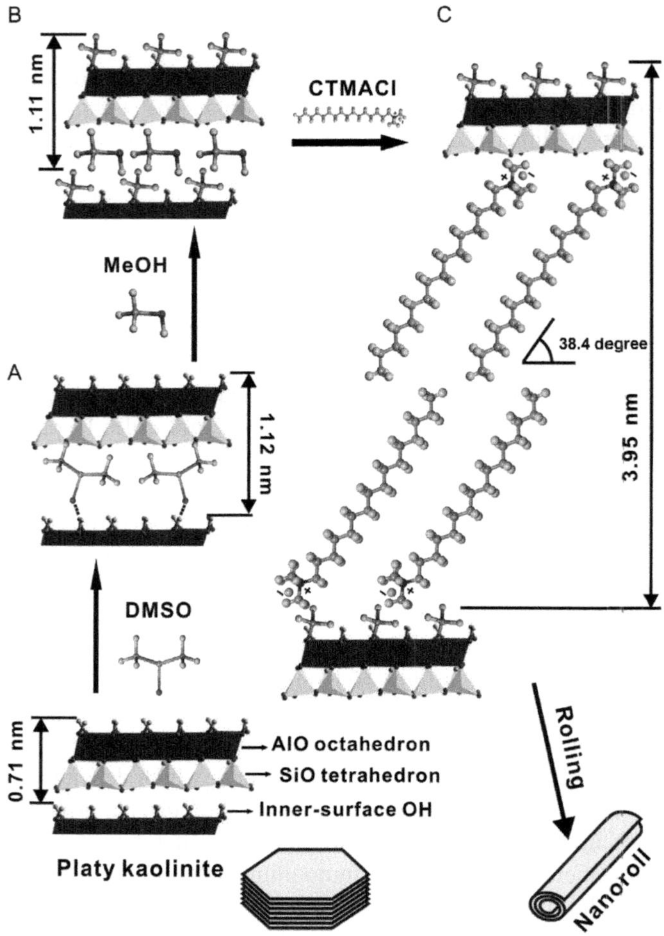

FIG. 17.5 The structure of kaolinite intercalated with hexadecyltrimethylammonium chloride (CTMACl) and the scheme of halloysite-like nanotube formation More details can be found in the original paper. *Reprinted with permission from Yuan et al. (2013). Copyright 2015 Elsevier.*

A more detailed investigation regarding the CTMACl kaolinite intercalate showing a nanotubular morphology was continued by Yuan et al. (2013). These authors focused on the effect of temperature on the efficiency of nanotube formation and calcination aspects (ie, thermal stability and accompanied changes of textural properties). An increase of intercalation temperature up to 80°C significantly enhanced the rolling of layers, whereas the transformation was only partial at lower temperatures (30°C). This was reflected by the values of S_{BET} and V_{tot}, which reached a maximum for the intercalation temperature of about 80°C (Fig. 17.7). The given possible explanation of this

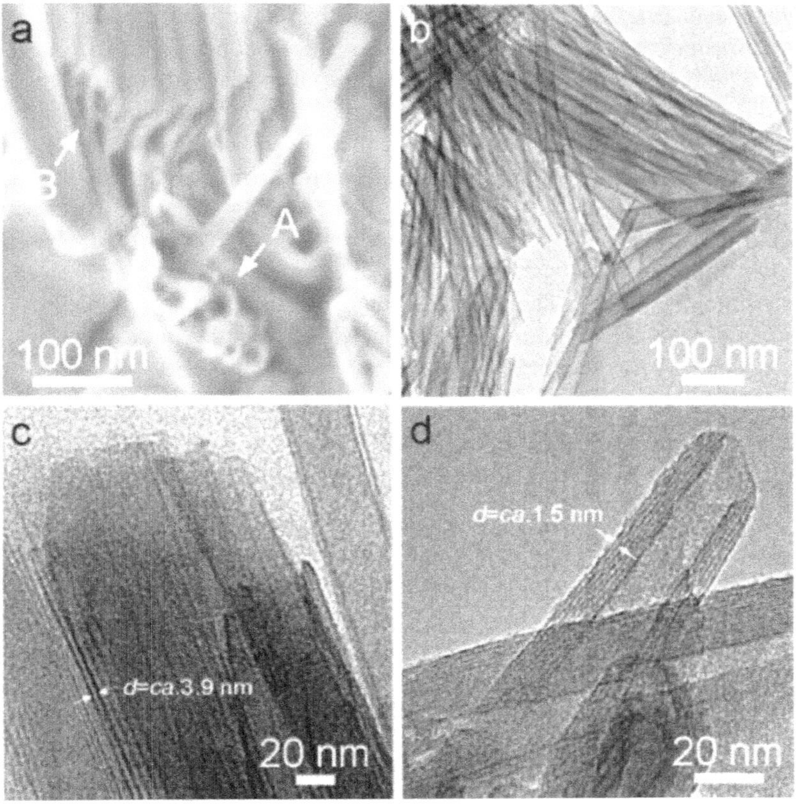

FIG. 17.6 TEM microphotographs of kaolinite nanotubes obtained via the one-step procedure. More details can be found in the original paper. *Reprinted with permission from Kuroda et al. (2011). Copyright 2015 American Chemical Society.*

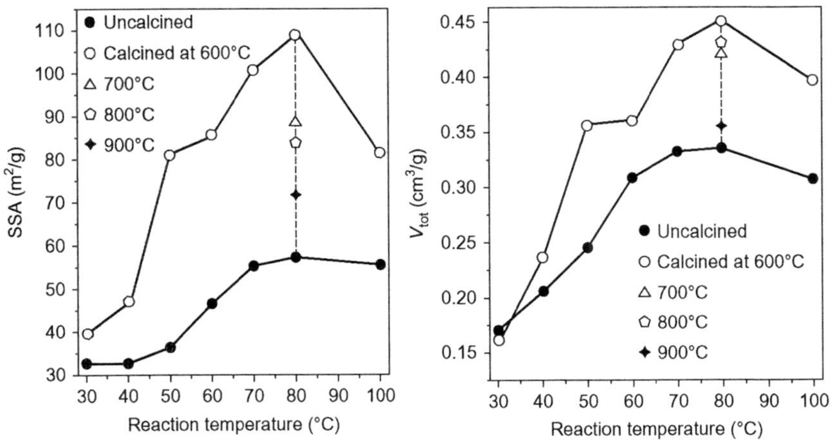

FIG. 17.7 S_{BET} and V_{tot} changes of formed halloysite-like nanotubes vs intercalation temperature. *Reprinted with permission from Yuan et al. (2013). Copyright 2015 Elsevier.*

phenomenon was based on the assumption that higher temperatures enhance the diffusion of organic guest molecules into the interlayer space. Thus, delamination and further curling of layers takes place to a larger extent. High-temperature treatment (600–800°C) of nanotubular particles did not cause significant disturbance of the morphology and changes of their texture. A clear morphology alteration and deformation accompanied with a S_{BET} decrease was observed at 900°C which was probably connected with recrystallisation and phase segregation of SiO_2 and Al_2O_3. Generally, the destruction of halloysite-like layers took place at temperatures lower than observed for natural halloysite. This was attributed to the much lower thickness of synthetic nanotube walls in comparison with those of natural tubes, which promotes dehydroxylation.

17.5 DELAMINATION AND ROLLING PROCEDURES INVOLVING THE USE OF POLYMERS AND IONIC LIQUIDS

A special treatment procedure which deserves attention was described by Letaief and Detellier (2009). The intention of these researchers was to achieve complete delamination of a kaolinite structure in a polymer matrix. The chemical route involved the use of sodium polyacrylate (PANa), which is an anionic polyelectrolyte (Fig. 17.8A). In order to maximise the interaction between PANa and kaolinite interlayer surfaces, a positive charge was generated by a combination of intercalation, grafting and quaternarisation. The treatment started with intercalation of urea, followed by interlayer grafting with the use of triethanolamine; the aminoalcohol forms Al–O–C covalent bonds through inner-surface OH groups (Letaief and Detellier, 2007). The amine nitrogen localised in the centre of the molecule can be quaternised in the reaction with iodomethane. As a result, the molecule (and subsequently the mineral) become positively charged. Afterwards, the reaction with anionic PANa was performed by mechanical grinding. Such treatment caused a very extensive delamination, which was confirmed by XRD. The recorded XRD pattern did not show any reflections confirming disruption in layer stacking. The analysis of PANa and triethanolamine-modified kaolinite physical mixture enabled to exclude the effect of dilution by the polymer. Most important, the SEM observations revealed the formation of very thin particles composed of up to several layers. The applied procedure enhanced the curling and folding of 1:1 layers similar to the results of experiments by Gardolinski and Lagaly (2005b). Letaief et al. (2011) investigated the possibility of using three different ionic liquids of imidazolium- and pyrrolidium-type with a vinyl group in the structure. As a precursor for in situ polymerisation of ionic liquids, the intercalation compounds with urea and DMSO were used. The spontaneous polymerisation was shown by ^{13}C CP-MAS NMR spectroscopy. The study showed that the use of urea intercalate and subsequent polymerisation induced a very efficient exfoliation and delamination of kaolinite particles (Fig. 17.8B). In turn, the use of DMSO precursors caused only the

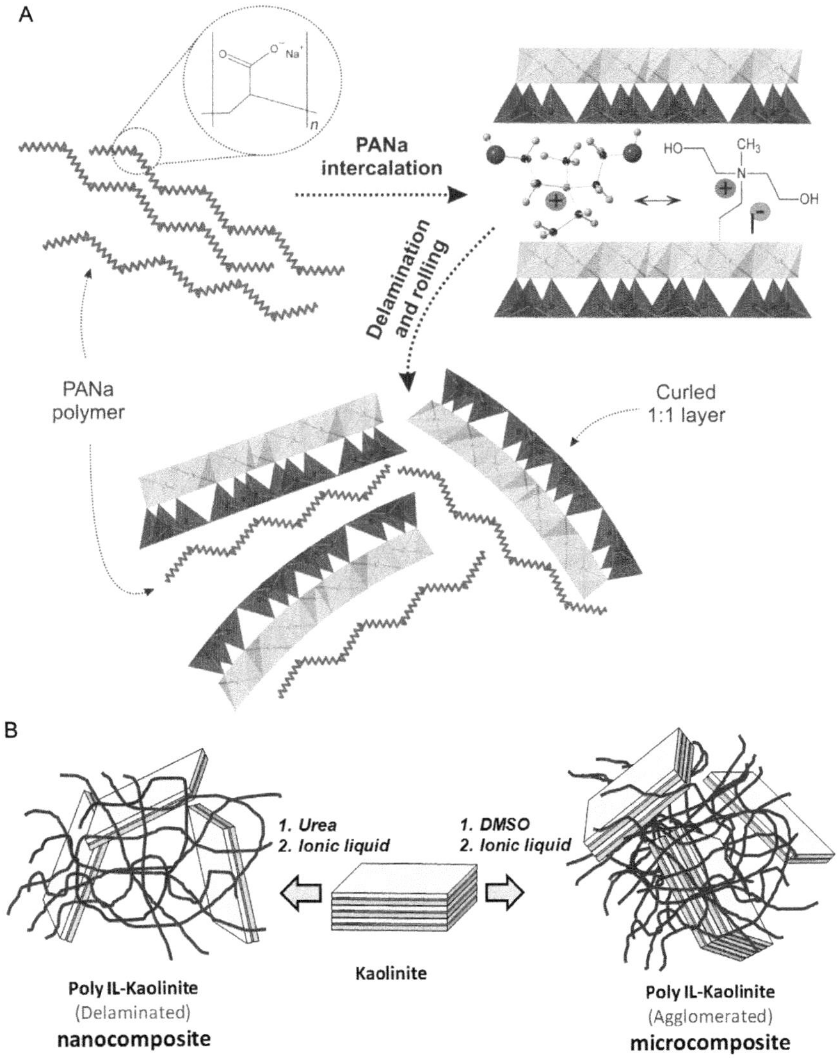

FIG. 17.8 A scheme showing the polymer-induced delamination procedure with the use of (A) sodium polyacrylate PANa and (B) ionic liquid. *Panel (A): Adapted with permission from Letaief and Detellier (2009). Copyright 2015 American Chemical Society. Panel (B): Reprinted with permission from Letaief et al. (2011). Copyright 2015 American Chemical Society.*

formation of polymer microcomposites. The delaminated particles showed signs of layer folding in SEM microphotographs (Fig. 17.9). It is worth underlining that in contrast to the procedure involving PANa, the nanomaterials obtained with the use of ionic liquids contained no more than 30% organic materials.

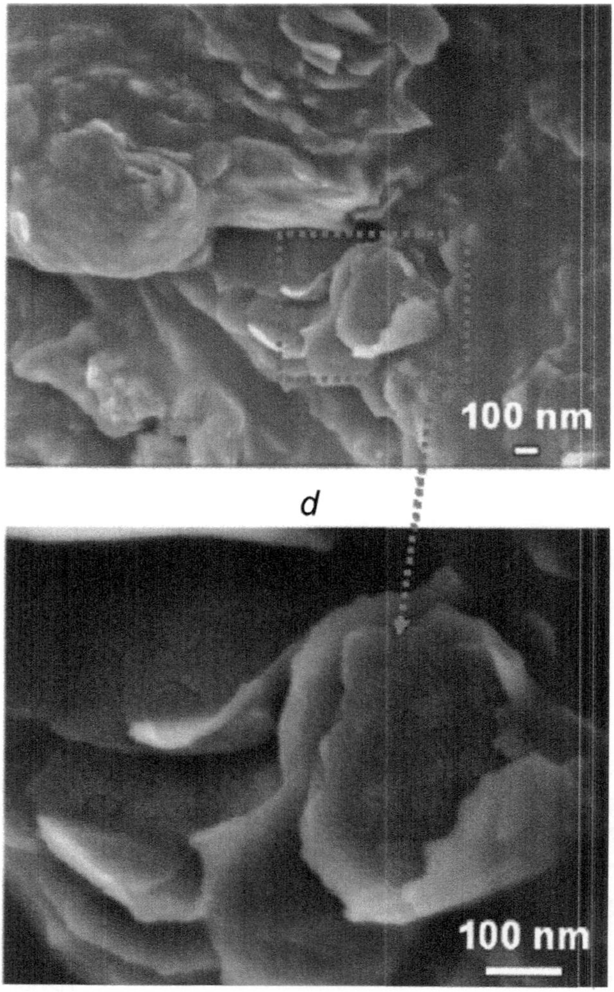

FIG. 17.9 SEM microphotographs of curled kaolinite particles formed through ionic liquid polymerisation. *Reprinted with permission from Letaief et al. (2011). Copyright 2015 American Chemical Society.*

17.6 CONCLUDING REMARKS

The acquired knowledge concerning transformation of kaolinite morphology enables an appropriate selection of treatment in order to obtain halloysite-like nanotubes. The following synthetic routes were summarised in this chapter: one-step and two-step techniques, as well as polymer-induced delamination/rolling. A comparison of research data indicates that the two-step intercalation/deintercalation shows the highest efficiency in terms of nanotubular particle yield. In all cases, the layer folding was due to weakening of interlayer hydrogen bonds by intercalation of neutral, primary alkylamines and quaternary ammonium salts or the insertion of polymer chains either directly or through in situ polymerisation. The transformation of morphology by the reported treatment did not cause significant structural perturbations, and the initial aluminosilicate framework was preserved. A 1:1 aluminium silicate exhibiting this type of nanotubular morphology are of broad interest as supports for catalytic, adsorption and drug delivery.

REFERENCES

Avila, L.R., de Faria, E.H., Ciuffi, K.J., Nassar, E.J., Calefi, P.S., Vicente, M.A., Trujillano, R., 2010. New synthesis strategies for effective functionalization of kaolinite and saponite with silylating agents. J. Colloid Interf. Sci. 341, 186–193.

Barrios, J., Plançon, A., Cruz, M.I., Tchoubar, C., 1977. Qualitative and quantitative study of stacking faults in a hydrazine treated kaolinite—relationship with the infrared spectra. Clays Clay Miner. 25, 422–429.

Bates, T.F., Hildebrand, F.A., Swineford, A., 1950. Morphology and structure of endellite and halloysite. Am. Mineral. 35, 463–484.

Bogatyrev, B.A., Matveyeva, L.A., Zhukov, V.V., Magazina, L.O., 1997. Kaolinite and halloysite synthesis in gibbsite-silica solution system under normal conditions. Geokhimiya 8, 851–862.

Busenberg, E., 1978. The products of the interaction of feldspars with aqueous solutions at 25°C. Geochim. Cosmochim. Acta 42, 1679–1683.

Chen, X., Sun, X., Li, Y., 2002. Self-assembling vanadium oxide nanotubes by organic molecular templates. Inorg. Chem. 41, 4524–4530.

Cho, H.D., Mermut, A.R., 1992. Evidence for halloysite formation from weathering of ferruginous chlorite. Clays Clay Miner. 40, 608–619.

Churchman, G.J., Aldridge, L.P., Carr, R.M., 1972. The relationship between the hydrated and dehydrated states of a halloysite. Clays Clay Miner. 20, 241–246.

Churchman, G.J., Whitton, J.S., Claridge, G.G.C., Theng, B.K.G., 1984. Intercalation method using formamide for differentiating halloysite from kaolinite. Clays Clay Miner. 32, 241–248.

da Silva, T.H., de Souza, T.F.M., Ribeiro, A.O., Ciuffi, K.J., Nassar, E.J., Silva, M.L.A., Henrique de Faria, E., Calefi, P.S., 2014. Immobilization of metallophthalocyanines on hybrid materials and in-situ synthesis of pseudo-tubular structures from an aminofunctionalized kaolinite. Dyes Pigm. 100, 17–23.

Detellier, C., Schoonheydt, R.A., 2014. From platy kaolinite to nanorolls. Elements 10, 201–206.

Dong, W., Li, W., Yu, K., Krishna, K., Song, L., Wang, X., Wang, Z., Coppens, M.-O., Feng, S., 2003. Synthesis of silica nanotubes from kaolin clay. Chem. Commun. 7, 1302–1303.

Du, M., Guo, B., Jia, D., 2010. Newly emerging applications of halloysite nanotubes: a review. Polym. Int. 59, 574–582.

Elbokl, T., Detellier, C., 2006. Aluminosilicate nanohybrid materials. Intercalation of polystyrene in kaolinite. J. Phys. Chem. Solids 67, 950–955.

Fialips, C.I., Petit, S., Decarreau, A., Beaufort, D., 2000. Influence of synthesis pH on kaolinite "crystallinity" and surface properties. Clays Clay Miner. 48, 173–184.

Gardolinski, J.E.F.C., 2005. Interlayer Grafting and Delamination of Kaolinite. PhD Thesis, Christian-Albrechts University, Kiel, Germany.

Gardolinski, J.E.F.C., Lagaly, G., 2005a. Grafted organic derivatives of kaolinite: I. Synthesis, chemical and rheological characterization. Clay Miner. 40, 537–546.

Gardolinski, J.E.F.C., Lagaly, G., 2005b. Grafted organic derivatives of kaolinite: II. Intercalation of primary n-alkylamines and delamination. Clay Miner. 40, 547–556.

Gastuche, M.C., Delvigne, J., Fripiat, J.J., 1954. Chemical alteration of kaolinites. In: Trans. Fifth Int. Congr. Soil Sci. vol. 2, pp. 439–456.

Hem, J.D., Lind, C.J., 1974. Kaolinite synthesis at 25°C. Science 184, 1171–1173.

Hendricks, S.B., 1938. On the crystal structure of the clay minerals: dickite, halloysite and hydrated halloysite. Am. Mineral. 23, 295–301.

Hope, E.W., Kittrick, J.A., 1964. Surface tension and the morphology of halloysite. Am. Mineral. 49, 859–866.

Huertas, F.J., Fiore, S., Huertas, F., Linares, J., 1999. Experimental study of the hydrothermal formation of kaolinite. Chem. Geol. 156, 171–190.

Inagaki, S., Fukushima, Y., Kuroda, K., 1993. Synthesis of highly ordered mesoporous materials from a layered polysilicate. Chem. Commun. 8, 680–682.

Joussein, E., Petit, S., Churchman, J., Theng, B., Righi, D., Delvaux, B., 2005. Halloysite clay minerals—a review. Clay Miner. 40, 383–426.

Kittrick, J.A., 1970. Precipitation of kaolinite at 25°C and 1 atm. Clays Clay Miner. 18, 261–267.

Kohyama, N., Fukushima, K., Fukami, A., 1978. Observation of the hydrated form of tubular halloysite by an electron microscope equipped with an environmental cell. Clays Clay Miner. 26, 25–40.

Komori, Y., Sugahara, Y., Kuroda, K., 1999. Intercalation of alkylamines and water into kaolinite with methanol kaolinite as an intermediate. Appl. Clay Sci. 15, 241–252.

Komori, Y., Enoto, H., Takenawa, R., Hayashi, S., Sugahara, Y., Kuroda, K., 2000. Modification of the interlayer surface of kaolinite with methoxy groups. Langmuir 16, 5506–5508.

Kresge, C.T., Leonowicz, M.E., Roth, W.J., Vartuli, J.C., Beck, J.S., 1992. Ordered mesoporous molecular sieves synthesized by a liquid-crystal template mechanism. Nature 359, 710–712.

Kuroda, Y., Ito, K., Itabashi, K., Kuroda, K., 2011. One-step exfoliation of kaolinites and their transformation into nanoscrolls. Langmuir 27, 2028–2035.

La Iglesia, A., Martin-Vivaldi, J.L., 1975. Synthesis of kaolinite by homogeneous precipitation at room temperature: I. Use of anionic resins in (OH) form. Clays Clay Miner. 10, 399–405.

La Iglesia, A., Van Oosterwyck-Gastuche, M.C., 1978. Kaolinite synthesis. I. Crystallization conditions at low temperatures and calculation of thermodynamic equilibria. Application to laboratory and field observations. Clays Clay Miner. 26, 397–408.

Letaief, S., Detellier, C., 2007. Functionalized nanohybrid materials obtained from the interlayer grafting of aminoalcohols on kaolinite. Chem. Commun. 25, 2613–2615.

Letaief, S., Detellier, C., 2009. Clay-polymer nanocomposite material from the delamination of kaolinite in the presence of sodium polyacrylate. Langmuir 25, 10975–10979.

Letaief, S., Leclercq, J., Liu, Y., Detellier, C., 2011. Single kaolinite nanometer layers prepared by an in situ polymerization–exfoliation process in the presence of ionic liquids. Langmuir 27, 15248–15254.

Levard, C., Rose, J., Masion, A., Doelsch, E., Borschneck, D., Olivi, L., Dominici, C., Grauby, O., Woicik, J.C., Bottero, J.Y., 2008. Synthesis of large quantities of single-walled aluminogermanate nanotube. J. Am. Chem. Soc. 130, 5862–5863.

Li, X., Yang, H., 2014. Morphology-controllable Li_2SiO_3 nanostructures. CrystEngComm 16, 4501.

Li, Y.D., Li, X.L., He, R.R., Zhu, J., Deng, Z.X., 2002. Artificial lamellar mesostructures to WS_2 nanotubes. J. Am. Chem. Soc. 124, 1411–1416.

Li, X., Liu, Q., Cheng, H., Zhang, S., Frost, R.L., 2015. Mechanism of kaolinite sheets curling via the intercalation and delamination process. J. Colloid Interf. Sci. 444, 74–80.

Lin, H.P., Mou, C.Y., 1996. Tubules-within a tubule hierarchical order of mesoporous molecular sieves in MCM-41. Science 273, 765–768.

Liu, Q., Li, X., Guo, P., Cheng, H., Ji, L., Zhang, S., 2014. Formation of nanoscrolls and preparation of kaolinite intercalation compound with alkylamines. J. Chin. Ceram. Soc. 42, 1064–1069.

Ma, R., Bando, Y., Sasaki, T., 2004. Directly rolling nanosheets into nanotubes. J. Phys. Chem. B 108, 2115–2119.

Matusik, J., Kłapyta, Z., 2013. Characterization of kaolinite intercalation compounds with benzylalkylammonium chlorides using XRD, TGA/DTA and CHNS elemental analysis. Appl. Clay Sci. 83–84, 433–440.

Matusik, J., Gaweł, A., Bielańska, E., Osuch, W., Bahranowski, K., 2009. The effect of structural order on nanotubes derived from kaolin-group minerals. Clays Clay Miner. 57, 452–464.

Matusik, J., Stodolak, E., Bahranowski, K., 2011a. Synthesis of polylactide/clay composites using structurally different kaolinites and kaolinite nanotubes. Appl. Clay Sci. 51, 102–109.

Matusik, J., Wisła-Walsh, E., Gaweł, A., Bielańska, E., Bahranowski, K., 2011b. Surface area and porosity of nanotubes obtained from kaolin minerals of different structural order. Clays Clay Miner. 59, 116–135.

Matusik, J., Gaweł, A., Bahranowski, K., 2012a. Grafting of methanol in dickite and intercalation of hexylamine. Appl. Clay Sci. 56, 63–67.

Matusik, J., Scholtzová, E., Tunega, D., 2012b. Influence of synthesis conditions on the formation of a kaolinite-methanol complex and simulation of its vibrational spectra. Clays Clay Miner. 60, 227–239.

Matusik, J., Kłapyta, Z., Olejniczak, Z., 2013. NMR and IR study of kaolinite intercalation compounds with benzylalkylammonium chlorides. Appl. Clay Sci. 83–84, 426–432.

McDonald, A., Scott, B., Villemure, G., 2009. Hydrothermal preparation of nanotubular particles of a 1:1 nickel phyllosilicate. Microporous Mesoporous Mater. 120, 263–266.

Nakagaki, S., Wypych, F., 2007. Nanofibrous and nanotubular supports for the immobilization of metalloporphyrins as oxidation catalysts. J. Colloid Interf. Sci. 315, 142–157.

Nakagaki, S., Machado, G., Halma, M., Dossantosmarangon, A., Defreitascastro, K., Mattoso, N., Wypych, F., 2006. Immobilization of iron porphyrins in tubular kaolinite obtained by an intercalation/delamination procedure. J. Catal. 242, 110–117.

Niu, J., Qiang, Y., Li, X., Liu, Z., Zhang, S., Feng, P., Ou, X., 2014. Morphology and orientation of curling of kaolinite layer in hydrate. Appl. Clay Sci. 101, 215–222.

Olejnik, S., Aylmore, L.A.G., Posner, A.M., Quirk, J.P., 1968. Infrared spectra of kaolin mineral-dimethyl sulfoxide complexes. J. Phys. Chem. 72, 241–249.

Ouyang, J., Zhou, Z., Zhang, Y., Yang, H., 2014. High morphological stability and structural transition of halloysite (Hunan, China) in heat treatment. Appl. Clay Sci. 101, 16–22.

Pan, J., Yao, H., Xu, L., Ou, H., Huo, P., Li, X., Yan, Y., 2011. Selective recognition of 2,4,6-trichlorophenol by molecularly imprinted polymers based on magnetic halloysite nanotubes composites. J. Phys. Chem. C 115, 5440–5449.

Parham, W.E., 1969. Formation of halloysite from feldspar: low temperature, artificial weathering versus natural weathering. Clays Clay Miner. 17, 13–22.

Robertson, I.D.M., Eggleton, R.A., 1991. Weathering of granitic muscovite to kaolinite and halloysite and of plagioclase-derived kaolinite to halloysite. Clays Clay Miner. 39, 113–126.

Robson, H.E., 1979. Synthetic halloysites. Google Patents. US 4150099.

Singh, B., 1996. Why does halloysite roll? A new model. Clays Clay Miner. 44, 191–196.

Singh, B., Gilkes, R.J., 1992. An electron optical investigation of the alteration of kaolinite to halloysite. Clays Clay Miner. 40, 212–229.

Singh, B., Mackinnon, D.R., 1996. Experimental transformation of kaolinite to halloysite. Clays Clay Miner. 44, 825–834.

Sugahara, Y., Satokawa, S., Kuroda, K., Kato, C., 1988. Evidence for the formation of interlayer polyacrylonitrile in kaolinite. Clays Clay Miner. 36, 343–348.

Sugahara, Y., Satokawa, S., Kuroda, K., Kato, C., 1990. Preparation of a kaolinite–polyacrylamide intercalation compound. Clays Clay Miner. 38, 137–143.

Sun, D., Li, B., Li, Y., Yu, C., Zhang, B., Fei, H., 2011. Characterization of exfoliated/delamination kaolinite. Mater. Res. Bull. 46, 101–104.

Thompson, J.G., Cuff, C., 1985. Crystal structure of kaolinite: dimethylsulfoxide intercalate. Clays Clay Miner. 33, 490–500.

Tunney, J.J., Detellier, C., 1996a. Aluminosilicate nanocomposite materials. Poly(ethylene glycol)-kaolinite intercalates. Chem. Mater. 8, 927–935.

Tunney, J.J., Detellier, C., 1996b. Chemically modified kaolinite. Grafting of methoxy groups on the interlamellar aluminol surface of kaolinite. J. Mater. Chem. 6, 1679–1685.

Van Oosterwyck-Gastuche, M.C., La Iglesia, A., 1978. Kaolinite synthesis. II. A review and discussion of the factors influencing the rate process. Clays Clay Miner. 26, 409–417.

Wada, K., 1961. Lattice expansion of kaolin minerals by treatment with potassium acetate. Am. Mineral. 46, 78–91.

Weiss, A., 1963. Ein Geheimnis des chinesischen Porzellans. Angew. Chem. 75, 755–762.

White, R.D., Bavykin, D.V., Walsh, F.C., 2012. Spontaneous scrolling of kaolinite nanosheets into halloysite nanotubes in an aqueous suspension in the presence of GeO_2. J. Phys. Chem. C 116, 8824–8833.

Yuan, P., Tan, D., Aannabi-Bergaya, F., Yan, W., Fan, M., Liu, D., He, H., 2012. Changes in structure, morphology, porosity, and surface activity of mesoporous halloysite nanotubes under heating. Clays Clay Miner. 60, 561–573.

Yuan, P., Tan, D., Annabi-Bergaya, F., Yan, W., Liu, D., Liu, Z., 2013. From platy kaolinite to aluminosilicate nanoroll via one-step delamination of kaolinite: effect of the temperature of intercalation. Appl. Clay Sci. 83–84, 68–76.

Chapter 18

From Molecular Precursor to Imogolite Nanotubes

A. Thill*
Laboratoire Interdisciplinaire sur l'Organisation Nanometrique et Supramoleculaire, NIMBE, CEA, CNRS, University Paris-Saclay, Paris, France
Corresponding author: e-mail: antoine.thill@cea.fr

18.1 INTRODUCTION

Imogolite is a natural, nanotubular clay mineral that was discovered in 1962 in Japanese volcanic soil by Yoshinaga and Aomine (1962). All the earliest studies were performed on natural nanotubes purified from soil samples, and natural purified imogolite is still used even in recent publications, especially in Japan. Farmer et al. (1977) published a recipe to synthesise imogolite. They explained that imogolite can be obtained from the hydrolysis of an aluminium salt in the presence of $Si(OH)_4$ after heat treatment in acidic conditions. This seminal paper defined the basis for the production of synthetic imogolite. Nowadays, most of the publications on imogolite have used a synthetic version of these nanoparticles. The recipe initially proposed by these authors is the one most often used. Several alternative synthesis pathways have been proposed, and much progress has been made to optimise the yield and purity of the obtained imogolite. The formation mechanism of imogolite nanotubes has also been studied.

In this chapter, the types of recipes that have been used to prepare imogolite nanotubes are summarised and the current understanding of their formation process is reviewed.

18.2 SUMMARY OF SYNTHESIS METHODS

18.2.1 Main Chemical Recipes

We can distinguish three main types of chemical recipes that produce imogolite or imogolite-like nanotubes based on the type of chemical precursors used. The first type uses the neutralisation of an acidic aluminium salt solution in the presence of a silicon source. In this kind of recipe, hydroxyls are slowly added to

Developments in Clay Science, Vol. 7. http://dx.doi.org/10.1016/B978-0-08-100293-3.00018-2

the initial solution containing aluminium and silicon to reach a required pH or OH/Al ratio. The second type of synthesis introduced by Farmer et al. (1983) uses the neutralisation of an acidic aluminium salt solution by a sodium silicate solution. In this case, the sodium silicate solution is the source of both silicon and hydroxyls. The OH/Al ratio, however, can be adjusted by using acids or bases. The last recipes proposed by Farmer et al. (1983) use organic sources for both aluminium and silicon. In this case, the precipitation occurred directly under low-pH conditions fixed by the added acid (HCl, $HClO_4$, HNO_3 and CH_3COOH). A few exotic methods also will be described.

18.2.2 Aluminium Salts and Silicon Sources

A schematic diagram of the synthesis pathway is presented in Fig. 18.1. The aluminium salts used most often are $AlCl_3$ and $Al(ClO_4)_3$, with $Al(NO_3)_3$ less frequently used. The initial solution has a very low pH. At pH < 3, Al is stable in the form of hydrated ions or very small polycations (Bottero et al. 1980).

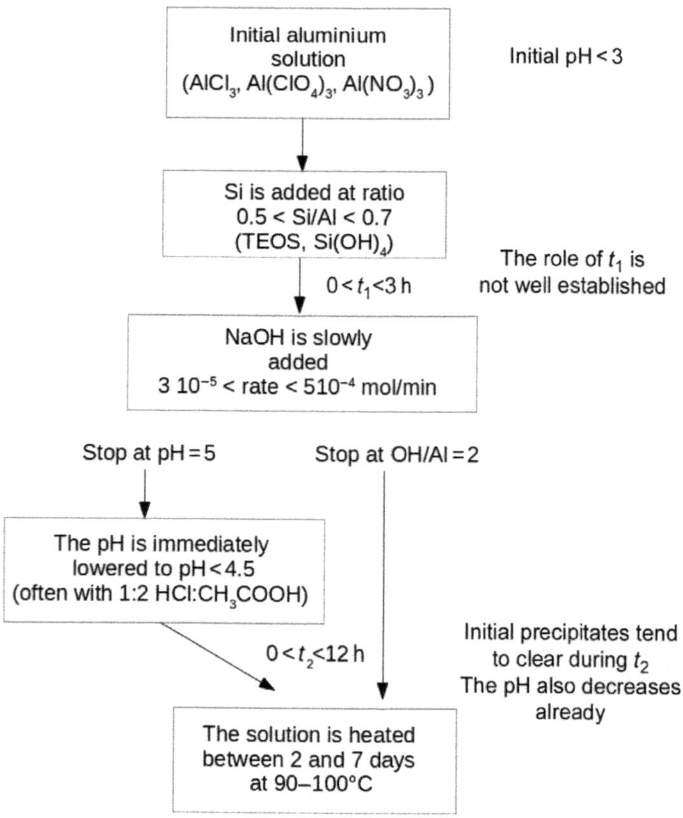

FIG. 18.1 Schematic reaction pathway for the synthesis of imogolite from salt precursors.

Silicon is added to this acidic solution, most often using tetraethoxysilane (TEOS). In the initial paper by Farmer et al. (1977), the precise nature of the silicon source is not given. Some authors used the hydrolysis of a diluted solution of TEOS to produce $Si(OH)_4$ at concentrations below the amorphous silica solubility (about 2 mM). Recently, Abidin et al. (2008) used dialysis to produce 1.6 mM of $Si(OH)_4$ solution obtained from the dissolution of silica gels.

The ratio between Si and Al is often slightly higher than 0.5, the stoichiometric ratio of imogolite. It has been shown that in the concentration range used for this kind of imogolite synthesis, an excess of silicon prevents the formation of aluminium hydroxide (boehmite or gibbsite) (Farmer et al. 1977). Afterward, the introduction of the silicon source, a time t_1 is sometimes allowed for the solution to age. For example, Barrett et al. (1991) let the solution stir for 45 min to allow the full hydrolysis of the TEOS precursor to occur. Other authors do not allow any ageing time after the introduction of the silicon (or germanium) source. For example, according to Farmer et al. (1983) introducing TEOS into a 10-mM $Al(ClO_4)_3$ solution and immediately increasing the pH to 4.5 resulted in a satisfactory product. Allowing the full hydrolysis of an alkoxide precursor at low pH is probably only possible at concentrations below the solubility of amorphous silica. Indeed, at low pH, if the concentration of Si exceeds the solubility of amorphous silica, then polymerisation of silicon occurs, which makes the silicon tetrahedra unavailable for bonding with the aluminium clusters.

Generally, the solution is hydrolyzed using a very slow rate of base addition, but the rates are not always given precisely. Inoue and Huang (1985) used an injection speed of 1.5×10^{-2} molOH/molAl/min. Su et al. (1992) gave an injection rate of 3.4×10^{-2} molOH/molAl/min. Denaix (1993) studied the titration of both 0.1-M and 10^{-3}-M solutions at an Si/Al ratio of 2. She used an even smaller injection speed of 1.6×10^{-3} molOH/molAl/min under vigorous stirring. A too-fast injection of hydroxyls would lead to concentration gradient or localised high pH in the synthesis solution, which could favour the irreversible precipitation of aluminium hydroxide. Therefore, a too-fast hydrolysis may result in loss of purity and the formation of gibbsite or boehmite. This generally gives a whitish colour to the nanotube dispersion.

After the hydrolysis step, some authors leave the solution at rest for a time noted as t_2 in Fig. 18.1. Farmer et al. (1983) waited 24 h after the hydrolysis step before the heating stage. Denaix et al. (1999) waited 1 h for 10^{-3} M and 24 h for 0.1-M synthesis. Mukherjee et al. (2005) waited under stirring for 3 h before heating. To our knowledge, there have been no systematic observations dedicated to understanding the effect of this ageing period before heating. The solutions were then heated to temperatures between 90°C and 100°C for 2–7 days. Mukherjee et al. (2007) studied the influence of the heating temperature from 65°C to 95°C. They showed that the growth kinetic is strongly dependent on heating. After 300 h at 95°C, the nanotubes almost finished growing, whereas at 65°C, almost no nanotube growth was observed.

Wada (1987) studied a synthesis of imogolite at 25°C; it took them 7 years to obtain imogolite at this temperature. Very few experiments used higher temperature and hydrothermal reactions. Indeed, at higher temperatures, imogolite is no longer stable, and it decomposes to boehmite and kaolinite (Farmer et al., 1979).

18.2.3 Aluminium Salts and Sodium Silicates

Farmer et al. (1983) were the first to propose the preparation of a concentrated dispersion of proto-imogolite (i.e., small nanoparticles, at the expense of which imogolite nanotubes occur) that uses inexpensive reagents. These authors mixed an $Al(NO_3)$ salt with a freshly prepared sodium silicate in the presence of an $HClO_4$ acid in a molar ratio Al:Si:$HClO_4$ of 1.8:1:1.5 until a clear dispersion was obtained. This dispersion is precipitated at pH 7. The precipitate is immediately redispersed in $HClO_4$. It is diluted at a concentration of about 10 mM in aluminium to obtain the maximum yield.

Inoue and Huang (1985) prepared orthosilicic acids by passing Na metasilicate through an H^+ saturated resin column. After ion exchange, the concentration was maintained below 2 mM to avoid the polymerisation of Si. This solution is then added to an aluminium salt at an Si/Al ratio of 0.5. Ohashi et al. (2004) mixed directly Na_4SiO_4 at 0.1 M with $AlCl_3$ at 0.15 M. The resulting solution was neutralised with NaOH 1 M up to pH 6. After desalinisation, the solution pH is reduced with HCl to less than 4 and autoclaved for 2 days at 100°C. Nakanishi et al. (2007) created almost exactly the same preparation with Na_4SiO_4 and $AlCl_3$ at a lower Si/Al ratio of 0.5. The main difference was the absence of desalinisation. In this case, they did not obtain imogolite, but rather something closer to allophane. Nakanishi et al. (2008) explored the effect of various dialysis times of the concentrated mixture of Na_4SiO_4 and $AlCl_3$. Before dialysis, the OH/Al ratio was changed to 2.4 using 1 M of NaOH. These authors observed that allophane is obtained without dialysis or for dialysis times of less than 48 h. Between 48 and 108 h of dialysis, imogolite is the only formed product. For longer dialysis times, boehmite starts to form instead of imogolite.

Kuroda and Kuroda (2008) performed a similar synthesis by mixing 20 mM of Na_4SiO_4 with 30 mM of $AlCl_3$ solution. The pH of the mixture is brought up to 6.7 with 1 M of NaOH. The obtained precipitate is centrifuged and redispersed in water three times. Finally, the washed precipitate is redispersed in HNO_3 and heated for 5 days at 100°C under reflux.

18.2.4 Aluminium and Silicon Alkoxides

In a patent, Farmer (1979) described several protocols to produce imogolite at relatively high concentrations using a mixture of aluminium-*sec*-butoxide (ASB) and tetraethoxysilane (TEOS). Typically, the two compounds are added together in a glovebox under nitrogen at a molar Si/Al ratio of 0.5

and immediately mixed with an acid solution (HCl, $HClO_4$ or HNO_3). Barrett et al. (1991) used this protocol with an Al concentration of 70 mM, Si/Al ratio of 0.5 and $Al/HClO_4=2$. The alkoxides are slowly added to an $HClO_4$ solution at 80°C under stirring and stirred for 4 h. The resulting solution is left to stand, and a white precipitate settled. This precipitate, identified by Barrett et al. (1991) as aluminium oxide, is removed and the remaining supernatant is refluxed for 120 h. Qi et al. (2008) made almost exactly the same synthesis with an $Si:Al:HClO_4$ ratio of 1:2:1 at an aluminium concentration of 50 mM. The solution became clear after being stirred for 5 h at 70°C and then autoclaved at 100°C for 3 days. Yucelen et al. (2011) prepared a 1:2:1 $Si:Al:HClO_4$ solution with an aluminium concentration of 0.1 M. The mixture was stirred at 25°C for 18 h before heating. Using ^{27}Al nuclear magnetic resonance (NMR), they studied the evolution of the aluminium local environment during the ageing period. These authors showed that important evolutions occurred during the first 24 h. Liz-Marzan and Philipse (1994) performed the same synthesis with a 1:2:1 $Si:Al:HClO_4$ ratio and an aluminium concentration of 50 mM. The alkoxides were mixed and homogenised by stirring for 20 min in open air. The solution was then stirred at 70°C for 5 h and filtrated before being heated at 95°C for 3 days. Koenderink et al. (1999) compared two heating procedures (autoclave and reflux) and concluded that the same product is obtained. However, they showed that after 3 days of heating, the product still contained many proto-imogolites. Bonelli et al. (2009) used a molar ratio of $1+\varepsilon$:2:1 $Si:Al:HClO_4$. A slight excess ε of silicon was used to prevent the formation of aluminium hydroxide. The 75 mM Al concentration solution was mixed at 20°C for 18 h and then diluted at 20 mM before being heated at 100°C for 4 days. Bottero et al. (2011) used this protocol to prepare hybrid nanotubes in a single step. They replaced TEOS with triethoxymethylsilane with the exact same protocol.

Using alkoxides as precursors, Barrett et al. (1991) showed that it is possible to form imogolite by heating for 5–10-h at 100°C and then letting it stand for 1–3 weeks at room temperature.

18.2.5 Other Recipes

Garcia-Hernandez and Rodriguez-Rodriguez (1990) studied the alteration of a basaltic glass coming from a volcanic ash deposit by a 0.1-mM ammonium oxalate solution. The basaltic glass is grounded between 0.05 and 2 mm and packed in 11 successive Teflon columns. The authors observed the formation of a gel at the output of the eighth column. They linked the formation of imogolite with the precipitation of iron oxides. Imogolite is no longer formed if the solution is circulated in a closed loop on a single column. In this case, only allophane was obtained.

Hongo et al. (2013) used rice husk ash (RHA) as a silicon source for imogolite synthesis. The RHA was first mixed with 3 N of NaOH at 80°C for

24 h. This mixture was then filtrated and the filtrate mixed with $Al(NO_3)_3$. The pH of this solution was adjusted to pH 5 and a precipitate was obtained. The precipitate was collected by centrifugation and redispersed in water acidified with acetic acid at pH 4. This solution was refluxed at 100°C for 4 days.

18.3 INFLUENCE OF SOME IMPORTANT PARAMETERS: CONCENTRATION, AL/SI AND OH/AL RATIOS

18.3.1 Concentration Range

18.3.1.1 Al/Si-Based Imogolite Synthesis

The synthesis of imogolite is known to be easier at low concentrations. Indeed, if the concentration of Si is below the solubility of amorphous silica (about 2 mM), then silicon is mainly present in solution as $Si(OH)_4$ and is directly available for bonding with aluminium polycations. If the silicon concentration used is 2 mM, then a maximum of about 0.4 g/L of imogolite could be produced. If the concentration is increased above this limit, then silicon polymerisation may occur. This is observed, for example, by NMR. Denaix et al. (1999) showed that before the heating stage beside the main −78 ppm peak corresponding to the imogolite structure, a large peak with an average chemical shift of −100 ppm was observed corresponding to the presence of silicon atoms with various degrees of polymerisation. These polymerised species decreased considerably upon heating.

Levard et al. (2009) showed that it is possible to obtain imogolite from decimolar aluminium concentration. However, this concentration increase also corresponds with a strong decrease of the nanotube growth kinetics. It took 2 months at 0.1 M to obtain imogolite nanotubes. This kinetic effect has also been observed by Farmer (1979). Indeed, when the yield was qualitatively discussed based on the volume of gel collected after centrifugation, the maximum gel volume was obtained at intermediate concentrations. Further increase of the concentration finally decreased the imogolite yields. Fig. 18.2 shows the gel volumes collected by Farmer (1979) from synthesis, with initial silicon concentrations ranging between 5 and 75 mM.

The best gel volume is obtained when precursor solution is gradually added each day to reach a final concentration of 65 mM. With this procedure, a final gel volume of 6.9 cm^3 for 10 cm^3 sample is obtained, whereas only about 3 cm^3 is obtained with a starting concentration of 65 mM. Ackerman et al. (1993) noted that seeding the precursor solution with previously formed imogolite considerably accelerated the synthesis, with the heating period shrinking from 5 to 2 days.

18.3.1.2 Al–Ge-Based Imogolite Synthesis

Wada and Wada (1982) replaced silicon with germanium and showed that nanotubes were also formed, but they were shorter in length and had a larger

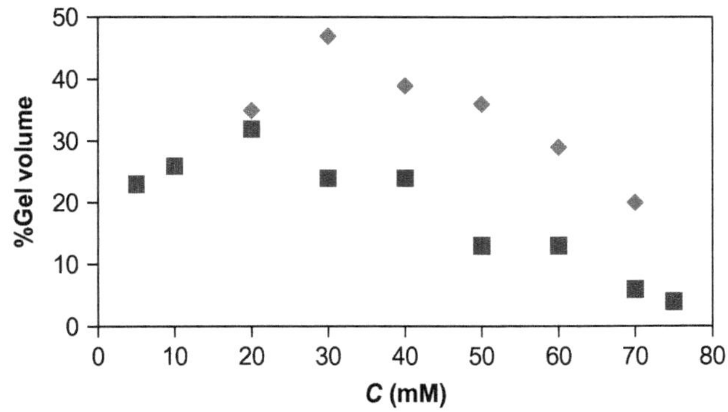

FIG. 18.2 Percentage of gel as a function of aluminium concentration after 1 day (*squares*) and 4 days (*diamonds*) of heating. *Adapted from Farmer patent 1979 no: GB 2 025 384 A.*

diameter. Levard et al. (2008) showed that it is possible to increase the concentration in the synthesis of Al–Ge imogolite up to a 1-M concentration of aluminium. Unlike standard Al–Si imogolite, only nanotubes were obtained in this case. This is explained by the fact that germanium oxide has a much higher solubility and do not tend to polymerise. Later, Maillet et al. (2010) showed that the concentration used in the synthesis may change the global shape of the Al–Ge nanotubes. Indeed, at low concentrations, double-walled (DW) Al–Ge imogolite nanotubes are formed. If the concentration is higher than 0.5 M, then single-walled (SW) nanotubes are obtained. The transition from SW to DW was explained later by Thill et al. (2012a). This shape transition is attributed to the competition between the electrostatic attractive energy between asymmetric proto-imogolite structures and the curvature energy. Upon increasing the concentration, it is supposed that a strong screening of the electrostatic attraction hinder the stability of the DW structure. Amara et al. (2013) showed that changing the hydrolysis protocol for these nanotubes by the use of the thermal decomposition of urea instead of the slow injection of NaOH allowed the formation of very long nanotubes, even at 0.1 M concentrations in aluminium. Finally, the synthesis of Al–Ge nanotubes is much easier and more convenient than standard imogolite. Applying the Amara et al. (2013) protocol, a concentration of about 10 g/L is obtained in a single step. After dialysis, such samples directly displayed a strong birefringence. Obtaining the same properties for standard imogolite would require several concentration steps (including precipitation, centrifugation and redispersion). Al–Ge imogolite synthesis, however, suffers from cost and toxicity problems (Van den Brule et al., 2014).

18.3.2 Al/Si Ratio

Farmer (1979) studied the effect of the Al/Si ratio upon the yield, measured as gel volume. At low concentrations (i.e., 10 mM in aluminium), the gel yield increased continuously with the Si/Al ratio up to the maximum ratio tested of 0.6. At high concentrations (i.e., 100 mM in aluminium), a maximum gel yield after 4 days was obtained for Si/Al = 0.4.

A slight excess of silicon is used very often in order to prevent the formation of aluminium hydroxides (Farmer et al., 1977, 1983; Clark and McBride, 1984; Wada, 1987; Barret et al., 1991; Hu et al., 2004; Mukherjee et al., 2005; Kuroda and Kuroda, 2008; Ookawa et al., 2008; Yang et al., 2008; Bac et al., 2009). The typical Si/Al ratios are between 0.5 and 0.66. For a classical synthesis protocol starting from salt precursors, Barrett et al. (1991) showed that the optimum Si/Al ratio was 0.56. In the case of synthesis starting from alkoxides, imogolite was always the sole product, provided that Si/Al was more than 0.5. Aluminium hydroxide impurities are generally formed as much larger particles and their presence is readily noticed by the whitish appearance of the dispersion. Infrared spectroscopy is a good tool to assess the presence of impurities in the synthetic imogolite. For example, a peak at 1060–1070 cm^{-1} is sometimes associated with the presence of aluminium hydroxides. In the synthesis where the Si/Al ratio is slightly more than 0.5 (Clark and McBride, 1984; Wada, 1987; Kuroda and Kuroda, 2008; Arancibia-Miranda et al., 2011, 2013), no marked peak at 1060–1070 cm^{-1} was observed. In some syntheses of imogolite with a Si/Al ratio of 0.5, a strong 1060–1070 cm^{-1} peak is obtained (Liz-Marzan and Philipse, 1994; Denaix et al., 1999; Koenderink et al., 1999). This peak is sometimes very weak (McBride et al., 1984; Inoue and Huang, 1985) and is not observed in some cases (Clark and McBride, 1984; Abidin et al., 2007; Arancibia-Miranda et al., 2011). A slight increase of the Si/Al ratio seems to be a good way to prevent the formation of side minerals, but it is not the only possibilities. The reasons for the formation of side products at an Si/Al ratio of 0.5 are not very clear. For example, Abidin et al. (2008) made the same synthesis by only modifying the heating stage. In the first case, the solution was heated on a hot plate under reflux; and in the other case, it was autoclaved. The autoclaved sample did not show the 1068-cm^{-1} peak, whereas with the hot plate heating, this peak was strong.

18.3.3 OH/Al Ratio

Many synthesis protocols used the hydrolysis of a mixture of aluminium and silicon by the slow addition of hydroxyls. In the original protocol proposed by Farmer et al. (1977), the hydroxyl addition is performed up to a pH of 5, the resulting solution being immediately acidified by the addition of HCl and CH_3COOH. This mysterious acidification stage was later discarded in many synthesis protocols. Denaix (1993) made titration curves of a 2-mM $HClO_4$ solution as a function of the injected hydroxyls. The evolution of pH as a function of the OH/Al ratio is reported in Fig. 18.3.

The pH is almost constant at about 4.7 up to an OH/Al ratio of about 2, and then it increases more rapidly after OH/Al = 2, with a steep increase at

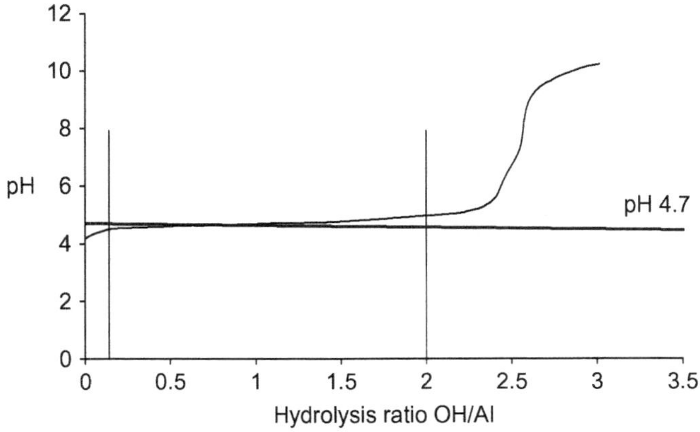

FIG. 18.3 Evolution of pH as a function of added hydroxyls for a 2-mM $AlClO_4$ solution with Si/Al = 0.5. *Figure reproduced from Denaix (1993).*

OH/Al = 2.5. In the recipe of Farmer et al. (1977), the targeted pH after acidification is 4.5. Going directly to pH 4.5 would probably result in an ill-defined hydrolysis ratio and a lack of reproducibility. Therefore, going to pH 5 and then acidifying the solution constitute a way to have a reproducible OH/Al ratio when using pH measurements. Many syntheses are directly based on the ratio of added hydroxyls. The value used most often is OH/Al = 2. In this case, the acidification step is generally not performed. There may be a more subtle reason for the two-step procedure. Indeed, the relative polymerisation kinetic of Al and Si is very important, and the two procedures may result in contrasted situations. Arancibia-Miranda et al. (2011) compared the two procedures. The final products are very similar. However, the characteristic doublet near 1000 cm^{-1} in the infrared spectra occurs very early in the two-step procedures and only at the end in the method without acidification.

Denaix (1993) used liquid NMR on mixtures of 0.1 M $Al(ClO_4)_3$ and 0.05 M TEOS with various OH/Al ratios. The obtained product is closer to allophane than imogolite in this case. The same synthesis using a 2-mM aluminium solution gives imogolite. Yucelen et al. (2011) did the same measurement using alkoxide precursors with 0.1 M of ASB.

Denaix did not observe any tetrahedral aluminium species. Aluminium was present in the form of octahedral $Al(H_2O)_6^{3+}$. The quantity of $Al(H_2O)_6^{3+}$ decreases as the OH/Al ratio increases, and more polymerised forms of aluminium are obtained. At OH/Al = 1, isolated octahedral aluminium still represents 50% of the total aluminium and 20% at OH/Al = 2. The proportion of isolated aluminium octahedra is constant with time and seems to depend only on the hydrolysis ratio. This result demonstrates that the polymerisation of aluminium is driven by the hydrolysis OH/Al ratio. Levard et al. (2011) followed the effect of the hydrolysis ratio R = OH/Al during the synthesis of Al–Ge imogolites and showed that at $R < 1.5$, a chemical shift at about 66 ppm is observed and accounts for 7% ($R = 0.5$) to 2% ($R = 1.5$) of the total aluminium. This contribution is generally associated with tetrahedral aluminium, often due to the formation of Keggin Al_{13} polycations. This also was observed by Yucelen et al. (2011) before the heating stage for the synthesis of Al–Si imogolite from organic precursors.

From $R = 2$–3, Al is only in octahedral coordination, with a chemical shift at about 6 ppm. Using EXAFS measurements at the Ge K edge, Levard et al. (2011) showed that the average number of Al atoms in the second coordination shell of Ge increases with R up to 2, where it reaches its final value of about 6. In the same way, the average Al/Si ratio in the obtained solid reaches a final value of 2 at $R = 2$. Before $R = 2$, Levard et al. (2011) showed that the imogolite structure contained Ge lacunae (more than 20% of Ge sites). The obtained nanotubes are also SW at $R = 1.5$ (Thill et al., 2012a), whereas at $R = 2$, most of the nanotubes are DW. This work demonstrates that the normal structure of Al–Ge imogolite (without lacuna and with correct composition) is DW. The SW structure is associated with the presence of lacunae in the Ge layer, which may change the electrostatic interaction between proto-imogolite surfaces or change the mechanical properties of the walls.

18.4 IMOGOLITE FORMATION MECHANISM

In this section, for pedagogic purposes, pictures of three-dimensional (3D) printed plastic models will be used to describe chemical reactions between Al in octahedral sites and Si in tetrahedral sites. The plastic models are linked using spherical magnets illustrating oxo or hydroxo bridges.[1] The unoccupied sites of the plastic models correspond to aquo ligands.

18.4.1 Aluminium Hydrolysis and Condensation

$Al(H_2O)_6^{3+}$ forms an unhydrolysed cation at pH < 3. At higher pH values, it is hydrolysed following this reaction:

$$[Al(H_2O)_6]^{3+} + RH_2O \leftrightarrow [Al(OH)_R(H_2O)_{6-R}]^{(3-R)+} + RH_3O^+$$

Once hydrolysed, the aluminium cations are no longer stable and condensation reaction will occur through olation reactions. The coordinated hydroxyl groups being able to produce a nucleophilic attack on an aquo ligand of another metallic cation:

1. 3D printing files to reproduce these plastic models are available upon request to Thill.

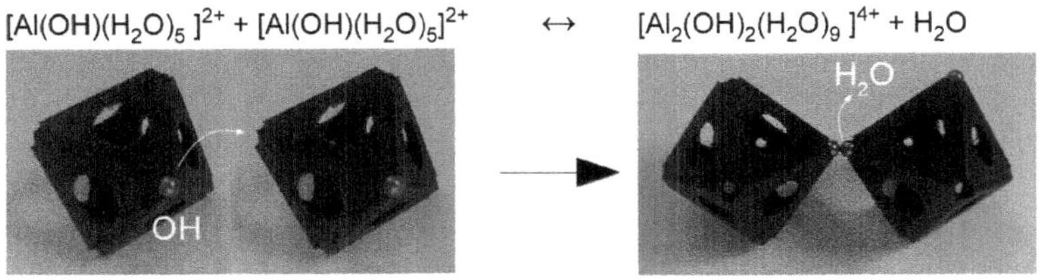

After this first olation, further intramolecular olation reaction is very favourable, and an aluminium dimer linked through two Al–OH–Al bonds can form:

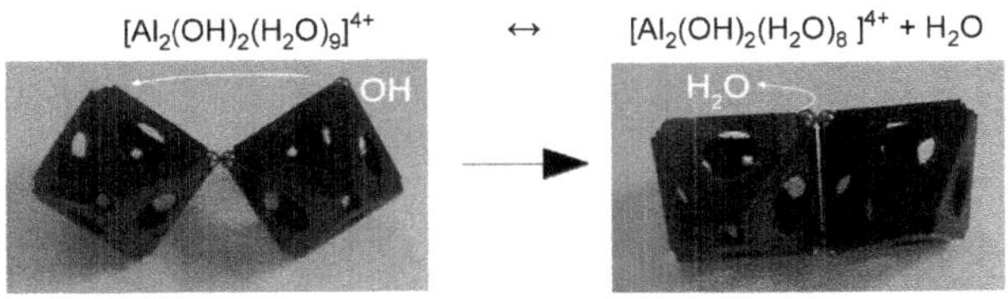

When aluminium is the sole metallic cation, the condensation reactions will produce three dominating polycations $[Al_2(OH)_2(H_2O)_8]^{4+}$, $[Al_3(OH)_4(H_2O)_{10}]^{5+}$ and the Keggin ion $[Al_{13}O_4(OH)_{24}(H_2O)_{12}]^{7+}$ (Brinker and Scherer, 1989). Bottero et al. (1980) studied the hydrolysis and condensation of Al using NMR and small-angle X-ray scattering (SAXS). The proportions of the dominating polycation as a function of the hydrolysis ratio according to their results are given in Fig. 18.4.

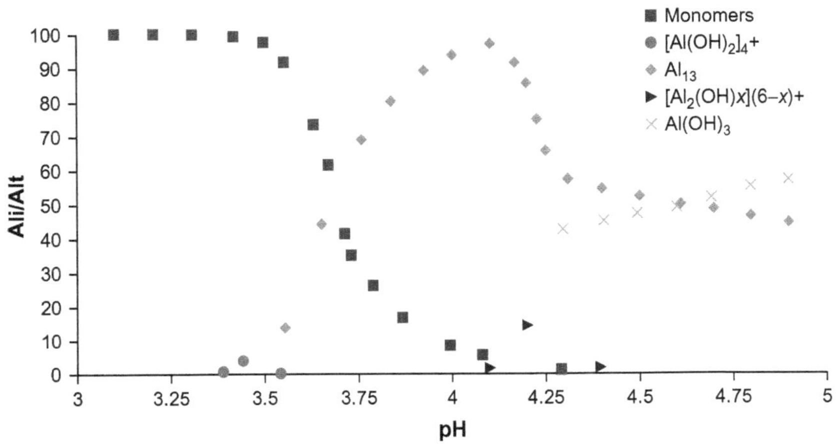

FIG. 18.4 Aluminium speciation as a function of a hydrolysis ratio of a 10^{-3} M $AlCl_3$ solution. *Figure adapted from Bottero et al. (1980).*

In NMR, only very small oligomers are observed with chemical shifts around 4 ppm, and Al_{13} is directly formed significantly at $R=1$ (Akitt and Elders, 1988). Upon further hydrolysis and condensation, $Al(OH)_3$ will form. This initial precipitate is an amorphous gel that comes from the flocculation of Al_{13} polycations. With time, the NMR signal of Al_{13} polycation will decrease and crystalline phase will appear (Axelos et al., 1985). The restructuring of the $Al(OH)_3$ gel is easier if the neutralisation is performed directly and without heating.

18.4.2 Silicon Hydrolysis and Condensation

Monosilicic acid is soluble and stable in aqueous solution at 25°C so long as the concentration is below about 1.6 mM. Above this concentration, the monosilicic acid will polymerise through an oxolation reaction:

$$2[Si(OH)_4] \leftrightarrow [Si_2O(OH)_6] + H_2O$$

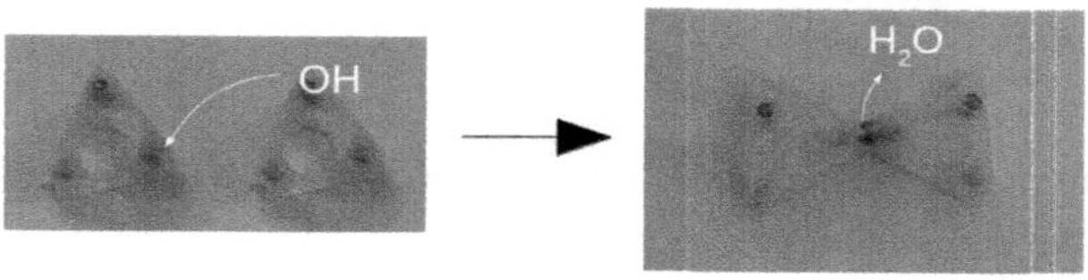

The pKa of $[Si(OH)_4]$ is 9.9, and upon polymerisation, it tends to decrease. The small oligomers are thus more easily deprotonated and efficient to bond to another $[Si(OH)_4]$ molecule. Already at the stage of a linear tetramer, intramolecular oxolation is favoured and cyclic tetramers are formed. The polymerisation rate is proportional to the concentration of hydroxyls and is at its minimum at pH 2, which is the isoelectric point (IEP) of amorphous silica. The condensation of silica continues until the formation of particles in the nanometer range. At pH below 7, these particles are not stable and they tend to aggregate and form gels.

A solution of monomeric silicic acid can be prepared through the dissolution of silica gel. This has been used by Inoue and Huang (1985) to prepare imogolite. The dissolution of metasilicate hexahydrate in acidic solution at low temperatures can also be used (Iller, 1979). Doing so will produce a significant amount of salt, which may hinder the formation of imogolite. To prevent excess salt, a solution of orthosilicate can by exchanged with an H^+ exchange resin with special care to maintain an acidic pH. This type of method has been proposed by Farmer et al. (1983). One of the most often used methods is the hydrolysis of tetraethylsilicate (TEOS). TEOS is not miscible with water and often the hydrolysis reaction of TEOS is performed in an ethanol/water mixture. This is not the case for imogolite synthesis, where TEOS is generally directly added in an acidic water dispersion containing an aluminium salt. It is hydrolysed through a nucleophilic attack by water molecules. In excess water and in the presence of an acid, the hydrolysis occurs rapidly. Bernards et al. (1991) compared the hydrolysis and condensation reaction of tetramethoxysilane (TMOS), TEOS and tetrapropoxysilane (TPOS), and showed that the hydrolysis rate occurs in the order TMOS > TEOS > TPOS and that the dimerisation followed the reverse order (TPOS > TEOS > TMOS). According to their results, the use of TMOS could be interesting for imogolite synthesis, as it will quickly create –Si(OH) groups to react with aluminium polycations. The reduced dimerisation kinetic is also interesting for maintaining monomeric silicon atoms required for the imogolite structure.

18.4.3 From Polycation to Proto-Imogolite

When $Si(OH)_4$ is present in solution, it involves an inhibition of the growth of hydroxyaluminium (HA). Wada and Wada (1980) studied the incorporation of silicon in HA as a function of the OH-Al and Si/Al ratios. The presence of $Si(OH)_4$ did not change the hydrolysis and condensation, but it changes the pH of the solution. It also modifies the reaction times with molybdate ions as soon as NaOH is added. This tends to demonstrate that the Si reacts at very early stages with HA ions to form hydroxyaluminosilicate (HAS) ions. Up to an OH/Al ratio of 2, the proportion of Si incorporated in the HAS cations increased linearly in the same trend as the proportion of Al in polymer form. These authors suggest that the reaction mainly proceed between silicic acid and Al–OH–Al bridges. The different solutions did not show any precipitation or turbidity, even if they were supersaturated with respect to gibbsite and no clear morphologies could be identified by electronic microscopy.

Exley et al. (2002) studied the formation of HAS at pH 6.2 with an Al/Si ratio varying from 8:1 to 1:4. They identified two distinct types of HAS, which they called HAS_A and HAS_B. HAS_A had an ideal Si/Al ratio of 0.5, and its structure is dominated by Si–O–Al linkage (Q^3(3Al)) to Al in octahedral geometry. HAS_B had an ideal Si/Al ratio of 1. They kept part of the structural characteristics of HAS_A but with a new contribution of a significant Si–O–Si framework Q^4(1-2Al) and Q^3(1-2Al) and the presence of tetrahedral aluminium sites. The formation of HAS_A or HAS_B is controlled by the Si/Al ratio of the precipitation solution. HAS_A tended to form in solution where Al was present in excess, whereas HAS_B formed when $Si(OH)_4$ dominated. In Fig. 18.5, the different reactions between the dominating initial polycation $[Al_2(OH)_2(H_2O)_8]^{4+}$ and $Si(OH)_4$ are displayed.

HAS_A and HAS_B are amorphous materials. The infrared spectra of the HAS cations is similar to that of allophane or proto-imogolite. According to Exley et al. (2002), they share similar local structures, but they probably have nothing to do with proto-imogolites. Proto-imogolites could be viewed as a stable type of HAS_A that requires some internal restructuring through thermal

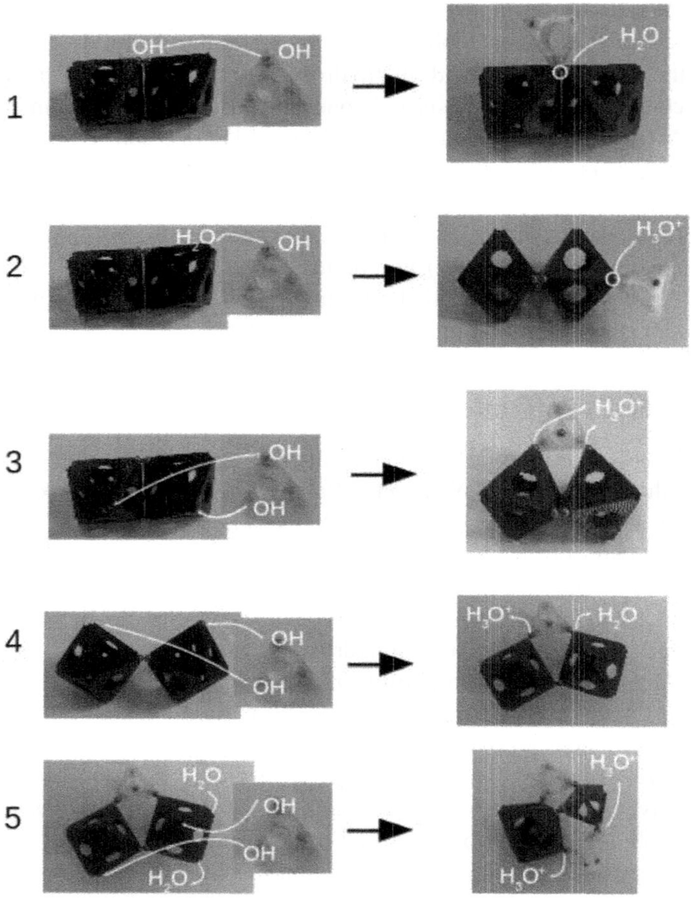

FIG. 18.5 Possible reaction scheme proposed by Exley et al. (2002) for the formation of HAS_A (schemes 1–4) and HAS_B (scheme 5).

activation to form. Both Denaix (1993) and Exley et al. (2002) presented the polycation $[Al_2SiO_2(OH)_2(H_2O)_6]^{2+}$ (structure 3 of Fig. 18.5) as an important species in the further condensation reactions at very early stages of condensation. It is interesting to note that structure 4 of Fig. 18.5 readily gives structure 3 upon intramolecular olation. Denaix (1993) observed an NMR peak at −74 ppm, which was attributed to this structure 3 and was interpreted as the smallest polycation at which condensation can lead to proto-imogolite. Upon further hydrolysis of $[Al_2SiO_2(OH)_2(H_2O)_6]^{2+}$, the cation $[Al_2SiO_2(OH)_3(H_2O)_5]^+$ has a strong negative partial charge of $\delta_O = -0.36$ on the oxo bridge, while the newly added hydroxo group has a weaker negative partial charge of $\delta_{OH} = -0.12$. It is thus reasonable to propose a further condensation through a nucleophilic attack of the oxo bridge on an aquo group and a further intramolecular condensation with the hydroxo group, as shown in Fig. 18.6. This reaction would lead to a new edge-sharing bond between two aluminium atoms $[Al_4Si_2O_4(OH)_{10}(H_2O)_8]^{2+}$. There are 3 possible edges for this new bond and 2 orientations of the polycation attaching to this edge, giving a total of 12 possible structures.

When two 'proto-imogolite-compatible' structures are chosen among the 12 possible tetramers, their condensation can lead to a dense polycation with the formation of a ring of 6 edge-sharing aluminium. It can be further stabilised by oxolation with the hydroxo group of the central silicon tetrahedral, as shown in Fig. 18.7A. Similar reactions between randomly selected tetramers proved more difficult and generally do not allow further intramolecular condensation. This additional intramolecular oxolation is probably a key point explaining the enhanced stability of the proto-imogolite structures. In Fig. 18.7B, the condensation between two randomly selected tetramers leads to a less dense structure with Q^3(2Al) silicon tetrahedra. On this structure, further condensation is still possible on the edge with a dotted line.

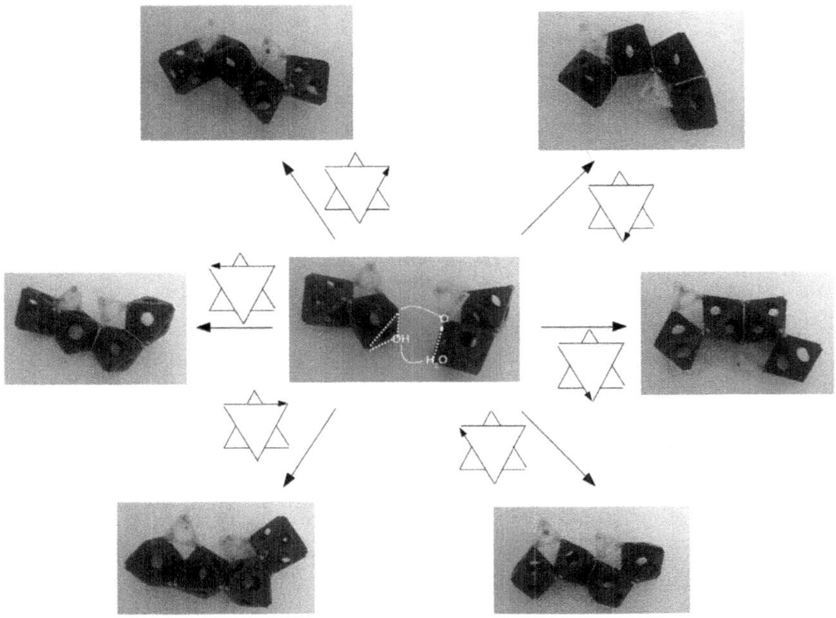

FIG. 18.6 Possible condensations between two $[Al_2SiO_2(OH)_3(H_2O)_5]^+$ cations. The *triangle* shows the three edges on which the condensation can occur, and the *arrow* shows the position of the nucleophilic attack of the oxo ligand. This scheme shows six possible structures arising from this reaction. Six other structures are possible with the second edge of the condensing polycation. The *bottom-right* structure is the only one to be compliant with an imogolite structure. All other polycations require internal restructuring with bonds breaking to make an imogolite structure.

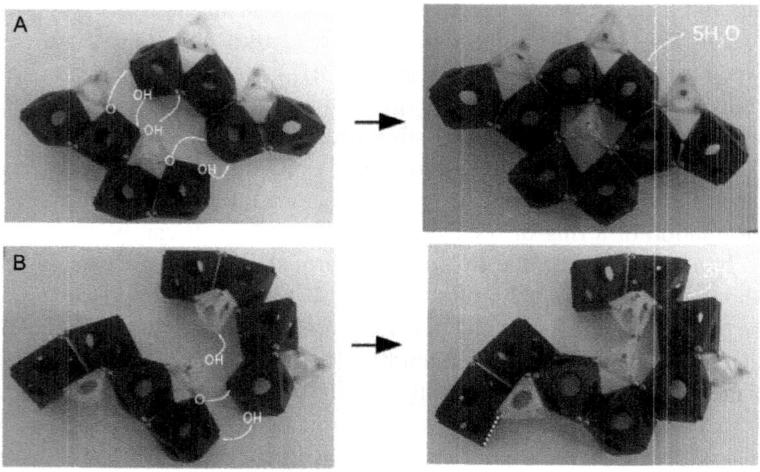

FIG. 18.7 (A) Condensation between two 'proto-imogolite compatible tetramer'. The obtained cation is further stabilised by the Q3(3Al) silicon tetrahedra responsible for the characteristic NMR signal at −78 ppm. (B) Condensation between two randomly selected tetramers. Their condensation leads to a more open structure, on which further condensation is possible on the *dotted line*.

The octamer $[Al_8Si_4O_9(OH)_{18}(H_2O)_{10}]^{4+}$, can condense further into proto-imogolite structures upon hydrolysis of aquo ligands. Already at this stage, this polycation shows a curvature. This succession of possible reaction pathways is mostly presented here for pedagogic purposes. It illustrates the fact that obtaining the imogolite structure directly is not very likely. The formation of imogolite-like structures will probably require internal restructuring, as suggested by Mukherjee et al. (2005). The real situation is also more complex because of the presence of alcohol and complexing anions.

Yucelen et al. (2011) used electro-spray coupled to mass spectroscopy to assess the possible structures at the early stages of hydrolysis and condensation. They showed that a wide range of possible polycations are present in their spectra. Most important, the silicon in these polycations may not be fully hydrolysed and thus still contain some $-Si-OC_2H_5$ groups. It is also observed that the polycations contain adsorbed anions. Using molecular dynamic simulation, Yucelen et al. (2011) showed that these anions play an important role in the early curvature of the proto-imogolite. This could explain the role of the different acid types in the control of the diameter of the final imogolite nanotubes (Yucelen et al., 2011).

18.5 PROTO-IMOGOLITE SHAPE AND INTERACTION

Very early on, it was supposed that proto-imogolite was in fact a piece of an imogolite nanotube. Farmer et al. (1979) said: 'Similarly, proto-imogolite can be presumed to incorporate orthosilicate groups attached to fragments of gibbsitelike sheets of aluminium hydroxide'. The observation of this proto-imogolite is difficult, however, as it is obtained generally at very low concentrations (about millimolars of silicon); and many in situ techniques, such as liquid NMR, SAXS are not sensitive enough to obtain some precise structural insights. More recently, Mukherjee et al. (2005) talked about oligomeric precursors, which they assumed to be in thermodynamic equilibrium with monomeric chemical precursors of aluminium and silicon. However, later, Mukherjee et al. (2007) used the term 'amorphous nanoparticles' to qualify the proto-imogolites. The term 'amorphous' is correct if it describes a structure with no long-range order, and therefore no well-defined diffraction pattern. But according to the early assumption of Farmer et al. (1979), proto-imogolites are not amorphous. They grow with a well-defined local structure similar to the final imogolite structure, but they are nano-objects with a size of about 5 nm. The amorphous states and the internal restructuring are probably important at early stages, as described in the previous section.

The discovery of very concentrated synthesis of Al–Ge imogolite nanotubes by Levard et al. (2008) opened new opportunities to explore in situ the structures of proto-imogolite. Using X-ray adsorption spectroscopy, inductively coupled plasma mass spectroscopy and SAXS, Levard et al. (2008) showed that a small piece of imogolite nanotube containing about 200 Al atoms with 20% lacunae of Ge tetrahedra and a curvature slightly less than that of the final nanotube could consistently explain the different experimental observations. Later, Thill et al. (2012a) published the first cryo-TEM images of proto-imogolite from these high-concentration Al–Ge nanotube syntheses (Fig. 18.8). The supposed shape of a roof-tile 5-nm nanotube piece is consistent with the cryo-TEM observations.

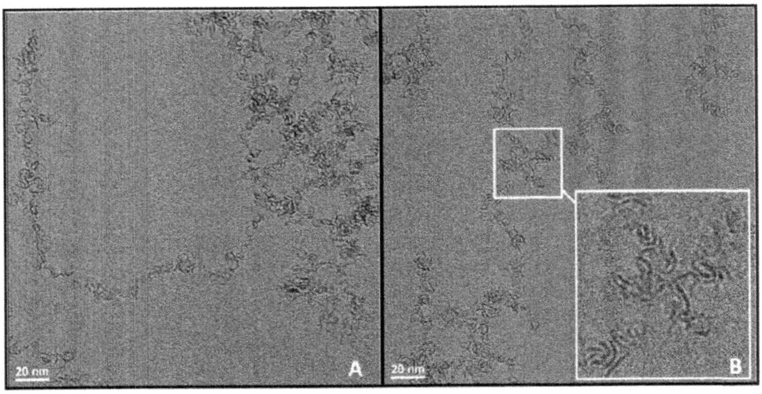

FIG. 18.8 Cryo-TEM images of the dispersions obtained for an initial aluminium perchlorate concentration of 0.5 M and a hydrolysis ratio of OH/Al = 0.5. (A) and (B) are two areas of the same sample. The inset in image (B) displays a detailed view of the objects.

The asymmetric shape of the proto-imogolite is very important to understand the growth of imogolite nanotubes. Arancibia-Miranda et al. (2011) showed that following the IEP of imogolite during their synthesis, a significant shift towards higher pH is observed. They interpreted this result by the evolution of the accessibility of Si–OH groups upon the formation of imogolite nanotubes at the expense of proto-imogolite. In proto-imogolite, the Si-OH groups are accessible, while in final nanotubes, they are confined in the internal cavity. The decrease of the total accessible Si–OH groups explained the increase of the IEP. Thill et al. (2012b) showed that proto-imogolite experiences an electrostatic interaction between the internal surface covered with Si–OH groups, which tends to be negatively charged, and the external surface covered with (Al–OH–Al) groups, which tends to be positively charged (Gustafsson, 2001). This interaction leads to early staking of the proto-imogolite as observed in cryo-TEM pictures (Fig. 18.8).

A competition exists between this electrostatic attractive energy and the curvature energy of the proto-imogolite. Thill et al. (2012b) showed that in the case of Al–Si imogolite, the curvature energy profile is so stiff that the proto-imogolite cannot grow up to the final nanotube with curvatures very far from their minimum energy. This is not the case for Al–Ge imogolite, whose energy profile is smoother (Konduri et al., 2006). In this case, the minimum energy corresponds to DW nanotubes, which have a significant attractive energy gain with not much of an increase in curvature energy. This theory has been rationalised and generalised for MW nanotubes and spirals (Thill et al., 2012b).

Amara et al. (2015) showed that when methyltriethoxygermanium is used instead of tetraethoxygermanium, then only SW nanotubes formed whatever the Si-Ge ratio was. This is easily explained by the fact that Ge–CH_3 groups can no longer create electrostatic interactions with the external proto-imogolite surface. It is also interesting to note that synthesis with methyltriethoxysilane instead of TEOS allowed nanotubes to be obtained more easily than with classical synthesis (Bottero et al., 2011). This is probably because without mutual electrostatic attraction and with more limited possible polymerisation, imogolite structure formation is greatly favoured.

Several authors have showed that the proto-imogolite shape and interaction are very important in the shape of the obtained nanostructure. Abidin et al. (2007) showed that the introduction of an alkali or alkalimetal cation in the form of chloride can change the obtained product from imogolite towards allophane. This transition occurs at an Na-Al ratio of 30 or a Ca-Al ratio of 5. Abidin et al. (2007) made simulations on a proto-imogolite structure with an Si/Al ratio of 1.17. They showed that the model changes shape upon the dissociation of one of the three hydroxyl of –$Si(OH)_3$ edge groups. Bac et al. (2009) made an experimental work on the synthesis of Al–Ge nanosphere (allophane-like) or nanotubes (imogolite-like). They showed that the selection of the type of nanostructure is performed by the pH. They first add NaOH up to a pH of 13. Then the pH is reduced with HCl to 4 or 9. Upon acidification at pH 4, nanotubes formed, whereas at pH 9, nanospheres are obtained. They explained this selection by the self-assembly of oligomers, which is influenced by the pH and salt concentration. Konduri et al. (2006) used a simplified mechanical model for imogolite and used harmonic potentials for the stretching of Si–O and Al–O bonds. The bending of O–Al–O and O–Si–O bonds was not considered. Applying the same methods for the proto-imogolite shapes of increasing size is interesting, as it helps one to understand the kinetic control introduced by the experiments of Bac et al. (2009).

Indeed, if the association of Si tetrahedra with the two-dimensional (2D) network of gibbsitelike Al octahedra occurs at the very early stages, then very small proto-imogolite structures like the one considered by Abidin et al. (2007), with six Al and one Si, can form. Using the harmonic potentials shown in Table 18.1, the equilibrium energy of such structures was computed using a Monte Carlo (MC) algorithm.

For the smallest proto-imogolite containing one Si tetrahedra and six Al octahedra, the structure contains $1\times(6+4)+6\times(12+6)=118$ harmonic oscillators. Six bending angles corresponding to Al–O–Al edge-sharing octahedra are not free to move. Three Si–O–Al bonds also reduced the number of free oscillators. The number of freely moving oscillators, thus, is 109. The kinetic energy of such a structure, thus, is $E_c=\frac{1}{2}kT\times 109=54kT$. The same calculation can be performed for proto-imogolites of increasing size. Proto-imogolite structures containing 1, 4, 9, 16, 25, 36, 49, 64, 100, 144 and 196 Si tetrahedral have been considered. The energy of the proto-imogolite containing 1 and 25 atoms as a function of the number of MC steps is shown in Fig. 18.9. The initial structures start from high-energy configurations, and after about 4000 MC steps per atom, they relaxed to a constant energy.

For the smallest structure, the average relaxed energy is close to the E_c limits. This structure behaves as a network of an unconstrained bonded spring network where all springs oscillate around their energy minimum. As the proto-imogolite grows, this quickly becomes no longer the case.

TABLE 18.1 Harmonic Potentials Used for MC Relaxation of Proto-Imogolite Structures

	Equilibrium distance/angle	Bond strength
Si–O	1.65 Å	5.89×10^5 J mol^{-1} Å^{-2}
Al–O	2.15 Å	2.54×10^5 J mol^{-1}Å^{-2}
O–Si–O	109.47 degree	10^5 J mol^{-1} degree^{-2}
O–Al–O	90 degree	10^5 J mol^{-1} degree^{-2}

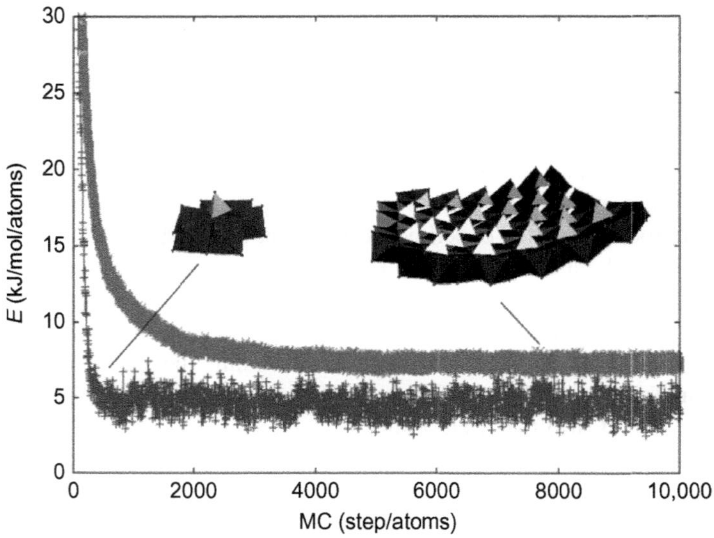

FIG. 18.9 Energy of proto-imogolite structures containing 1 and 25 silicon atoms. The smallest proto-imogolite relaxed to $E_c = 4.32$ kJ/mol/atoms, which is very close to the kinetic energy of the fully relaxed structure ($E_c = 4.19$ kJ/mol/atoms). The large proto-imogolite has a relaxed energy of 7.64 kJ/mol/atoms, which is much higher than its fully relaxed kinetic energy ($E_c = 4.7$ kJ/mol/atoms).

For a proto-imogolite containing 25 tetrahedra, the equilibrium energy is $7.64\ \mathrm{kJ\,mol^{-1}\,atoms^{-1}}$, which is already far from the E_c limit of $4.7\ \mathrm{kJ\,mol^{-1}\,atoms^{-1}}$. This increase of the equilibrium energy is explained by the boundary effects. Indeed, all tetrahedra aim at having the same local curvature as the smallest fully relaxed structure. But when the proto-imogolite gets bigger, the border effects prevent this relaxation. It is like trying to give a spherical curvature to a sheet of paper; at some point, you will need to cleverly cut the paper. Even if some average energy increase is observed for the small proto-imogolite, they still look like pieces of spheres, and one can imagine that the association of such objects would lead to allophane-like structures. However, for the largest proto-imogolite, it is obvious that the curvature starts to depend strongly on the direction. Such transition is shown in Fig. 18.10 as a function of the energies of the proto-imogolites. The association of such proto-imogolites may lead to an imogolite structure.

This transition between allophane and imogolite structures gives a change in slope of the average energy of the proto-imogolite as a function of its size. This change occurs for proto-imogolites having about 100 Si atoms (240 Al atoms). This is the typical size observed by Levard et al. (2008) using SAXS for Al–Ge proto-imogolites.

It is thus very important to understand the growth kinetic of aluminium with respect to the reaction kinetic of silicon. In a two-step process, with the first being the formation of proto-imogolites and then followed by internal

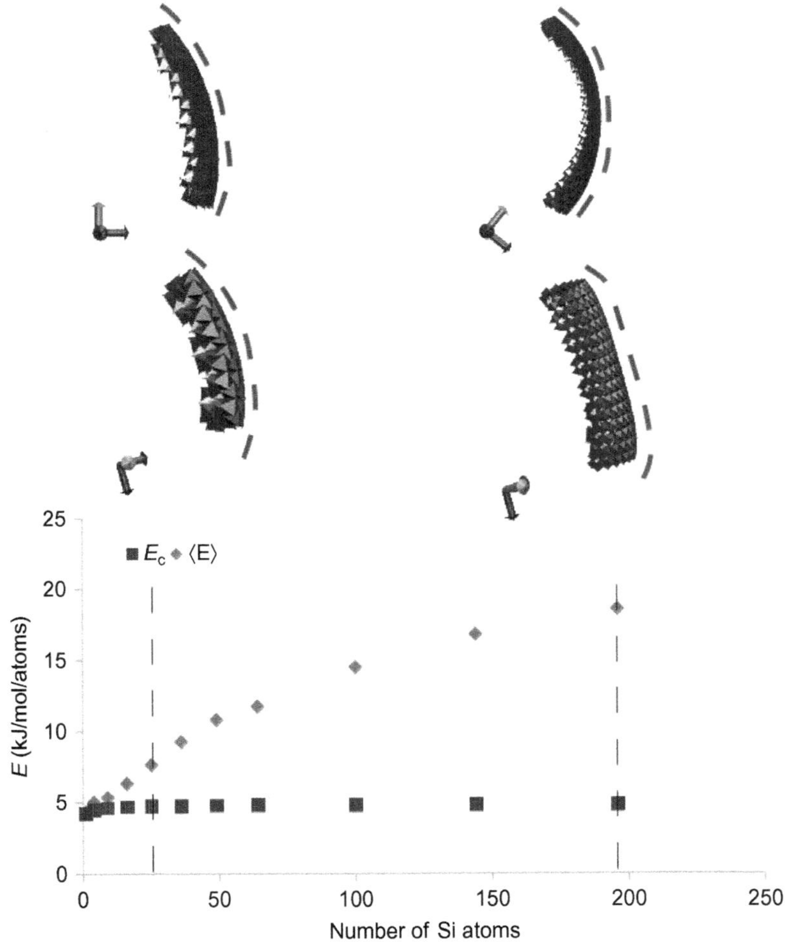

FIG. 18.10 Average energy $\langle E \rangle$ compared to the theoretical fully relaxed kinetic energy of proto-imogolite containing 1, 4, 9, 16, 25, 36, 49, 64, 100, 144 and 196 Si atoms. For the proto-imogolite containing 25 and 196 Si atoms, two views of the relaxed structures are given with a 90-degree rotation of the z-axis. For the small proto-imogolite, the structure appears curved in both directions. For the largest proto-imogolite, the relaxed structure has almost no curvature along the z-axis.

reorganisation and self-assembly, the size distribution of proto-imogolites could be an important control parameter, especially for the transition between allophane and imogolite structures.

Proto-imogolite is a very important product created by the imogolite formation process. Depending on the particular physicochemical conditions, the obtained objects can be hollow spheres or nanotubes with one or several walls. However, up to now, very few detailed studies are available for this stage of the reaction.

18.6 IMOGOLITE GROWTH MECHANISM (KINETIC MODELS)

The growth kinetic of germanium-based imogolite was followed by dynamic light scattering (DLS) and ultraviolet (UV)–vis spectroscopy at different temperatures by Mukherjee et al. (2007). Based on their results, they proposed a simple two-stage kinetic model, with the first step accounting for the formation of amorphous nanoparticles from the atomic precursors of aluminium and germanium. Then, these amorphous nanoparticles transformed into fully formed nanotubes through internal rearrangement. From the fits of the growth kinetics, they obtained final sizes for the nanotubes of 17.6 nm and for the amorphous nanoparticles of 5.9 nm. They also give activation energy values for the formation of nanotubes from precursors (181 kJ/mol) and amorphous nanoparticles (110 kJ/mol).

Yang et al. (2008) followed the growth of imogolite nanotubes using a special TEM grid preparation technique (developed by Yang and Su, 2007), which allowed the observation of well-dispersed individual nanotubes. They showed that the number of nanotube increases during the whole observation period of 120 h. Unlike their number, the nanotube length reaches a maximum value at about 100 nm after 40 h of heating. Yang et al. (2008) measured the length distribution of nanotubes. Very few nanotubes are observed with a length of up to 500 nm. The size distribution has a characteristic asymmetric shape with a lot of small nanotubes and fewer and fewer long nanotubes. Yang et al. (2008) observed that the growth occurred at the expense of proto-imogolite and small nanotubes. They also showed that the final nanotubes can continue to grow if they are added to a precursor solution. It means that the nanotube end stays open and active towards further growth.

Maillet et al. (2011) followed the growth kinetic of germanium-based imogolite nanotubes by SAXS from the end of hydrolysis to 120 h of heating. They first observed that the initial solution at the end of the hydrolysis step already displays a scattering signal. This means that nanometric structures are already formed at this stage. The shape of the scattering corresponds to proto-imogolites having a roof-tile shape (Levard et al. 2010). The integration of the scattered intensity multiplied by the square of the scattering vector gives a value Q, which is only a function of the volume fraction and contrast of the scattering objects in solution. They showed that Q is constant throughout the reaction. This can be interpreted by the complete formation of the proto-imogolite species at the end of the hydrolysis stage. When the heating stage begins, the volume fraction of scattering solids is already at its maximum value and the structural modifications occurs only through internal restructuring or dissolution precipitation at constant solubility. Upon heating, the scattering signal slowly transforms in two ways: (i) an oscillation appears and stays at the same scattering vector q value throughout the synthesis, and this oscillation is linked to the characteristic diameter of the nanotubes; and (ii) the second transformation is linked to the increase of the average length of the nanotubes, inducing an increase in the low-angle intensity. By using atomic force microscopy (AFM), Maillet et al. (2011) found that the length distribution of the nanotubes have the same asymmetric shape as the one measured by Yang et al. (2008); and they proposed a kinetic model to quantitatively explain the evolution of the length of the nanotubes. Their model was based on the hypothesis that nanotubes are formed through the internal restructuring of proto-imogolites, and the growth is due to tube–tube collisions. Once formed, the small nanotube pieces can assemble into longer nanotubes. These authors introduced collision kernels inspired from the formation of rigid polymers, which took into account the one-dimensional (1D) shape of the growing imogolite and introduced a linkage probability corresponding to the number of collisions actually needed for the formation of a longer nanotube. The computation of the model revealed that it is necessary to consider the collision between fully formed and already grown nanotubes to reproduce the asymmetric shape of the length distribution. If only the growth of the nanotubes from small proto-imogolites is considered, then the shape of the length distribution is no longer asymmetric towards long nanotubes.

The fitting parameter corresponds to the bonding probability of 5×10^{-7} for SW nanotubes and 10^{-7} for DW nanotubes. These very small linkage probabilities mean that more than 10 million collisions are required to actually link two nanotubes. This explains the observed very slow kinetic despite the relatively high concentrations (0.25 M and 0.5 M in aluminium). This development is linked to three factors. First, the collision kernels quickly go to zero when the length of the nanotube increases. It is easy to understand that the probability of having a properly oriented collision between 1D objects is very low when the objects are long. Second, even for these rare properly aligned collisions, only 1 out of millions actually leads to a new bond between the nanotubes. The third factor is less intuitive. When an object grows in only one dimension, then the average distance $\langle D \rangle$ between the objects increases more slowly than the objects' average size $\langle L \rangle$. Therefore, at a certain point of this growth, the nanotubes get longer than their average distance. This obviously hinders the growth mechanism involving tip–tip collisions. This is the same phenomenon as the transition from a diluted to semidiluted regime in polymers. For a synthesis of an Al–Si imogolite containing 12 Si atoms per 0.425 nm performed with an initial Si concentration of 2 mM, this transition occurs for an average nanotube length of about 150 nm. As the average nanotube length is generally larger than this value, a significant part of the growth occurs in a semidiluted regime. Maillet et al. (2011) took this transition into account through the use of an effective concentration for collision kernels in the semidiluted regime.

Yucelen et al. (2013) also proposed a detailed kinetic model. They considered a four-step model, with k_1 being the rate of transformation of precursors (P) into amorphous nanoparticles (A); k_2 being the rate of transformation of (A) into the smallest nanotubes; k_3 being the growth of nanotubes through the incorporation of P at the tube end; and k_4 being the tip–tip collision between nanotubes. The Yucelen et al. (2013) model also allows for explaining quantitatively the growth and length distribution shapes of the nanotubes. The main difference with the Maillet et al. (2011) model comes from the fact that the k_4 rate has no length dependence. This means that two very long nanotubes have the same collision rate as two short ones. They also refined the early-stage process by introducing a k_3 rate for the direct incorporation of proto-imogolites into fully formed nanotubes. In the Maillet et al. (2011) model, proto-imogolites are first transformed into nanotubes before they can be added to the nanotube end. Despite the small divergence in the hypotheses, these works have converged towards a common general description of the imogolite growth. Indeed, both Yucelen et al. (2013) and Maillet et al. (2011) concluded that the early assumption by Farmer et al. (1979) that nutrient addition as the sole growth mechanism is not sufficient. Rather, it is necessary to consider tip–tip collisions between fully formed nanotubes to account for the shape of the length distribution. Details about the early process (k_1, k_2 and k_3 in the Yucelen et al. (2013) model) would certainly benefit from more detailed kinetic studies of the early stages of imogolite formation.

18.7 CONCLUDING REMARKS

The general mechanism for the formation of imogolite nanotube is now well accepted. In Fig. 18.11, a schematic diagram of the proposed reaction pathway is presented. In the first step of the reaction, the coprecipitation of Al and Si gives essentially an amorphous precipitate that can fit into the definition of HAS_A proposed by Exley et al. (2002). This precipitate is probably far from an imogolite-like structure despite the fact that it has similar infrared spectra as proto-imogolites. It is probable that internal restructuring occurs during ageing or at the beginning of the heating stage. This restructuring leads to the formation of proto-imogolites that are pieces of imogolite nanotubes. Their average size, the presence of complexing anions and their mutual interactions are probably very important to the selection of the obtained nanostructures. In proper physicochemical conditions (i.e., low concentration, low pH, low salinity, and an Si/Al ratio between 0.5 and 0.6), nanotubes are obtained at the expense of proto-imogolites upon heating. From these conditions, some recipes still allow for obtaining nanotubes, but allophanes with eventual mixtures of aluminium hydroxide and amorphous silicon often are obtained. The success of an imogolite synthesis can be assessed by simple visual inspection. The dispersion has to be water clear. It forms a gel as the pH or the salinity increases. When concentrated at volume fractions of 1% or more, the dispersion is birefringent when observed between crossed polarisers.

It is fascinating to see that starting from the most abundant elements of the Earth's crust (silicon and aluminium), it is possible to self-assemble perfectly defined nanotubes with a diameter controlled at the angstrom level. It is

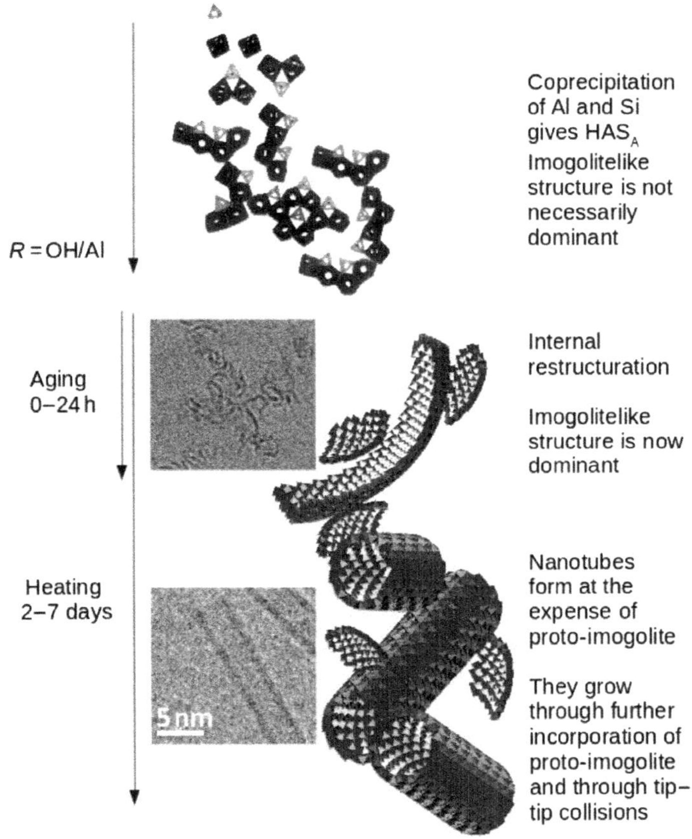

FIG. 18.11 Schematic reaction pathway to obtain imogolite.

significant to realise that the very large community working on carbon nanotubes has failed to achieve an equivalent success upon structural control of the synthesis despite major research efforts over time.

Recent studies have allowed a deep understanding of this phenomenon. Researchers have also considerably increased the types of available imogolite-like structures. It is possible to obtain short or long nanotubes, as well as to significantly modify the diameter and to control wall composition and the surface properties through relatively simple chemical recipes. This opens a research field towards a whole family of 'geo- inspired' nanostructures. The initial steps of the synthesis are still unexplored to a large extent. In particular, the parameters controlling the shape and interaction of proto-imogolites are not fully established. This is probably an interesting area for researchers to study in order to gain even better control over the synthesis and to explore even more structural modifications of imogolite-like structures.

REFERENCES

Abidin, Z., Matsue, N., Henmi, T., 2007. Differential formation of allophane and imogolite: experimental and molecular orbital study. J. Computer-Aided Mater. Des. 14, 5–18.

Abidin, Z., Matsue, N., Henmi, T., 2008. A new method for nano tube imogolite synthesis. Jpn. J. Appl. Phys. 47, 5079–5082.

Ackerman, W.C., Smith, D.M., Huling, J.C., Kim, Y.-W., Bailey, J.K., Brinker, C.J., 1993. Gas/vapor adsorption in imogolite: a microporous tubular aluminosilicate. Langmuir 9, 1051–1057.

Akitt, J.W., Elders, J.M., 1988. Multinuclear magnetic resonance studies of hydrolysis of aluminum (III) 8. Base hydrolysis monitored at very high magnetic field. J. Chem. Soc. Dalton Trans. 5, 1347–1355.

Amara, M.-S., Paineau, E., Bacia-Verloop, M., Krapf, M.-E., Davidson, P., Belloni, L., Levard, C., Rose, J., Launois, P., Thill, A., 2013. Single-step formation of micron long $(OH)_3Al_2O_3GeOH$ imogolite-like nanotubes. Chem. Commun. 49, 11284–11286.

Amara, M.-S., Paineau, E., Rouzière, S., Guiose, B., Krapf, M.E.M., Taché, O., Launois, P., Thill, A., 2015. Hybrid tunable-diameter metal-oxide nanotubes for organic molecules trapping. Chem. Mater. 27, 1488–1494.

Arancibia-Miranda, N., Escudey, M., Molina, M., Garcia-Gonzales, M.T., 2011. Use of isoelectric point and pH to evaluate the synthesis of a nanotubular aluminosilicate. J. Non Cryst. Solids 357, 1750–1756.

Arancibia-Miranda, N., Escudey, M., Molina, M., Garcia-Gonzales, M.T., 2013. Kinetic and surface study of single-walled aluminosilicate nanotubes and their precursors. Nanomaterials 3, 126–140.

Axelos, M., Tchoubar, D., Bottero, J.Y., Fiessinger, F., 1985. Détermination par DPAX de la structure fractale d'agrégats obtenus par collage d'amas. Etude de deux solutions d'hydroxyde d'aluminum Al(OH)x avec x = 2.5 et 3. J. Phys. France 46, 1587–1593.

Bac, B.H., Song, Y., Kim, M.H., Lee, Y.B., Kang, I.M., 2009. Surface-modified aluminogermanate nanotube by OPA: synthesis and characterization. Inorg. Chem. Commun. 12, 1045–1048.

Barrett, S.M., Budd, P.M., Price, C., 1991. The synthesis and characterization of imogolite. Eur. Polym. J. 27, 609–612.

Bernards, T.N.M., Vanbommel, M.J., Boonstra, A.H., 1991. Hydrolysis-condensation processes of the tetra-alkoxysilanes TPOS, TEOS and TMOS in some alcoholic solvents. J. Non Cryst. Solids 134, 1–13.

Bonelli, B., Bottero, I., Ballarini, N., Passeri, S., Cavani, F., 2009. IR spectroscopic and catalytic characterization of the acidity of imogolite-based systems. J. Catal. 264, 15–30.

Bottero, J.Y., Cases, J.M., Fiessinger, F., Poirier, J.E., 1980. Studies of hydrolyszed aluminum chloride solutions. 1. Nature of aluminum species and composition of aqueous solutions. J. Phys. Chem. 84, 2933–2939.

Bottero, I., Bonelli, B., Ashbrook, S.E., Wright, P.A., Zhou, W., Tagliabue, M., Armandi, M., Garrone, E., 2011. Synthesis and characterization of hybrid organic/inorganic nanotubes of the imogolite type and their behaviour towards methane adsorption. Phys. Chem. Chem. Phys. 13, 744–750.

Brinker, C.J., Scherer, G.W., 1989. Sol-Gel Science, the Physics and Chemistry of Sol-Gel Processing. Academic press, Harcourt Bace Jovanovich Publishers, Boston.

Clark, C.J., McBride, M.B., 1984. Cation and anion retention by natural and synthetic allophane and imogolite. Clays and Clay Miner. 32, 291–299.

Denaix, L., 1993. Synthèse et propriétés d'aluminosilicates non lamellaires: l'imogolite et les allophanes. PhD thesis, University Paris VI, 21 January 1993.

Denaix, L., Lamy, I., Bottero, J.Y., 1999. Structure and affinity towards Cd^{2+}, Cu^{2+}, Pb^{2+} of synthetic colloidal amorphous aluminosilicates and their precursors. Colloids Surf. A 158, 315–325.

Exley, C., Schneider, C., Doucet, F.J., 2002. The reaction of aluminum with silicic acidic solution: an important mechanism in controlling the biological availability of aluminum. Coord. Chem. Rev. 228, 127–135.

Farmer, V.C., 1979. Synthesising imogolite. UK Patent GB 2025384A.

Farmer, V.C., Fraser, A.R., Tait, J.M., 1977. Synthesis of imogolite: a tubular aluminum silicate polymer. J. Chem. Soc. Chem. Commun. 13, 462–463.

Farmer, V.C., Fraser, A.R., Tait, J.M., 1979. Characterization of the chemical structure of natural and synthetic aluminosilicate gels and sols by infrared spectroscopy. Geochim. Cosmochim. Acta 43, 1417–1420.

Farmer, V.C., Adams, M.J., Fraser, A.R., Palmieri, F., 1983. Synthetic imogolite: properties, synthesis, and possible applications. Clay Miner. 18, 459–472.

Garcia-Hernandez, J.E., Rodriguez-Rodriguez, A., 1990. Altération expérimentale des cendres basaltiques dans l'oxalate d'ammonium 10^{-4}M. Science du sol 28, 193–201.

Gustafsson, J.P., 2001. The surface chemistry of imogolite. Clay Clay Miner. 49, 73–80.

Hongo, T., Sugiyama, J., Yamazaki, A., Yamasaki, A., 2013. Synthesis of imogolite from rice husk ash and evaluation of its acetaldehyde adsorption ability. Ind. Eng. Chem. Res. 52, 2111–2115.

Hu, J., Kannangara, K., Wilson, M.A., Reddy, N., 2004. The fused silicate route to protoimogolite and imogolite. J. Non Cryst. Solids 347, 224–230.

Iler, R.K., 1979. The Chemistry of Silica: Solubility, Polymerization, Colloid and Surface Properties, and Biochemistry of Silica. John Wiley and Sons, New York.

Inoue, K., Huang, P.M., 1985. Influence of citric acid on the formation of short range ordered aluminosilicates. Clays Clay Miner. 33, 312–322.

Koenderink, G.H., Kluijtmans, S.G.J.M., Philipse, A.P., 1999. On the synthesis of colloidal imogolite fibers. J. Colloid Interface Sci. 216, 429–431.

Konduri, S., Mukherjee, S., Nair, S., 2006. Strain energy minimum and vibrational properties of single-walled aluminosilicate nanotubes. Phys. Rev. B 74, 033401.

Kuroda, Y., Kuroda, K., 2008. Layer-by-layer assembly of imogolite nanotubes and polyelectrolytes into core-shell particles and their conversion to hierarchically porous spheres. Sci. Technol. Adv. Mater. 9, 025018.

Levard, C., Rose, J., Masion, A., Doelsch, E., Borschneck, D., Olivi, L., Dominici, C., Grauby, O., Woicik, J.C., Bottero, J.Y., 2008. Synthesis of large quantity of single-walled aluminogermanate nanotube. J. Am. Chem. Soc. 130, 5862–5863.

Levard, C., Masion, A., Rose, J., Doelsch, E., Borschneck, D., Dominici, C., Ziarelli, F., Bottero, J.Y., 2009. Synthesis of imogolite from decimolar concentration at low temperature and ambient pressure: a promising route for inexpensive nanotubes. J. Am. Chem. Soc. 131, 17080–17081.

Levard, C., Rose, J., Thill, A., Masion, A., Doelsch, E., Maillet, P., Spalla, O., Olivi, L., Cognigni, A., Ziarelli, F., Bottero, J.Y., 2010. Formation and growth mechanism of imogolite-like aluminogermanate nanotubes. Chem. Mater. 22, 2466–2473.

Levard, C., Masion, A., Rose, J., Doelsch, E., Borschneck, D., Olivi, L., Chaurand, P., Dominici, C., Ziarelli, F., Thill, A., Maillet, P., Bottero, J.Y., 2011. Synthesis of Ge-imogolite: influence of the hydrolysis ratio on the structure of the nanotubes. Phys. Chem. Chem. Phys. 13, 14516–14522.

Liz-Marzan, L., Philipse, A.P., 1994. Synthesis of platinum nanoparticles in aqueous host dispersions of inorganic (imogolite) rods. Colloids Surf. A Physicochem Eng. Asp. 90, 95–109.

Maillet, P., Levard, C., Larquet, E., Mariet, C., Spalla, O., Menguy, N., Masion, A., Doelsch, E., Rose, J., Thill, A., 2010. Evidence of double-walled Al-Ge imogolite-like nanotubes. A cryo-TEM and SAXS investigation. J. Am. Chem. Soc. 132, 1208–1209.

Maillet, P., Levard, C., Spalla, O., Masion, A., Rose, J., Thill, A., 2011. Growth kinetic of single and double-walled aluminogermanate imogolite-like nanotubes: an experimental and modelling approach. Phys. Chem. Chem. Phys. 13, 2682–2689.

McBride, M.B., Farmer, C.V., Russel, J.D., Tait, J.M., Goodman, B.A., 1984. Iron substitution in aluminosilicate sols synthesized at low pH. Clay Miner. 19, 1–8.

Mukherjee, S., Bartlow, V.M., Nair, S., 2005. Phenomenology of the growth of single-walled aluminosilicate and aluminogermanate nanotubes of precise dimensions. Chem. Mater. 17, 4900–4909.

Mukherjee, S., Keesuk, K., Nair, S., 2007. Short, highly ordered, single-walled mixed oxide nanotubes assemble from amorphous nanoparticles. J. Am. Chem. Soc. 129, 6820–6826.

Nakanishi, R., Wada, S.I., Suzuki, M., Maeda, M., 2007. Heat-induced gelation of hydroxy-aluminosilicate synthesized by instantaneous mixing of sodium silicate and aluminum chloride. J. Fac. Agric. Kyushu Univ. 52, 147–151.

Nakanishi, R., Suzuki, M., Inukai, K., Maeda, M., Wada, S.I., 2008. Effects of dialysis on the formation of imogolite from highly concentrated hydroxyaluminum-silicate solution. Trans. Mater. Res. Soc. Jpn 33, 493–496.

Ohashi, F., Tomura, S., Akaku, K., Hayashi, S., Wada, S.I., 2004. Characterization of synthetic imogolite nanotubes as gas storage. J. Mater. Sci. 39, 1799–1801.

Ookawa, M., Takata, Y., Suzuki, M., Inukai, K., Maekawa, T., Yamaguchi, T., 2008. Oxidation of aromatic hydrocarbons with H_2O_2 catalyzed by a nano-scale tubular aluminosilicate, Fe-containing imogolite. Res. Chem. Intermed. 34, 679–685.

Qi, X., Yoon, H., Lee, S.H., Yoon, J., Kim, S.J., 2008. Surface-modified imogolite by 3-APS-OsO4 complex: Synthesis, characterization and its application in the dehydroxylation of olefins. J. Ind. and Eng. Chem. 14, 136–141.

Su, C., Harsh, J.B., Bertsch, P.M., 1992. Sodium and chloride sorption by imogolite and allophanes. Clays and Clay Miner. 40, 280–286.

Thill, A., Maillet, P., Guiose, B., Spalla, O., Belloni, L., Chaurand, P., Auffan, M., Olivi, L., Rose, J., 2012a. Physico-chemical control over the single- or double-wall structure of aluminogermanate imogolite-like nanotubes. J. Am. Chem. Soc. 134, 3780–3786.

Thill, A., Guiose, B., Bacia-Verloop, M., Geertsen, V., Belloni, L., 2012b. How the diameter and structure of $(OH)_3Al_2O_3Si_xGe_{1-x}(OH)$ imogolite nanotubes are controlled by an adhesion versus curvature competition. J. Phys. Chem. C 116, 26841–26845.

Van den Brule, S., Beckers, E., Chaurand, P., Liu, W., Ibouraadaten, S., Palmai-Pallag, M., Uwambayinema, F., Yakoub, Y., Avellan, A., Levard, C., Haufroid, V., Marbaix, E., Thill, A., Lison, D., Rose, J., 2014. Nanometer-long Ge-imogolite nanotubes cause sustained lung inflammation and fibrosis in rats. Part. Fiber Toxicol. 11, 67.

Wada, S.I., 1987. Imogolite synthesis at 25°C. Clay Miner. 35, 379–384.

Wada, S.I., Wada, K., 1980. Formation, composition and structure of hydroxy-aluminosilicate ions. J. Soil Sci. 31, 457–467.

Wada, S.I., Wada, K., 1982. Effects of substitution of germanium for silicon in imogolite. Clay Clay Miner. 30, 123–128.

Yang, H., Su, Z., 2007. Individual dispersion of synthetic imogolite nanotubes via droplet evaporation. Chin. Sci. Bull. 52, 2301–2303.

Yang, H., Wang, C., Su, Z., 2008. Growth mechanism of synthetic imogolite nanotubes. Chem. Mater. 20, 4484–4488.

Yoshinaga, N., Aomine, S., 1962. Imogolite in some andosoils. Soil Sci. Plant Nutr. 8 (3), 22–29.

Yucelen, G.I., Choudhury, R.P., Vyalikh, A., Scheler, U., Beckham, H.W., Nair, S., 2011. Formation of single-walled aluminosilicate nanotubes from molecular precursors and curved nanoscale intermediates. J. Am. Chem. Soc. 133, 5397–5412.

Yucelen, G.I., Kang, D.Y., Schmidt-Krey, I., Beckham, H.W., Nair, S., 2013. A generalized kinetic model for the formation and growth of single-walled metal oxides nanotubes. Chem. Eng. Sci. 90, 200–212.

Chapter 19

Imogolite-Like Family

N. Arancibia-Miranda[a,b,1] and M. Escudey[a,b]
[a]*Center for the Development of Nanoscience and Nanotechnology, CEDENNA, Santiago, Chile*
[b]*Facultad de Química y Biología, Universidad de Santiago de Chile, Santiago, Chile*
[1]*Corresponding author: e-mail: nicolas.arancibia@usach.cl*

19.1 INTRODUCTION

In the world of nanotechnology, nanotubular materials are considered key pieces for the development of new functional materials. They have a huge development potential in areas such as electronics, biotechnology, sensing, separations, energy storage management and catalysis (Bursill et al., 2000; Tenne, 2006; Bottero et al., 2010; Kijima, 2010).

Significant advances in this type of nanostructures were made following the discovery by Iijima (1991) of carbon nanotubes and by Tenne et al. (1992) of disulphide molybdenum nanotubes for which protocols of synthesis, mechanisms of growth, physicochemical properties and potential applications have been intensively explored. However, both nanotubes present a number of problems in their technological development and in the quality of the final products (Charlier, 2002; Kijima, 2010). Some examples of the difficulties impeding the full development of these nanoparticles (Sloan et al., 1999; Rao and Nash, 2003; Tenne, 2006; Yucelen et al., 2012a,b) include:

- insufficient control of the nanotubes dimensions (radius and length),
- extreme and expensive preparation conditions,
- presence of contaminants (heavy metals) and
- structural defects, polydispersion and presence of multi-layered nanotubes.

In this context, imogolite emerges as an alternative nanotube whose synthesis has low technical requirements and high reproducibility (Yoshinaga and Aomine, 1962; Cradwick et al., 1972; Wilson et al., 2007). Imogolite was the first synthetic metal oxide nanotube obtained in a laboratory by Farmer et al. (1977) and later by other authors (Barrett et al., 1991; Bursill et al., 2000). This synthesis was patented in 1981 (Farmer, 1981). In this patent, the possible addition of other ions, such as germanium (Ge), titanium (Ti), chromium (Cr) and iron (Fe), and replacing Si or Al with an isomorphic

Developments in Clay Science, Vol. 7. http://dx.doi.org/10.1016/B978-0-08-100293-3.00019-4

substitution were described. Wada and Wada (1982) synthesised an aluminogermanate nanotube with an imogolite local structure by coprecipitation. More recently, Bottero et al. (2010) demonstrated that it is possible to obtain by coprecipitation imogolite-like nanotubes where the internal Si–OH groups were replaced by Si–CH_3 groups. Since these seminal publications, many modifications happened and/or were tested in laboratories to obtain new materials sharing a common imogolite local structure. These materials can also adopt many different shapes and can lead to nanospheres, single-walled (SW) nanotubes, double-walled (DW) nanotubes, nanoscrolls, or a combination. Based on all those findings, all the new nanomaterials with different shapes are now described as a whole imogolite-like family sharing an imogolite local structure with a theoretical general formula defined as

$$(OH)_{3-h}R_2Al_{2-m}X_mO_3Si_{1-n}Y_n(OH)_{1-k}R_{1k},$$

where X = Ga, In, Fe(III), Cr(III); Y = Ge, Ti, As, P, C and R_1 and $R_2 = CH_3$, C_2H_5, C_2H_2, C_6H_6, ... This list, for X, Y and R, is probably not exhaustive.

Because of the attractive chemical and structural flexibilities that considerably expand their physicochemical properties beyond those of imogolite (Zanzottera et al., 2012; Amara et al., 2014, 2015; Boyer et al., 2014), a great deal of research has been done to understand the mechanisms of formation for such nanoparticles (Levard et al., 2010; Maillet, 2010; Yucelen et al., 2011; Thill et al., 2012b). Indeed, their high specific surface area (SSA) and the coupling of different surface reactivity (both, inside and outside) make these zero-dimensional (allophane-like) and one-dimensional (imogolite-like) materials excellent candidates for technological applications. These include clay polymer nanocomposites, gas fuel storage, absorbent, exchange material for a heat pump system, humidity-controlling material, antidwelling material and systems of controlled delivery of drugs and material for the removal of trace elements in water (Liz-Marzan and Philipse, 1994; Ohashi et al., 2004; Bac et al., 2009; Levard et al., 2009a; Guerra et al., 2011; Baroña et al., 2013; Amara et al., 2014; Arancibia-Miranda et al., 2014, 2015; Avellan et al., 2014).

This chapter discusses the different theoretical predictions made for the existence and properties of the imogolite-like family, the strategies that were successfully applied to obtain several members of this family, and the detailed formation mechanism and properties of some members of this family. Finally, some perspectives and the prospects of not-yet discovered members of the family will be reported.

19.2 THEORETICAL PREDICTIONS CONCERNING THE EXTENT OF THE IMOGOLITE-LIKE FAMILY

Inspired by the successful synthesis of Ge-imogolite carried out by Wada and Wada (1982), Alvarez-Ramirez (2007, 2009) performed theoretical research on imogolite isomorphic structures in which Si was replaced by elements of group IV (C, Ge and Sn) and Al by Ga from group III. These new imogolites showed differences in their electronic and mechanical properties compared with the traditional imogolite. Similar studies carried out by Duarte et al. (2012) and Guimarães et al. (2013) suggest that it is possible to find derivatives of imogolite by replacing the silicate with different oxyanions such as phosphate (PO_4^{3-}), phosphite (PO_3^{3-}), arsenate (AsO_4^{3-}) or arsenite (AsO_3^{3-}) (Fig. 19.1).

From these theoretical considerations, the authors proposed the synthesis of a new type of imogolite without major modifications based on the partial or total replacement of the silicate by arsenite (Imo-As-ite). In contrast, to obtain aluminophosphate (Imo-P-ate), aluminophosphite (Imo-P-ite) and aluminoarsenate (Imo-Al-As-ate) imogolites, it is critical to carry out the synthesis at pH values less than or equal to the pK_a of the corresponding acid, so that the predominant species is the fully protonated oxyanion (Duarte et al., 2012; Guimarães et al., 2013). The nanotube diameter was sensitive to the substituting oxyanion, establishing the following ascending sequence: Imo-P-ite = Imo-As-ite < Imo-P-ate < Imo-Al-As-ate, with values of 11.5, 11.5, 12.0 and 13.1 Å, respectively. A significant effect on the electronic properties were predicted for these nanotubes, when P(V) and As(V) was used since the calculated band gaps were 10 and 7 eV for Imo-P-ate and Imo-As-ate,

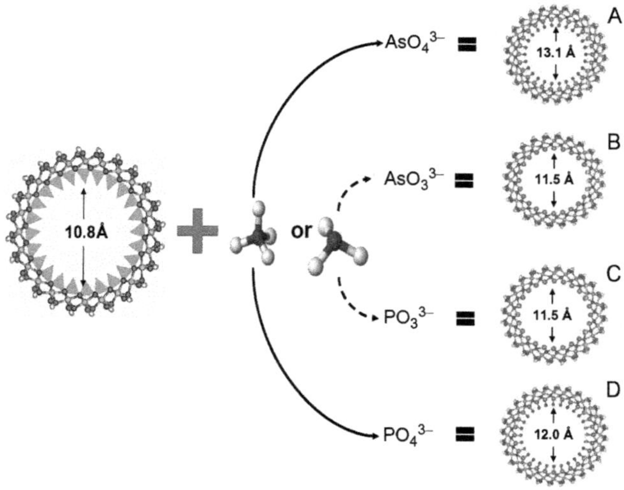

FIG. 19.1 Optimised zigzag imogolite-like nanotubes: (A) (13.0) aluminophosphate NT; (B) (12.0) aluminophosphite NT; (C) (14.0) aluminoarsenate NT and (D) (12.0) aluminoarsenite NT. *White atoms* are H, *red* (*dark grey* in the print version) are O, *blue* (*light grey* in the print version) are Al, *orange* (*grey* in the print version) are P, and *lilac* (*grey* in the print version) are As. *Adapted from Guimarães et al. (2013).*

respectively, while semiconductors with band gaps features around 5 eV were estimated for Imo-P-ite and Imo-As-ite (Guimarães et al., 2013). Possible effects on solvation, hydrophobicity and adsorption capacities could be expected in these nanotubes, as a consequence of the substitution of Si(IV) for P(III) and As(III), due to the absence of the internal –OH (Fig. 19.1).

Konduri et al. (2007) conducted a detailed work on aluminogermanate nanotubes to understand the factors controlling their dimension using different spectroscopic techniques such as X-ray photoelectron spectroscopy, vibrational spectroscopy, energy-dispersive X-ray spectroscopy, X-ray diffraction (XRD) and computational modelling. The model showed variations and displacement in the minimum energy as a function of the nanotube radius, as the Si/Ge ratio was modified. The model takes into account the rigidity and length of the Si–O, Ge–O and Al–O bonds. It allowed them to explain the increase in diameter, but the prediction of the model was a linear relationship between the proportion of Ge and the diameter. This was not fully consistent with the experimental observations showing that the diameter almost stops increasing before reaching a full replacement of Si in the structure (Fig. 19.2). This inconsistency was due to the lack of consideration of DW structures occurring at high Ge content. This was latterly considered by Maillet (2010). These authors observed that changes in the synthesis conditions allowed to obtain SW or DW nanotubes. DW nanotubes predominated

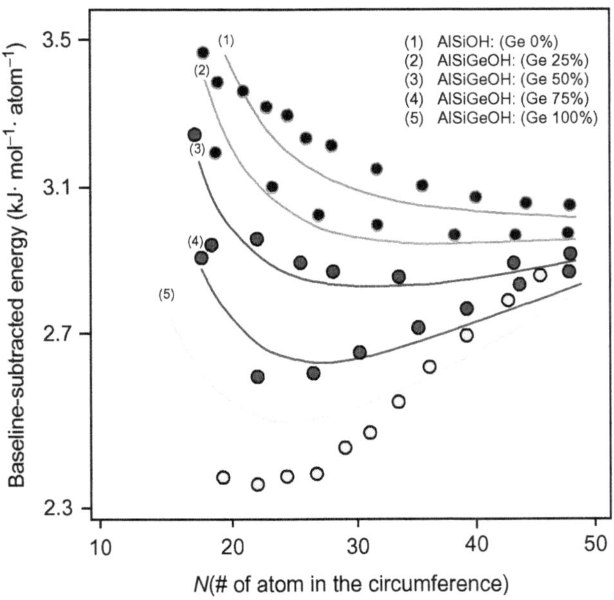

FIG. 19.2 Total energy per atom at 298 K of AlSiGeOH nanotubes as a function of the number of aluminium atoms (N) in the nanotube circumference. *Adapted from Konduri et al. (2007).*

at low concentrations of Ge (0.25 M) and show a slightly greater diameter than SW nanotubes. The coupling between internal and external elastic walls through attractive electrostatic energies was not considered in the model proposed by Konduri et al. (2007) (Fig. 19.2). This coupling explains the formation and predominance of DW or SW nanotubes.

Ab initio studies suggested that the structural functionalisation of imogolite with small percentages of Fe (5–10%) could reduce the value of the band gap between 4.6–2.6 and 4.2–1.0 eV, respectively. Fe inclusion into imogolite-like structures produces a shift of the Fermi energy and the overlapping of Fe electronic states to the original imogolite-like states at the top, the middle and the bottom of the gap region, conferring on Fe-imogolite some semiconductor properties, which would expand its application in the fields of physics, specifically in the design of sensors (Alvarez-Ramírez, 2009).

Another theoretical study was done by González et al. (2014) using classical molecular dynamics simulations. They studied the formation process of these imogolite nanotubes, starting from a perfect imogolite layer and allowing it to self-roll at different temperatures. These authors reported a lot of different conformations including binocular-like ones and nanoscrolls, depending on the temperature and the number of gibbsite units considered for each scenario. They concluded that the kinetic factors and agglomeration are issues that play a significant role in the synthesis of imogolite (and probably for other imogolite-like nanotubes). Also, their simulations suggested that the rolling of aluminosilicate layer seems to be a valid mechanism for the formation of imogolite.

19.3 SYNTHESIS STRATEGIES TO SUCCESSFULLY OBTAIN NEW IMOGOLITE-LIKE FAMILY MEMBERS

There are two different strategies that were successfully applied to obtain new compositions and shapes for imogolite-like nanoparticles. The first strategy used a modification of the reactants in a classical coprecipitation route. It is the way aluminogermanate nanotubes were initially obtained by Wada and Wada (1982) using $GeCl_4$ as a germanium precursor instead of tetraethyl orthosilicate (TEOS) for silicon. The modification of the composition in a coprecipitation synthesis route most often accompanies a change in shape. In the case of aluminogermanate, the obtained imogolite nanotubes were shorter and have a larger diameter. The second strategy, using postsynthesis modifications, has been applied to incorporate Fe(III) in imogolite by isomorphic substitution (Shafia et al., 2015) or to incorporate Ni, Cd and Cu in the external wall of imogolite through specific site adsorption (Levard et al., 2009a; Arancibia-Miranda et al., 2015). It is also possible to modify the surfaces through grafting molecules on the internal Si–OH groups or outer Al–OH groups without modifying the overall imogolite structure (Bottero et al., 2010; Zanzottera et al., 2012; Kang et al., 2014). Up to now, these postsynthesis modifications mainly allowed superficial composition modification.

19.3.1 Examples of Imogolite-Like Nanoparticles Obtained Through Coprecipitation

Multiple coprecipitation protocols for obtaining imogolite have been developed (Suzuki et al., 2000; Levard et al., 2009b; Yucelen et al., 2011, 2012a,b). Differences involve (i) the concentration of reagents (from millimolar up to decimolar concentrations); (ii) the type of acids (or Al salts) and (iii) the ageing time and temperature (Wada, 1987; Suzuki et al., 2000; Levard et al., 2009b; Yucelen et al., 2012a,b; Lee et al., 2014). It is interesting to note that these modifications already allow for obtaining very different shapes while keeping the same overall composition. However, the most significant change related to the incorporation of different elements in imogolite structure during synthesis is the wide range of possible imogolite-like structures. In the following sections, the generalised methods of synthesis of three types of imogolite, aluminogermanates (Ge-imogolite), Fe-containing imogolite (Fe-imogolite) and oxy-imogolite are described.

19.3.1.1 General Coprecipitation Synthesis

A generalised synthesis protocol to obtain imogolite-like nanoparticles with formula $(OH)_{3-h}R_{2h}Al_{2-m}X_mO_3Si_{1-n}Y_n(OH)_{1-k}R_{1k}$ consists in the coprecipitation of aluminium/X and silicon/Y monomers in an acidic aqueous solution ($pH < 5$). When the initial pH is very low (ie, when starting from an aluminium salt, $pH < 2.0$), the pH is raised by the slow addition of NaOH. When the precipitation starts from fully organic precursors (ie, aluminium sec butoxide + Si/Y alkoxides), the pH is fixed by an acid (HCl, $HClO_4$, HNO_3, CH_3COOH). This initial coprecipitation leads to dispersed proto-imogolites, which are undergoing a process of ageing at 95°C, over 120 h in normal pressure conditions (1.0 atm) (Farmer et al., 1977). The introduction of OH/R_1 groups is possible by mixing tetraethoxide and a triethoxide initially containing a $Si–R_1$ group. Composition variation is not the only possibility for changing the obtained nanoparticles. Indeed, factors such as temperature, pH and mainly the initial concentration of the reagents used can inhibit the formation of imogolite, generating a wide range of by-products, such as gibbsite, bayerite or amorphous silica (Barrett et al., 1991; Denaix et al., 1999; Arancibia-Miranda et al., 2011). Different conditions can also change the shape of the nanoparticle (SW/DW or SW/spheres), being reported in various studies in imogolite and imogolite-like nanoparticles (Bac et al., 2009; Thill et al., 2012a).

Modifications in the synthesis procedures developed for similar structures of imogolite, such as Ge-imogolite, Fe-imogolite and Imo-As-ite, focus mainly on concentration and percentage of replacement of the starting reagents, keeping factors such as temperature and ageing time similar to the standard synthesis of imogolite method (Wada and Wada, 1982; Mukherjee et al., 2007; Ookawa et al., 2008; Maillet et al., 2010b; Thill et al., 2012a,b). However, despite considering similar conditions to what was reported for imogolite, the synthesis of these similar structures is not absent from the formation of by-products (Ookawa et al., 2008; Levard et al., 2011).

19.3.1.1.1 Organic Grafting Inside Imogolite

A modification of the inner surface of imogolite is possible by the partial replacement of TEOS in the bare imogolite with an organosiloxane reagent (R'–Si–[O–R]$_3$) during their synthesis process (Bottero et al., 2010; Kang et al., 2014). A first example of this new type of imogolite was obtained by replacing TEOS with triethoxymethylsilane (Bottero et al., 2010). An internally methylated, imogolite was obtained successfully. Such hybrid imogolite favours the adsorption of methane compared to bare imogolite due to increased hydrophobicity of the lumen. The degree of substitution in these modifications should be controlled to reduce potential steric hindrance, taking into consideration molecular interactions that could inhibit the formation of the imogolite. Recently, a study conducted by Kang et al. (2014), allowed obtaining an amine-functionalised nanotubes aluminosilicate using aminomethyltriethoxy-silane (AMTES) as a starting reagent. The design of a special organosiloxane reagent (AMTES) was needed because of the limited internal diameter of the imogolite. Indeed, classically used larger precursors, such as aminopropyltriethoxysilane, do not allow successful internal functionalisation (Bottero et al., 2010; Kang et al., 2014). The obtained nanotube was characterised in detail (structure and morphology) using conventional and cryogenic transmission electron microscopy (TEM/cryo-TEM), XRD, nitrogen physisorption, elemental analysis, vibrational spectroscopy and solid-state nuclear magnetic resonance (solid-state NMR). Such functionalisation gives amine-functionalised nanotubes aluminosilicates having a length of 60 nm, almost three times less than the imogolite. The internal analysis of amine-functionalised nanotubes aluminosilicates, using solid-state ^{13}C NMR and FT-Raman spectroscopy, showed that almost 15% of the Si–OH surface groups are replaced by the aminomethyl group ($\equiv$Si—CH_2NH_2); a value close to the 20% theoretically expected. The presence of the aminomethyl groups increases the selectivity in the adsorption of CO_2 compared to bare imogolite when molecular pairs CO_2/N_2 and CO_2/CH_4 were evaluated. Such selectivity is desired for material aiming at the selective removal of greenhouse gases.

19.3.1.2 Tested Ratio of Elemental Substitution

As discussed earlier, the possibility of successful synthesised imogolite isostructures by adding various elements (Ge, Fe, As and P), depends on a series of experimental factors, including the concentration and ratio of the starting reagents, its subsequent hydrolysis, the percentage of replacement and the chemical nature of the substituent element (Farmer, 1981; Wada and Wada, 1982; Maillet et al., 2010a; Levard et al., 2011).

Taking a generalised notation for the imogolite-like composition

$$(OH)_{3-h}R_{2h}Al_{2-m}X_mO_3Si_{1-n}Y_n(OH)_{1-k}R_{1k},$$

up to now, the combinations have been tested as listed in Table 19.1.

19.3.2 Examples of Imogolite-Like Nanoparticles Obtained Through Postmodifications

In this type of modification, the process starts from an already obtained imogolite nanotube using classical synthesis routes. The imogolite nanotubes are then modified by isomorphic substitution, site-specific adsorption or grafting of organic molecules.

19.3.2.1 Isomorphic Substitution vs Fe_2O_3 Clusters (Shafia et al., 2015)

Recent research in imogolite nanotubes doped with Fe, obtained by two different synthesis pathways: (i) a direct method with incorporation during synthesis of Fe^{3+} (Fe-0.025-IMO and Fe-0.050-IMO) and (ii) another postsynthesis loading was evaluated to see how Fe^{3+} (Fe-L-IMO) can be incorporated into the structure of the imogolite (Shafia et al., 2015). A detailed spectroscopic characterisation of the obtained samples suggest that the Fe is incorporated into octahedral sites in the outer surface of the imogolite through isomorphic substitutions, which preferably occur in the samples synthesised by direct methods, with low Fe concentrations ($x=0$, 0.025). This is a common process in most natural aluminosilicates. However, increases in the concentration of Fe in this kind of synthesis results in combined processes of isomorphic substitution and Fe_2O_3 cluster formation, indicating that the imogolite structure can support a limited amount of Fe.

The isomorphic substitution processes and cluster formation were observed only in samples doped with Fe postsynthesis and those obtained by direct methods, where the substitution of Fe was higher, as the case of Fe-0.050-IMO (Shafia et al., 2015).

The incorporation of Fe in low concentrations (Fe-0.025-IMO), through a direct synthesis, increased the distance between two aligned nanotubes (cell parameter *a*), evidence of an increase in the diameter of the obtained imogolite nanotubes. The magnetic parameters are also altered by the effect of the incorporation of Fe in the imogolite structure. The imogolite (bare) has a diamagnetic behaviour, which changes to paramagnetic when iron is incorporated into the structure. A detailed study of the magnetic properties of the samples (Fe-0050-IMO and Fe-L-IMO) showed that these materials possess a ferro and ferrimagnetic behaviour, which is associated with the nanoparticles of hematite and magnetite, formed on the surface, respectively. Another effect of the presence of Fe in the imogolite was the variation in the values of the band gap. For example, with a decrease in this parameter

TABLE 19.1 Composition of Different Isostructures of Proposed Imogolite

X (*m*)	Y (*n*)	R_1 (*k*)	R_2 (*h*)	Products obtained	References
0	Ge (0–1)	0	0	Nanotubes successfully obtained for all ratios tested	Wada and Wada (1982)
Ga (ns)	Ge (0–1)	0	0	Substitution of Al for Ga was tested without success	Wada and Wada (1982)
0	0	CH_3 (1)	0	Successful synthesis of AlSi nanotubes with internal CH_3 groups	Bottero et al. (2010)
Fe (0.05)	0	0	0	Successful synthesis of Fe-containing imogolite. First trimetallic nanotube	Ookawa et al. (2006)
0	0	0	P-HEMA (0.05–0.10–1)	Successful synthesis of AlSi nanotubes with external 2-acidphosphoxyethyl methacrylate groups	Yamamoto et al. (2007)
0	Ge (1)	0	OPA (0.20–1)	Successful synthesis of AlGe nanotubes with external $CH_3(CH_2)_{17}PO(OH)_2$ groups	Bac et al. (2009)
0	Ge (0–1)	0	0	From $n=0$–0.5, SW nanotubes with increasing diameter	Thill et al. (2012b)
				For $n>0.5$, SW/DW mixture at constant average diameter	
0	As(III) (0–1)	0	0	Theoretical proposal of oxy-imogolite	Duarte et al. (2012)
	As(V) (0–1)				
	P(III) (0–1)				
	P(V) (0–1)				

0	0	CH_2NH_2	0	Successful synthesis of AlSi nanotubes with internal $\equiv Si—CH_2NH_2$ groups	Kang et al. (2014)
Fe (0.01)	Ge (0–1)	0	0	Successful synthesis of Al–Fe/Ge nanotubes. Second trimetallic nanotube	Avellan et al. (2014)
0	Ge (0–1)	CH_3 (0 or 1)	0	Only SW nanotubes are successfully obtained	Amara et al. (2015)
Fe (0.025–0.050)	0	0	0	Successful synthesis of Fe-containing imogolite	Shafia et al. (2015)

(semiconductor behaviour) in samples of 'Fe-0025-IMO' (band gap, =2.8 eV) and 'Fe-L-IMO' (band gap, =2.4 eV), nearly two units less than the imogolite (band gap =4.6 eV), which is considered an insulating material (Alvarez-Ramírez, 2009).

19.3.2.2 Ni Adsorption at a Specific Surface Sites (Levard et al., 2009a)

Imogolite is naturally found mainly in soils of volcanic origin, such as Andisols (Yoshinaga and Aomine, 1962; Farmer et al., 1977; Gustafsson, 2001), and a few research studies were focused on its role in adsorption of trace elements (contaminants and nutrients) (Clark and McBride, 1984; Arai et al., 2006; Guerra et al., 2011; Levard et al., 2012). In this context, Levard et al. (2009a) used Ge-imogolite as an imogolite model to study the removal capacity and the type of interaction with Ni. It was found that adsorption of Ni was significantly lower (twice) compared to that reported for other aluminium oxides (γ-alumina) at similar conditions of pH. Indeed, despite its large SSA (228 $m^2\ g^{-1}$), it was determined that only 0.11 Ni atoms are adsorbed per square nanometre.

This can be explained by two facts: (i) Ge-imogolite had a positive external surface charge in the conditions of the Ni adsorption and (ii) the use of X-ray scattering (XAS) determined that Ni is coordinated with six oxygen atoms, probably occupying a few vacancies or structural defects of Ge-imogolite (Levard et al., 2009a). A similar phenomenon could occur in the adsorption of Cd and Cu in imogolite, recently studied by Arancibia-Miranda et al. (2015).

19.3.2.3 Grafting of Organic Molecules

One of the main advantages that metal oxide nanotubes, compared to carbon nanotubes, is their easy functionalisation due to the hydroxyl groups present on the surface (Kang et al., 2014). Kang et al. (2014) presented a protocol for the postfunctionalisation of the internal surface of SW imogolite using vacuum impregnation of dried powder. On the outer surface, the predominant aluminol groups of the imogolite have a high affinity towards the phosphate group. Post-synthesis functionalisation is facilitated by using alkyl phosphoric acid derivatives (Yamamoto et al., 2007; Bac et al., 2009) allowing for strengthening the mechanical and optical properties of polymers (Yamamoto et al., 2007). Bac et al. (2009) successfully applied the same chemical method to modify the surface of Ge-imogolite with octadecylphosphonic acid (OPA), obtaining a nanohybrid dispersed in hydrophobic solvents, such as chloroform.

These investigations show that both imogolite and Ge-imogolite can be relatively easily functionalised using organic derivatives of Si or Ge or alkyl phosphoric acids (Bac et al., 2009; Amara et al., 2015). Internal changes are generally successful when the reagent is incorporated at the beginning of the synthesis, while the changes to the external surface are performed on the nanotube already formed to avoid possible complication with Al^{3+}, since it shows a high affinity for organic compounds (Inoue and Huang, 1984).

19.4 EXAMPLES OF IMOGOLITE-LIKE FAMILY NANOPARTICLES

Aluminogermanate nanoparticles and Fe-containing imogolite will be reviewed in detail next, followed by unpublished results on oxy-imogolite. (The case of hybrid imogolite is the subject of detailed sections in Chapters 12 and 23.)

19.4.1 Aluminogermanate Nanotubes/Nanospheres

The replacement of silicate (O_3SiOH) by germanate (O_3GeOH) conducted by Wada and Wada (1982) was the first important modification in the synthesis of imogolite, which successfully gave rise to similar but new structures. Solutions containing Ge and Si with Ge/(Ge + Si) atomic ratios of 0.2, 0.5 and 1.0 were added dropwise under constant stirring to a solution of $AlCl_3$. Sodium hydroxide (0.1 mol L^{-1}) was then added to the solutions, and they were stirred vigorously to adjust the pH to a value of about 5.0. The solutions were then immediately reacidified by the addition of 1 mmol HCl and 2 mmol CH_3COOH per mL. The final pH values were about 4.5, and the solutions were heated at 95°C–100°C. The aged solutions were dialysed or centrifuged. The obtained products were structurally characterised using XRD, electron diffraction patterns, differential thermal analysis curves, infrared (IR) spectra and TEM (Wada and Wada, 1982). The resulting products appeared to be nanotubes of shorter length than imogolite, but with an external diameter of 3.3 nm. The increase in diameter was attributed to the differences in the bond distances between the Ge–O in comparison to the Si–O (Wada and Wada, 1982; Pohl et al., 1996; Mukherjee et al., 2005; Konduri et al., 2007; Levard et al., 2008). This replacement also resulted in a significant decrease in the length of the nanotubes, reaching values of 20–50 nm (Wada and Wada, 1982; Mukherjee et al., 2007) instead of several hundred nanometres for imogolite.

Since then, aluminogermanates nanotubes have been the most developed and studied among all analogous imogolite-like structures. Their synthesis is highly reproducible and flexible in comparison to standard synthesis (Wada and Wada, 1982; Mukherjee et al., 2005; Konduri et al., 2007; Thill et al., 2012b).

Thanks to the simplicity of the synthesis, various techniques of local and semilocal characterisation have been used to better understand the formation mechanism of aluminogermanates. It also served as a model that can be extrapolated to better understand what also happens with imogolite

(Mukherjee et al., 2005; Konduri et al., 2007; Levard et al., 2010; Maillet, 2010; Thill et al., 2012b). This was the original motivation of Wada and Wada (1982) who wanted to better understand the IR signals of imogolite. The replacement of Si by Ge goes with a displacement of the IR bands coherent with the mass difference between Si and Ge. Due to these experiments, the bands appearing at 990 and 939 cm^{-1} were associated to the stretching Si–O bounds. In the case of aluminogermanates, these are shifted at 910 and 810 cm^{-1} (Wada and Wada, 1982) as expected.

The first approaches in the search for the critical steps that control the mechanism of formation and growth of Ge-imogolite were investigated at the Georgia Institute of Technology by the group of Sankar Nair (Mukherjee et al., 2005, 2007; Konduri et al., 2007; Yucelen et al., 2011) under similar conditions of imogolite synthesis. Using multiple techniques of structural analysis, aluminogermanates nanotubes begin to form within the first 24 h of synthesis, maintaining their structure almost unchanged throughout the process, which is consistent with the results obtained by Wilson et al. (2001) and Hu et al. (2004). It was also established that the average length of Ge-imogolite is approximately 20 nm instead of several hundred nanometres, and that the external diameter is 3.3 nm (Wilson et al., 2001; Hu et al., 2004; Mukherjee et al., 2005).

In this same line of investigation, Mukherjee et al. (2007) determined that the key structure that allows the development of Ge-imogolite is comprised of a set or aggregate of amorphous nanoparticles called 'proto-imogolite', whose dimensions are close to 6 nm. They come from monomers or chemically stable oligomers controlled by the pH of the medium formed in the early stages of synthesis. Then, proto-imogolites self-assemble during the ageing period to form short nanotubes (Mukherjee et al., 2007; Yang et al., 2008; Yucelen et al., 2011; Arancibia-Miranda et al., 2013). Mukherjee et al. (2007) claim that this process is mainly under thermodynamic control.

Significant progress has been made by the Nair group to understand the formation of imogolite using Ge-imogolite as a model. Further progress was difficult due to the limitations of in situ analytical techniques, which require high concentrations of reagent. An important modification to the aluminogermanates synthesis was introduced by Levard et al. (2008), who discovered that the concentration of the starting reagents can be increased 100 times without preventing nanotube formation. Due to this discovery, the study of Ge-imogolite synthesis in situ was possible through local-scale analyses (XAS at the Ge–K edge and ^{27}Al NMR) and a semi-local-scale technique such as small-angle X-ray scattering (SAXS) (Levard et al., 2008; Maillet, 2010) (Fig. 19.3).

The most important results of these studies suggest that in the first 24–48 h of ageing, the Al atoms are hydrolysed and polymerised, forming a structure similar to gibbsite. On this gibbsite-like structure, three –OH are replaced with the covalent linking with each germanate groups. This process was

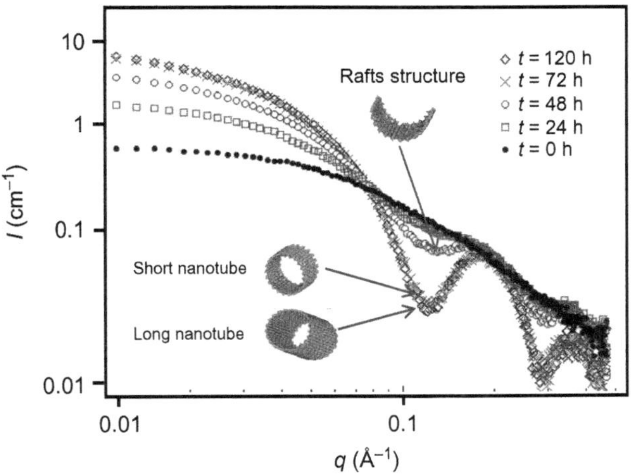

FIG. 19.3 Small-angle X-ray scattering (SAXS) intensities measured at $t=0$, 24, 48, 72 and 120 h of ageing. *Adapted from Levard et al. (2010).*

determined and confirmed by extended X-ray absorption fine structure spectroscopy (EXAFS) and NMR analyses (Levard et al., 2008, 2010).

SAXS analysis indicated that the formation of nanotubes was identifiable from 24 h ageing. Nanotubular structure development was clearly observed through small oscillations in the scattering curves from the second day of synthesis (Fig. 19.3).

A structural precursor model consisting of a roof-tile-shaped particle with an imogolite local structure was proposed. It contains an average of 200 Al atoms that gives an average size of 5 nm, similar to the observations performed by Mukherjee et al. (2005). Close to 26% of Ge vacancies was estimated to be present at this stage, which may induce a different curvature compared to the final nanotubes (Levard et al., 2010) (Fig. 19.4).

Investigations carried out by Maillet (2010) showed that modifying the Al concentration induces a change in the nanotube structure. It was observed by SAXS and cryo-TEM that for Al concentrations of 0.25 M, the obtained material had unexpected oscillations at a large scattering vector, revealing an additional structuration at shorter distances. These oscillations are explained by a DW structure confirmed with cryo-TEM images (Maillet et al., 2010a,b). When the concentration is increased above 0.5 M in Al, the DW structure is progressively changed into a SW structure (Fig. 19.5).

The DW aluminogermanate diameter is greater than the SW, reaching values of 4.0 ± 0.1 nm, while the inner nanotube diameter does not exceed 2.4 ± 0.1 nm. The separation distance between both nanotubes was 0.27 nm, which is usually associated with the typical value of a layer of water, which in turn prevents the formation of covalent bonds (Maillet et al., 2010a,b).

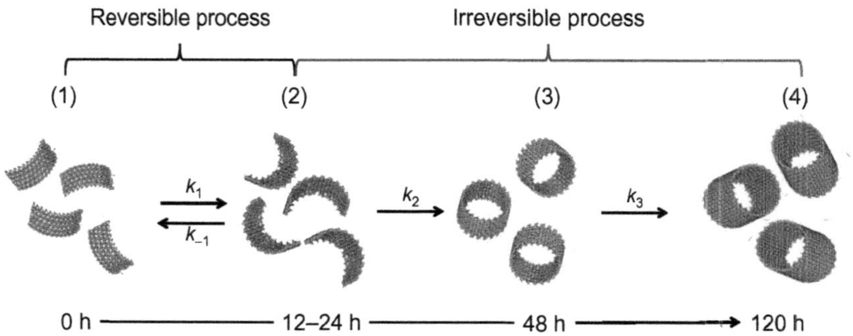

FIG. 19.4 Proposed growth mechanisms of Ge-imogolite nanotubes. *Light grey symbols* represent Al octahedra and *dark grey symbols* represent Ge tetrahedra.

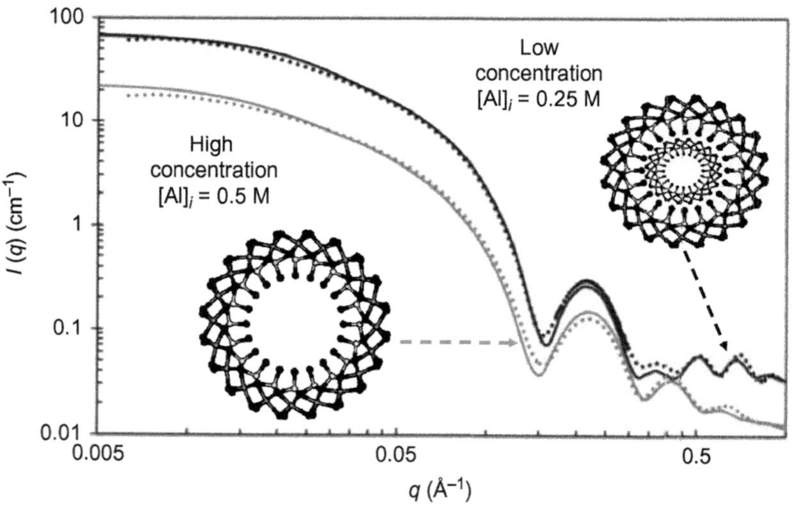

FIG. 19.5 SAXS curves of 0.5 and 0.25 M Al–Ge imogolite-like NTs. Experimental data are plotted using *dots*, and the theoretical scattering curve of the represented structure is plotted using an *unbroken line. Adapted from Maillet (2010).*

Research conducted on SW and DW aluminogermanate imogolite-like nanotubes, using analytical techniques and semilocal scale techniques such as SAXS, have allowed to develop a new hypothesis about kinetic mechanisms and the formation of imogolites, in which this process occurs in two general steps: (i) the topological transformation of roof-tile-shaped proto-imogolite into short nanotube sections and (ii) nanotube growth through successful collisions (tip–tip collisions). In the case of the DW imogolite-like aluminogermanate, the same amount of proto-imogolite, generates a smaller number of nanotubes. The efficiency of the tip–tip collisions is also reduced by the DW structure (Levard et al., 2010; Maillet, 2010) (Figs 19.3 and 19.4).

The conditions that favour a type of nanotube synthesis, either SW or DW, was studied in depth by Thill et al. (2012b) where the transition from SW to DW upon variation of the Al concentration was analysed from a physico-chemical point of view. The results indicated that the formation of one nanotube type or another is strongly related to variations of elastic (ΔE_c) and electrostatic energies (ΔE_e) (Thill et al., 2012b). Once the formation and rearrangements of a given number of precursors is finished (roof-tile-shaped), DW formation occurs when the attractive energy ΔE_e loss is greater than the gain of elastic energy ($\Delta E_e + \Delta E_c < 0$). This condition is fulfilled at low ionic strength (ie, a low concentration of reactants). On the contrary, SW nanotubes formation is favoured at high concentrations of reactant (0.5–0.75 M). The increase in ionic strength reduces the electrostatic attractive energy between proto-imogolites. The electrostatic energy, then, is less than the gain in curvature energy ($\Delta E_e + \Delta E_c > 0$), and the DW structure is no longer stable (Thill et al., 2012b) (Fig. 19.6).

Levard et al. (2011) and Thill et al. (2012b) showed that the hydrolysis of the initial reagents is a critical factor influencing the type of nanotube (SW or DW). When the hydrolysis ratio (OH)/(Al) is low (ratio = 1.75) mainly SW nanotubes were obtained. In these conditions, defects and lacunas appear in the imogolite wall. The consequence of this phenomenon is a modification of the precursor surface charge and mechanical properties (Fig. 19.6).

The transition between SW and DW aluminogermanates was also assessed by varying the ratio of Si/Ge, to obtain nanotubes of the general formula $(OH)_3Al_2O_3Si_xGe_{1-x}OH$ ($0 \leq x \leq 1$). Thill et al. (2012a) observed that the

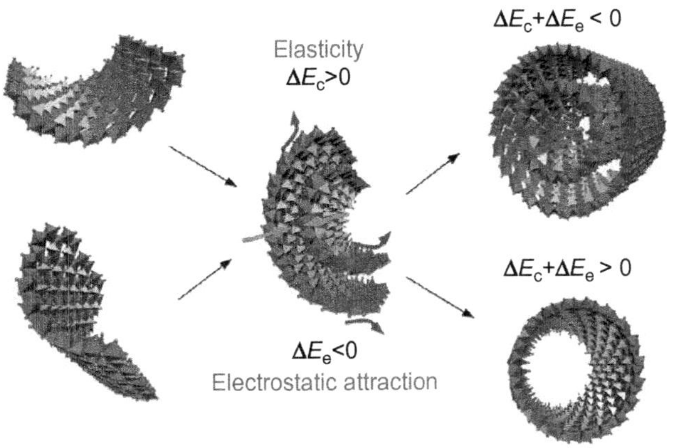

FIG. 19.6 Qualitative schematic scenario for the proto-imogolite transformation into SW or DW imogolite nanotubes. *Reprinted (adapted) with permission from Thill et al. (2012b). Copyright (2012) American Chemical Society.*

addition of Si in small amounts during the synthesis induces a transition between the DW and SW structures. In those solutions that contain a mixture of precursor structures, the average diameter reported for the final nanotube remained almost constant. The experimental intensities scattered and the scattering models showed a proper fit when x is between 0.25 and 1, wherein the SW nanotubes predominate. For $x<0.25$, values that correspond to the SW/DW transition region, the adjustment quality decreases. In the case of DW nanotubes, no significant changes were observed in the outer diameter, reporting a value of about 4.2 nm. However, SW nanotubes showed a variation in the outer diameter of between 2 nm at $x=1$ and 3.5 nm at $x=0.5$. For $x<0.5$, the outer diameter evolution was barely visible. The cryo-TEM analysis showed the presence of a few nanoscrolls. The incorporation of Si atoms in the structure of DW Ge-imogolite generates a decrease of the imogolite layer curvature, which could initially increase the rolling of nanoscrolls.

19.4.2 Iron-Containing Imogolite

The incorporation of Fe in the synthesis of imogolite gave rise to a new analogous structure, Fe-imogolite, as explained earlier. Its synthesis was developed by Suzuki et al. (2000) although the first approximation of this structural change was carried out by McBride et al. (1984). The method considers, as in the preparation of aluminogermanates, the use of solutions in decimolar concentrations (Ookawa et al., 2006).

- Briefly, 100 mL of a solution 0.15 mol L^{-1} containing $FeCl_3$ and $AlCl_3$, in an atomic ratio $Fe/(Fe+Al)=0.05$, is added dropwise by stirring in 100 mL of aqueous solution of Na_4SiO_4 (0.1 M), maintaining the stirring for 1.5 h. The precursors are obtained by the slow addition of 0.1 M NaOH solution to reach a pH value close to 5.5. To remove excess salt (NaCl), the precursors are washed with Milli-Q water and separated by centrifugation (three times). Finally, they are dispersed in 2.0 L of water. Subsequently, 40 mL of 0.1 M HCl are added, and the dispersion is stirred for 18 h to ensure homogeneity and complete dispersion of the precursors. The solution is then subjected to an ageing process at 95°C for 7 days. This aged solution is then either dialysed or centrifuged.

This new structure has similar dimensions to imogolite, but with more catalytic activity (Ookawa et al., 2008). Recently, Avellan et al. (2014) reported the successful use of Fe^{2+} as a partial replacement of Al^{3+}, being incorporated as a starting reactant during the synthesis of aluminogermanates and obtaining an isostructural of imogolite and the first trimetallic nanotube belonging to the Ge-imogolite family.

Unpublished results performed by the authors of this chapter, related to this imogolite analogous structure, indicate that in order to obtain

Fe-imogolite as the major product, the replacement percentage of Al^{3+} by Fe^{3+} may not exceed 10%. Above this limit, iron tends to form by-products as described in previous sections. These results are consistent with those reported by McBride et al. (1984) and Shafia et al. (2015).

Attempts to modify imogolite with Fe atoms have focused mainly on structural substitution of Al^{3+}, which has been successfully performed in aluminosilicates and aluminogermanates using salts of Fe^{3+} and Fe^{2+}, respectively, as starting reagents, partially replacing Al^{3+} (Ookawa et al., 2006, 2008; Avellan et al., 2014). Magnetite-imogolite binary systems (Ferro-imogolite) have also been prepared with Fe oxide as a coating, which allows one to substantially enhance the removal of the As from aqueous matrices (Arancibia-Miranda et al., 2014).

This imogolite isostructure has not been studied as much as the aluminogermanate nanotubes. However, the most significant work in Fe-imogolite has been carried out by Suzuki et al. (2000) who successfully developed a new synthetic procedure to get a Fe-doped imogolite. The replacement of Al by structural Fe does not significantly modify the dimensions of Fe-imogolite in comparison to imogolite, which has also been reported by the addition of Fe^{2+} in Ge-imogolite (Avellan et al., 2014). A comparative study between imogolite and Fe-imogolite, prepared with different percentages of Fe^{3+} [molar ratio of $x = Fe/(Al + Fe) = 0.1$ and 0.05] was performed by Ookawa et al. (2006, 2008). For $x = 0.1$, it was not possible to find imogolite in the XRD and FTIR spectra. Very similar results have also been reported by McBride et al. (1984). However, with $x < 0.05$, the XRD and FTIR spectra were similar to imogolite. The three reflections in the low-angle-region characteristics of nanotubes are observed (Ookawa, 2012). IR bands assigned to Fe-imogolite appear in similar regions as the ones reported for imogolite, with a doublet around 1000 cm^{-1} (Ookawa et al., 2006, 2008; Ookawa, 2012).

The dimensions of Fe-imogolite were estimated using atomic force microscopy (AFM), reaching 100–400 nm in length and an outer diameter between 2.2 and 2.4 nm (Ookawa, 2012). In addition, the TEM results indicated that Fe-imogolite have a tubular structure with a length that varies between 100 and 1000 nm, with an abundance of associated globular structures (allophane and Fe oxides). Mössbauer spectroscopy analysis shows that Fe oxides are mainly goethite and ferrihydrite (Shafia et al., 2015).

UV–vis, X-ray absorption near edge structure spectroscopy and EXAFS were carried out to confirm the location of Fe by comparing the results with those obtained with an imogolite where a Fe^{3+} solution (named 'Fe/imogolite') and Fe_2O_3 were previously adsorbed. The comparison of results revealed that Fe in Fe-imogolite was not tetrahedrally coordinated. Moreover, the fact that the EXAFS data reveals a different state compared to the Fe adsorbed on Fe/imogolite tends to demonstrate that Fe was coordinated in an octahedron site of the crystalline structure of imogolite. The results are consistent with those of Ookawa et al. (2006, 2008) and Ookawa (2012).

The mechanism explaining why the formation of this type of imogolite is not possible with a replacement of more than 10% of Al^{3+} has not been proposed yet. Avellan et al. (2014) synthesised, in strict anoxic conditions, a series of aluminogermanates with Fe^{2+} at molar ratios $n_{Fe}/n_{(Al+Fe)}=0$, 0.02, 0.05 and 0.1. When the concentration of Fe^{2+} was greater (0.05 and 0.1), solid globular structures of a large size (100–500 nm) occurred, possibly attributed to coprecipitated Fe (oxy)hydroxides. The presence of these by-products was not determined in samples prepared at lower concentrations of Fe (Fig. 19.7).

The material obtained was analysed using AFM. The images confirmed the presence of nanotubes in all $n_{Fe}/n_{(Al+Fe)}$ ratios. The diameters were similar in samples 0, 0.02, but they increased with the increasing concentration of Fe. Nanotube lengths vary between 30 and 150 nm for each sample. The XRD patterns were similar to those reported for Ge-imogolite. Using EXAFS, Fe was determined to be Fe(III) in an octahedral coordination. The change of oxidation state $Fe^{2+} \rightarrow Fe^{3+}$ is attributed to the loss of the anoxic condition during the process of ageing (Avellan et al., 2014).

The difference in size between FeO_6 and AlO_6 octahedral, potentially causes lattice strain, deficit of charge in the initial nucleation stage (due to the substitution of an Al^{3+} by Fe^{2+}), or both. This could be a clue as to why it is not possible to obtain a complete replacement of Al by Fe. Avellan et al. (2014) have highlighted the importance of eliminating by-products in order to boost this material, from an engineering point of view. A simple and effective alternative to eliminate the unwanted material produced during the synthesis of Fe-imogolite is to perform a selective dissolutive treatment for Fe oxides known as 'Na-citrate/dithionite/bicarbonate treatment', which is widely used to remove free iron oxides from the inorganic fraction of soil (Escudey et al., 1986; Pizarro et al., 2003).

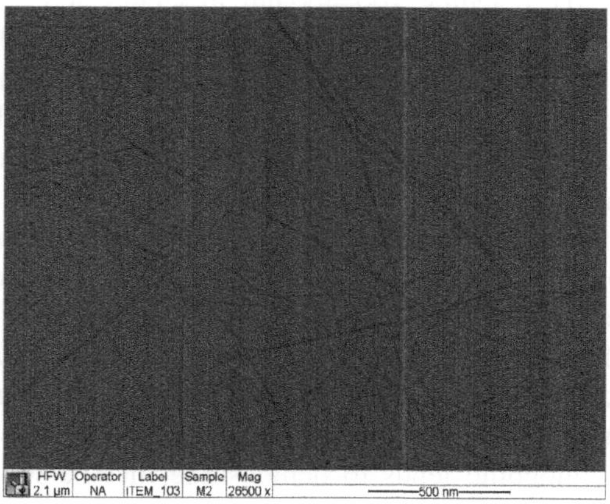

FIG. 19.7 TEM image of Fe-imogolite and cluster iron oxide (substitution 10% iron). *Arancibia-Miranda (unpublished work).*

This variant of imogolite has catalytic properties superior to those observed for the aluminosilicate without functionalising (Ookawa et al., 2008). The Fe-imogolite may act as a catalyst for different reactions, mainly the oxidation of organic compounds such as benzaldehyde, cyclohexane, toluene and chlorobenzenes, generating epoxies, ketonic and alcoholic derivatives and highlighting the conversion of benzene to phenol (Ookawa et al., 2006, 2008; Ookawa, 2012).

Ookawa et al. (2008) examined the effect of solvent on the catalytic properties of Fe-imogolite in the oxidation of benzene. These authors find that it is possible to detect phenol by analysis of GC-mass only when the solvent is acetonitrile, while in solvents such as acetic acid, 2-propanol and propionic acid, this reaction is inhibited. However, this behaviour remains unclear.

Other aromatic hydrocarbons (eg, chlorobenzene, toluene, benzaldehyde and phenol) have been oxidised with H_2O_2, using Fe-imogolite as a catalyst (Ookawa et al., 2008). In general, with the products formed in those aromatic hydrocarbons that have a side chain of the type –OH or a group, –Cl, the oxidation occurs in the ring at *ortho* and *para* positions, while in toluene and benzaldehyde, it was observed that the side-chain and ring can be oxidised (Fig. 19.8).

19.4.3 Oxy-imogolite

In the case of oxy-imogolite synthesis, it is important to consider pK_a and solubility as critical factors that can influence the rate of formation and amount

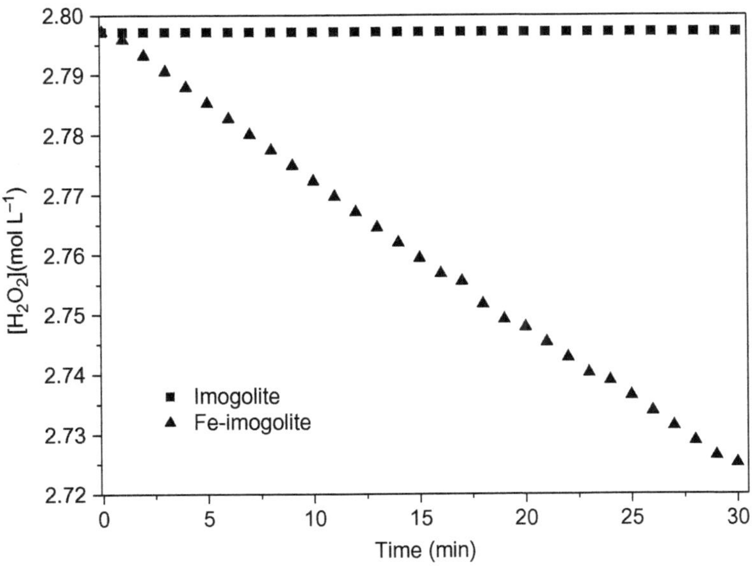

FIG. 19.8 H_2O_2 concentration as a function of the reaction time at 25°C and pH 5.8. *Arancibia-Miranda (unpublished work).*

of obtained products. The pK_a value will determine the conditions for the synthesis of nanotubes (Duarte et al., 2012; Guimarães et al., 2013). Unpublished research performed by the authors of this chapter has focused on developing isomorphic structures similar to imogolite, based on the total or partial replacement of silicate, by oxyanions such as H_3PO_4, H_3PO_3, H_3AsO_3, H_3AsO_4 and H_3SeO_4. An overview of the synthesis of oxy-imogolite, is shown here:

- TEOS and sodium arsenite were added dropwise to a stirred solution of 5 mmol L^{-1} $AlCl_3$ solution until the Al/(Si+$As^{(III)}$) ratio was 2, using a total percentage of 10–25% arsenite. It was allowed to stand for 45 min with vigorous stirring. Then a 1×10^{-2} M NaOH solution was added at a rate of 0.5 mL min^{-1} until an Al:(Si+$As^{(III)}$):OH ratio of 2:1:4 was obtained, reaching a pH of 5.0, and the mixture was stirred vigorously for 4 h and then heated at 95°C for 5 days. To achieve the synthesis, 0.1 M solution of NH_4OH was added until the pH reached 8.5, yielding a transparent gel, which was then separated and washed with deionised water by centrifuging at 9000 rpm for 30 min until the specific conductivity was reduced to less than 0.78 dS m^{-1}. Finally, this aged solution was either dialysed or centrifuged.

The microscopic analysis of nanotubes prepared with a substitution of 10% in weight of Si by As, called 'Imo-As_{10}-ite', indicated that there are differences in dimensions in comparison to imogolite. The external average diameter determined for Imo-As_{10}-ite was 3.2 ± 0.4 nm, whereas imogolite has a diameter of 2.1 ± 0.2 nm. These values were determined based on the measurement of various samples at different scales ($n = 25$). A decrease in length was observed for Imo-As_{10}-ite in comparison to imogolite, suggesting that substitutions involving an increase in diameter affect the one-dimensional growth of the nanotube. Similar results have been reported for aluminogermanates (Wada and Wada, 1982; Mukherjee et al., 2005; Levard et al., 2008). The pH reaction in Imo-As_{10}-ite synthesis was measured every 12 h, showing that the process of condensation of the precursors (oxolation mechanism), caused a significant decrease in pH values compared to imogolite, where a greater variation of this parameter was observed between 24 h of ageing, with a difference of two pH units (Yang et al., 2008; Arancibia-Miranda et al., 2011).

This experiment is in progress in order to confirm and quantify the possible substitution with As and other oxyanions as phosphate (PO_4^{3-}), phosphite (PO_3^{3-}), arsenate (AsO_4^{3-}), arsenite (AsO_3^{3-}), selenate (SeO_4^{3-}) and selenite (SeO_3^{3-}).

19.5 POSSIBLE PROJECTION

Like zeolites or metal-organic frameworks, imogolite-like structures may become a very rich family of new microporous nanodispersed materials in the near future. Their very high SSAs and the existence of two surfaces with clearly differentiated physicochemical properties make them potentially useful in many applications.

This family is far from being explored in full, and many possibilities are foreseen both in terms of structure and compositions. The possibility of developing systems for controlled drug delivery and nanobiosensors could promote this imogolite-like nanoparticles from a medical point of view. However, many open questions remain about the behaviour of this type of material in a biological environment, and a long research path is ahead.

Considering that the imogolite-like structure has two types of surface, high SSA, a pH-dependent surface charge and a synthesis process that allows one to obtain large quantities of nanoparticles belonging to the imogolite-like family, there may be a number of useful applications in analytical chemistry, agriculture and environmental science. Their ease of structural modification with other elements such as Fe and surface functionalisation with organic compounds makes them promising candidates as catalysts in multiple industrial reactions.

The lack of existing information about Fe-imogolite could be an advantage for the development of fundamental science studies that seek to explain this nanotube formation phenomena and the impossibility of obtaining a nanotube via a 100% replacement of Al by Fe. However, the most obvious projections point to their implementation and evaluation as a catalyst in oxidation reactions or degradation of polycyclic aromatic, dyes, pesticides and volatile aldehyde compounds. In regard to this, significant progress concerning their production cost will be needed. The preparation of sensors based on Fe-imogolite or hybrid imogolite for the determination of As or organic pollutants could be evaluated in the near future. The synthesis and characterisation of the different types of oxy-imogolite, proposed by Guimarães et al. (2013), are closer to fundamental research and will enhance the general knowledge at the frontier of the imogolite-like family.

19.6 CONCLUDING REMARKS

In this chapter, the existing members of the imogolite-like family were reviewed in detail. The extent of the structure and composition that the imogolite-like family allows to be achieved with the incorporation of elements like Ge, Fe and As (III) confers them new physicochemical properties, expanding the range of areas where they can find useful applications, such as in physics, biology and environmental science. However, there are still some grey areas that need to be studied in depth, which have sometimes been theoretically predicted but not yet experimentally achieved, such as the incorporation of other elements such as Cr, P, Se and Ti. There is a clear need to improve the synthesis protocols, decreasing the generation of by-products and costs, in addition to determining whether some of these isostructures are present in nature.

The exploration of the imogolite-like family and their newly obtained properties is still in its infancy. For example, the photocatalytic properties of Fe-imogolite and the physicochemical properties of hybrid imogolite have barely been studied. These areas are worthy of examination and research in the near future.

REFERENCES

Alvarez-Ramirez, F., 2007. *Ab initio* simulation of the structural and electronic properties of aluminosilicate and aluminogermanate nanotubes with imogolite-like structure. Phys. Rev. B. 76, 125421–125436.

Alvarez-Ramírez, F., 2009. First principles studies of Fe-containing aluminosilicate and aluminogermanate nanotubes. J. Chem. Theory Comput. 5, 3224–3231.

Amara, M.S., Rouzière, R., Paineau, E., Bacia-Verloop, M., Thill, A., Launois, P., 2014. Hexagonalization of aluminogermanate imogolite nanotubes organized into closed-packed bundles. J. Phys. Chem. C 118, 9299–9306.

Amara, M.S., Paineau, E., Rouzière, R., Guiose, B., Krapf, M.-E.M., Taché, O., Launois, P., Thill, A., 2015. Hybrid, tunable-diameter, metal oxide nanotubes for trapping of organic molecules. Chem. Mater. 27, 1488–1494.

Arai, Y., McBeath, M., Bargar, J.R., Joye, J., Davis, J.A., 2006. Uranyl adsorption and surface speciation at the imogolite–water interface: self-consistent spectroscopic and surface complexation models. Geochim. Cosmochim. Acta 70, 2492–2509.

Arancibia-Miranda, N., Escudey, M., Molina, M., García-González, M.T., 2011. Use of isoelectric point and pH to evaluate the synthesis of a nanotubular aluminosilicate. J. Non-Cryst. Solids 357, 1750–1756.

Arancibia-Miranda, N., Escudey, M., Molina, M., García-González, M.T., 2013. Kinetic and surface study of single-walled aluminosilicate nanotubes and their precursors. Nanomaterials 3, 126–140.

Arancibia-Miranda, N., Escudey, M., Pizarro, C., Denardin, J.C., García-González, M.T., Fabris, J.D., Charlet, L., 2014. Preparation of magnetic single-walled aluminosilicate nanotube: a structural and magnetic study. Mater. Res. Bull. 51, 145–152.

Arancibia-Miranda, N., Silva-Yumi, J., Escudey, M., 2015. Effect of cations in the background electrolyte on the adsorption kinetics of copper and cadmium and the isoelectric point of imogolite. J Hazard. Mater. 299, 675–684.

Avellan, A., Levard, C., Kumar, N., Rose, J., Olivi, L., Thill, A., Chaurand, P., Borschneck, D., Masion, A., 2014. Structural incorporation of iron into Ge–imogolite nanotubes: a promising step for innovative nanomaterials. RSC Adv. 4, 49827.

Bac, B.H., Song, Y., Kim, M.H., Lee, Y.-B., Kang, I.M., 2009. Surface-modified aluminogermanate nanotube by OPA: synthesis and characterization. Inorg. Chem. Commun. 12, 1045–1048.

Baroña, G.N.B., Lim, J., Choi, M., Jung, B., 2013. Interfacial polymerization of polyamide-aluminosilicate SWNT nanocomposite membranes for reverse osmosis. Desalination 325, 138–147.

Barrett, S., Budd, P.M., Price, C., 1991. The synthesis and characterization of imogolite. Eur. Polym. J. 7, 609–612.

Bottero, I., Bonelli, B., Ashbrook, S., Wright, P., Zhou, W., Tagliabue, M., Armandia, M., Garrone, E., 2010. Synthesis and characterization of hybrid organic/inorganic nanotubes of the imogolite type and their behaviour towards methane adsorption. Phys. Chem. Chem. Phys. 13, 744–750.

Boyer, M., Paineau, E., Bacia-Verloop, M., Thill, A., 2014. Aqueous dispersion state of amphiphilic hybrid aluminosilicate nanotubes. Appl. Clay Sci. 96, 45–49.
Bursill, L.A., Peng, J.L., Bourgeois, L.N., 2000. Imogolite: an aluminosilicate nanotube material. Philos. Mag. A 80, 105–117.
Charlier, J.C., 2002. Defects in carbon nanotubes. Acc. Chem. Res. 35, 1063–1069.
Clark, C.J., McBride, M.B., 1984. Chemisorption of Cu(II) and Co(II) on allophane and imogolite. Clay Clay Miner. 32, 300–311.
Cradwick, P.D.G., Farmer, V.C., Russell, J.D., Masson, C.R., Wada, K., Yoshinaga, N., 1972. Imogolite—a hydrated aluminium silicate of tubular structure. Nat. Phys. Sci. 240, 187–189.
Denaix, L., Lamy, I., Bottero, J.Y., 1999. Structure and affinity towards Cd^{2+}, Cu^{2+} Pb^{2+} of synthetic colloidal amorphous aluminosilicates and their precursors. Colloids Surf. A 158, 315–325.
Duarte, H.A., Lourenço, M.P., Heine, T., Guimarães, L., 2012. Clay mineral nanotubes: stability, structure and properties, stoichiometry and materials science—when numbers matter. In: Innocenti, Alessio (Ed.), InTech. ISBN: 978-953-51-0512-1. http://dx.doi.org/10.5772/34459. http://www.intechopen.com/books/stoichiometry-and-materials-science-when-numbers-matter/clay-mineral-nanotubes-stability-structure-and-properties-.
Escudey, M., Galindo, G., Ervin, J., 1986. Effect of iron oxide dissolution treatment on the isoelectric point of allophanic soils. Clay Clay Miner. 34, 108–110.
Farmer, V.C., 1981. Possible roles of a mobile hydroxy-aluminium orthosilicate complex (protoimogolite) and other hydroxyaluminium and hydroxy-iron species in pod-zolization. In: Migrations Organominérales dans les Sols Tempérés. 303. Colloques Internationaux du C.N.R.S, Paris, France, pp. 275–279.
Farmer, V.C., Fraser, A.R., Tait, J.M., 1977. Synthesis of imogolite: a tubular aluminium silicate polymer. J. Chem. Soc. Chem. Commun. 13, 462–463.
González, R.I., Ramírez, R., Rogan, J., Valdivia, J.A., Muñoz, F., Valencia, F., Ramírez, M., Kiwi, M., 2014. Model for self-rolling of an aluminosilicate sheet into a single-walled imogolite nanotube. J. Phys. Chem. C 118 (48), 28227.
Guerra, D., Batista, A., Viana, R., Airoldi, C., 2011. Adsorption of rubidium on raw and MTZ- and MBI-imogolite hybrid surfaces: an evidence of the chelate effect. Desalination 275, 107–111.
Guimarães, L., Nunes, Y., Lourenço, M.P., Duarte, H., 2013. Imogolite-like nanotubes: structure, stability, electronic and mechanical properties of the phosphorous and arsenic derivatives. Phys. Chem. Chem. Phys. 14, 4303–4309.
Gustafsson, J.P., 2001. The surface chemistry of imogolite. Clay Clay Miner. 49, 73–80.
Hu, J., Kannangara, G.S.K., Wilson, M.A., Reddy, N., 2004. The fused silicate route to protoimogolite and imogolite. J. Non-Cryst. Solids 347, 224–230.
Iijima, S., 1991. Helical microtubules of graphitic carbón. Nature 354, 1–2.
Inoue, K., Huang, R.M., 1984. Influence of citric acid on the natural formation of imogolite. Nature 308, 58–60.
Kang, D.-Y., Brunelli, N.A., Yucelen, G.I., Venkatasubramanian, A., Zang, J., Leisen, J., Hesketh, P.J., Jones, C.W., Nair, S., 2014. Direct synthesis of single-walled aminoaluminosilicate nanotubes with enhanced molecular adsorption selectivity. Nat. Commun. 5, 1–9.
Kijima, T., 2010. Inorganic and Metallic Nanotubular Materials: Recent Technologies and Applications. Springer, Berlin. 117.
Konduri, S., Mukherjee, S., Nair, S., 2007. Controlling nanotube dimensions: correlation between composition, diameter, and internal energy of single-walled mixed oxide nanotubes. ACS Nano 5, 393–402.
Lee, H., Jeon, Y., Lee, Y., Lee, S.-U., Takahara, A., Sohn, D., 2014. Thermodynamic control of diameter-modulated aluminosilicate nanotubes. J. Phys. Chem. C 118, 8148–8152.

Levard, C., Rose, J., Masion, A., Doelsch, E., Borschneck, D., Olivi, L., Dominici, C., Grauby, O., Woicik, J.C., Bottero, J.-Y., 2008. Synthesis of large quantities of single-walled aluminogermanate nanotube. J. Am. Chem. Soc. 130, 5862–5863.

Levard, C., Doelsch, E., Rose, J., Masion, A., Basile-Doelsch, I., Proux, O., Hazemann, J.L., Borschneck, D., Bottero, J.-Y., 2009a. Role of natural nanoparticles on the speciation of Ni in andosols of la Reunion. Geochim. Cosmochim. Acta 73, 4750–4760.

Levard, C., Masion, A., Rose, J., Doelsch, E., Borschneck, D., Dominici, C., Ziarelli, F., Bottero, J.Y., 2009b. Synthesis of imogolite fibers from decimolar concentration at low temperature and ambient pressure: a promising route for inexpensive nanotubes. J. Am. Chem. Soc. 131, 17080–17081.

Levard, C., Rose, J., Thill, A., Masion, A., Doelsch, E., Maillet, P., Spalla, O., Olivi, L., Cognigni, A., Ziarelli, F., 2010. Formation and growth mechanisms of imogolite-like aluminogermanate nanotubes. Chem. Mater. 22, 2466–2473.

Levard, C., Masion, A., Rose, J., Doelsch, E., Borschneck, D., Olivi, L., Chaurand, P., Dominici, C., Ziarelli, F., Thill, A., Maillet, P., Bottero, J.Y., 2011. Synthesis of Ge-imogolite: influence of the hydrolysis ratio on the structure of the nanotubes. Phys. Chem. Chem. Phys. 13, 14516–14522.

Levard, C., Doelsch, E., Basile-Doelsch, I., Abidin, Z., Miche, H., Masion, A., Rose, J., Borschneck, D., Bottero, J.-Y., 2012. Structure and distribution of allophanes, imogolite and proto-imogolite in volcanic soils. Geoderma 184, 100–108.

Liz-Marzan, L., Philipse, A.P., 1994. Synthesis of platinum nanoparticles in aqueous host dispersions of inorganic (imogolite) rods. Colloids Surf. A 90, 95–109.

Maillet, P., 2010. Structure et croissance de nanotubes de Ge-imogolite simple et double-paroi. Physics. Université Pierre et Marie Curie—Paris VI, 2010. French.

Maillet, P., Levard, C., Spalla, O., Masion, A., Rose, J., Thill, A., 2010a. Growth kinetic of single and double-walled aluminogermanate imogolite-like nanotubes: an experimental and modeling approach. Phys. Chem. Chem. Phys. 3, 2682–2689.

Maillet, P., Levard, C., Larquet, E., Mariet, C., Spalla, O., Menguy, N., Masion, A., Doelsch, E., Rose, J., Thill, T., 2010b. Evidence of double-walled Al-Ge imogolite-like nanotubes. A cryo-TEM and SAXS investigation. J. Am. Chem. Soc. 132, 1208–1209.

McBride, M.B., Farmer, V.C., Russell, J.D., Tait, J.M., Goodman, B.A., 1984. Iron substitution in aluminosilicate sols synthesized at low pH. Clay Miner. 19, 1–8.

Mukherjee, S., Bartlow, V.M., Nair, S., 2005. Phenomenology of the growth of single-walled aluminosilicate and aluminogermanate nanotubes of precise dimensions. Chem. Mater. 17, 4900–4909.

Mukherjee, S., Kim, K., Nair, S., 2007. Short, highly ordered, single-walled mixed-oxide nanotubes assemble from amorphous nanoparticles. J. Am. Chem. Soc. 129, 6820–6826.

Ohashi, F., Tomura, S., Akaku, K., Hayashi, S., Wada, S.-I., 2004. Characterization of synthetic imogolite nanotubes as gas storage. J. Mater. Sci. 39, 1799–1801.

Ookawa, M., 2012. Synthesis and characterization of fe-imogolite as an oxidation catalyst, clay minerals in nature—their characterization, modification and application. In: Valaskova, Marta (Ed.), InTech, ISBN: 978-953-51-0738-5. http://dx.doi.org/10.5772/48171. http://www.intechopen.com/books/clay-minerals-in-nature-their-characterization-modification-and-application/synthesis-and-characterization-of-fe-imogolite-as-an-oxidation-catalyst.

Ookawa, M., Inoue, Y., Watanabe, M., Suzuki, M., Yamaguchi, T., 2006. Synthesis and characterization of Fe containing imogolite. Clay Sci. 2, 280–284.

Ookawa, M., Takata, Y., Suzuki, M., Inukai, K., Maekawa, T., Yamaguchi, T., 2008. Oxidation of aromatic hydrocarbons with H_2O_2 catalyzed by a nano-scale tubular aluminosilicate, Fe-containing imogolite. Res. Chem. Intermed. 34, 679–685.

Pizarro, C., Escudey, M., Fabris, J.D., 2003. Influence of organic matter on the iron oxide mineralogy of volcanic soils. Hyperfine Interact. 148, 53–59.

Pohl, P.I., Faulon, J.-L., Smith, D.M., 1996. Pore structure of imogolite computer models. Langmuir 12, 4463–4468.

Rao, C.N.R., Nash, M., 2003. Inorganic nanotubes. Dalton Trans. 1, 1–24.

Shafia, E., Esposito, S., Manzoli, M., Chiesa, M., Tiberto, P., Barrera, G., Menard, G., Allia, P., Freyria, F.S., Garrone, E., Bonelli, B., 2015. Al/Fe isomorphic substitution versus Fe_2O_3 clusters formation in Fe-doped aluminosilicate nanotubes (imogolite). J. Nanopart. Res. 17, 336.

Sloan, J., Hutchison, J.L., Tenne, R., Feldman, Y., Srilina, T., Homyonfer, M., 1999. Defect and ordered tungsten oxides encapsulated inside 2H–WX_2(X = S and Se) fullerene-related structures. J. Solid State Chem. 144, 100–117.

Suzuki, M., Ohashi, F., Inukai, K., Maeda, M., Tomura, S., 2000. Synthesis of allophane and imogolite from inorganic solution—influence of co-existing ion concentration and titration rate on forming precursor. J. Clay Sci. Soc. Japan 40, 1–14.

Tenne, R., 2006. Inorganic nanotubes and fullerene-like nanoparticles. Nat. Nanotechnol. 1, 103–111.

Tenne, R., Margulis, L., Genut, M., Hodes, G., 1992. Polyhedral and cylindrical structures of tungsten disulfide. Nature 360, 444–446.

Thill, A., Guiose, B., Bacia-Verloop, M., Geertsen, V., Belloni, L., 2012a. How the diameter and structure of $(OH)_3Al_2O_3Si_xGe_{1-x}OH$ imogolite nanotubes are controlled by an adhesion versus curvature competition. J. Phys. Chem. C 116, 26841–26849.

Thill, A., Maillet, P., Guiose, B., Spalla, O., Belloni, L., Chaurand, P., Auffan, M., Olivi, L., Rose, J., 2012b. Physico-chemical control over the single- or double-wall structure of aluminogermanate imogolite-like nanotubes. J. Am. Chem. Soc. 134, 3780–3786.

Wada, S.-I., 1987. Imogolite synthesis at 25°C. Clay Clay Miner. 35, 379–384.

Wada, S.-I., Wada, K., 1982. Effects of substitution of germanium for silicon in imogolite. Clay Clay Miner. 30, 123–128.

Wilson, M.A., Lee, G.S.H., Taylor, R.C., 2001. Tetrahedral rehydration during imogolite formation. J. Non-Cryst. Solids 296, 172–181.

Wilson, M.A., Tran, N.H., Milev, A.S., Kannangara, G.S.K., Volk, H., Lu, G.Q.M., 2007. Nanomaterials in soils. Geoderma 146, 291–302.

Yamamoto, K., Otsuka, H., Takahara, A., 2007. Preparation of novel polymer hybrids from imogolite nanofiber. Polym. J. 39, 1–15.

Yang, H., Wang, C., Su, Z., 2008. Growth mechanism of synthetic imogolite nanotubes. Chem. Mater. 20, 4484–4488.

Yoshinaga, N., Aomine, S., 1962. Imogolite in some Ando soils. Soil Sci. Plant Nutr. (Tokyo) 8, 114.

Yucelen, G.I., Choudhury, R.P., Vyalikh, A., Scheler, U., Beckham, H.W., Nair, S., 2011. Formation of single-walled aluminosilicate nanotubes from molecular precursors and curved nanoscale intermediates. J. Am. Chem. Soc. 133, 5397–5412.

Yucelen, G.I., Choudhury, R.P., Leisen, J., Nair, S., Beckham, H.W., 2012a. Defect structures in aluminosilicate single-walled nanotubes: a solid-state nuclear magnetic resonance investigation. J. Phys. Chem. C 116, 17149–17157.

Yucelen, G.I., Kang, D.-Y., Guerrero-Ferreira, R., Wright, E.R., Beckham, H.W., Nair, S., 2012b. Shaping single-walled metal oxide nanotubes from precursors of controlled curvature. Nano Lett. 12, 827–832.

Zanzottera, C., Vicente, A., Armandi, M., Fernandez, C., Garrone, E., Bonelli, B., 2012. Physico-chemical properties of imogolite nanotubes functionalized on both external and internal surfaces. J. Phys. Chem. C 116, 23577–23584.

Part IV

Applications of Nanosized Tubular Clay Minerals

Chapter 20

An Overview on the Safety of Tubular Clay Minerals

M.-C. Jaurand*
Institut National de la Santé et de la Recherche Médicale (INSERM), UMR-S 116, 27 rue Juliette Dodu, Paris, France
Université Paris Descartes, Labex Immuno-Oncology, Sorbonne Paris Cité, Faculté de Médecine, Paris, France
**Corresponding author: e-mail: marie-claude.jaurand@inserm.fr*

20.1 INTRODUCTION

20.1.1 The HARN Paradigm

The high-aspect ratio nanoparticle (HARN) paradigm was stated as a legacy of our knowledge of the action mechanism of asbestos, and further toxicological investigations developed with carbon nanotubes (CNT). The historical and scientific context is summarised here.

The adverse effects of asbestos that is induction of lung fibrosis (asbestosis), pleural fibrosis (pleural plaques) and lung cancer have been known since the early 20th century. Further, studies were stimulated by the report of an association between asbestos exposure and a rare, very severe neoplasm called 'malignant pleural mesothelioma', a cancer arising from the transformation of mesothelial cells, in the population of North Western Cape Province in South Africa, where asbestos was mined (Wagner et al., 1960). Although human exposure occurs by inhalation, the lung and pleura are not the only organs or tissues where cancers are observed. Asbestos exposure is also found to be associated with significant enhancement of ovarian and laryngeal cancers (Straif et al., 2009).

Among the particles present in the environment, fibres pose greater human health problems in comparison with other particles because of their form. Their elongated shape makes long fibres easier to be deposited in the lung. Deposition of particles in the respiratory tract is dependent on their aerodynamic diameter; the upper size limit is about 5 μm for inhaled particles that reach the lung alveoli. Higher-diameter particles are deposited or retained in

Developments in Clay Science, Vol. 7. http://dx.doi.org/10.1016/B978-0-08-100293-3.00020-0

the upper airways and the bronchial tract (Heyder, 2004). For fibres, the aerodynamic diameter is mostly dependent on the diameter, and long fibres (with a length of dozens of micrometres) can reach the alveolar region (Baron, 1996). Suzuki et al. (2005) reported that in malignant mesothelioma, the majority (89.4%) of fibres detected in the lung and mesothelial tissues had lengths shorter than or equal to 5 μm. These authors reported values of 0.62 ± 3.72 μm as the geometric mean length (range 0.07–82.4 μm) and 0.06 ± 2.17 μm of the geometric mean diameter (range 0.01–6.5 μm). Asbestos fibres are composed of very small fibrils, which associate to form bundles. Chrysotile is a phyllosilicate, which has a cylindrical crystal habit and occurs as fibrils 20–40 nm in diameter. Fibrils of amphiboles are generally thicker (about 50–150 nm) (Hamade et al., 2015). The most frequent types of amphiboles used in industry in the past were amosite and crocidolite. Asbestos fibres may naturally occur in the micrometre or millimetre size, or even larger.

Following the findings of the adverse health effects of asbestos fibres, other fibres present in the environment were found to induce lung cancer and pleural mesothelioma in humans, including erionite, an aluminium silicate fibrous found in volcanic and sedimentary rocks; and amphiboles found in different geographical areas (such as fluoro-edenite in Italy and winchite in Libby, Montana) (Meeker et al., 2003; Carbone et al., 2012; Jean et al., 2014). Furthermore, experimental studies carried out with artificial fibres (including refractory ceramic fibres and some glass fibres) demonstrated a carcinogenic potency in experimental animals, entailing their classification as 2B (possibly carcinogenic to humans) by the International Agency for Research on Cancer (IARC, 2002). This body of data stimulated strong questioning in the area of the eventual toxicity of other elongated particles. This concern was particularly acute, considering the development of nanotechnologies and nanosized materials, including the manufacturing of several sorts (simple-, double- or multiwalled) CNT (Service, 2003). CNT have been compared to asbestos in terms of their similarities to asbestos fibrils. First, CNT are in the nanometre range, although smaller single-walled (SW) CNT may also exist, and form hollow cylinders that may be compared to chrysotile asbestos fibrils.

Early studies alerted on the toxicological effects of CNT, demonstrating similar results for in vivo cell responses (inflammation, formation of granulomas, frustrated phagocytosis) and carcinogenic potency (mesotheliomas in rodents following intraperitoneal injection). Further studies documented their cytotoxicity and genotoxicity. A recent IARC monograph documents the possible carcinogenic effect on humans of one largely tested multiwalled (MW) CNT (Grosse et al., 2014). The results of studies developed with asbestos and CNT, as well as with other nanoparticles such as nanowires or nanorods, led to the HARN paradigm, which is applicable to elongated particles of a high-aspect ratio (with length/diameter ≥ 3 or 5). This means that HARN particles may have a pathologic potential in human beings and that the effects are modulated by physicochemical properties, especially dimensions and the state of aggregation. The expected effects concern inflammation and the response to stress in the form of oxidative and cellular stresses, including mitosis impairment. Discussions of the comparison of the effects of asbestos and CNT can be found in several reviews (eg, Jaurand et al., 2009; Shvedova et al., 2009; Donaldson et al., 2013).

Being aware of these issues and of the huge development of nanotechnologies, working groups and organisations have proposed definitions for the nanoscale and for nanomaterials and also tried to determine whether HARN should raise the same concerns as asbestos fibres did (Tran et al., 2008). The challenging questions to emerge recently are whether nanosized tubular particles such as halloysite and imogolite fall into the category of HARN with potential toxicity, and if so, to what extent.

20.1.2 Specific Industrial Applications of Asbestos and Carbon Nanotubes—The Different Spin-Off Effects in Terms of Toxicological Issues

In general, the field of applications of particles defines the types of health concern involved. Briefly, asbestos has still been used in many countries, mainly in buildings and other construction projects, and has been present in numerous products for consumers. Humans are exposed by inhalation; primary targets are the respiratory airways, lung and pleura, but translocation may account for the incidence of cancers in other tissues, as has been found in the case of exposure to asbestos. Potential health risks occur in production industries and in use of products for consumers. Environmental exposures have been described more recently (Maynard et al., 2004; Maynard and Kuempel, 2005; Tsai et al., 2009; Bello et al., 2010; Johnson et al., 2010; Cena and Peters, 2011; Lee et al., 2015).

CNT differ from asbestos because they are engineered nanoparticles produced to be pristine or functionalised. The production of CNT is lower than that of asbestos, and the uses appear to be more specific, but CNT have found widespread application in fields ranging from microelectronics to medicine (Ali-Boucetta and Kostarelos, 2013; Wang et al., 2014b). Human exposure to CNT can therefore occur by inhalation, but also via other routes as well, and the target organs are not limited to respiratory airways, lung and serosa, but may concern all other organs and tissues as well. CNT may pose health risks in production industries for users and consumers as well as patients. Present studies have investigated the ecological life cycle of CNT and raised concerns about the release of CNT into the environment, which may directly or indirectly lead to human exposure. In that area, a number of ecotoxicological studies are available, but they have not been considered in this discussion.

The situation is even more complicated for nanotubular clay minerals since the range of their applications is very wide; their commercial production is high, especially with regard to industrial and agricultural applications. Moreover, they are used and engineered for numerous medical purposes (Lecouvet, 2015; Vahedi and Pasbakhsh, 2015a). From research data, toxicology-related studies schematically encompass the search for biocompatibility as a reinforcement of material or carriers, but few studies have focused on the evaluation of the toxicity of the mineral form in the occupational or general environment.

20.2 POTENTIAL HEALTH INJURY RESULTING FROM INDUSTRIAL USES OF HALLOYSITE AND IMOGOLITE

20.2.1 Uses of Halloysite and Imogolite

Halloysite and imogolite are natural minerals used in a large number of applications (Vahedi and Pasbakhsh, 2015a); similar to CNT, they are tubular nanostructures that can be also synthesised and functionalised (Zanzottera et al., 2012; Vahedi and Pasbakhsh, 2015b). Briefly, these natural clay minerals are used for their catalytic and adsorption properties and as loading agents in polymers. Halloysite is used in a wider range of applications than imogolite, including various medical functions such as drug carriers and reinforcing agents. Therefore, considering the great increase in industrial applications of nanosized tubular clay minerals, the health issues raised cover a variety of toxicological fields. They concern the hazard and risk assessment associated with occupational and environmental exposures, biocompatibility and information on the traceability and control of ecological life cycles.

20.2.2 Spin-Off in Occupational, Environmental and Therapeutic Issues

Attempts to define halloysite and imogolite safety, then, should focus on large research areas: epidemiological studies, ecotoxicological studies, biodistribution and toxicokinetics at the body, tissue and cell levels (ie, cytotoxicity, genotoxicity and modification of cell functions). So far, no epidemiological study exists to qualify the exposures to halloysite and imogolite, and the health risks remain undefined. The number of studies concerning the potential toxicity of these nanosized tubular clay minerals is rather low. The present discussion will summarise the studies that have been carried out on the toxicity of halloysite and imogolite. Some of them will be detailed to illustrate the present level of knowledge about their biological effects. Generally, toxicological studies aiming to evaluate risk by inhalation investigate cytotoxicity, genotoxicity and specific cell responses on cultured cells and lung injury. The published toxicological studies on nanotubular clay minerals concern more specific aspects (mainly their use for medical purposes). Schematically, the purpose of the investigations can be subdivided into three branches: (i) drug delivery, (ii) clay polymer nanocomposites (CPN) and (iii) safety.

20.3 THE BIOLOGICAL EFFECTS OF POTENTIAL INTEREST IN THE TOXICOLOGY OF NANOTUBULAR PARTICLES

20.3.1 Experimental Studies on the Biological Effects of Halloysite

The potential of halloysite in the fields of pharmacy and medicine to be used as a reinforcement of implants to increase cell adhesion or deliver chemicals was reviewed recently (Rabiskova, 2012; Lvov et al., 2014; Shutava et al., 2014). The objectives of the studies may be orthogonal in terms of toxicity analysis. For instance, cytotoxicity is expected to kill cancer cells, whereas maintenance of tissue integrity is needed to deliver antibiotics preventing from infection. Many studies have been carried out. The available data on the different uses, routes of exposure and biological end points of halloysite are summarised in Table 20.1.

20.3.1.1 Cytocompatibility and Drug Release

Resveratrol, a drug with potential in cancer chemoprevention and chemotherapy, was encapsulated inside halloysite and coated with two different couples of polyelectrolytes: PEI-PSS/PAH [polyethyleneimine (PEI), poly(sodium 4-styrene-sulfonate) (PSS)/poly(allylamine hydrochloride) (PAH)] or PRM/DXS [protamine salt (PRM) grad III/dextran sulphate sodium salt (DXS)]. Resveratrol-loaded, native and PEI-PSS/PAH-coated halloysite were not cytotoxic on a mammary cancer cell line (MCF-7), but resveratrol-loaded-PRM/DXS halloysite was slightly cytotoxic, a finding consistent with the release of resveratrol investigated in buffer solution (Vergaro et al., 2012).

In another study, pristine halloysite used as a control was loaded either with methotrexate, artemisinin, quercetin or taurolidine and tested on osteosarcoma cell lines. Results showed that the drug-loaded halloysite can have cytotoxicity, but the same was not true for unloaded halloysite (Sun and Mills, 2014).

Cell viability was investigated on seven cancer cell lines using pristine halloysite and several functionalised halloysites that were positively charged with triazolium salts and either unloaded or loaded with curcumin (Riela et al., 2014). This molecule exhibits antioxidant, anti-inflammatory, antiviral, antibacterial, antifungal and anticancer activities with potential to fight various diseases (Aggarwal et al., 2007). These effects are mediated through the regulation of various transcription and growth factors, inflammatory cytokines, protein kinases and other enzymes. Pristine halloysite or curcumin-loaded pristine halloysite were not cytotoxic, whereas functionalised halloysite exerted some variable cytotoxic effects against the different cell lines (Riela et al., 2014). Although these toxic effects increased for functionalised halloysite, they were loaded with curcumin, and the toxicity of these modified halloysites is difficult to determine.

TABLE 20.1 Summary of Studies Carried Out with Halloysite

Objective/ context	Mineral system tested	Goal	Drug load or bind	Biological/ toxicological assay	Results	References
CPN film	Halloysite	Dental regeneration	Polydioxanone (PDS II®)	Attachment and proliferation of fibroblasts	Biocompatibility; enhanced proliferation	Bottino et al. (2013)
CPN film	Halloysite	Elastomer for use in soft tissue engineering	Poly(glycerol sebacate) (PGS)	Cytotoxicity, measurement of lactate dehydrogenase (LDH) level of mouse fibroblasts	Satisfactory biocompatibility; cytotoxicity at concentrations PGS/halloysite 10% and 20 wt.%	Chen et al. (2011)
Drug release	Dextrin-coated halloysite closed with dextrin	Cancer cell therapy	Triazole dye brilliant green (BG)	Uptake of halloysite, cell morphology, drug delivery in A549 and Hep3b cells	Drug delivery	Dzamukova et al. (2015)
Drug carrier	Incorporation into adhesive therapeutical dental resin	Inhibition of dentin degradation	Doxycycline	Antibacterial activity; anti–matrix metalloproteinase (MMP)	Inhibition of bacterial growth. Protection against MMP degradation	Feitosa et al. (2014)
Not specifically defined	Halloysite	Evaluate toxicity	None	Cytotoxicity and proteomic profiling of Caco-2/HT29-MTX cells	Absence of cytotoxicity in spite of elevated proinflammatory cytokine release	Lai et al. (2013)

Drug carrier	DNA-wrapped halloysite	Cancer therapy	Doxorubicin	Cell morphology A549 cells	Sustained release	Lee et al. (2013)
CPN film	Halloysite	Multiple uses	Chitosan	Morphology and viability of fibroblasts	Cytocompatibility	Liu et al. (2012a)
CPN film	Halloysite bounded to poly(ʟ-lactide) (PLLA)	Adhesivity	Poly(ʟ-lactide)	Adhesion and proliferation of mouse M3T3 fibroblasts	Promotion of adhesion and cell proliferation	Luo et al. (2013)
Drug carrier	f-Halloysite (coated with triazolium salts)	Enhanced drug release	Cardanol	Kinetics of release; cell viability of cell lines HA22T/VGH, HepG2 and Hep3b	Cytotoxicity of cardanol-loaded f-halloysite	Massaro et al. (2015)
Drug carrier	Halloysite/ poly(lactic-*co*-glycolic acid) composite nanofibres (HNTs/ PLGA)	Antibiotic activity	Tetracycline hydrochloride	Cytotoxicity on mouse fibroblasts	Cytocompatibility; sustained drug release	Qi et al. (2012)
CPN	Halloysite	Improvement polymer properties	Poly(lactic-*co*-glycolic acid)	Adhesion and proliferation of mouse fibroblasts	Cytocompatibility; proliferation of mouse fibroblasts	Qi et al. (2012)
Drug carrier	Halloysite and f-halloysite (positively charged with triazolium salts)	Cancer cell therapy	Curcumin	Viability of 7 cancer cell lines	Cytotoxicity	Riela et al. (2014)

Continued

TABLE 20.1 Summary of Studies Carried Out with Halloysite—Cont'd

Objective/ context	Mineral system tested	Goal	Drug load or bind	Biological/ toxicological assay	Results	References
Drug carrier	Unloaded or loaded halloysite	Cancer cell therapy	Antitumor drugs (methotrexate, artemisinin), antioxidant (quercetin) and antimicrobial (taurolidine)	Viability of osteosarcoma cells	No cytotoxicity of unloaded halloysite. Inhibition of proliferation with halloysite/ methotrexate; cell growth arrest and apoptosis with the other molecules	Sun and Mills (2014)
CPN	Halloysite	Mutagenic effect	None	Bacteria strains; Ames test	No mutagenicity	Taylor et al. (2014)
Not specifically defined	Untreated and f-halloysite (APTES) (fluorescently labelled)	Mechanism of action, toxic potency	None	Viability of human epithelial adenocarcinoma cell line (HeLa) and human breast cancer cell line (MCF-7)	Dose-dependent decrease in cell viability. Halloysite uptake	Vergaro et al. (2010)
Drug carrier	f-Halloysite (polyelectrolytes coating)	Cancer therapy	Resveratrol	Cytotoxicity on a cancer cell line (MCF-7)	No cytotoxicity of halloysite alone or polyelectrolyte-coated halloysite. Cytotoxicity of resveratrol-loaded-coated halloysite (MTT assay)	Vergaro et al. (2010)

Drug carrier	Acid- and heat-treated halloysite	Adsorption/release of cationic drugs	Ofloxacin	No biologic system	–	Wang et al. (2014a)
siRNA release	f-Halloysite	Cancer cell therapy	Antisurvivin	Morphological and biological study of f-halloysite-treated cells	Knocking down of the gene	Wu et al. (2014)
Drug carrier	Halloysite-doped membrane	Bone regeneration	Metronidazole	Cytotoxicity on L929 cells (MTT assay) antibacterial activity	Cytotoxicity (MTT assay) prevention of infection	Xue et al. (2015)
CPN film	Halloysite modified with APTES in poly(caprolactone)/gelatin microfibres	Antimicrobial	Metronidazole	Cytotoxicity in L929 fibroblast cells (MTT assay). Antibacterial activity against *F. nucleatum*	No cytotoxicity. Antibacterial effect	Xue et al. (2015)
Protection of drug toxicity	Halloysite	Reduction of toxic effects	Zearalenone	In vivo study in pregnant pigs	Reduction of pathological effects	Yin et al. (2015)
Protection of drug toxicity	Halloysite modified with the surfactant stearyldimethyl-benzylammonium chloride (MHNT)	Reduction of toxic effects	Zearalenone	In vitro: drug adsorption from simulated biologic fluids (gastric, intestinal)—in vivo (MHNT in diet of rats): reduced toxic and oestrogenic effects	Decreased toxic and estrogenic effects zearalenone (changes in oxidative stress biomarkers and organ weights)	Zhang et al. (2014)

Cytotoxicity was often assessed by the MTT (3-(4,5-dimethylthiazol-2-yl)-2,5-diphenyltetrazolium bromide) assay. It must be noted that this assay reflects the activity of mitochondria by the reduction of MTT into formazan, measured by densitometry. Then, a decrease in the optical density may be due to loss of cell viability or/and cell growth inhibition. The measurement of lactate dehydrogenase (LDH) levels in the extracellular medium demonstrates a leakage of the cell membrane, attesting loss of cell viability.

Dzamukova et al. (2015) developed a sophisticated model to control the release of drugs within the cells. This system is based on the loading of halloysite with a model anticancer drug (triazole dye brilliant green, BG) and coating of halloysite with dextrin (DX) at the tube end. DX is a molecule cleavable by intracellular glycosyl hydrolases for controlled release inside cells. The effects were studied in two cell lines: a human lung carcinoma epithelial cell line (A549), and a hepatocellular carcinoma cell line (Hep3b). Several end points were investigated; ie, cell morphology and cellular uptake (by optical and electron microscopy), cell viability and cytoskeleton integrity. Differences were observed in the uptake of DX-halloysite. Nanotubes were concentrated in perinuclear regions in A549 cells, but they were randomly distributed through and accumulated in the cell membrane of Hep3b cells. There was no modification of cell viability compared to cell cultures without DX-halloysite, and no change in cytoskeleton organisation. Exposure of A549 cells to BG-loaded DX-halloysite demonstrated a decreased viability that depends on halloysite concentrations when compared to cells cultured without halloysite. However, the effect of BG-loaded halloysite was similar to that of BG-loaded DX-halloysite. In contrast, BG-loaded halloysite were more toxic than BG-loaded DX-halloysite in Hep3b cells (Dzamukova et al., 2015).

Massaro et al. (2015) investigated the toxicity of functionalised halloysite loaded with cardanol, a possible anticancer molecule, on three hepatoma cell lines. Free cardanol had no effect on cell viability, but the cardanol functionalised halloysite was strongly cytotoxic to the three cell types.

Other studies aimed at developing tools preventing bacterial infection that may occur while using membranes to improve tissue regeneration and cell attachment. In one study, membranes were made with halloysite modified with (aminopropyl)triethoxysilane (APTES) and further included at several concentrations (0, 1, 2, 5, 10 and 20 mass%) in a polymer solution of polycaprolactone and gelatin (PG). The cytotoxicity of the membranes was determined by the viability and proliferation of murine fibrosarcoma cells (L929). Furthermore, these halloysites were loaded with metronidazole, an antibacterial agent, and tested against the bacteria *F. nucleatum*. The results demonstrated that L929 cells adhered well to all the membranes tested, without toxicity. The growth of *F. nucleatum* was inhibited by PG-metronidazole and metronidazole-loaded PG-halloysite, while drug-free PG-halloysite samples did not show inhibition (Xue et al., 2015).

20.3.1.2 Biocompatibility of Clay Polymer Nanocomposites

Halloysite is also used to reinforce biomaterials devoted to replacing or sustaining natural tissues. Chen et al. (2011) developed synthetic biomaterials available for clinical use that can replace soft tissues such as muscle and connective tissue. These authors synthesised a new CPN system made of poly(glycerol sebacate) (PGS) at 0, 1, 3, 5, 10 and 20 mass% filled with halloysite nanotubes in order to improve the mechanical properties of the CPM.

Cell viability was not modified in comparison with control cultures (culture medium without PGS or halloysite/PGS). PGS alone and the halloysite/1–5 mass% PGS nanocomposites were not cytotoxic. However, statistically significant cytotoxicity was observed in the CPN with 10 and 20 mass% PGS (Chen et al., 2011).

Bionanocomposite films made of chitosan and halloysite were developed for applications in bone tissue engineering and drug carrier systems (Liu et al., 2012a). Viability was assessed on a mouse embryo fibroblast cell line (NIH/3T3) by the bromure de 3-(4,5-dimethylthiazol-2-yl)-2,5-diphenyltetrazolium bromide (MTT) assay that measures mitochondrial activity. Although no cytotoxicity was found, the proliferation of fibroblasts was reduced as compared with cultures in tissue culture plates, possibly due to a different cell attachment to chitosan and halloysite/chitosan as compared with culture plates (Liu et al., 2012a).

Human dental pulp-derived fibroblast cells were seeded on normal untreated culture plates, on polydioxanone (PDS II®) (PDS), or on a nanocomposite scaffold of halloysite and PDS, at four concentrations of halloysite (0.5–10%) (Bottino et al., 2013). Proliferation assays demonstrated a significant enhancement of fibroblast proliferation in comparison with the control culture on tissue culture plates. Proliferation was not significantly modified in the CPN scaffolds in comparison with in cultures made on PDS alone, suggesting a biocompatibility of the bionanocomposite (Bottino et al., 2013).

Other studies were related to research in tooth restoration. Feitosa et al. (2014) reported biological activity, cytocompatibility and bonding characteristics to dentin of a doxycycline (DOX)–encapsulated halloysite-modified adhesive. The end points concerned cytotoxicity evaluated on human dental pulp stem cells (HDPSC), the antimicrobial activity of synthesised adhesives and DOX-containing eluates against *Streptococcus mutans* and inhibition of metalloproteinase-mediated casein cleavage of metalloproteinase-mediated casein cleavage (Feitosa et al., 2014). None of the adhesives, either unmodified halloysite, loaded halloysite or loaded halloysite+DOX-loaded, were cytotoxic to HDPSC. A significant growth inhibition of *Streptococcus* was observed with loaded halloysite with DOX adhesive added. Moreover, DOX-containing eluates promoted greater inhibition of metalloproteinase activity compared to the control.

20.3.1.3 Conventional Toxicity Studies

There have been a number of studies designed to evaluate potential health risk effects. Lai et al. (2013) aimed at investigating the toxicity of halloysite on intestinal epithelial cells. They used a model of Caco-2/HT29-MTX co-culture simulating the large intestine. Caco-2 and HT29-MTX adenocarcinoma cell lines are derived from intestinal absorptive and mucus-secreting goblet cell types, respectively. The clay contained 95% halloysite and the rest consisted of kaolin and silica. The biological assays consisted of measurement of cell viability, membrane integrity and permeability, as well as chemokine, cytokine and proteomic analysis. Exposure was performed at concentrations of 1, 10 and 100 μg/mL. The results demonstrated (i) significant proinflammatory effects at the lowest concentrations and (ii) significant changes in protein expression at the highest concentrations (Lai et al., 2013). Proteomic analyses revealed significant quantitative expression changes in halloysite-treated cells. At 1 μg/mL, 35 proteins were upregulated and 32 were downregulated, among 4081 proteins analysed. More proteins were upregulated at 10 μg/mL, but less were downregulated. At 100 μg/mL, 415 of 4081 proteins were differentially expressed, most of them (404) being upregulated (Lai et al., 2013). These results demonstrate a different dose-dependent response of the halloysite-treated cells. Lai et al. (2013) consider that halloysite do not appear to have toxic effects at moderate levels of exposure, but exposure at higher concentrations induced cell responses associated with injury and oxidant stress, stimulation of cell growth and proliferation pathways that warrant further investigation.

Human cancer cells, HeLa and MCF-7, were exposed to halloysite (either untreated or functionalised with APTES) or coated using a layer-by-layer technique via alternate adsorption of polycations and polyanions, and fluorescently labelled with fluorescein isothiocyanate. All samples induced a dosedependent cytotoxicity with an IC_{50} (50% of viable cells in comparison with untreated cells) at a concentration of about 100 mg/mL. Halloysite localisation followed by fluorescence detection indicated that halloysite were internalised by the cells (Vergaro et al., 2010).

20.3.2 Experimental Studies on the Biological Effects of Imogolite

A few studies have been performed on the biological effects of imogolite. The summarised results with further details are given in Table 20.2. Rotoli et al. (2014) assessed the cytotoxicity and genotoxicity of single-walled imogolite on murine macrophages and human airway epithelial cells. The results were compared to those obtained with SW and multiwalled CNT. Viability and gene expression were investigated on two murine macrophage cell lines (RAW264.7 and MH-S a continuous cell line established following transformation of alveolar macrophages with simian virus 40) (Mbawuike and Herscowitz, 1989). Gene studies concerned the expression of *Nos2* (inducible nitric oxide synthesis) and *Arg1* (arginase 1), markers of classical or alternative macrophage activation. Genotoxicity was studied on RAW264.7 and epithelial cancer cells A549 (comet assay, an indicator of DNA breakage and micronucleus assay, an indicator or chromosome damage). Epithelial barrier integrity was evaluated from the trans-epithelial electrical resistance (TEER) using a human lung carcinoma cell line Calu-3. RAW264.7 cells internalised imogolite which reduced the cell viability in a dose-dependent manner, but to

TABLE 20.2 Summary of Studies Carried Out with Imogolite

Objective/context	Mineral system tested	Goal	Drug load or bind	Biological/toxicological assay	Results	References
CPN	Imogolite	Biocompatibility	None	Effects on mouse osteoblast-like cells cultured on imogolite	Good biocompatibility between the cells and imogolite scaffold; enhancement effect of proliferation and osteoblastic differentiation	Ishikawa et al. (2010)
Toxicity	Single wall imogolite	Study of adverse effects	None	In vitro toxicity on macrophages and epithelial cells (viability, epithelial barrier integrity, genotoxicity)	No or low cytotoxicity; no genotoxicity (comet assay); no alteration in membrane permeability	Rotoli et al. (2014)
Toxicity	Ge-imogolite short tubes (ST, ca. 6 nm)	Study of adverse effects	None	In vitro toxicity on fibroblasts (viability, genotoxicity)	No cytotoxicity; genotoxicity	Liu et al. (2012b)
Toxicity	Ge-imogolite long tubes (LT, ca. 50 nm)	Study of adverse effects	None	In vitro toxicity on fibroblasts (viability, genotoxicity)	No cytotoxicity; low genotoxicity	Liu et al. (2012b)

Continued

TABLE 20.2 Summary of Studies Carried Out with Imogolite—Cont'd

Objective/context	Mineral system tested	Goal	Drug load or bind	Biological/toxicological assay	Results	References
Toxicity	Double wall Ge-imogolite	Study of adverse effects	Ge-substituted	Intra-tracheal injection in rats. Biopersistence of Ge-imogolites and their localisation in the lung. Acute inflammation and genotoxicity.	Persistent dose-dependent inflammation. Fibrotic response. Genotoxicity (micronuclei). Higher than crocidolite	van den Brule et al. (2014)
Toxicity	Single wall Ge-imogolite	Study of adverse effects	Ge-substituted	Intratracheal injection in rats. Biopersistence of Ge-imogolites and their localisation in the lung. Acute inflammation and genotoxicity	Persistent dose-dependent inflammation. Fibrotic response. Genotoxicity (micronuclei). Higher than crocidolite	van den Brule et al. (2014)

Cytotoxicity was often assessed by the MTT [3-(4,5-dimethylthiazol-2-yl)-2,5-diphenyltetrazolium bromide] assay. It should be noted that this assay reflects the activity of mitochondria, by the reduction of MTT into formazan, measured in densitometry. A decrease in the optical density may be due to loss of cell viability, cell growth inhibition or both. The measurement of lactate dehydrogenase (LDH) levels in the extracellular medium demonstrates a leakage of the cell membrane, attesting loss of cell viability.

a lesser extent than SW and multiwalled CNT, on a per mass basis. A genotoxicity was observed in A549 cells in the comet assay but not in RAW264.7 cells. TEER was not significantly reduced by imogolite but significantly with SW and multiwalled CNT, leading the authors to admit a relative inertness of imogolite samples. Although a hypothesis is made as to the particle parameters that may account for the lack of an adverse effect (extreme hydrophilicity and/or the presence of an external alumina layer in the absence of free radical release) further research are necessary to confirm these hypotheses. Moreover, attention should be paid to the fact that physicochemical properties of imogolite are determined in vitro, in acellular conditions and that the biological medium can modify these properties. One example is the modification of zeta potential in culture medium (Rotoli et al., 2014). It must be reminded that a biological medium contains several types of biological molecules. Interactions may occur with DNA, proteins and phospholipids (Jaurand, 1989). After inhalation, interactions may occur with proteins, mucopolysaccharides and phospholipids in the alveolar surfactant.

Imogolite-like nanotubes (see Chapter 19) were also tested for their cytotoxic and genotoxic effects towards human dermal fibroblasts (Liu et al., 2012b). These synthesised aluminogermanate nanotubes (Ge-imogolite) particles were of similar compositions and with various sizes and shapes (PR are proto-imogolite, ST and LT are short and long nanotubes, respectively). End points were Ge-imogolite uptake, cell viability, comet and micronucleus assays for genotoxicity. Internalisation into human fibroblasts was only observed for the small particles PR and ST samples. None of the samples showed a statistically significant diminution of cell viability. The three types of particles induced statistically significant dose-dependent DNA and chromosome damage, and the two short samples (PR and ST) were more active, likely related to oxidative stress. In contrast, this process seems to be a minor factor with LT (Liu et al., 2012b).

The in vivo effects of synthesised Ge-imogolite, substituting Si by Ge, were investigated in rats exposed by intratracheal instillation (van den Brule et al., 2014). Two samples, SW (70-nm length) and double-walled (DW, 62-nm length) Ge-imogolite were injected at doses of 0.02–2 mg/rat. Other samples, crocidolite asbestos and the hard metal particles WC-Co, were used as positive controls. Several toxicity parameters were studied, including biopersistence and localisation in the lung, lung inflammation and genotoxicity (micronuclei in isolated type II pneumocytes) (van den Brule et al., 2014). The results showed that Ge-imogolite is biopersistent in the lung, induces sustained inflammation and promotes genotoxicity in lung cells. Lung histology revealed the presence of granulomas with Ge-imogolite in the same zone and fibrosis. Overall, the inflammatory and fibrotic responses induced by SW and DW Ge-imogolite were more severe (on a mass-dose basis) than those induced by crocidolite. These results obtained on Ge-imogolite call for future similar studies in the case of imogolite.

20.3.3 Outline of Studies Carried Out so Far and Perspectives

Studies carried out with halloysite nanotubes mainly investigated the biocompatibility, cytotoxicity and effects related to drug delivery. Experiments were performed on different sorts of living cells. Several end points were determined, depending on the context of halloysite use: bacterial toxicity, cytotoxicity, genotoxicity, proteomic profiling and physiological mammalian cell function. Finally, the number of studies is limited, and most of the existing studies employed toxicity assays focusing on cell viability; only one animal study concerned the effect on mycotoxin toxicity in pigs. As opposed to the aim of most studies carried out with halloysite, the toxicity of imogolite nanotubes has been tested in a context of general health concern, more precisely as studies were carried out with CNT. Cytotoxicity and genotoxicity were both investigated in cell cultures, and one animal study was performed in rats to assess the effects on the lung. Nevertheless, the number of studies has been limited.

In the future, toxicological studies should be developed to assess the biological effects of nanotubular clay minerals. The end points developed in asbestos and CNT toxicology should be examined with the aim of understanding the action mechanism of the observed effects. However, the large range of uses of nanotubular clay minerals must focus on relevant approaches consistent with the fields of application. In production, safety should concern both producers and consumers, just as for other minerals or engineered particles. Toxicological studies involve the analysis of cell viability, response to stress, genotoxicity, biopersistence, toxicokinetics and anatomo-pathological changes. Recently, the development of 'omics'-type methodologies that allow deeper information of the changes induced in gene and protein expression in response to HARN-exposed organs and cells are of great interest. As far as possible, the assessment of the biological effects should be performed in pertinent assays, and on target cells, tissues and organs. Regarding the great number of halloysite or imogolite nanotubes that may be constructed, an effort should be made to develop more sophisticated in vitro assays, to mimic the tissues and to reduce animal experiments. With nanotubular clay minerals, it will be needed to define the possible impacts considering the mode of exposure, biodistribution, toxicokinetics and possible interactions with cells and tissues. Some mechanistic points should also be considered, such as particle uptake and interactions with the components' biological media. As for asbestos, the formation of a protein corona can be present at the nanoparticle surface (Lanone et al., 2013). The presence of nanoparticles in the cells is a key point of interaction with cell components, especially for DNA and chromosome damage. Phagocytosis, diffusion and endocytosis are three uptake mechanisms for nanoparticles. With CNT, as it might be with nanotubular clay minerals, the uptake is modulated by the degree of dispersion, the formation of supramolecular complexes and the nanotube length (Raffa et al., 2010).

20.4 DETERMINANTS OF NANOPARTICLE TOXICITY AND MECHANISMS

20.4.1 Determinants of Nanoparticle Toxicity

This section briefly summarises the asbestos parameters determining the adverse health effects of asbestos. First, the elongated shape and dimensions (length, diameter) are important since they influence the fibre deposition in the lung following inhalation, as discussed earlier. Fibre dimensions also affect the clearance mechanism; short fibres may be cleared by macrophages following the phagocytic process (fibre uptake and internalisation), while fibres too long to be internalised by these cells are not cleared (or are only partially cleared), which may provoke the so-called 'frustrated phagocytosis'; macrophage damage can persist and accumulate in the lung. As far as inhalation is concerned, once deposited in the lung asbestos fibres can migrate to other sites in the body (Oberdorster et al., 2004, 2009). A translocation to the pleura is of particular importance to account for the fibrogenic and carcinogenic health effects of asbestos at the pleural level (Broaddus et al., 2011). The physicochemical properties of asbestos may also play a role related to the presence of surface contaminants. Metals, especially iron, may be involved in the generation of reactive oxygen species (ROS) and reactive nitrogen species (RNS), and polycyclic aromatic hydrocarbons have a carcinogenic potency. Moreover, the surface may interact with the biological milieu and biological molecules may be adsorbed at the fibre surface (Jaurand, 1989; Jaurand et al., 2009). A role of fibre dimensions accounting for the carcinogenic potential of asbestos is also found in epidemiological studies (Loomis et al., 2010; Hamra et al., 2014). The chemical stability and biopersistence of fibres in biological medium is another issue, as it may modulate both the fibre properties and duration of retention in the lung.

20.4.2 Mechanism of Action

Toxicological studies have demonstrated that asbestos fibres induce several responses in mammalian cells. Although the mechanisms of fibrogenesis and carcinogenesis by asbestos are not fully elucidated, the importance of fibre dimensions is demonstrated. A number of in vitro and in vivo studies have demonstrated that long and thin fibres are more toxic than short fibres (Sanchez et al., 2009). The definitions of 'short' and 'long' are a matter of discussion. To make a long story short, a cutoff at 5 μm has been chosen for technical and regulatory reasons related to the measurement of asbestos in air samples, but not for toxicological reasons (Boulanger et al., 2014). However, it is clear from animal and cell studies that fibres less than 5 μm of length are less fibrogenic and carcinogenic for the lung than longer fibres (Boulanger et al., 2014). This does not imply an absence of reactivity of nanoparticles that may induce damage in areas of the body such as the lungs or cardiovascular system (Donaldson and Poland, 2012; Donaldson et al., 2013).

The main toxicological features are reported in different studies (eg, Donaldson and Poland, 2012; Donaldson et al., 2013; Jean et al., 2014; Kane et al., 2014), so they are only briefly summarised here. Asbestos fibres induce the production of ROS and RNS, as well as of inflammatory factors in asbestos-exposed cells. Inflammation is also observed in the lung after exposure of rodents by inhalation. A genotoxic effect has also been demonstrated, both in vitro and in vivo. Genotoxicity results from the DNA-damaging effect of ROS, and from the formation of abnormal mitoses likely resulting from the physical interaction between fibres and mitotic chromosomes or penetration of the nucleus during interphase. It must be noted that both epithelial and mesothelial cells, not only macrophages, have the capability to internalise asbestos fibres (Jaurand et al., 1979). Several in vitro studies have illustrated the mitotic impairment in asbestos-treated cells by the morphological observation of abnormal chromosome segregation and binucleated/multinucleated cells in asbestos-treated (Cortez and Machadosantelli, 2008; Cortez et al., 2011; Huang et al., 2011). Genotoxicity also can be mediated by mitochondria-derived reactive species (Huang et al., 2012). Another parameter that may modulate the fibre toxicity is their persistence for a long duration of time in the tissues, so-called biopersistence. This makes the fibres able to exert sustained and prolonged effects on the affected tissues (Kane et al., 2014).

20.4.3 Comparison of Halloysite, Imogolite and Related Nanotube Characteristics

The different physicochemical characteristics of halloysite and imogolite will be found in chapters 4 and 9 in this book. Halloysite and imogolite are both short nanotubular clay minerals with a high-aspect ratio. From a simple viewpoint, both their dimensions in the nanometre range and elongated shapes, their high-aspect ratio and, to a lesser extent, their structure (hollow cylinders) justify their inclusion in the group of HARN. These particles appear to have different ranges of dimensions, with halloysite thicker and longer than imogolite (Table 20.3). From a theoretical point of view, the small length of individual nanotubes might limit some forms of toxicity. In the experiments reported here, they differ by the treatment they receive; halloysite are used as natural particles that undergo numerous treatments as a response to the objectives of the applications. In contrast, imogolite were synthesised and not significantly modified. Moreover, their functionalisation and state of aggregation modify their properties, needing to further determine the structure and effects relationships. The limited toxicological data available do not allow for reaching clear-cut conclusions on their pathogenicity. This is particularly evident considering the type of studies carried out with halloysite and imogolite.

TABLE 20.3 Dimensions of Halloysite and Imogolite Used in Some Studies

	Outer diameter (nm)	Inner diameter or [wall thickness] (nm)	Length (nm)
Halloysite			
Lai et al. (2013)	50–100	10–20	200–2000
Massaro et al. (2015)	50	15	—
Riela et al. (2014)	50	15	—
Ge-imogolite			
PR imogolite (Liu et al., 2012b)	—	[0.5]	5
ST imogolite[a] (Liu et al., 2012b)	4.3	[0.6]	6±4
LT imogolite[a] (Liu et al., 2012b)	4.3	[0.6]	51±40
SW imogolite[b] (van den Brule et al., 2014)	3.8	—	70
DW imogolite[b] (van den Brule et al., 2014)	4.3	—	62
Crocidolite (van den Brule et al., 2014)	330	—	2500

[a]*Number of particles/g:* 3.5×10^{18} *(ST);* 9.8×10^{17} *(LT).*
[b]*Number of particles/g:* 6.2×10^{17} *(SW);* 4.05×10^{17} *(DW).*

20.5 CONCLUDING REMARKS AND PERSPECTIVES

In this chapter, an assessment of the studies related to the biological effects of halloysite and imogolite was summarised. Most studies found in the literature concerned halloysite, but a few were related to imogolite. Different toxicological approaches were developed with halloysite and imogolite because of the different goals of the studies. Halloysite were employed with the objectives to cure a disease, repair tissues or prevent injury, whereas imogolite were considered as potential toxicants and tested accordingly. Consequently, a variety of halloysite samples were investigated, adding a complexity level to defining the effects of the native particles. Despite the small number of toxicological studies, the results suggest different reactivities between halloysite and imogolite in the biological environment. When tested unmodified, halloysite does not seem to exert significant cytotoxic effects but seems capable of delivering toxic molecules when needed. The effects resulting from other uses remain unknown. In contrast, Ge-imogolite appears to show some toxicity, causing inflammation and DNA damage. However, with imogolite, this is based on a few in vitro studies and one in vivo study carried out according to the principles of toxicological assays. It has so far not been possible to identify the parameters of the nanotubular clay minerals that play a role in the effect observed.

This review emphasises a number of debatable points concerning the state of the nanotubes, the goals of the experiments and the resulting acquired knowledge relative to their toxic potential for humans. It also suggests developing studies to better define the possible issues related to the uses of these nanomaterials. In the future, additional research should be done to clarify the potential risk of exposure to nanotubular clay minerals by both producers and consumers. One important basic issue is the definition of the perimeter of the uses of nanosized tubular clay minerals, as well as the populations potentially exposed, both in the occupational and general environments. This knowledge will allow the definition of possible routes of exposure. So far, toxicity by inhalation is the major health concern. However the possibility of exposure via other routes should be considered when industrial or commercial applications of nanotubular clay minerals concern other routes of exposure. Consequently, toxicokinetics and particle distribution in the body should be investigated. Halloysite and imogolite share similarities but also show varying characteristics and uses. When associated with other HARN, especially CNT and asbestos, this world is very diverse in its nature, uses and production. This suggests that specific approaches should be designed to investigate health effects of a given product in the context of its uses. A flowchart considering first the type of exposure, then dose-dependent possibilities of interactions with cells and tissues, could help experts to design relevant tests and pertinent end points.

REFERENCES

Aggarwal, B.B., Sundaram, C., Malani, N., Ichikawa, H., 2007. Curcumin: the Indian solid gold. Adv. Exp. Med. Biol. 595, 1–75.

Ali-Boucetta, H., Kostarelos, K., 2013. Pharmacology of carbon nanotubes: toxicokinetics, excretion and tissue accumulation. Adv. Drug Deliv. Rev. 65, 2111–2119.

Baron, P.A., 1996. Application of the thoracic sampling definition to fiber measurement. Am. Ind. Hyg. Assoc. J. 57, 820–824.

Bello, D., Wardle, B.L., Zhang, J., Yamamoto, N., Santeufemio, C., Hallock, M., Virji, M.A., 2010. Characterization of exposures to nanoscale particles and fibers during solid core drilling of hybrid carbon nanotube advanced composites. Int. J. Occup. Environ. Health 16, 434–450.

Bottino, M.C., Yassen, G.H., Platt, J.A., Labban, N., Windsor, L.J., Spolnik, K.J., Bressiani, A.H., 2013. A novel three-dimensional scaffold for regenerative endodontics: materials and biological characterizations. J. Tissue Eng. Regen. Med. 9, E116–E123.

Boulanger, G., Andujar, P., Pairon, J.C., Billon-Galland, M.A., Dion, C., Dumortier, P., Brochard, P., Sobaszek, A., Bartsch, P., Paris, C., Jaurand, M.C., 2014. Quantification of short and long asbestos fibers to assess asbestos exposure: a review of fiber size toxicity. Environ. Health 13, 59.

Broaddus, V.C., Everitt, J.I., Black, B., Kane, A.B., 2011. Non-neoplastic and neoplastic pleural endpoints following fiber exposure. J. Toxicol. Environ. Health B Crit. Rev. 14, 153–178.

Carbone, M., Ly, B.H., Dodson, R.F., Pagano, I., Morris, P.T., Dogan, U.A., Gazdar, A.F., Pass, H.I., Yang, H., 2012. Malignant mesothelioma: facts, myths, and hypotheses. J. Cell. Physiol. 227, 44–58.

Cena, L.G., Peters, T.M., 2011. Characterization and control of airborne particles emitted during production of epoxy/carbon nanotube nanocomposites. J. Occup. Environ. Hyg. 8, 86–92.

Chen, Q.Z., Liang, S.L., Wang, J., Simon, G.P., 2011. Manipulation of mechanical compliance of elastomeric PGS by incorporation of halloysite nanotubes for soft tissue engineering applications. J. Mech. Behav. Biomed. Mater. 4, 1805–1818.

Cortez, B.A., Machadosantelli, G.M., 2008. Chrysotile effects on human lung cell carcinoma in culture: 3-D reconstruction and DNA quantification by image analysis. BMC Cancer 8, 181.

Cortez, B.A., Quassollo, G., Caceres, A., Machado-Santelli, G.M., 2011. The fate of chrysotile-induced multipolar mitosis and aneuploid population in cultured lung cancer cells. PLoS One 6, e18600.

Donaldson, K., Poland, C.A., 2012. Inhaled nanoparticles and lung cancer—what we can learn from conventional particle toxicology. Swiss Med. Wkly. 142, w13547.

Donaldson, K., Poland, C.A., Murphy, F.A., Macfarlane, M., Chernova, T., Schinwald, A., 2013. Pulmonary toxicity of carbon nanotubes and asbestos—similarities and differences. Adv. Drug Deliv. Rev. 65, 2078–2086.

Dzamukova, M.R., Naumenko, E.A., Lvov, Y.M., Fakhrullin, R.F., 2015. Enzyme-activated intracellular drug delivery with tubule clay nanoformulation. Sci. Rep. 5, 10560.

Feitosa, S.A., Palasuk, J., Kamocki, K., Geraldeli, S., Gregory, R.L., Platt, J.A., Windsor, L.J., Bottino, M.C., 2014. Doxycycline-encapsulated nanotube-modified dentin adhesives. J. Dent. Res. 93, 1270–1276.

Grosse, Y., Loomis, D., Guyton, K.Z., Lauby-Secretan, B., El Ghissassi, F., Bouvard, V., Benbrahim-Tallaa, L., Guha, N., Scoccianti, C., Mattock, H., Straif, K., 2014. Carcinogenicity of fluoro-edenite, silicon carbide fibres and whiskers, and carbon nanotubes. Lancet Oncol. 15, 1427–1428.

Hamade, A.K., Long, C.M., Valberg, P.A., 2015. Naturally occuring mineral fibers. In: Harbison, R.D. (Ed.), Hamilton & Hardy's Industrial Toxicology. John Wiley & Sons, Hoboken, New Jersey.

Hamra, G.B., Loomis, D., Dement, J., 2014. Examining the association of lung cancer and highly correlated fibre size-specific asbestos exposures with a hierarchical Bayesian model. Occup. Environ. Med. 71, 353–357.

Heyder, J., 2004. Deposition of inhaled particles in the human respiratory tract and consequences for regional targeting in respiratory drug delivery. Proc. Am. Thorac. Soc. 1, 315–320.

Huang, S.X., Jaurand, M.C., Kamp, D.W., Whysner, J., Hei, T.K., 2011. Role of mutagenicity in asbestos fiber-induced carcinogenicity and other diseases. J. Toxicol. Environ. Health B Crit. Rev. 14, 179–245.

Huang, S.X., Partridge, M.A., Ghandhi, S.A., Davidson, M.M., Amundson, S.A., Hei, T.K., 2012. Mitochondria-derived reactive intermediate species mediate asbestos-induced genotoxicity and oxidative stress-responsive signaling pathways. Environ. Health Perspect. 120, 840–847.

IARC, 2002. Man-made vitreous fibres. IARC Monogr. Eval. Carcinog. Risks Hum. 81, 1–381.

Ishikawa, K., Akasaka, T., Yawaka, Y., Watari, F., 2010. High functional expression of osteoblasts on imogolite, aluminosilicate nanotubes. J. Biomed. Nanotechnol. 6, 59–65.

Jaurand, M.C., 1989. Particulate-state carcinogenesis. A survey of recent studies on the mechanism of action of fibres. In: Bignon, J., Peto, J., Saracci, R. (Eds.), Non-Occupational Exposure to Mineral Fibres, 90. IARC Scientific Publication, Lyon, pp. 54–73.

Jaurand, M.C., Kaplan, H., Thiollet, J., Pinchon, M.C., Bernaudin, J.F., Bignon, J., 1979. Phagocytosis of chrysotile fibers by pleural mesothelial cells in culture. Am. J. Pathol. 94, 529–538.

Jaurand, M.C., Renier, A., Daubriac, J., 2009. Mesothelioma: do asbestos and carbon nanotubes pose the same health risk? Part. Fibre Toxicol. 6, 16.

Jean, D., Le Pimpec-Barthes, F., Andujar, P., Gregoire, M., Pairon, J., Scherpereel, A., Galateau-Sallé, F., Jaurand, M.C., 2014. Thoracic neoplasia–mesothelioma. In: McManus, L.M., Mitchell, R.N. (Eds.), Pathobiology of Human Disease: A Dynamic Encyclopedia of Disease Mechanisms. Elsevier, Amsterdam, Netherlands; Oxford, UK and Waltham, MA USA.

Johnson, D.R., Methner, M.M., Kennedy, A.J., Steevens, J.A., 2010. Potential for occupational exposure to engineered carbon-based nanomaterials in environmental laboratory studies. Environ. Health Perspect. 118, 49–54.

Kane, A., Jean, D., Knuutila, S., Jaurand, M.C., 2014. Malignant mesothelioma: mechanism of carcinogenesis. In: Anttila, S. (Ed.), Occupational Cancers. Springer-Verlag, London.

Lai, X., Agarwal, M., Lvov, Y.M., Pachpande, C., Varahramyan, K., Witzmann, F.A., 2013. Proteomic profiling of halloysite clay nanotube exposure in intestinal cell co-culture. J. Appl. Toxicol. 33, 1316–1329.

Lanone, S., Andujar, P., Kermanizadeh, A., Boczkowski, J., 2013. Determinants of carbon nanotube toxicity. Adv. Drug Deliv. Rev. 65, 2063–2069.

Lecouvet, B., 2015. Natural mineral nanotubes as nanoreactors used in industrial and agricultural applications. In: Pasbakhsh, P., Churchman, G.J. (Eds.), Natural Mineral Nanotubes. Properties and Applications. Apple Academic Press Inc., Oakville, ON, Canada; Wareton, NJ, USA.

Lee, Y., Jung, G.E., Cho, S.J., Geckeler, K.E., Fuchs, H., 2013. Cellular interactions of doxorubicin-loaded DNA-modified halloysite nanotubes. Nanoscale 5, 8577–8585.

Lee, J.S., Choi, Y.C., Shin, J.H., Lee, J.H., Lee, Y., Park, S.Y., Baek, J.E., Park, J.D., Ahn, K., Yu, I.J., 2015. Health surveillance study of workers who manufacture multi-walled carbon nanotubes. Nanotoxicology 9, 802–811.

Liu, M., Zhang, Y., Wu, C., Xiong, S., Zhou, C., 2012a. Chitosan/halloysite nanotubes bionanocomposites: structure, mechanical properties and biocompatibility. Int. J. Biol. Macromol. 51, 566–575.

Liu, W., Chaurand, P., Di Giorgio, C., De Meo, M., Thill, A., Auffan, M., Masion, A., Borschneck, D., Chaspoul, F., Gallice, P., Botta, A., Bottero, J.Y., Rose, J., 2012b. Influence of the length of imogolite-like nanotubes on their cytotoxicity and genotoxicity toward human dermal cells. Chem. Res. Toxicol. 25, 2513–2522.

Loomis, D., Dement, J., Richardson, D., Wolf, S., 2010. Asbestos fibre dimensions and lung cancer mortality among workers exposed to chrysotile. Occup. Environ. Med. 67, 580–584.

Luo, E., Song, G., Li, Y., Shi, P., Hu, J., Lin, Y., 2013. The toxicity and pharmacokinetics of carbon nanotubes as an effective drug carrier. Curr. Drug Metab. 14, 879–890.

Lvov, Y., Aerov, A., Fakhrullin, R., 2014. Clay nanotube encapsulation for functional biocomposites. Adv. Colloid Interface Sci. 207, 189–198.

Massaro, M., Colletti, C.G., Noto, R., Riela, S., Poma, P., Guernelli, S., Parisi, F., Milioto, S., Lazzara, G., 2015. Pharmaceutical properties of supramolecular assembly of co-loaded cardanol/triazole-halloysite systems. Int. J. Pharm. 478, 476–485.

Maynard, A., Kuempel, E., 2005. Airborne nanostructured particles and occupational health. J. Nanopart. Res. 7, 587–614.

Maynard, A.D., Baron, P.A., Foley, M., Shvedova, A.A., Kisin, E.R., Castranova, V., 2004. Exposure to carbon nanotube material: aerosol release during the handling of unrefined single-walled carbon nanotube material. J. Toxicol. Environ. Health 67, 87–107.

Mbawuike, I.N., Herscowitz, H.B., 1989. MH-S, a murine alveolar macrophage cell line: morphological, cytochemical, and functional characteristics. J. Leukoc. Biol. 46, 119–127.

Meeker, G.P., Bern, A.M., Brownfield, I.K., Lowers, H.A., Sutley, S.J., Hoefen, T.M., Vance, J.S., 2003. The composition and morphology of amphiboles from the Rainy Creek Complex, near Libby, Montana. Am. Mineral. 88, 1955–1969.

Oberdörster, G., Sharp, Z., Atudorei, V., Elder, A., Gelein, R., Kreyling, W., Cox, C., 2004. Translocation of inhaled ultrafine particles to the brain. Inhal. Toxicol. 16, 437–445.

Oberdorster, G., Elder, A., Rinderknecht, A., 2009. Nanoparticles and the brain: cause for concern? J. Nanosci. Nanotechnol. 9, 4996–5007.

Qi, R., Cao, X., Shen, M., Guo, R., Yu, J., Shi, X., 2012. Biocompatibility of electrospun halloysite nanotube-doped poly(lactic-co-glycolic acid) composite nanofibers. J. Biomater. Sci. Polym. Ed. 23, 299–313.

Rabiskova, M., 2012. Halloysite—interesting nanotubular carrier for drugs. Ceska Slov. Farm. 61, 255–260.

Raffa, V., Ciofani, G., Vittorio, O., Riggio, C., Cuschieri, A., 2010. Physicochemical properties affecting cellular uptake of carbon nanotubes. Nanomedicine (Lond.) 5, 89–97.

Riela, S., Massaro, M., Colletti, C.G., Bommarito, A., Giordano, C., Milioto, S., Noto, R., Poma, P., Lazzara, G., 2014. Development and characterization of co-loaded curcumin/triazole-halloysite systems and evaluation of their potential anticancer activity. Int. J. Pharm. 475, 613–623.

Rotoli, B.M., Guidi, P., Bonelli, B., Bernardeschi, M., Bianchi, M.G., Esposito, S., Frenzilli, G., Lucchesi, P., Nigro, M., Scarcelli, V., Tomatis, M., Zanello, P.P., Fubini, B., Bussolati, O., Bergamaschi, E., 2014. Imogolite: an aluminosilicate nanotube endowed with low cytotoxicity and genotoxicity. Chem. Res. Toxicol. 27, 1142–1154.

Sanchez, V.C., Pietruska, J.R., Miselis, N.R., Hurt, R.H., Kane, A.B., 2009. Biopersistence and potential adverse health impacts of fibrous nanomaterials: what have we learned from asbestos? Wiley Interdiscip. Rev. Nanomed. Nanobiotechnol. 1, 511–529.

Service RF, 2003. American Chemical Society meeting. Nanomaterials show signs of toxicity. Science 300, 243.

Shutava, T.G., Fakhrullin, R.F., Lvov, Y.M., 2014. Spherical and tubule nanocarriers for sustained drug release. Curr. Opin. Pharmacol. 18, 141–148.

Shvedova, A.A., Kisin, E.R., Porter, D., Schulte, P., Kagan, V.E., Fadeel, B., Castranova, V., 2009. Mechanisms of pulmonary toxicity and medical applications of carbon nanotubes: two faces of Janus? Pharmacol. Ther. 121, 192–204.

Straif, K., Benbrahim-Tallaa, L., Baan, R., Grosse, Y., Secretan, B., El Ghissassi, F., Bouvard, V., Guha, N., Freeman, C., Galichet, L., Cogliano, V., 2009. A review of human carcinogens—part C: metals, arsenic, dusts, and fibres. Lancet Oncol. 10, 453–454.

Sun, L., Mills, D.K., 2014. Halloysite nanotube-based drug delivery system for treating osteosarcoma. Conf. Proc. IEEE Eng. Med. Biol. Soc. 2014, 2920–2923.

Suzuki, Y., Yuen, S.R., Ashley, R., 2005. Short, thin asbestos fibers contribute to the development of human malignant mesothelioma: pathological evidence. Int. J. Hyg. Environ. Health 208, 201–210.

Taylor, A.A., Aron, G.M., Beall, G.W., Dharmasiri, N., Zhang, Y., McLean, R.J., 2014. Carbon and clay nanoparticles induce minimal stress responses in gram negative bacteria and eukaryotic fish cells. Environ. Toxicol. 29, 961–968.

Tran, C.L., Hankin, S.M., Ross, B., Aitken, R.J., Jones, A.D., Donaldson, K., Stone, V., Trantra, R., 2008. An Outline Scoping Study to Determine Whether High Aspect Ratio Nanoparticles (HARN) Should Raise the Same Concerns as do Asbestos Fibres: Report on Project CB0406. IOM, Research Consulting Services, Edinburgh, UK. http://www.iom-world.org.

Tsai, S.J., Hofmann, M., Hallock, M., Ada, E., Kong, J., Ellenbecker, M., 2009. Characterization and evaluation of nanoparticle release during the synthesis of single-walled and multiwalled carbon nanotubes by chemical vapor deposition. Environ. Sci. Technol. 43, 6017–6023.

Vahedi, V., Pasbakhsh, P., 2015a. Applications of natural mineral nanotubes in nanocomposites. In: Pasbakhsh, P., Jock Churchman, G. (Eds.), Natural Mineral Nanotubes. Properties and Applications. Apple Academic Press Inc., Oakville, ON, Canada; Wareton, NJ, USA.

Vahedi, V., Pasbakhsh, P., 2015b. Functionalization and compatibilization of the halloysite nanotubes. In: Pasbakhsh, P., Jock Churchamn, G. (Eds.), Natural Mineral Nanotubes. Properties and Applications. Apple Academic Press Inc., Oakville, ON, Canada; Wareton, NJ, USA.

van den Brule, S., Beckers, E., Chaurand, P., Liu, W., Ibouraadaten, S., Palmai-Pallag, M., Uwambayinema, F., Yakoub, Y., Avellan, A., Levard, C., Haufroid, V., Marbaix, E., Thill, A., Lison, D., Rose, J., 2014. Nanometer-long Ge-imogolite nanotubes cause sustained lung inflammation and fibrosis in rats. Part. Fibre Toxicol. 11, 67.

Vergaro, V., Abdullayev, E., Lvov, Y.M., Zeitoun, A., Cingolani, R., Rinaldi, R., Leporatti, S., 2010. Cytocompatibility and uptake of halloysite clay nanotubes. Biomacromolecules 11, 820–826.

Vergaro, V., Lvov, Y.M., Leporatti, S., 2012. Halloysite clay nanotubes for resveratrol delivery to cancer cells. Macromol. Biosci. 12, 1265–1271.

Wagner, J.C., Sleggs, C.A., Marchand, P., 1960. Diffuse pleural mesothelioma and asbestos exposure in the North Western Cape Province. Br. J. Ind. Med. 17, 260–271.

Wang, Q., Zhang, J., Mu, B., Fan, L., Wang, A., 2014a. Facile preparation of magnetic 2-hydroxypropyltrimethyl ammonium chloride chitosan/Fe3O4/halloysite nanotubes microspheres for the controlled release of ofloxacin. Carbohydr. Polym. 102, 877–883.

Wang, W., Zhu, Y., Liao, S., Li, J., 2014b. Carbon nanotubes reinforced composites for biomedical applications. Biomed. Res. Int. 2014, 518609.

Wu, H., Shi, Y., Huang, C., Zhang, Y., Wu, J., Shen, H., Jia, N., 2014. Multifunctional nanocarrier based on clay nanotubes for efficient intracellular siRNA delivery and gene silencing. J. Biomater. Appl. 28, 1180–1189.

Xue, J., Niu, Y., Gong, M., Shi, R., Chen, D., Zhang, L., Lvov, Y., 2015. Electrospun microfiber membranes embedded with drug-loaded clay nanotubes for sustained antimicrobial protection. ACS Nano 9, 1600–1612.

Yin, S., Meng, Q., Zhang, B., Shi, B., Shan, A., Li, Z., 2015. Alleviation of zearalenone toxicity by modified halloysite nanotubes in the immune response of swine. Food Addit. Contam. Part A Chem. Anal. Control Expo. Risk Assess. 32, 87–99.

Zanzottera, C., Vicente, A., Celasco, E., Fernandez, C., Garrone, E., Barbara, B., 2012. Physicochemical properties of imogolite nanotubes functionalized on both external and internal surfaces. J. Phys. Chem. C 116, 7499–7560.

Zhang, Y., Gao, R., Liu, M., Yan, C., Shan, A., 2014. Adsorption of modified halloysite nanotubes in vitro and the protective effect in rats exposed to zearalenone. Arch. Anim. Nutr. 68, 320–335.

Chapter 21

Halloysite Polymer Nanocomposites

J. Huang, Z.H. Tang, X.H. Zhang and B.C. Guo*
South China University of Technology, Guangzhou, China
Corresponding author: e-mail: psbcguo@scut.edu.cn

21.1 INTRODUCTION

Halloysite has sparked considerable interest in the fabrication of halloysite polymer nanocomposites due to its unique characteristics that combine excellent mechanical properties, tubular microstructure, biocompatibility, flame retardancy and a versatile surface chemistry. To date, halloysite has been incorporated into various polymer matrices, including thermoplastics, thermosets and elastomers, to form halloysite polymer nanocomposites. The incorporation of halloysite into polymers gives rise to multiple effects on the pure polymer matrix. The mechanical performance, thermal stability and flame retardancy of clay polymer nanocomposites (CPN) can be improved compared to those of pure polymers. Moreover, halloysite can fine-tune the crystallization of the resulting CPN. For ease of processing, because halloysite is available in large quantities and can be dispersed in water and many organic solvents, halloysite polymer nanocomposites can be produced by various approaches, such as solution mixing, melt blending, in situ polymerization and spinning. The dispersion of halloysite in CPN is primarily determined by the surface characteristics and processing approaches. Generally, solution mixing and in situ polymerization can result in a CPN with uniform dispersion, whereas spinning leads to an orientation of the halloysite in the CPN.

To maximize the potential of halloysite in CPN, two factors (namely, the dispersion of the halloysite and the interfacial interactions between the halloysite and polymer matrices) play vital roles in determining the final performance of the CPN. Overall, halloysite is hydrophilic, but it is more hydrophobic than nanosilica and other layered silicates because of its lower density of surface hydroxyl groups at the outer surface, so the tube–tube interaction of halloysite is relatively weak. Therefore, it is expected that halloysite can be readily dispersed in many polymers even without modification.

Developments in Clay Science, Vol. 7. http://dx.doi.org/10.1016/B978-0-08-100293-3.00021-2

Halloysite can interact with polar polymers such as polyvinyl alcohol (PVA), chitosan and carboxylated rubbers via hydrogen bonding to strengthen the interfacial adhesion. However, halloysite interacts poorly with nonpolar polymers such as the polyolefins. To address this issue, various strategies such as physically wrapping halloysite with surfactants or polar macromolecules as compatibilizers, covalently grafting halloysite with coupling agents or polymer chains and charge transferring between conjugated molecules with halloysite have been explored as ways to enhance interfacial adhesion.

In this chapter, processing methods for preparing halloysite polymer nanocomposites are summarized. Next, the various interactions between halloysite and polymers are elaborated. The multiple effects of halloysite on CPN will then be discussed, with final concluding remarks.

21.2 PROCESSING METHODS FOR THE FABRICATION OF HALLOYSITE POLYMER NANOCOMPOSITES

Processing methods have significant effects on the dispersion status of halloysite nanotubes in the resulting halloysite polymer nanocomposites. Due to relatively weak internanotube interactions, halloysite can be better dispersed in polymers than carbon nanotubes or nanosilica. Various methods, such as solution mixing, melt mixing, in situ polymerization and electrospinning, can be used to fabricate halloysite polymer nanocomposites. However, the fabrication methods should be selected in accordance with the intrinsic properties of the matrix and the target application of the CPN. Because halloysite is generally hydrophilic, for water-soluble polymers, solution mixing in water is the first choice. For thermoplastics and rubbers, the solution method is not favourable because it uses organic solvents. For those applications, melt mixing is highly desired due to its convenient processing and compatibility with available industrial facilities. In situ polymerization is a better choice for achieving well-defined structures and strong interfacial bonds between halloysite and polymers. Layer-by-layer assembly (LbL) and electrophoretic deposition (EPD) are useful for fabricating multilayered functional films. Melt spinning or electrospinning are usually employed to fabricate fibre CPN. In the fabrication of rubber-based halloysite composites, latex co-coagulation is a promising method due to the environmentally friendly nature of the process and the availability of latex products for most rubbers. The pros and cons of these processing methods are examined in detail next.

21.2.1 Solution Mixing

As already stated, halloysite is generally hydrophilic and can easily be dispersed in water with vigorous stirring or by ultrasound. Therefore, the CPN can be prepared via directly mixed halloysite with water-soluble polymers in aqueous solutions. This process is simple and environmentally friendly.

Halloysite has also been identified as a nontoxic and biocompatible material (Kommireddy et al., 2005; Vergaro et al., 2010b), and most of the bionanocomposite films that are based on halloysite and polymers, such as PVA (Liu et al., 2007a; Zhou et al., 2010; Fujii et al., 2014), polyethylene glycol (PEG) (Cavallaro et al., 2012), polylactic acid (PLA) (De Silva et al., 2013a), chitosan (Liu et al., 2012c, 2013b; De Silva et al., 2013b), alginate (Liu et al., 2015) and starch (He et al., 2012), are fabricated by this method. Generally, it is not necessary to modify the surface of halloysite due to the strong interaction between the halloysite and the water-soluble polymer (as in hydrogen bonding, for example). However, the interaction can be enhanced via treating the halloysite with either alkalis (Vahedi and Pasbakhsh, 2014; Zeng et al., 2014) or acids (Fujii et al., 2014), by increasing the amount of hydroxyls on the surface of the halloysite or by increasing the specific surface area (SSA) of the halloysite. Hydroxylated halloysite (h-Hal) exhibits less transmittance and better colloidal stability after equilibrium than purified halloysite (p-Hal) (Fig. 21.1). Moreover, it has been reported that the addition of a small amount of acid in the halloysite chitosan aqueous solution could result in an electrostatic attraction between them, which further enhances their interactions (Liu et al., 2012c).

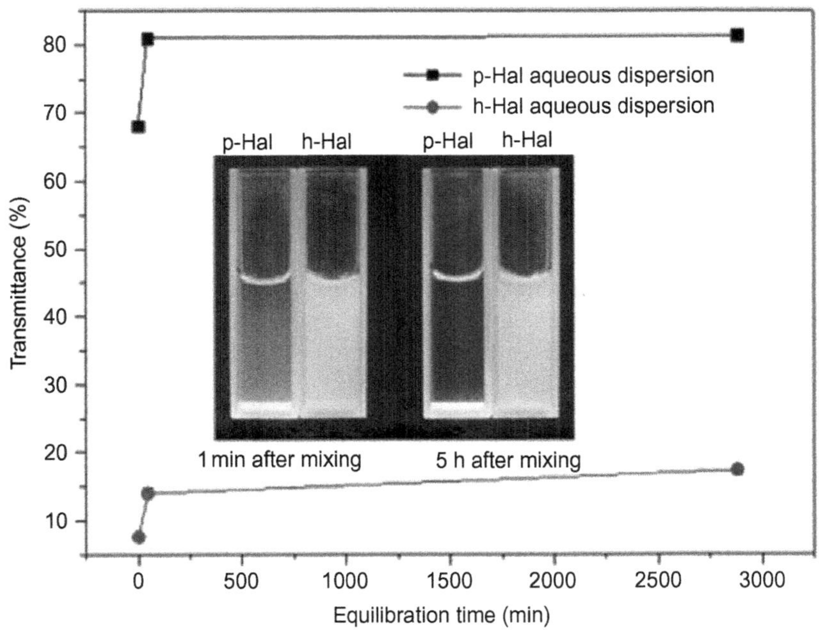

FIG. 21.1 Transmittance vs time for p-Hal and h-Hal aqueous dispersion (1.0 mg mL^{-1}). The *inset* shows digital photographs of the samples at different stages in the testing. *Reprinted with permission from Zeng et al. (2014). Copyright 2014 Elsevier Ltd.*

The polarity of halloysite enables it to be dispersed readily in some polar solvents; thus, solution mixing can also be carried out in organic solutions. For example, halloysite and ethylene-vinyl acetate have been dispersed in tetrahydrofuran to prepare the CPN (Bidsorkhi et al., 2015). When the mixing temperature is high enough, the nonpolar solvents can also be used as a mixing medium. For example, a halloysite polypropylene (PP) nanocomposite was prepared in xylene at 120°C by stirring. Specifically, halloysite epoxy resin nanocomposites are usually prepared by first dispersing halloysite directly in epoxy resin and then pouring the mixture with a hardener into moulds to cure. To obtain good dispersion state in the halloysite epoxy mixture, a surface treatment with various silanes is generally essential (Liu et al., 2007b, 2008b). In addition, it is worth noting that ionic liquids, a new class of 'green' solvent, have been used in the fabrication of halloysite cellulose nanocomposites, which provides a new method for solution-mixing processing (Hanid et al., 2014).

Generally, postprocessing is necessary after homogeneous solution mixing. Casting or coagulation is the most common process to obtain the final CPN. However, halloysite has been observed to aggregate during the drying process of the as-cast film, whereas aggregation was suppressed in the coagulation method due to the fast coalescence of the particles (Liu et al., 2007a). In addition, halloysite alginate gel beads, which can be applied in the removal of toxic dyes, can be obtained using calcium ions as the flocculent (Liu et al., 2012a; Cavallaro et al., 2013). Freeze drying can be used to prepare porous halloysite polymer nanocomposite scaffolds for tissue engineering (Liu et al., 2013b, 2015), and electrospinning can be used to fabricate nanofibres, which will be detailed further in Section 21.2.4 later in this chapter.

21.2.2 Melt Mixing

Although solution mixing is simple and convenient, it is largely limited to water-soluble polymers, and the removal of solvent is usually time consuming and environmentally unfriendly when organic solvents are used. In view of these disadvantages, melt processing is a good alternative that is particularly useful for thermoplastics and rubber-based nanocomposites. Furthermore, melt processing is a more favourable approach for producing halloysite polymer nanocomposites industrially because it is always easy to integrate into standard industrial facilities. Various moulding processes can be utilized, including extrusion, injection moulding, compression moulding and microinjection compression moulding (Wang et al., 2015). However, high temperatures, as well as the strong shearing force, may induce an unexpected polymer degradation and oxidation that may damage the polymeric properties. As a result, small amounts of antioxidants and thermal stabilizers are usually added during processing. Furthermore, the anisotropy of the tubular structure will be oriented under the strong shearing force that would also largely and positively affect the performance of the CPN, especially for crystalline polymers.

During the melt-mixing process, the wetting of the halloysite by the polymer chains is determined by interfacial compatibility. The surface polarity of halloysite allows its easy and favourable dispersion in polar polymer matrices, but halloysite needs to be surface-modified using silane or polymer grafts to improve its dispersibility in nonpolar polymers (Du et al., 2006; Jia et al., 2009). Certainly, further improvement in the interaction between halloysites and polar polymers has been achieved when the halloysites were modified (Guo et al., 2009a). For clay rubber nanocomposites in particular, if the CPN is cured by sulphur, then the addition of an organosilane is the most common and effective approach for improving interfacial compatibility (Pasbakhsh et al., 2010; Rooj et al., 2010; Raman et al., 2013). When the CPN is cured by peroxide, unsaturated carboxylic acids such as acrylic acid (Guo et al., 2008) and sorbic acid (Guo et al., 2009b) can be used as interfacial modifiers. The dispersion of halloysite has been significantly improved by the inclusion of methacrylic acid (MAA) into halloysite styrene butadiene rubber (SBR) nanocomposites (Fig. 21.2).

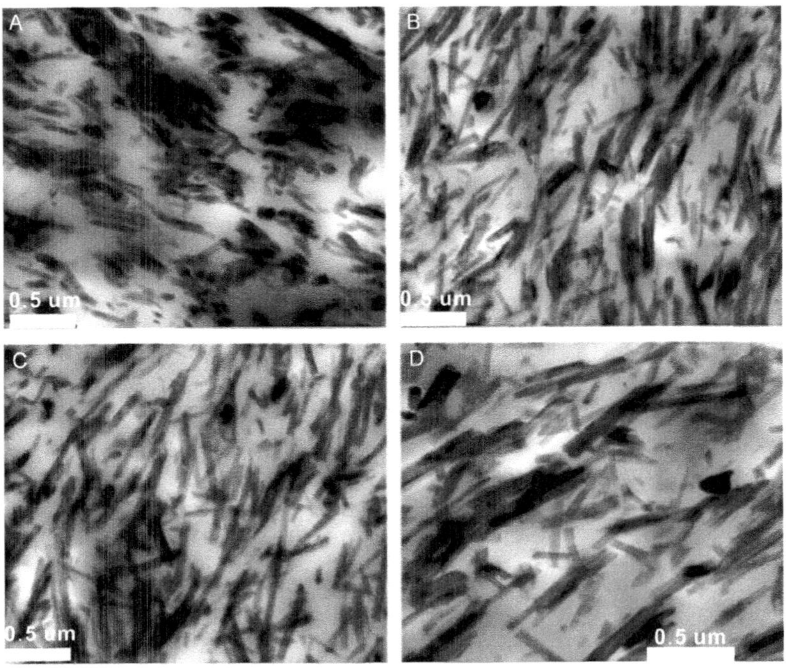

FIG. 21.2 Transmission electron microscope (TEM) images of halloysite SBR nanocomposites with different MAA loading. (A) SBR, (B) SBR2A, (C) SBR7A and (D) SBR15A. *Reprinted with permission from Guo et al. (2008). Copyright 2008 Elsevier Ltd.*

The mixing routes or methods also have an influence on halloysite dispersion. For instance, a two-step master batch dilution process can be used to obtain a more uniform dispersion of halloysite in the polymer matrix than that obtained from direct mixing. The mixing temperature can also affect the mixing homogeneity (Prashantha, 2011; Fernández et al., 2013; Marini et al., 2014). Moreover, the water-assisted process is known to be useful for improving clay mineral dispersibility during extrusion moulding when preparing the CPN, in which water plays multiple roles depending on the nature of the components. First, water acts as a solvent and reduces the melt viscosity of the blend. Second, water molecules can diffuse between the silicate layers, increasing the interlayer distance. Third, water can alter the chemistry of the system during extrusion (Yu et al., 2005; Lecouvet et al., 2014). Green melt blending has been successfully used to prepare halloysite polymer nanocomposites by Lecouvet et al. (2011b, 2013a, 2014). For example, a well-dispersed halloysite PP nanocomposite has been prepared using water-assisted extrusion. Injected water can hydrolyse anhydride functions of maleic anhydride-grafted PP (Plb) into carboxylic diacid functions, enabling the formation of hydrogen bonds between the carboxylic diacid functions of Plb and halloysite surface in the high-compression zone. Then, good dispersion of the halloysite nanotubes was achieved. In the terminal plastification zone, although hydrogen bonds disappeared with water degassing, the filler remained well dispersed in the polymer matrix. The synergistic effect of Plb and water injection on the halloysite dispersion is visible in TEM images (Fig. 21.3; Lecouvet et al., 2011b). Furthermore, another study revealed that

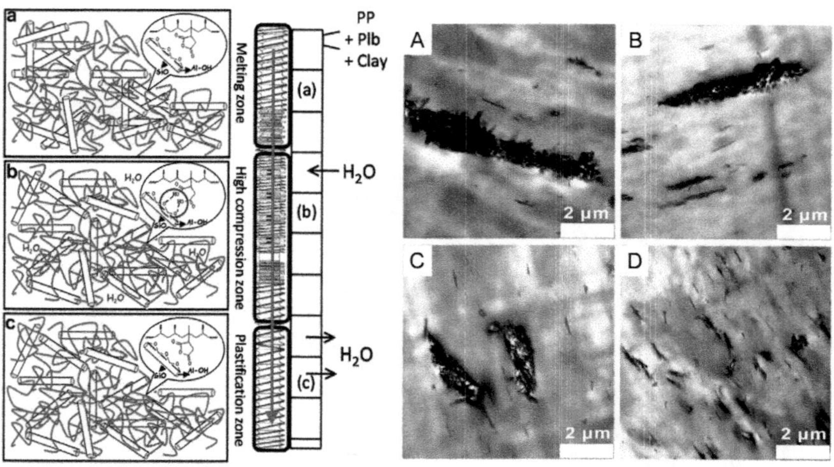

FIG.21.3 Water-assisted extrusion process for Plb-H-W nanocomposites (*left*) and TEM micrographs of (A) PP-H8, (B) PP-H8-W, (C) PP-Plb-H8 and (D) PP-Plb-H8-W nanocomposites at a magnification of 2300 (*right*). *Reprinted with permission from Lecouvet et al. (2011b). Copyright 2011 Elsevier Ltd.*

high-pressure water penetration during injection moulding produced high shearing and cooling rates that strongly oriented the halloysite in the PP. Unfortunately, the orientation of the halloysite did not strongly enhance its nucleating efficiency for PP (Wang and Huang, 2013).

21.2.3 In Situ Polymerization

The in situ polymerization fabrication method begins with the dispersion of halloysite in monomers or a solution of monomers, followed by in situ polymerization of the monomers. Because the monomers interact with the halloysite surface and cause a uniform dispersion, a high level of dispersion of halloysite can be achieved via in situ polymerization. The in situ technique enables covalent bonding between functionalized halloysite and polymer chains, which facilitates an improvement in the compatibility between the filler and polymer matrix. In situ polymerization can be conducted by various reactions, such as free radical polymerization, emulsion polymerization, atom transfer radical polymerization (ATRP) and plasma polymerization.

The oxidative polymerization of aniline or pyrrole in the presence of halloysite has been reported. This method allows the coating of the resulted polymer onto the halloysite nanotubes in a noncovalent fashion. Because raw halloysite exhibits good dispersibility in aniline and pyrrole, CPN prepared by this method show a uniform dispersion of halloysite. Luca and Thomson (2000) reported the intercalation and polymerization of aniline within halloysite via the exposure of copper(II)-doped halloysite film to an aniline vapour. Zhang et al. (2008a) prepared polyaniline (PANI)–coated halloysite by an in situ soapless emulsion polymerization of anilinium chloride adsorbed onto a halloysite surface. The anilinium chloride–adsorbed halloysite was obtained by dispersing halloysite into an acidic aqueous solution of aniline. The conductivity of the halloysite PANI nanocomposites increased with increasing amounts of the hydrochloric acid dopant added during the emulsion polymerization. Murali et al. (2014) modified halloysite by the in situ polymerization of aniline, and asymmetric mixed matrix membranes were synthesized by varying the proportion of the modified halloysite loading of the polysulfone polymer, which exhibited superior gas permeability and separation properties compared to those of a pristine polysulfone membrane under the same conditions.

Coaxial nanotubular halloysite polypyrrole (PPy) nanocomposites were synthesized by the in situ chemical oxidative polymerization method based on the Lewis acid–base type between the basic amine-functionalized halloysite and acidic N–H bonds in pyrrole (Fig. 21.4). The resulting CPN exhibited superior conductivity and a weak temperature dependence of the conductivity, suggesting their potential application in electrode materials for supercapacitors (Yang et al., 2010).

In situ radical polymerization in the presence of halloysite in various situations, such as dispersion, solution and bulk, has been reported. Dong et al.

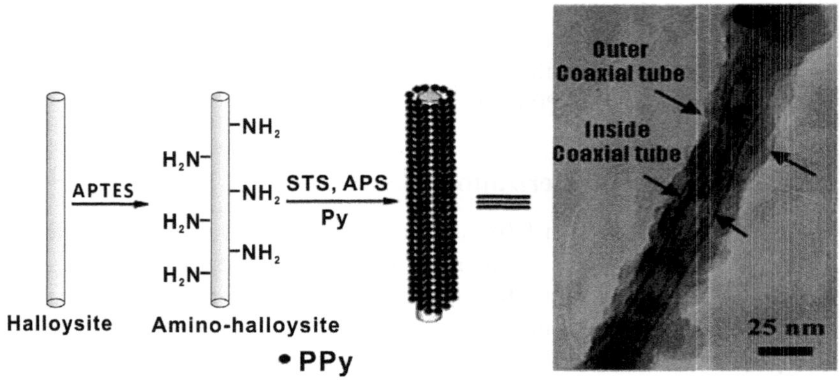

FIG. 21.4 Schematic of the synthesis process of halloysite PPy nanocomposites with a coaxial tubular morphology. *Reprinted with permission from Yang et al. (2010). Copyright 2010 Elsevier Ltd.*

(2012) fabricated halloysite poly(sodium acrylate-*c*-acrylamide) (PAA-*c*-AM) nanocomposites using an inverse dispersing polymerization method. The PAA-*c*-AM copolymer reacted with and grafted onto the hydroxyls of halloysite during the polymerization process. In addition, in situ radical polymerization can be performed in bulk. According to Zhao and Liu (2008), halloysite polystyrene (PS) nanocomposites could be prepared simply by the bulk polymerization of styrene with a silane-modified halloysite. Because the silane (KH-570) modified halloysite bears C=C bonds, PS chains can be covalently coupled onto halloysite surfaces during radical polymerization. Lin et al. (2011) prepared halloysite PS nanocomposites by an emulsion polymerization of styrene in an aqueous emulsion of halloysite and styrene using sodium dodecyl sulphate (SDS) as a surfactant. SDS was an excellent dispersing agent for halloysite in an aqueous solution, which could adsorb onto the halloysite surface and form a bilayer SDS coat to stabilize the halloysite. The bilayer coat of SDS prevented the adsorption of styrene onto the halloysite surface during polymerization, leading to the formation of PS nanospheres separating individual halloysite materials in the CPN. Because of the uniform dispersion of the halloysite, the impact strength of the resulting halloysite PS nanocomposites was 300% higher than that of the neat PS. The halloysite polyacrylamide nanocomposite gels were synthesized through the radical polymerization of acrylamide in the presence of halloysite in an aqueous system. Due to hydrogen bonding interactions between the halloysite and polyacrylamide, the halloysite was uniformly dispersed in the gels and acted as multifunctional cross-linking points for the polyacrylamide to form mechanically strong organic/inorganic hybrid hydrogels (Liu et al., 2012b).

The abovementioned radical polymerization was conducted in an uncontrolled fashion. The preparation of the CPN also can be performed in a controlled polymerization such as ATRP and reversible addition-fragmentation chain transfer (RAFT) polymerization, which permits the formation of well-defined structures in the CPN. ATRP is a promising strategy to graft well-defined polymer chains onto halloysite surfaces to prepare corresponding CPN. Prior to the polymerization reaction, the ATRP initiator is first introduced onto the halloysite surface by the surface modification of the halloysite. Li et al. (2008) reported the synthesis of a halloysite PS nanocomposite via a surface-initiated ATRP onto the halloysite. The halloysite was first subjected to modification with 3-aminopropyltrimethoxysilane, and then the silanized halloysite was further reacted with 2-bromoisobutyryl bromide to immobilize the initiator. Afterwards, the halloysite PS nanocomposite with a core–shell coaxial structure was produced by grafting PS (Fig. 21.5). In particular, the composition and shell thickness could be easily fine-tuned by adjusting the monomer structure and polymerization time (Li et al., 2008). Furthermore, the halloysite PS nanocomposite could serve as a general template for fabricating other CPN by sulphonating the PS layer and compounding with metallic compounds, polymers and inorganic materials (Li et al., 2009).

Moreover, halloysite grafted with 2-methacryloyloxyethyl phosphorylcholine was synthesized via reverse ATRP (Wang et al., 2014). A halloysite nanocomposite with a hyperbranched polymer shell was synthesized by a surface-initiated self-condensing vinyl (co)polymerization of 2-((bromoacetyl)oxy)ethyl acrylate via the ATRP technique. The initiator, 2-bromoisobutyric acid, was immobilized onto the halloysite surface via the electrostatic force between them (Mu et al., 2008).

The amphiphilic brushes of poly(4-vinylpyridine)-block-polystyrene were grafted onto the halloysite via a surface RAFT living polymerization. Before the polymerization, the R group in the RAFT agent *S*-1-dodecyl-*S*′-(*R*, *R*′-dimethyl-*R*″-acetic acid) trithiocarbonates was anchored. To demonstrate the amphiphilicity of the modified halloysite, the Pickering emulsification behaviour of the resulted CPN in a water/soybean oil biphase mixture was studied. The modified halloysite could emulsify the water/soybean oil biphase

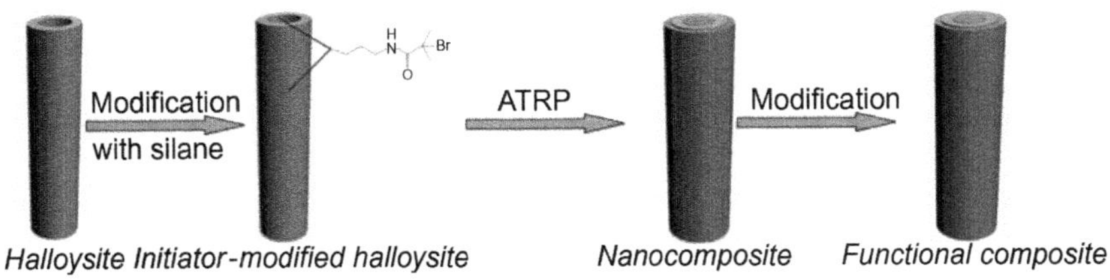

FIG. 21.5 An illustration of the synthesis of halloysite polymer nanocomposite nanotubes via surface-initiated ATRP onto the halloysite. *Reprinted with permission from Li et al. (2008). Copyright 2008 John Wiley & Sons, Inc.*

mixture. The effects of the microstructure of the amphiphilic brushes, such as the hydrophilic/hydrophobic segment size and sequence, on the emulsification performance have been documented (Hou et al., 2014).

Plasma polymerization can be used to coat a polymer onto the surface of halloysite. Using the energy from a plasma generated by an electric discharge, the monomer gases are activated and form electrons, ions and radicals, which are able to polymerize and form a deposit on the surface of the halloysite. The plasma polymerizations of various monomers, such as thiophene, pyrrole and styrene, were used to modify the hydrophilic surfaces of the halloysite. After modification, the dispersion of the halloysite and the interfacial affinity between the halloysite and polymer improved (Shamsi et al., 2010; Poikelispaa et al., 2013).

21.2.4 Electrospinning and Melt Spinning

Electrospinning is an efficient and economical method to prepare three-dimensional fibrous structures that have a high porosity, very small pore size and very large surface-to-volume ratio (Reneker and Chun, 1996; Frenot and Chronakis, 2003). Biodegradable PLA (or its copolymer or blend) and PVA have good electrospinning properties. The incorporation of halloysite does not significantly influence the fibre morphology except for an increase in the fibre diameter. The inclusion of halloysite led to the improved mechanical strength and biocompatibility of the CPN (Liao et al., 2010; Qi et al., 2012; Zhao et al., 2013). It can be seen from SEM images that the uniform nanofibres were maintained even though the content of halloysite was as high as 30%. More details can be seen on a single electrospun halloysite PVA nanocomposite fibre; the halloysite appears to be embedded in the fibre matrix and is oriented along the fibre axis (Fig. 21.6; Liao et al., 2010). This may be due to the elongation

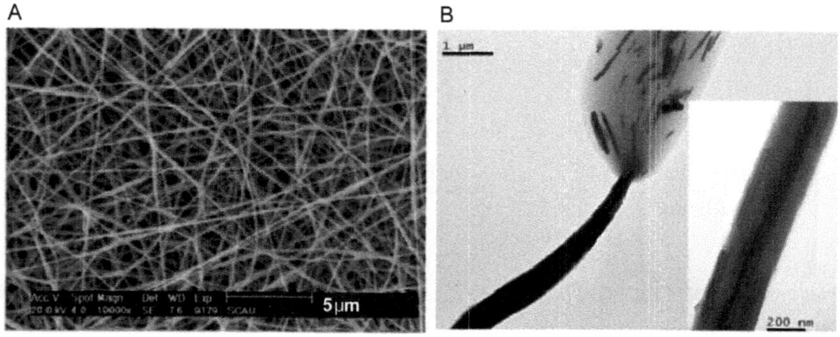

FIG. 21.6 (A) SEM micrograph of the 30% halloysite contents of halloysite PVA nanocomposite fibres. (B) TEM micrograph taken from a single electrospun halloysite PVA nanocomposite fibre.

and shear force that are imposed during the electrospinning process, which breaks the interactions among the nanotubes (Zhao et al., 2013). Moreover, the cross-linking of the fibres did not change the morphology significantly, and the porous characteristics of the mat were retained (Liao et al., 2010).

Electrospinning is often greatly influenced by both material parameters, such as solution viscosity, solution concentration, surface tension and electrical conductivity; and processing factors, including a high voltage setting, feed rate of solution, collector type and working distance. Among all these parameters, solution viscosity is the critical key for determining the fibre morphology, owing to its direct link with the number of entanglements of the polymeric chains (Li and Wang, 2013). Generally, the incorporation of halloysite increases the solution viscosity, resulting in the production of a uniform fibre. Moreover, the fibre diameter usually increased with the addition of halloysite. This may be attributed to two factors: (i) The solution viscosity increases with the halloysite content; and (ii) negatively charged halloysite would result in a decrease of the surface charge density of the spinning jet, leading to a larger fibre diameter (Qi et al., 2012; Haroosh et al., 2013). Moreover, silane-treated halloysite has a better dispersion in the fibre CPN (Haroosh et al., 2013).

Halloysite loaded with a drug can also spin with a polymer. For example, halloysite loaded with tetracycline hydrochloride, a model drug, was mixed with poly(lactic-*co*-glycolic acid) (PLGA) for subsequent electrospinning to form drug-loaded nanocomposite nanofibrous mats. This double-container drug delivery system (both the halloysite and PLGA polymer are drug carriers) beneficially reduces the burst release of the drug, and at the same time, the introduction of halloysite can significantly improve the tensile strength of the polymer nanofibrous mats (Qi et al., 2010).

For thermoplastics like PP, melt spinning is a common means of preparing fibres. Lin (2011) first prepared halloysite PP nanofibres via a master batch method and showed that halloysite loading had a significant influence on the spinnability of the CPN. The maximum loading for the halloysite during spinning with PP was 1.5 mass%. Above that level, no continuous fibres could be obtained. This was mainly due to poor interactions between the halloysite and PP, leading to aggregates in the fibre. Despite the relatively low content of halloysite, the mechanical properties of the halloysite PP nanocomposite fibres were greatly improved. For example, for fibre containing 1 mass% of halloysite, the breaking strength of the fibre increased by 18.6%.

21.2.5 Layer-by-Layer Self-assembly

Multilayered, thin films can be simply and flexibly prepared using an LbL self-assembly technique. This technique is commonly based on electrostatic attraction through which oppositely charged components are alternately deposited. In addition to electrostatic interaction, other weak interactions such as hydrogen bonding have been used to drive the LbL technique (Stockton and Rubner, 1997; Wang et al., 1997). Negatively charged halloysite can be used with LbL with polycations such as poly(ethyleneimine) (PEI), poly(allylaminehydrochloride) (PAH) and polydimethyldiallyl ammonium chloride (PDDA). Much of this work was summarized by Lvov et al. (2008) and Lvov and Abdullayev (2013). For instance, a halloysite PEI assembly thin film was obtained via the LbL technique (Lvov et al., 2002). Those authors showed that the multilayered assembly could not have occurred solely by the sequential immersion of the substrate in the halloysite solution or by the alternate immersion of the substrate in the solutions of both the negative halloysite and sodium poly(styrene sulphonate) (PSS). This result illustrated that an alternate adsorption of the oppositely charged components was necessary for a successful assembly. They also showed that the nanotubes in the layer were not parallel but rather were randomly oriented, and the layer density was rather low, ranging from 0.48 to 0.08 g cm^{-3}. The halloysite polycation layer thickness was found to be 54 nm, which was close to the halloysite diameter. Moreover, other components, including silica and alcohol dehydrogenase, were successfully introduced. Therefore, the assembly of the nanotube/nanosphere superlattices, in addition to an assembly of the alcohol dehydrogenase and halloysite multilayers, were accomplished (Fig. 21.7A). Another interesting work involved the three-dimensional assembly of halloysite nanotubes. Yeast cells were selected as biomimetic microtemplates, then PAH, halloysite, PAH and PSS were deposited, one after the other, on the cell surface. Finally, the cells and polymer multilayer were thermally

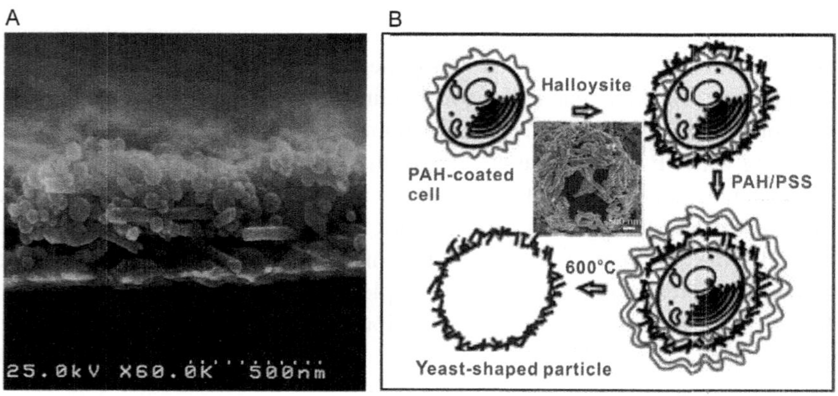

FIG. 21.7 (A) SEM cross-sectional view of a PEI/PSS/PEI+(halloysite/PEI)$_2$+(PEI/silica/PEI/halloysite)$_3$ multilayer. (B) Schematic of the 3D-halloysite fabrication process. The *inset* is an SEM photo. *Panel (A): Reprinted with permission from Lvov et al. (2002). Copyright 2002 Elsevier Ltd. Panel (B): Reprinted with permission from Konnova et al. (2013). Copyright 2013 the Royal Society of Chemistry.*

decomposed, and 3D assembled halloysite formed (Fig. 21.7B). This development has a promising application in cell-controlled growth (Konnova et al., 2013).

Different surfaces can act as the template and will experience different effects. These LbL-assembled nanoparticle films would be utilized to modify the surfaces of implantable materials with the aim of improving their biocompatibility (Kommireddy et al., 2005). Moreover, the structural reinforcement of microvascular networks was realized by employing the LbL technique to deposit films containing halloysite nanotubes onto scaffold surfaces, followed by matrix infiltration and scaffold removal (Olugebefola et al., 2014). Furthermore, organized multilayers of halloysite PDDA deposited on lignocellulose wood microfibres via the LbL technique were used to prepare paper handsheets that had a 30–50% higher porosity and had a tensile strength index close to that of the control sample (Lu et al., 2007).

In addition, halloysite can be used as a depositional template for forming multilayered organic shells by LbL self-assembly for the controlled release of macromolecules (Shchukin and Möhwald, 2007; Shchukin et al., 2008; Veerabadran et al., 2009; Vergaro et al., 2012; Shutava et al., 2014). The shell of polycation/polyanion multilayers on the halloysite provides a diffusion barrier, thereby slowing the release. For instance, a large-molecular-mass PAH (Mm 70,000)/PSS (Mm 70,000) polyelectrolyte shell effectively decreased the release rate of a dexamethasone drug from 6 to 15 h. However, the release profiles were significantly affected by the type of polyelectrolytes, their molecular mass, the degree of ionization (Veerabadran et al., 2009) and the pH of the solution (Shchukin and Möhwald, 2007). For example, for alkaline or acidic pH, an increase in the release of benzotriazole in an aqueous solution would occur in all nanocontainers composed of silica or halloysite with poly(diallyldimethylammonium chloride)/PSS, PAH/PSS or PAH/PMA as polyelectrolyte shells (Shchukin and Möhwald, 2007).

21.2.6 Electrophoretic Deposition

EPD has also been used for the deposition of CPN containing halloysite in several studies (Wang et al., 2011; Deen et al., 2012; Deen and Zhitomirsky, 2014). It is a process in which charged colloidal particles or polymer macromolecules, dispersed in a liquid medium, migrate under an electric field and are deposited onto the electrode surface (Fig. 21.8A). It allows the fabrication of uniform films of controlled thickness and offers many processing advantages, including high deposition rates and the potential for making depositions on substrates with complex shape. Moreover, EPD is well suited for the fabrication of CPN films. For example, halloysite, chitosan and halloysite chitosan–hydroxyapatite (HA) nanocomposite films have been fabricated by EPD. It took only 100 s to form a stable halloysite chitosan film at a deposition voltage of 5 V when the solution was composed of 0.1 g L^{-1}

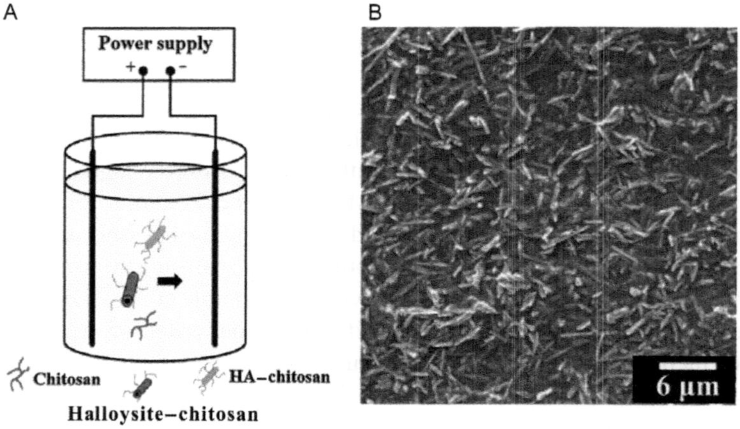

FIG. 21.8 (A) Schematic of the EPD cell; (B) SEM image of a CPN film prepared from 0.5 g L^{-1} chitosan solutions containing 0.6 g L^{-1} halloysite. *Reprinted with permission from Deen et al. (2012). Copyright 2012 Elsevier Ltd.*

chitosan and 0.1 g L^{-1} halloysite. SEM images (Fig. 21.8B) show that the films were continuous, crack-free and without significant agglomerations of halloysite. An investigation of the film cross sections showed that the film thickness varied from 0.1 to 5 μm as the deposition time varied from 1 to 10 min and the deposition voltages varied from 3 to 20 V. Furthermore, multilayered films containing chitosan, halloysite and HA with functional gradients have been prepared using EPD. The number of layers and the thicknesses of individual layers can be changed by varying the deposition time. Moreover, the layered structure obtained in this work allows the possible combination of the functional properties of different materials (Deen et al., 2012).

21.2.7 Latex Coagulation

Latex coagulation is a simple but effective way to prepare rubber-based nanocomposites filled with hydrophilic nanoparticles, especially for those that are difficult to disperse directly, such as graphite oxide. In this process, nanoparticles are first dispersed in aqueous solution and then mixed with rubber latex under stirring or ultrasonically. Subsequently, a flocculent such as calcium chloride and acetic acid, according to the nature and application of the CPN, is used for coagulation (Varghese and Karger-Kocsis, 2003; Jia et al., 2006). Further processing, such as mechanical mixing and curing the coagulation compound, is similar to that of ordinary rubber compounds. Compared with mechanical mixing directly, a latex coagulation pretreatment enables more dispersible nanoparticles and improves the mechanical performance of

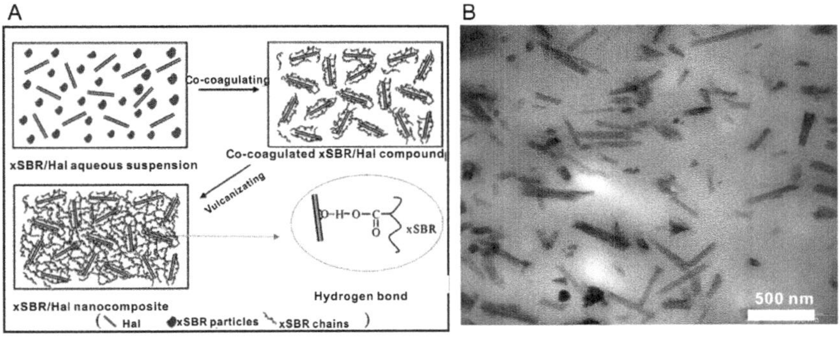

FIG. 21.9 (A) Schematic of the preparation process of halloysite (Hal) xSBR nanocomposites. (B) TEM photo of halloysite xSBR nanocomposites. *Reprinted with permission from Du et al. (2008a). Copyright 2008 Elsevier Ltd.*

the CPN. Halloysite has a wide range of colloidal stability, and rubbers filled with halloysite can be prepared by latex coagulation to homogeneously disperse them. For example, Du et al. (2008a) prepared halloysite carboxylated butadiene-styrene rubber (xSBR) nanocomposites via latex coagulation (Fig. 21.9A). Halloysite nanotubes with tubular structure were uniformly dispersed in the matrix (Fig. 21.9B). In addition, the strong interfacial bonding between the halloysite and xSBR matrix also contributed to the good dispersibility. Another example concerns a hybrid consisting of halloysite and graphene coagulating with a SBR emulsion. The high loading of the halloysite (40 phr) led to an aggregation in the matrix. However, with the addition of tannic acid–functionalized graphene, the dispersion of the halloysite greatly improved (Tang et al., 2013b).

21.3 INTERFACE INTERACTION IN HALLOYSITE POLYMER NANOCOMPOSITES

It has been well documented that interfacial adhesion plays a critical role in determining the overall performance of polymer CPN. Generally, the interfacial modification or pretreatment of the filler also contributes to the improved dispersion state of the resulting CPN, which also accounts for the performance enhancement. Therefore, interfacial modification is a prerequisite for maximizing the performance of the resulting CPN. Although halloysite is chemically similar to kaolinite, its modification method is somewhat special due to the 1D and lumen structures. Different strategies have been developed to improve interfacial interactions in the halloysite polymer nanocomposites.

Broadly, the modification methods for halloysite can be categorized into covalent and noncovalent methods, which are chemically similar to those for other clay minerals. However, there are some special requirements for the modification of halloysite. Geometrically, the location of the functional groups is different from those of layered clay minerals. The silanols (with relatively lower densities) are mostly located on the outer side, and the aluminols are located in the inner side and at the ends. In addition, tubular halloysite has a special lumen structure. This lumen structure may be useful for special nanotemplating or for the formation of tubular micelle via the selective modification of the inner and outer sides of the lumen. Interfacial interactions occurring in the halloysite polymer nanocomposites are discussed next.

21.3.1 Covalent Bonding

Covalent bonding is a strong interfacial interaction in CPN, effectively strengthening the transfer of interfacial stress. As the reactive groups on the surface of halloysite are only hydroxyl groups, the covalent linkage between halloysite and polymer is usually constructed by the reactions of the hydroxyls. Because the outside of a halloysite is composed of siloxane and silanols, condensation between the hydrolysed silanes and the surface hydroxyl groups of the halloysite is the most common way to create a covalent linkage with halloysite, through which a functionality can be anchored onto the halloysite via covalent bonding. It should be noted that the aluminols on the edges and inner side of the lumen also react towards various silanes.

The functional groups introduced by silanization are deliberately chosen for their subsequent reaction with the polymer matrix (Du et al., 2006; Liu et al., 2008b; Guo et al., 2009a; Zhang et al., 2014; Tian et al., 2015). Du et al. (2006) first introduced amino groups onto halloysite via silanization, and then created covalent interfaces between the modified halloysite and maleic anhydride-grafted PP, which greatly improved the reinforcing efficiency of the halloysite towards PP. If glycidyl groups were anchored, the modified halloysite exhibited an improved affinity with epoxy resin due to the co-curing reaction in the interface (Liu et al., 2008b). Bis(3-triethoxysilylpropyl) tetrasulphide is a well-known sulphide-containing silane in the rubber industry. If it is used to modify the interface between halloysite and rubber, a covalent sulphide-based linkage is created in the halloysite rubber nanocomposites (Rooj et al., 2010). Tian et al. (2015) found that PEI was covalently linked to amino-functionalized halloysite via a Schiff base reaction. In this way, a halloysite PEI nanocomposite with abundant amino groups was fabricated, and such a CPN could very efficiently remove chromium. To increase the silanization efficiency, an acid activation of the surface to increase the silanol content and the removal of organic impurities by using hydrogen peroxide can be considered.

Surface-initiated polymerization on the surface of inorganics generates inorganic–organic nanocomposites with well-defined structures. Generally, the preanchoring of the initiator by silanization on the surface is an essential step for such a procedure. If the halogen species are anchored, ATRP can be performed. If thiocarbamate species are used, RAFT can be realized. Li et al. (2008, 2009) immobilized an ATRP initiator using a two-step method. Then, ATRP of styrene or acrylonitrile is performed with the modified halloysite as the initiator. Accordingly, the halloysite PS and halloysite polyacrylonitrile nanocomposites were prepared. When the halloysite was removed using hydrofluoric acid, the polymeric nanotubes and nanowires were derived. The CPN could be further converted into nanocarbons via carbonization. The direct casting of the solution of the halloysite PS nanocomposites resulted in nonwoven fabrics, which exhibited unique superhydrophobicity. The antifouling poly(2-methacryloyloxyethylphosphorylcholine) can also be formed on halloysite with a similar ATRP process (Wang et al., 2014). It should be noted that if double bonds are introduced via silanization, the subsequent free radical copolymerization with vinyl monomers can also yield covalent bonding between the halloysite and the matrix, although the structure is not well defined due to the formation of the 'free' polymers (Zhang et al., 2013a). If thiocarbamate is grafted onto halloysite via a two-step procedure, then the RAFT polymerization can be performed on the modified halloysite. Through this method, Hou et al. (2014) prepared the amphiphilic brushes of poly (4-vinylpyridine)-block-polystyrene and polystyrene-block-poly(4-vinylpyridine) on the halloysite via a surface RAFT living polymerization.

As a type of silicate, the surface silanols and/or aluminols are also reactive to other species to form covalent linkages. Liu et al. (2007b) confirmed the direct covalent interface between cyanate ester resin and halloysite via the addition of cyanate on surface silanols and aluminols. The reaction mechanisms were proposed and substantiated by various characterizations. Specifically, the formation of the iminocarbonate consisting of an aluminol moiety was confirmed (Fig. 21.10). The cyanate ester can co-cure with epoxy resin. Therefore, a covalent interface resulted in the halloysite epoxy resin nanocomposites, which was very important for lowering the thermal expansion of the resin. It was also claimed that the surface aluminols were reactive to the carboxyl of L-lactic acid to form an aluminium lactate by the condensation

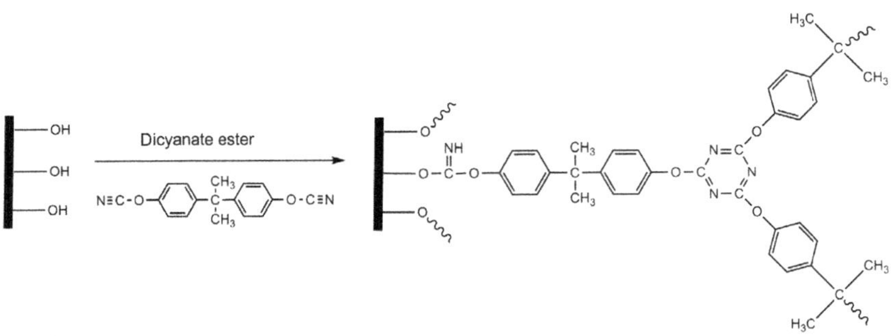

FIG. 21.10 Reaction mechanism between the silanols/aluminols and cyanate ester.

reaction. Therefore, high-performance halloysite PLA nanocomposites were prepared (Xu et al., 2015). Conjugated polyfluorenes with –C≡C–H end groups was covalently grafted onto the azide groups modified halloysite via 'click' chemistry (Zhang et al., 2015).

21.3.2 Hydrogen Bonding

Due to the presence of oxygenic groups on the surfaces of halloysite, they can form direct hydrogen bonding with many polar groups such as carboxylates, hydroxyls and aminos. Therefore, if the matrices possess such groups, interfacial hydrogen bonding will result. Actually, there are many commercially available polymers that can readily form hydrogen bonds with halloysite (Liu et al., 2007a, 2013b; Du et al., 2008a; Guo et al., 2009c; Zhou et al., 2010). The reported matrices with hydrogen bonding capability include PVA, polyamide (PA) and xSBR. For example, a halloysite PVA bionanocomposite with low cytotoxicity was prepared by the facial aqueous solution blending of PVA and halloysite dispersion. Due to the hydrogen bonding between the halloysite and PVA, a very uniform dispersion state of halloysite resulted. The mechanical properties of the PVA improved significantly by the incorporated halloysite. In addition, the incorporation of halloysite induced changes in the nanotopography and surface chemistry of the PVA films. On the halloysite-dominant surface, cells exhibited a significantly higher level of adhesion than that of the neat PVA film. The halloysite PVA films favoured fibroblast attachment and growth when the halloysite content was not very high (<7.5 mass%) (Zhou et al., 2010).

Due to its excellent aqueous dispersion stability, halloysite is very suitable for the preparation of rubber-based nanocomposites by co-coagulating rubber latex and halloysite dispersion. For example, carboxylated rubber latex has been used to coagulate with halloysite dispersion. The obtained halloysite rubber nanocomposite exhibited a very high reinforcing efficiency. With the inclusion of only 5 phr of halloysite, the modulus (at 100% elongation) is almost double that of a neat xSBR vulcanizate. The striking reinforcing efficiency is attributable to the strong interfacial hydrogen bonding between the pendent carboxylic groups on the rubber chains and the surfacial hydroxyls on the halloysite (Du et al., 2008a).

Many natural macromolecules contain diverse oxygenic and nitrogenous groups. In addition, they can be dissolved readily in water. Consequently, many halloysite nanocomposites based on natural polymers have been reported (Shamsi and Geckeler, 2008; De Silva et al., 2013b; Lee et al., 2013; Liu et al., 2013b, 2015). These systems could find various biomedical applications, such as halloysite chitosan nanocomposites (Liu et al., 2013a), halloysite alginate nanocomposites (Liu et al., 2015) for scaffolds in tissue engineering and halloysite DNA nanocomposites for the controlled release of drugs (Lee et al., 2013). The good dispersibility of halloysite in the matrices and the hydrogen bonding interface contributed to the high reinforcement in those systems, which is important in some scaffolds.

Low-molecular substances could also be utilized as interfacial modifiers of halloysite polymer nanocomposites. Du et al. (2007, 2010a) reported a series of low-molecular substances as hydrogen bond acceptors or donors (Table 21.1). If this type of substance was introduced into a halloysite PP nanocomposite, the inorganic–organic hybrids bridged by hydrogen bonding exhibited a much-improved efficiency in enhancing the mechanical properties of the halloysite PP nanocomposites. If the low-molecular substances can react further with the matrix, the interfacial adhesion can definitely improve. Guo et al. (2008, 2009b, 2010b) and Lei et al. (2011) developed a series of novel interfacial modifiers for halloysite rubber nanocomposites. On the one hand, these modifiers can form hydrogen bonds with the halloysite and can therefore absorb onto the halloysite. On the other hand, they can be chemically grafted onto the rubber chains via radical reaction. Consequently, a much-improved reinforcing efficiency has been realized in these halloysite rubber nanocomposites. These modifiers are incorporated by traditional two-roll milling, so they have great industrial promise.

21.3.3 Interfacial Charge Transferring

Many clay minerals exhibit the capability of absorbing organics via an electron transfer mechanism due to the presence of metal atoms with unoccupied orbitals such as aluminium, iron and transition metals. Such behaviour offers opportunities for enhancing the interfacial properties between halloysite and polymers or organics via electron transfer interactions or by altering the matrix behaviour.

Liu et al. (2009c) grafted *N*-cyclohexyl-2-benzothiazole sulphonamide (CBS) onto PP chains and prepared halloysite PP-graft nanocomposites. Due to electron transfer between the CBS and halloysite, the CPN exhibited much-improved mechanical properties. Although the interactions between the halloysite and CBS may involve multiple mechanisms such as Lewis acid–base interaction, hydrogen bonding, ionic bonding and electron transfer, the electron transfer has been characterized in detail. An example of this type of unique compatibility is illustrated in Fig. 21.11.

Many fluorescent whitening agents have conjugated structures, which are potentially capable of reacting with halloysite via electron transfer (Liu et al., 2008a,c). Liu et al. (2008c) demonstrated the in situ formation of hybrid fibrils consisting of halloysite and 2,5-bis(2-benzoxazolyl) thiophene (BBT) in a PP matrix. It has been confirmed that the formation of the hybrid fibrils was due to the combination of the electron transferral between the BBT and halloysite and shearing during the processing. The formation of hybrid fibrils led to a much-improved crystallinity of the PP matrix, which predominately accounted for the much-improved mechanical properties. In another example,

TABLE 21.1 Organics Used by Du et al. (2010a) to Form Hydrogen Bonds

Organics	Abbreviation	Molecular structures	Hydrogen-bonding functionalities
Melamine	MEL	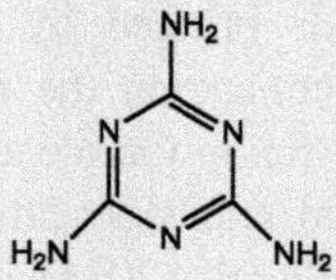	N, NH_2
Diphenyl guanine	DPG	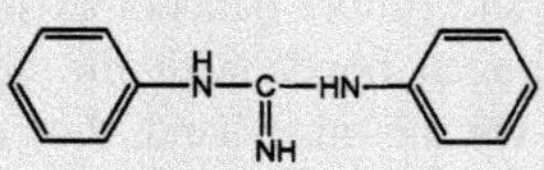	NH
2,4,6-Trimercapto-*s*-triazine	TCY	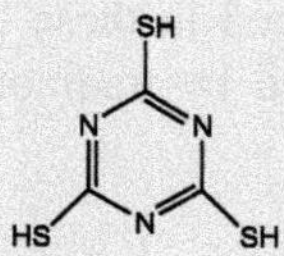	N, SH
Melamine cyanuric acid	MCA	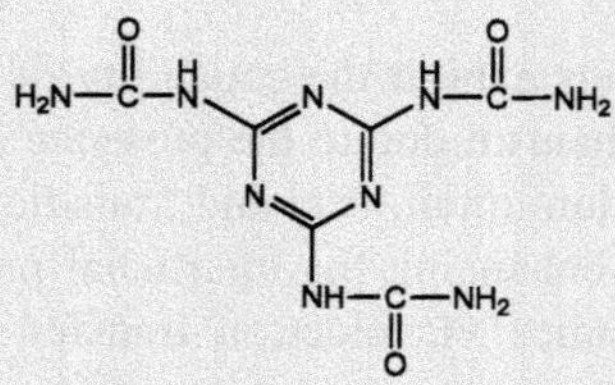	O, N, NH, NH_2
Tri-(2-hydroxyethyl) isocyanurate	THEIC	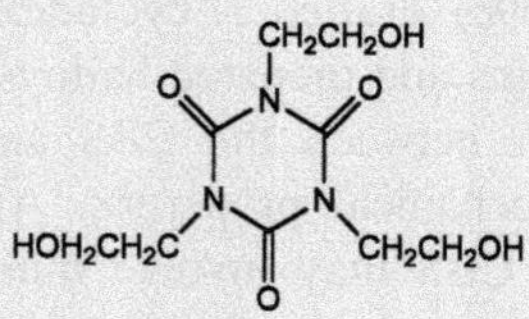	O, OH
Beta cyclodextrin	Beta-CD	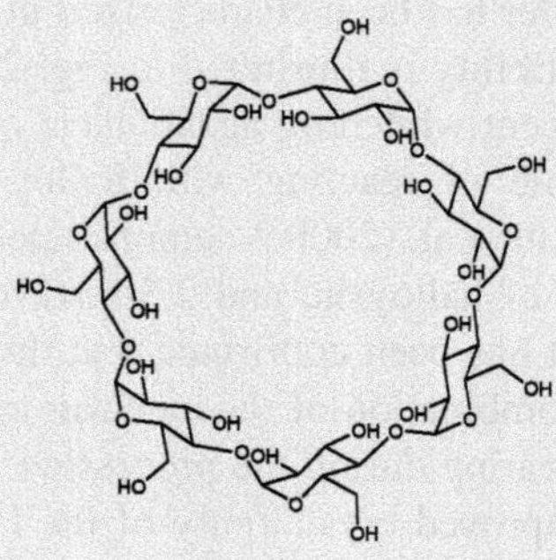	OH

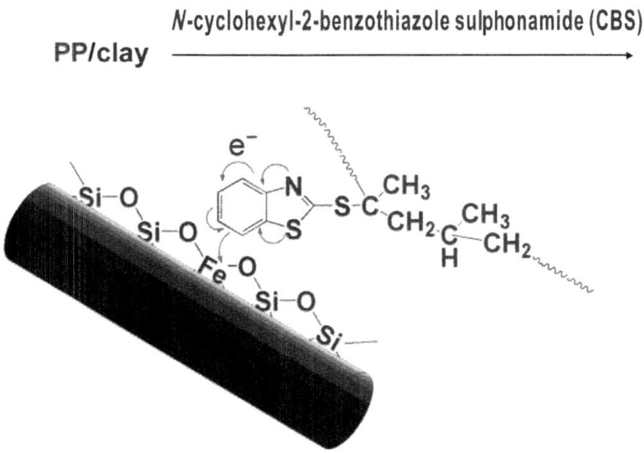

FIG. 21.11 Compatibility between halloysite and PP via the electron transfer mechanism. *Reprinted with permission from Liu et al. (2009c). Copyright 2009 Elsevier Ltd.*

2,2′-(1,2-ethenediyldi-4,1-phenylene) bisbenzoxazole was introduced into halloysite PP nanocomposites. The microsized aggregates were formed due to electron transfer between the halloysite and the fluorescent whitening agent. It has been shown that the cracking of the aggregates of halloysite during fracture plays an important role in improving the mechanical properties of halloysite PP nanocomposites.

21.3.4 Complexation

The lumen surfaces of halloysite possess different chemistries. The inner surface is mainly aluminols, and the outer surface is mainly silanols. Thus, halloysite could be selectively hydrophobized in the inner surface by octadecylphosphonic acid via a complexation mechanism (Yah et al., 2012a). This can be realized based on the fact that octadecylphosphonic acid can bind to alumina sites at the nanotube lumen and cannot bind the outer siloxane surface of the nanotube. The characterizations confirmed the binding within the halloysite lumen through bidentate and tridentate P–O–Al linkages. If the silanization of the outer surface with a hydrophilic silane is subsequently performed, then a bifunctionalized tube results. Due to the hydrophilic outer surface and the hydrophobilized inner surface, the selectively modified halloysite nanotube is actually a novel type of inorganic micelle. Such an architecture has been claimed to enable various applications, ranging from water purification to drug immobilization and controlled release.

The substance-bearing catechol group can bind with inorganics and aluminols via a complexation mechanism. Therefore, the selective modification of the lumen can be realized in another way. A dopa biomimetic initiator is utilized to treat the halloysite, and the inner surface and outer surface both adsorb the initiator molecules. However, the interaction mechanism is definitely different. The binding of the initiator on the inner surface via complexation is much stronger than that achieved on the outer surface, which is due to hydrogen bonding. Subsequently, the initiator adsorbed on the outer surface can be removed by rinsing and sonication. Accordingly, the initiator is selectively anchored onto the inner surface of halloysite. Then, the surface-initiated polymerization via ATRP is performed on the selectively modified halloysite (Yah et al., 2012b). Moreover, Chao et al. (2013, 2014) showed that the dopamine could self-polymerize to adhere to the surface of halloysite via the complexation between the halloysite and the catechols in polydopamine. Unlike the processes that have been described earlier, the rinsing step is not needed because the resulting modified halloysite was used to immobilize the enzyme on the halloysite, and the more immobilization sites, the better.

21.4 MULTIPLE EFFECTS OF HALLOYSITE ON POLYMER NANOCOMPOSITES

21.4.1 Mechanical Reinforcement

Tubular halloysite consists of natural nanotubes with combined high mechanical properties and aspect ratio. In addition, tubular halloysite can be easily individualized due to its having a rodlike structure and being free of entanglement since it is in the form of carbon nanotubes. Further, it is much cheaper and more naturally abundant than carbon nanotubes. Therefore, halloysite may serve as a promising reinforcement for various polymers.

The mechanical properties of halloysite have been theoretically and experimentally studied. The Young's moduli for zigzag and armchair single-walled (SW) nanotubes have been calculated in the range of 230–340 GPa. The modelled stiffness for halloysite is comparable to that of imogolite (~175–390 GPa), gallium sulphide (~270 GPa) and chrysotile nanotubes (125 GPa). The Young's moduli seem to have an asymptotic behaviour with the increase of the SW halloysite radii. The Young's moduli of zigzag nanotubes are normally approximately 20–30 GPa less than the respective armchair nanotubes (Guimarães et al., 2010). Lecouvet et al. (2013b) directly measured the elastic modulus of halloysite by an atomic force microscope in three-point blending mode. The modulus, which had an average of 140 GPa, was derived from the recorded force curve for the halloysite nanotube, with outer diameters ranging from 50 to 160 nm. The elastic modulus increased with decreasing outer diameters, with a steep jump when it is less than 50 nm. The observed size dependence of the elastic modulus was attributed to surface tension in the thinner nanotubes and a nonnegligible contribution of shear deformation of the larger tubes (Lecouvet et al., 2013b). Lu et al. (2011) measured the Young's modulus of a SW halloysite nanotube using a TEM with a bending stage. Their study showed that the halloysite nanotube was surprisingly flexible and could be bent to almost 90 degrees without fracture or reduction in the cross-sectional area (Lu et al., 2011).

In comparison with platelike clay minerals such as montmorillonite, halloysite is individualized much more easily due to the much lower interunit interaction and the rodlike geometry. The interlayer interaction in montmorillonite is primarily by ionic interaction, which is much stronger than the hydrogen bonding and van der Waals force, which account for the internanotube interaction in halloysite. The much smaller interunit contact area for halloysite also contributes to its higher dispersibility. Therefore, under normal polymer processing conditions, the dispersion of halloysite is expected to be much better than that achieved in CPN with platelike clay minerals.

In view of the 1D geometry, halloysite is expected to be useful in fabricating CPN with anisotropic properties. During the processing of the CPN, the halloysite can be oriented by shearing and by the melt flow. Spinning is a typical form of processing with unidirectional shearing. As discussed previously, the halloysite nanotubes are highly oriented in the CPN fibres, which affords them great strength (Haroosh et al., 2013).

Although halloysite seems to be dispersed in polymer matrices under normal processing conditions, the actual dispersion depends on the properties of the matrix. Due to the potential hydrogen bonding in the interface, halloysite exhibits much better dispersibility in polar matrices, including polar rubbers and engineering plastics. The reinforcing efficiency for bio-based matrices is especially high due to the presence of abundant polar groups in those polymers (Liu et al., 2012c, 2013b; Nakamura et al., 2013; Wu et al., 2013; Fujii et al., 2014).

The reinforcement of polymers by halloysite has been extensively studied. The reported systems include polyolefins (Du et al., 2006, 2007, 2008b; Liu et al., 2008c; Jia et al., 2009; Du et al., 2010a; Khunova et al., 2013); PS (Shamsi et al., 2010); epoxy resin (Liu et al., 2008b; Alamri and Low, 2012; Vahedi and Pasbakhsh, 2014; Zhang et al., 2014); PA (Guo et al., 2009a; Handge et al., 2010; Lecouvet et al., 2011a); polyurethane (PU) (Jing et al., 2015); polyvinyl chloride (Liu et al., 2011); polymethyl methacrylate (Jing et al., 2015); polyvinylidene fluoride (PVDF) (Tang et al., 2013a) and various rubbers (Du et al., 2008a; Guo et al., 2008, 2009b).

Guo et al. (2009a) prepared PA6 nanocomposites with halloysite via a twin-screw extruder. Before the extrusion, the halloysite is pretreated with silane to further increase the reinforcing efficiency. In such halloysite PA6 nanocomposites, the halloysite is mostly individualized, leading to a significantly improved modulus and strength. The heat distortion temperature (HDT) of PA6 is greatly improved by the incorporation of silane-modified halloysite. For example, when 10 phr (approximately 9 mass%) of halloysite is incorporated, an increment above 48°C in HDT is achieved. The large improvement in HDT is attributable to the effective load transference from the matrix to halloysite via strong interfacial interactions.

Much research has been done on the toughening of brittle epoxy resins by incorporating small amounts of halloysite (Ye et al., 2007, 2011a,b; Deng et al., 2008). For example, the incorporation of 2.3 mass% halloysite in epoxy led to a fourfold greater impact strength without sacrificing the flexural modulus, strength and thermal stability. These authors proposed that the impact energy was dissipated via the formation of damage zones with a large number of microcracks in front of the main crack. The microcracks were stabilized by nanotube bridging. Nanotube bridging, pullout and breaking were also observed (Fig. 21.12) and proposed as the major energy-dissipating events (Ye et al., 2007). These authors also found that the fracture toughness of the halloysite particle-modified epoxies was markedly increased; the largest improvements were in the critical stress intensity factor (K_{IC}) (up to 50%) and in the critical strain energy rate (G_{IC}) (127%) (Deng et al., 2008). Halloysite has also been employed as a toughness modifier for unpolyester resin and linear low-density polyethylene (LLDPE) (Albdiry et al., 2013; Pedrazzoli et al., 2015).

In CPN with polymer blend as the matrix, the selective location of nanoparticles in one specific site or phase has a profound effect on the mechanical performance of the CPN. Lin et al. (2013) utilized halloysite to modify a model PP/PET blend (90/10). The mechanical performance of the blend was optimized by the selective dispersion of the nanotubular halloysite into the interfacial region of the blends via a two-step process. The selective dispersion of the halloysite is illustrated in Fig. 21.13. Compared with the control, the overall mechanical properties of the blend were substantially improved, which was attributed to the formation of the unique morphology. When halloysite was blended into PLA that was toughened with PA11 to create PLA-based compounds with enhanced ductility, but without sacrificing stiffness and strength, a salamilike structure formed, which revealed a halloysite-rich, tubularlike (fibrillar) PA11 phase. Compared to neat PLA, this led to a remarkable improvement in the tensile and impact properties. The toughening mechanism was attributed to the combined effect of the resistance to crack propagation and the load-bearing capacity of the fibrillar structure (Rashmi et al., 2015).

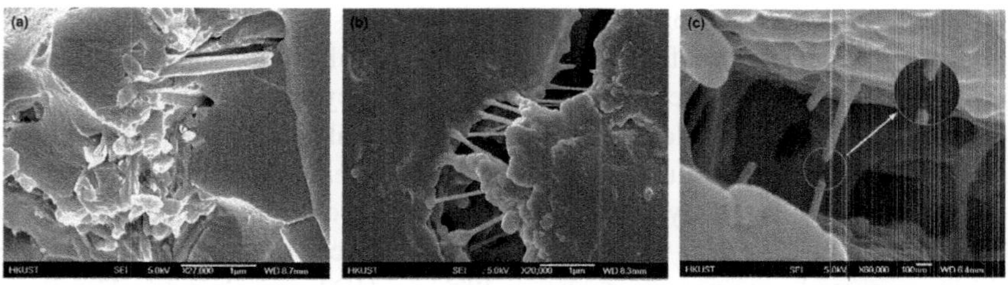

FIG. 21.12 Fracture surfaces of halloysite epoxy nanocomposites: (A) nanotube debonding/pullout, (B) nanotube bridging and (C) nanotube fracture. *Reprinted with permission from Ye et al. (2007). Copyright 2007 Elsevier Ltd.*

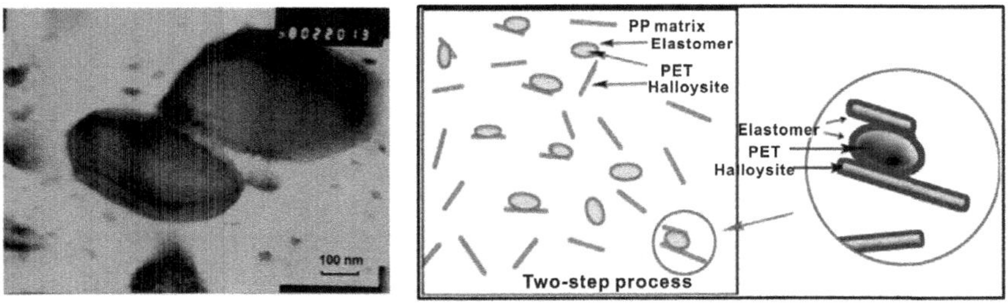

FIG. 21.13 TEM photo and schematic representation for the sample with selective location of halloysite (the *blue* (*dark grey* in the print version) *layer* is the compatibilizer SEBS-g-MAH). *Reprinted with permission from Lin et al. (2013). Copyright 2013 John Wiley & Sons, Inc.*

21.4.2 Thermal Stability and Flame Resistance

Halloysite greatly enhances the thermal stability and flame retardancy of polymers. The following causes contribute to the improvement in thermal stability. First, as a natural inorganic clay mineral, halloysite has a higher thermal stability than that of the organic polymer. An overview of the thermal stability of halloysite was presented in Chapter 6. Moreover, halloysite acts as a thermal barrier that protects the polymer from making contact with fire and a mass transport barrier that slows the escape of volatile products during the process of degradation. The hollow nanotube structure can trap polymer degradation products, delaying mass transport and further improving thermal stability.

The dispersion and interfacial interaction are crucial to realize high thermal resistance because the barrier effects deeply depend on an even dispersion and strong interfacial interaction. Jia et al. (2009) found that a PE graft could effectively improve the thermal stability and flame retardancy of halloysite LLDPE nanocomposites. The temperature at a 5% mass loss increased from 455°C to 464°C when 5 phr (parts per one hundred parts of the polymer matrix) of the PE graft were added. At the same time, the heat release rates (HRR) decreased from ~900 kW m^{-2} for neat LLDPE to ~400 kW m^{-2} for the modified halloysite LLDPE nanocomposites (30 phr of halloysite). Du et al. (2006) showed that that the temperature at a 5% mass loss of PP increased by 7°C with the inclusion of 10 phr of halloysite, and increased by 23°C when silane-modified halloysite was used. The HRR of the neat PP and halloysite PP nanocomposites are compared (Fig. 21.14). An enhanced thermal stability was obtained by adding halloysite in chitosan, PVA (Qiu and Netravali, 2013) and ethylene-propylene-diene monomer (EPDM) (Ismail et al., 2008). It is necessary at this point to emphasize that the addition of halloysite can also depress the coefficient of thermal expansion (CTE) due to the aluminosilicate nature of halloysite and the interfacial interactions

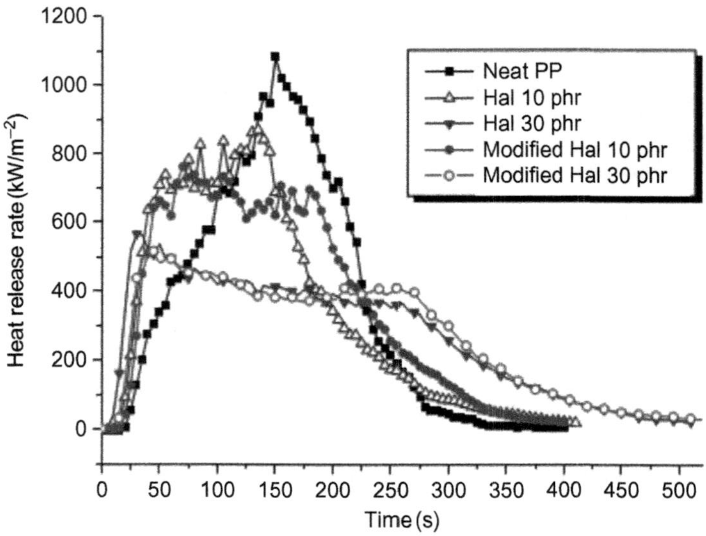

FIG. 21.14 Heat-release rate of neat PP and halloysite PP nanocomposites. *Reprinted with permission from Du et al. (2006). Copyright 2006 Elsevier Ltd.*

between the halloysite and matrix, which may restrict the relaxation mobility of the polymer chain segment at the inorganic polymer, especially at higher temperatures (Yano et al., 1997).

Concerning the mechanisms of enhanced flame retardancy by halloysite, in addition to the abovementioned heat and mass barrier effects, the demonstrated existence of iron oxides in the silicate fillers (Liu et al., 2013b) that serve as flame-retardant additives during the process of degradation (Zhu et al., 2001; Kashiwagi et al., 2002) is another source of improved thermal stability in CPN. Furthermore, halloysite decomposes and releases ~13.1% of crystal water in the temperature range 430°C–570°C, which may dilute and cool flammable gases. Therefore, the combination of cooling, dilution, barrier and radical trapping effects grant CPN great thermal stability and flame resistance. Attia et al. (2014) found that with the inclusion of halloysite into xSBR, the flammability properties of xSBR decreased greatly; HRR decreased from ~1423 to 885 kW m^{-2} when 30 phr halloysite were added. As a consequence, the emission of CO and CO_2 gases also reduced, indicating that the barrier effects dominated the depression of the flammability. Superior flame retardancy obtained by adding halloysite into other polymers such as SBR (Guo et al., 2010a), PA6 (Marney et al., 2008; Ounoughene et al., 2015), soy protein (Nakamura et al., 2013) and epoxy (Liu et al., 2007b) matrices have also been demonstrated. It is notable that halloysite enhances thermal stability and flame resistance, as well as improving the mechanical properties of polymers, which may be advantageous in the design of flame-retardant CPN.

21.4.3 Crystallization Behaviour

The incorporation of inorganic particles into a semipolymer generally influences the crystallization process of CPN because the particles can act as a nucleating agent for heterogeneous nucleation. As expected, halloysite is demonstrated to have a significant effect on the crystallization of semipolymers such as PP, PA6, PVA, PLA, PVDF and polycaprolactone (PCL) by acting as heterogeneous nucleation agents.

Ning et al. (2007) observed that well-dispersed halloysite could serve as a nucleation agent in a PP matrix, leading to an enhancement of the overall crystallization rate and the nonisothermal crystallization temperature. However, halloysite could increase the activation energy of crystallization due to the restriction on the motion of PP chains imposed by halloysite (Liu et al., 2012b). Similarly, the nucleating effects of halloysite in a PP matrix have also been reported by Du et al. (2010b) and Wang and Huang (2013). These observations suggest that halloysite serves as a nucleation agent and facilitates the crystallization of CPN, similar to other nanosized inorganics such as silica and montmorillonite. In particular, Liu et al. (2009a) reported that halloysite had a dual nucleating ability for α- and β-isotactic polypropylene (iPP) crystals and that the formation of β-iPP crystals is due to the unique surface characteristics of the halloysite. According to the dimensional lattice-matching theory, a lattice matching between the *c*-axis periodicity of iPP (6.5 Å) and a corresponding distance in the substrate crystal face of the nucleating agent is the main reason for inducing the β-iPP polymorph. Halloysite nanotubes have a unique morphology among other common nanoparticles, such as nanosilica, and other clay minerals, such as montmorillonite. They are multiwalled inorganic nanotubes and are rolled by 15–20 aluminosilicate layers. The 001 basal reflection of the halloysite nanotubes is 0.73 nm, which is fairly close to the length of the *c*-axis corresponding to the period of the 31 helix iPP, indicating good epitaxy (Fig. 21.15). In addition, both the outer surfaces and the ends of halloysite nanotubes may provide nucleation surfaces for iPP crystals. In particular, the defects on the surfaces with appropriate sizes and the ends of halloysite nanotubes may provide numerous nucleation surfaces when they are close to the dimension of the *c*-axis of PP. Thus, those CPN with high halloysite nanotube content (above 10 phr) provided enough sites for the nucleation and development of the β-iPP crystal, and this led to the formation of high β-iPP content under appropriate kinetics conditions. Moreover, the cooling rate in a nonisothermal crystallization process or the crystallization temperature in an isothermal crystallization process has a large influence on the β-crystal nucleating ability of halloysite nanotubes. Liu et al. (2009b) also observed that a surface treatment of halloysite with BBT led to a significant increase in the crystallinity of PP, which was attributed to the formation of microfibrils consisting of halloysite and BBT in the nanocomposites. The remarkably improved crystallinity was

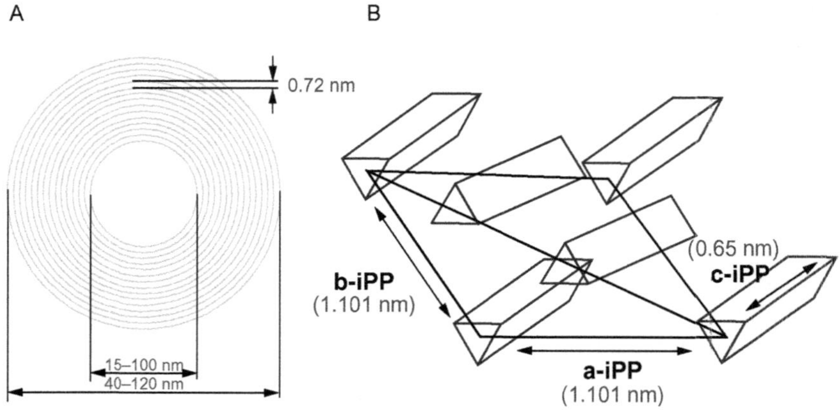

FIG. 21.15 Schematics of the crystalline structures of halloysite (A) and b-iPP (B). *Reprinted with permission from Liu et al. (2009a). Copyright 2009 Elsevier Ltd.*

responsible for the substantially increased tensile and flexural properties of the CPN. In addition, the processing methods had effects on the crystallization behaviour of PP. In halloysite PP nanocomposites prepared by solution mixing, it was found that both the crystallization onset temperature and the crystallization peak temperature increased consistently with the halloysite content (Liu et al., 2010b).

In halloysite PA6 nanocomposites, the halloysite could act as nucleating agents and accelerate the crystallization rate. A kinetic analysis indicated that the addition of halloysite resulted in a larger fold-surface-free energy of halloysite PA6 nanocomposites, which was restricted at higher halloysite loading. The halloysite content exerted a large effect on the crystallinity of the halloysite PA6 nanocomposites; the crystallinity reached a maximum with 5 phr of halloysite, and the percentage γ-phase crystals of total crystals increased with the halloysite content. The authors argued that halloysite played multiple roles that affected the crystallization behaviour of the PA6. First, halloysite served as a nucleating agent, which resulted in a higher crystallization temperature. Second, the strong hydrogen bonding between the halloysite and PA6 chains restricted the mobility of the PA6 chains. Finally, the halloysite favoured the formation of the γ-PA crystals (Guo et al., 2009a,c). The nucleation effects of halloysite in PA6 and PA12 have also been reported by other authors. It was found that the crystallization temperature and crystallinity of PA were improved with the addition of halloysite (Lecouvet et al., 2011a; Prashantha et al., 2011).

The crystallization temperature of a halloysite-filled PVA nanocomposite initially increased with halloysite loading due to its nucleation effect. However, the overloading of halloysite decreased the improvement in the crystallization temperature due to the aggregation of halloysite at high loading (Liu et al., 2007a). In halloysite-filled PVDF nanocomposites, halloysite could also increase the crystallization temperature of PVDF due to its nucleating effect. However, the enthalpy of crystallization significantly decreased with the addition of halloysite. Moreover, halloysite promoted the formation of γ-phase crystals, and the fraction of the γ-phase crystals increased to a maximum at 10-phr halloysite loading and then decreased as the halloysite loading further increased (Tang et al., 2013a). Wang and Huang (2014) reported that halloysite could induce the formation of the electroactive β- and γ-phases in a PVDF matrix. This could be explained by the fact that the synergistic effect of electrical and hydrogen interaction between halloysite and PVDF made the CH_2 groups of the PVDF align along the halloysite surfaces and the fluorine atoms align at the opposite side of the molecular chains, a development that facilitates the formation of long trans-conformations at the surfaces of the halloysite (Fig. 21.16).

Liu et al. (2013a) reported that the cold crystallization temperature for PLA shifted to a lower temperature and that the crystallinity increased with the addition of halloysite as a nucleating agent. Lee and Chang (2013) observed that the crystallinities and melting temperatures were reduced in the halloysite PCL nanocomposites compared to in the neat PCL ones because the restricted mobility of the PCL chains caused a reduction in the PCL crystal size and imperfections in the crystals. In addition, the crystallization

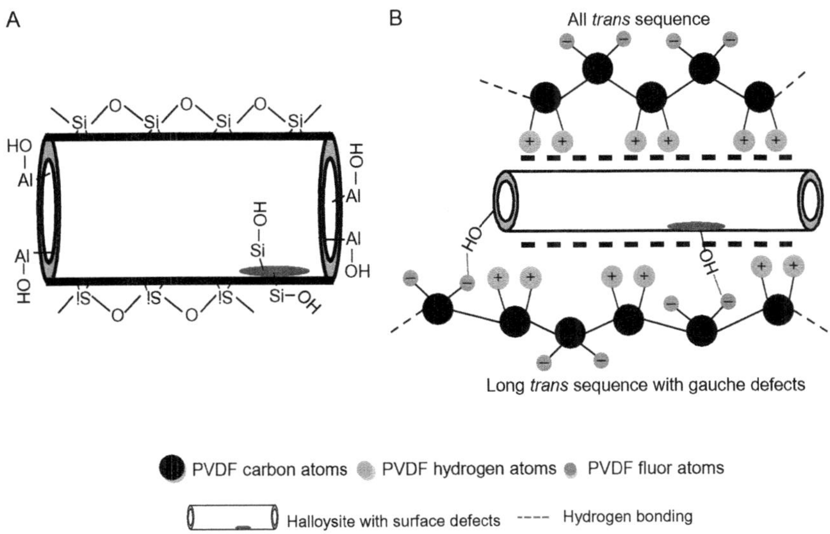

FIG. 21.16 Schematics of (A) the molecular structure at the surface and edges of halloysite and (B) the formation mechanisms of β- and γ-phase crystals in halloysite PVDF nanocomposites. *Reprinted with permission from Wang and Huang (2014). Copyright 2014 Elsevier Ltd.*

temperature and crystallization rate increased with increasing halloysite content due to the nucleation effects of the halloysite.

21.4.4 Dielectric Property

The dielectric property is the capacity of a material to store or dissipate energy under the application of an electric field. It is determined by monitoring the relaxation processes of the polymer chains and provides information about the molecular dynamics, which is an attractive prospect for researchers. To conduct research that will allow the wide use of halloysite, it is necessary to understand the dielectric property of CPN that contain halloysite.

Due to the polar nature of halloysite, the inclusion of halloysite usually brings about favourable changes in the dielectric characteristics of the polymer matrix. Thakur et al. (2014) synthesized PVDF with different contents of clay minerals (kaolinite and halloysite) by the solution-casting method to investigate the electroactive β-phase formation mechanism and the dielectric properties of CPN films. As expected, the value of the dielectric constant of the CPN films increased almost linearly with the content of both clay minerals. The dielectric constant of CPN decreased with an increase in frequency for both of the loadings (Fig. 21.17). This may be explained by the Maxwell–Wagner–Sillars interfacial polarization mechanism (Lopes et al., 2013). When an electric field is applied to the samples, the charge carriers from the impurities (ie, the clay minerals) or from the electrode can move and accumulate at the interface between the clay mineral and polymer matrix due to the difference in electrical conductivity between the clay mineral and polymer. The movement and accumulation of the charge carriers in the interface might cause a large polarization, as well as a large dielectric constant. Because the charge carriers have low mobility within the bulk of CPN, as both the clay minerals and PVDF are highly resistive at room temperature and atmospheric pressure, their movement and accumulation process need to take a relatively long time. As a result, the interfacial polarization in CPN decreased with increasing frequency, and a maximum tangent loss of the samples was observed at low frequencies rather than at high frequencies. Moreover, compared to a normal PVDF, a 15 mass% addition of kaolinite increased ε_r approximately 4.5 times, whereas the addition of the same amount of halloysite caused a dramatic increase in ε_r of approximately 33 times. This is because the high electrical conductivity difference between the halloysite and PVDF and the larger SSA of halloysite led to an increased probability for charge carriers to accumulate at the interface of the halloysite and PVDF matrix, resulting in larger interfacial polarizations, as well as larger dielectric constants, than in kaolinite PVDF films. Similar phenomena have been observed in halloysite hydroxypropyl cellulose (Cavallaro et al., 2011) and highly dispersed halloysite PA-11 nanocomposites (Prashantha et al.,

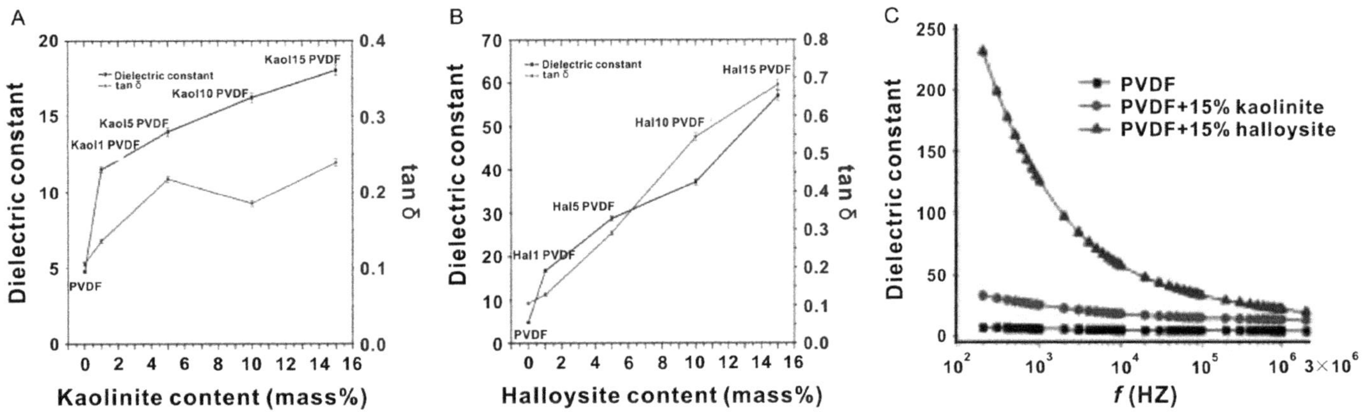

FIG. 21.17 Clay mineral content dependence of dielectric constants and tangent losses at 10 kHz: (A) Kaolinite PVDF thin films, (B) Halloysite PVDF thin films and (C) Frequency dependence of dielectric properties of pure PVDF, kaolinite PVDF thin films and halloysite PVDF thin films. *Reprinted with permission from Thakur et al. (2014). Copyright 2014 Elsevier Ltd.*

2013). However, the influence of halloysite content on the dielectric constants and electrical conductivity of halloysite hydroxypropyl cellulose is much smaller.

Halloysite LLDPE nanocomposite compounds are being used in the cable industry, where the dielectric properties are essential. According to the findings of Čermák et al. (2015), the polarization indexes, which are related to the rate of change of the polarization current, increased in accordance with the halloysite content and positively improved the polarizations' current characteristics. Based on the results from broadband dielectric spectroscopy, a small influence of halloysite content on the dissipation factor in the industrial frequency range and at working temperatures has been observed. It is worth noting that the thermal treatment of halloysite causes a notable reduction in the dissipation factor and much stronger relaxations. This behaviour is possibly caused by the rearrangement of the structure of the halloysite or by the evaporation of water or organic compounds from the halloysite. Further detailed analyses of pure halloysite in the delivered state and its thermally treated modifications will help clarify the exact physical processes that produce changes in the dielectric behaviour (Čermák et al., 2015).

21.4.5 Wettability Properties

Wettability is a fundamental property of solid surfaces that plays important roles in daily life, industry and agriculture (Zhang et al., 2008b; Liu et al., 2010a). It is strongly influenced by surface microstructures—the rougher a surface is, the poorer its wettability is (Cassie and Baxter, 1944; Khorasani et al., 2005; He et al., 2010). Halloysite, a naturally occurring hydrophilic clay mineral, has a complex influence on the wettability of polymers. Liu et al. (2010b) incorporated halloysite into PP for tailoring the surface microstructures of the CPN. Micropapillalike hybrid spherulites with nanostructures were formed on the halloysite PP nanocomposite surfaces. The contact angles of the CPN increase with halloysite content, but the content is less than 60%. Superhydrophobicity, with a maximum water contact angle of nearly 170 degrees and sliding angle of about 2 degrees, was seen in the halloysite PP nanocomposites due to the rough surfaces that arose from the halloysite nucleation ability for PP (Fig. 21.18). However, overloading halloysite caused a slightly decreased hydrophobicity of the CPN, although they also have a greater surface roughness. This was because halloysite aggregated on the CPN surfaces with high halloysite content, which was detrimental for the hydrophobicity of the surfaces due to the hydrophilic character of these aggregates. Overall, the idea for using nanoparticles for tailoring the microstructures and wettability of polymer surfaces is simple but flexible, so they can also be adopted in other systems. Cavallaro et al. (2011) prepared halloysite HPC and halloysite pectin nanocomposites. Decreases in the contact angles

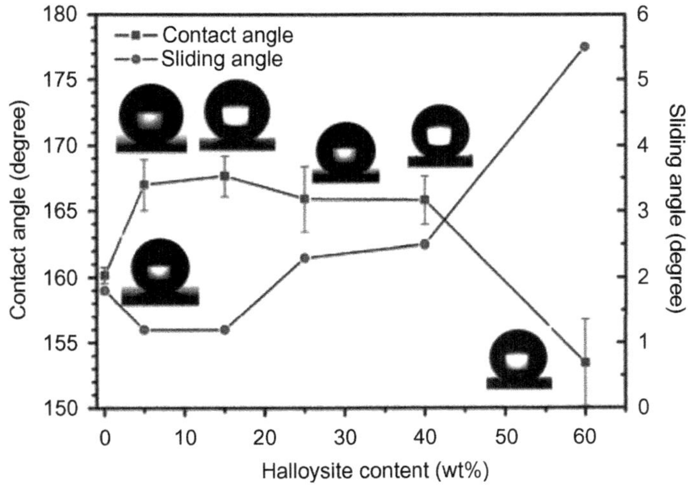

FIG. 21.18 Contact angles and sliding angles of the PP and halloysite PP nanocomposite surfaces. *Reprinted with permission from Liu et al. (2010b). Copyright 2010 Elsevier Ltd.*

and drop volume and increases in the contact area indicated improved hydrophilicity. By comparing the volume drop and the contact area data, Cavallaro et al. concluded that both spreading and water absorption occurred during the experiment indicating the hydrophilic nature of halloysite.

21.4.6 Biocompatibility and Related Properties

As a naturally occurring clay mineral, halloysite has been demonstrated to be biocompatible. No poisonous elements have been detected in halloysite. Vergaro et al. (2010b) studied the biocompatibility of halloysite and demonstrated that halloysite is nontoxic in cells up to concentrations of 75 $\mu g\ mL^{-1}$. Therefore, halloysite is promising for fabricating various biomedical CPN such as tissue-engineering scaffolds.

One of the most critical problems for tissue engineering is the establishment of a scaffold that is capable of supporting a three-dimensional tissue formation (Chen et al., 2002; Hollister, 2005). Tissue-engineering scaffolds are structures made of artificial or natural substances that are used to shape cell growths. To achieve this goal, scaffolds must meet special requirements, such as high porosity with adequate pore size, biodegradability and mechanical strength to support the body (Hutmacher, 2000). The biocompatibility of halloysite has been reported widely (eg, Levis and Deasy, 2003; Kelly et al., 2004; Abdullayev and Lvov, 2010; Vergaro et al., 2010a; Yuan et al., 2012; Aguzzi et al., 2013), as has the high modulus and strength of the hollow nanotubular structure. Both of these characteristics make halloysite an ideal

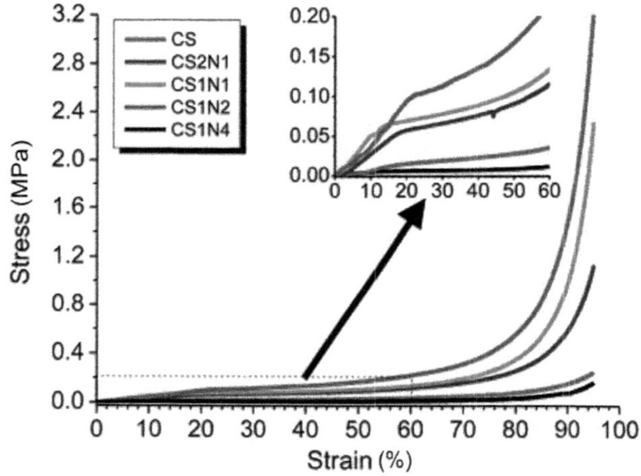

FIG. 21.19 Compressive stress–strain curves of chitosan and halloysite chitosan nanocomposite scaffolds. The *inset* shows an expansion of the strain range 0–60%, showing the changes in the compressive modulus of the samples. *Reprinted with permission from Liu et al. (2013b). Copyright 2013 the Royal Society of Chemistry.*

reinforcing filler for fabricating tissue-engineering scaffolds. Halloysite has been added to biodegradable polymers such as chitosan (Liu et al., 2013b), PVA (Zhou et al., 2010), PLGA (Qi et al., 2012) and PLA (Haroosh et al., 2013) to fabricate tissue-engineering scaffolds. Liu et al. (2013b) fabricated halloysite chitosan nanocomposites by combining solution-mixing and freeze-drying techniques. The determination of the mechanical properties demonstrated that the CPN scaffolds exhibited significant enhancements in compressive strength and the compressive modulus when compared with those of the pure chitosan scaffold (Fig. 21.19). The halloysite chitosan nanocomposites scaffolds exhibited a highly porous structure, and the halloysite had nearly no effect on the pore structure and porosity of the scaffolds. The good cell attachment and viability of the NIH3T3-E1 mouse fibroblasts cultured on halloysite chitosan nanocomposites indicated that the CPN were cytocompatible even when the loading of the halloysites was 80%. Transparent PVA and halloysite PVA bionanocomposite films were prepared by Zhou et al. (2010). With the inclusion of 7.5 mass% halloysite, the tensile strength increased by ~90%, and on a nanotube-dominant surface, the cells exhibited a significantly higher level of adhesion than on a neat PVA film (Fig. 21.20). The in vitro fibroblast response demonstrated that both the neat PVA and PVA halloysite nanocomposite films were biocompatible, and the PVA halloysite films were favourable for the attachment and growth of fibroblasts below 7.5 mass% of the incorporated halloysite. Therefore, the halloysite polymer nanocomposites exhibited great potential for applications in tissue engineering.

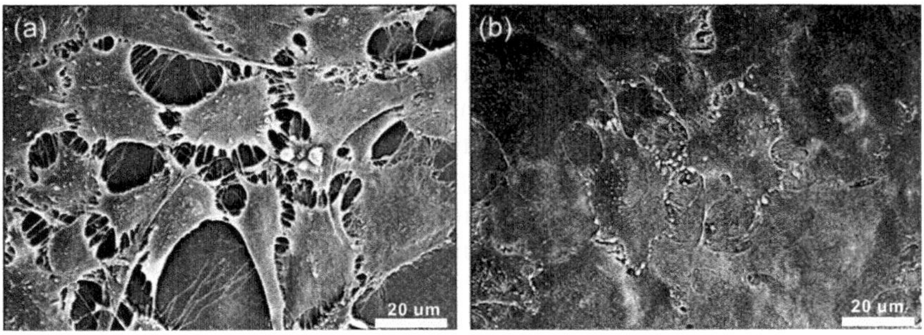

FIG. 21.20 Scanning electron microscope (SEM) images of osteoblasts (MC3T3-E1) cultured on neat PVA and PVA/halloysite bionanocomposite films on day 4: (A) PVA and (B) 7.5%. *Reprinted with permission from Zhou et al. (2010). Copyright 2010 John Wiley & Sons, Inc.*

Because the empty tubular lumens are capable of providing halloysite with space for encapsulating a range of active agents, halloysite provides additional functions, such as drug delivery (Levis and Deasy, 2002; Lee et al., 2013), antibiotics (Wei et al., 2012; Zhang et al., 2013b; Gorrasi, 2015), anticorrosion (Abdullayev and Lvov, 2010), crack self healing (Wei et al., 2012) and antioxidant (Fu et al., 2015) to CPN that can be synergistically combined with enhanced tensile, impact and adhesive strengths.

21.5 CONCLUDING REMARKS

As a naturally occurring nanotubular clay mineral, halloysite exhibits many unique structural features for fabricating polymer-based nanocomposites, including 1D nanoreinforcement, lumen structure and distinctly different inside and outside surface chemistries. Accordingly, halloysite polymer nanocomposites possess many distinct properties, such as high mechanical performance, enhanced thermal stability and flame retardancy, specially regulated crystallization and controlled organic–inorganic nanoassembly. These processing methods are especially crucial for the morphological and organization control of halloysite polymer nanocomposites. The interfacial designs and modifications are prerequisites for optimizing the performance of the CPN and for the realization of the target functions. Since the surface chemistry, as well as the unique structure of halloysite, have been well understood, there are abundant opportunities for the design of functional halloysite polymer nanocomposites. Some interesting developing directions for the functions of halloysite polymer nanocomposites, including nanotemplating and controlled release with the lumen structure, nanotubular micelles via selective modification, high-performance membranes and fibres by nanoassembling, will be a great focus in the future.

ABBREVIATIONS

ATRP	atom transfer radical polymerization
BBT	2,5-bis(2-benzoxazolyl) thiophene
CBS	N-cyclohexyl-2-benzothiazole sulphonamide
CPN	clay polymer nanocomposite
EPD	electrophoretic deposition
EPDM	ethylene-propylene-diene monomer
HA	hydroxyapatite
HDT	heat distortion temperature
HPC	hydroxypropylcellulose
LbL	layer-by-layer self-assembly
LLDPE	linear low-density polyethylene
MAA	methacrylic acid
PA	polyamides
PAA-*c*-AM	poly(sodium acrylate-*c*-acrylamide)
PAH	poly(allylaminehydrochloride)
PANI	polyaniline
PCL	polycaprolactone
PDDA	polydimethyldiallyl ammonium chloride
PEI	poly(ethyleneimine)
PLA	polylactic acid
PLGA	poly(lactic-*co*-glycolic acid)
PP	polypropylene
PPy	polypyrrole
PS	polystyrene
PSS	poly(styrene sulphonate)
PVA	polyvinyl alcohol
RAFT	reversible addition-fragmentation chain transfer
SBR	styrene butadiene rubber
SDS	sodium dodecyl sulphate
SSA	specific surface area
xSBR	carboxylated butadiene-styrene rubber

REFERENCES

Abdullayev, E., Lvov, Y., 2010. Clay nanotubes for corrosion inhibitor encapsulation: release control with end stoppers. J. Mater. Chem. 20, 6681–6687.

Aguzzi, C., Viseras, C., Cerezo, P., Salcedo, I., Sanchez-Espejo, R., Valenzuela, C., 2013. Release kinetics of 5-aminosalicylic acid from halloysite. Colloids Surf. B Biointerfaces 105, 75–80.

Alamri, H., Low, I.M., 2012. Microstructural, mechanical, and thermal characteristics of recycled cellulose fiber-halloysite-epoxy hybrid nanocomposites. Polym. Compos. 33, 589–600.

Albdiry, M.T., Ku, H., Yousif, B.F., 2013. Impact fracture behaviour of silane-treated halloysite nanotubes-reinforced unsaturated polyester. Eng. Fail. Anal. 35, 718–725.

Attia, N.F., Hassan, M.A., Nour, M.A., Geckeler, K.E., 2014. Flame-retardant materials: synergistic effect of halloysite nanotubes on the flammability properties of acrylonitrile-butadiene-styrene composites. Polym. Int. 63, 1168–1173.

Bidsorkhi, H.C., Adelnia, H., Heidar Pour, R., Soheilmoghaddam, M., 2015. Preparation and characterization of ethylene-vinyl acetate/halloysite nanotube nanocomposites. J. Mater. Sci. 50, 3237–3245.

Cassie, A.B.D., Baxter, S., 1944. Wettability of porous surfaces. Trans. Faraday Soc. 40, 546–551.

Cavallaro, G., Donato, D.I., Lazzara, G., Milioto, S., 2011. Films of halloysite nanotubes sandwiched between two layers of biopolymer: from the morphology to the dielectric, thermal, transparency, and wettability properties. J. Phys. Chem. C 115, 20491–20498.

Cavallaro, G., De Lisi, R., Lazzara, G., Milioto, S., 2012. Polyethylene glycol/clay nanotubes composites. J. Therm. Anal. Calorim. 112, 383–389.

Cavallaro, G., Gianguzza, A., Lazzara, G., Milioto, S., Piazzese, D., 2013. Alginate gel beads filled with halloysite nanotubes. Appl. Clay Sci. 72, 132–137.

Čermák, M., Kadlec, P., Kruliš, Z., Polanský, R., 2015. Dielectric analysis of halloysite nanotubes LLDPE nanocomposite compounds. In: 31st International Conference of Polymer Processing Society, At Jeju, South Korea.

Chao, C., Liu, J.D., Wang, J.T., Zhang, Y.W., Zhang, B., Zhang, Y.T., Xiang, X., Chen, R.F., 2013. Surface modification of halloysite nanotubes with dopamine for enzyme immobilization. ACS Appl. Mater. Interfaces 5, 10559–10564.

Chao, C., Zhang, B., Zhai, R., Xiang, X., Liu, J.D., Chen, R.F., 2014. Natural nanotube-based biomimetic porous microspheres for significantly enhanced biomolecule immobilization. ACS Sustainable Chem. Eng. 2, 396–403.

Chen, G., Ushida, T., Tateishi, T., 2002. Scaffold design for tissue engineering. Macromol. Biosci. 2, 67–77.

De Silva, R., Pasbakhsh, P., Goh, K., Chai, S.P., Chen, J., 2013a. Synthesis and characterisation of poly(lactic acid)/halloysite bionanocomposite films. J. Compos. Mater. 48, 3705–3717.

De Silva, R., Pasbakhsh, P., Goh, K., Chai, S.P., Ismail, H., 2013b. Physico-chemical characterisation of chitosan/halloysite composite membranes. Polym. Test. 32, 265–271.

Deen, I., Zhitomirsky, I., 2014. Electrophoretic deposition of composite halloysite nanotube–hydroxyapatite–hyaluronic acid films. J. Alloys Compd. 586, S531–S534.

Deen, I., Pang, X., Zhitomirsky, I., 2012. Electrophoretic deposition of composite chitosan–halloysite nanotube–hydroxyapatite films. Colloid Surf. A Physicochem. Eng. Asp. 410, 38–44.

Deng, S.Q., Zhang, J.N., Ye, L., Wu, J.S., 2008. Toughening epoxies with halloysite nanotubes. Polymer 49, 5119–5127.

Dong, F., Wang, J.H., Wang, Y.S., Ren, S.P., 2012. Synthesis and humidity controlling properties of halloysite/poly(sodium acrylate-acrylamide) composite. J. Mater. Chem. 22, 11093–11100.

Du, M.L., Guo, B.C., Jia, D.M., 2006. Thermal stability and flame retardant effects of halloysite nanotubes on poly(propylene). Eur. Polym. J. 42, 1362–1369.

Du, M.L., Guo, B.C., Liu, M.X., Jia, D.M., 2007. Formation of reinforcing inorganic network in polymer via hydrogen bonding self-assembly process. Polym. J. (Tokyo, Jpn.) 39, 208–212.

Du, M.L., Guo, B.C., Lei, Y.D., Liu, M.X., Jia, D.M., 2008a. Carboxylated butadiene-styrene rubber/halloysite nanotube nanocomposites: interfacial interaction and performance. Polymer 49, 4871–4876.

Du, M.L., Guo, B.C., Cai, X.J., Jia, Z.X., Liu, M.X., Jia, D.M., 2008b. Morphology and properties of halloysite nanotubes reinforced polypropylene nanocomposites. E-Polymers 8, 1490–1503.

Du, M.L., Guo, B.C., Liu, M.X., Cai, X.J., Jia, D.M., 2010a. Reinforcing thermoplastics with hydrogen bonding bridged inorganics. Phys. B Condens. Matter. 405, 655–662.

Du, M.L., Guo, B.C., Wan, J.J., Zou, Q.L., Jia, D.M., 2010b. Effects of halloysite nanotubes on kinetics and activation energy of non-isothermal crystallization of polypropylene. J. Polym. Res. 17, 109–118.

Fernández, A., Muniesa, M., González, J., 2013. Characterisation and processing of reinforced PA 6 with halloysite nanotubes (HNT) for injection molding. J. Mech. Eng. 59, 183–192.

Frenot, A., Chronakis, I.S., 2003. Polymer nanofibers assembled by electrospinning. Curr. Opin. Colloid Interface Sci. 8, 64–75.

Fu, Y., Zhao, D., Yao, P., Wang, W., Zhang, L., Lvov, Y., 2015. Highly aging-resistant elastomers doped with antioxidant-loaded clay nanotubes. ACS Appl. Mater. Interfaces 7, 8156–8165.

Fujii, K., Nakagaito, A.N., Takagi, H., Yonekura, D., 2014. Sulfuric acid treatment of halloysite nanoclay to improve the mechanical properties of PVA/halloysite transparent composite films. Compos. Interfaces 21, 319–327.

Gorrasi, G., 2015. Dispersion of halloysite loaded with natural antimicrobials into pectins: characterization and controlled release analysis. Carbohydr. Polym. 127, 47–53.

Guimarães, L., Enyashin, A.N., Seifert, G., Duarte, H.A., 2010. Structural, electronic, and mechanical properties of single-walled halloysite nanotube models. J. Phys. Chem. C 114, 11358–11363.

Guo, B., Lei, Y., Chen, F., Liu, X., Du, M., Jia, D., 2008. Styrene–butadiene rubber/halloysite nanotubes nanocomposites modified by methacrylic acid. Appl. Surf. Sci. 255, 2715–2722.

Guo, B., Zou, Q., Lei, Y., Jia, D., 2009a. Structure and performance of polyamide 6/halloysite nanotubes nanocomposites. Polym. J. (Tokyo, Jpn.) 41, 835–842.

Guo, B., Chen, F., Lei, Y., Liu, X., Wan, J., Jia, D., 2009b. Styrene-butadiene rubber/halloysite nanotubes nanocomposites modified by sorbic acid. Appl. Surf. Sci. 255, 7329–7336.

Guo, B., Zou, Q., Lei, Y., Du, M., Liu, M., Jia, D., 2009c. Crystallization behavior of polyamide 6/halloysite nanotubes nanocomposites. Thermochim. Acta 484, 48–56.

Guo, B., Chen, F., Lei, Y., Jia, D., 2010a. Tubular clay composites with high strength and transparency. J. Macromol. Sci. B. Phys. 49, 111–121.

Guo, B., Liu, X., Zhou, W., Lei, Y., Jia, D., 2010b. Adsorption of ionic liquid onto halloysite nanotubes: mechanism and reinforcement of the modified clay to rubber. J. Macromol. Sci. Part B Phys. 49, 1029–1043.

Handge, U.A., Hedicke-Hochstotter, K., Altstadt, V., 2010. Composites of polyamide 6 and silicate nanotubes of the mineral halloysite: influence of molecular weight on thermal, mechanical and rheological properties. Polymer 51, 2690–2699.

Hanid, N.A., Wahit, M.U., Guo, Q., Mahmoodian, S., Soheilmoghaddam, M., 2014. Development of regenerated cellulose/halloysites nanocomposites via ionic liquids. Carbohydr. Polym. 99, 91–97.

Haroosh, H.J., Dong, Y., Chaudhary, D.S., Ingram, G.D., Yusa, S., 2013. Electrospun PLA: PCL composites embedded with unmodified and 3-aminopropyltriethoxysilane (ASP) modified halloysite nanotubes (HNT). Appl. Phys. A 110, 433–442.

He, S.J., Wang, Y.Q., Feng, Y.P., Liu, Q.S., Zhang, L.Q., 2010. The preparation of an elastomer/silicate layer nanocompound with an exfoliated structure and a strong ionic interfacial interaction by utilizing an elastomer latex containing pyridine groups. Nanotechnology 21, 115601.

He, Y., Kong, W., Wang, W., Liu, T., Liu, Y., Gong, Q., Gao, J., 2012. Modified natural halloysite/potato starch composite films. Carbohydr. Polym. 87, 2706–2711.

Hollister, S.J., 2005. Porous scaffold design for tissue engineering. Nat. Mater. 4, 518–524.

Hou, Y.F., Jiang, J.Q., Li, K., Zhang, Y.W., Liu, J.D., 2014. Grafting amphiphilic brushes onto halloysite nanotubes via a living RAFT polymerization and their pickering emulsification behavior. J. Phys. Chem. B 118, 1962–1967.

Hutmacher, D.W., 2000. Scaffolds in tissue engineering bone and cartilage. Biomaterials 21, 175–189.

Ismail, H., Pasbakhsh, P., Fauzi, M.N.A., Abu Bakar, A., 2008. Morphological, thermal and tensile properties of halloysite nanotubes filled ethylene propylene diene monomer (EPDM) nanocomposites. Polym. Test. 27, 841–850.

Jia, Q., Wu, Y., Xu, Y., Mao, H., Zhang, L., 2006. Combining in-situ organic modification of montmorillonite and the latex compounding method to prepare high-performance rubber-montmorillonite nanocomposites. Macromol. Mater. Eng. 291, 218–226.

Jia, Z., Luo, Y., Guo, B., Yang, B., Du, M., Jia, D., 2009. Reinforcing and flame-retardant effects of halloysite nanotubes on LLDPE. Polym. Plast. Technol. Eng. 48, 607–613.

Jing, H., Higaki, Y., Ishikawa, T., White, K.L., Otsuka, H., Takahara, A., 2015. Polyurethane nanocomposites reinforced with surface modified halloysite nanotubes. Sci. Adv. Mater. 7, 974–980.

Kashiwagi, T., Grulke, E., Hilding, J., Harris, R., Awad, W., Douglas, J., 2002. Thermal degradation and flammability properties of poly(propylene)/carbon nanotube composites. Macromol. Rapid Commun. 23, 761–765.

Kelly, H.M., Deasy, P.B., Ziaka, E., Claffey, N., 2004. Formulation and preliminary in vivo dog studies of a novel drug delivery system for the treatment of periodontitis. Int. J. Pharm. 274, 167–183.

Khorasani, M.T., Mirzadeh, H., Kermani, Z., 2005. Wettability of porous polydimethylsiloxane surface: morphology study. Appl. Surf. Sci. 242, 339–345.

Khunova, V., Kristof, J., Kelnar, I., Dybal, J., 2013. The effect of halloysite modification combined with *in situ* matrix modifications on the structure and properties of polypropylene/halloysite nanocomposites. Express Polym. Lett. 7, 471–479.

Kommireddy, D.S., Ichinose, I., Lvov, Y.M., Mills, D.K., 2005. Nanoparticle multilayers: surface modification for cell attachment and growth. J. Biomed. Nanotechnol. 1, 286–290.

Konnova, S.A., Sharipova, I.R., Demina, T.A., Osin, Y.N., Yarullina, D.R., Ilinskaya, O.N., Lvov, Y.M., Fakhrullin, R.F., 2013. Biomimetic cell-mediated three-dimensional assembly of halloysite nanotubes. Chem. Commun. (Camb.) 49, 4208–4210.

Lecouvet, B., Gutierrez, J.G., Sclavons, M., Bailly, C., 2011a. Structure-property relationships in polyamide 12/halloysite nanotube nanocomposites. Polym. Degrad. Stab. 96, 226–235.

Lecouvet, B., Sclavons, M., Bourbigot, S., Devaux, J., Bailly, C., 2011b. Water-assisted extrusion as a novel processing route to prepare polypropylene/halloysite nanotube nanocomposites: structure and properties. Polymer 52, 4284–4295.

Lecouvet, B., Sclavons, M., Bourbigot, S., Bailly, C., 2013a. Thermal and flammability properties of polyethersulfone/halloysite nanocomposites prepared by melt compounding. Polym. Degrad. Stab. 98, 1993–2004.

Lecouvet, B., Horion, J., D'haese, C., Bailly, C., Nysten, B., 2013b. Elastic modulus of halloysite nanotubes. Nanotechnology 24, 105704.

Lecouvet, B., Sclavons, M., Bourbigot, S., Bailly, C., 2014. Towards scalable production of polyamide 12/halloysite nanocomposites via water-assisted extrusion: mechanical modeling, thermal and fire properties. Polym. Adv. Technol. 25, 137–151.

Lee, K.S., Chang, Y.W., 2013. Thermal, mechanical, and rheological properties of poly(epsilon-caprolactone)/halloysite nanotube nanocomposites. J. Appl. Polym. Sci. 128, 2807–2816.

Lee, Y., Jung, G.E., Cho, S.J., Geckeler, K.E., Fuchs, H., 2013. Cellular interactions of doxorubicin-loaded DNA-modified halloysite nanotubes. Nanoscale 5, 8577–8585.

Lei, Y., Tang, Z., Zhu, L., Guo, B., Jia, D., 2011. Functional thiol ionic liquids as novel interfacial modifiers in SBR/HNTs composites. Polymer 52, 1337–1344.

Levis, S., Deasy, P., 2002. Characterisation of halloysite for use as a microtubular drug delivery system. Int. J. Pharm. 243, 125–134.

Levis, S.R., Deasy, P.B., 2003. Use of coated microtubular halloysite for the sustained release of diltiazem hydrochloride and propranolol hydrochloride. Int. J. Pharm. 253, 145–157.

Li, Z., Wang, C., 2013. Effects of working parameters on electrospinning. In: Li, Z., Wang, C. (Eds.), One-Dimensional Nanostructures. Springer, Berlin Heidelberg, China, pp. 15–28.

Li, C., Liu, J., Qu, X., Guo, B., Yang, Z., 2008. Polymer-modified halloysite composite nanotubes. J. Appl. Polym. Sci. 110, 3638–3646.

Li, C., Liu, J., Qu, X., Yang, Z., 2009. A general synthesis approach toward halloysite-based composite nanotube. J. Appl. Polym. Sci. 112, 2647–2655.

Liao, R., Guo, B., Zhao, J., Zhou, W., 2010. Electrospun polyvinyl alcohol/halloysite nanotubes composite nanofibers. In: The Second International Conference on Advanced Textile Materials and Manufacturing Technology. Hangzhou, China, pp. 63–67.

Lin, T., 2011. Modifications of Polypropylene Fiber and FIBC Model Blend with Halloysite Nanotubes. Master D, South China University of Technology.

Lin, Y., Ng, K.M., Chan, C., Sun, G., Wu, J., 2011. High-impact polystyrene/halloysite nanocomposites prepared by emulsion polymerization using sodium dodecyl sulfate as surfactant. J. Colloid Interface Sci. 358, 423–429.

Lin, T., Zhu, L., Chen, T., Guo, B., 2013. Optimization of mechanical performance of compatibilized polypropylene/poly(ethylene terephthalate) blends via selective dispersion of halloysite nanotubes in the blend. J. Appl. Polym. Sci. 129, 47–56.

Liu, M., Guo, B., Du, M., Jia, D., 2007a. Drying induced aggregation of halloysite nanotubes in polyvinyl alcohol/halloysite nanotubes solution and its effect on properties of composite film. Appl. Phys. A 88, 391–395.

Liu, M., Guo, B., Du, M., Cai, X., Jia, D., 2007b. Properties of halloysite nanotube-epoxy resin hybrids and the interfacial reactions in the systems. Nanotechnology 18, 455–703.

Liu, M., Guo, B., Du, M., Jia, D., 2008a. The role of interactions between halloysite nanotubes and 2,2′-(1,2-ethenediyldi-4,1-phenylene) bisbenzoxazole in halloysite reinforced polypropylene composites. Polym. J. (Tokyo, Jpn.) 40, 1087–1093.

Liu, M., Guo, B., Du, M., Lei, Y., Jia, D., 2008b. Natural inorganic nanotubes reinforced epoxy resin nanocomposites. J. Polym. Res. 15, 205–212.

Liu, M., Guo, B., Zou, Q., Du, M., Jia, D., 2008c. Interactions between halloysite nanotubes and 2,5-bis(2-benzoxazolyl) thiophene and their effects on reinforcement of polypropylene/halloysite nanocomposites. Nanotechnology 19, 205709.

Liu, M., Guo, B., Du, M., Chen, F., Jia, D., 2009a. Halloysite nanotubes as a novel β-nucleating agent for isotactic polypropylene. Polymer 50, 3022–3030.

Liu, M., Guo, B., Du, M., Zou, Q., Jia, D., 2009b. Influence of hybrid fibrils of 2,5-bis (2-benzoxazolyl) thiophene and halloysite nanotubes on the crystallization behaviour of polypropylene. J. Phys. D Appl. Phys. 42, 75306–75314.

Liu, M., Guo, B., Lei, Y., Du, M., Jia, D., 2009c. Benzothiazole sulfide compatibilized polypropylene/halloysite nanotubes composites. Appl. Surf. Sci. 255, 4961–4969.

Liu, K., Yao, X., Jiang, L., 2010a. Recent developments in bio-inspired special wettability. Chem. Soc. Rev. 39, 3240–3255.

Liu, M., Jia, Z., Liu, F., Jia, D., Guo, B., 2010b. Tailoring the wettability of polypropylene surfaces with halloysite nanotubes. J. Colloid Interface Sci. 350, 186–193.

Liu, C., Luo, Y., Jia, Z., Zhong, B., Li, S., Guo, B., Jia, D., 2011. Enhancement of mechanical properties of poly(vinyl chloride) with polymethyl methacrylate-grafted halloysite nanotube. Express Polym. Lett. 5, 591–603.

Liu, L., Wan, Y., Xie, Y., Zhai, R., Zhang, B., Liu, J., 2012a. The removal of dye from aqueous solution using alginate-halloysite nanotube beads. Chem. Eng. J. 187, 210–216.

Liu, M., Li, W., Rong, J., Zhou, C., 2012b. Novel polymer nanocomposite hydrogel with natural clay nanotubes. Colloid Polym. Sci. 290, 895–905.

Liu, M., Zhang, Y., Wu, C., Xiong, S., Zhou, C., 2012c. Chitosan/halloysite nanotubes bionanocomposites: structure, mechanical properties and biocompatibility. Int. J. Biol. Macromol. 51, 566–575.

Liu, M., Zhang, Y., Zhou, C., 2013a. Nanocomposites of halloysite and polylactide. Appl. Clay Sci. 75–76, 52–59.

Liu, M., Wu, C., Jiao, Y., Xiong, S., Zhou, C., 2013b. Chitosan–halloysite nanotubes nanocomposite scaffolds for tissue engineering. J. Mater. Chem. B. 1, 2078.

Liu, M., Dai, L., Shi, H., Xiong, S., Zhou, C., 2015. In vitro evaluation of alginate/halloysite nanotube composite scaffolds for tissue engineering. Mater. Sci. Eng. C 49, 700–712.

Lopes, A.C., Costa, C.M., Serra, R.S.I., Neves, I.C., Ribelles, J.L.G., Lanceros-Mendez, S., 2013. Dielectric relaxation, ac conductivity and electric modulus in poly(vinylidene fluoride)/NaY zeolite composites. Solid State Ion. 235, 42–50.

Lu, Z., Eadula, S., Zheng, Z., Xu, K., Grozdits, G., Lvov, Y., 2007. Layer-by-layer nanoparticle coatings on lignocellulose wood microfibers. Colloid Surf. A Physicochem. Eng. Asp. 292, 56–62.

Lu, D., Chen, H., Wu, J., Chan, C.M., 2011. Direct measurements of the Young's modulus of a single halloysite nanotube using a transmission electron microscope with a bending stage. J. Nanosci. Nanotechnol. 11, 7789–7793.

Luca, V., Thomson, S., 2000. Intercalation and polymerisation of aniline within a tubular aluminosilicate. J. Mater. Chem. 10, 2121–2126.

Lvov, Y., Abdullayev, E., 2013. Functional polymer–clay nanotube composites with sustained release of chemical agents. Prog. Polym. Sci. 38, 1690–1719.

Lvov, Y., Price, R., Gaber, B., Ichinose, I., 2002. Thin film nanofabrication via layer-by-layer adsorption of tubule halloysite, spherical silica, proteins and polycations. Colloids Surf. A Physicochem. Eng. Asp. 198, 375–382.

Lvov, Y.M., Shchukin, D.G., Mohwald, H., Price, R.R., 2008. Halloysite clay nanotubes for controlled release of protective agents. ACS Nano 2, 814–820.

Marini, J., Pollet, E., Averous, L., Bretas, R.E.S., 2014. Elaboration and properties of novel biobased nanocomposites with halloysite nanotubes and thermoplastic polyurethane from dimerized fatty acids. Polymer 55, 5226–5234.

Marney, D.C.O., Russell, L.J., Wu, D.Y., Nguyen, T., Cramm, D., Rigopoulos, N., Wright, N., Greaves, M., 2008. The suitability of halloysite nanotubes as a fire retardant for nylon 6. Polym. Degrad. Stab. 93, 1971–1978.

Mu, B., Zhao, M.F., Liu, P., 2008. Halloysite nanotubes grafted hyperbranched (co)polymers via surface-initiated self-condensing vinyl (co)polymerization. J. Nanopart. Res. 10, 831–838.

Murali, R.S., Padaki, M., Matsuura, T., Abdullah, M.S., Ismail, A.F., 2014. Polyaniline *in situ* modified halloysite nanotubes incorporated asymmetric mixed matrix membrane for gas separation. Sep. Purif. Technol. 132, 187–194.

Nakamura, R., Netravali, A.N., Morgan, A.B., Nyden, M.R., Gilman, J.W., 2013. Effect of halloysite nanotubes on mechanical properties and flammability of soy protein based green composites. Fire Mater. 37, 75–90.

Ning, N.Y., Yin, Q.J., Luo, F., Zhang, Q., Du, R.N., Fu, Q., 2007. Crystallization behavior and mechanical properties of polypropylene/halloysite composites. Polymer 48, 7374–7384.

Olugebefola, S.C., Hamilton, A.R., Fairfield, D.J., Sottos, N.R., White, S.R., 2014. Structural reinforcement of microvascular networks using electrostatic layer-by-layer assembly with halloysite nanotubes. Soft Matter. 10, 544–548.

Ounoughene, G., Le Bihan, O., Chivas-Joly, C., Motzkus, C., Longuet, C., Debray, B., Joubert, A., Le Coq, L., Lopez-Cuesta, J.M., 2015. Behavior and fate of halloysite nanotubes (HNTs) when incinerating PA6/HNTs nanocomposite. Environ. Sci. Technol. 49, 5450–5457.

Pasbakhsh, P., Ismail, H., Fauzi, M.N.A., Bakar, A.A., 2010. EPDM/modified halloysite nanocomposites. Appl. Clay Sci. 48, 405–413.

Pedrazzoli, D., Pegoretti, A., Thomann, R., Kristof, J., Karger-Kocsis, J., 2015. Toughening linear low-density polyethylene with halloysite nanotubes. Polym. Compos. 36, 869–883.

Poikelispaa, M., Das, A., Dierkes, W., Vuorinen, J., 2013. Synergistic effect of plasma-modified halloysite nanotubes and carbon black in natural rubberbutadiene rubber blend. J. Appl. Polym. Sci. 127, 4688–4696.

Prashantha, K., 2011. Processing and characterization of halloysite nanotubes filled polypropylene nanocomposites based on a masterbatch route: effect of halloysites treatment on structural and mechanical properties. Express Polym. Lett. 5, 295–307.

Prashantha, K., Schmitt, H., Lacrampe, M.F., Krawczak, P., 2011. Mechanical behaviour and essential work of fracture of halloysite nanotubes filled polyamide 6 nanocomposites. Compos. Sci. Technol. 71, 1859–1866.

Prashantha, K., Lacrampe, M.F., Krawczak, P., 2013. Highly dispersed polyamide-11/halloysite nanocomposites: thermal, rheological, optical, dielectric, and mechanical properties. J. Appl. Polym. Sci. 130, 313–321.

Qi, R., Guo, R., Shen, M., Cao, X., Zhang, L., Xu, J., Yu, J., Shi, X., 2010. Electrospun poly(lactic-co-glycolic acid)/halloysite nanotube composite nanofibers for drug encapsulation and sustained release. J. Mater. Chem. 20, 10622.

Qi, R., Cao, X., Shen, M., Guo, R., Yu, J., Shi, X., 2012. Biocompatibility of electrospun halloysite nanotube-doped poly(lactic-co-glycolic acid) composite nanofibers. J. Biomater. Sci. Polym. Ed. 23, 299–313.

Qiu, K., Netravali, A.N., 2013. Halloysite nanotube reinforced biodegradable nanocomposites using noncrosslinked and malonic acid crosslinked polyvinyl alcohol. Polym. Compos. 34, 799–809.

Raman, V.S., Rooj, S., Das, A., Stöckelhuber, K.W., Simon, F., Nando, G.B., Heinrich, G., 2013. Reinforcement of solution styrene butadiene rubber by silane functionalized halloysite nanotubes. J. Macromol. Sci. Part A: Pure Appl. Chem. 50, 1091–1106.

Rashmi, B.J., Prashantha, K., Lacrampe, M.F., Krawczak, P., 2015. Toughening of poly(lactic acid) without sacrificing stiffness and strength by melt-blending with polyamide 11 and selective localization of halloysite nanotubes. Express Polym. Lett. 9, 721–735.

Reneker, D.H., Chun, I., 1996. Nanometre diameter fibres of polymer, produced by electrospinning. Nanotechnology 7, 216–223.

Rooj, S., Das, A., Thakur, V., Mahaling, R.N., Bhowmick, A.K., Heinrich, G., 2010. Preparation and properties of natural nanocomposites based on natural rubber and naturally occurring halloysite nanotubes. Mater. Des. 31, 2151–2156.

Shamsi, M.H., Geckeler, K.E., 2008. The first biopolymer-wrapped non-carbon nanotubes. Nanotechnology 19, 075604.
Shamsi, M.H., Luqman, M., Basarir, F., Kim, J.S., Yoon, T.H., Geckeler, K.E., 2010. Plasma-modified halloysite nanocomposites: effect of plasma modification on the structure and dynamic mechanical properties of halloysite-polystyrene nanocomposites. Polym. Int. 59, 1492–1498.
Shchukin, D.G., Möhwald, H., 2007. Surface-engineered nanocontainers for entrapment of corrosion inhibitors. Adv. Funct. Mater. 17, 1451–1458.
Shchukin, D.G., Lamaka, S.V., Yasakau, K.A., Zheludkevich, M.L., Ferreira, M.G.S., Mohwald, H., 2008. Active anticorrosion coatings with halloysite nanocontainers. J. Phys. Chem. C 112, 958–964.
Shutava, T.G., Fakhrullin, R.F., Lvov, Y.M., 2014. Spherical and tubule nanocarriers for sustained drug release. Curr. Opin. Pharmacol. 18, 141–148.
Stockton, W.B., Rubner, M.F., 1997. Molecular-level processing of conjugated polymers. 4. Layer-by-layer manipulation of polyaniline via hydrogen-bonding interactions. Macromolecules 30, 2717–2725.
Tang, X., Hou, M., Zou, J., Truss, R., 2013a. Poly(vinylidene fluoride)/halloysite nanotubes nanocomposites: the structures, properties, and tensile fracture behaviors. J. Appl. Polym. Sci. 128, 869–878.
Tang, Z., Wei, Q., Lin, T., Guo, B., Jia, D., 2013b. The use of a hybrid consisting of tubular clay and graphene as a reinforcement for elastomers. RSC Adv. 3, 17057.
Thakur, P., Kool, A., Bagchi, B., Das, S., Nandy, P., 2014. Enhancement of β phase crystallization and dielectric behavior of kaolinite/halloysite modified poly(vinylidene fluoride) thin films. Appl. Clay Sci. 99, 149–159.
Tian, X.K., Wang, W.W., Wang, Y.X., Komarneni, S., Yang, C., 2015. Polyethylenimine functionalized halloysite nanotubes for efficient removal and fixation of Cr (VI). Microporous Mesoporous Mater. 207, 46–52.
Vahedi, V., Pasbakhsh, P., 2014. Instrumented impact properties and fracture behaviour of epoxy/modified halloysite nanocomposites. Polym. Test. 39, 101–114.
Varghese, S., Karger-Kocsis, J., 2003. Natural rubber-based nanocomposites by latex compounding with layered silicates. Polymer 44, 4921–4927.
Veerabadran, N.G., Mongayt, D., Torchilin, V., Price, R.R., Lvov, Y.M., 2009. Organized shells on clay nanotubes for controlled release of macromolecules. Macromol. Rapid Commun. 30, 99–103.
Vergaro, V., Abdullayev, E., Lvov, Y.M., Leporatti, S., Zeitoun, A., Cingolani, R., Rinaldi, R., 2010a. Halloysite clay nanotubes: characterization, biocompatibility and use as drug carriers. Nanotech 3, 395–396.
Vergaro, V., Abdullayev, E., Lvov, Y.M., Zeitoun, A., Cingolani, R., Rinaldi, R., Leporatti, S., 2010b. Cytocompatibility and uptake of halloysite clay nanotubes. Biomacromolecules 11, 820–826.
Vergaro, V., Lvov, Y.M., Leporatti, S., 2012. Halloysite clay nanotubes for resveratrol delivery to cancer cells. Macromol. Biosci. 12, 1265–1271.
Wang, B., Huang, H., 2013. Effects of halloysite nanotube orientation on crystallization and thermal stability of polypropylene nanocomposites. Polym. Degrad. Stab. 98, 1601–1608.
Wang, B., Huang, H., 2014. Incorporation of halloysite nanotubes into PVDF matrix: nucleation of electroactive phase accompany with significant reinforcement and dimensional stability improvement. Compos. A: Appl. Sci. Manuf. 66, 16–24.
Wang, L., Wang, Z., Zhang, X., Shen, J., Chi, L.F., Fuchs, H., 1997. A new approach for the fabrication of an alternating multilayer film of poly(4-vinylpyridine) and poly(acrylic acid) based on hydrogen bonding. Macromol. Rapid Commun. 18, 509–514.

Wang, Y., Deen, I., Zhitomirsky, I., 2011. Electrophoretic deposition of polyacrylic acid and composite films containing nanotubes and oxide particles. J. Colloid Interface Sci. 362, 367–374.

Wang, Z., Wang, H., Liu, J., Zhang, Y., 2014. Preparation and antifouling property of polyethersulfone ultrafiltration hybrid membrane containing halloysite nanotubes grafted with MPC via RATRP method. Desalination 344, 313–320.

Wang, B., Huang, H., Guan, W., 2015. Squeeze flow-induced hierarchical morphology in micro-injection compression molded thin-walled plates of PP/SAN blend and its nanocomposite with halloysite nanotubes. Polym. Plast. Technol. Eng. 54, 814–821.

Wei, W., Abdullayev, E., Hollister, A., Mills, D., Lvov, Y.M., 2012. Clay nanotube/poly (methylmethacrylate) bone cement composites with sustained antibiotic release. Macromol. Mater. Eng. 297, 645–653.

Wu, W., Cao, X., Zhang, Y., He, G., 2013. Polylactide/halloysite nanotube nanocomposites: thermal, mechanical properties, and foam processing. J. Appl. Polym. Sci. 130, 443–452.

Xu, W., Luo, B., Wen, W., Xie, W.J., Wang, X.Y., Liu, M.X., Zhou, C.R., 2015. Surface modification of halloysite nanotubes with L-lactic acid: an effective route to high-performance poly(L-lactide) composites. J. Appl. Polym. Sci. 132, 41451–41459.

Yah, W.O., Takahara, A., Lvov, Y.M., 2012a. Selective modification of halloysite lumen with octadecylphosphonic acid: new inorganic tubular micelle. J. Am. Chem. Soc. 134, 1853–1859.

Yah, W.O., Xu, H., Soejima, H., Ma, W., Lvov, Y., Takahara, A., 2012b. Biomimetic dopamine derivative for selective plymer modification of halloysite nanotube lumen. J. Am. Chem. Soc. 134, 12134–12137.

Yang, C., Liu, P., Zhao, Y.Q., 2010. Preparation and characterization of coaxial halloysite/polypyrrole tubular nanocomposites for electrochemical energy storage. Electrochim. Acta 55, 6857–6864.

Yano, K., Usuki, A., Okada, A., 1997. Synthesis and properties of polyimide-clay hybrid films. J. Polym. Sci. A Polym. Chem. 35, 2289–2294.

Ye, Y., Chen, H., Wu, J., Ye, L., 2007. High impact strength epoxy nanocomposites with natural nanotubes. Polymer 48, 6426–6433.

Ye, Y., Chen, H., Wu, J., Chan, C., 2011a. Evaluation on the thermal and mechanical properties of HNT-toughened epoxy/carbon fibre composites. Compos. B Eng. 42, 2145–2150.

Ye, Y., Chen, H., Wu, J., Chan, C., 2011b. Interlaminar properties of carbon fiber composites with halloysite nanotube-toughened epoxy matrix. Compos. Sci. Technol. 71, 717–723.

Yu, Z.Z., Hu, G.H., Varlet, J., Dasari, A., Mai, Y.W., 2005. Water-assisted melt compounding of nylon-6/pristine montmorillonite nanocomposites. J. Polym. Sci. B Polym. Phys. 43, 1100–1112.

Yuan, P., Southon, P.D., Liu, Z., Kepert, C.J., 2012. Organosilane functionalization of halloysite nanotubes for enhanced loading and controlled release. Nanotechnology 23, 375705.

Zeng, S., Reyes, C., Liu, J., Rodgers, P.A., Wentworth, S.H., Sun, L., 2014. Facile hydroxylation of halloysite nanotubes for epoxy nanocomposite applications. Polymer 55, 6519–6528.

Zhang, L., Wang, T., Liu, P., 2008a. Polyaniline-coated halloysite nanotubes via in-situ chemical polymerization. Appl. Surf. Sci. 255, 2091–2097.

Zhang, X., Shi, F., Niu, J., Jiang, Y., Wang, Z., 2008b. Superhydrophobic surfaces: from structural control to functional application. J. Mater. Chem. 18, 621–633.

Zhang, J.H., Zhang, D.H., Zhang, A.Q., Jia, Z.X., Jia, D.M., 2013a. Poly (methyl methacrylate) grafted halloysite nanotubes and its epoxy acrylate composites by ultraviolet curing method. J. Reinf. Plast. Compos. 32, 713–725.

Zhang, Y., Chen, Y., Zhang, H., Zhang, B., Liu, J., 2013b. Potent antibacterial activity of a novel silver nanoparticle-halloysite nanotube nanocomposite powder. J. Inorg. Biochem. 118, 59–64.

Zhang, J.H., Jia, Z.X., Jia, D.M., Zhang, D.H., Zhang, A.Q., 2014. Chemical functionalization for improving dispersion and interfacial bonding of halloysite nanotubes in epoxy nanocomposites. High Perform. Polym. 26, 734–743.

Zhang, H.L., Zhu, X.Y., Wu, Y.G., Song, H.Z., Ba, X.W., 2015. High-efficiency grafting of halloysite nanotubes by using p-conjugated polyfluorenes via "click" chemistry. J. Mater. Sci. 50, 4387–4395.

Zhao, M., Liu, P., 2008. Halloysite nanotubes/polystyrene (HNTs/PS) nanocomposites via *in situ* bulk polymerization. J. Therm. Anal. Calorim. 94, 103–107.

Zhao, Y.L., Wang, S.G., Guo, Q.S., Shen, M.W., Shi, X.Y., 2013. Hemocompatibility of electrospun halloysite nanotube- and carbon nanotube-doped composite poly(lactic-co-glycolic acid) nanofibers. J. Appl. Polym. Sci. 127, 4825–4832.

Zhou, W.Y., Guo, B.C., Liu, M.X., Liao, R.J., Rabie, A.B.M., Jia, D.M., 2010. Poly (vinylalcohol)/halloysite nanotubes bionanocomposite films: properties and in vitro osteoblasts and fibroblasts response. J. Biomed. Mater. Res. A 93A, 1574–1587.

Zhu, J., Uhl, F.M., Morgan, A.B., Wilkie, C.A., 2001. Studies on the mechanism by which the formation of nanocomposites enhances thermal stability. Chem. Mater. 13, 4649–4654.

Chapter 22

Halloysite for Controllable Loading and Release

E. Abdullayev[a] and Y. Lvov[b,*]

[a]*Ennis-Flint Science and Technology Center, Thomasville, NC, United States*

[b]*Institute for Micromanufacturing, Louisiana Tech University, Ruston, LA, United States*

[*]*Corresponding author: e-mail: hovsan@gmail.com*

22.1 INTRODUCTION

Halloysite is an alumosilicate clay mineral with a submicron-sized tubular morphology that has an external diameter of 50–200 nm, lumen diameter of 5–30 nm and length of 0.5–2 μm (Bates et al., 1950). Chemically, halloysite is similar to kaolinite (Joussein et al., 2005). Due to its tubular shape with an empty lumen, halloysite may be used as a 'green' and environmentally available container for encapsulation of the active agents that benefit the medical and chemical industries. Research activities in this area have gained significant attention in recent years due to the obvious benefits of halloysite, such as elongated shape, empty lumen, ease of mixing with a broad range of polymers, biocompatibility, environmental friendliness and availability (Yelleswarapu et al., 2010). Incorporating nanotubes loaded with active agents within the polymer matrix provides smart clay polymer nanocomposite (CPN) material with a specific functionality apart from improved mechanical properties, benefiting from controlled release of the active agents such as self-healing, improved corrosion protection, drug and antibiotic release and mould growth resistance. As such, halloysite resembles 'ceramic bones' within a polymer matrix loaded with functional chemicals, like real bones loaded with marrow (Abdullayev and Lvov, 2013; Lvov and Abdullayev, 2013).

Some properties of well-known inorganic siliceous materials, which are being studied as controlled-release containers along with halloysite, are listed in Table 22.1. Halloysite holds a unique position among these materials, however, due to its small particle size and large pores. The small size makes halloysite suitable for applications in bio-CPN and tissue engineering, while the large pores allow for the immobilisation of biomolecules like enzymes and proteins within halloysite pores. The combination of biopolymers and

Developments in Clay Science, Vol. 7. http://dx.doi.org/10.1016/B978-0-08-100293-3.00022-4

TABLE 22.1 Comparison of Halloysite and Other Siliceous Materials (Bates et al., 1950; Mahani and Kazemeini, 2003; Joussein et al., 2005; Horcajada et al., 2006; Aguzzi et al., 2007; Abbasov et al., 2013; Abdullayev et al., 2013; Datt et al., 2013; Lvov and Abdullayev, 2013; Rives et al., 2014; Tan et al., 2014)

	Halloysite nanotubes	Diatomaceous Earth	Platy clay minerals	Zeolites
Chemical composition	$Al_2Si_2O_5(OH)_4$	Mainly SiO_2, below 10% Al_2O_3 and Fe_2O_3	Al_2O_3, SiO_2, MgO, H_2O and other metal oxides	Al_2O_3, SiO_2, Na_2O, K_2O and traces of other metal oxides
Morphology	Tubular	Thousands of shapes representing ancient algae	Multilayers stacked with ionic and hydrogen bonds	Bulk microporous material
Size	Length 0.5–2.0 μm; width 50–200 nm; pores 5–30 nm	5–100 μm; pores: 0.1–1.0 μm	5–20 μm in length and width, thickness of individual layers: 0.7–2.5 nm	Generally several microns, depending on grinding level; pores: 0.5–2.0 nm
Specific surface area, m^2/g	20–150	16–70	10–200	100–900
Number of publications on controlled release[a] overall/last 5 years	94/72	19/11	Montmorillonite 257/138 Bentonite 117/39 Hydrotalcite 100/40 Kaolinite 68/24 Vermiculite 37/9	251/118

[a]*Publications were searched from Scopus using the keywords 'controlled release', 'sustained release' and 'drug delivery', along with the name of the template as of 7 May 2015.*

halloysite generates new materials with properties that match the advantages to be derived from natural resources (namely, environmental friendliness and low cost). In addition, halloysite has an easily modifiable surface chemistry that differs at the external surface and inner pores (exterior silica, interior aluminium oxide), allowing for the selective modification of each of them (Yuan et al., 2008; Yah et al., 2012a,b; Massaro et al., 2015). All these advantages have attracted the attention of researchers all over the world, as indicated by the increase in studies performed in this area during the past several years. Close to 80% of the research papers on halloysite applications have been published in the last 5 years, as opposed to 20–40% of papers looking into other materials. These studies include controlled delivery in protective coatings (Fix et al., 2009; Abdullayev et al., 2013), bone implants (Wei et al., 2012), teeth fillers (Kelly et al., 2004), cosmetics (Suh et al., 2011), artificial tissue scaffolds and wound-healing materials (Zhou et al., 2009; Qi et al., 2010; Nitya et al., 2012; Liu et al., 2013). Halloysite functionalised with selected proteins was used as a target-specific coating to capture leukaemic and epithelial cancer cells (Hughes and King, 2010). This chapter attempts to summarise the results of the recent research activities focusing on halloysite as a controlled-release nanocontainer. Special attention was given to applications in bionanocomposites, drug delivery and protective coatings.

22.2 ASPECTS OF HALLOYSITE STRUCTURE RELATED TO CONTROLLED-RELEASE APPLICATIONS

Halloysite occurs in nature as rolled alumosilicate layers containing tetrahedral SiO_4 and octahedral $AlO_2(OH)_4$ sheets with intercalated water molecules between adjacent layers (Fig. 22.1). Pure halloysite (10 Å) has the chemical formula of $Al_2Si_2O_5(OH)_4{\cdot}nH_2O$, where nH_2O stands for intercalated water. After drying in mild conditions, it irreversibly loses intercalated water, transforming to halloysite (7 Å), where $n=0$. Commercially available halloysite is generally halloysite (7 Å), due to the relative instability of halloysite (10 Å) (Joussein et al., 2005; Abdullayev et al., 2011). In this section, only aspects of the halloysite structure related to the loading and controlled release of active agents will be described (for a detailed description of the halloysite structure, see Chapter 2).

The loading efficiency of active agents is closely related to the size of the halloysite lumen, and hence by its effective pore volume. Churchman et al. (1995) studied the porosity of halloysite samples mined from various deposits in New Zealand and Australia using nitrogen adsorption and desorption isotherms and obtained specific surface areas (SSA) in the range of 30–94 m^2/g. Samples with small halloysite particles had abundant pores in the 5–15 nm range (Fig. 22.1B and C). Apart from cylindrical tube lumens, they also reported slit-shaped pores formed between layers due to their deformation upon drying (Churchman et al., 1995).

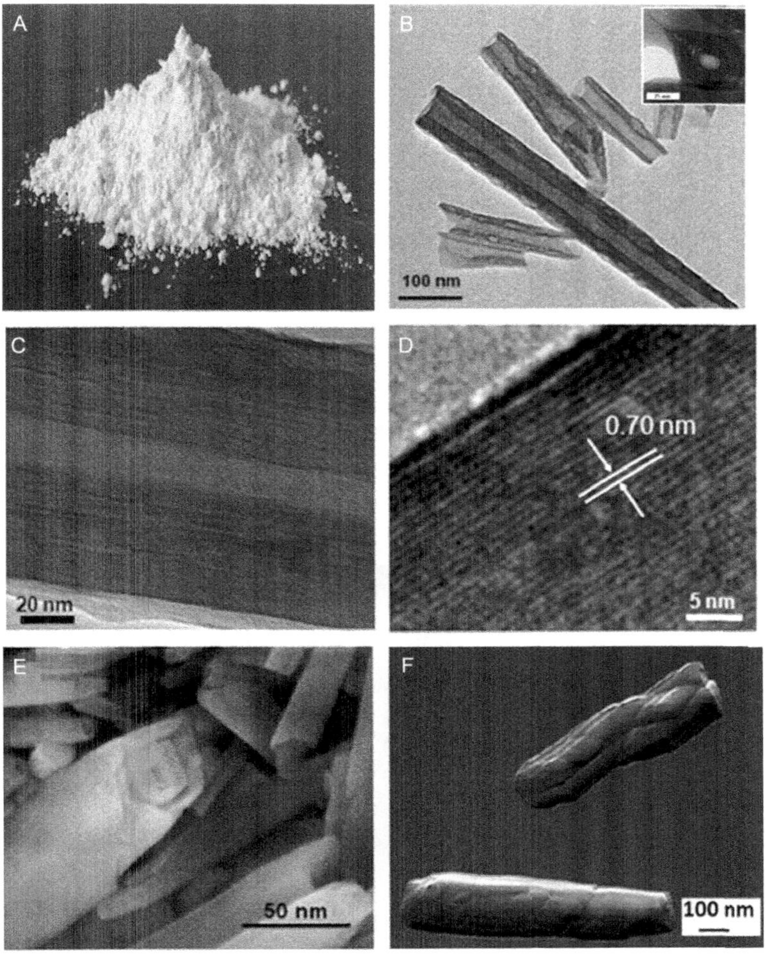

FIG. 22.1 Halloysite images: (A) halloysite powder, (B–E) transmission electron microscopy (TEM) images of halloysite tubes including tube cross-section, (E and F) scanning electron microscopy (SEM) and atomic force microscopy (AFM) images of the clay mineral nanotubes. *Reprinted with permission from Abdullayev et al. (2011). Copyright 2011 American Chemical Society.*

One of the important features of halloysite is its different surface chemistries at the inner and outer sides of the tubes: silica sheets correspond to the external surfaces of the tubes, while aluminium oxide makes up the inner (lumen) surface chemistry. Aluminium and silicon oxides have different ionisation properties and surface charges, which is evident from the zeta-potentials of their colloids in water (Fig. 22.2A). Alumina has a positive charge up to pH 8.5, while silica is negative above pH values of 1.5. Due to silica on the external surfaces, halloysite has a negative surface charge

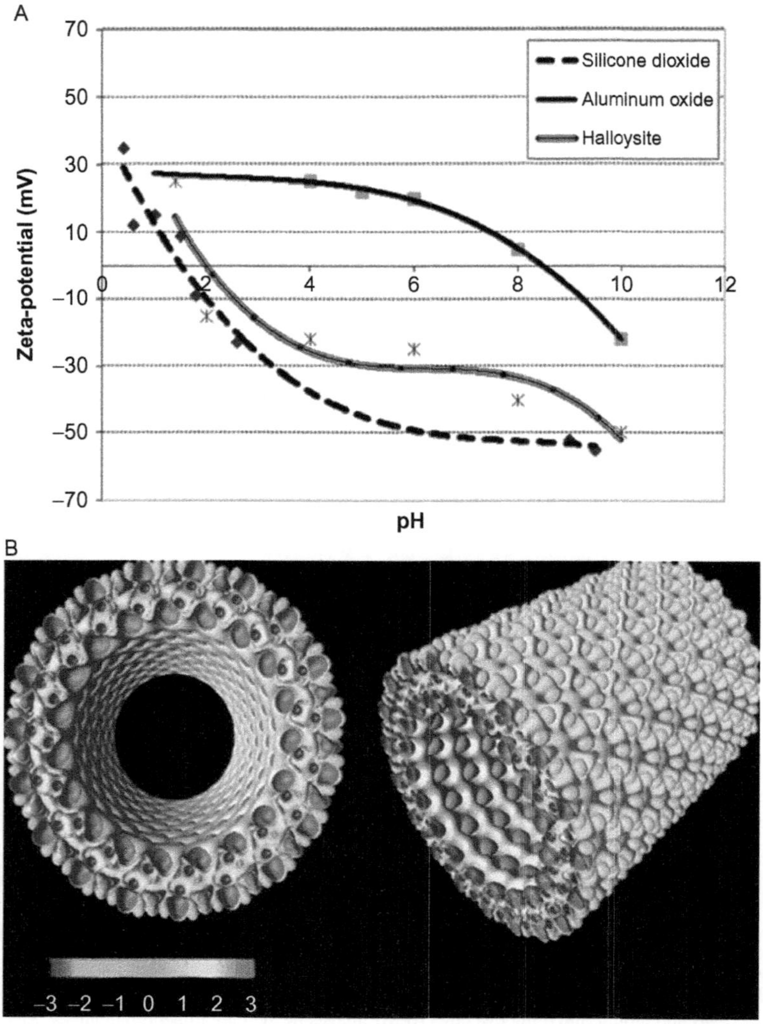

FIG. 22.2 (A) Zeta-potential for halloysite nanotubes, silica and alumina nanoparticles (compounds of similar size and surface chemistry). (B) Distribution of surface charge on single-walled hypothetical halloysite nanotubes calculated by self-consistent charge density calculations. *Panel (A): Reprinted with permission from Vergaro et al. (2010). Copyright 2010 American Chemical Society. Panel (B): Reprinted with permission from Guimarães et al. (2010). Copyright 2010 American Chemical Society.*

in a neutral environment; however, inner lumens (ie, aluminium oxide) are positively charged (Veerabadran et al., 2007). The surface charges of the halloysite faces were further confirmed by self-consistent charge density calculations of single-walled (SW) halloysite tubules provided by Guimarães et al. (2010)

(Fig. 22.2B). Difference in surface charges allows for the selective modification of inner and external faces of halloysite nanotubes and selective loading of negatively charged molecules within halloysite internal lumens (Lvov and Abdullayev, 2013).

Cytocompatibility is the prerequisite for the halloysite usage in drug-release applications. A large amount of work has been done in this area, showing the biocompatible nature of halloysite, even though approval of this material by the US Food and Drug Administration (FDA) for medical applications has yet to be achieved. In vitro cell toxicity tests on breast cancer and human dermal fibroblast cells showed that raw halloysite is 50 times less toxic than sodium chloride (Veerabadran et al., 2007). Penetration and accumulation of halloysite within MCF-7 and HeLa cell membranes did not prevent their proliferation, as evidenced by confocal microscopy research. Halloysite nanotubes were found safe for a model soil nematode (*Caenorhabditis elegans*) at concentrations below 1 mg/mL (Vergaro et al., 2010; Fakhrullina et al., 2015). Chapter 20 gives a detailed overview of halloysite safety issues.

22.3 LOADING AND RELEASE CHARACTERISTICS OF UNMODIFIED HALLOYSITE NANOTUBES

Halloysite nanotubes were successfully loaded with diverse types of active agents, including pharmaceutical agents, antibacterials, corrosion inhibitors and mould growth inhibitors, made of simple organic and inorganic substances, as well as enzymes and proteins with relatively large molecular dimensions. Halloysite entraps molecules in three different ways: (i) adsorption to the external and internal walls of the tubes, (ii) loading into lumen and (iii) intercalation of substance between layers. The most important of these is the lumen loading, as this provides the highest loading capability and controlled release. Release from halloysite lumen typically lasts from several hours to days, depending on molecular mass, structure and solubility of the active agent in the release medium (Abdullayev and Lvov, 2013; Lvov and Abdullayev, 2013).

22.3.1 Loading Active Agents Within Halloysite Lumens

Empty lumens of the halloysite correspond to about 20% of the total volume, which makes them suitable for loading with about 10–15% of the active agents within internal lumens. There are some studies claiming even higher loading efficiency, which could be associated with the external adsorption of active agents along with the lumen loading.

A general procedure for the halloysite lumen loading, shown in Fig. 22.3, is as follows. Halloysite is mixed as a dry powder with a saturated (highly

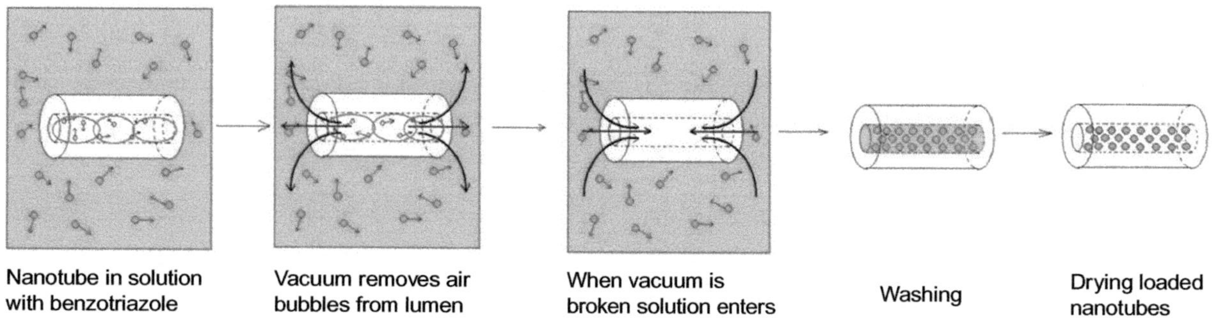

FIG. 22.3 Loading halloysite nanotubes with benzotriazole. *Reprinted with permission from Abdullayev et al. (2009). Copyright 2009 American Chemical Society.*

concentrated) solution of the active agent. Solvents are generally chosen among liquids of low viscosity that dissolve active agents and wet halloysite walls. Water, acetone and ethanol are among the most commonly used solvents. The dispersion of halloysite is stirred and transferred to a jar for evacuation with a vacuum pump, which then should be kept in a vacuum for 10–30 min, allowing air bubbles to be removed from halloysite pores, which is generally indicated by a slight fizzing of the solution. Once the vacuum is broken, solution enters the lumen. This is generally repeated three to four times to increase the loading efficiency. After loading, the halloysite nanotubes are washed to remove unloaded substance and dried (Abdullayev et al., 2009).

The concentration gradient of the active agent between halloysite pores and external solution is important for reaching maximum loading efficiency. This is usually maintained by drying of the solution while loading (generally in a moderate vacuum). On the other hand, drying speed has to be relevant to the diffusion rate of the active agents. Therefore, volatile and low-viscosity solvents like acetone or ethanol are preferable in order to increase the process speed. Water is generally used for medicines, enzymes and proteins when the stability of the active agent in the abovementioned solvent is questionable, or in cases when solubility is not sufficiently high (Abdullayev and Lvov, 2011).

Some of the active agents successfully encapsulated by halloysite nanotubes are presented in Table 22.2. The loading efficiency of the unmodified halloysite nanotube lies in the range of 5–13% for low-molecular-mass substances and about 1–2% for enzymes. Loading of low-molecular-mass actives increased up to 20% by various surface functionalisation methods. It should be noted that these values are less than the typical loading efficiency reported for layered clay minerals and synthetic zeolites (20–30% range) (Horcajada et al., 2006; Aguzzi et al., 2007; Datt et al., 2013; Rives et al., 2014; Tan et al., 2014). However, halloysite provides complete isolation of the loaded active agent from surrounding media, which is critical for both sustained and controlled release. Therefore, halloysite is the template of choice when the benefits of the nanosized container and slow release are more important than the loading efficiency.

High thermal stability is one of the main advantages of the halloysite nanotube as a carrier. Tubular morphology is preserved at temperatures as high as 900°C, even though crystalline structures transform slightly above 450°C because they lose structural water molecules (Yuan et al., 2012b). The high thermal stability of halloysite was beneficial for loading halloysite with various metals through the thermal decomposition of loaded metal-organic salts. One of these metals is silver that was loaded into halloysite pores as silver acetate and decomposed at 400°C (Abdullayev et al., 2011). A similar approach was used for the preparation of iron- and cobalt-supported halloysite nanotubes through the thermal decomposition of corresponding metal acetyl acetonates at 500°C (Fig. 22.4) (Abdullayev and Lvov, 2013). More information on the thermal behaviour of halloysite is presented in Chapter 7.

TABLE 22.2 Some of the Active Agents Encapsulated by Halloysite Nanotubes

Template	Loaded active agent	Loading efficiency (%)	References
Medicines			
Halloysite	Ofloxacin (361.3 Da)	4.5	Wang et al. (2014)
Alkali-modified halloysite	Ofloxacin	6.3	Wang et al. (2013)
Halloysite	Dexamethasone (392.5 Da)	3–5	Veerabadran et al. (2007)
Halloysite	Nifedipine (346.3 Da)	3–5	Veerabadran et al. (2007)
Halloysite	Furosemide (330.7 Da)	3–5	Veerabadran et al. (2007)
Halloysite	Ibuprofen (206 Da)	11.7	Tan et al. (2013)
Halloysite functionalised with γ-aminopropyltriethoxysilane (APTES)	Ibuprofen	14.8	Tan et al. (2013)
Halloysite coated with polybutyl cyanoacrylate	Diltiazem HCl (414 Da)	18–19	Levis and Deasy (2003)
Halloysite coated with chitosan	Diltiazem HCl	5–8	Levis and Deasy (2003)
Halloysite	Diphenhydramine HCl (255 Da)	5.9	Hemmatpour et al. (2015)
Halloysite coated with poly(*N,N*-dimethylaminoethyl methacrylate)	Diphenhydramine HCl	4.5	Hemmatpour et al. (2015)
	Diclofenac sodium salt (318 Da)	3.5	Hemmatpour et al. (2015)
Halloysite	Curcumin	1.5	Riela et al. (2014)
Halloysite functionalised with butyl triazolium salt	Curcumin	12.5	Riela et al. (2014)
Halloysite	Resveratrol	5	Vergaro et al. (2012)

Continued

TABLE 22.2 Some of the Active Agents Encapsulated by Halloysite Nanotubes—Cont'd

Template	Loaded active agent	Loading efficiency (%)	References
Corrosion inhibitors			
Halloysite	BTA (119 Da)	4	Abdullayev et al. (2013)
Halloysite	Mercaptobenzimidazole (150 Da)	25	Abdullayev et al. (2013)
Halloysite	Mercaptobenzothiazole (167 Da)	19	Abdullayev et al. (2013)
Antibacterials			
Halloysite	Gentamicin (477.6 Da)	11.0	Wei et al. (2012)
Halloysite	Brilliant green (482.6 Da)	2.8	Wei et al. (2014)
Halloysite with copper-benzotriazole stoppers	Brilliant green	6–15	Wei et al. (2014)
Cosmetic additives			
Halloysite	Glycerin	19	Suh et al. (2011)
Dyes			
Halloysite	Orange II (350.3 Da)	5.9	Yuan et al. (2012a,b)
APTES functionalised with halloysite	Orange II	7.8	Yuan et al. (2012a,b)
Halloysite	Methylene blue (319.9 Da)	4–9	Zhao and Liu (2008)
Halloysite	Neutral red (288.8 Da)	3–5	Luo et al. (2010)
Halloysite	Rhodamine 6G (479.0 Da)	4.4	Zhao et al. (2012)
Halloysite	Chromeazurol S (605.3 Da)	3.9	Zhao et al. (2012)
80% Halloysite-chitosan nanocomposite	Malachite green	2.9	Peng et al. (2015)
Enzymes			
Halloysite	α-Amylase	1.54	Zhai et al. (2010)
Halloysite	Urease	1.17	Zhai et al. (2010)

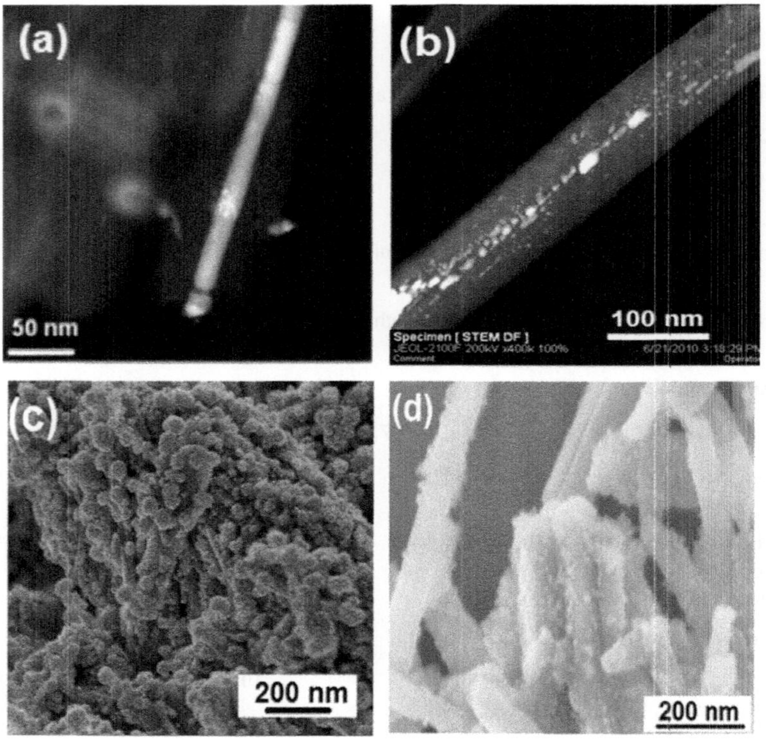

FIG. 22.4 (A and B) Transmission electron microscope image of halloysite samples loaded with silver and SEM images of iron (C) and cobalt (D) nanoparticles seeded on halloysite walls. *Reprinted with permission from Abdullayev et al. (2011); Abdullayev and Lvov (2013). Copyright 2013 Royal Society of Chemistry.*

22.3.2 Release Characteristics of Unmodified Halloysite Nanotubes

Halloysite nanotubes have been extensively studied as controlled-release nanocontainers over the last decade for a wide variety of substances, and various release kinetics models have been proposed. In this section, a focus on release behaviour of active agents will be reported as previously described in Table 22.2.

22.3.2.1 Inorganic Salts

Release of the inorganic salts from unmodified halloysite nanotubes takes place very fast, with complete release achieved within 1–2 h. This is related to the greater mobility and solubility of the inorganic ions in release medium and weaker interaction with halloysite nanotube walls. Therefore, halloysite

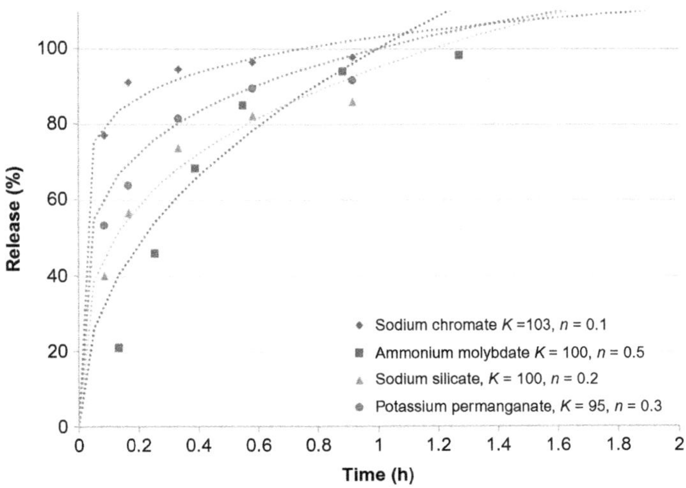

FIG. 22.5 Release profiles of ammonium molybdate ($(NH_4)_6Mo_7O_{24}$), potassium permanganate ($KMnO_4$), sodium silicate (Na_2SiO_3) and sodium chromate (Na_2CrO_4) from unmodified halloysite. *Dashed lines* indicate power function release profile with *K*- and *n*-values indicated near the names of the active agents.

was not studied for the loading and release of inorganic substances, though there are a few studies describing the release kinetics of various salts. In Fig. 22.5, release kinetics curves of ammonium molybdate, potassium permanganate, sodium metasilicate and sodium chromate are shown. Release experiments were conducted in deionised (DI) water under magnetic stirring. At certain time intervals, small volumes (about 100 μL) of release media was collected and centrifuged in order to determine the concentration of released ions. Fresh water was added to replenish collected liquid. The concentrations of the ions were determined by their characteristic absorption peaks in UV–vis regions using the Beer–Lambert law. Release profiles were fit to power function (also called the *Korsmeyer–Peppas model*); $R_t = k \cdot t^n$, where R_t is the amount of the released inorganic substance at time t, and k and n are experimental constants determined from the best fit. This model was developed by Korsmeyer et al. (1983) for release from cylindrical and bottle-shaped pores with small openings, in which k is related to the diffusion speed of the leaking substance and n is related to the diffusion mechanism and geometry of the pores. When $n = 0.5$, diffusion is Fickian, and when $n = 1.0$, zero diffusion follows zero-order kinetics (linear release).

22.3.2.2 *Corrosion Inhibitors*

Halloysite has been actively studied for loading and release containers for organic corrosion inhibitors, especially aromatic triazole and imidazole derivatives that are primarily used to protect copper and aluminium alloys and

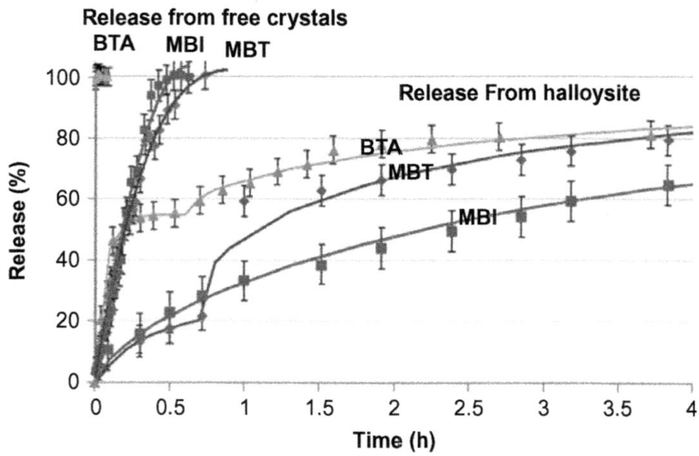

FIG. 22.6 Release profiles of corrosion inhibitors; benzotriazole (BTA), 2-mercaptobenzimidazole (MBI) and 2-mercaptobenzothiazole (MBT) from free crystals and halloysite tubes. *Reprinted with permission from Abdullayev et al. (2013). Copyright 2013 American Chemical Society.*

partially steels. The release profiles of three corrosion inhibitors [benzotriazole (BTA), 2-mercaptobenzimidazole and 2-mercaptobenzothiazole] from free crystals and halloysite nanotubes are illustrated in Fig. 22.6. Inhibitor release from halloysite nanotubes is retarded significantly, extending over 30 h. Calculations based on the release profile predict that 99% of the release of 2-mercaptobenzimidazole is achieved within 35 h, which is 50 times slower than the release from its microcrystals. BTA and 2-mercaptobenzothiazole released from halloysite nanotubes in two stages; initial rapid release of externally adsorbed inhibitors, followed by slow release from internal lumens (Abdullayev et al., 2013). A two-stage release profile was also reported for 5-aminosalicylic acid release from halloysite (Aguzzi et al., 2013). Closer inspection of the release curves indicates that the lumen loading increases in the order of benzotriazole <2-mercaptobenzothiazole < 2-mercaptobenzimidazole (Abdullayev et al., 2013).

22.3.2.3 *Antiseptics and Antibiotics*

Several antiseptic agents have been encapsulated by halloysite nanotubes, including silver, brilliant green and iodine. Brilliant green release profiles from halloysite nanotubes coated with copper-BTA release stoppers are described in Fig. 22.7. Release from free microcrystals was also shown for comparison. Brilliant green release was significantly extended after loading within halloysite lumens and further extended with release stoppers (release stoppers will be discussed later in this chapter). Brilliant green is completely dissolved in 15 min, while only 70% of the antiseptic is released from

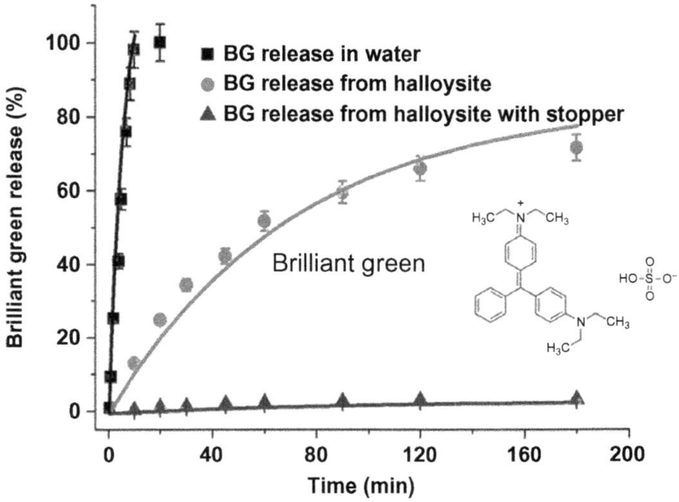

FIG. 22.7 Brilliant green (BG) release curves in deionised water at room temperature. *Reprinted with permission from Wei et al. (2014). Copyright 2014 Royal Society of Chemistry.*

halloysite lumens within 3 h. First-order kinetics has been used to describe release kinetics as follows: $R = M_{\infty}\left(1 - e^{-kt}\right)$, where M_{∞} is the total amount of the active agent loaded within halloysite lumens and k is the release rate constant (Wei et al., 2014).

Gentamicin, a commonly used antibiotic in hip replacement surgery, was encapsulated within halloysite nanotubes. The purpose of this study was to effectively incorporate gentamicin within acrylic bone cements commonly used as bone implants. Loading was achieved by vacuum cycling, as described earlier in this chapter. Without washing the loaded halloysite nanotubes, gentamicin comprised 15% of the total mass, and it was reduced to 11% after two rinses with DI water, which roughly corresponds to the inter lumen loading capacity. The 4% of lost gentamicin may be associated with the negatively charged outer surface of the nanotubes. Gentamicin, having an isoelectric point of over 8.0, possesses a small positive charge at pH 7 and may adsorb on the outer surface of halloysite. Release kinetics was studied in water under magnetic stirring. Gentamicin release from virgin halloysite nanotubes lasted over 48 h and was further extended to 2 weeks with halloysite-poly(methylmethacrylate) bone cement nanocomposites (Wei et al., 2012).

22.3.2.4 *Medicines*

Drug delivery systems are designed to improve therapeutic efficacy by maintaining therapeutic drug concentrations at the site of action. Halloysite nanotubes can show several advantages in drug delivery applications due to its submicron size and 10 nm pores that are suitably fitted for loading medicines with a broad molecular range.

A number of anti-inflammatory drugs have been effectively encapsulated within halloysite lumens, including 5-aminosalicylic acid (Aguzzi et al., 2013), dexamethasone (Veerabadran et al., 2007) and ibuprofen (Veerabadran et al., 2007; Tan et al., 2013) (Fig. 22.8). All these drugs are poorly soluble in neutral water; hence, loading within halloysite lumens provides not only sustained release, but also their effective dispersion in aqueous media.

Release behaviour of 5-aminosalicylic acid from halloysite nanotubes have been studied in detail by Aguzzi et al. (2013) in three different release media; purified water, 0.1 M HCl and pH 6.8 buffer solution. These authors show that release kinetics takes place in two stages: namely, fast desorption of externally adsorbed molecules and slow drug release from nanotube lumens. Desorption rate increased in acidic and buffer solutions as compared to purified water, which was ascribed to the increased ionic strength of the release medium.

Release profiles of three poorly soluble drugs from halloysite nanotubes are shown in Fig. 22.9. Encapsulation in halloysite reduced drug-release time by more than 20 times. Only 70% of dexamethasone was released within 4 h. In the case of furosemide and nifedipine, the release rate was even lower (50% and 40%, respectively). Release profiles from the naked drug crystals, also studied for comparison, show that complete dissolution in water took place in 15, 30 and 40 min for furosemide, dexamethasone and nifedipine, respectively. Release curves were fit to the Korsmeyer–Peppas model ($R_t = k^*t^n$, described in Section 22.3.2.1, earlier in this chapter) with the k parameter varying from 18–26 and n ranging from 0.5–0.8 (shown in Fig. 22.9). Furosemide has the lowest k-value, indicating the strongest interaction with the halloysite walls compared to dexamethasone and nifepidine. This interaction can be ascribed to the coordination of sulfonate groups in its structure around Al^{3+} ions of halloysite or due to the presence of the acidic COOH group, which increases its negative charge. The drug-release mechanism fits the nanopore-controlled regime for nifedipine

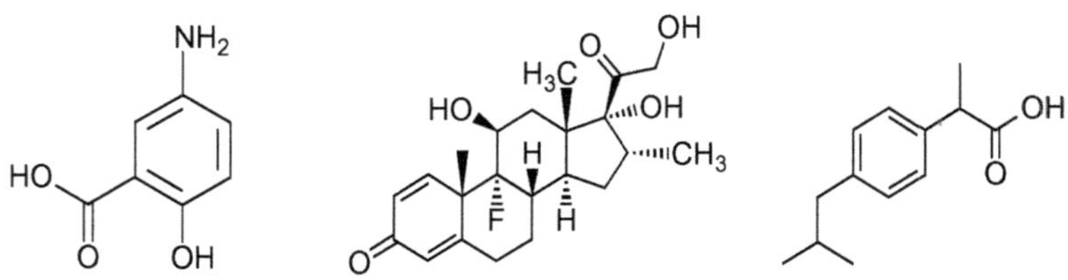

FIG. 22.8 Chemical formulas of 5-aminosalicylic acid (153.1 Da), dexamethasone (392.5 Da) and ibuprofen (206.3 Da).

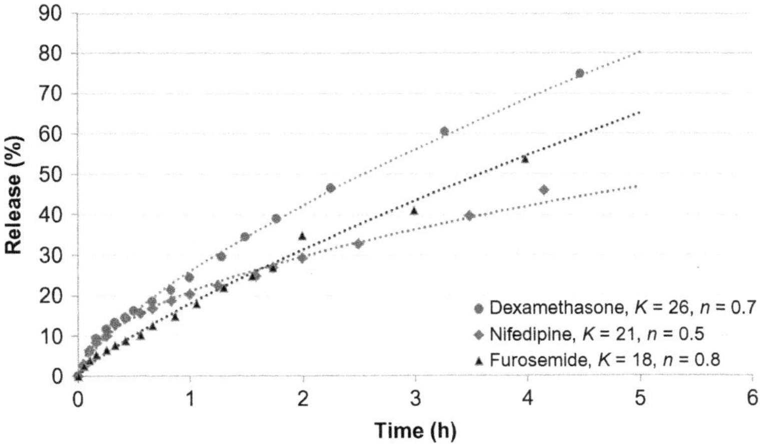

FIG. 22.9 Release of three poorly water-soluble drugs from halloysite nanotubes in water. Drugs were loaded into halloysite by using 50% ethanol-water solution. *Dashed lines* indicate power function release profile with *K*- and *n*-values indicated near the names of the active agents.

and approaches zeroth-order kinetics for furosemide and dexamethasone (Veerabadran et al., 2007).

Loading the anti-anginal drug diltiazem hydrochloride within halloysite lumens showed promise for oral administration of this drug. The loading efficiency was about 25.6% by weight, and release lasted about 1 h from raw halloysite nanotubes. The release was further retarded to 7 h by coating halloysite with chitosan. This formulation has been proposed for treating periodontitis disease in dogs (Levis and Deasy, 2003; Kelly et al., 2004).

22.3.2.5 Proteins and Enzymes

Halloysite is one of the few containers capable of encapsulating globular proteins with diameters ranging from 2–6 nm. The advantage of having a large tubule diameter is clear in this case. The release rate of loaded proteins is rather slow; only 5% of catalase and glucose oxidase was released within 5 h (Fig. 22.10). Proteins have lower diffusion coefficients, as well as stronger interactions with the lumen walls; this is believed to be the main reason behind their slow release (Lvov and Price, 2008).

Halloysite pores were shown to be efficient nanocages for the protection of enzymes allowing longtime preservation of their biocatalytic activities. Glucose oxidase/halloysite film preserved its biocatalytic activity over 50 days in a phosphate-buffered saline solution (Abdullayev et al., 2012a; Abdullayev and Lvov, 2014). Another study claims 90% retention of enzymatic activity of urease and α-amylase enzymes immobilised in halloysite lumen during 15 days of storage (Zhai et al., 2010).

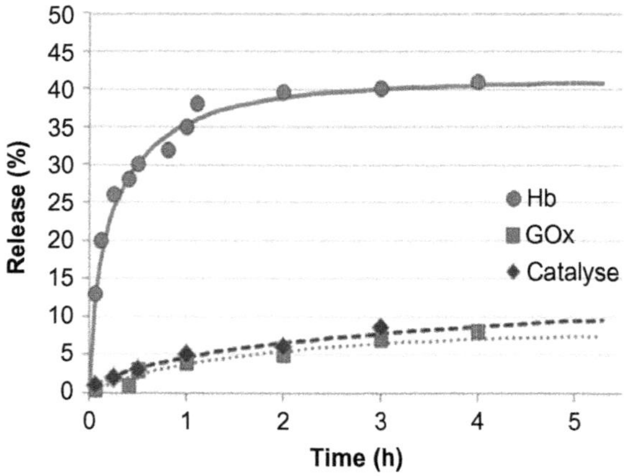

FIG. 22.10 Release curves for haemoglobin (*circles*), glucose oxidase (*squares*) and catalase (*diamonds*) in water at pH 6.5. *Reproduced with permission from Abdullayev and Lvov (2013). Copyright 2013 Royal Society of Chemistry.*

22.4 STRATEGIES FOR CONTROLLING THE LOADING CAPACITY AND RELEASE RATE OF ACTIVE AGENTS

Halloysite nanotubes were proven to be effective in the encapsulation of a wide variety of active agents with typical loading efficiencies below 15% and release rates extending to 10–30 h. Enhanced loading efficiency and extended release rates is possible by the chemical modification and surface functionalisation of halloysite nanotubes. These methods not only help in improving loading and release behaviour, but also allow for rendering nanotube lumens suitable for each specific target active agent to obtain maximum chemical stability, such as favourable pH within nanotube lumens for acid- or base-sensitive actives. Reactive aluminol groups within halloysite lumen and nanotube edges, as well as OH groups at nanotube defects, are useful for surface functionalisation, while negative surface charge allows for building polyelectrolyte shells. Selective etching aluminium of oxide from halloysite lumens allowed for increasing the pore volume and loading efficiency.

22.4.1 Selective Lumen Etching

Selective etching of aluminium oxide from halloysite lumens is one of the most effective ways to improve loading efficiency. Aluminium oxide from halloysite lumen can be leached with strong acids like sulfuric and hydrochloric acids. External silicone dioxide sheets remain intact with acid treatment due to the high acid stability of SiO_2. However, care must be taken not to collapse the tubular structure by overetching the aluminium oxide. Typically, up to 40%

removal of aluminium oxide is possible without degrading the tubular structure. Such treatment allowed for increasing the lumen diameter from the original 15 nm up to 25 nm, resulting in 30–40% loading efficiency (Fig. 22.11). This is higher than most of the siliceous templates, and comparable with polymeric capsules (Abdullayev et al., 2012b).

Belkassa et al. (2013) reported an expansion of the lumen diameter of the Algerian halloysite by heat treatment at 600°C, followed by hydrochloric acid etching. SSA increased up to 503 m^2/g from the original 60.5 m^2/g. Maximum loading efficiency was achieved by treating halloysite nanotubes with a 0.5 M hydrochloric acid solution and was found to be equal to 70 mg/g with crystal violet dye. Wang et al. (2014) studied the loading efficiency of ofloxacin on acid-etched and heat-treated halloysites. Loading efficiency was slightly reduced from 44.7 to 37.9 mg/g with acid etching, which was attributed to breaking nanotubes. This is not surprising, considering that hydrochloric acid solutions with concentrations of up to 12 M were used. None of these works used vacuum cycling for loading active agents; this may cause some of the internal pores to be inaccessible for the active agent.

22.4.2 Formation of Halloysite Nanotube Release Stoppers

Halloysite nanotubes allow synthesising release stoppers to control the release rate of the loaded active agents. Fig. 22.12A shows an example of such stoppers. Caps were formed on halloysite tube endings by melting and crystallisation of starch at the halloysite tube ends at 200°C. This method was not very practical due to the requirement of such a high temperature (Lvov and Abdullayev, 2013). Other methods include formation of thin films, precipitate or both at the tube endings by the interaction of triazole and imidazole derivatives with transition metal salts (Fig. 22.12B) (Wei et al., 2014) and polyelectrolyte complexation through layer-by-layer (LbL) nanoassembly (Fig. 22.17) (Lvov et al., 2008).

The first practical approach for synthesising halloysite nanotube end caps was demonstrated by the interaction of the leaking active agent, BTA, with copper (II) sulphate to form copper-BTA film at the tube endings, which acts as a barrier that controls the release rate via varying film thicknesses and porosity (Fig. 22.13) (Abdullayev et al., 2009; Abdullayev and Lvov, 2010).

The benzotriazole release rate was controlled by a variation of $CuSO_4$ concentrations used for the formation of the release stopper. Copper sulphate solutions with six different concentrations were used to obtain controllable BTA release, as shown in Fig. 22.14A. The benzotriazole release rate was continuously decreased by using higher concentrations of $CuSO_4$ solution for stopper formation. For example, 75% of BTA was released within 6 h from halloysite nanotubes treated with 0.04 mM copper sulphate solution, while for nanotubes treated with 4.0 and 8.0 mM $CuSO_4$ solution, these values dropped to 50% and 30%, respectively. Excellent correlation between the

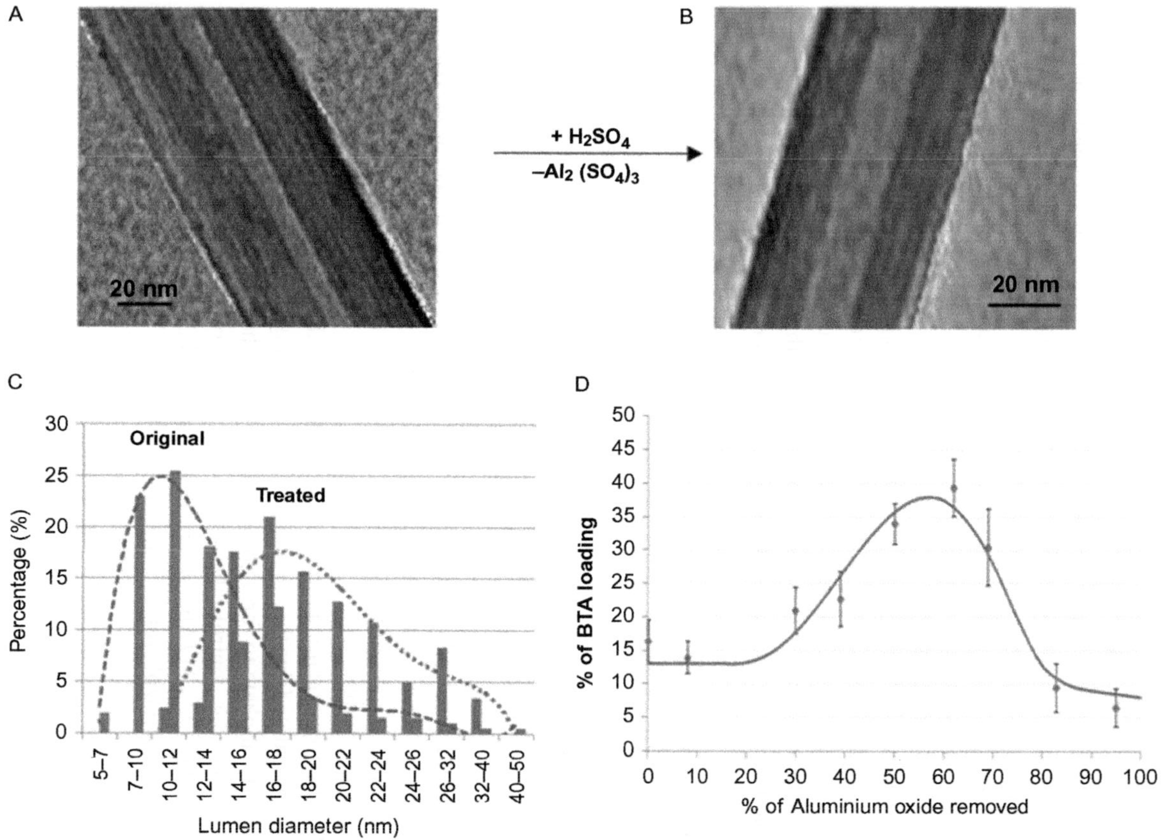

FIG. 22.11 TEM of halloysite nanotubes before (A) and after (B) treatment with H_2SO_4. (C) Histogram of tube lumen diameters based on 200 measurements of TEM images. (D) Benzotriazole (BTA) loading efficiency within halloysite lumen at various levels of aluminium oxide removal. Average lumen diameter increases from 10 to 20 nm, allowing loading efficiency to go up to 40%. *Reproduced with permission from Abdullayev et al. (2012a). Copyright 2012 American Chemical Society.*

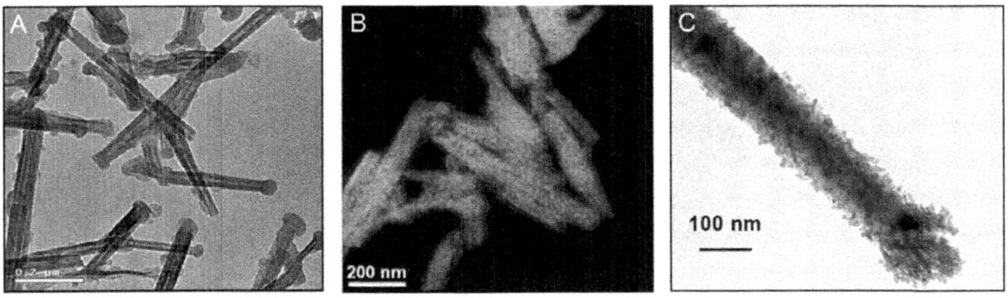

FIG. 22.12 (A) Halloysite nanotube with end stoppers of starch complex. (B) Oxygen (*green*; *light grey* in the print version) and nitrogen (*brown*; *dark grey* in the print version) elemental maps of the halloysites covered with copper-mercaptobenzimidazole taken by scanning transmission electron microscope-electron dispersive X-ray analysis (STEM-EDX). (C) Polymeric shell synthesised on halloysite surface by means of LbL nanoassembly of polyethyleneimine and 7-nm silica nanoparticles. Silica-covered external surface and tube endings acted as a release stopper for a loaded active agent. *Reprinted with permission from Lvov and Abdullayev (2013). Copyright 2013 Elsevier; Joshi et al. (2013). Copyright 2013 American Chemical Society; Lvov et al. (2008). Copyright 2008 American Chemical Society.*

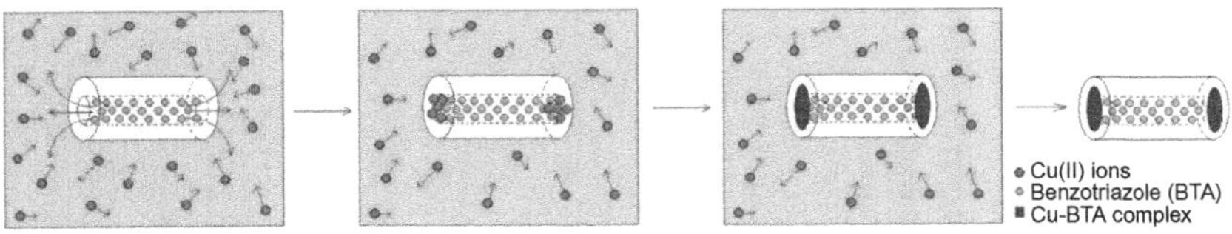

FIG. 22.13 Illustration of stopper formation at halloysite tube endings by interaction of leaking benzotriazole (BTA) at Cu(II) ions. *Reprinted with permission from Abdullayev et al. (2009). Copyright 2009 American Chemical Society.*

BTA release rate and the concentration of the Cu (II) ions indicates that permeability of the nanotube end caps can be fine-tuned to obtain the desired release profile (Abdullayev and Lvov, 2010).

The amount of BTA at the nanotube ending that is available for stopper formation is another parameter influencing the BTA release rate. This was tuned by washing loaded halloysite with DI water prior to processing with Cu(II) ions. Stoppers were synthesised on the nanotubes after two and four water-washing cycles and tested for the BTA release rate. Tubes washed two times with water showed a slower release than those washed four times before exposure to the $CuSO_4$ solution (Fig. 22.14B). Evidently, four washing cycles significantly reduced the amount of BTA at the tube ends, resulting in weaker stopper formation (Abdullayev et al., 2009).

Fig. 22.15 shows a modified version of the abovementioned approach for release stopper formation to include a broader range of actives, including consecutive adsorption of BTA and Cu(II) onto the halloysite surface after being

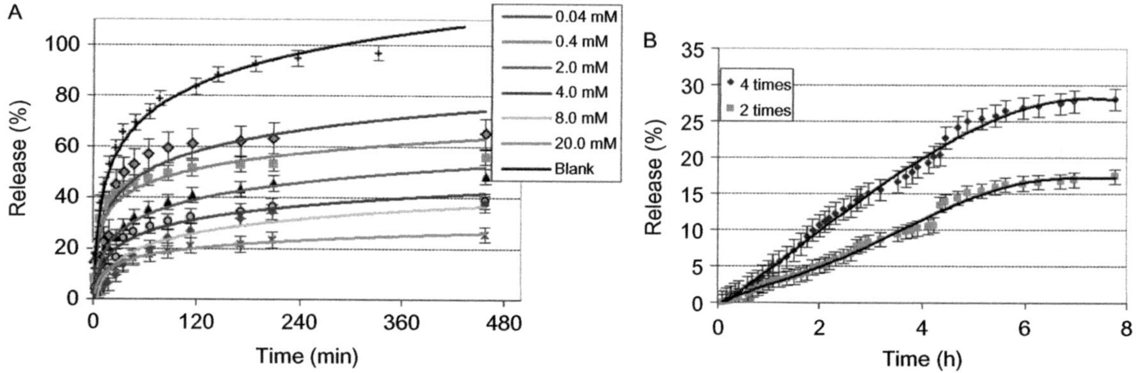

FIG. 22.14 (A) Release profiles of benzotriazole from halloysite washed with $CuSO_4$ solution of different concentrations in the 0.04–20.0 mM range. Release of benzotriazole from untreated halloysite is also shown (labelled 'Blank'). (B) Benzotriazole release profiles from nanotube lumen treated with two and four water-washing cycles to reduce benzotriazole concentrations prior to Cu-BTA stopper formation. *Reprinted with permission from Abdullayev et al. (2009). Copyright 2009 American Chemical Society.*

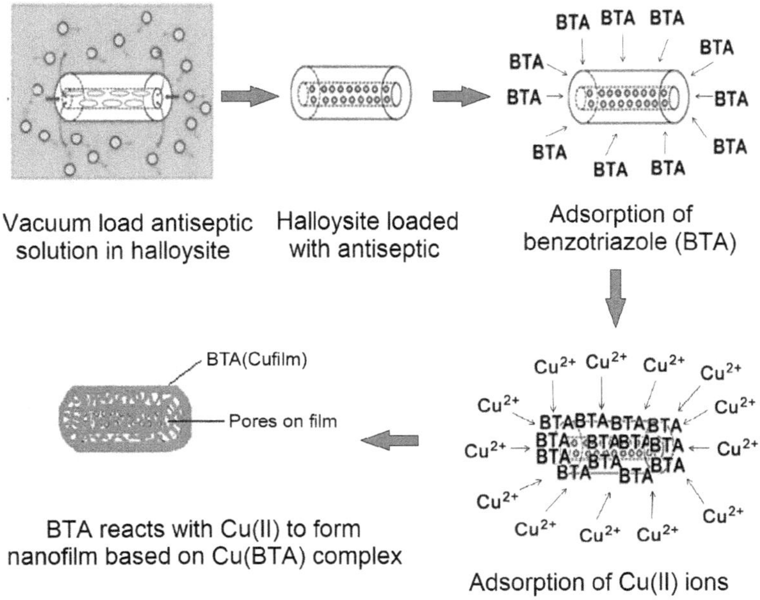

FIG. 22.15 Formation of halloysite release stoppers with sequential adsorption of benzotriazole and Cu(II) ions. *Reprinted with permission from Wei et al. (2014). Copyright 2014 Royal Society of Chemistry.*

loaded with the active agent. Brilliant green, a well-known antiseptic agent, was encapsulated within halloysite pores using this approach. The brilliant green release rate was significantly decreased by encapsulating halloysite nanotubes with copper-BTA coating (Fig. 22.15). The desired release rate can be obtained by varying the concentration of the metal ions at the external bulk solution (Wei et al., 2014).

Copper-BTA film formation was studied in detail by varying the concentrations of BTA and $CuSO_4$. Brilliant green release curves treated at different complexation conditions are presented in Fig. 22.16. BTA concentrations varied in the 41–165 mM range at two fixed concentrations of $CuSO_4$: 4.0 and 80.0 mM. At 80 mM, the $CuSO_4$ concentration release rate gradually decreases upon the increase of BTA level (Fig. 22.15A). Release stoppers formed at higher BTA concentrations lead to more intact film, and hence tighter encapsulation (Wei et al., 2014).

Surprisingly, the opposite phenomenon was observed at a Cu(II) concentration of 4.0 mM: brilliant green release rate was highest at the 165 mM BTA concentration (Fig. 22.16B). This might be related to the faster film formation at higher concentrations of BTA, which results in higher loading efficiency (brilliant green leaks from halloysite pores during encapsulation

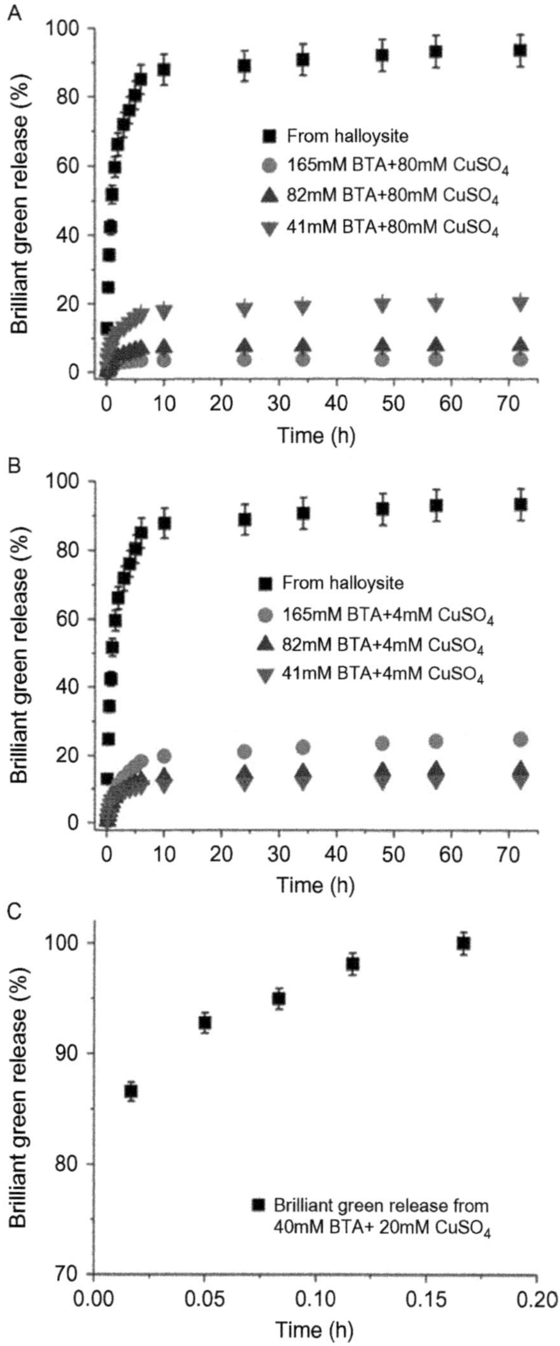

FIG. 22.16 Brilliant green release with different tube encapsulations: halloysite with a BTA-Cu stopper, which has various concentrations of benzotriazole (BTA) and 80 mM $CuSO_4$ (A); halloysites with BTA-Cu stoppers, which have various concentrations of BTA and 4 mM $CuSO_4$ (B). Brilliant green release from microcrystals BTA-Cu encapsulated without halloysite tubes (much faster release) (C). *Reprinted with permission from Wei et al. (2014). Copyright 2014 Royal Society of Chemistry.*

as tubes are not sealed yet, causing loading efficiency to be reduced). Indeed, at the 4.0 mM $CuSO_4$ concentration, loading efficiency increased from 7.8% to 10.8% by increasing BTA concentration from 41 to 165 mM. Higher loading efficiency causes a higher concentration gradient between inner pores and external solution, which is the major driving force for drug release. Encapsulation using copper-BTA film has been attempted on bare brilliant green crystals, with no success (Fig. 22.16C) (Wei et al., 2014).

22.4.3 Polymeric Shells on Halloysite Nanotubes

Halloysite nanotubes have a negative external surface charge in their native form due to exterior SiO_2 sheets, making them suitable for adsorption of polyelectrolytes. Adsorption of common polycations like polyethyleneimine (PEI), poly(allylamine hydrochloride) (PAH) and poly(diallyldimethylammonium chloride) (PDDA) renders halloysite zeta-potential positive, rather than neutral. This charge reversal allows sequential adsorption of polycations and polyanions on the halloysite surface and builds polymeric shells by means of LbL nanoassembly. Halloysite zeta-potential alteration by sequential adsorption of PEI–PAA and PAH–PSS is shown in Fig. 22.17B. Surface charges typically range from ±30 to 60 mV, which is characteristic of stable colloidal particles and indicate a strong interaction between polyelectrolytes and halloysite. Apart from polyelectrolytes, small nanoparticles were effectively deposited on the halloysite surface (Fig. 22.17C). Such shells provide a release barrier for loaded active agents. The thickness of the barrier can be controlled by the thickness of the polyelectrolyte shell (ie, the number of polyelectrolyte layers) (Shutava et al., 2014).

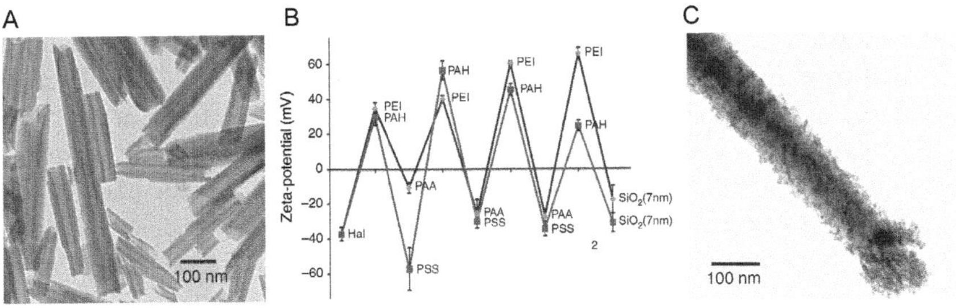

FIG. 22.17 (A) TEM of raw halloysite (B) Zeta-potential alternation in the processes of LbL assembly of polyelectrolyte multilayer shells on a halloysite nanotube (*PAA*, polyacrylic acid; *PSS*, polystyrene sulfonate; *PEI*, polyethyleneimine; *PAH*, polyallylamine hydrochloride); (C) TEM image of (PAH/PSS/(PAH/silica)$_2$) nanotube coating. *Reproduced with permission from Shutava et al. (2014). Copyright 2014 Elsevier; Lvov et al. (2008). Copyright 2008 American Chemical Society.*

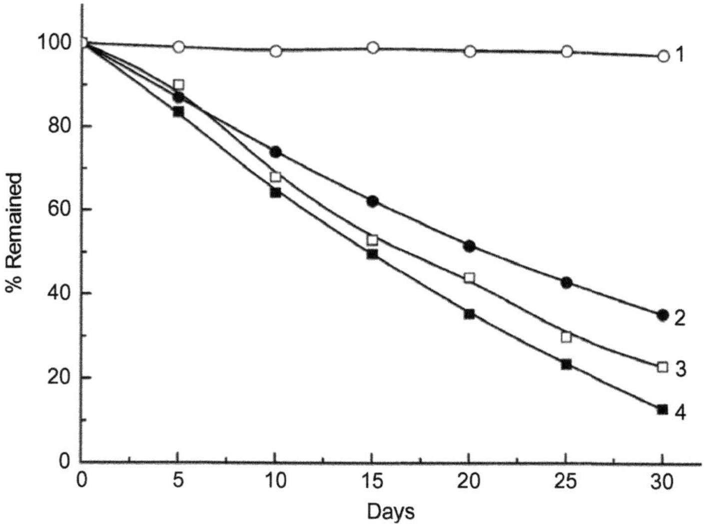

FIG. 22.18 Release of the 2-mercaptobenzothiazole in aqueous solution from halloysite nanotubes; (1) halloysite coated with PAH/PSS shell at pH 6.5; (2) unmodified halloysite nanotube at pH 6.5; (3) halloysite coated with PAH/PSS shell at pH 10; (4) unmodified halloysite nanotube at pH 10. *Reprinted with permission from Shchukin et al. (2008). Copyright 2008 American Chemical Society.*

PSS/PAH bilayers were used to obtain pH-responsive release of 2-mercaptobenzothiazole from halloysite nanotubes in sol–gel anticorrosive coatings (Fig. 22.18). The polyelectrolyte shell covered the halloysite nanotube ends, providing a controlled release. About 10% of the loaded active agent was released from halloysite nanotubes without polyelectrolyte shells after 6 days of storage, whereas polyelectrolyte shells terminated the spontaneous release of the 2-mercaptobenzothiazole at neutral pH. The release was triggered at pH 10 due to the neutralisation of the PAH charge, leading to the formation of the defects between polyelectrolyte multilayers (PAH is a weak polyelectrolyte with pK_a of 9.6). The pH-responsive release of active agents is crucial in the preparation of self-healing anticorrosive coating capable of delivering corrosion inhibitors at coating defect sites at the time of corrosion (Shchukin et al., 2008).

Several attempts have been made to synthesise polymeric shells using in situ polymerisation on the halloysite surface. Levis and Deasy (2003) studied diltiazem hydrochloride release from halloysite nanotubes coated with in situ polymerised poly(alkyl cyanoacrylate) derivatives. Release profiles of the drug-loaded halloysite coated with poly(alkyl cyanoacrylate) derivatives or *n*-butyl, and/or isobutyl and octyl cyanoacrylate, are shown in Fig. 22.19. Polymer shells were synthesised by stirring halloysite in a 5% solution of corresponding monomers in heptane for 8 h (polymerisation was

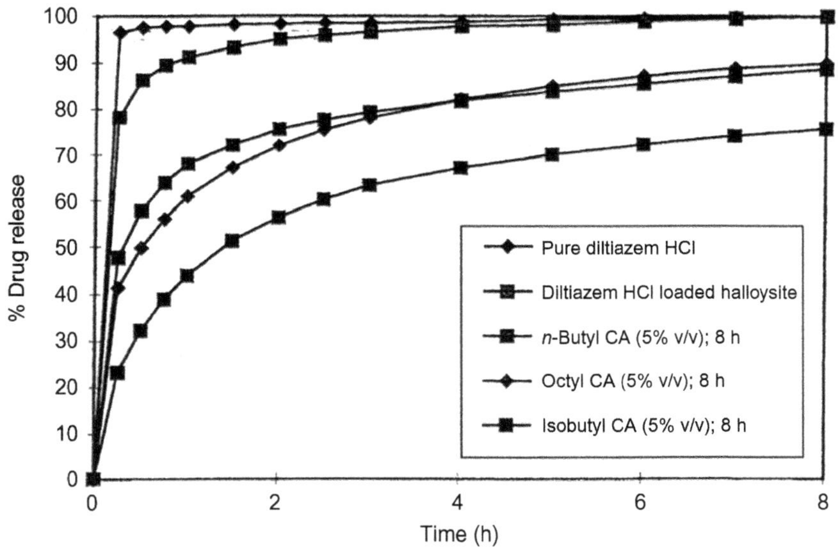

FIG. 22.19 Release profiles of diltiazem hydrochloride from halloysite nanotubes coated with poly-*n*-butyl, poly-*iso*-butyl or poly-octyl cyanoacrylate derivatives in aqueous solution at 37°C and pH 6.8. *Reprinted with permission from Levis and Deasy (2003). Copyright 2003 Elsevier.*

initiated by residual moisture on halloysite surface). Poly(iso-butyl cyanoacrylate) provided the greatest retardation in release rate, with about 70% release in 8 h, as opposed to 80% for other cyanoacrylate derivatives and 100% for uncoated halloysite. Researchers have ascribed this to the highest molecular mass of the *iso*-butyl cyanoacrylate shell, as confirmed by gel permeation chromatography analysis [with *M* being approximately equal to 720,000 for poly(*iso*-butyl cyanoacrylate) vs 550,000 for poly-*n*-butyl and 570,000 for poly-octyl coatings]. Drug-loading efficiencies were 17.83%, 18.75% and 21.46% for poly-*n*-butyl, poly-*iso*-butyl and poly-octyl cyanoacrylate coatings, respectively.

Urea-formaldehyde (UF) films were synthesised on halloysite surfaces by in situ polycondensation of urea and formaldehyde. Polymers were used with the aim of retarding the release rate of corrosion inhibitors in self-healing metal coatings. The performance of the polymer shell was compared with the copper-triazole and imidazole stoppers, as described in the previous section. Release profiles of BTA and 2-mercaptobenzothiazole from halloysite pores covered with UF polymer and copper-inhibitor (inhibitor = BTA and 2-mercaptobenzothiazole) stoppers are shown in Fig. 22.20. The Cu(II)-inhibitor complexation was found to be more efficient than the UF polymer for retarding inhibitor release (Joshi et al., 2013).

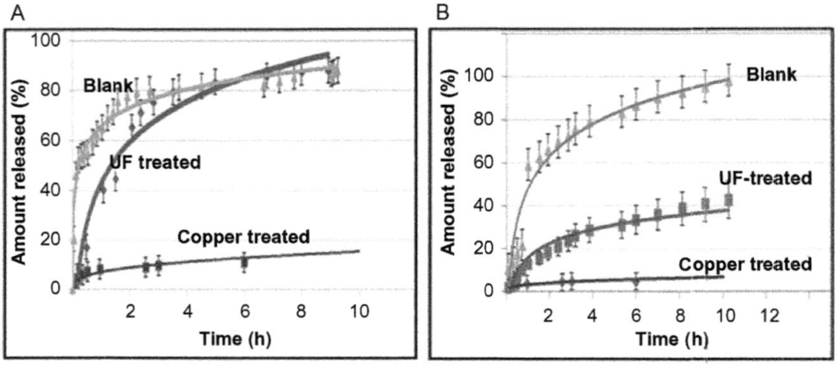

FIG. 22.20 Release time reduction for (A) benzotriazole and (B) 2-mercaptobenzothiazole from untreated halloysite nanotubes and halloysite coated with urea-formaldehyde (UF) polymer and copper (II)-based release stoppers. *Reprinted with permission from Joshi et al. (2013). Copyright 2013 American Chemical Society.*

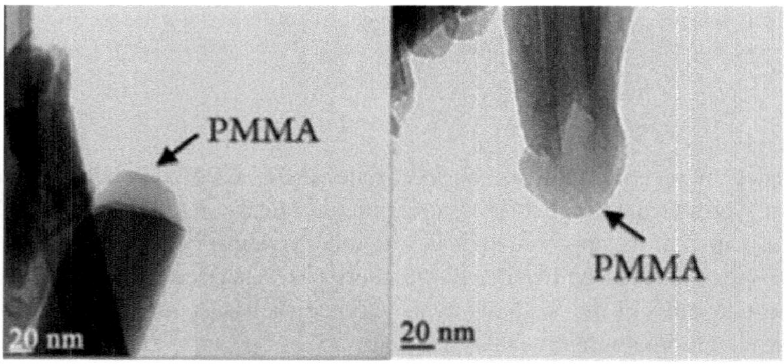

FIG. 22.21 TEM images of halloysite nanotubes with poly(methylmethacrylate) (PMMA) plugs synthesized by atomic transfer radical polymerisation. *Reprinted with permission from Yah et al. (2012a). Copyright 2012 American Chemical Society.*

Yah et al. (2012b) synthesised poly(methylmethacrylate) (PMMA) plugs at halloysite tube endings by in situ surface-initiated atomic transfer radical polymerisation (SI-ATRP) of methyl methacrylate. Halloysite was modified with 2-bromo-*N*-[2-(3,4-dihydroxyphenyl)ethyl]-isobutyryl amide, which also served as an atomic transfer radical polymerisation (ATRP) initiator (Fig. 22.21). Wang et al. (2008) synthesised poly(methylmethacrylate)-poly (*N*-isopropylacrylamide) brushes on halloysite nanotubes using reverse ATRP. However, release behaviours of active agents have not been studied in any of these previous works.

Halloysite nanotubes were coated with two different lengths of poly(*N*,*N*-dimethylaminoethyl methacrylate) (PDMAEMA) to obtain pH-responsive release of drugs, and the modified halloysites were denoted 'halloysite-PDMAEMA-1' and 'halloysite-PDMAEMA-2' (Hemmatpour et al., 2015). Halloysite was functionalised with 3-aminopropyltriethoxysilane and α-bromoisobutyryl bromide prior to grafting with PDMAEMA by SI-ATRP. PDMAEMA chain length was varied by means of the monomer-to-initiator ratio. Grafting with polymer on halloysite was confirmed by Fourier transform infrared (FTIR) spectroscopy, X-ray photoelectron spectroscopy (XPS) and thermal gravimetric analysis. PDMAEMA grafting of halloysite nanotubes led to the sustained release of diphenhydramine hydrochloride and diclofenac sodium salt. Initial burst release of medicines was lower in halloysite-PDMAEMA-1. Diphenhydramine hydrochloride showed pH-responsive release due to the differing packing density of PDMAEMA in acidic and basic environments. PDMAEMA chains are protonated in acidic media, resulting in repulsion between the grafted polymer chains and stretched chain conformation. Such chain conformation lowers the packing density and accelerates drug release. The opposite effect was observed in the basic environment (Hemmatpour et al., 2015).

22.4.4 Surface Functionalisation of Halloysite for Controlled Release

Halloysite nanotube external surfaces and inner lumen walls have been modified with a variety of surface-modifying agents through interaction with reactive aluminol and silanol groups (at surface defects) and ionic and hydrogen bonding. These surface modifiers include silane-coupling agents, alkyl phosphonic acid derivatives, quaternary ammonium salts, fatty acid salts, polymers, ionic liquids and various metal and metal oxide nanoparticles. The most common purpose of halloysite surface modification involves its compatibilisation with polymer matrices to obtain enhanced dispersion (Du et al., 2010). However, surface modification was also shown to increase the loading and release properties of active agents. Next, the surface functionalisation from the perspective of controlled release of active agents will be discussed.

Having different inner- and outer-wall chemistries, the inner lumens of halloysite nanotubes have been selectively modified with octadecyl phosphonic acid and dopamine derivatives to render them hydrophobic (Yah et al., 2012a). The grafting of octadecyl phosphonic acid on the halloysite inner lumens took place by means of formation of Al–O–P bonds, as demonstrated by XPS and FTIR studies. This reaction involved stepwise formation of monodentate, bidentate and tridentate phosphonate bonds with aluminol groups. Phosphonate did not react with silica surfaces, primarily due to the weak nature of Si–O–P bonds that readily hydrolyse in aqueous solution

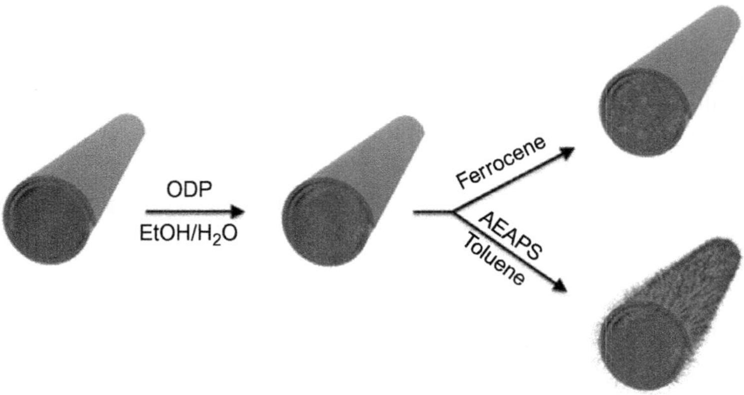

FIG. 22.22 Schematic illustration of bifunctionalisation of the silica-alumina oxide surface of halloysite by octadecyl phosphonic acid (ODP) and subsequent silylation with 3-(2-aminoethyl) aminopropyltrimethoxy silane (AEAPS) and loading with ferrocene. *Reproduced with permission from Yah et al. (2012a). Copyright 2012 American Chemical Society.*

(Fig. 22.22). Further reactions of halloysite-ODP with AEAPS resulted in functionalising external surfaces with amine groups, as evidenced by nuclear magnetic resonance (NMR) studies. This modification allows the bifunctionalisation of external and inner surfaces with desired surface chemistries to enhance the loading selectivity towards active agents of choice (Yah et al., 2012a).

Halloysite with such hydrophobic lumen provided enhanced selectivity towards the adsorption of hydrophobic agents as demonstrated by ferrocene derivatives. Halloysite behaved like reverse micelles with hydrophilic exterior and hydrophobic interior walls. The mass of ferrocene released from halloysite- and octadecyl phosphonic acid (ODP)–modified halloysite lumen (Fig. 22.23A) shows that ODP-modified halloysite contained four times more ferrocene than unmodified halloysite, indicating its higher adsorption capacity. The release profiles were described by the Higuchi equation with the square root of time dependence of release rate; $R_t = k_H\sqrt{t}$, where k_H is the release rate constant. R_t vs $\sqrt{t}$ plots showed a two-step release for halloysite-ODP, reflecting desorption of the outside surface and from the tube lumen. No lumen release was observed for unmodified halloysite, indicating that ferrocene molecules were adsorbed only to the external surface of halloysite by weak van der Waals forces. Data about unmodified halloysite did not fit the Higuchi equation well (Yah et al., 2012a).

Silane-coupling agents are amongst the most commonly used surface modifiers for halloysite. Yuan et al. (2008) conducted the first detailed study of the mechanism of APTES grafting on halloysite walls. APTES was grafted to both silica and alumina surfaces, despite the lower abundance

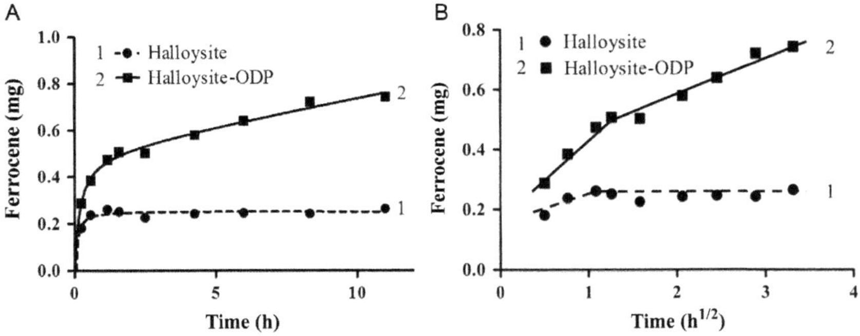

FIG. 22.23 Release profile (A) and Higuchi square root of time plots (B) for release of ferrocene from halloysite (*circles, dashed line*) and halloysite functionalised with octadecyl phosphonic acid (halloysite-ODP) (*squares, solid line*). *Reproduced with permission from Yah et al. (2012a). Copyright 2012 American Chemical Society.*

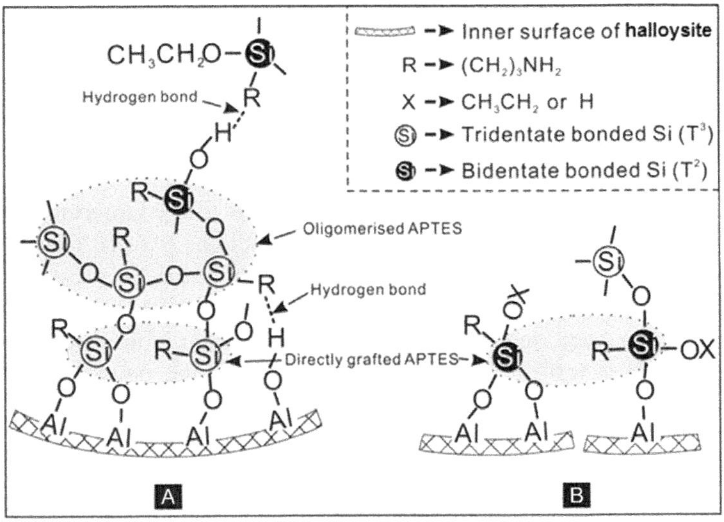

FIG. 22.24 Schematic representation of the proposed mechanism of the formation of a cross-linked network (A) and of the grafting between bidentate bonded Si and AlOH groups (B). *Reprinted with permission from Yuan et al. (2008). Copyright 2008 American Chemical Society.*

of external-surface hydroxyl groups. This was explained by the involvement of hydrogen bonding and oligomerisation of silane on halloysite surfaces, in addition to direct grafting to surface hydroxyl groups at external defect sites (Fig. 22.24). Coupling efficiency reduced with thermal treatment and increased with vacuum evacuation of the reaction mixture. The first effect was ascribed to the removal of surface moisture that leads to APTES oligomerisation, and the latter was ascribed to the removal of air bubbles from halloysite pores, which allowed silane to interact with internal aluminol groups (Yuan et al., 2008). Such silane treatment was found to increase the loading efficiency of ibuprofen by about 25% (11.8% for raw halloysite vs 14.8% for APTES-functionalised halloysite) (Tan et al., 2013). Modification of halloysite lumen by APTES promoted the loading of Ibuprofen due to the electrostatic attraction between aminopropyl groups of APTES and the carboxyl groups of ibuprofen. A similar phenomenon was also observed with Orange II dye (Yuan et al., 2012a).

The halloysite surface was functionalised with positively charged triazolium salts (Fig. 22.25) and used as a nanocontainer for the sustained release of the anticancer drug curcumin. Curcumin was adsorbed to both the internal lumen and external surface of the functionalised sample due to the reversal of the external surface charge from negative to positive. As a result, curcumin-loading efficiency increased from 1.5% for raw halloysite to 11.0% for methyl and 12.5% for butyl triazolium derivatives (shown as 3a and 3b in Fig. 22.25, respectively) (Riela et al., 2014).

The release kinetics of curcumin is shown in Fig. 22.26. The functionalised halloysite sample showed pH-dependent release kinetics. The release of curcumin was faster at pH 1 than with pH 5 and pH 8, with the latter two showing only a small amount of drug released. This is related to the differing solubilities of curcumin in acidic and basic environments and the pH dependence of curcumin–triazole interactions. At pH 1 both triazolium salt and curcumin are positively charged, causing electrostatic repulsions that accelerate drug release. The curcumin release rates at pH 1 from functionalised and raw halloysite were similar. The release mechanism was characterised by two different models: the power law $R_t = kt^n$ and first-order kinetics $R_t = 1 - e^{-kt}$. First-order kinetics showed better correlation with the experimental data as compared to the power law (Riela et al., 2014).

The first detailed study on the functionalisation of halloysite walls by cationic and anionic surfactants was performed by Ng et al. (2011) with XPS and time-of-flight secondary ion mass spectrometry (ToF-SIMS) characterisation. Halloysite was functionalised with model cationic and anionic surfactants [benzalkonium chloride (BAC) and sodium dodecyl sulphate (SDS), respectively]. Halloysite modified with SDS readily dispersed in aqueous solvent contrary to the BAC-modified sample. This is the indication that the different faces of halloysite were hydrophobised with surfactants (the external face in the case of BAC and the internal pores in the case of SDS). This conclusion is further supported by XPS surface characterisation of samples, showing that the halloysite external surface with the SDS-modified sample had very little carbon content. BAC-modified halloysite,

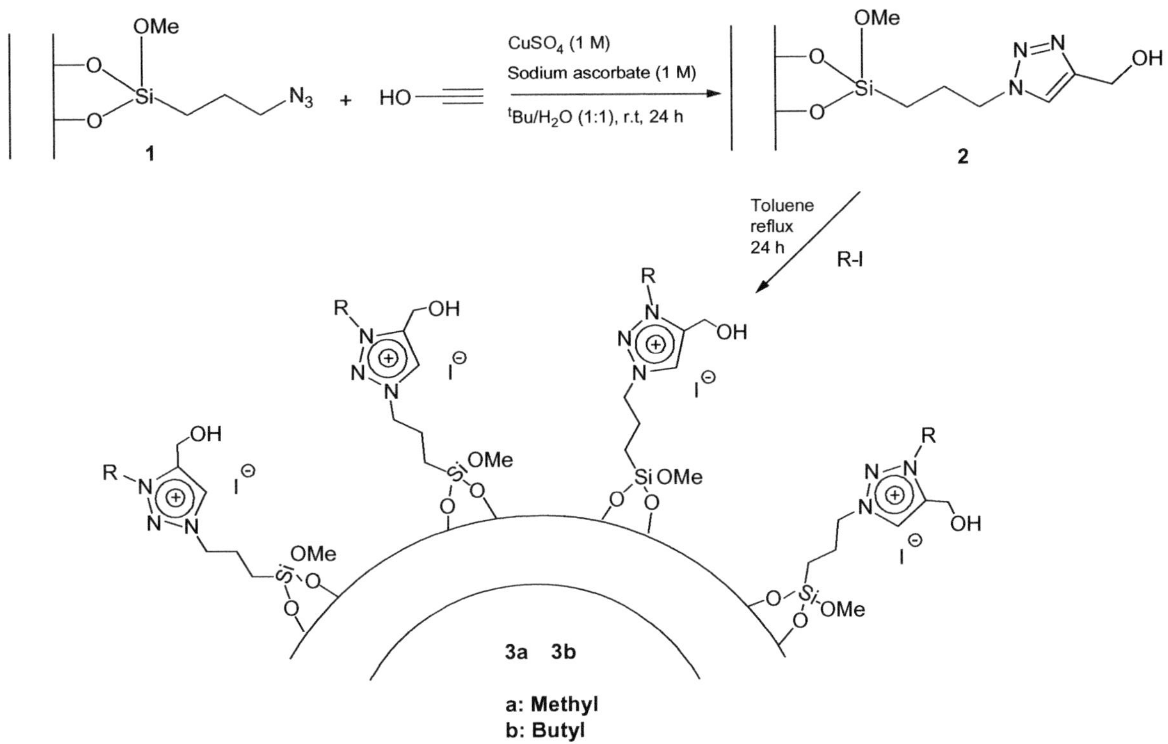

FIG. 22.25 Halloysite surface functionalization with triazolium salts. Structure 1: halloysite modified with 3-azidopropyl trimethoxysilane, structure 2: halloysite modified with 1N-propylsilyl, 4-hydroxymethyl, 1,2,3-triazole, structures 3a and 3b: halloysite modified with and 1N-propylsilyl, 3N-methyl (or butyl) 4-hydroxymethyl 1,2,3-triazolium iodide. *Reproduced with permission from Riela et al. (2014). Copyright 2014 Elsevier.*

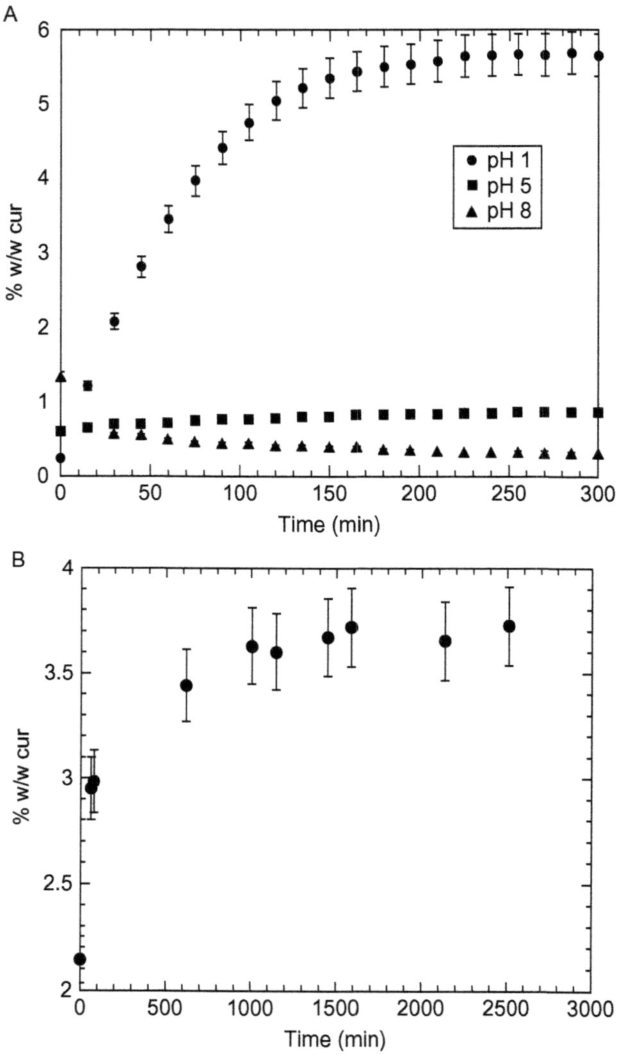

FIG. 22.26 Amount of curcumin released from (A) halloysite functionalised with structure 3b as a function of time for three different pH values and (B) powdered halloysite at pH 1. *Reprinted with permission from Riela et al. (2014). Copyright 2014 Elsevier.*

on the other hand, showed significant levels of carbon and nitrogen on the external surface (Ng et al., 2011).

The internal lumen of halloysite was hydrophobised with anionic perfluoro C5–C7 alkyl surfactants (Cavallaro et al., 2014). Modified halloysites showed enhanced colloidal stability due to the neutralisation of the internal alumina surface and causing the overall surface ζ-potential to increase. Nanotubes with hydrophobic cores were promising as oxygen nanoreservoirs in aqueous media. Aqueous dispersion of raw and modified halloysite was saturated with oxygen and left for equilibration. Foams were not generated with fluorinated halloysite dispersion during oxygen bubbling and stirring, contrary to the raw halloysite dispersion and pure water, indicating that oxygen is loaded within halloysite lumens. Oxygen concentrations were recorded by time at two different conditions: in steady solution and under vigorous stirring (1250 rpm). In both cases, the oxygen reduction rate was significantly retarded in dispersions with halloysite. This is the indication of oxygen slowly releasing from halloysite lumens (Fig. 22.27). About 50% of O_2 was released from raw halloysite after 6 h, while the modified nanotubes showed an oxygen loss of about 10%. Remarkably, this enhancement in oxygen entrapment ability was generated by the presence of only 0.86% of fluoroalkyl surfactant, a concentration that is two orders of magnitude less than its critical micellar concentration (Cavallaro et al., 2014).

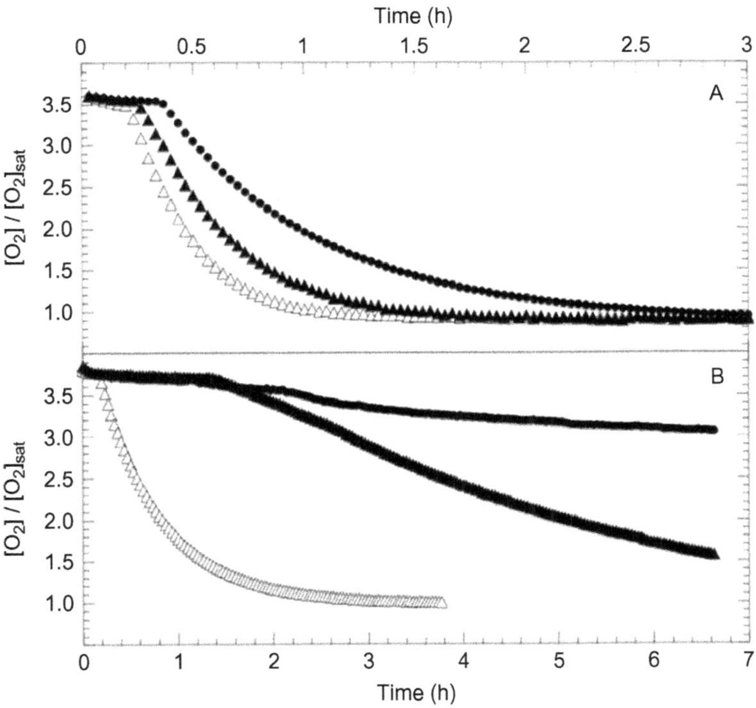

FIG. 22.27 Oxygen concentration as a function of time in pure water (*open triangles*) and in aqueous dispersions of raw halloysite (*solid triangles*) and halloysite with fluorinated lumen (*solid circles*). Collected under magnetic stirring at 1250 rpm (A) and under static conditions (B). *Reprinted with permission from Cavallaro et al. (2014). Copyright 2014 Elsevier.*

22.5 POTENTIAL APPLICATIONS OF HALLOYSITE AS A CONTROLLED-RELEASE NANOCONTAINER

22.5.1 Functional Halloysite Polymer Nanocomposites and Coatings

Halloysite has been used as a capsule for loading several paint additives, including corrosion inhibitors, biocides, mould growth inhibitors and flame retardants. Corrosion inhibitors loaded within halloysite lumens provide sustained release for self-healing coatings. Anticorrosive coatings were proposed to protect various copper (Abdullayev et al., 2013), aluminium (Shchukin et al., 2008; Fix et al., 2009) and steel alloys (Joshi et al., 2013). The addition of halloysite loaded with BTA, 2-mercaptobenzothiazole and 2-mercaptobenzimidazole into oil-based urethane paint significantly improved its corrosion protection for copper alloys. Copper strips were exposed to 30 g/L of aqueous sodium chloride solution for 6 months. The original paint showed very poor performance, as shown by the accumulation of green patina underneath the paint on strips coated with ordinary paint, while strips coated with halloysite-polyurethane nanocomposite coating showed no evidence of corrosion (Joshi et al., 2013; Abdullayev and Lvov, 2014). Corrosion inhibitors loaded in halloysite lumens were slowly released from lumens and inhibited corrosion by interacting with exposed metal parts.

Halloysite studies for aluminium-based metal protection primarily involve sol–gel type coatings. Most of the studies were directed towards the protection of 2024 aluminium alloy, which is commonly used in aviation products. In Fig. 22.28, localised corrosion current densities on sol–gel coated and artificially scratched aluminium strips (2024 alloy) are shown. The activity of the anodic corrosion current at sol–gel film is very high and rapidly increases over several hours, indicating a fast-pitting corrosion. Anodic current is completely suppressed on modified sol–gel coating with halloysite nanotubes, which clearly indicates efficient corrosion protection (Abdullayev et al., 2009).

Halloysite nanotube loaded with silver has been applied as antimicrobial additives in coatings to prevent bacterial infection. This coating suppressed the growth of both Gram-positive (*Staphylococcus aureus*) and Gram-negative (*Escherichia coli*) bacteria on the paint surface. A clear zone of inhibition around painted samples (as shown in Fig. 22.29B) indicates Ag^+ ions slowly leaking from the tubes and killing bacteria. Contrary to the silver oxide nanoparticles, halloysite core-shell structures provided better colour stability for the coating (Figs 22.29C and D) (Abdullayev et al., 2011).

Antiageing styrene-butadiene rubber nanocomposite was prepared using halloysite nanotubes loaded with antioxidant 4010NA (Fig. 22.30A, *N*-isopropyl-*N*′-phenyl-*p*-phenylenediamine). 4010NA is a model chemical antioxidant added to elastomers to extend their oxidational ageing by binding with oxygen. However, antioxidants often diffuse towards the surface of rubber,

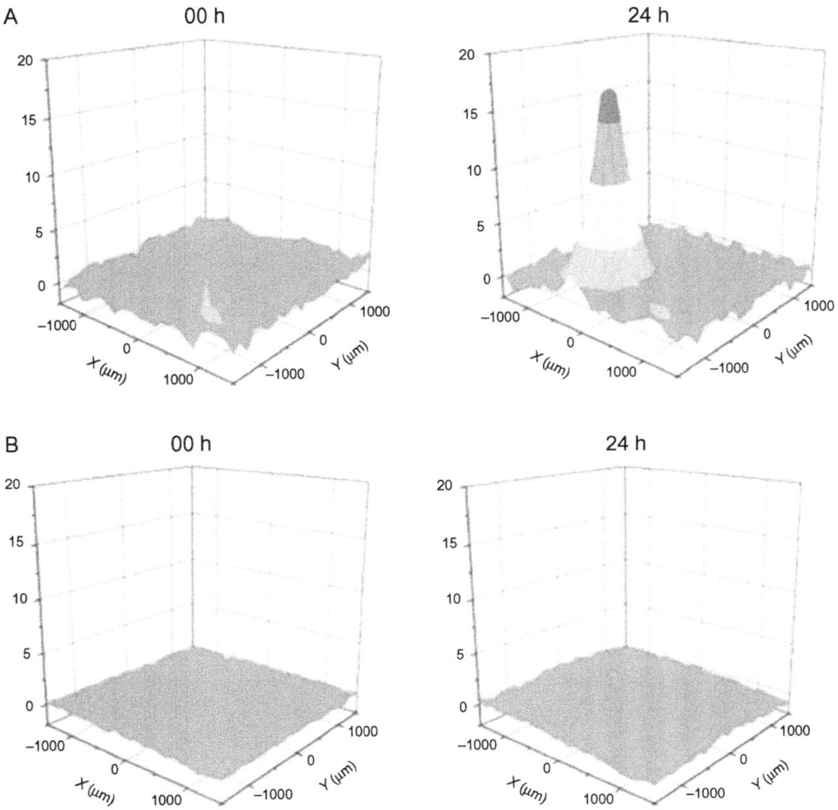

FIG. 22.28 Corrosion current densities from metal strips made from 2024 Al alloy. Strips were coated with the usual sol–gel coating (A) and with sol–gel containing halloysite (B). *Reproduced with permission from Lvov et al. (2008). Copyright 2008 American Chemical Society.*

causing 'blooming' that results in the formation of microdefects, loss of antioxidant protection and environmental pollution. The optimal loading level of 4010NA in rubber reached 3.2% by weight after its loading within halloysite lumen, i.e. threefold increase over blooming threshold concentration of free antioxidant. This nanocomposite provided 9 months of sustained antioxidant release at 25°C without causing 'blooming' defects. (Fig. 22.30D) (Fu et al., 2015).

Styrene-butadiene rubber with 27 mass% halloysite loaded with 4010NA shows a noticeable increase in ageing resistance. The ageing resistance of the nanocomposite were studied by comparing the tensile strength, elongation at break and shore hardness before and after thermal-oxidative ageing at 90°C and 120°C for 24, 72 and 168 h. Halloysite rubber nanocomposite containing 2.2% of encapsulated and 1.0% of free antioxidant resulted in the best oxidation protection after 168 h thermal ageing at 90°C, yielding only a 2%

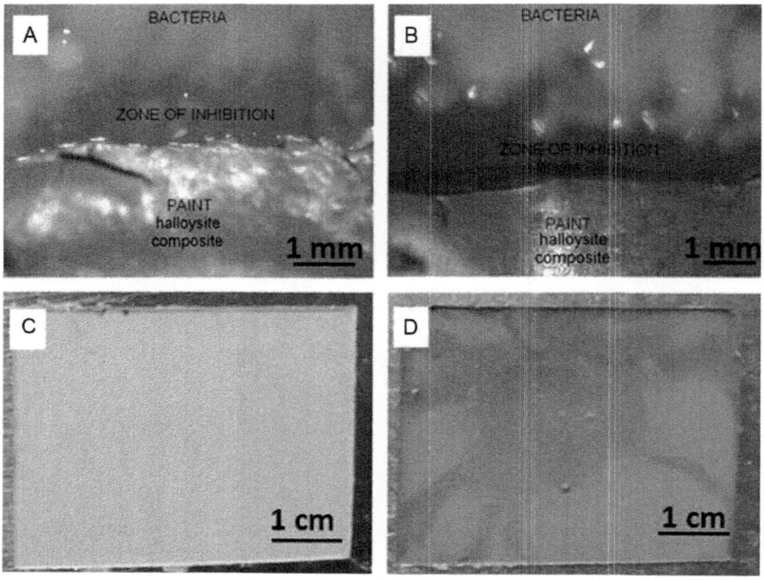

FIG. 22.29 Images of the solvent-based blue paint doped with 5% halloysite loaded with silver after 1 week of exposure to *S. aureus* (A) and *E. coli* (B). Images of acrylic latex paint with 5% halloysite loaded with silver (C) and with 5% silver oxide (D). *Reproduced with permission from Abdullayev et al. (2011). Copyright 2011 American Chemical Society.*

decrease in elongation break without losing tensile strength and elasticity. For comparison, original styrene-butadiene rubber (SBR) after such thermal treatment became brittle and crashed during mechanical testing. Antioxidant blooming was analysed by nitrogen content on rubber surfaces using XPS, and halloysite SBR nanocomposite showed significantly less blooming than the control sample with free antioxidants. The amount of nitrogen (which correlates with antioxidant concentrations) on the surface of the control sample increases to 13.7 atomic% in 7 days of ageing at 90°C, and then decreases to 12.0% after 15 days, indicating a significant amount of antioxidant migration to the surface. The maximum nitrogen amount on the surface of halloysite SBR nanocomposites was 8.4% after 7 days of ageing, which decreased to 4.8% afterwards. Blooming resistance was further confirmed by SEM imaging of the surface (Fu et al., 2015).

22.5.2 Polymer Scaffolds for Tissue Engineering and Wound Healing

Halloysite nanotubes show excellent compatibility with biopolymers commonly used for tissue engineering, such as polylactic acid, polycaprolacton

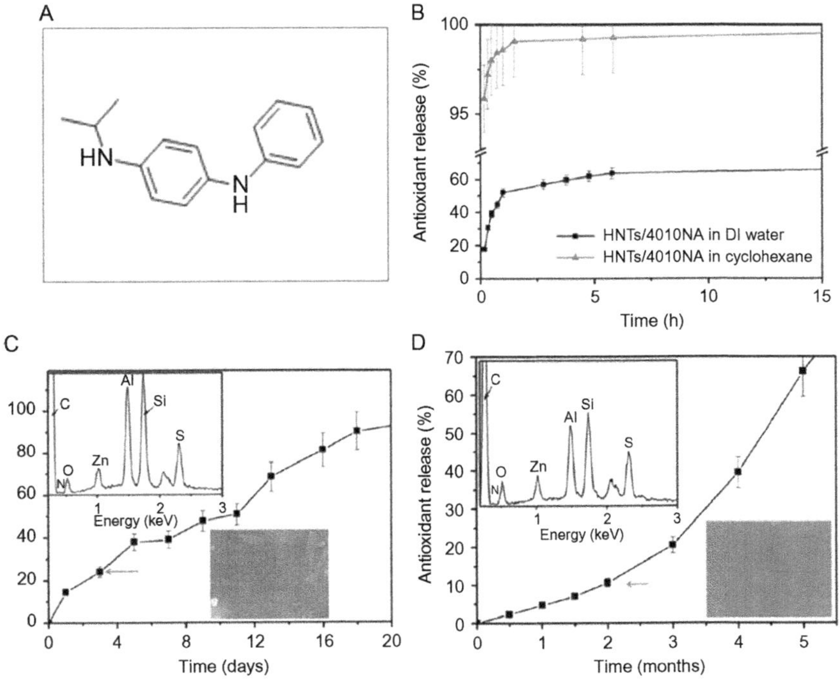

FIG. 22.30 (A) Structure of antioxidant 4010NA. Antioxidant release from (B) loaded halloysite in water and cyclohexane; halloysite rubber composite at (C) 90°C and (D) 25°C. The inset images are the SEM images and the energy-dispersive X-ray spectra of the selected area. *Reprinted with permission from Fu et al. (2015). Copyright 2015 American Chemical Society.*

(Lvov and Abdullayev, 2013), chitosan and polyvinyl alcohol (PVA). The advantages of halloysite in these systems include enhanced cell attachment and growth, improved mechanical properties and encapsulation of certain medications within scaffolds for controllable release.

Halloysite loaded with brilliant green as an antiseptic agent was used as additive in electrospun polycaprolacton (PCL) scaffolds. Brilliant green release from these hybrid scaffolds was extended to 5 days. About 20% of the initial drug burst was observed, which was associated with the release of molecules adsorbed on the external surface, and this phenomenon is typical (Lvov and Abdullayev, 2013). In another study, halloysite-PCL scaffolds were explored for bone regeneration. In vitro bioactivity of a scaffold was tested in simulated body fluid, and enhanced bioactivity was observed by the addition of halloysite, as evidenced by mineral deposition. Halloysite-PCL fibres were covered by an apatite-like layer in 21 days. This hybrid scaffold exhibited excellent attachment, spreading and proliferation of human mesenchymal stromal cells (hMSC) and osteoblast cells, indicating superior performance for bone tissue engineering (Nitya et al., 2012).

The 10% halloysite-PVA scaffolds significantly improved cell attachment and growth. Osteoblastlike cells grew on halloysite-PVA films better than neat PVA, as shown by filopodia extensions on 10% halloysite-PVA nanocomposites on the fourth day of cell seeding. Halloysite imparted nanoscale roughness on PVA films, causing enhanced osteoblast adhesion (Zhou et al., 2009). Similar results were observed on surfaces modified by halloysite and poly(dimethyldiallyl ammonium chloride) using LbL nanoassembly. Human dermal fibroblast cells grew faster on substrates coated with halloysite than on analogous substrates with montmorillonite and silica nanoparticles. Both nanotopographic features and surface chemistry were responsible for cell adhesion. SEM revealed nanoscale roughness caused by halloysite nanotubes (Kommireddy et al., 2005). In addition, external silica surfaces have been found to improve cell bioactivity by stimulating osteoblasts and osteoblastlike cells to secrete type I collagen and other biochemical markers (Zhou et al., 2009).

Halloysite-chitosan scaffolds were prepared by mixing halloysite and chitosan in aqueous acetic acid solution and freeze drying. The addition of halloysite improved mechanical and thermal properties such as compressive strength, modulus, thermal stability and water uptake. Scaffolds preserved their porous structure even at 80% halloysite addition and doubled the bioactivity for growing NIH 3T3 cells (Liu et al., 2013).

Hybrid material based on halloysite-chitosan-poloxamer nanocomposite loaded with tetracycline was proposed for healing animal periodontitis. Halloysite was loaded with an antibiotic and mixed with acetate buffer solution containing 0.2% chitosan, 20% poloxamer, 0.5% polyethylene glycol and 1% octylcyanoacrylate. In vivo antibacterial research was performed by using a syringe to inject this CPN into surgical pockets artificially created in dogs' mouths. This CPN was capable of stopping bacterial infection for up to 2.5 months, while syringeability was retained over 9 months (Kelly et al., 2004).

Porous scaffolds based on halloysite-chitosan nanocomposite have been proposed recently for wound healing. A CPN with 67% halloysite exhibited 89% increase in blood clotting and enhanced early-stage wound healing in Sprague-Dawley rats (Liu et al., 2015). A halloysite alginate nanocomposite scaffold was prepared by mixing in solution and freeze drying. Incorporation of halloysite significantly increased the fibroblast cell attachment, mechanical strength and stability of this scaffold against enzymatic degradation (Liu et al., 2014).

Nanocomposite hydrogels of oligo(trimethylene carbonate)-poly(ethylene glycol)-oligo(trimethylene carbonate) diacrylate (TPT) were prepared by the photopolymerisation of TPT in the presence of halloysite. Sodium alginate (Alg) was added to construct an interpenetrating network with ionic cross-linked and chemical cross-linked moieties to obtain desirable mechanical properties and tunable drug-release behaviour. Hydrogels exhibited a three-dimensional porous structure with an irregular shape, with dimensions in the

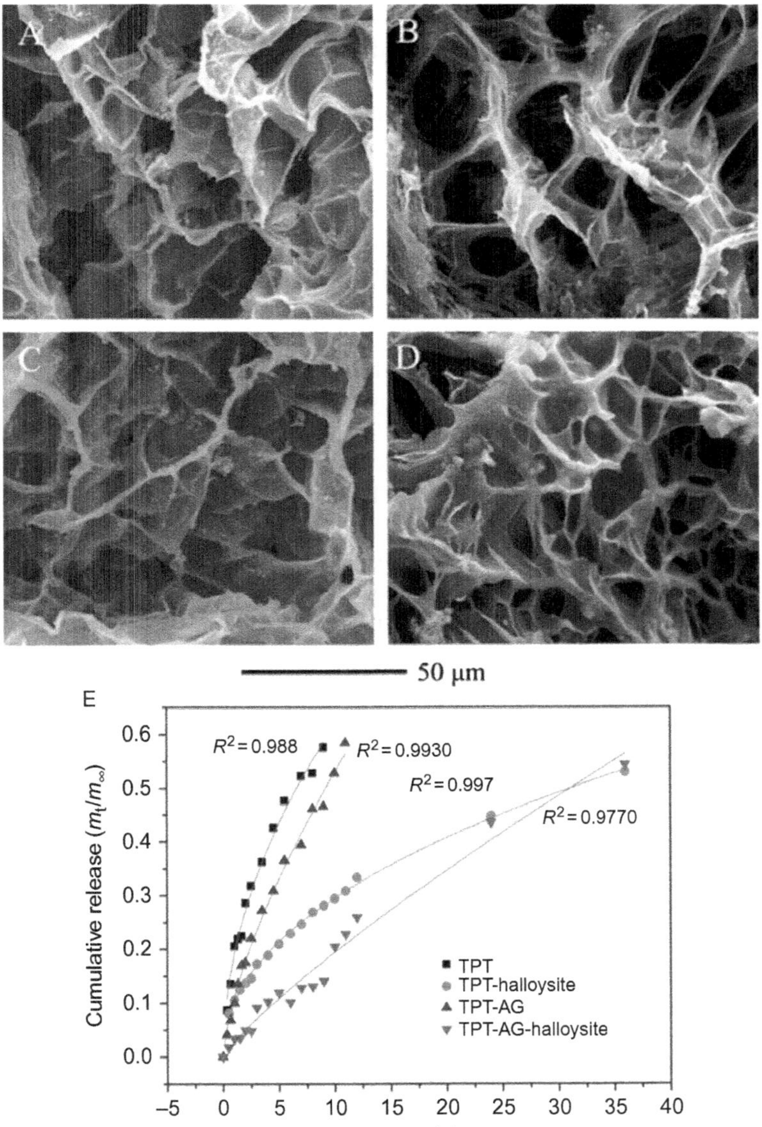

FIG. 22.31 SEM of (A) TPT, (B) TPT-halloysite, (C) TPT-AG and (D) TPT-AG-halloysite hydrogels. 2% of halloysite was added to composite hydrogels. (E) Cumulative release of BSA from the hydrogels in PBS at pH 7.4. *Reprinted with permission from Tu et al. (2013). Copyright 2013 Elsevier.*

range of 20–40 μm (Fig. 22.31A–D). No obvious phase separation was observed by the addition of Alg, indicating great miscibility of TPT and Alg. The mechanical properties of the hydrogel were greatly enhanced; compression modulus and fracture toughness were increased fivefold with the addition of 2% halloysite (Tu et al., 2013).

The drug-release behaviour of the hydrogels was examined using bovine serum albumin (BSA). BSA release was significantly retarded by the incorporation of halloysite within scaffolds. About 63–68% of loaded active was released from the original scaffolds, which dropped to 25–30% upon the addition of only 2% halloysite. Complete release lasted over 100 h for halloysite-TPT scaffolds, as opposed to 36 h with only TPT (Fig. 22.31E) (Tu et al., 2013).

Halloysite-poly(hydroxyethyl methacrylate)-sodium hyaluronate nanocomposite hydrogels were prepared by free-radical redox polymerisation of 2-hydroxyethyl methacrylate and *N,N*-methylene bis-acrylamide. Halloysite was loaded with 5-fluorouracil, an anticancer drug, and incorporated into scaffold in situ during polymerisation. The hydrogels exhibited pH-sensitive swelling behaviour, allowing controllable drug delivery triggered by acidic environments. In vitro release at 37°C in stimulated gastric fluid (pH 1.2) for 2 h, followed by a stimulated intestinal medium (pH 7.4), showed that drug release effectively stops when the pH of medium is increased. These hydrogels were proposed for treating colon cancer (Rao et al., 2014).

Thermo-responsive hydrogel was constructed based on halloysite/poly (*N*-isopropylacrylamide) nanocomposite for thermally induced drug delivery of medicines. Hydrogel shrunk to about 20% of the original volume above ~40°C, blocking the pores necessary for drug release. Methylene blue release was conducted at solution temperatures of 23°C and 44°C, demonstrating a steady drug release from embedded halloysite lumens at below the lower temperature (23°C), and effectively no release after shrinking at 44°C. Drug release was reinstated by lowering the temperature. This suggests that the drug release from hydrogel can be controlled by manipulating the environmental temperature (Xi et al., 2013).

Electrospun scaffolds for guided bone/tissue regeneration were developed by electrospinning drug-loaded halloysite-poly(caprolactone)/gelatin nanocomposites to provide sustained drug delivery. Scaffolds were collected on a rotating collector lined with aluminium foil. The mechanical strength of the halloysite based nanocomposite scaffolds doubled in the direction of rotation and halved in the perpendicular direction as compared to the raw polymer scaffolds, indicating the isotropic nature of the material. Metronidazole (MNA), a model antibiotic used against anaerobic bacteria, was encapsulated into scaffolds and studied with the bacteria *Fusobacterium nucleatum*. Three different scaffolds were designed containing 20% free MNA (PG-MNA), 5% MNA loaded within halloysite and 20% free+5% halloysite-loaded MNA (PG-MNA/halloysite-MNA). The free drug released from scaffolds very quickly, with almost the entire release reached in about 4 days, whereas

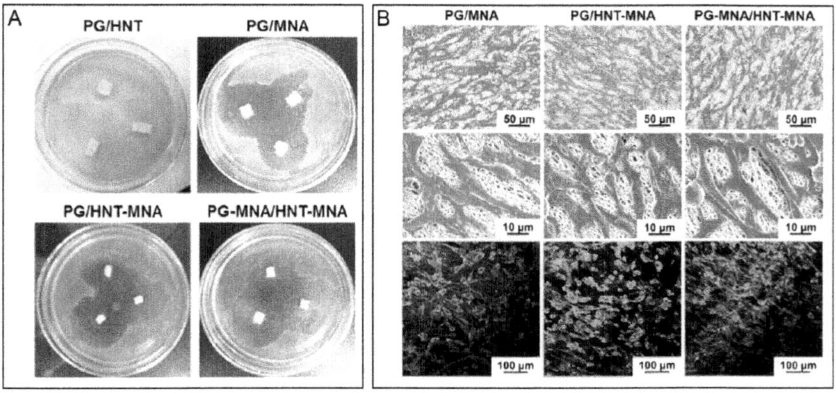

FIG. 22.32 (A) Inhibition of bacterial growth on agar plates of the drug-loaded nanomicro fibre membranes after incubation for 1 day under anaerobic conditions at 37°C. (B) Scanning electron micrographs and confocal fluorescent images of L929 fibroblasts proliferated on the scaffolds after 5 days. *Reprinted with permission from Xue et al. (2015). Copyright 2015 American Chemical Society.*

the halloysite-loaded drug provided sustained release over 2 weeks, although drug release was extremely low during the initial 2 days. The combination of free and halloysite-loaded MNA allowed fast initial release followed by slow release over an extended time. The sustained release of metronidazole from the membranes prevented the colonisation of anaerobic *Fusobacteria,* as shown by the zone of inhibition in agar plates (Fig. 22.32A), while L929 fibroblasts could still adhere to and proliferate on scaffolds (Fig. 22.32B) (Xue et al., 2015).

22.5.3 Bone Cements and Implants

Bone cements and bone graft substitutes are one of the rapidly growing medical practices in the world. Currently, there is a tremendous need to make commercial bone cements and more functional, bio-instructional and longer-lasting products available. Several studies performed in this area show great promise for the application of halloysite nanotubes. Halloysite offers several benefits; first, most bone cements include antibiotics, added to prevent bacterial infection during surgery. However, direct addition of antibiotics leads to weakening of the cement—in particular, a loss of mechanical strength. The encapsulation of antibiotics within halloysite lumens helps to eliminate such effects. In addition, sustained release of the antibiotics from the current bone cement varieties is available for only a few hours. Halloysite retards the release rate up to several weeks, allowing protection against bacterial infection throughout the treatment. Finally, mixing halloysite with bone cements has been shown to significantly increase mechanical and adhesive properties, which extends the lifetime of the product.

The influence of halloysite addition to PMMA-based bone cement, one of the most commonly used materials in prostheses, was studied by Wei et al. (2012). Gentamicin was loaded into halloysite lumens to provide antibacterial protection. Tensile strength of the CPN increased from 18.0 to 26.1 MPa with the addition of 5% halloysite, while flexural strength was slightly reduced. On the other hand, addition of free gentamicin reduced mechanical properties. An optimal halloysite level of 5% was suggested to provide optimal mechanical properties, which corresponded to 0.75% of gentamicin, considering the 15% loading efficiency. Gentamicin release from such CPN lasted over 10 days. Fig. 22.33 presents a zone of inhibition formed around PMMA cement after being doped with halloysite after 1 week, while bacteria readily grew around the original PMMA cement. These data clearly suggest that the

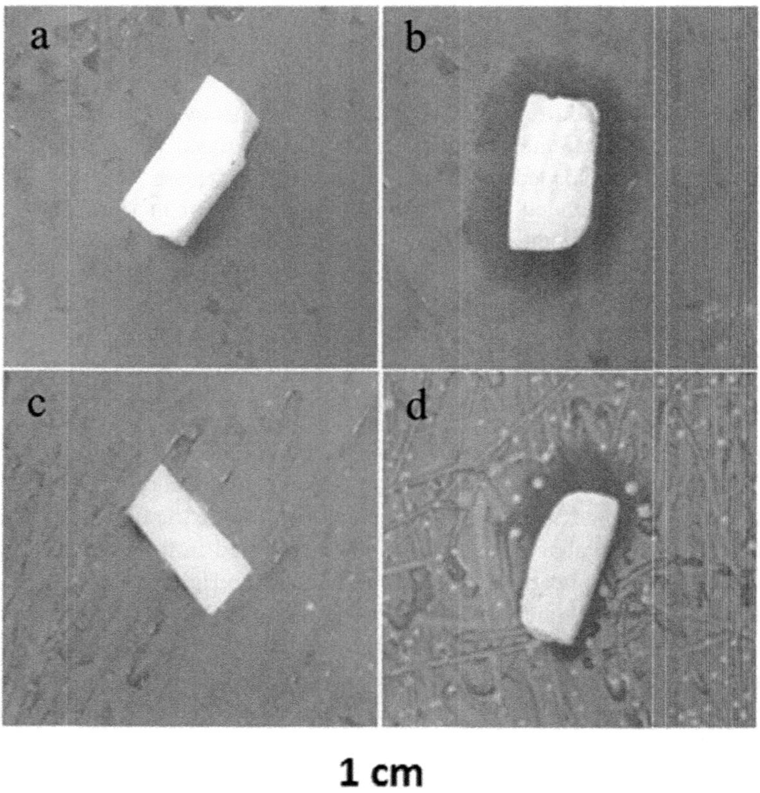

FIG. 22.33 Images of *E. coli* (A and B) and *S. aureus* (C and D) around original poly(methylmethacrylate) (PMMA) bone cement (*left images*), and PMMA+halloysite loaded with gentamicin (*right images*) after 1 week of the bacterial culture growth. *Reprinted with permission from Wei et al. (2012). Copyright 2012 John Wiley & Sons.*

release of antibiotic gentamicin from the halloysite-PMMA nanocomposite protects bacterial infection over 1 week without compromising the mechanical properties of the implant (Wei et al., 2012).

A significant increase of PMMA cement adhesion to bone was possibly the most significant finding in this research study. Adhesion of the cement was tested by sticking two pieces of cow femur bone together with bone cement and pulling them apart with a tensile tester. The maximum force required to break the halloysite-PMMA nanocomposite sample was twice that of the original PMMA cement sample and the sample with free gentamicin (600 N vs 290 N and 260 N, respectively). In addition, bones stuck with halloysite-PMMA cement failed within the bone (ie, by breaking it), whereas the other two samples failed at the cement-bone interface. More precise measurement of the adhesive strength revealed a fivefold increase in adhesive strength with 10% halloysite. Entanglement of halloysite tubes within bone pores is believed to be the cause of this rapid improvement of adhesion (Wei et al., 2012).

Halloysite was also shown to be efficient in calcium phosphate bone cement, which is commonly used due to its excellent biocompatibility and osteoconductivity. These formulations are made of various calcium phosphate salts, which upon mixing with water form a workable and self-setting putty. The growth of crystallites within this putty helps it retain its shape and mechanical strength. The addition of 10% halloysite improved the yield strength of beta tricalcium phosphate-based bone cement by over 10 times (0.33 MPa for original cement vs 3.5 MPa for halloysite-calcium phosphate cement). Elongation at the break was improved by 4 times. This was associated with halloysite-induced nonisotropic crystallisation of calcium phosphate, resulting in crystalline fibres of 50 microns length and 0.5 micron diameter (Abdullayev, 2014).

22.5.4 Cosmetics

Halloysite was proposed for the encapsulation of various active agents commonly used in cosmetic creams, including humectants, vitamins, fragrances, antimicrobials, antioxidants and soothing agents. With these substances, halloysite elongates the time range for their useful action. In addition, halloysite was found to be effective in cleansing the skin of oils and unwanted toxins. Commercial nail polish utilising natural halloysite nanotubes is already available (Riedlinger and Corkery, 2007; Rawtani and Agrawal, 2012).

The first study on the use of halloysite in cosmetics was directed towards the encapsulation and long-lasting release of active agents. Encapsulation of active cosmetic ingredients within halloysite lumens significantly improves the dermatological effects of these products, allowing effective skin treatment. According to US patent application 2007/0202061 (Riedlinger and Corkery, 2007) consistent release of glycerol and vitamins C and E from halloysite lumens has been achieved within 4 h.

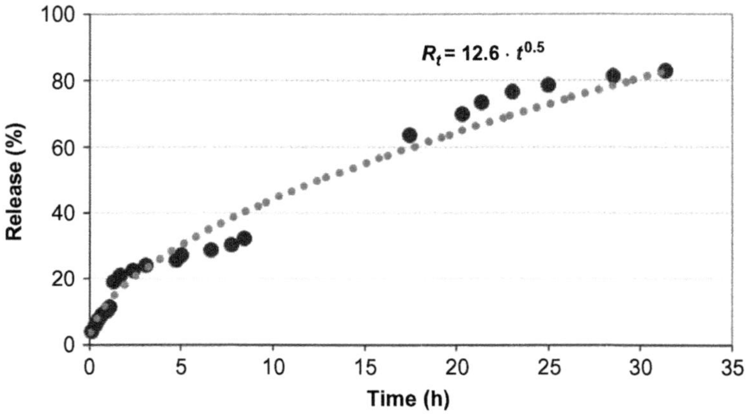

FIG. 22.34 Glycerol release profile from halloysite nanotubes.

Loading and release characteristic of glycerol, a well-known skin humidifying agent, was studied by Suh et al. (2011). A loading efficiency of 19% was achieved by loading glycerol from 40% aqueous solution using a vacuum pump. The release profile fits the Korsmeyer–Peppas model with pore-controlled diffusion kinetics. Complete release exceeded 30 h, which provides a long-lasting humidifying effect (Fig. 22.34). Halloysite was shown to be beneficial for skin even without active agents. Due to its adsorption capacity, halloysite removes unwanted toxins and unpleasant oils from the skin, and as such, serves as a hypoallergenic skin cleanser. Halloysite also performs as a gentle exfoliator, drawing dead skin cells away from the surface to leave it fresh, young and healthy (Rawtani and Agrawal, 2012).

22.5.5 Biomedical Devices

Halloysites loaded with proteins and enzymes are promising for use in various biomedial devices. Urease-loaded halloysite was used as a nanoreactor for the biomimetic synthesis of calcium carbonate. The urease-catalysed reaction occurred exclusively inside halloysite nanotubes, as no reaction product was found on the outer surface of the nanotubes (Shchukin et al., 2005). Similarly, halloysite, functionalised with hematin at the tube exterior, allowed the polymerisation of aniline to cover the halloysite surface (Tierrablanca et al., 2010). Urease and α-amylase, immobilised in halloysite lumens, retained 90% of their enzymatic activity after 15 days of storage (Zhai et al., 2010).

Halloysite nanotubes functionalised with P-selectin protein were used to fabricate a microfluidic bioanalytical device for detecting and capturing circulating tumour cells from the bloodstream. Thanks to the increased surface roughness of microfluidic channels caused by a high aspect ratio, halloysite tubule devices showed significantly better performance in capturing Colo205 colon carcinoma and KG1a leukaemic cancer cells. Tall structures comprising halloysite nanoneedles presented more P-selectin protein to the bloodstream, capturing cancer cells far from the surface (Hughes and King, 2010).

Surfaces coated with liposome and halloysite have been investigated with regard to the capture of MCF-7 and Colo205 cancer cells. These surfaces showed excellent selectivity towards cancer cells by repelling healthy blood cells due to the perfusion of neutrophils. This provides a unique capability to deliver anticancer drugs to cancerous cells by reducing chemotherapeutic drug dosages and toxic nonspecific effects (Mitchell et al., 2012). In another study, halloysite nanotubes functionalised with negatively charged surfactant-sodium dodecanoate were utilised for enhanced tumour cell capture while negating leukocyte adhesion from the bloodstream. Interestingly, surface functionalisation with the positive surfactant decyltrimethylammonium bromide promoted leukocyte adhesion while abolishing cancer cell attachment (Mitchell et al., 2015).

22.6 CONCLUDING REMARKS

Halloysite nanotubes are among the most promising materials for the encapsulation of chemically and biologically active agents. A wide variety of active agents with diverse molecular sizes and structures have been effectively encapsulated within halloysite lumens and provided extended release periods in aqueous environments. These agents include corrosion inhibitors, biocides, medicines, cosmetic agents, proteins and enzymes. In addition, due to differing surface chemistries on the inner and external walls, halloysite nanotubes were selectively functionalised with various surface modifiers, including silane-coupling agents, phosphonates, surfactants and polymers to enhance loading efficiency and controlled release. Selective alumina etching from halloysite lumens enlarged the space in halloysite internal lumens, causing more agents to be loaded within nanotubes.

In addition to loading active agents, halloysite nanotubes show excellent compatibility with a variety of polymers with moderate and high polarities. Unlike other clay minerals, halloysite are a submicron size, with few hydroxyl groups at the external surface, causing fewer aggregates in polymer blends. This simplifies mixing halloysite with polymers during compounding.

Halloysite nanotubes loaded with corrosion inhibitors and fungicides find applications in protective coatings. Preliminary studies have shown that such functional coatings provide extended corrosion protection for copper and aluminium alloys, even in extremely harsh environments. Encapsulation of medical agents provides the potential for the preparation of functional polymer composite materials for tissue engineering and bone implants. Bionanocomposite scaffolds using halloysite and polymers, such as PCL, PVA and chitosan, were reported. Halloysite was found to significantly improve mechanical properties and enhance cell adhesion. Sustained drug delivery from such scaffolds allows for the treatment of various diseases. Finally, halloysite nanotubes were shown to be very effective in artificial bone cements and dental fillers. Addition of halloysite into bone cement significantly improves its mechanical properties, which is crucial for bone implants. In addition, active agents that need to be embedded into bone cement (such as growth hormone or antibiotics) can effectively be encapsulated to provide longtime pharmaceutical activity without suffering drug overdoses.

Studies on halloysite nanotubes are still heavily focused on polymers and coatings in spite of all the abovementioned benefits. Lack of sufficient safety data from large animal studies with halloysite seems to be the main hurdle in its application for controlled drug delivery. Therefore, medical applications featuring skin-controlled release, such as cosmetic creams and wound-healing scaffolds, seems to be the most promising in the short term.

ABBREVIATIONS

3T3 cells	a cell culture produced by transferring (T) primary mouse embryonic fibroblast cells every 3 days (the first '3'), and inoculating at the rigid density of 3×10^5 cells per 20-cm^2 dish (the second '3') continuously
4010NA	*N*-isopropyl-*N*′-phenyl-4-phenylenediamine
AEAPS	3-(2-aminoethyl)aminopropyltrimethoxy silane
Alg	sodium alginate
APTES	(3-aminopropyl)triethoxysilane
ATRP	atomic transfer radical polymerisation
BAC	benzalkonium chloride
BG	brilliant green
BiBB	α-bromoisobutyryl bromide
BSA	bovine serum albumin
BTA	benzotriazole
Colo205	a type of human colon cancer cell
DI water	deionised water
DPH	diphenhydramine
DS	diclofenac sodium
FDA	Food and Drug Administration
FTIR	Fourier transform infrared spectroscopy
hMSC	human mesenchymal stromal cells
HNT	Halloysite nanotubes
KG1a	type of leukaemic cancer cell
LbL	layer by layer
MBI	2-mercaptobenzimidazole
MBT	2-mercaptobenzothiazole
MCF-7	Michigan Cancer Foundation-7
MNA	metronidazole

NIH	National Institutes of Health (USA)
NMR	nuclear magnetic resonance
ODP	octadecyl phosphonic acid
PAA	poly(acrylic acid)
PAH	poly(allylaminehydrochloride)
PBS	phosphate-buffered saline
PCL	poly(caprolactone)
PDDA	poly(diallyldimethylammoniumchloride)
PDMAEMA	poly(*N*,*N*-dimethylaminoethyl methacrylate)
PEI	polyethyleneimine
PG	polycaprolactone/gelatin scaffolds
PLGA	poly(lactic-*co*-glycolic acid)
PMMA	poly(methylmethacrylate)
PSS	poly(styrenesulfonate)
PVA	polyvinyl alcohol
SBR	styrene-butadiene rubber
SDS	sodium dodecyl sulphate
SEM	scanning electron microscopy
SI-ATRP	surface-initiated atomic transfer radical polymerisation
TEM	transmission electron microscopy
ToF-SIMS	time-of-flight secondary ion mass spectrometry
TPT	oligo(trimethylene carbonate)-poly(ethylene glycol)-oligo (trimethylene carbonate) diacrylate
UF	urea-formaldehyde
UV–vis	ultraviolet visible
XPS	X-ray photoelectron spectroscopy

REFERENCES

Abbasov, V., Mammadova, T., Andrushenko, N., Hasankhanova, N., Lvov, Y., Abdullayev, E., 2013. Halloysite–Y-zeolite blends as novel mesoporous catalysts for the cracking of waste vegetable oils with vacuum gasoil. Fuel 117, 552–555.

Abdullayev, E., 2014. Medical and health applications of halloysite nanotubes. In: Churchman, G.J., Pasbakhsh, P. (Eds.), Natural Mineral Nanotubes Properties and Applications. Apple Academic Press and CRC Press, Boca Raton, Florida, USA, pp. 422–437. Chapter 22.

Abdullayev, E., Lvov, Y., 2010. Clay nanotubes for corrosion inhibitor encapsulation: release control with end stoppers. J. Mater. Chem. 20, 6681–6687.

Abdullayev, E., Lvov, Y., 2011. Halloysite clay nanotubes for controlled release of protective agents. J. Nanosci. Nanotechnol. 11, 10007–10026.

Abdullayev, E., Lvov, Y., 2013. Halloysite clay nanotubes as a ceramic "skeleton" for functional biopolymer composites with sustained drug release. J. Mater. Chem. B 1, 2894–2903.

Abdullayev, E., Lvov, Y., 2014. Halloysite tubule nanoreactors in industrial and agricultural applications. In: Churchman, G.J., Pasbakhsh, P. (Eds.), Natural Mineral Nanotubes Properties and Applications. Apple Academic Press, Ontario, Canada, pp. 364–383. Chapter 19.

Abdullayev, E., Shchukin, D., Price, R., Lvov, Y., 2009. Halloysite tubes as nanocontainers for anticorrosion coating with benzotriazole. ACS Appl. Mater. Interf. 1, 1437–1443.

Abdullayev, E., Sakakibara, K., Okamoto, K., Wei, W., Ariga, K., Lvov, Y., 2011. Natural tubule clay template synthesis of silver nanorods for antibacterial composite coating. ACS Appl. Mater. Interf. 2011 (3), 4040–4046.

Abdullayev, E., Joshi, A., Wei, W., Zhao, Y., Lvov, Y., 2012a. Economically viable fabrication method of nanocomposite materials from linear arrays of metallic nanoparticles and nanorods on tubular halloysite templates. In: Proc. of ASME 2012 International Mechanical Engineering Congress and Exposition, Micro- and Nano-Systems Engineering and Packaging, vol. 9, Parts A and B, Houston, Texas, USA, November 9–15, 2012, Paper No. IMECE2012-87006, pp. 1033–1042.

Abdullayev, E., Joshi, A., Wei, W., Zhao, Y., Lvov, Y., 2012b. Enlargement of halloysite clay nanotube lumen by selective etching of aluminum oxide. ACS Nano 6, 7216–7226.

Abdullayev, E., Abbasov, V., Tursunbayeva, A., Portnov, V., Ibrahimov, H., Mukhtarova, G., Lvov, Y., 2013. Self-healing coatings based on halloysite clay polymer composites for protection of copper alloys. ACS Appl. Mater. Interf. 5, 4464–4471.

Aguzzi, P.C., Viseras, C., Caramella, C., 2007. Use of clays as drug delivery systems: possibilities and limitations. Appl. Clay Sci. 36, 22–36.

Aguzzi, C., Viseras, C., Cerezo, P., Salcedo, I., Sánchez-Espejo, R., Valenzuela, C., 2013. Release kinetics of 5-aminosalicylic acid from halloysite. Coll. Surf. B Biointerfaces 105, 75–80.

Bates, T., Hilderbrand, F., Swineford, A., 1950. Morphology and structure of endellite and halloysite. Am. Miner. 35, 463–485.

Belkassa, K., Bessaha, F., Khelifa, K.M., Gener, I.B., Comparot, J.D., Khelifa, A., 2013. Physicochemical and adsorptive properties of a heat-treated and acid-leached Algerian halloysite. Coll. Surf. A Physicochem. Eng. Asp. 421, 26–33.

Cavallaro, G., Lazzara, G., Milioto, S., Palmisano, G., Parisi, F., 2014. Halloysite nanotube with fluorinated lumen: non-foaming nanocontainer for storage and controlled release of oxygen in aqueous media. J. Coll. Interf. Sci. 417, 66–71.

Churchman, G.J., Davy, T.J., Aylmore, L.A.G., Gilkes, R.J., Self, P.G., 1995. Characteristics of fine pores in some halloysites. Clay Miner. 30, 89–98.

Datt, A., Burns, E.A., Dhuna, N.A., Larsen, S.C., 2013. Loading and release of 5-fluorouracil from HY zeolites with varying SiO_2/Al_2O_3 ratios. Micropor. Mesopor. Mater. 167, 182–187.

Du, M., Guo, B., Jia, D., 2010. Newly emerging applications of halloysite nanotubes: a review. Polym. Int. 59, 574–595.

Fakhrullina, G.I., Akhatova, F.S., Lvov, Y.M., Fakhrullin, R.F., 2015. Toxicity of halloysite clay nanotubes in vivo: a *Caenorhabditis elegans* study. Environ. Sci. Nano 2, 54–59.

Fix, D., Andreeva, D., Möhwald, H., Lvov, Y., Shchukin, D., 2009. Application of inhibitor loaded halloysite nanotubes in active anticorrosive coatings. Adv. Funct. Mater. 19, 1720–1727.

Fu, Y., Zhao, D., Yao, P., Wang, W., Zhang, L., Lvov, Y., 2015. Highly aging-resistant elastomers doped with antioxidant-loaded clay nanotubes. ACS Appl. Mater. Interf. 7, 8156–8165.

Guimarães, L., Enyashin, A.N., Seifert, G., Duarte, H.A., 2010. Structural, electronic, and mechanical properties of single-walled halloysite nanotube models. J. Phys. Chem. C 114, 11358–11363.

Hemmatpour, H., Asl, V.H., Mamaqani, H.R., 2015. Synthesis of pH-sensitive poly (N,N-dimethylaminoethyl methacrylate)-grafted halloysite nanotubes for adsorption and controlled release of DPH and DS drugs. Polymer 65, 143–153.

Horcajada, P., Alvarez, C.M., Rámila, A., Pariente, J.P., Regí, M.V., 2006. Controlled release of Ibuprofen from dealuminated faujasites. Solid State Sci. 8, 1459–1465.

Hughes, A., King, M., 2010. Use of naturally occurring halloysite nanotubes for enhanced capture of flowing cells. Langmuir 26, 12155–12159.

Joshi, A., Abdullayev, E., Vasiliev, A., Volkova, O., Lvov, Y., 2013. Interfacial modification of clay nanotubes for the sustained release of corrosion inhibitors. Langmuir 29, 7439–7448.

Joussein, E., Petit, S., Churchman, J., Theng, B., Righi, D., Delvaux, B., 2005. Halloysite clay minerals—a review. Clay Miner. 40, 383–426.

Kelly, H., Deasy, P., Ziaka, E., Claffey, N., 2004. Formulation and preliminary in vivo dog studies of a novel drug delivery system for the treatment of periodontitis. Int. J. Pharm. 274, 167–183.

Kommireddy, D., Ichinose, I., Lvov, Y., Mills, D., 2005. Nanoparticle multilayer: surface modification for cell attachment and growth. J. Biomed. Nanotechnol. 1, 286–290.

Korsmeyer, R.W., Gurny, R., Doelker, E., Buri, P., Peppas, N.A., 1983. Mechanisms of solute release from porous hydrophilic polymers. Int. J. Pharm. 15, 25–35.

Levis, S., Deasy, P., 2003. Use of coated microtubular halloysite for the sustained release of diltiazem hydrochloride and propranolol hydrochloride. Int. J. Pharm. 253, 145–157.

Liu, M., Wu, C., Jiao, Y., Xiong, S., Zhou, C., 2013. Chitosan-halloysite nanotubes nanocomposite scaffolds for tissue engineering. J. Mater. Chem. B. 1, 2078–2089.

Liu, M., Shen, Y., Ao, P., Dai, L., Liu, Z., Zhou, C., 2014. The improvement of hemostatic and wound healing property of chitosan by halloysite nanotubes. RSC Adv. 4, 23540–23553.

Liu, M., Dai, L., Shi, H., Xiong, S., Zhou, C., 2015. In vitro evaluation of alginate/halloysite nanotube composite scaffolds for tissue engineering. Mater. Sci. Eng. C 49, 700–712.

Luo, P., Zhao, Y., Zhang, B., Liu, J., Yang, Y., Liu, J., 2010. Study on the adsorption of Neutral Red from aqueous solution onto halloysite nanotubes. Water Res. 44, 1489–1497.

Lvov, Y., Abdullayev, E., 2013. Functional polymer-clay nanotube composites with sustained release of chemical agents. Prog. Polym. Sci. 38, 1690–1719.

Lvov, Y., Price, R., 2008. Halloysite nanotubules a novel substrate for the controlled delivery of bioactive molecules. In: Ruiz-Hitzky, E., Ariga, K., Lvov, Y. (Eds.), Bio-Inorganic Hybrid Nanomaterials. Wiley, London, Berlin, pp. 440–478.

Lvov, Y.M., Shchukin, D.G., Möhwald, H., Price, R.R., 2008. Halloysite clay nanotubes for controlled release of protective agents. ACS Nano 2, 814–820.

Mahani, H., Kazemeini, M., 2003. Treatment of diatomaceous earth to obtain its catalyst support. Scientia Iranica 10, 350–356.

Massaro, M., Colletti, C.G., Noto, R., Riela, S., Poma, P., Guernelli, S., Parisi, F., Milioto, S., Lazzara, G., 2015. Pharmaceutical properties of supramolecular assembly of co-loaded cardanol/triazole-halloysite systems. Int. J. Pharm. 478, 476–485.

Mitchell, M.J., Castellanos, C.A., King, M.R., 2012. Nanostructured surfaces to target and kill circulating tumor cells while repelling leukocytes. J. Nanomater. 2012, 1–10 Article ID 831263.

Mitchell, M.J., Castellanos, C.A., King, M.R., 2015. Surfactant functionalization induces robust, differential adhesion of tumor cells and blood cells to charged nanotube-coated biomaterials under flow. Biomaterials 56, 179–186.

Ng, K.M., Ting, Y., Lau, R., Chan, C.M., Weng, L.T., Wu, J., 2011. Surface studies of halloysite nanotubes by XPS and ToF-SIMS. Surf. Interf. Anal. 43, 795–802.

Nitya, G., Nair, G.T., Mony, U., Chennazhi, K.P., Nair, S.V., 2012. In vitro evaluation of electrospun PCL/nanoclay composite scaffold for bone tissue engineering. J. Mater. Sci. Mater. Med. 23, 1749–1761.

Peng, Q., Liu, M., Zheng, J., Zhou, C., 2015. Adsorption of dyes in aqueous solutions by chitosan-halloysite nanotubes composite hydrogel beads. Micropor. Mesopor. Mater. 201, 190–201.

Qi, R., Guo, R., Shen, M., Cao, X., Zhang, L., Xu, J., Yu, J., Shi, X., 2010. Electrospun poly(lactic-co-glycolic acid)/halloysite nanotube composite nanofibers for drug encapsulation and sustained release. J. Mater. Chem. 20, 10622–10629.

Rao, K.M., Nagappan, S., Seo, D.J., Ha, C.S., 2014. pH sensitive halloysite-sodium hyaluronate/poly(hydroxyethyl methacrylate) nanocomposites for colon cancer drug delivery. Appl. Clay Sci. 97–98, 33–42.

Rawtani, D., Agrawal, Y.K., 2012. Multifarious applications of halloysite nanotubes: a review. Rev. Adv. Mater. Sci. 30, 282–295.

Riedlinger, M.D., Corkery, R.W., 2007. Cosmetic skincare applications employing mineral-derived tubules for controlled release. US Patent Application 2007/0202061 A1, assigned to NaturalNano Inc.

Riela, S., Massaro, M., Colletti, C.G., Bommarito, A., Giordano, C., Milioto, S., Noto, R., Poma, P., Lazzara, G., 2014. Development and characterization of co-loaded curcumin/triazole- halloysite systems and evaluation of their potential anticancer activity. Int. J. Pharm. 475, 613–623.

Rives, V., Arco, M., Martín, C., 2014. Intercalation of drugs in layered double hydroxides and their controlled release: a review. Appl. Clay Sci. 88–89, 239–269.

Shchukin, D., Price, R., Lvov, Y., 2005. Biomimetic synthesis of vaterite in the interior of clay nanotubules. Small 1, 510–513.

Shchukin, D.G., Lamaka, S.V., Yasakau, K.A., Zheludkevich, M.L., Ferreira, M.G.S., Möhwald, H., 2008. Active anticorrosion coatings with halloysite nanocontainers. J. Phys. Chem. C 112, 958–964.

Shutava, T.G., Fakhrullin, R.F., Lvov, Y.M., 2014. Spherical and tubule nanocarriers for sustained drug release. Curr. Opin. Pharmacol. 18 (2014), 141–148.

Suh, Y., Kil, D., Chung, K., Abdullayev, E., Lvov, Y., Mongayt, D., 2011. Natural nanocontainer for the controlled delivery of glycerol as a moisturizing agent. J. Nanosci. Nanotechnol. 11, 661–665.

Tan, D., Yuan, P., Bergaya, F.A., Yu, H., Liu, D., Liu, H., He, H., 2013. Natural halloysite nanotubes as mesoporous carriers for the loading of ibuprofen. Micropor. Mesopor. Mater. 179, 89–98.

Tan, D., Yuan, P., Bergaya, F.A., Liu, D., He, H., 2014. High-capacity loading of 5-fluorouracil on the methoxy-modified kaolinite. Appl. Clay Sci. 100, 60–65.

Tierrablanca, E., García, J.R., Roman, P., Silva, R.C., 2010. Biomimetic polymerization of aniline using hematin supported on halloysite nanotubes. Appl. Catal. A Gen. 381, 267–273.

Tu, J., Cao, Z., Jing, Y., Fan, C., Zhang, C., Liao, L., Liu, L., 2013. Halloysite nanotube nanocomposite hydrogels with tunable mechanical properties and drug release behavior. Comp. Sci. Technol. 85, 126–130.

Veerabadran, N., Price, R., Lvov, Y., 2007. Clay nanotubes for encapsulation and sustained release of drugs. NANO 2, 215–222.

Vergaro, V., Abdullayev, E., Lvov, Y.M., Cingolani, R., Rinaldi, R., Leporatti, S., 2010. Cytocompatibility and uptake of halloysite clay nanotubes. Biomacromolecules 11, 820–828.

Vergaro, V., Lvov, Y.M., Leporatti, S., 2012. Halloysite clay nanotubes for resveratrol delivery to cancer cells. Macromol. Biosci. 12, 1265–1271.

Wang, L.P., Wang, Y.P., Pei, X.W., Peng, B., 2008. Synthesis of poly(methyl methacrylate)-*b*-poly(*N*-isopropylacrylamide) (PMMA-*b*-PNIPAM) amphiphilic diblock copolymer brushes on halloysite substrate via reverse ATRP. React. Funct. Polym. 68, 649–655.

Wang, Q., Zhang, J., Wang, A., 2013. Alkali activation of halloysite for adsorption and release of ofloxacin. Appl. Surf. Sci. 287, 54–61.

Wang, Q., Zhang, J., Zheng, Y., Wang, A., 2014. Adsorption and release of ofloxacin from acid- and heat-treated halloysite. Coll. Surf. B Biointerfaces 113, 51–58.

Wei, W., Abdullayev, E., Hollister, A., Lvov, Y., Mills, D., 2012. Clay nanotube/poly(methyl methacrylate) bone cement composite with sustained antibiotic release. Macromol. Mater. Eng. 301, 645–653.

Wei, W., Minullina, R., Abdullayev, E., Fakhrullin, R., Mills, D., Lvov, Y., 2014. Enhanced efficiency of antiseptics with sustained release from clay nanotubes. RCS Adv. 4, 488–494.

Xi, L., Xiaojie, J., Rui, X., Mingyue, J., Jie, W., Liangyin, C., 2013. Halloysite nanotube composited thermo-responsive hydrogel system for controlled-release. Chinese J. Chem. Eng. 21, 991–998.

Xue, J., Niu, Y., Gong, M., Shi, R., Chen, D., Zhang, L., Lvov, Y., 2015. Electrospun microfiber membranes embedded with drug-loaded clay nanotubes for sustained antimicrobial protection. ACS Nano 9, 1600–1612.

Yah, W.O., Takahara, A., Lvov, Y.M., 2012a. Selective modification of halloysite lumen with octadecylphosphonic acid: new inorganic tubular micelle. J. Am. Chem. Soc. 134, 1853–1859.

Yah, W.O., Xu, H., Soejima, H., Ma, W., Takahara, T., Lvov, Y., 2012b. Biomimetic dopamine derivative for selective polymer modification of halloysite nanotube lumen. J Am. Chem. Soc. 134, 12134–12137.

Yelleswarapu, C.S., Gu, G., Abdullayev, E., Lvov, Y., Rao, D.V.G.L.N., 2010. Nonlinear optics of nontoxic nanomaterials. Optics Commun. 283, 438–441.

Yuan, P., Southon, P.D., Liu, Z., Green, M.E.R., Hook, J.M., Antill, S.J., Kepert, C.J., 2008. Functionalization of halloysite clay nanotubes by grafting with γ-aminopropyltriethoxysilane. J. Phys. Chem. C 112, 15742–15751.

Yuan, P., Southon, P.D., Liu, Z., Kepert, C.J., 2012a. Organosilane functionalization of halloysite nanotubes for enhanced loading and controlled release. Nanotechnology 23, 1–5. Article number 375705.

Yuan, P., Tan, D., Bergaya, F.A., Yan, W., Fan, M., Liu, D., He, H., 2012b. Changes in structure, morphology, porosity, and surface activity of mesoporous halloysite nanotubes under heating. Clay Clay Miner. 60, 561–573.

Zhai, R., Zhang, B., Liu, L., Xie, Y., Zhang, H., Liu, J., 2010. Immobilization of enzyme biocatalyst on natural halloysite nanotubes. Catal. Commun. 12, 259–263.

Zhao, M., Liu, P., 2008. Adsorption behavior of methylene blue on halloysite nanotubes. Micropor. Mesopor. Mater. 112, 419–424.

Zhao, Y., Abdullayev, E., Vasiliev, A., Lvov, Y., 2012. Halloysite nanotubule clay for efficient water purification. J. Coll. Interf. Sci. 406, 121–129.

Zhou, W., Guo, B., Liu, M., Liao, R., Bakr, A., Rabie, M., Jia, D., 2009. Poly(vinyl alcohol)/halloysite nanotubes bionanocomposite films: properties and in vitro osteoblasts and fibroblasts response. J. Biomed. Mater. Res. A 2009, 1574–1581.

Chapter 23

Halloysite for Adsorption and Pollution Remediation

J. Matusik*

Faculty of Geology, Geophysics and Environmental Protection, AGH University of Science and Technology, Krakow, Poland

Corresponding author: e-mail: jakub_matusik@wp.pl

23.1 INTRODUCTION

The prevention of environmental pollution caused by inorganic and organic toxic (and very often cancerogenic) chemical compounds is very important. The remediation and immobilization of contaminants applies to drinking water, wastewater and soil solutions, among other substances. The potentially hazardous inorganic substances include heavy metals (Alloway, 2013). In turn, the organic pollutants show a variety of structures and chemical activities (eg, dyes and pigments, hydrocarbons and their chlorinated derivatives and polycyclic aromatic hydrocarbons). Many of these chemical compounds are not required for the development of living organisms and may cause serious health problems, even at low or trace concentrations.

The proposed remediation techniques involve precipitation of highly insoluble phases, membrane filtration and electrolytic reduction (Wang et al., 2003). However, adsorption-based approaches are in particular applied because the materials used for treatment are of low cost and very often may be reused after appropriate regeneration (Hua et al., 2012). The adsorbents represent a group of structurally different materials, with the most important including functionalized silica, activated carbons, zeolites, metal oxides, biomass, polymer-based materials and raw or modified clay minerals (Bailey et al., 1999; Babel and Kurniawan, 2003; Wang et al., 2003; Saad et al., 2007; Abou Taleb et al., 2008; Bhattacharyya and Gupta, 2008; Hua et al., 2012; Lee and Tiwari, 2012; Qu et al., 2013).

Halloysites, especially those exhibiting nanotubular morphology, are promising as a nanomaterial used for adsorption and pollution remediation (Zhao et al., 2013). Its several derivatives are known, which are inorganic organohybrids designed on a molecular scale (Pan et al., 2011; Matusik,

Developments in Clay Science, Vol. 7. http://dx.doi.org/10.1016/B978-0-08-100293-3.00023-6

2014; Matusik and Wścisło, 2014). The known synthesis procedures allow to perform grafting and intercalation of several organic compounds. This helps to control and adjust adsorption properties of the clay mineral for a specific pollutant (either polar/apolar or positively/negatively charged). The grafted materials with covalently attached organic molecules and halloysite polymer nanocomposites are particularly important. This is connected with their stability in aqueous solutions, which makes them promising candidates for water treatment (Tian et al., 2015).

This chapter discusses the possibility of using halloysite and its derivatives for the remediation of polluted waters and wastewaters, including organic and inorganic contaminants. A literature review has been made, and selected experimental approaches are summarized, with an emphasis on the mechanisms of pollutant removal.

23.2 SURFACE PROPERTIES OF RAW HALLOYSITE AND ADSORPTION MECHANISMS

The reported research on halloysite confirmed that the adsorption of ions and charged molecules on pure kaolin-group minerals may be explained by two mechanisms (Fig. 23.1). The first, involving ion exchange, is related to permanently charged sites present at particle faces. The generated negative charge is due to isomorphic substitutions of Al^{3+} for Si^{4+} in the tetrahedral sheet, Fe^{2+} for Al^{3+} in the octahedral sheet or both. As the substitution usually does not exceed ~1% (Newman et al., 1994; Joussein et al., 2005) this first

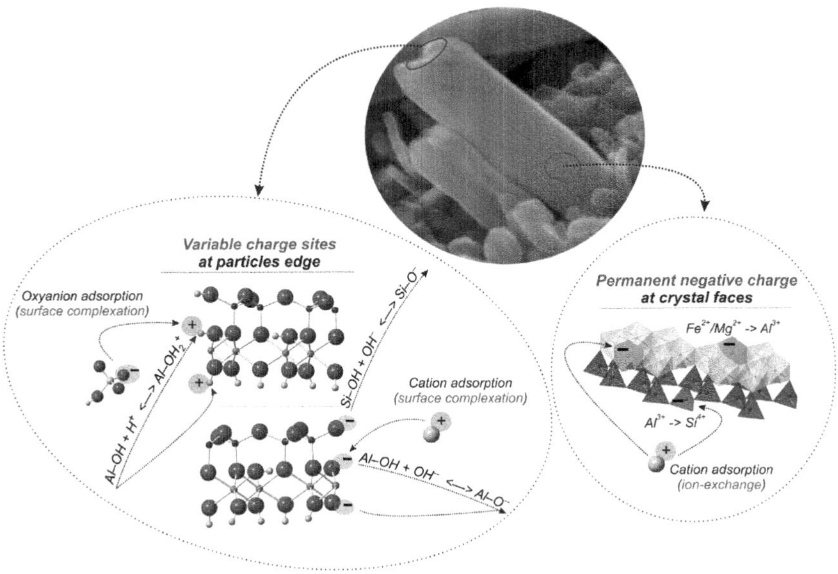

FIG. 23.1 Possible adsorption mechanisms on raw halloysite (7 Å).

mechanism is of minor importance in adsorption processes. The second mechanism is connected with variable charge sites found at particles edge and involves surface complexation. The nature of charge is strongly affected by the pH of solution in which the clay mineral is dispersed (Brady et al., 1996; Tombácz and Szekeres, 2006). Under acidic or alkaline conditions the edge aluminol Al–OH and/or silanol Si–OH groups may undergo protonation or deprotonation (proton donor–acceptor reactions) which induces positive or negative charge, respectively. The interlayer space of pure, unmodified halloysite is not directly available for ions and molecules with exception of selected molecules, eg, dimethyl sulphoxide (DMSO) which has a large dipole moment and destroys interlayer hydrogen bonding during intercalation. More details about the properties of halloysite are reported in Chapters 2 and 4.

23.3 FUNCTIONALIZATION OF HALLOYSITE STRUCTURE FOR THE IMPROVEMENT OF ADSORPTION PROPERTIES

23.3.1 The Effect of Calcination and Acid Activation

The common, inexpensive surface modification methods of kaolin-group minerals, including halloysite, involve thermal treatment (calcination) and acid activation. Calcination of halloysite leads to structural changes similar to kaolinite (Smith et al., 1993; Yuan et al., 2012b). In the 600–850°C temperature range the structure undergoes dehydroxylation which affects largely the coordination of aluminium in the octahedral sheet. Further heating leads to formation of nanocrystalline (<5 nm) γ-Al_2O_3 (1000°C) which with increasing temperature up to 1400°C reacts with silica forming aluminium-rich mullite which is finally converted to mullite—$3Al_2O_3 \cdot 2SiO_2$. From the view point of adsorption the calcination is often performed at dehydroxylation temperature. The generation of active centres is connected with reduced coordination of aluminium. More details regarding the thermal behaviour of halloysite are reported in Chapter 7. Although thermal treatment enhances adsorption properties this is not always accompanied with an increase of specific surface area (SSA). For example, the SSA of 64 m^2/g measured for halloysite sample from Djebel Debagh, Guelma (Algeria) did not change significantly after calcination at 600°C for 2 h (60.5 m^2/g) (Mellouk et al., 2009; Belkassa et al., 2013). The large increase of SSA which favours adsorption can be noticed for acid-activated halloysites (Zhang et al., 2012; Belkassa et al., 2013). The Algerian halloysite treated with HCl at concentrations in the 0.5–7.0 mol/L at 70°C for 4 h showed a SSA increase even up to 503.3 m^2/g (Belkassa et al., 2013). In turn, a SSA increase from 47.8 to 267.1 m^2/g was noticed for halloysite from Hebei Province (China) after treatment with 3 mol/L H_2SO_4 at 90°C for 13 h (Zhang et al., 2012). However prolonged treatment times or too high acid concentration leads to aggregation of partially amorphous clay mineral particles, which lowers the SSA and overall porosity and subsequently may inhibit adsorption. The traditional use of inorganic acids in the first stage removes mineral impurities and promotes octahedral sheet dissolution with simultaneous release of Al and relative increase of SiO_2 content. Afterwards, the etching of the, more stable in acidic conditions, tetrahedral sheet starts and the formed SiO_2 precipitate fills the particles lumen (White et al., 2012). The affected and/or partially destroyed surface layers showed increased reactivity. The acid treatment influences strongly the zeta potential of halloysite surfaces which is crucial in adsorption phenomena (Joo et al., 2013). The zeta potential values were found to be equal to: −3.5, −32.4 and −44.8 mV at pH 2, 7 and 11, respectively. The pH and corresponding zeta potential affect the stability of halloysite particles which may agglomerate, thus lowering SSA and changing pore size distribution. The acid activation can be done selectively due to differences in chemical nature of Al and Si sheets. The H_2SO_4 acid was used for an enlargement of halloysite lumen by dissolving the octahedral sheet (Abdullayev et al., 2012). Such treatment increased the lumen diameter (from 15 to 25 nm) and SSA (from 40 to 250 m^2/g). As a result a four-fold adsorption increase of benzotriazole was reported. The acid-activated halloysite shows high adsorption capacity towards methylene blue (MB) dye of 61.35 mg/g (Zhang et al., 2012). The dispersion of acid-activated halloysite in aqueous solution rapidly decreases the pH which negatively affects the cations adsorption. Thus, efficient adsorption requires the control and adjustment of pH.

23.3.2 Raw and Interlayer-Grafted Halloysite for the Removal of Cations

The halloysite structure with adsorption centres and its specific morphology enable to use the clay mineral as heavy metals scavenger. A comprehensive report on Zn(II) adsorption phenomena on raw halloysites of different origin was given by Wada and Kakuto (1980). The authors underlined the high selectivity of Zn(II) ions to adsorb on halloysite active sites. The selected halloysites showed a broad range of cation-exchange capacity (CEC) from 7.7 to 56.7 meq/100 g and different Fe_2O_3 content (0.11–3.76%). A significant effect of halloysite morphology was found, that is the selectivity of Zn(II) adsorption was the highest for spherical halloysite, medium for the platy halloysite, and was found the lowest for tubular halloysites. In turn, no correlation between the adsorption and iron content was reported. It was concluded that the Zn(II) adsorbs preferentially to Ca(II) ions on halloysites and most of the adsorbed Zn(II) is present on the cation-exchange sites of the clay mineral. Moreover, the experiments indicated that charge sites show high affinity for H^+ which are mostly responsible for the attraction and binding of Zn(II). The authors discussed the role of pH-dependent cation-exchange sites at the particles edges. The deprotonated hydroxyls may act as ligands which complex Zn(II) in preference to Ca(II). The stability of the formed Zn(II)-clay mineral monodentate or bidentate complex depends greatly on the local chemical environment in which the nature of OH groups was influenced by sterical effect of adjacent cation-exchange sites. Different Zn(II) adsorption behaviour of halloysites suggested a crucial influence of the genesis and geological occurrence on the active sites. The adsorption of Zn(II) on halloysite showing nanotubular morphology was also studied by Dong et al. (2012). The authors examined the influence of solution characteristics (pH, ionic strength, temperature and competitive effect of other ions) on the adsorption. The results showed a domination of cation-exchange mechanism and outer-sphere complexation at low pH. In turn, at high pH the removal was also due to inner-sphere complexation and precipitation. The detailed analysis also showed a significant influence of the background electrolyte, ie, competitive ions on the Zn(II) removal efficiency (Fig. 23.2). The presence of Na^+, K^+, Ca^{2+} and Mg^{2+} inhibited the Zn(II) removal. The effect was much stronger when divalent cations were present which competed for binding sites with Zn(II). The authors also pointed that the decrease of Zn(II) removal in the presence of divalent cations is due to their influence on surface potential. It becomes less negative and this hampers the Zn(II) adsorption. The anionic background electrolyte showed a much more pronounced effect on the Zn(II) removal efficiency. The Cl^- and SO_4^{2-} inhibited strongly the Zn(II) adsorption which can be attributed to the formation of soluble Zn(II)-Cl or Zn(II)-SO_4 complexes. These being electronegative do not have the affinity to adsorb. On the other hand, CO_3^{2-} ions had a positive effect on the Zn(II) adsorption as at $pH < 6.5$, the anion may induce the formation of ternary or multinary complexes, eg, –O–Zn–CO_3, –O–CO_3–Zn, –O–Zn–CO_3–Zn–. At higher pH values the simultaneous precipitation of $ZnCO_3$ may occur which increases the removal efficiency to 100%. The performed thermodynamic experiments,

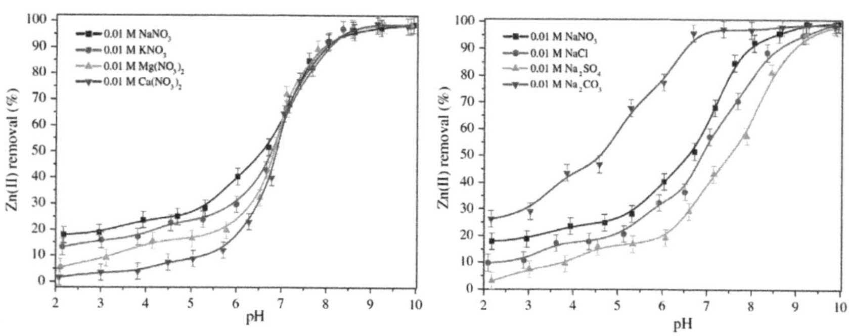

FIG. 23.2 *Left*: influence of coexisting electrolyte cations on the removal of Zn(II) by halloysites as a function of pH values. $T = 293$ K, $C_{(Zn)\ initial} = 1.54 \times 10^{-4}$ mol/L, $m/V = 1.0$ g/L. *Right*: influence of coexisting electrolyte anions on the removal of Zn(II) by halloysites as a function of pH values. $T = 293$ K, $C_{(Zn)\ initial} = 1.54 \times 10^{-4}$ mol/L, $m/V = 1.0$ g/L. *Reprinted with permission from Dong et al. (2012). Copyright 2012 Springer.*

which allowed drawing temperature-dependent adsorption isotherms, showed that the adsorption process was endothermic and spontaneous.

The usability of pure halloysite from New Zealand for the highly efficient removal of Ag(I) ions was also shown (Kiani, 2014). The experimentally obtained maximal adsorption capacity equal to 109.79 mg/g was higher than the values reported for bentonite (80.0 mg/g) and comparable to adsorption capacity of polyacrylonitrile nanofibre mats (105.7 mg/g). The investigation allowed to estimate the energy of Ag(I) adsorption on halloysite which was equal to 7.28 kJ/mol and corresponded to physical nonspecific adsorption. The binding took place mainly at the deprotonated edge sites where Ag(I) ions competed with protons especially at low pH. Due to relatively weak interaction of halloysite surface, the regeneration using HCl was very efficient and reached almost 100%, which enables the material to be reused. In other studies, a halloysite intercalated with Na^+, Pb^{2+} or $NH_4{}^+$ acetate was tested for the removal of Cu(II) ions (Mellouk et al., 2009, 2011). The nanotubular clay mineral had a CEC of 19.2 meq/100 g, SSA of 64 m^2/g and an isoelectric point at pH 2.5. The intercalation of acetates improved the removal efficiency and a 2.2-fold increase of adsorption capacity was noticed in contrast to the pure halloysite. The adsorption enhancement was attributed to cation exchange connected with the acetate presence and cation complexation at the particles edges.

The experimental procedures leading to the synthesis of kaolinite interlayer-grafted derivatives were shown to be effective also for halloysite due to structural similarities of the clay minerals. The exposed inner surface hydroxyls of nanotubular halloysite were reacted with diethanolamine $(CH_2CH_2OH)_2NH$ (DEA) and triethanolamine $(CH_2CH_2OH)_3N$ (TEA) which may play a role of complexing agents (Fig. 23.3) (Matusik and Wścisło, 2014).

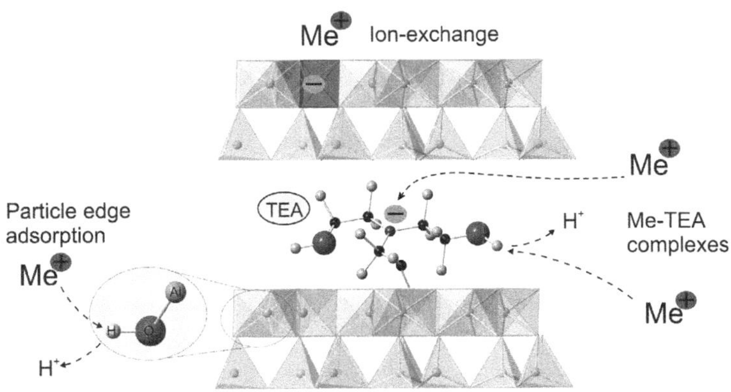

FIG. 23.3 The approximate structure of halloysite interlayer grafted with triethanolamine (TEA) and possible adsorption mechanisms in contact with aqueous solution containing cations (Me). *Graphical abstract reprinted with permission from Koteja and Matusik (2015). Copyright 2015 Elsevier.*

For the research a halloysite sample from Dunino deposit (SW Poland) was chosen. The synthetic route consisted of two successive steps. Firstly, an intercalation with DMSO was carried out at room temperature (22°C). Secondly, the formed intercalation compound was refluxed with DEA or TEA at 180°C under argon flow for 24 h (Letaief and Detellier, 2007). The obtained dispersion was centrifuged, washed with isopropanol and subsequently with water and finally dried at 65°C. The extensive washing allowed to ensure that only aminoalcohol molecules which are covalently attached to the octahedral sheet through Al–O–C bonds were left. The formation of Al–O–C bonds affected the dehydroxylation temperature of the halloysite due to change of the local environment of the octahedra. The tailored materials were resistant to decomposition during prolonged washing in aqueous solutions as attested by X-ray diffraction (XRD) and Fourier transform infrared spectroscopy (FTIR) methods. Moreover, they did not show swelling properties and had a fixed interlayer space equal to 3.1 Å (DEA) or 3.6 Å (TEA). The interlayer space value was exclusively controlled by the size of the aminoalcohol molecules which formed a monolayer. The introduced DEA and TEA molecules possess metal-binding properties due to presence of amine nitrogen with a lone electron pair behaving as a Lewis base in relation to metals acting as Lewis acids (Sen and Dotson, 1970; Tauler et al., 1985; Karadag et al., 2001; Matusik and Wścisło, 2014). The binding of metals through deprotonated OH groups which terminate the alkyl chain of aminoalcohols is also possible (Whitmire et al., 1999). Based on the earlier studies of Letaief and Detellier (2007) it was assumed that the aminoalcohols were attached through Al–O–C bonds to the octahedral sheet involving one of their OH groups. Thus, the formulas of the modified materials calculated on the basis of nitrogen content were: $Al_2Si_2O_5(OH)_{3.93}(DEA)_{0.07}$ and $Al_2Si_2O_5(OH)_{3.90}(TEA)_{0.10}$, respectively. The small content of the grafted molecules [~16% (DEA derivative) and ~25% (TEA derivative) in relation to maximal content] could be connected with the steric hindrance effect (ie, blocking of the interlayer by DEA or TEA). The applied chemical treatment induced the formation of mesopores and macropores in the case of the TEA derivative, and in turn, part of the mesopores was blocked in the case of the DEA derivative. The textural parameters revealed the mesoporous character of the synthesized materials, with the content of micropores not exceeding 7% of the total pore volume (V_{TOT}). The grafting of TEA led to the exfoliation of particles to a small extent, in contrast to the DEA grafting, where exfoliation did not take place and probably some of the particles formed aggregates. This was attested by changes in SSA, which increased for the TEA derivative to 57.0 m^2/g and decreased to 47.37 m^2/g for the DEA derivative, in comparison to the 49.52 m^2/g initial value of raw halloysite. The applied grafting procedure did not improve the CEC and it also did not affect the point of zero charge (PZC).

The performed adsorption tests revealed that the modification significantly improved the halloysite adsorption capacity with respect to Pb(II), Cd(II), Zn(II) and Cu(II). Equilibrium and kinetic experiments connected the removal enhancement with a two-step gradual migration of the cations into the halloysite interlayer space, followed by their complexation by amine nitrogen. In general, the equilibrium was achieved after about 4 min. The Weber–Morris plots enabled the identification of two stages of the process due to the presence of two linear trend lines (Weber and Morris, 1963). The two-step process consisted of external mass transfer, which was followed by migration of the cations into the interlayer space. On the other hand, in the case of pure halloysite, the adsorption took place rapidly (in the first minute of the reaction) due to surface complexation, ion exchange or both. The pH of the system increased due to the competitive behaviour of protons, in contrast to a system with raw halloysite, where the pH decreased. The latter was attributed to a domination of ion-exchange and surface complexation mechanisms, and the adsorption efficiency was mainly influenced by the metals' hydrolysis constants. In the case of grafted materials, in particular, the Cu(II) ions were preferentially removed from solution, as they readily form complexes with N-donor ligands (Brannon et al., 1971; Whitmire et al., 1999). This was very visible in multielement systems where the adsorption on grafted materials followed the sequence Cu(II) > Pb(II) ≫ Zn(II) ≈ Cd(II) (Fig. 23.4). The Zn(II) and Cd(II) adsorption was significantly hampered. The kinetic data for these elements indicated a one-step rapid adsorption. In turn, the kinetics for Cu(II) and Pb(II) showed a gradual increase of adsorption, suggesting a binding in the interlayer space of halloysite. The formation of Cu(II)-aminoalcohol complexes was confirmed using UV–vis spectroscopy. Moreover, the decomposition of grafted aminoalcohol after Cu(II) took place at different temperatures and affected the shape of the exothermic peak in comparison to pure grafted derivatives. The equilibrium pH in the multielement system was always below the precipitation pH limit of metal phases. Thus, the concurrent precipitation mechanism was excluded.

Experiments with the use of real wastewater showed that the materials' adsorption properties were not hampered due to the presence of organic media and other ions. Analogous to synthetic aqueous solutions, the same conclusions were drawn, including the preferential Cu(II) adsorption. The desorption experiments revealed a complex binding strength of the metal cations and showed that a complete regeneration of the materials was not possible when using ammonium acetate as a mild extractant.

The mathematical description of the equilibrium adsorption data showed the best agreement when using the Langmuir model. The theoretically obtained adsorption capacity was close to the experimental data with a very high correlation coefficient. The kinetic data in the whole time range followed the pseudo-second order equation with a high linear correlation coefficient ($R^2 > 0.98$).

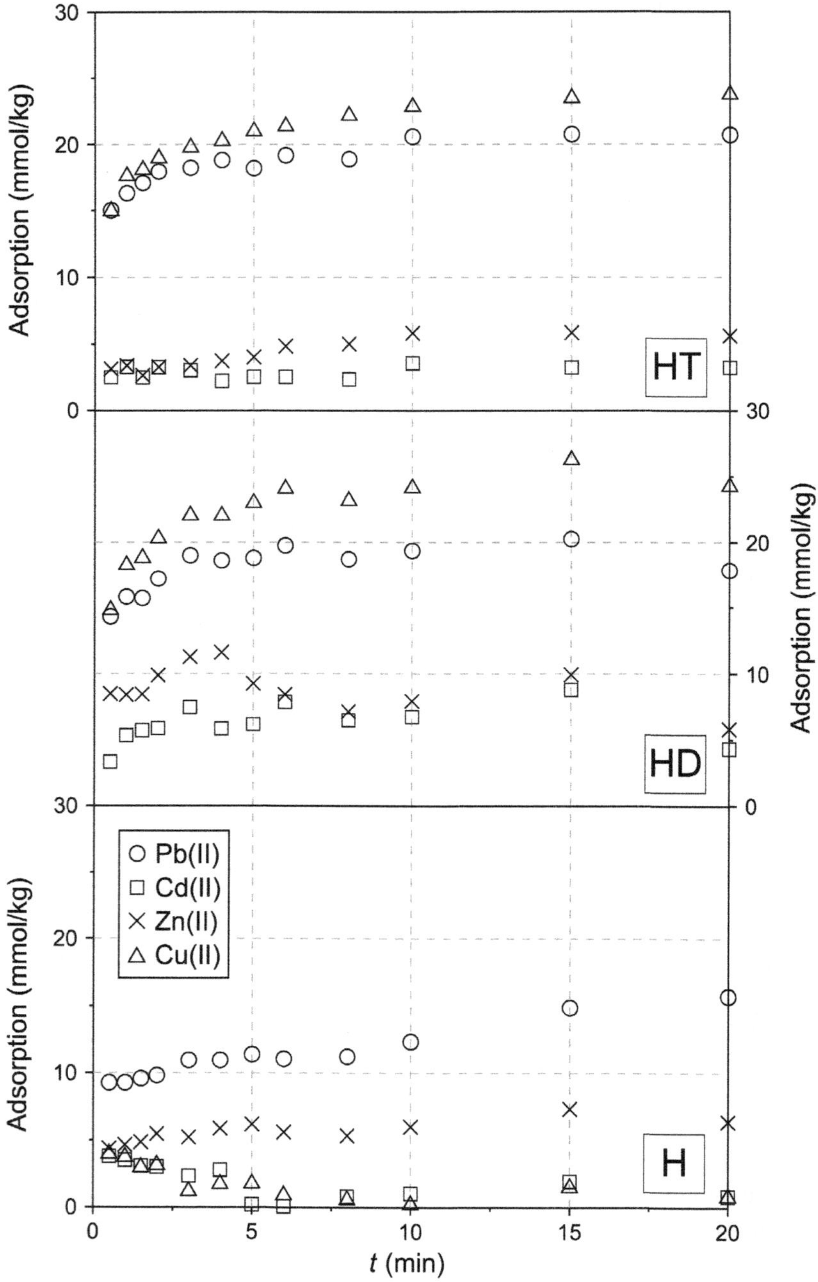

FIG. 23.4 Kinetics of Pb(II), Cd(II), Zn(II) and Cu(II) uptake in the multielement system for the H (pure halloysite), HD (DEA grafted halloysite) and HT samples (TEA-grafted halloysite). *Reprinted with permission from Matusik and Wścisło (2014). Copyright 2014 Elsevier.*

23.3.3 Surface and Interlayer-Modified Halloysite for the Removal of Anions

The adsorption and removal of anions is less frequently studied due to preferential affinity of halloysite to bind cations. In theory, the surface of raw halloysite does not have the capability to attract anions in high amounts. However, the research literature reports modification procedures regarding halloysite structures that enable the enhancement of the adsorption capacity of oxyanion contaminants. One of the methods for improving oxyanion adsorption is an introduction of surfactants, which could be either intercalated or attached to the clay mineral surface (Haggerty and Bowman, 1994; Li and Bowman, 2001; Bowman, 2003; Xi et al., 2010). The raw halloysite was treated with an aqueous solution of hexadecyltrimethylammonium bromide (HDTMA-Br) in different amounts in relation to its CEC (Xi et al., 2010). In the studied experimental conditions, the HDTMA-Br was exclusively present on the clay mineral surface, as the XRD pattern of halloysite did not show changes after treatment. Moreover, the introduced molecules blocked the internal porosity, as N_2 adsorption/desorption measurements indicated a decrease in SSA from 26.5 to ~17.0 m^2/g. The presence of quaternary ammonium bromide, especially in amounts exceeding the CEC, enhanced the clay mineral's ability to remove nitrates. A twofold to threefold increase, however, was not as high as for modified bentonite (increase of more than 10-fold), where the HDTMA-Br molecules were also intercalated into the interlayer space. A similar modification approach for the halloysite in the Henan province, China, was studied by Wang et al. (2010) in terms of Cr(VI) removal. The surface modification of halloysite nanotubes by HDTMA-Br increased the adsorption capacity to ~132 mmol/kg. The authors emphasized the fast kinetics of Cr(VI) uptake, with equilibrium achieved after 1 h as compared to 3–48 h equilibrium times that previously had been reported for organic polymer-modified and zeolite adsorbents (Wang et al., 2010). The adsorption of N(V) and Cr(VI) oxyanions by HDTMA-Br modified halloysite mainly involved electrostatic interaction of the oxyanion with an adsorbed organic cation.

A synthetic route leading to significant improvement in oxyanion adsorption by halloysite was based on interlayer grafting and subsequent quaternization (Matusik and Bajda, 2013; Matusik, 2014). The procedure was based on the treatments reported for kaolinite, which was used as a size-selective sensor of cyanide, iodide and other anions (Letaief et al., 2008; Dedzo et al., 2012; Dedzo and Detellier, 2013). The synthesis consisted of three subsequent stages. First, the halloysite from Polish Dunino deposits was pre-intercalated with DMSO. Second, the pre-intercalates were refluxed with TEA for 2 h at 180°C under argon flow. The third step included the methylation reaction with iodomethane (CH_3I) at 42°C. The key methylation reaction involved the

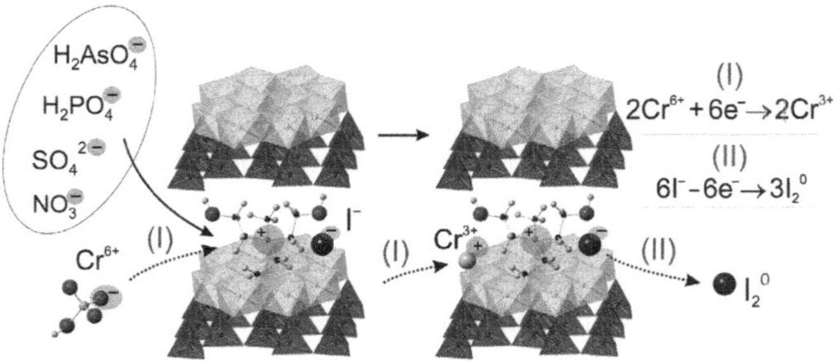

FIG. 23.5 The approximate structure of quaternized interlayer-grafted halloysite and its interaction with oxyanions involving ion exchange and Cr(VI) reduction. *Reprinted with permission from Matusik and Bajda (2013). Copyright 2013 Elsevier.*

attachment of the CH_3 group to the interlayer-grafted TEA molecule and was confirmed by the ^{13}C CP-MAS NMR spectroscopy. This treatment was designed to form positively charged centres associated with ammonia nitrogen of ammonium salt molecules formed in the interlayer space (Fig. 23.5). The generated positive charge was compensated by mobile iodide anions, which could be ion-exchanged easily. Subsequent detailed investigations indicated a significant increase in oxyanions: Cr(VI), As(V), P(V), S(VI) and N(V) adsorption capacity by the modified halloysite (Matusik and Bajda, 2013; Matusik, 2014).

In the case of Cr(VI) removal, the adsorption maximum (initial pH 5) for the pure halloysite (SSA: 50.6 m^2/g) reached only ~12.0 mmol/kg. In turn, the modification led to an approximately eightfold increase of the adsorption capacity of up to ~95 mmol/kg (Fig. 23.6). It is worth pointing out that the chemical treatment enabled the access of the chromate to the interlayer space by increasing the d_{001} value up to ~10.8 Å. Kinetic studies confirmed that the adsorption on the interlayer-quaternized halloysite is a pseudo-second order reaction and that over 90% of the finally immobilized chromium was removed after 2 min of the reaction. The adsorption equilibrium was achieved after ~20 min, indicating that the nanotubular morphology of halloysite inhibited the adsorption rate.

The analysis of solution chemistry before and after adsorption experiments, coupled with the FTIR and X-ray photoelectron spectroscopy (XPS) analyses of solid, enabled proposing a two-step removal of Cr(VI). First, an ion exchange of Cr(VI) → I^- takes place, followed by Cr(VI) to Cr(III) reduction by iodide, which releases iodine into solution (Fig. 23.5). The formation of Cr(III) is desirable from an environmental point of view, as Cr(III) is much less toxic than Cr(VI). The pH-dependent ratio of the removed Cr(VI) to released iodide was related to the presence of either the $HCrO_4^-$ or CrO_4^{2-} ionic form and confirmed initial ion-exchange mechanism.

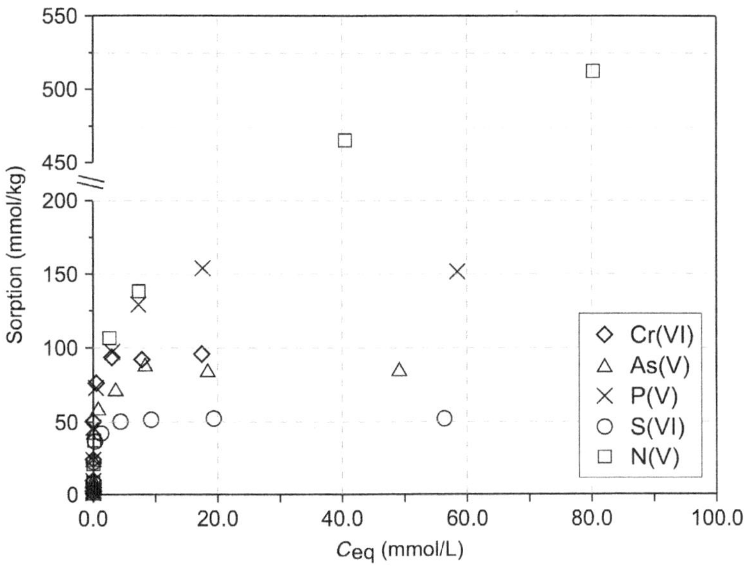

FIG. 23.6 Adsorption isotherms of selected oxyanions on the quaternized interlayer grafter halloysite. *Adapted with permission from Matusik (2014). Copyright 2014 Elsevier.*

The obtained interlayer-quaternized halloysite also showed improved anion adsorption properties towards As(V), P(V), S(VI) and N(V) (Fig. 23.6). However, the improvement was less than for the material derived from kaolinite of high structural order (Matusik, 2014). The adsorption isotherms for all the anions follow the Langmuir model, with a high correlation coefficient and good fit to the experimental maximum adsorption capacity. In analogy to Cr(VI), the studied anions were mainly ion exchanged with mobile iodide ions, which was confirmed through analysis of solution chemistry and ratio of removed oxyanion to released iodide.

After adsorption, the d_{001} value of the modified halloysite was unchanged, as attested by the XRD pattern confirming that swelling did not occur and only the grafted organic molecules controlled the basal spacing. Thus, the interlayer penetration by larger molecules may not be possible, giving rise to selective adsorption properties of the quaternized halloysite. Moreover, the modification does not alter or destroy the halloysite structure and morphology.

Another protocol for improving Cr(VI) adsorption involved a formation of halloysite nanocomposite with polypyrrole (PPy) polymer through in situ polymerization technique (Fig. 23.7) (Ballav et al., 2014). Although the polymer itself is showing ion-exchange properties, an addition of nanofiller (eg, halloysite nanotubes prevents agglomeration induced by π–π stacking). This enhances its adsorption properties and enables its usage as efficient oxyanion CPN (clay polymer nanocomposite) adsorbent. The reported maximal

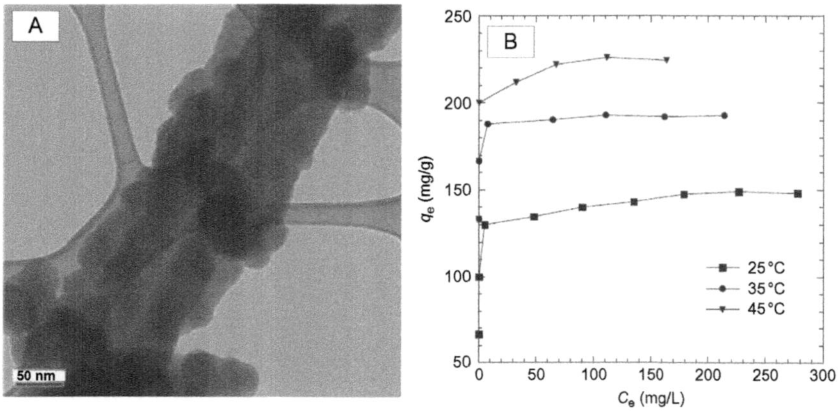

FIG. 23.7 (A) Transmission electron microscope (TEM) images of polypyrrole-coated halloysite, (B) adsorption isotherms of Cr(VI) onto the polypyrrole-coated halloysite. TEM images of polypyrrole-coated halloysite and its adsorption isotherms towards Cr(VI). *Reprinted with permission from Ballav et al. (2014). Copyright 2014 Elsevier.*

adsorption capacity is exceptionally high, reaching ~2870 mmol Cr(VI)/kg at pH 2 and room temperature (Fig. 23.7). The performed XPS analyses showed that the Cr(VI) removal involves its reduction to Cr(III), induced by electron-rich organic PPy molecules. The conducted desorption studies indicated the possibility of halloysite PPy nanocomposite regeneration and its further reuse for up to three times. Moreover, the adsorbent showed very high (~99%) Cr(VI) removal efficiency from chromium-contaminated groundwater and mine wastewater. Interestingly, the accompanying SO_4^{2-} and Cl^- ions did not hamper the Cr(VI) removal, suggesting that the material has high selectivity for chromium (Ballav et al., 2014). A CPN approach was also tested using polyethylenimine (PEI) for the functionalization of halloysite nanotubes (Tian et al., 2015). The procedure enabled to obtain a nanomaterial adsorbent with a relatively low SSA of 30.9 m^2/g, but again with a very high Cr(VI) adsorption capacity, equal to 1240 and 1970 mmol/kg measured at 25°C and 55°C, respectively. The reported possible removal mechanism included (i) Cr(VI) electrostatic attraction by protonated amine groups and (ii) reduction of Cr(VI) to Cr(III) through electron transfer from aldehyde and amine groups. Improvement of Cr(VI) adsorption was also achieved by using raw halloysite nanotubes grafted with the silane-coupling agent *N*-β-aminoethyl-γ-aminopropyl trimethoxysilane (Luo et al., 2011). The protonated amino groups of the grafted silane attracted Cr(VI) oxyanions, and the maximum adsorption capacity reached ~716 mmol/kg (Fig. 23.8). The experiments confirmed a fast removal, with 95% uptake after ~5 min. In contrast to other adsorbents, the silane-grafted halloysite showed greater removal efficiency at lower temperatures.

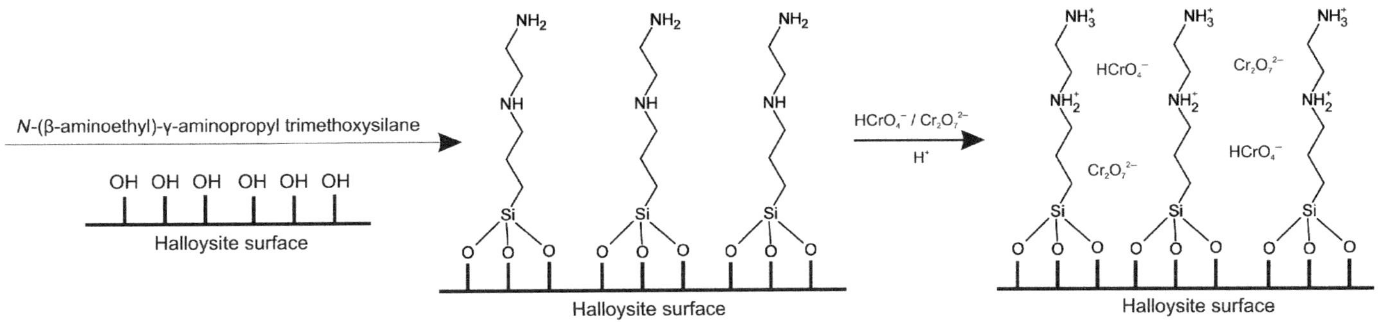

FIG. 23.8 Schematic diagram illustrating Cr(VI) adsorption onto halloysite nanotubes surface grafted with silane-coupling agent. *Adapted with permission from Luo et al. (2011). Copyright 2011 American Chemical Society.*

23.3.4 Raw Halloysite and Halloysite-Based Composites for the Removal of Organic Pollutants

Apart from the common usage of nanotubular halloysite as an adsorbent of cationic and anionic inorganic species, the clay mineral is often used in the treatment of aqueous media contaminated with organic compounds. The hydrated halloysite (10 Å) has the ability to form intercalation compounds with various organic molecules (Carr and Chih, 1971). The intercalation proceeds most efficiently when the molecules are polar and contain hydroxyl –OH and amine $-NH_2$ functional groups capable of forming hydrogen bonds in the halloysite interlayer space. As dehydration of halloysite (10 Å) takes place easily at room temperature in the absence of water, a halloysite (7 Å) is more often used in adsorption experiments. The nanotubular halloysite and its modifications, as well as its nanocomposites with other minerals, were used for the uptake of organic acids, dyes and simple aromatic hydrocarbons, among others.

Pristine halloysite nanotubes (Hebei province, China) were tested for the removal of a commonly found cationic MB dye from aqueous solutions in batch experiments (Zhao and Liu, 2008). The maximal adsorption capacity reached 84.32 mg/g and was achieved after about 30 min. The removal was the highest at high pH values, which was attributed to extensive particle edge deprotonation. The thermodynamic experiments indicated that the adsorption capacity increased with decreasing temperature of the solution. The formed temporary charge attracted the cationic dye. The authors underlined the fact that the sedimentation rate of halloysite dispersion with adsorbed MB proceeds very fast. In contrast, the dispersion of pure halloysite remains stable for a much longer time. This behaviour of the adsorbent is desired from the viewpoint of applications due to the simplicity of water filtration after dye removal. The same research team tested the halloysite from Hebei province for the uptake of dye used in biological research—neutral red (NR), which forms cations in aqueous solution (Luo et al., 2010). The reported maximal dye uptake at room temperature (298 K) reached 54.85 mg/g and increased with increasing temperature to 59.24 mg/g (308 K) and to 65.45 mg/g (318 K). Thus, the observed temperature effect was the opposite of that noticed for the MB, with the adsorption being an endothermic process. The plots of adsorption capacity vs time$^{1/2}$ allowed one to distinguish the two-step nature of NR dye removal. This was connected to (i) intraparticle diffusion in the first stage and (ii) equilibrium achievement in the second stage. The reported removal of methyl violet (MV) by nanotubular halloysite from Henan province was equal to 105.26 mg/g at 298 K and reached 113.64 mg/g at 318 K (Liu et al., 2011). The adsorption values were higher than those reported for other substances, such as sepiolite (10.24 mg/g) and baggase fly ash (26.25 mg/g). Although the adsorption was about five times less than activated carbon (526.32 mg/g), the authors emphasized the low cost of halloysite (200$/tonne, with 4000$/tonne for activated carbon in the year 2011). This makes the inexpensive halloysite a promising candidate to use for dye removal from aqueous media. Moreover, it was shown that adsorbent regeneration was possible by calcination at 300°C. This treatment led to dye combustion without destroying the aluminium silicate framework; however, after six calcination/adsorption cycles, the regeneration efficiency decreased to 60%. Kiani et al. (2011) showed that halloysite nanotubes (China Clays Ltd., New Zealand) can efficiently remove malachite green (MG) through physical adsorption, with a maximal capacity of 99.6 mg/g achieved at pH 9.5. In agreement with earlier observations, the adsorbed dye led to halloysite particle aggregation and favoured adsorbent sedimentation.

An interesting study showing the capability of halloysite to remove not only cationic but also anionic dyes was shown by Zhao et al. (2013). The research revealed that adsorption of rhodamine 6G (R6G) and chrome azurol S (CS) reaches 43.6 and 38.7 mg/g, respectively, and it is higher for halloysite in comparison to kaolinite. The affinity to both positively and negatively charged dyes resulted from the pH-dependent charge generated by the halloysite Al_2O_3 and SiO_2 surfaces. Such phenomenon was defined by the authors as 'bivalent adsorbancy' (Zhao et al., 2013). The regeneration of halloysite-based filters was possible up to 50 times through dye combustion. The removal of R6G increased with temperature with the process being endothermic and was attributed to electrostatic dye-silanol surface interaction. In contrast, a decrease of CS adsorption was noticed, indicating an exothermic process. This was attributed to the CS hydrogen bonding through its carboxyl and hydroxyl groups to aluminol groups of the nanotube interior (Zhao et al., 2013). For the adsorption, these authors also used halloysite surfaces modified with selected surfactants and complexing agents. This treatment gave positive results in the case of CS adsorption. In particular, modification with dioctadecyldimethylammonium bromide (DODAB) led to a sixfold increase of the anionic dye adsorption. This was due to CS interaction with quaternary ammonium nitrogen showing a positive charge. The halloysite treatment with organic molecules did not improve cationic R6G removal, which was highest when a raw clay mineral was used (Zhao et al., 2013).

Yuan et al. (2012a) showed that grafting an organosilane γ-aminopropyltriethoxysilane (APTES) on halloysite nanotubes increased the loading of anionic acid orange II dye (AOII) (Fig. 23.9). Under appropriate pH conditions, the protonated amine groups NH_3^+ of covalently bonded APTES inside halloysite lumen had strong affinity to the AOII.

An interesting approach for halloysite nanotubes used for dye adsorption was shown by Xie et al. (2011) and Duan et al. (2012). The halloysite nanotubes were covered with nanocrystalline Fe_3O_4 (3–10 nm in size) through precipitation. Experimentally, the halloysite was dispersed in a solution containing iron chloride ($FeCl_3 \cdot 6H_2O$) and iron sulphate ($FeSO_4 \cdot 7H_2O$) at high pH (in the 9–10 range), which favoured iron oxide precipitation. The

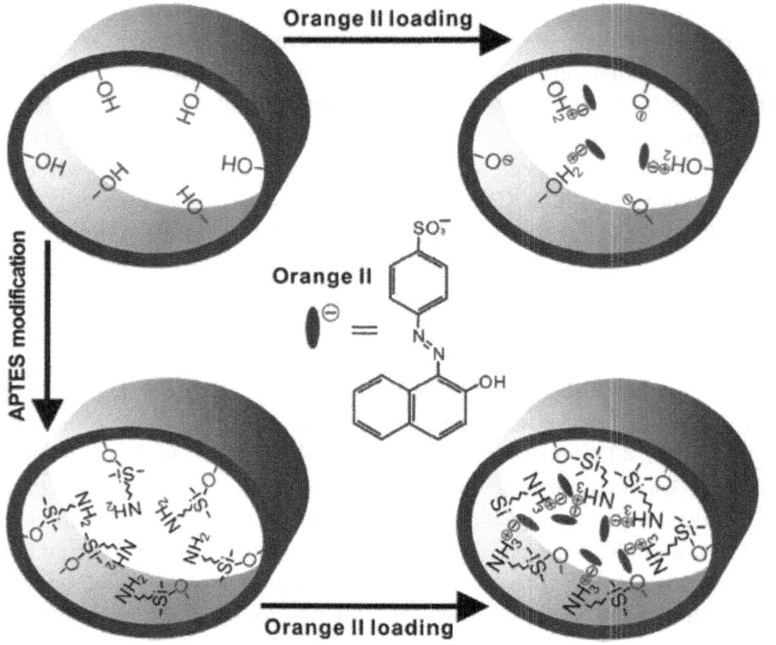

FIG. 23.9 Schematic representation of the amine modification and orange II loading of the halloysite nanotube. *Reprinted with permission from Yuan et al. (2012a). Copyright 2012 ©IOP Publishing.*

formed halloysite iron nanocomposite showed high adsorption capacity towards several dyes, including MV, MB and NR, with a lower adsorption capacity towards methyl orange (MO) (Xie et al., 2011; Duan et al., 2012). However, the most striking feature of the used halloysite iron nanocomposite was the possibility of its simple separation from the adsorbate liquid using its magnetic properties (Fig. 23.10). The external magnetic field attracted the halloysite iron nanocomposite, which could be regenerated by calcination and further used for dye adsorption. The nanocomposite composed of halloysite nanotubes and magnetic iron oxides was also used to support the preparation of sophisticated magnetic molecularly imprinted polymers (MMIP) (Pan et al., 2011). The obtained MMIPs showed selective recognition of a model phenolic compound: 2,4,6-trichlorophenol. The organic can be selectively removed from polluted samples with a simultaneous magnetic separation of the used adsorbent.

Dye adsorption was also improved through the preparation of halloysite complexes with natural biopolymers. One of them involves the use of chitosan—a cationic polymeric carbohydrate whose adsorption properties are governed by the presence of amine and hydroxyl groups, allowing electrostatic interaction with dyes (Peng et al., 2015). A synergetic effect of the prepared

FIG. 23.10 *Left*: TEM micrograph of halloysite-Fe_3O_4 composite, *right*: magnetic separation after adsorption onto halloysite-Fe_3O_4 composite. *Reprinted with permission from Xie et al. (2011); Duan et al. (2012). Copyright 2012 Elsevier.*

halloysite chitosan hydrogel beads (2.29–2.75 mm in diameter) was observed, and subsequently, the adsorption capacity of MB and MC dyes was enhanced. The adsorption of MB and MC equalled 217.39 and 243.90 mg/g, respectively. Meanwhile, the adsorption capacity of the best halloysite-based nanocomposites reached 270.27 mg/g (MB dye) and 294.18 mg/g (MC dye). The scanning electron microscopy (SEM) observations allowed these authors to conclude that the microstructure of halloysite-based nanocomposites was much rougher than in the case of pure chitosan. Furthermore, the loosened microstructure of hydrogel beads, which possess large SSA enables efficient molecular diffusion of adsorbate and its further immobilization. Alginate salts of alginic acid were also used to prepare the halloysite biopolymer nanocomposite. Their adsorption capacity is connected with abundant carboxyl groups having an affinity for metal binding (Liu et al., 2012). Although pure alginate can act as an excellent adsorbent, its application is limited because of its low rigidity, low mechanical strength and dense gel layer, which hinder ion diffusion. The prepared organic/inorganic hybrid beads showed an improved MB dye adsorption capacity, with a maximum equal to 222 mg/g at 298 K. The dynamic experiment in a column system revealed that 10 g of the hybrid beads are able to treat 29.7 L of MB aqueous solution (55 mg/L concentration) with a removal efficiency higher than 90% (Liu et al., 2012). It is worth mentioning that hybrid morphology did not change during experiments, in contrast to pure alginate, showing its increased stability.

Halloysite was used for the adsorption of selected organic molecules; eg, aromatic naphthalene (Lee and Kim, 2002), 5-aminosalicylic acid (Viseras et al., 2008) and chloroanilines (Szczepanik et al., 2014). The modification of halloysite surface by HDTMA increased naphthalene adsorption. The reported mechanism involved the organic partitioning of condensed aromatic rings in the hydrophobic environment of alkyl chains. Lee and Kim (2002) proved that the amount of adsorbed HDTMA is less important than its conformation and configuration, which strongly affect the partitioning effect. The adsorption properties are often studied in the context of medical applications (ie, drug delivery). A strong interaction between halloysite surfaces and 5-aminosalicylic acid (to cite one example) was reported (Viseras et al., 2008). Thus, the clay mineral can act as a promising support for drugs. The acid-activated halloysite has a strong affinity to remove chloroanilines such as 3-chloroaniline, 4-chloroaniline and 3,4-chloroaniline from aqueous solutions (Szczepanik et al., 2014).

The halloysite with an adsorbed dye can serve as a functionalized nanomaterial that induces the degradation of complex organic compounds. For example, a clay mineral-based hybrid photosensitizer was obtained by surface modification of halloysite by Rose Bengal dye (Bielska et al., 2015). The hybrid material can generate singlet oxygen, which significantly enhances the photodegradation of a common pesticide—4-*n*-nonylphenol. The pesticide biodegradation kinetics are very slow, but the identified photodecomposition products are expected to be much more susceptible to environmental degradation with the help of bacteria.

23.4 CONCLUDING REMARKS

Halloysite possesses interesting adsorption properties and is capable for efficient removal of inorganic (eg, heavy metals) and organic (eg, dyes) contaminants. Its low cost and environmental friendliness are its main advantages over other adsorbents. The unique morphology of the clay mineral, which could be found in nature, makes it a promising candidate for selective adsorption processes. Moreover, the complex morphology directly affects the SSA and V_{TOT}, which are relatively high, as for a raw clay mineral adsorbent. Apart from classical modification techniques (eg, calcination and acid activation), new trends can be observed that significantly evolve. This includes the possibility of modification through grafting and intercalation processes, as well as the formation of complexes with polymers and other minerals (eg, oxides). Such modifications lead to an intended adjustment and control of adsorption properties towards selected pollutants.

REFERENCES

Abdullayev, E., Joshi, A., Wei, W., Zhao, Y., Lvov, Y., 2012. Enlargement of halloysite clay nanotube lumen by selective etching of aluminum oxide. ACS Nano 6, 7216–7226.

Abou Taleb, M.F., Mahmoud, G.A., Elsigeny, S.M., Hegazy, E.-S.A., 2008. Adsorption and desorption of phosphate and nitrate ions using quaternary (polypropylene-g-N, N-dimethylamino ethylmethacrylate) graft copolymer. J. Hazard. Mater. 159, 372–379.

Alloway, B.J., 2013. Heavy Metals in Soils: Trace Metals and Metalloids in Soils and Their Bioavailability. Springer, Netherlands. http://dx.doi.org/10.1007/978-94-007-4470-7.

Babel, S., Kurniawan, T.A., 2003. Low-cost adsorbents for heavy metals uptake from contaminated water: a review. J. Hazard. Mater. 97, 219–243.

Bailey, S.E., Olin, T.J., Bricka, R.M., Adrian, D.D., 1999. A review of potentially low-cost sorbents for heavy metals. Water Res. 33, 2469–2479.

Ballav, N., Choi, H.J., Mishra, S.B., Maity, A., 2014. Polypyrrole-coated halloysite nanotube clay nanocomposite: synthesis, characterization and Cr(VI) adsorption behaviour. Appl. Clay Sci. 102, 60–70.

Belkassa, K., Bessaha, F., Marouf-Khelifa, K., Batonneau-Gener, I., Comparot, J.-d., Khelifa, A., 2013. Physicochemical and adsorptive properties of a heat-treated and acid-leached Algerian halloysite. Colloids Surf. A 421, 26–33.

Bhattacharyya, K., Gupta, S., 2008. Adsorption of a few heavy metals on natural and modified kaolinite and montmorillonite: a review. Adv. Colloid Interf. Sci. 140, 114–131.

Bielska, D., Karewicz, A., Lachowicz, T., Berent, K., Szczubiałka, K., Nowakowska, M., 2015. Hybrid photosensitizer based on halloysite nanotubes for phenol-based pesticide photodegradation. Chem. Eng. J. 262, 125–132.

Bowman, R.S., 2003. Applications of surfactant-modified zeolites to environmental remediation. Micropor. Mesopor. Mater. 61, 43–56.

Brady, P.V., Cygan, R.T., Nagy, K.L., 1996. Molecular controls on kaolinite surface charge. J. Colloid Interf. Sci. 183, 356–364.

Brannon, D.G., Morrison, R.H., Hall, J.L., Humphrey, G.L., Zimmerman, D.N., 1971. Spectra and bonding for copper(II)-aminoalcohol complexes I. The IR spectra of complexes of mono-, di- and triethanolamine. J. Inorg. Nucl. Chem. 33, 981–990.

Carr, R.M., Chih, H., 1971. Complexes of halloysite with organic compounds. Clay Miner. 9, 153–166.

Dedzo, G.K., Detellier, C., 2013. Ionic liquid-kaolinite nanohybrid materials for the amperometric detection of trace levels of iodide. Analyst 138, 767–770.

Dedzo, G.K., Letaief, S., Detellier, C., 2012. Kaolinite–ionic liquid nanohybrid materials as electrochemical sensors for size-selective detection of anions. J. Mater. Chem. 22, 20593–20601.

Dong, Y., Liu, Z., Chen, L., 2012. Removal of Zn(II) from aqueous solution by natural halloysite nanotubes. J. Radioanal. Nucl. Chem. 292, 435–443.

Duan, J., Liu, R., Chen, T., Zhang, B., Liu, J., 2012. Halloysite nanotube-Fe_3O_4 composite for removal of methyl violet from aqueous solutions. Desalination 293, 46–52.

Haggerty, G.M., Bowman, R.S., 1994. Sorption of chromate and other inorganic anions by organo-zeolite sorption of chromate and other inorganic anions by organo-zeolite. Environ. Sci. Technol. 28, 452–458.

Hua, M., Zhang, S., Pan, B., Zhang, W., Lv, L., Zhang, Q., 2012. Heavy metal removal from water/wastewater by nanosized metal oxides: a review. J. Hazard. Mater. 211–212, 317–331.

Joo, Y., Sim, J.H., Jeon, Y., Lee, S.U., Sohn, D., 2013. Opening and blocking the inner-pores of halloysite. Chem. Commun. 49, 4519.

Joussein, E., Petit, S., Churchman, J., Theng, B., Righi, D., Delvaux, B., 2005. Halloysite clay minerals—a review. Clay Miner. 40, 383–426.

Karadag, A., Yilmaz, V.T., Thoene, C., 2001. Di- and triethanolamine complexes of Co(II), Ni(II), Cu(II) and Zn(II) with thiocyanate: synthesis, spectral and thermal studies. Crystal structure of dimeric Cu(II) complex with deprotonated diethanolamine $[Cu_2(\mu\text{-dea})_2(NCS)_2]$. Polyhedron 20, 635–641.

Kiani, G., 2014. High removal capacity of silver ions from aqueous solution onto halloysite nanotubes. Appl. Clay Sci. 90, 159–164.

Kiani, G., Dostali, M., Rostami, A., Khataee, A.R., 2011. Adsorption studies on the removal of malachite green from aqueous solutions onto halloysite nanotubes. Appl. Clay Sci. 54, 34–39.

Koteja, A., Matusik, J., 2015. Di- and triethanolamine grafted kaolinites of different structural order as adsorbents of heavy metals. J. Colloid Interf. Sci. 455, 83–92.

Lee, S.Y., Kim, S.J., 2002. Adsorption of naphthalene by HDTMA modified kaolinite and halloysite. Appl. Clay Sci. 22, 55–63.

Lee, S.M., Tiwari, D., 2012. Organo and inorgano-organo-modified clays in the remediation of aqueous solutions: an overview. Appl. Clay Sci. 59–60, 84–102.

Letaief, S., Detellier, C., 2007. Functionalized nanohybrid materials obtained from the interlayer grafting of aminoalcohols on kaolinite. Chem. Commun. 25, 2613–2615.

Letaief, S., Tonle, I.K., Diaco, T., Detellier, C., 2008. Nanohybrid materials from interlayer functionalization of kaolinite. Application to the electrochemical preconcentration of cyanide. Appl. Clay Sci. 42, 95–101.

Li, Z., Bowman, R.S., 2001. Retention of inorganic oxyanions by organo-kaolinite. Water Res. 35, 3771–3776.

Liu, R., Zhang, B., Mei, D., Zhang, H., Liu, J., 2011. Adsorption of methyl violet from aqueous solution by halloysite nanotubes. Desalination 268, 111–116.

Liu, L., Wan, Y., Xie, Y., Zhai, R., Zhang, B., Liu, J., 2012. The removal of dye from aqueous solution using alginate-halloysite nanotube beads. Chem. Eng. J. 187, 210–216.

Luo, P., Zhao, Y., Zhang, B., Liu, J., Yang, Y., Liu, J., 2010. Study on the adsorption of neutral red from aqueous solution onto halloysite nanotubes. Water Res. 44, 1489–1497.

Luo, P., Zhang, J., Zhang, B., Wang, J., Zhao, Y., Liu, J., 2011. Preparation and characterization of silane coupling agent modified halloysite for Cr(VI) removal. Ind. Eng. Chem. Res. 50, 10246–10252.

Matusik, J., 2014. Arsenate, orthophosphate, sulfate, and nitrate sorption equilibria and kinetics for halloysite and kaolinites with an induced positive charge. Chem. Eng. J. 246, 244–253.

Matusik, J., Bajda, T., 2013. Immobilization and reduction of hexavalent chromium in the interlayer space of positively charged kaolinites. J. Colloid Interf. Sci. 398, 74–81.

Matusik, J., Wścisło, A., 2014. Enhanced heavy metal adsorption on functionalized nanotubular halloysite interlayer grafted with aminoalcohols. Appl. Clay Sci. 100, 50–59.

Mellouk, S., Cherifi, S., Sassi, M., Marouf-Khelifa, K., Bengueddach, A., Schott, J., Khelifa, A., 2009. Intercalation of halloysite from Djebel Debagh (Algeria) and adsorption of copper ions. Appl. Clay Sci. 44, 230–236.

Mellouk, S., Belhakem, A., Marouf-Khelifa, K., Schott, J., Khelifa, A., 2011. Cu(II) adsorption by halloysites intercalated with sodium acetate. J. Colloid Interf. Sci. 360, 716–724.

Newman, R.H., Childs, C.W., Churchman, G.J., 1994. Aluminium coordination and structural disorder in halloysite and kaolinite by ^{27}Al NMR spectroscopy. Clay Miner. 29, 305–312.

Pan, J., Yao, H., Xu, L., Ou, H., Huo, P., Li, X., Yan, Y., 2011. Selective recognition of 2,4,6-trichlorophenol by molecularly imprinted polymers based on magnetic halloysite nanotubes composites. J. Phys. Chem. C 115, 5440–5449.

Peng, Q., Liu, M., Zheng, J., Zhou, C., 2015. Adsorption of dyes in aqueous solutions by chitosan–halloysite nanotubes composite hydrogel beads. Micropor. Mesopor. Mater. 201, 190–201.

Qu, X., Alvarez, P.J.J., Li, Q., 2013. Applications of nanotechnology in water and wastewater treatment. Water Res. 47, 3931–3946.

Saad, R., Belkacemi, K., Hamoudi, S., 2007. Adsorption of phosphate and nitrate anions on ammonium-functionalized MCM-48: effects of experimental conditions. J. Colloid Interf. Sci. 311, 375–381.

Sen, B., Dotson, R.L., 1970. Characterization and studies of some triethanolamine complexes of transition and representative metals. J. Inorg. Nucl. Chem. 32, 2707–2716.

Smith, M.E., Neal, G., Trigg, M.B., Drennan, J., 1993. Structural characterization of the thermal transformation of halloysite by solid state NMR. Appl. Magn. Reson. 4, 157–170.

Szczepanik, B., Słomkiewicz, P., Garnuszek, M., Czech, K., 2014. Adsorption of chloroanilines from aqueous solutions on the modified halloysite. Appl. Clay Sci. 101, 260–264.

Tauler, R., Casassas, E., Rainer, M.J.A., Rode, B.M., 1985. The complex formation of Cu(II) with triethanolamine in aqueous solution. Inorg. Chim. Acta 105, 165–170.

Tian, X., Wang, W., Wang, Y., Komarneni, S., Yang, C., 2015. Polyethylenimine functionalized halloysite nanotubes for efficient removal and fixation of Cr (VI). Micropor. Mesopor. Mater. 207, 46–52.

Tombácz, E., Szekeres, M., 2006. Surface charge heterogeneity of kaolinite in aqueous suspension in comparison with montmorillonite. Appl. Clay Sci. 34, 105–124.

Viseras, M.T., Aguzzi, C., Cerezo, P., Viseras, C., Valenzuela, C., 2008. Equilibrium and kinetics of 5-aminosalicylic acid adsorption by halloysite. Micropor. Mesopor. Mater. 108, 112–116.

Wada, K., Kakuto, Y., 1980. Selective adsorption of zinc on halloysite. Clay Clay Miner. 28, 321–327.

Wang, Y.H., Lin, S.H., Juang, R.S., 2003. Removal of heavy metal ions from aqueous solutions using various low-cost adsorbents. J. Hazard. Mater. 102, 291–302.

Wang, J.H., Zhang, X., Zhang, B., Zhao, Y.F., Zhai, R., Liu, J.D., Chen, R.F., 2010. Rapid adsorption of Cr(VI) on modified halloysite nanotubes. Desalination 259, 22–28.

Weber Jr., W.J., Morris, J.C., 1963. Kinetics of adsorption on carbon from solution. J. Sanit. Eng. Div. 89, 31–59.

White, R.D., Bavykin, D.V., Walsh, F.C., 2012. The stability of halloysite nanotubes in acidic and alkaline aqueous suspensions. Nanotechnology 23, 065705.

Whitmire, K.H., Hutchison, J.C., Gardberd, A., Edwards, C., 1999. Triethanolamine complexes of copper. Inorg. Chim. Acta 294, 153–162.

Xi, Y., Mallavarapu, M., Naidu, R., 2010. Preparation, characterization of surfactants modified clay minerals and nitrate adsorption. Appl. Clay Sci. 48, 92–96.

Xie, Y., Qian, D., Wu, D., Ma, X., 2011. Magnetic halloysite nanotubes/iron oxide composites for the adsorption of dyes. Chem. Eng. J. 168, 959–963.

Yuan, P., Southon, P.D., Liu, Z., Kepert, C.J., 2012a. Organosilane functionalization of halloysite nanotubes for enhanced loading and controlled release. Nanotechnology 23, 375705.

Yuan, P., Tan, D., Aannabi-Bergaya, F., Yan, W., Fan, M., Liu, D., He, H., 2012b. Changes in structure, morphology, porosity, and surface activity of mesoporous halloysite nanotubes under heating. Clay Clay Miner. 60, 561–573.

Zhang, A.B., Pan, L., Zhang, H.Y., Liu, S.T., Ye, Y., Xia, M.S., Chen, X.G., 2012. Effects of acid treatment on the physico-chemical and pore characteristics of halloysite. Colloids Surf. A 396, 182–188.

Zhao, M., Liu, P., 2008. Adsorption behavior of methylene blue on halloysite nanotubes. Micropor. Mesopor. Mater. 112, 419–424.

Zhao, Y., Abdullayev, E., Vasiliev, A., Lvov, Y., 2013. Halloysite nanotubule clay for efficient water purification. J. Colloid Interf. Sci. 406, 121–129.

Chapter 24

Imogolite Polymer Nanocomposites

W. Ma, Y. Higaki and A. Takahara*
Institute for Materials Chemistry and Engineering, and International Institute for Carbon-Neutral Energy Research (WPI-I2CNER), Kyushu University, Fukuoka, Japan
**Corresponding author: e-mail: takahara@cstf.kyushu-u.ac.jp*

24.1 INTRODUCTION

Through the combination of several components at the molecular level, the fabrication of nanohybrids is one of the most effective approaches for producing materials with advanced properties that had been unavailable in the past. Organic–inorganic nanohybrids formed by dispersing inorganic nanostructures in organic matrices are attracting a great deal of research interest; the complementation and synergy of the rigid inorganic and the soft organic ingredients are capable of giving the resultant nanohybrids a number of novel features (Sanchez et al., 2005). Well-dispersed inorganics in the nanohybrids are generally used as reinforcing fillers that provide high strength or stability for the organics. Since the invention of clay polymer nanocomposites (CPN) by Toyota Central Research Laboratory (Okada et al., 1987; Okada and Usuki, 2006), extensive worldwide research on organic–inorganic nanohybrids has been conducted and several products, such as nylon–clay (Usuki et al., 1993, 1995) and rubber–silica (Mackenzie et al., 1992), are already used in the fields of the packaging industry, sporting goods and automobiles (Nicole et al., 2010) Discovery or synthesis of new nanomaterials, such as carbon nanotubes and graphene have been providing diverse candidate nanofillers for the construction of nanohybrids (Iijima and Ichihashi, 1993). Carbon nanotubes even made a boom for nanohybrid preparation (Spitalsky et al., 2010). Against such a background, clay minerals have survived the rapid advance in material science. They still keep their vitality due to the versatile characteristics of an attractive structure, tunable surface chemistry and 'green' features (ie, easily available and biocompatible) that are in great demand for addressing several severe global climate issues.

Developments in Clay Science, Vol. 7. http://dx.doi.org/10.1016/B978-0-08-100293-3.00024-8

In addition to commonly used lamellar clay minerals, such as kaolinite and montmorillonite, clay minerals can form tubular structures, which are garnering research interest due to their unique hollow and layered structures, high porosity and large specific surface area (SSA). Halloysite and imogolite are naturally occurring tubular aluminosilicate clay minerals. Halloysite, ($Al_2Si_2O_5(OH)_4 \cdot nH_2O$), with a length of 500–1000 nm, diameter of ca. 50 nm and lumen of 15 nm, is mined from abundant mineral deposits (Lvov et al., 2008). The dimensions of imogolite ($Al_2SiO_3(OH)_4$) are considerably smaller than those of halloysite. It has a length of approximately 100 nm, external diameter of about 2 nm, and internal diameter of about 1 nm (Cradwick et al., 1972; Mukherjee et al., 2005; Yang et al., 2008); therefore, it can provide a much greater SSA than halloysite. Imogolite presents a SiOH-exposed interior and AlOH-exposed exterior surfaces, and it is positively charged in acidic conditions (Gustafsson, 2001; Guimares et al., 2007). Typical applications of imogolite involve its use as a gel former (Inoue et al., 2006; Nakano et al., 2010; Jiravanichanun et al., 2012), for gas storage (Ohashi et al., 2004; Kuroda and Kuroda, 2011), as a coagulant and thickener (Farmer et al., 1983) and as a nanofiller for polymers (Yamamoto et al., 2005a,b; Ma et al., 2011a,b,c; Shikinaka et al., 2011, 2015). Natural imogolite was discovered in certain volcanic soil deposits in Japan (Yoshinaga and Aomine, 1962), and then in many other volcanic and podzolic soils throughout the world. However, mining is not an economic way to access imogolite because the concentration of natural imogolite is very low. As an alternative, this nanomaterial can be obtained by chemical synthesis under mild conditions (Farmer et al., 1977). Recent progress in synthetic protocols has resulted in several promising methods for producing imogolite (or imogolite-like) nanotubes in large quantities (Levard et al., 2008, 2009). These efforts may potentially stimulate new interest in imogolite research and open up possibilities for large-scale applications.

The applications of imogolite as a building block for nanodevices (Yang et al., 2007; Kuroda and Kuroda, 2008) and nanohybrids (Yamamoto et al., 2002, 2007; Jiravanichanun et al., 2008, 2009; Yah et al., 2010, 2011a,b; Ma et al. 2012a,b) have been demonstrated in previous studies. This chapter reviews the utility of imogolite for the fabrication of organic–inorganic nanohybrids, including (i) the surface modification of imogolite nanotubes based on the robust affinity between the phosphate group of the organic molecule and the aluminol (AlOH) group of imogolite external surface, (ii) imogolite-polymer nanocomposites and (iii) imogolite hybrid hydrogels. The direct mixing of imogolite with organic polymers is not a suitable way to prepare CPN because imogolite nanotubes are likely to build bundles due to their rigid structure and high surface energy, and the bundles may grow into even larger bundles or random aggregates. Several effective approaches for dispersing imogolite as small-sized bundles, or even individual nanotubes in polymer matrices, are introduced as well.

24.2 SURFACE MODIFICATION FOR IMOGOLITE POLYMER NANOCOMPOSITES

The surface energy of imogolite nanotubes can be reduced by the introduction of an organic layer, and this can greatly improve the dispersibility of imogolite in organic solvents, as well as in various polymer matrices. The metal–oxygen–phosphorus (M–O–P) interaction plays an important role in the surface modification of imogolite nanotubes. Silane coupling agent is a popular device for modifying inorganic surfaces. The external surfaces of imogolite have been modified with (γ-aminopropyl) triethoxysilane (Johnson and Pinnavaia, 1990). However, the instability of organosilane on the outer surface of imogolite has been reported. Johnson and Pinnavaia (1991) suggested the use of organophosphates or organophosphoric acids as coupling agents for the modification of the external AlOH surface of imogolite. Recently, Shikinaka et al. (2015) reported a successful modification of imogolite surface through interaction between Al–OH and aldehyde groups of benzaldehyde. The strong affinity between octadecylphosphonic acid and the imogolite surface was reported (Yamamoto et al., 2001). More recently, an approach for anchoring alkyl chains on an imogolite surface from aqueous solution was developed (Ma et al., 2011b) by converting the water-insoluble alkyl phosphate into the corresponding water-soluble ammonium salt. The adsorption of molecules on the inorganic surface from aqueous solution is particularly interesting because (i) imogolite nanotubes are dispersible only in aqueous

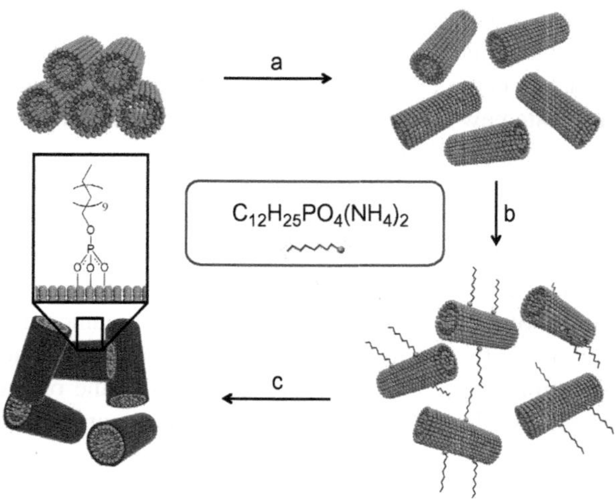

FIG. 24.1 Schematic illustration of dodecylphosphate chemisorbing onto the surface of individually dispersed imogolite nanotubes: (A) dispersion of freeze-dried imogolite powder into weak acidic water by electrostatic repulsion; (B) chemisorption of dodecylphosphate onto the outer surface of individually dispersed imogolite nanotubes and (C) purifying and solidifying the product by centrifugation and freeze drying. *Reprinted with permission from Ma et al. (2011b). © 2011, The Chemical Society of Japan.*

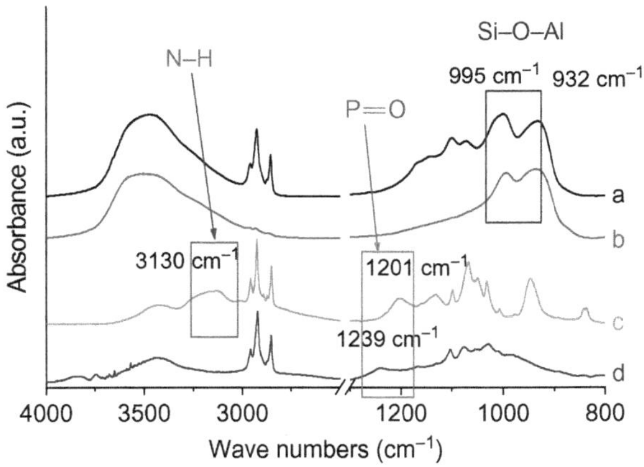

FIG. 24.2 FT-IR spectra of (A) imogolite–$DDPO_4$, (B) pristine imogolite, (C) $DDPO_4(NH_4)_2$ and (D) $DDPO_4H_2$.

medium and (ii) it is an environmentally friendly procedure. The details of this procedure are shown in Fig. 24.1. The ammonium salt of dodecylphosphate ($DDPO_4(NH_4)_2$) was precipitated from a 2-propanol solution of dodecylphosphoric acid ($DDPO_4H_2$) by the addition of ammonia.

The interaction between imogolite and $DDPO_4$ is confirmed by Fourier transform infrared (FT-IR) spectroscopy measurements. The FT-IR spectra of the imogolite modified with $DDPO_4$ (imogolite–$DDPO_4$), pristine imogolite, $DDPO_4(NH_4)_2$ and $DDPO_4H_2$ are shown in Fig. 24.2. For $DDPO_4H_2$, the absorption of the PO–H vibration around 3420 cm^{-1}, the absorption of P–O stretching vibration around 1030 cm^{-1}, and the absorption of P=O vibration at 1239 cm^{-1} were observed. The absorption spectrum of $DDPO_4(NH_4)_2$ shows some differences from those of $DDPO_4H_2$, as the absorption band around 1030 cm^{-1} split into several sharp absorptions and the P=O vibration shifts to 1201 cm^{-1} as well. The absorption around 3130 cm^{-1} is attributed to N–H vibration of the ammonium groups. For imogolite–$DDPO_4$, the absence of N–H absorption at 3130 cm^{-1} indicates that ammonium cations are absent on the surface of imogolite. The typical absorption bands of imogolite at 995, 935 and 560 cm^{-1} still exist, suggesting the Si–O–Al skeleton of imogolite nanotubes is intact, while the absorption at 995 cm^{-1} is strengthened by the coexistent absorption of the phosphate groups. The absorption of P=O at 1239 cm^{-1} for $DDPO_4H_2$ and 1201 cm^{-1} for $DDPO_4(NH_4)_2$ disappears from the spectrum of imogolite $DDPO_4$, presumably due to the condensation between the phosphate groups and the aluminol groups.

In order to obtain further insight into the molecular aggregation state of the imogolite $DDPO_4$, wide-angle X-ray diffraction (WAXD) measurements were carried out at the BL02B2 beam line of SPring-8 using the white powder product packed in a glass capillary. WAXD profiles for the freeze-dried imogolite and imogolite–$DDPO_4$ are shown in Fig. 24.3. Scattering vector q (nm^{-1}) is

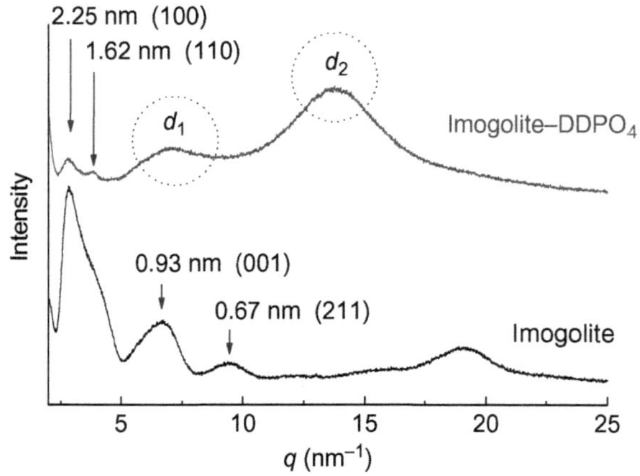

FIG. 24.3 WAXD profiles of pristine imogolite and imogolite–$DDPO_4$.

defined as $q = (4\pi/\lambda)\sin\theta$, where 2θ is the scattering angle. The d-value was calculated by d (nm) $= 2\pi/q$. The reflections at 2.25, 1.62, 0.93 and 0.67 nm for the freeze-dried imogolite are assigned to the lattice planes of hexagonally packed nanotubes and the scattering of individual nanotube (Wada et al., 1979). For imogolite–$DDPO_4$, the broad diffraction around $q = 13.8$ nm^{-1} (d_2) is probably attributed to the disordered grafted alkyl chains. The reflections at 2.25 and 1.62 nm suggest the presence of imogolite bundles. On the other hand, the intensity of the reflections at 2.25 and 1.62 nm significantly decreased compared with those of pristine imogolite, indicating the debundling or delamination of the imogolite bundles. Imogolite cylinders may interact through their Al–OH groups, and bundles of imogolite tubes still exist even in weak acidic water. When the dodecylphosphate attaches to the surface of those bundles, a core–shell structure is built with the imogolite bundles as the core and the dodecylphosphate layer as the shell. However, it is expected that only tightly packed bundles maintain during the modification process.

The above content demonstrated that the chemisorption of low-molecular-mass alkyl phosphonic chains on imogolite surface from an aqueous solution is insufficient to prevent nanotube aggregation. As a better alternative, a grafting of polymer chains from the nanostructure surface has been developed as a powerful technique for homogeneous dispersion in hybrid nanomaterials (Perruchot et al., 2001; Vestal and Zhang, 2002; Matsuno et al., 2004). Several strategies can be used to graft polymers from the inorganic surface, including grafting to, grafting through and grafting from (Wang et al., 2004). In many cases, grafting from, in which the polymer chains are grown in situ from the surface by means of surface-initiated polymerization, is preferred because it provides greater grafting density than the grafting-to and grafting-through approaches. The grafting-from process can be performed with various polymerization techniques, including ionic and free-radical polymerization (Jagur-Grodzinski, 2001). Free-radical polymerization is preferable to ionic processes on economic grounds because it is easier to perform and much less sensitive to the presence of water.

The grafting of poly(methyl methacrylate) (PMMA) on the surface of imogolite was achieved by a surface-initiated controlled radical polymerization technique called activators regenerated by electron transfer for atom transfer radical polymerization (ARGET ATRP) (Ma et al., 2011c) (see Fig. 24.4). ARGET ATRP is has been attracting interests due to the convenience it offers—eg, it can be carried out without strict deoxygenation and only needs a ppm-level of catalyst (Jakubowski et al., 2006; Matyjaszewski et al., 2006).

A water-soluble ATRP initiator carrying a phosphoric group, ammonium salt of 8-(2-bromo-2-methylpropanoyloxy) octyl phosphoric acid [BMPO-$PO_4(NH_4)_2$], was designed and synthesized for surface-initiated ARGET ATRP from imogolite nanotubes. Homogeneous modification of the imogolite surface was achieved by this water-soluble, surface-attachable initiator. In addition, the molecule is capable of providing the modified imogolite with adequate hydrophobicity because it contains a relatively long hydrophobic chain.

ARGET ATRP was carried out by using ascorbic acid as the reducing agent and anisole as the solvent. Polymeric products, including PMMA-grafted imogolite (imogolite-*g*-PMMA), were isolated by precipitation from methanol. PMMA was cleaved from the imogolite surface by HF/HCl for gel permeation chromatography (GPC) measurements. GPC data showed that grafted PMMA with molecular mass of $M_n = 26{,}600$ and 32,700, and the corresponding molecular mass distribution of $M_w/M_n = 1.22$ and 1.33, were obtained after polymerization times of 50 and 90 min, respectively.

Pristine imogolite cannot be dispersed in organic solvents. After modification with $BMPOPO_4$, the resulting modified imogolite becomes dispersable in various solvents. Unfortunately, the dispersions are neither homogenous nor

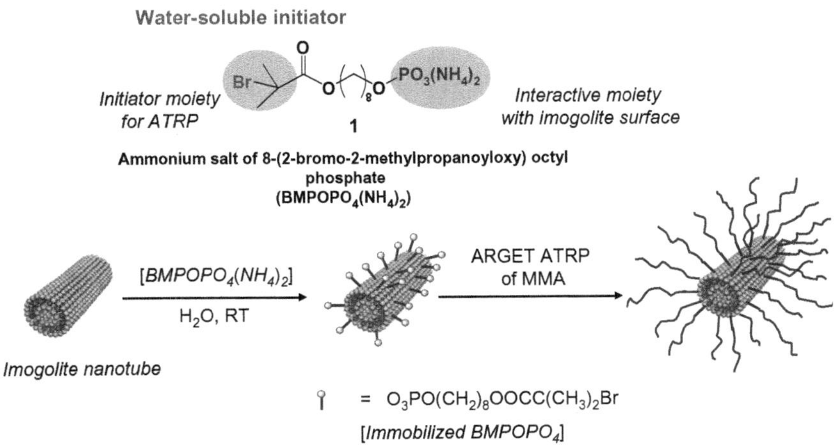

FIG. 24.4 Schematic representation for grafting PMMA onto an imogolite surface.

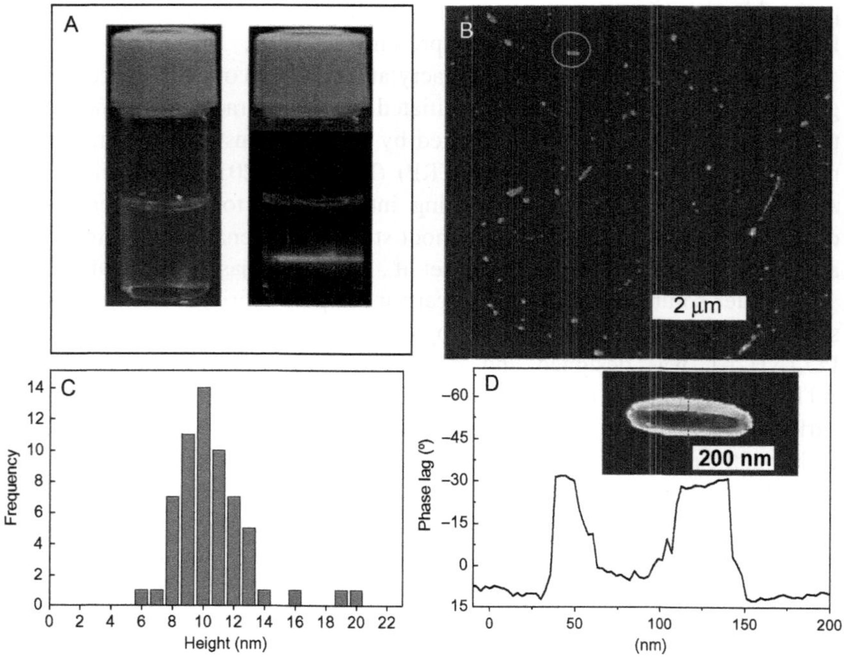

FIG. 24.5 (A) Photographs of THF dispersion of the imogolite-*g*-PMMA with a concentration of 10 mg mL^{-1} (the one on the right shows scattering of a *green* (*light grey* in the print version) *light* beam due to the Tyndall effect). (B) A AFM height image of the imogolite-*g*-PMMA (M_n=32,700, M_w/M_n=1.33). (C) Height distribution of the imogolite-*g*-PMMA as estimated by AFM. (D) A phase image (*inset*) and cross-sectional analysis of a PMMA-*g*-imogolite [marked with a *circle* in (B)]. *Reprinted with permission from Ma et al. (2011c). Copyright 2011 The Royal Society of Chemistry.*

stable. On the other hand, the imogolite-*g*-PMMA shows excellent dispersibility in organic solvents, such as THF, chloroform and toluene. As shown in Fig. 24.5A the homogenous dispersion of imogolite-*g*-PMMA in THF with a concentration of 10 mg mL^{-1} was stable for more than 2 months.

Morphology of imogolite-*g*-PMMA nanotubes was observed with atomic force microscopy (AFM) in a tapping mode using a sharp diamondlike carbon tip with curvature radius of 1 nm. A height image of the imogolite-*g*-PMMA with surface-modified PMMA chains of M_n=32,700 and M_n/M_w=1.33 is shown in Fig. 24.5B. Discrete nanostructures were randomly distributed on the mica surface and no aggregation was observed, indicating excellent dispersibility of imogolite-*g*-PMMA. The height distribution of this imogolite-*g*-PMMA is obtained from 60 AFM images (Fig. 24.5C). The obtained average height value is 10.6±2.5 nm, although there are still some objects with heights of more than 15 nm. If an object with the smallest height value (6 nm) is a single imogolite-*g*-PMMA nanotube, the ones with larger height values are attributed to bundles. The high-resolution phase image and the corresponding cross-section analysis in Fig. 24.5D indicate that the imogolite-*g*-PMMA consists of a hard middle part and a soft edge. This further confirms the core–shell structure of the imogolite-*g*-PMMA.

Imogolite is an insulator since it consists of wide-bandgap alumina and silica. Therefore, unlike carbon nanotube, imogolite lacks intrinsic semiconductivity, but thanks to its high SSA, abundant empty surface sites and high stability at ambient conditions, imogolite can be an interesting condensed phase for heterosupramolecules such as electron-emitting materials or water sensors in nanoelectronic devices. A more promising approach to obtain semiconducting aluminosilicate nanotubes is functionalization with conjugated molecules such as terthiophene with alkyl spacer consisting of $–CH_2CH_2–$ and terminal group of $P{=}O(OH)_2$ (Fig. 24.6). Thiophene oligomers have been extensively studied in recent years due to their excellent optical properties. They exhibit high quantum yields of photoluminescence and large coverage of fluorescence frequency in entire spectrum from visible ultraviolet (UV) to near-infrared (IR) in solution, as well as in the solid state (Barbarella et al., 1998a,b, 2001).

The preparation method of imogolite–terthiophene hybrid materials and their dispersion states are shown in Fig. 24.7. Imogolite solution was added dropwise to a THF solution of 2-(5″-hexyl-2,2′,5′,2″-terthiophen-5-yl)ethylphosphonic acid (HT3P) and stirred overnight in the dark. The HT3P-modified imogolite (imogolite-HT3P) was obtained by centrifugation and rinsed with fresh THF three times to remove weakly chemisorbed or nonchemisorbed HT3P. The precipitate was redispersed again in deionized water before freeze drying. Freeze-dried imogolite-HT3P was a cottonlike yellow solid. The same preparation method was used for 2-(5″-hexyl-2,2′:5′,2″-terthiophen-5-yl)ethylphosphonic acid 1,1-dioxide (HT3OP) to produce a cottonlike, pale brown solid of HT3OP-modified imogolite (imogolite-HT3OP). No precipitate was obtained for OH-group derivatives (HT3OH and HT3OOH) through the surface-modification procedure, indicating that the $P{=}O(OH)_2$ group is necessary for surface modification.

The conductivity of the pristine imogolite and imogolite–terthiophene hybrids was investigated by intensity–voltage (*I–V*) measurement. By coating the pristine imogolite and imogolite–terthiophene on a silicon wafer and connecting it by silver paste at the two ends, the conductivity was measured by source metre. The plots of current vs applied bias voltage give *I–V* curves in the voltage range −30 to 30 V, as shown in Fig. 24.8.

The average electrical conductance calculated from the forward bias region of pure imogolite *I–V* curve was 5.9 μS. It was speculated that the current flow was due to charge hopping on the hydrated imogolite surface. Oh et al. (2007) studied the *I–V* characteristic of imogolite and proposed that bound water molecules contribute to the surface conductivity of imogolite (Oh et al., 2007; Park et al., 2008). The current flow was attributed to the ability of OH groups on the imogolite surface to lose or gain a positive charge (proton) from water molecules, resulting in net surface change. The electrical

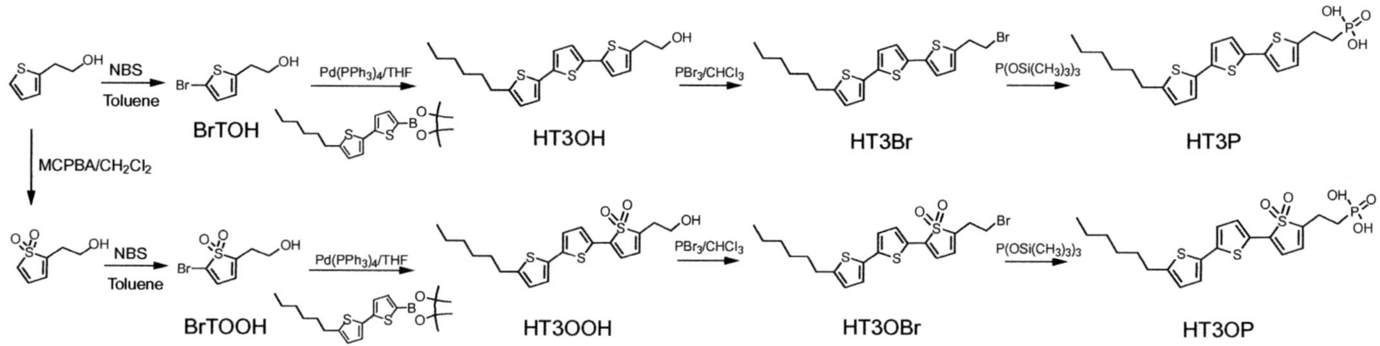

FIG. 24.6 Synthesis pathway for electron donating (HT3P) and accepting (HT3OP).

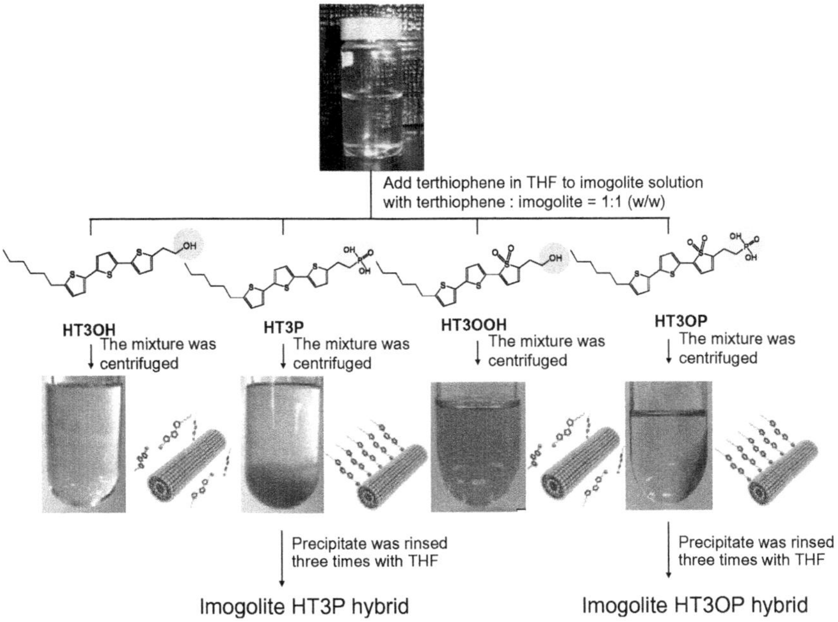

FIG. 24.7 Schematic illustration of the preparation method of imogolite–terthiophene hybrid materials and their dispersion states. *Reprinted with permission from Yah et al. (2011a). © 2011, The Chemical Society of Japan.*

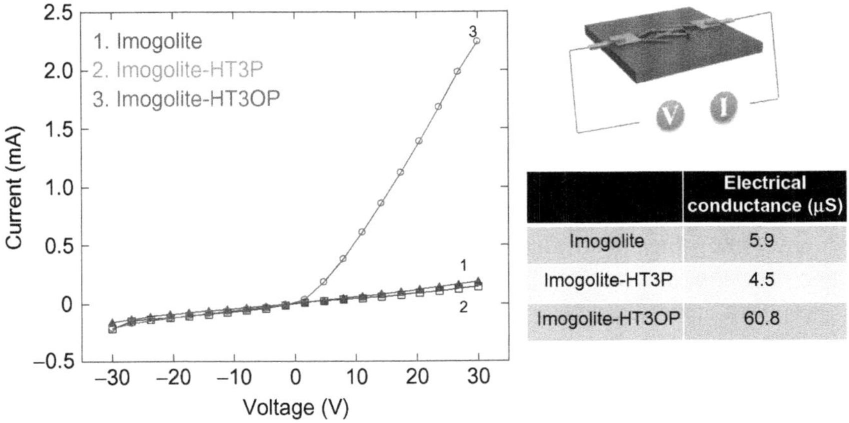

	Electrical conductance (μS)
Imogolite	5.9
Imogolite-HT3P	4.5
Imogolite-HT3OP	60.8

FIG. 24.8 Current–voltage (*I–V*) curves and conductivity of pristine imogolite, imogolite-HT3P and an imogolite-HT3OP hybrid.

conductance of imogolite-HT3OP improved to 60.8 μS, which is one order of magnitude greater than that of pristine imogolite. The imogolite-HT3OP showed a nonohmic characteristic in the *I–V* curve that indicates the disorder packing of HT3OP on the imogolite surface, where it behaved like a semiconductor or metal-semiconductor Schottky junction. The HT3OP can act as an electron acceptor when interacting with imogolite because the high electron affinity of S=O of HT3OP withdraws the negative charge from imogolite, resulting in the effective motion of positive charges on the imogolite surface. The introduction of HT3OP onto the imogolite surface amplified a p-type conductivity of imogolite that resembles the phenomenon of chemical doping of carbon nanotubes with alkaline metal (Mordkovich et al., 1996; Lee et al., 1997; Kong et al., 2000; Kim et al., 2008). On the other hand, the electrical conductance of imogolite HT3OP calculated from the forward bias region is 4.5 μS, which is less than it was before HT3P doping. Here, HT3P works as an electron donor when interacting with imogolite. The p-type conductivity of imogolite was reduced when a HT3P thiophene ring transfers a negative charge to imogolite, which restricts the effective motion of positively charged species on the imogolite surface.

24.3 IMOGOLITE POLYMER NANOCOMPOSITES BY THE SIMPLE BLENDING METHOD

Imogolite has an outer surface of hydrophilic Al–OH groups and can be dispersed in water under low acidic conditions. Therefore, it is very difficult to disperse imogolite in a hydrophobic polymer matrix. Unsuccessful attempts have been made to convert the hydroxylated surface of imogolite to a hydrophobic surface by a surface modification method using organosilanes (Johnson and Pinnavaia, 1990, 1991). Since alkyl phosphoric acid groups have been used for surface treatment of the Al-sheet, attempts have been made to modify the surface with alkyl phosphonic acid (Yamamoto et al., 2001). Yamamoto et al. (2001) have reported that modified imogolite nanofibres with alkylphosphonic acid can be dispersed in hydrophobic solvents such as hexane and chloroform while they aggregate in water. When focusing on the use of imogolite in CPN, it appeared that the surface modification of imogolite would improve its dispersibility into the hydrophobic polymer matrix. A novel imogolite polymer nanohybrid was prepared by free-radical polymerization of methyl methacrylate (MMA) in the presence of imogolite modified with phosphoric acid ester with a polymerizable methacrylate group, which grafts a PMMA chain from the imogolite surface. The preparation of an imogolite–PMAA nanohybrid (Yamamoto et al., 2005a) is illustrated in Fig. 24.9.

Imogolite (freeze-dried powder) was dispersed into weak acidic water (pH 5.0) using sonication. An aqueous solution of 2-acidphosphoxyethyl methacrylate (P-HEMA: Fig. 24.9) was added to the imogolite aqueous

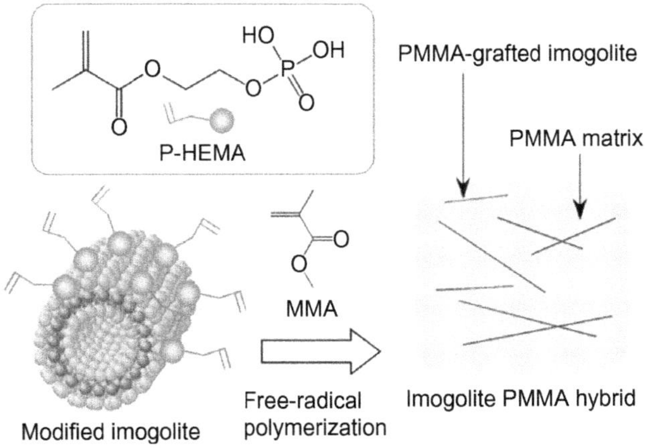

FIG. 24.9 Chemical structure of 2-acidphosphoxyethyl methacrylate (P-HEMA) and illustration of the preparation of polymer hybrid using imogolite modified with P-HEMA. *Reprinted with permission from Yamamoto et al. (2005a). © 2005, Elsevier B.V.*

dispersion under stirring to yield cloudy solution. After stirring for 24 h, the solution was centrifuged to collect the white precipitate of imogolite modified with P-HEMA (imogolite/P-HEMA). The precipitate was washed with water three times by centrifugation and decantation to remove the unreacted P-HEMA and then freeze-dried, resulting in a white powder of imogolite/P-HEMA.

The white powders of imogolite/P-HEMA and 2,2′-azo-bis-isobutyronitrile were mixed with MMA in a glass reactor. The molar ratio of the added monomer and initiator was 50:1. The mixture became cloudy due to the dispersed imogolite/P-HEMA. The free-radical polymerization was carried out by heating the mixture at 343 K after degassing by the freeze–pump–thaw cycle. After heating for 3 h, the obtained solid polymer was dissolved in chloroform and precipitated by a large amount of methanol to create a white powder of imogolite–PMMA hybrid. The yield of this hybrid was 89.6%. Hybrid films were then prepared by the solvent cast method from imogolite–PMMA solution. The 5 mass% concentration of imogolite–PMMA chloroform solution was kept in a glass Petri dish at room temperature for 12 h. After evaporating the solvent, the film was peeled off from the glass Petri dish and dried in a vacuum for 12 h.

The chemisorption of P-HEMA onto the imogolite surface was confirmed by FT-IR measurement. FT-IR spectra of pristine imogolite, P-HEMA and imogolite modified with P-HEMA are shown in Fig. 24.10. The spectrum of imogolite (Fig. 24.10, No. 5) indicated the characteristic absorption of the stretching vibration band ν(Si–O–Al) at 995 cm^{-1} and 935 cm^{-1}, and large broad peak of ν(OH) around 3450 cm^{-1} (Russell et al., 1969). The spectrum

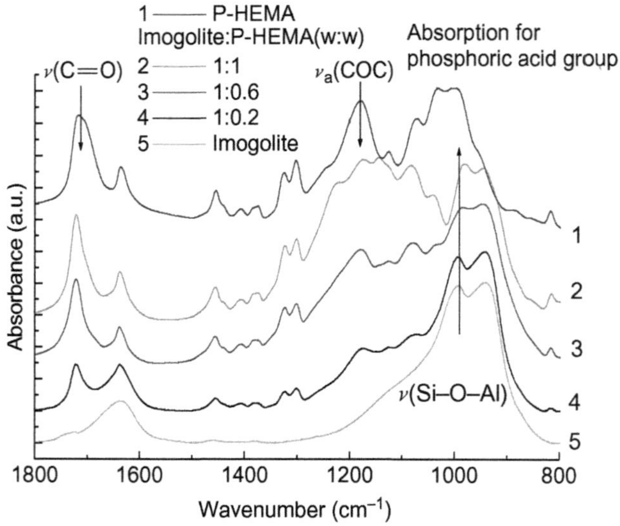

FIG. 24.10 FT-IR spectra of pristine imogolite, P-HEMA and imogolite modified with P-HEMA at various P-HEMA contents in the region from 1800 to 800 cm^{-1}. *Reprinted with permission from Yamamoto et al. (2005a). © 2005, Elsevier B.V.*

of P-HEMA (Fig. 24.10, No. 1) also showed a few characteristic absorption peaks, which are ascribed to the stretching vibration of carbonyl groups ν(C=O) at 1714 cm^{-1}, the asymmetric stretching vibration ν_{as}(COC) of the methacryloyloxy group at 1179 cm^{-1} and several peaks for coupled vibration of P–O stretching vibration at 1070, 1030 and 1000 cm^{-1} (Ando et al., 1996; Suzuki et al., 1997). In contrast, P-HEMA-chemisorbed imogolite with a weight ratio of 1:0.2 (Fig. 24.10, No. 4) showed adsorption peaks at 992 and 935 cm^{-1}, corresponding to the absorption of imogolite, and at 1722 and 1180 cm^{-1}, corresponding to ν(C=O) and ν_{as}(COC) of P-HEMA. These results provide evidence of P-HEMA adsorption onto the imogolite surface. The P-HEMA content in the imogolite/P-HEMA system (Fig. 24.10, Nos. 2 and 3) increased the intensity of absorption peaks at 1080 cm^{-1}, corresponding to the $\nu_{as}\left(PO_3{}^{2-}\right)$ band. Also, the $\nu_s\left(PO_3{}^{2-}\right)$ absorption peak generally appears around 980 cm^{-1} when a phosphoric acid group interacts with the substrate surface (Ando et al., 1996). The absorption around at 995 cm^{-1} broadened and shifted to 987 cm^{-1} for the imogolite/P-HEMA ratio of 1:0.6 and 982 cm^{-1} for the imogolite/P-HEMA ratio of 1:1 because of the extent of interaction between the phosphoric acid groups of P-HEMA and Al–OH groups of imogolite.

The fraction of imogolite and P-HEMA was estimated by thermogravimetric analysis (TGA) and elemental analysis. TGA measurement was carried out at a heating rate of 10 K min^{-1}. Approximately 1 mg of each sample was

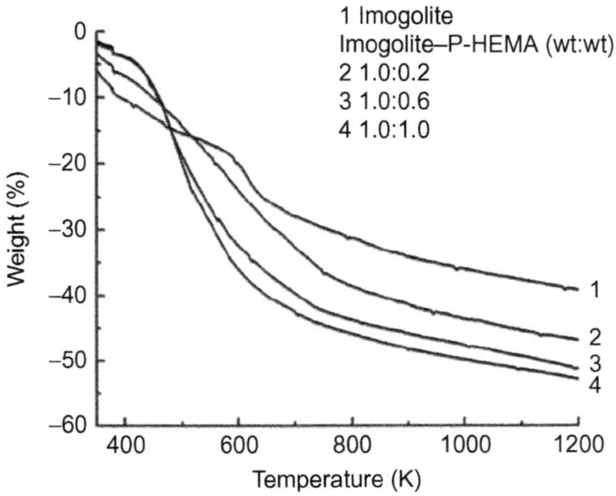

FIG. 24.11 TGA curves of imogolite and imogolite modified with P-HEMA at various levels of P-HEMA content. *Reprinted with permission from Yamamoto et al. (2005a). © 2005, Elsevier B.V.*

placed on a platinum pan. TGA curves of pristine imogolite and imogolite/P-HEMA with various mass contents are shown in Fig. 24.11. Two steps of mass loss were observed in the TGA curve of imogolite, which is similar to previous results (Donkai et al., 1992). These mass losses indicated the release of adsorbed water and structural water. On the other hand, the TGA curves of the modified imogolite showed a rapid mass loss from 430 K, reflecting the degradation of P-HEMA. The amounts of adsorbed P-HEMA on the surface of imogolite estimated from elemental analysis were 0.15 mg per 1 mg of imogolite (imogolite:P-HEMA = 1:0.2), 0.36 mg (1:0.6) and 0.51 mg (1:1), respectively. The coverage value of the imogolite surface was evaluated by the assumption of the complete monolayer modification of P-HEMA onto the imogolite nanotube surface. The coverage of the imogolite surface with P-HEMA was estimated to be 13.8%, 33.5% and 47.5%.

A TEM image of imogolite nanotubes modified with P-HEMA is shown in Fig. 24.12. The dispersibility of these nanotubes was found to be similar to the morphology of unmodified imogolite nanotubes dispersed into low acidic water, as the diameter of nanotubes around several nanometers or over 10 nm in both cases. The dispersibility of nanotubes also supports the successful surface modification of imogolite with P-HEMA.

The imogolite/P-HEMA with a large amount of adsorption (imogolite: P-HEMA = 1:1) was employed for in situ MMA polymerization to prepare the imogolite–PMMA nanohybrid. The polymerization of MMA was confirmed by IR and NMR measurement. The conversion of MMA in the polymerization condition was 95–98%, as estimated by NMR measurement.

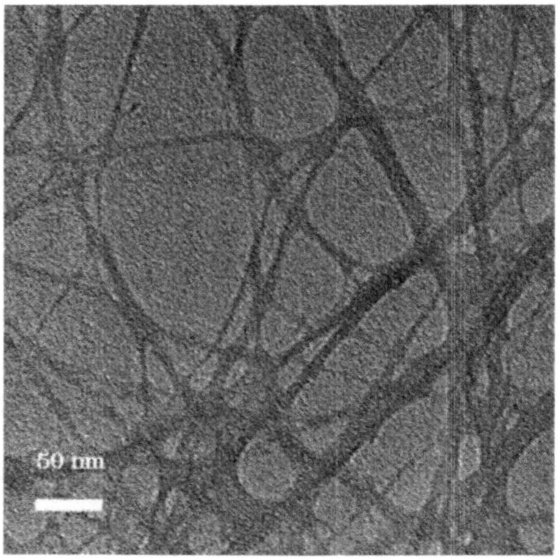

FIG. 24.12 TEM image of surface modified imogolite nanofibres modified with P-HEMA. The sample was prepared by depositing chloroform dispersion on a TEM grid. *Reprinted with permission from Yamamoto et al. (2005a). © 2005, Elsevier B.V.*

TABLE 24.1 Molecular Mass and Polydispersity Index of Soluble Polymer Contents in Imogolite–PMMA Hybrids

Run no.	Content of imogolite (mass%)	M_n	M_w/M_n	State of imogolite
1	0	58,800	2.50	–
2	1.0	65,100	2.55	Modified
3	1.0	65,600	2.54	Unmodified

The average molecular mass (M_n) and the polydispersity index (M_w/M_n) for the soluble polymer content were estimated by GPC and are summarized in Table 24.1. To confirm the side effects of the imogolite nanofibre on the polymerization, it was examined with both modified and unmodified imogolite. GPC was carried out at 313 K with THF as eluent, and using only the soluble part of the polymer hybrid through a 0.5-μm membrane filter. GPC results indicated that the presence of 1.0 mass% imogolite hardly influenced the MMA polymerization.

The surface grafting of PMMA chains on imogolite fibres was confirmed by IR measurement. The IR spectra of PMMA, imogolite–PMMA hybrids and unmodified imogolite powder recovered from the imogolite–PMMA

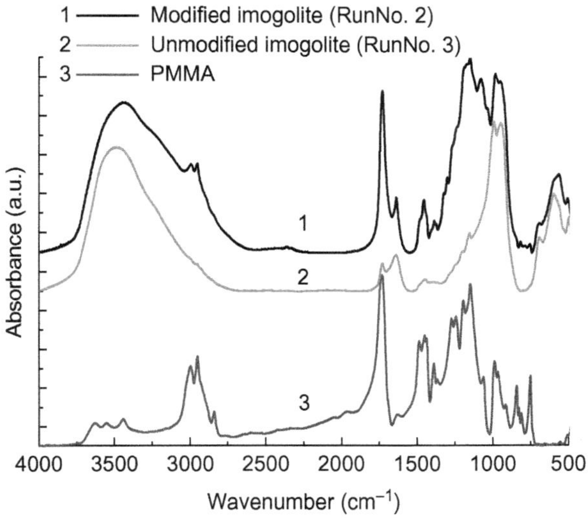

FIG. 24.13 FT-IR spectra of PMMA, PMMA-modified imogolite and unmodified imogolite recovered from PMMA/imogolite hybrid chloroform solution. *Reprinted with permission from Yamamoto et al. (2005a). © 2005, Elsevier B.V.*

hybrid chloroform solution are shown in Fig. 24.13. In the spectrum of PMMA, the stretching vibration bands of CH_2 and C=O at 2996, 2952 and 1733 cm^{-1} were clearly observed. The spectrum of the imogolite–PMMA hybrids after polymerization showed evidence of a PMMA chain attributed to the CH_2 and C=O absorption peaks at 2996, 2953 and 1727 cm^{-1} and retained the coupled vibration of the P–O stretching vibration at 1080 cm^{-1}. The C=O band peak of the imogolite–PMMA hybrids shifted slightly to the lower wavenumber side in comparison with the spectrum of PMMA, due to the interaction between the PMMA chains and imogolites. These results revealed that MMA polymerized and grafted via the methacryloyl group of adsorbed P-HEMA on the surface of imogolite. In addition, the presence of the absorption peak at 3450 cm^{-1} and the two peaks around 995–935 cm^{-1} demonstrates the retention of imogolite structure after free-radical polymerization.

The dispersion states of imogolite–PMMA hybrids and unmodified imogolite powders in water and chloroform interface are shown in Fig. 24.14. After sonication, the unmodified imogolite did not disperse in the chloroform phase, but it did in the aqueous phase. Unmodified imogolite could not disperse in chloroform because the PMMA chain desorbed from the imogolite surface and dissolved into the chloroform phase. Furthermore, unmodified imogolite was aggregated at the interface between water and chloroform after 3 h. On the other hand, imogolite–PMMA hybrids dispersed in the chloroform phase and did not precipitate in the solution even after several days. The stable

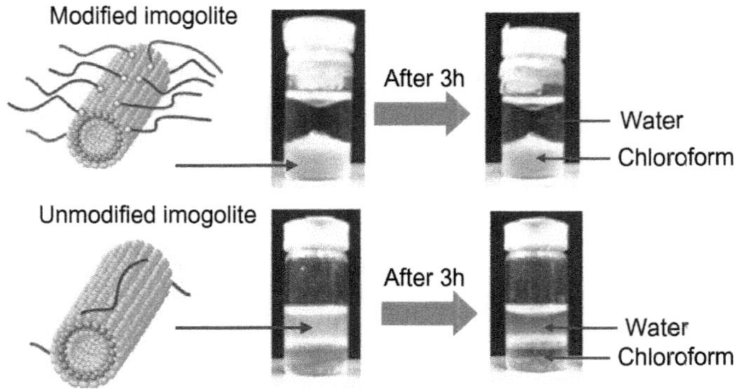

FIG. 24.14 Dispersion state of imogolite–PMMA hybrids and unmodified imogolite in the water/chloroform phase. *Reprinted with permission from Yamamoto et al. (2005a). © 2005, Elsevier B.V.*

dispersion of imogolite–PMMA hybrids was due to the high affinity for chloroform, which is a good solvent for PMMA.

PMMA is a typical transparent glassy polymer, and it is important to keep its optical properties even after the introduction of a reinforcing inorganic filler. The grafted PMMA chains on the imogolite surface provide a high affinity between imogolite nanofibres and PMMA matrix. A hybrid film including 1 mass% imogolite/P-HEMA was prepared to evaluate its optical property. For reference, the blend film of the same imogolite content was prepared from mixture of PMMA and unmodified imogolite. The transparency of PMMA, imogolite–PMMA hybrid and imogolite/PMMA blend films is shown in Fig. 24.15, and Table 24.2 summarizes the haze value of each film. Light transmission was measured with a UV–vis spectrometer at the region of the visible light wavelength from 350 to 700 nm. Haze measurements were carried out using a D65 light source, and the values for each sample were averaged in five measurements at different film locations. The imogolite–PMMA hybrid film retained the transparency of matrix PMMA, and showed more transparency than the imogolite/PMMA blend films. This difference in transparency can be attributed to the dispersibility of imogolite nanofibres in the polymer matrix. One of the factors to control the dispersibility is the affinity of the interface between PMMA matrix and the imogolite–PMMA hybrids.

The mechanical properties of the imogolite–PMMA hybrid film were investigated as well. Dynamic mechanical analysis measurements were carried out in the tensile mode at 11 Hz in nitrogen atmosphere. Tensile testing was carried out at a stretching rate of 1 mm/min. Fig. 24.16 shows the temperature dependence of the storage modulus (E') and loss tangent (tanδ) for PMMA, imogolite–PMMA hybrid and imogolite/PMMA blend films. The imogolite–PMMA hybrid and blend films contained 1 mass% of imogolite content. The dynamic storage modulus of the imogolite–PMMA hybrid film

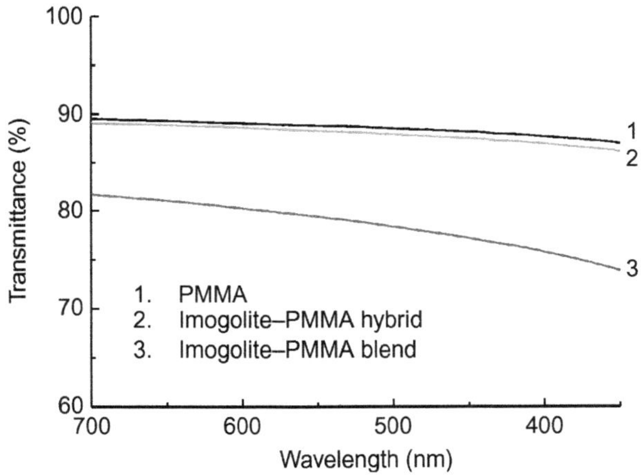

FIG. 24.15 Light transmittance of PMMA, imogolite–PMMA hybrid and imogolite/PMMA blend films in the visible light region.

TABLE 24.2 Haze Value and Transmittance of PMMA, Imogolite–PMMA Hybrid and Imogolite/PMMA Blend Films[a]

Sample	Haze value	Transmittance[b] (%)
PMMA	0.59 ± 0.11	88.6
Imogolite–PMMA (hybrid)	1.18 ± 0.07	88.0
Imogolite/PMMA (blend)	3.37 ± 0.56	78.4

[a]*The content of imogolite in polymer matrix was 1.0 mass%.*
[b]*Determined by light transmission measurement at* $\lambda = 500$ *nm.*

was approximately 1.5 times more than that of the PMMA film. In addition, the α_a-absorption temperature of the hybrid film was higher than that of the PMMA film. The tensile modulus, tensile strength and elongation of PMMA, imogolite–PMMA hybrid and imogolite/PMMA blend films evaluated by tensile testing are summarized in Table 24.3. The tensile modulus and the ultimate strength of the imogolite–PMMA hybrid were approximately 1.4 times more than those of the PMMA film. On the other hand, the tensile modulus and strength of the imogolite/PMMA blend film were similar to those of the PMMA film. As the dispersibility of imogolite in the PMMA matrix improved, the reinforcement effect of imogolite became remarkable in the case of imogolite–PMMA hybrid film. The elongation at the break of the imogolite–PMMA hybrid and the blend film was smaller than that of the matrix polymer film, but the hybrid film clearly exhibited a greater elongation at

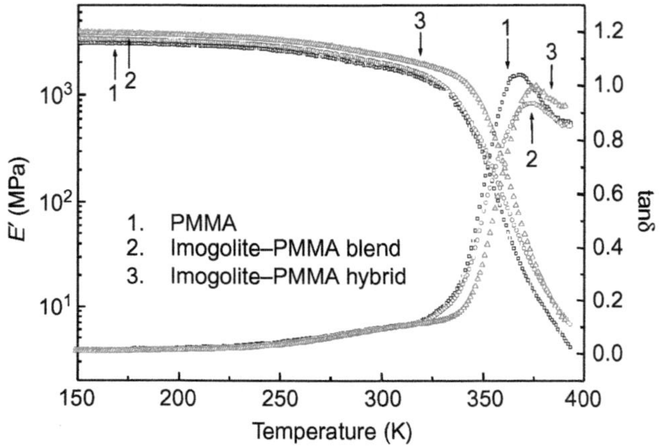

FIG. 24.16 Temperature dependence of dynamic storage modulus and loss tangent at 11 Hz for PMMA, imogolite–PMMA hybrid and imogolite/PMMA blend film with 1.0 mass% imogolite content. *Reprinted with permission from Yamamoto et al. (2005a). © 2005, Elsevier B.V.*

TABLE 24.3 Tensile Modulus, Strength and Elongation at Break of PMMA, Imogolite–PMMA Hybrid[a] and Imogolite/PMMA Blend[b] Film Evaluated by Tensile Testing

Sample	Modulus (GPa)	Strength (MPa)	Elongation (%)
PMMA	1.13±0.05	27.6±0.6	38.0±13.7
Imogolite–PMMA (hybrid)	1.55±0.05	38.5±0.8	21.0±8.9
Imogolite–PMMA (blend)	1.20±0.08	27.7±1.2	5.0±1.2

[a]*The content of imogolite in the polymer matrix was 1.0 mass%.*

the break than the blend film. The lack in the interfacial defect between imogolite–PMMA hybrids and matrix PMMA would lead to the large elongation on the stretching deformation. Shikinaka et al. (2015) also reported the successful dispersion of modified imogolite in aliphatic polyesters through modification of imogolite by benzaldehyde.

24.4 IMOGOLITE POLYMER NANOCOMPOSITES PREPARED BY IN SITU IMOGOLITE SYNTHESIS IN POLYMER SOLUTION

Imogolite has the advantage of being able to be prepared artificially like a carbon nanotube, making it potentially useful in various commercial applications. A synthetic pathway of imogolite was first described by Farmer et al.

(1977). It is synthesized from dilute solutions of aluminium chloride and monomeric orthosilicic acid, and several investigations concerning synthetic methods for imogolite have been reported (eg, Koenderink et al., 1999; Ohashi et al., 2004). Yamamoto et al. (2005b) introduced the synthetic method for an aluminosilicate nanofibre in polymer solution towards the preparation of a nanohybrid, and characterization of the imogolite/polymer hybrid prepared by in situ synthesis was reported.

Imogolite was synthesized essentially by the method described by Farmer et al. (1977). The white solid obtained by freeze drying of the dispersed solution was light and fluffy, similar to natural imogolite. The yield of the freeze-dried products was 67.3%.

The freeze-dried synthetic imogolite sample was characterized by IR measurement via the KBr method (Fig. 24.17, curve 4). The spectrum shows two sharp absorptions at 995 and 935 cm^{-1}, which were assigned to the Si–O–Al stretching vibration, and a large absorption at 3440 cm^{-1}, corresponding to the OH stretching vibration. The IR spectra in this region were not sufficiently distinctive to allow for the presence of other aluminosilicates such as allophane (spherical structure). In a previous study of the synthesis of imogolite, Wada et al. (1979) has indicated that the appropriate reaction time for the formation of imogolite is 4–5 days. Actually, the IR spectra of synthesized imogolite heated for 12–72 h showed adsorption peaks attributed to the by-products because of the remaining imogolite precursor. Consequently, the reaction time of the in situ synthesis in the polymer solution was set at 96 h.

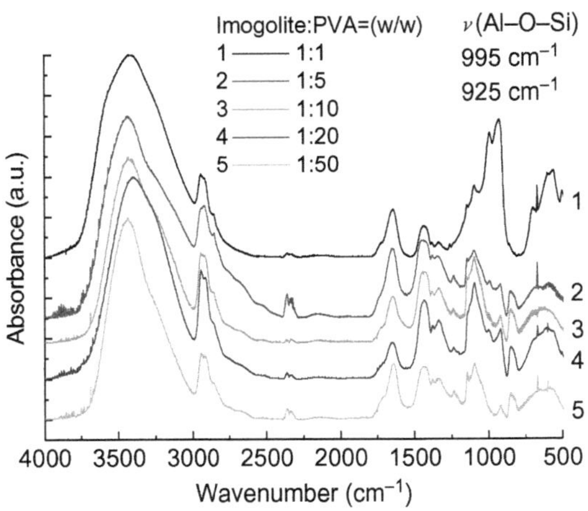

FIG. 24.17 FT-IR spectra of natural imogolite/PVA blend, in situ synthesized imogolite/PVA hybrid, PVA and synthesized imogolite. *Reprinted with permission from Yamamoto et al. (2005a). © 2005, Royal Society of Chemistry.*

The clay polymer nanohybrid was prepared by in situ synthesis of imogolite in the presence of poly(vinyl alcohol) (PVA; degree of polymerization: 630) in the reaction solution. The hydroxyl groups in PVA are supposed to interact with the imogolite surface. The dilute Al and Si solution with an adjusted pH was mixed with a PVA aqueous solution and stirred at 369 K for 96 h. The mixed solution was prepared with different imogolite/PVA mass-fraction ratios of 1:1, 1:5, 1:10, 1:20 and 1:50, respectively. After cooling, the mixed solution of imogolite and PVA was precipitated in ethanol, and the precipitate was filtered using a 450 nm Millipore filter and then rinsed with water and ethanol. The white product of in situ synthesized imogolite–PVA hybrid was then obtained. The cast film was prepared from the in situ synthesized imogolite/PVA solution without separation in order to investigate the optical and mechanical properties of the films. The concentration of imogolite/PVA aqueous solution was 5 mass%. This solution was kept in a hydrophobical surface-treated glass Petri dish at 323 K for 12 h and dried in a vacuum for 12 h. As reference samples, the PVA films and freeze-dried imogolite/PVA blend films were prepared in the same way. The thickness of these cast films were ca. 100 μm.

The IR spectra of PVA and the imogolite–PVA hybrids prepared by two different methods are shown in curves 1–3 of Fig. 24.17. One was prepared by the in situ method, and the other was prepared by the blending of PVA powder and freeze-dried natural imogolite. The mass ratio of the imogolite to PVA was 1:1 for each sample. The characteristic absorption peaks of PVA were appeared at 2945 cm^{-1}: $\nu_a(CH_2)$, 2915 cm^{-1}: $\nu_s(CH_2)$, 1430 cm^{-1}: $\delta(CH_2)$ and 1095 cm^{-1}: ν(C–O). (Krimm et al., 1956). The IR spectra of the hybrid with imogolite showed two sharp peaks corresponding to the Si–O–Al stretching vibration, which were the same as those shown by the synthesized imogolite (curve 4). The IR spectra of the natural imogolite/PVA and synthetic imogolite/PVA hybrids were quite similar to that of the PVA powder, except for the characteristic imogolite absorptions at 995 and 930 cm^{-1}. These results suggest that imogolite was successfully synthesized.

The molecular aggregation state of the synthesized imogolite–PVA hybrid was confirmed by WAXD measurements. Powder WAXD measurement was carried out at the BL02B2 beam line of SPring-8. Fig. 24.18 shows WAXD profiles of natural and synthetic imogolite and the imogolite–PVA hybrid. The imogolite:PVA mass ratio was 1:1. The WAXD pattern of imogolite consists of a number of broad diffractions. The diffractions at 1.83–1.36 and 0.79 nm for natural imogolite (Cradwick et al., 1972) and at 2.30–1.50 and 0.95 nm for synthetic imogolite (Wada et al., 1979) originated from the hexagonally packed nanotube and the scattering of individual nanotubes. Similar diffractions corresponding to the hexagonally packed nanotube have also been observed for a single-walled (SW) carbon nanotube (Maniwa et al., 2001). The PVA crystallite diffraction corresponding to the (101) lattice plane was

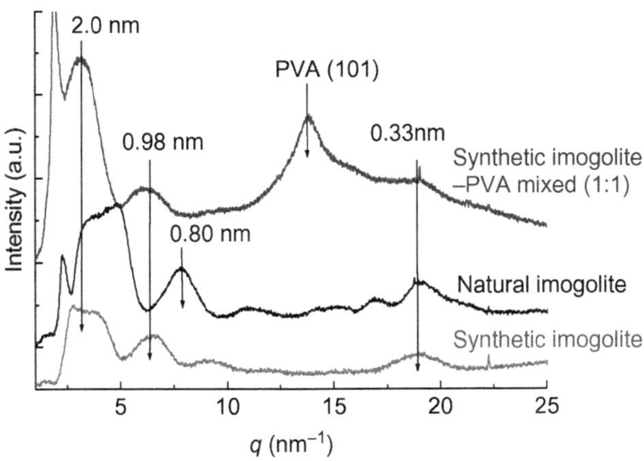

FIG. 24.18 WAXD profiles of natural and synthesized imogolite and in situ synthesized imogolite–PVA hybrids. *Reprinted with permission from Yamamoto et al. (2005a). © 2005, Royal Society of Chemistry.*

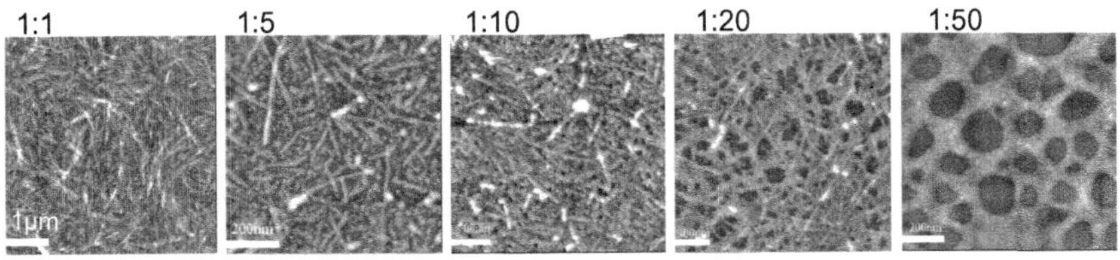

FIG. 24.19 Cyclic contact mode AFM height images of in situ synthesized imogolite–PVA hybrid film. Imogolite:PVA (w/w) = 1:1, 1:5, 1:10, 1:20 and 1:50. *Reprinted with permission from Yamamoto et al. (2005b). © 2005, Royal Society of Chemistry.*

observed at 0.45 nm. X-ray reflections for the synthetic imogolite were observed at a lower *q* region than that for natural imogolite, indicating that the diameter of the synthetic imogolite is slightly larger than that of the natural imogolite (Farmer et al., 1983). In the imogolite–PVA hybrids, the characteristic reflections at *d*-spacings of 2.0, 0.98, 0.33 nm and PVA's reflection were clearly observed. The absence of X-ray reflections from by-products suggests that imogolite can be synthesized in situ in PVA aqueous solution.

The morphology of in situ synthesized imogolite in the polymer matrix was observed by the cyclic contact mode AFM observation (Fig. 24.19). The sample for AFM observation was prepared by spin-coating of a 1 mass% dispersed aqueous solution of imogolite–PVA hybrid at 3000 rpm for 1 min on a silicon wafer, and the obtained film was then vacuum-dried for 12 h. The brighter areas in the cyclic contact mode AFM image correspond to the higher-height regions. The AFM images, except for the mass ratio of 1:50 for imogolite/PVA, showed the fibre morphology, which indicates formation of the imogolite bundles. The average length of the synthesized imogolite fibre prepared by the in situ method was several hundred nanometers, which is shorter than the length of typical natural and synthesized imogolite fibres. The AFM image of the imogolite/PVA mass ratio of 1:50 (Fig. 24.19E) showed no evidence of imogolite fibres. These results indicate that the PVA inhibits imogolite growth. The important factors in the formation of imogolite nanofibres are the pH, OH/Al, Si/Al molar ratio, temperature and additives in the solution (Wada et al., 1979; Inoue and Huang, 1986). In particular, additives such as organic ligands, which can interact with the aluminium ions, play a decisive role in the formation and nature of short-range ordered aluminosilicate (Inoue and Huang, 1986). The length of the imogolite nanofibres estimated from the AFM images were 689 ± 47 nm (imogolite:PVA = 1:1), 271 ± 21 nm (1:5), 224 ± 18 nm (1:10) and $178 \pm 9{:}3$ nm (1:20), respectively. In contrast, the length of natural imogolite nanofibre was several micrometres (Inoue and Huang, 1986). Therefore, the imogolite nanofibre growth in the axial direction might be impeded by the presence of PVA due to the interaction between imogolite and PVA. This intermolecular interaction was also indicated by the lyotropic mesophase in an imogolite and PVA mixture (Hoshino et al., 1992). Therefore, there is a critical concentration of PVA for the successful formation of synthesized imogolite.

Optical properties of hybrid films were measured by light transmission and haze value measurements. The light transmittance in the imogolite–PVA hybrid and blend films and PVA films is shown in Fig. 24.20. Imogolite–PVA

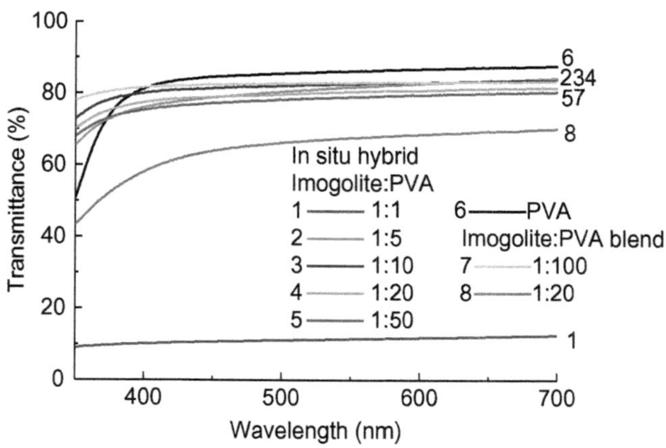

FIG. 24.20 Light transmittance of in situ synthesized imogolite–PVA hybrid, imogolite/PVA blend and PVA film. The magnitude of transmittance was normalized by the film thickness.

hybrid films prepared by the in situ synthesis method were optically transparent, and thus comparable to the imogolite/PVA blend films with the same quantity of freeze-dried imogolite powder. It appears to be difficult to redisperse the freeze-dried imogolite powder since imogolite fibres aggregate during the air-drying or freeze-drying process. However, imogolite nanofibres that were built up via in situ synthesis in a polymer solution exhibit fine dispersion because the adsorbed PVA chains on imogolite fibres during the synthetic process inhibit their aggregation. The imogolite–PVA hybrid films that contain large amounts of imogolite, up to a mass ratio of 1:5 (imogolite/PVA), could maintain fine optical transparency, not only because of the dispersibility of synthetic imogolite nanofibres in the PVA matrix, but also because of the crystallinity and crystallite size reduction of PVA crystallites due to the strong interaction between imogolite and PVA.

Table 24.4 summarizes the haze value of imogolite–PVA hybrid films. The appearance of in situ synthesized imogolite–PVA hybrid and blend films with the same imogolite content, which is imogolite:PVA = 1:20, is shown in Fig. 24.21. The haze value of the synthesized imogolite/PVA hybrid films was extremely low; in contrast, that of the imogolite/PVA blend films was high. The high haze value means that there are objects that scatter incident light. The aggregated imogolite fibres in the polymer matrix would be a main scattering bodies in the imogolite/PVA blend films. The pure PVA films shows slight haze originated from the microcrystallites. Light transmittance and haze measurements revealed that the in situ synthetic imogolite–PVA hybrids accomplish fine dispersion state compared with the film prepared by the simple blending of freeze-dried imogolite and PVA.

The temperature dependence of the dynamic storage modulus, E', of PVA, imogolite–PVA hybrid and blend films is shown in the top part of Fig. 24.22.

TABLE 24.4 Haze Value of the In Situ Synthesized Imogolite–PVA Hybrid, Imogolite/PVA Blend and PVA Film

Sample (mass ratio)	Haze value (%)
PVA	1.59±0.19
Imogolite:PVA 1:20	1.00±0.08
Imogolite:PVA 1:10	0.57±0.02
Imogolite:PVA 1:5	2.55±0.08
Imogolite:PVA 1:20[a]	29.94±3.60

[a]*Blend film.*

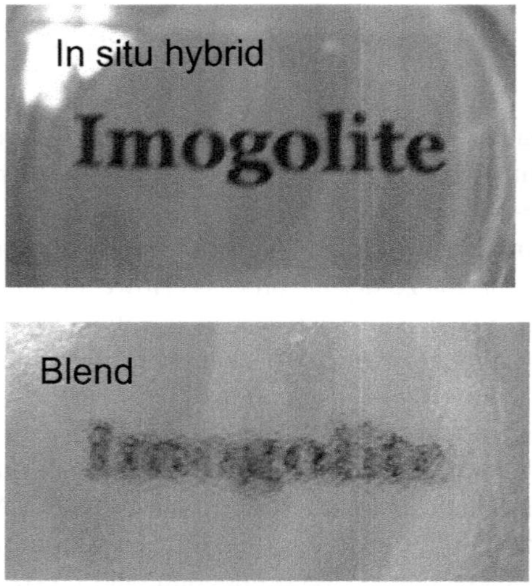

FIG. 24.21 Transparency of in situ synthesized imogolite–PVA hybrid and blend film with imogolite:PVA = 1:20. The thickness of the film is ca. 100 μm.

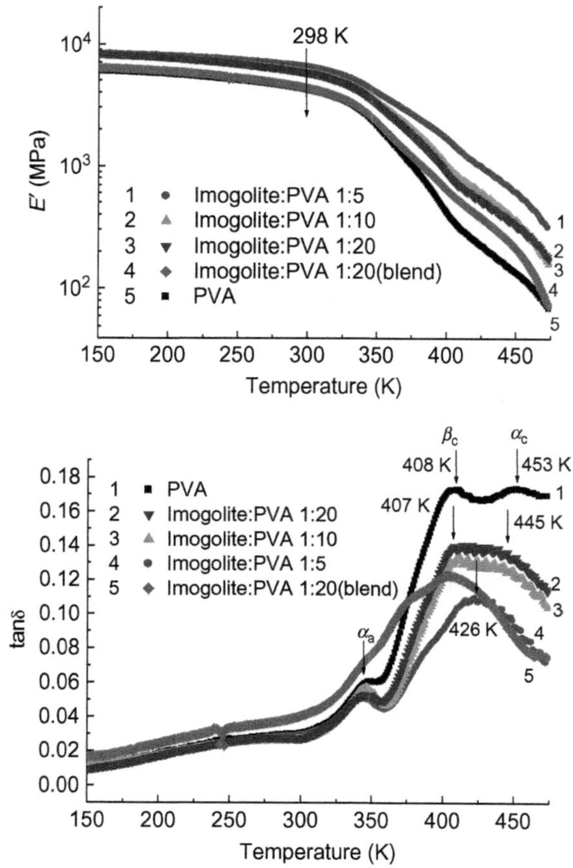

FIG. 24.22 Temperature dependence of storage modulus (E') and loss tangent (tanδ) of in situ synthesized imogolite–PVA hybrid, imogolite/PVA blend and PVA film. *Reprinted with permission from Yamamoto et al. (2005b). © 2005, Royal Society of Chemistry.*

TABLE 24.5 Room Temperature Modulus and Heat Distortion Temperature of In Situ Synthesized Imogolite–PVA Hybrid, Imogolite/PVA Blend and PVA Film Estimated from Dynamic Viscoelasticity Measurement

Sample	E' (GPa)[a]	HDT (K)
PVA	4.2	374
Imogolite:PVA 1:20	5.8	383
Imogolite:PVA 1:10	5.8	388
Imogolite:PVA 1:5	6.6	402
Imogolite:PVA 1:20[b]	4.3	379

[a]*Evaluated at 298 K.*
[b]*Blend film.*

Table 24.5 summarizes the modulus at room temperature and the heat distortion temperature (HDT) of the hybrid and blend films. The HDT value was estimated from the storage modulus data by determining the temperature at which the storage modulus drops to 25% of its value at room temperature (Shelley et al., 2001). E' at room temperature and HDT of the in situ synthesized imogolite–PVA hybrid films were found to be higher than that of the PVA film. Furthermore, E' and HDT values increased with imogolite content in the polymer matrix. In contrast, no change was observed in E' curves of the imogolite/PVA blend film and the PVA film. The temperature dependence of the mechanical loss tangent, tanδ, of imogolite/PVA hybrid films is shown on the bottom part of Fig. 24.22. Compared with the PVA film, changes in the α_c and β_c absorption peaks were observed in the in situ imogolite–PVA hybrid films and the imogolite/PVA blend films. Both the α_c and β_c peaks can be attributed to motions within the crystalline phase of PVA, while the α_c peak is assigned to translational modes of chain segments along the chain axis (McCrum et al., 1967). In contrast, the β_c peak is ascribed to an anharmonic twisting vibration around the chain axis. That is, the change in the α_c and β_c peaks can be attributed to the contrast in the molecular aggregation state in the crystalline phase of PVA (McCrum et al., 1967). The contrast in the crystalline aggregation states was also observed in the crystallinity.

DSC thermograms revealed the decrease in the heat of fusion and the melting temperature of the PVA crystal in the hybrid films (Fig. 24.23). Table 24.6 summarizes the melting temperature and crystallinity, calculated from the experimentally obtained heat of fusion, which was estimated from the value of pure PVA crystal as a standard (Tanigami et al., 1999). The magnitude of the heat of fusion of PVA ideal crystal was reported as $156\,J\,g^{-1}$. The crystallinity was evaluated under consideration of the mass fraction of

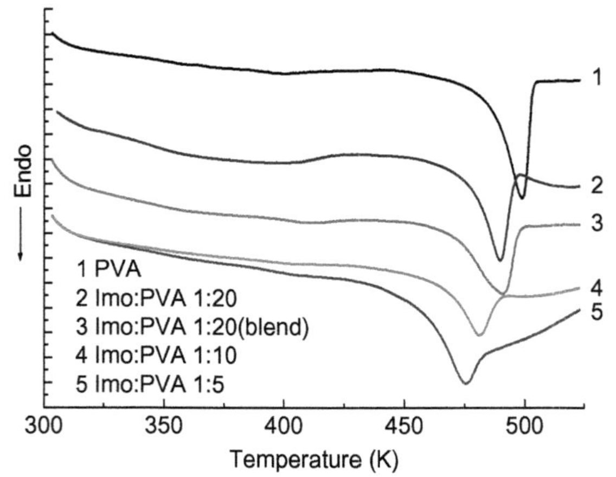

FIG. 24.23 DSC curves of the in situ synthesized imogolite–PVA hybrid, imogolite/PVA blend and PVA film. *Reprinted with permission from Yamamoto et al. (2005b). © 2005, Royal Society of Chemistry.*

TABLE 24.6 Crystallinity and Melting Temperature of PVA Calculated from Heat of Fusion for the In Situ Synthesized Imogolite–PVA Hybrid, Imogolite/PVA Blend and PVA Film

Sample	Crystallinity (%)	T_m (K)
PVA	50.9±5.8	487.4±0.3
Imogolite:PVA 1:20	32.2±3.1	476.9±1.3
Imogolite:PVA 1:10	27.8±1.8	470.9±1.4
Imogolite:PVA 1:5	23.2±1.8	467.2±2.3
Imogolite:PVA 1:20[a]	33.3±6.8	471.2±8.4

[a]*Blend film.*

PVA in the hybrids. The PVA crystallinity reduced in the imogolite–PVA hybrid film compared with the PVA film. Generally, the lower the crystallinity of polymer materials, the higher the intensity of α_a-absorption in tanδ curves (α_a-absorption is assigned to micro-Brownian motions of long chain segments in the amorphous regions of the polymer) (McCrum et al., 1967). However, even though the PVA crystallinity decreased in the in situ imogolite–PVA hybrid film, the α_a-absorption of the hybrid films was similar to that of the PVA films (Fig. 24.22). The growth of unstable PVA crystals that do not contribute to the α_a-absorption process would exist in the imogolite–PVA hybrid film because of the strong interaction between the hydroxyl groups of PVA and the hydrophilic imogolite surface. The imogolite–PVA hybrid prepared by the in situ method exhibited excellent mechanical properties due to the fine dispersion of imogolite in the polymer matrix. The improvement of mechanical properties of crystalline PVA films despite the decrease in crystallinity suggests the efficient reinforcing effect of imogolite nanotubes against the PVA matrix.

24.5 IMOGOLITE BIOPOLYMER NANOCOMPOSITES

Since imogolite is a natural hydrogel, it can be utilized for the immobilization of biological components. This section introduces the hybrid of imogolite and enzyme having phosphoric acid groups interacting with the imogolite surface (Inoue et al., 2006). By utilizing the immobilization of the enzyme that has a phosphoric acid group, it is possible to achieve high loading onto the imogolite fibre. Furthermore, immobilized enzyme activity can be expected to be maintained after the formation of the hybrid hydrogel because imogolite can build up a three-dimensional network structure in aqueous solution (Philips and Wierenga, 1998). Preparation of a hybrid hydrogel via immobilization of pepsin as a model enzyme with a phosphoric acid group and evaluation of its aggregation structure and enzyme activity was reported by Inoue et al. (2006).

The imogolite–pepsin hybrid hydrogel was prepared by adding an acidic (pH 3.1) pepsin solution (1 mg/mL) to an acidic (pH 3.1) imogolite solution (0.5 mg/mL). The mixture was incubated at 310 K by shaking at 120 rpm for 4 h. After that, it was centrifuged at 3000 rpm for 15 min and rinsed three times with pH 3.1 acetic acid solution to obtain the hybrid hydrogel. A photograph and a schematic representation of the network structure of the hybrid hydrogel are shown in Fig. 24.24. Pepsin has a phosphoric group at

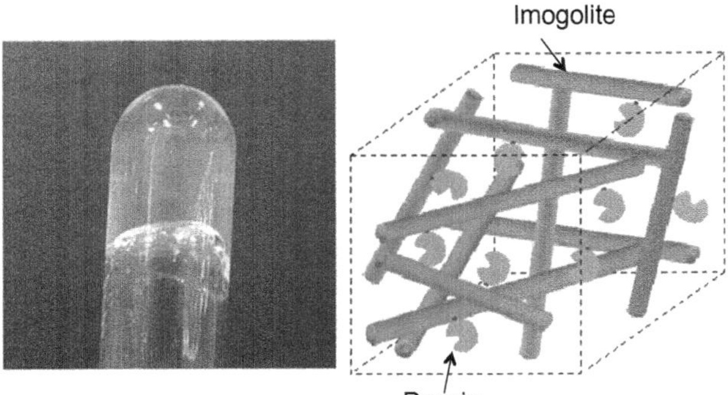

FIG. 24.24 Photograph and schematic representation of imogolite–pepsin hybrid hydrogel.

the serine of the 68th residue (Tang and Hartley, 1970; Sielecki et al., 1990). The Al–OH groups of imogolite are positively charged and dispersed under acidic conditions (pH below 5) in aqueous solution by the electrostatic repulsion among them, and pepsin is negatively charged at pH 3.1 because its isoelectric point is 1.0 (Tiselius et al., 1938). This hybrid hydrogel was formed due to the electrostatic interaction between imogolite and pepsin, as well as the interaction between Al–OH groups of the imogolite surface and the phosphoric group of pepsin.

The immobilization of pepsin was confirmed by IR measurement using the KBr method. IR spectra of imogolite, pepsin and the imogolite–pepsin hybrid hydrogel are shown in Fig. 24.25. The absorption peaks at 995 and 940 cm^{-1} can be attributed to the stretching vibration of Si–O–Al in imogolite, and those at 1647 and 1537 cm^{-1} to amide I and II bands in pepsin, respectively (Forato et al., 1998). The immobilization of pepsin by the interaction between pepsin and imogolite was indicated, since the absorption peaks of pepsin and imogolite were observed simultaneously and the intensity ratio of absorption peaks from imogolite at 995 and 940 cm^{-1} was changed before and after immobilization. The absorption peaks from amide I and II bands before and after immobilization suggest that pepsin was not denatured. The amount of immobilized pepsin was estimated by TGA measurement. The sample for TGA measurement was prepared by heating the freeze-dried hybrid hydrogel in toluene at 357 K, which is the azeotropic temperature between H_2O and toluene mixture, and residual toluene was removed by vacuum drying. The

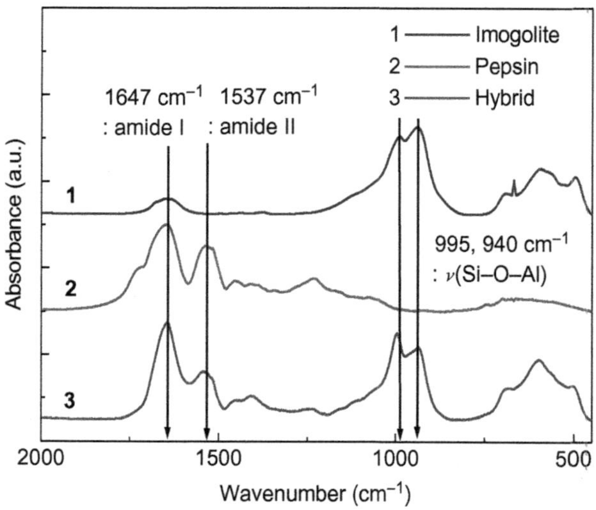

FIG. 24.25 FT-IR spectra of (1) imogolite, (2) pepsin and (3) imogolite–pepsin hybrid hydrogel.

maximum value was approximately 1.8 mg of pepsin per 1 mg of imogolite. Assuming a single-fibre dispersion of imogolite, this maximum value indicates that ca. 71% of the imogolite surface is covered with pepsin.

Confocal laser scanning microscopy (CLSM) and field-emission scanning electron microscopy (FE-SEM) were used to evaluate the dispersion state of pepsin in the hybrid hydrogel and the network structure of the hybrid hydrogel. The sample for CLSM observation was prepared with a hybrid hydrogel composed of prelabelled imogolite and pepsin. Imogolite and pepsin were labelled with fluorescein 5-isothiocyanate and adenosine 50-triphosphate, Alexa Fluor® 647 20-(or-30)-*O*-(*N*-(2-aminoethyl)-urethane) and hexa (triethylammonium) salt, respectively. For FE-SEM observation, the hybrid hydrogel was dehydrated by freeze drying and the specimen was prepared by precoating dehydrated hybrid hydrogel with OsO_4. Pepsin was found to be finely dispersed in the hybrid hydrogel, and the fluorescent images of pepsin and imogolite showed a similar morphology due to the immobilization of pepsin onto the imogolite surface. The FE-SEM image of the hybrid hydrogel with 99.7% water content is displayed in Fig. 24.26. The three-dimensional network structure of the hybrid hydrogel was observed. The average pore size of the hybrid hydrogel was 108 nm.

The enzyme activity of immobilized pepsin in the hybrid hydrogel was evaluated based on hydrolysis of haemoglobin at pH 3:1 (Fig. 24.27). The hybrid hydrogel and free-pepsin solution was mixed with 2.5 mL of 2.5 mass% haemoglobin solution and incubated at 310 K for 10 min. Next, 5 mL of 5 mass% trichloroacetic acid was added to the mixture and the reaction mixture was incubated at 310 K for 1 h. After centrifugation and filtration, the filtrate products were measured by UV–vis spectroscopy to estimate the reaction rate (ΔA280/min). This rate shows the change in absorbance at 280 nm; ΔA280/min of free pepsin was 0.183, and that of immobilized pepsin in the hybrid hydrogel was 0.048. Therefore, immobilized pepsin in the hybrid hydrogel retained ca. 26% of its enzyme activity compared with free pepsin

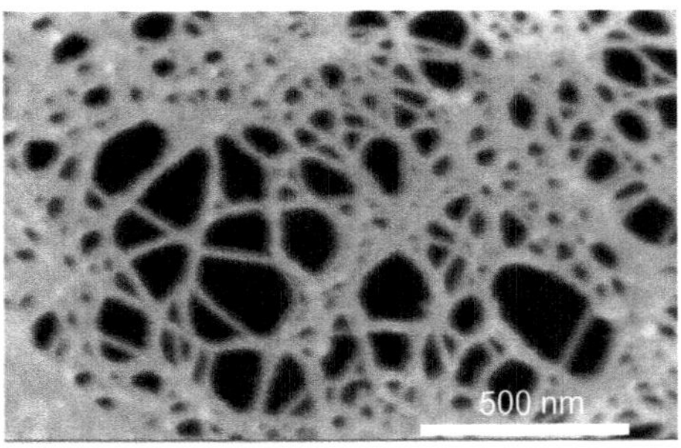

FIG. 24.26 FE-SEM image of imogolite–pepsin hybrid hydrogel.

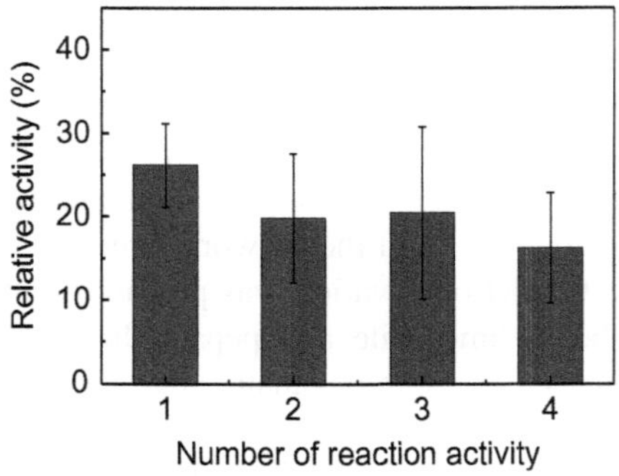

FIG. 24.27 The enzyme activity of immobilized pepsin in imogolite–pepsin hybrid hydrogel during repeated reactions at 410 K. *Reprinted with permission from Inoue et al. (2006).* © *2006, The Chemical Society of Japan.*

in aqueous solution. The enzyme activity of immobilized pepsin apparently decreased. Similar behaviour has been observed for the enzyme immobilized at the inner surface of halloysite (Shchukin et al., 2005). This decrease could be ascribed to the slow diffusion of substrate into the network structure of the hybrid hydrogel, as well as inhibition of the diffusion of substrate to the active site of immobilized pepsin due to the steric hindrance of the imogolite network. In the case of pepsin immobilized in the hybrid hydrogel, it can be easily recovered from the reaction system and can repeatedly react with the substrate. The change in enzyme activity was investigated in the repeated reaction. The enzyme activity of immobilized pepsin was maintained after four reaction cycles, with a slight decrease in activity as the number of reactions increased.

DNA is a very important biomolecule containing life information, and has also been studied for disease therapy. Free DNA is rapidly degraded, while DNA bound onto solid particles appears to be resistant to degradation (Alvarez et al., 1998). Therefore, an imogolite–DNA hybrid gel may have the potential to act as a reservoir for DNA storage and even control its release (Jiravanichanun et al., 2012). The driving force for the gel formation is attributed to the strong interaction between the phosphate groups on the outside of the DNA double helix and the aluminol groups on the imogolite surface.

The imogolite–DNA hybrid gels were prepared by mixing 1 mg/mL solutions of imogolite and DNA, which were adjusted to pH 4 by adding diluted HCl (Fig. 24.28). A series of hybrid hydrogels were prepared by varying the DNA/imogolite feed ratio, as given in Table 24.7. Adsorption of DNA on imogolite was immediately observed as a visual viscosity change after the solutions were mixed. The mixture was further shaken at 37 °C overnight. Then, all the mixtures were centrifuged to separate the hybrid gels and

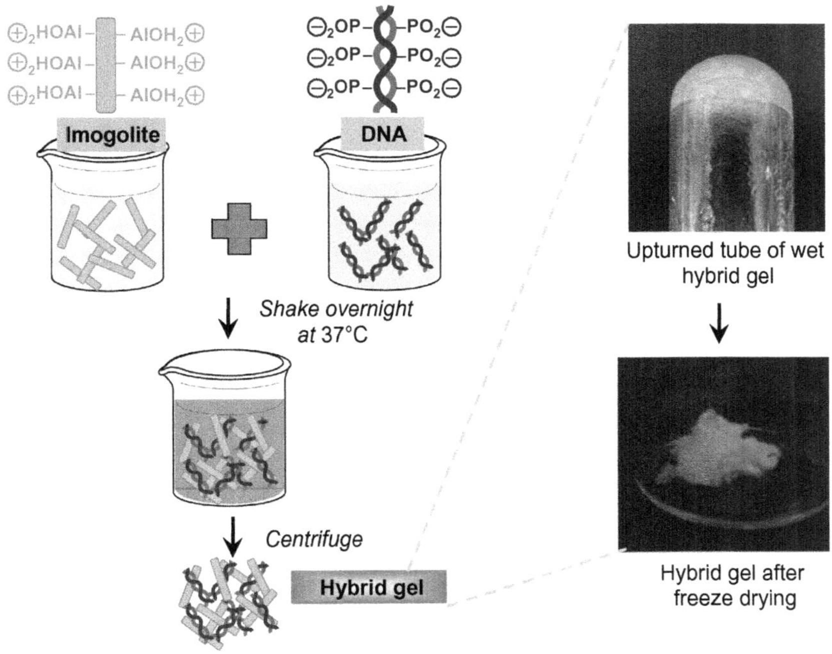

FIG. 24.28 Schematic representation for the preparation of imogolite–DNA hybrid gel. *Reprinted with permission from Jiravanichanun et al. (2012). © 2012, American Chemical Society.*

TABLE 24.7 Concentration of Imogolite and DNA for Hybrid Gel Preparation

Gel no.	1	2	3	4	5	6	7	8
DNA content (mass%)	9	17	29	38	50	63	71	83
Imogolite content (mass%)	91	83	71	62	50	37	29	17

supernatants. The collected hybrid gels were washed with water three times until no more DNA was desorbed in the supernatants.

The FT-IR spectra of a series of the hybrid gels with varied DNA/imogolite feed ratio are shown in Fig. 24.29. The characteristic peaks of imogolite were observed at 930 and 960 cm^{-1}, corresponding to the stretching of Al–O–Si groups. The prominent bands at 1090, 1236 and 1700 cm^{-1}

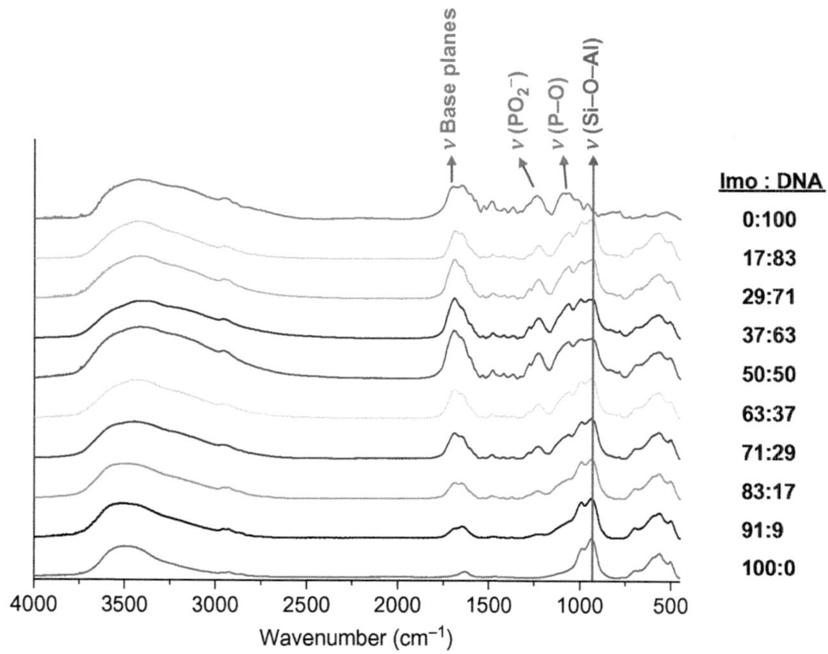

FIG. 24.29 FT-IR spectra of imogolite, DNA and imogolite–DNA hybrid gels.

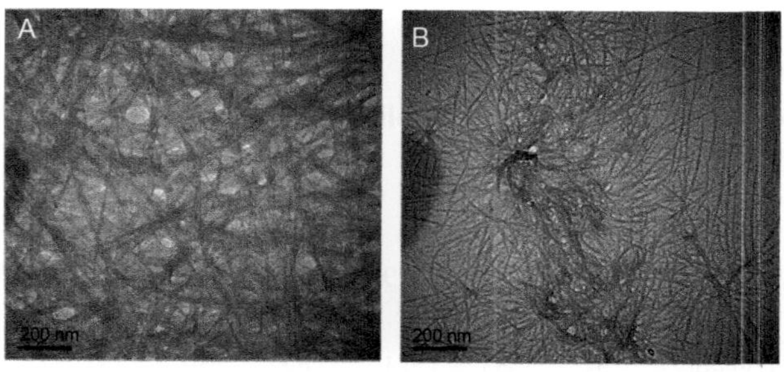

FIG. 24.30 TEM images of (A) imogolite and (B) imogolite–DNA hybrid gel. *Reprinted with permission from Jiravanichanun et al. (2012). © 2012, American Chemical Society.*

correspond to PO_2^- and P–O groups in DNA. These results indicate the existence of imogolite and DNA in the hybrid gels. When DNA feed content increased, there was an increase in the intensity of P–O and PO_2^- stretching bands. The FT-IR spectra provided an average DNA amount in the hybrid gels.

The morphology of freeze-dried imogolite and the imogolite–DNA hybrid gel was observed by TEM (Fig. 24.30). Imogolite bundles build a continuous network structure (Fig. 24.30A). Interestingly, imogolite presents a different morphology in the hybrid gels (Fig. 24.30B). The TEM image of the hybrid gel showed two characteristics of imogolite bundles: the fibrous morphology of imogolite bundles and an entanglement between imogolite fibres. A good dispersion of imogolite fibres is suggested so that DNA molecules wrap around a single imogolite; this leads to better fibre dispersion than that of pure imogolite. When DNA molecules bridged between imogolite fibres, imogolite fibres were curled and entangled, forming weblike structures. A good affinity of Al–OH groups of imogolite and phosphate groups of DNA is a driving force for DNA adsorption on imogolite. There was no significant difference in morphology for all hybrid gels.

WAXD measurement of freeze-dried imogolite, DNA and imogolite–DNA hybrid gels was carried out by a powder diffractometer at BL02B2 in SPring-8. Fig. 24.31 shows WAXD profiles of freeze-dried hybrid hydrogels. The WAXD pattern of DNA showed single diffuse diffraction at a *d* spacing of 5–30 nm. Imogolite exhibited reflections at *d* spacings of 1.53, 0.92 and 0.66 nm assigned to the diffraction from individual tubes

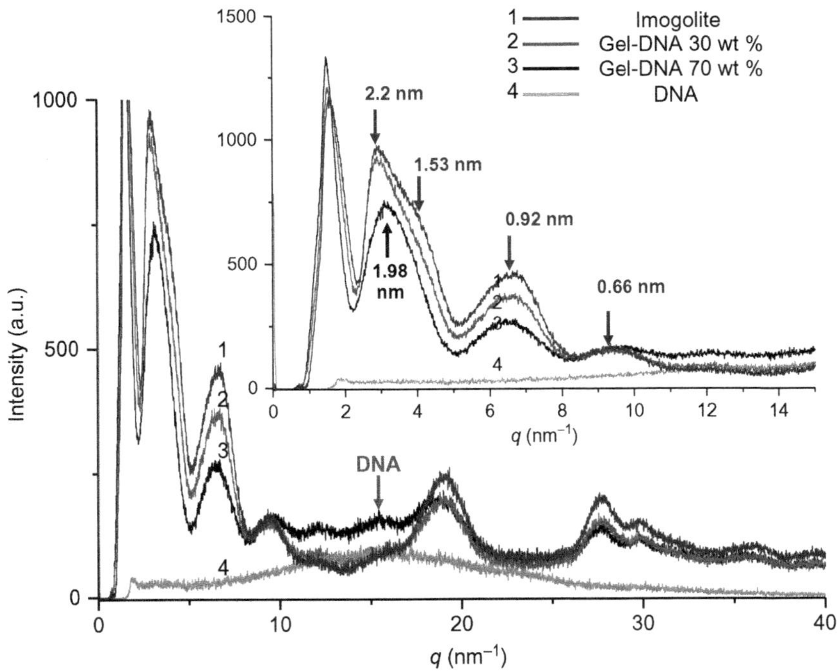

FIG. 24.31 WAXD profiles of freeze-dried imogolite, DNA and imogolite–DNA hybrid gels. *Reprinted with permission from Jiravanichanun et al. (2012). © 2012, American Chemical Society.*

(Wada et al., 1979) and a reflection at 2.2 nm associated with the centre-to-centre tube distance of imogolite bundles (Farmer et al., 1983). Two hybrid gels showed a WAXD pattern similar to imogolite, while there are some differences at $q=2$–5 and 11–$18\ nm^{-1}$. The shift of peak at the $q=2$–$5\ nm^{-1}$ region might suggest distortion of imogolite packing upon binding by DNA. This result agreed with TEM images. The high intensity of the peak at $q=11$–$18\ nm^{-1}$ indicated the difference of DNA content in the hybrid gels. The gel 7 with DNA feed content of 70 mass% clearly showed high intensity compared with other patterns.

The residues in supernatant represent the fraction of unbound DNA. The DNA/imogolite feed ratio dependence on the fraction of DNA in supernatants is shown in Fig. 24.32 [blue (dark grey in the print version) plots]. When DNA feed content was less than 50 mass%, DNA molecules were completely immobilized on imogolite to form the hybrid gels. In contrast, when DNA feed content was greater than 50 mass%, the amount of unbound DNA significantly increases. The yield of the imogolite–DNA hybrid gels is shown in Fig. 24.32 [green (grey in the print version) plots]. The amount of the obtained hybrid gels increased as the DNA feed content increased. The maximum amount of the hybrid gel was obtained at a DNA/imogolite feed ratio of 63/37 (gel 6). Above this point, the amount of obtained gels decreased because the amount of imogolite decreased.

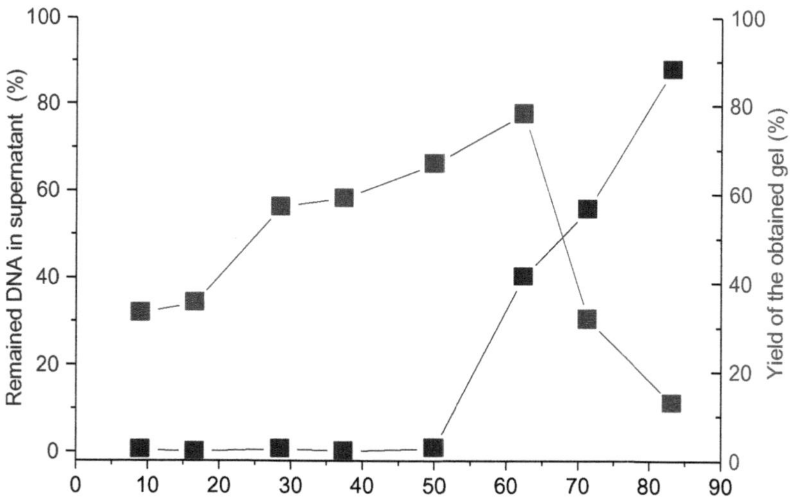

FIG. 24.32 Amount of DNA remaining in supernatant [*blue* (*dark grey* in the print version) *plots*] and yield of the obtained hybrid gels [*green* (*grey* in the print version) *plots*] as a function of DNA feed content. *Reprinted with permission from Jiravanichanun et al. (2012). © 2012, American Chemical Society.*

The amount of bound DNA on imogolite was estimated. A known mass of gels was dissolved in concentrated HCl for 1 week, neutralized with saturated $NaHCO_3$, integrated volume, and then the DNA amount was measured by UV–vis spectroscopy. The DNA content (mass%) in the hybrid gels was calculated by dividing the DNA amount in HCl solution by the initial mass of DNA. The bound DNA into imogolite in the hybrid gel and the UV–vis spectra of DNA adsorption for each hybrid gel are shown in Fig. 24.33. An increase in the DNA feed content led to an increase of DNA content in the hybrid gels. Gel 7 showed the highest DNA content in the hybrid gel (75%), which was derived from a DNA/imogolite feed ratio of 71/29. An imogolite–DNA mass ratio at 75/25 could be calculated for a 1:1 molar ratio. Above this point, DNA content decreased, which may be caused by electrostatic repulsion generated by the excessive negative charges of DNA.

The frequency dependence of dynamic viscoelasticity of the hybrid gels is shown in Fig. 24.34. Gels 1 and 3 were gel-like and highly viscous and did not fall down even after the vials were overturned. The lower modulus in gel 1 compared to gel 3 is probably due to the lower density of the cross-linking point in the hybrid network. Gels 5 and 7 were fluidlike and had a lower viscosity than gels 1 and 3 because of the incompleteness of the network structure. Photographs of the gel surface (upside view), and the smoothness of gels was arranged from rough to smooth: gel 3 > gel 1 > gel 5 > gel 7 (Fig. 24.35). The DNA content in the hybrid gels influenced the gels' viscosity and other physical properties. The aluminol groups on the external surface of imogolite can be both positively and negatively charged depending on the pH. Gels 1 and 3 have high imogolite content and high viscosity due to the entanglement of imogolite themselves and imogolite with DNA molecules. Gels 5 and 7, which have high DNA content and coverage of the

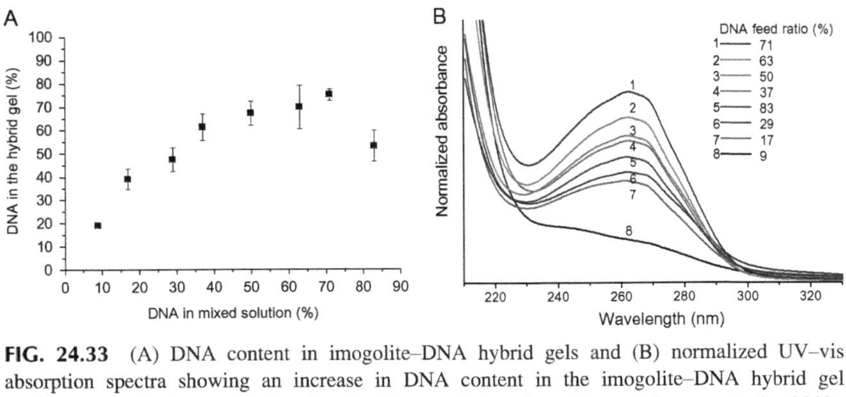

FIG. 24.33 (A) DNA content in imogolite–DNA hybrid gels and (B) normalized UV–vis absorption spectra showing an increase in DNA content in the imogolite–DNA hybrid gel against DNA feed content. *Reprinted with permission from Jiravanichanun et al. (2012). © 2012, American Chemical Society.*

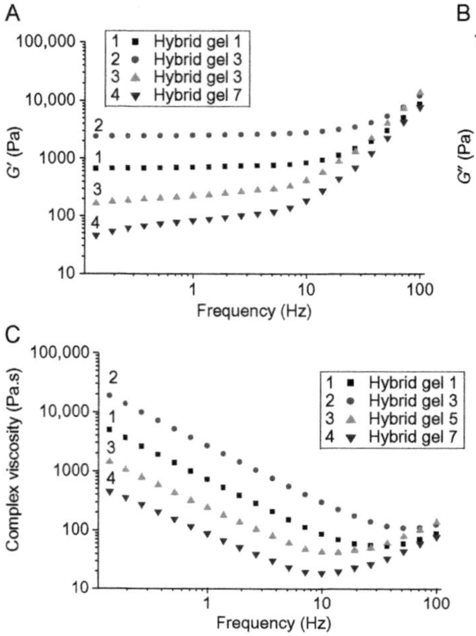

FIG. 24.34 (A) Storage, (B) loss moduli and (C) complex viscosity (G', G'', and η^*, respectively) as a function of frequency for imogolite–DNA hybrid gels 1 (*black square*), 3 [*red circle* (*grey* in the print version)], 5 [*green triangle* (*light grey* in the print version)] and 7 [*blue inverted triangle* (*dark grey* in the print version)]. *DNA/imogolite feed ratio for each gel is shown in Table 24.7. Reprinted with permission from Jiravanichanun et al. (2012). © 2012, American Chemical Society.*

imogolite–DNA network surface by DNA molecules, induced repulsion between small aggregates and led to low viscous liquidlike gels.

The surface charge of imogolite depends on pH and has point of zero charge (PZC) of about pH 6 (Tsuchida et al., 2005). Theoretically, DNA would be released from imogolite at pH values higher than PZC because Al–OH_2 groups are deprotonated. DNA was not released from the hybrid gel at pH 4 because the surface of imogolite was positively charged. DNA release in the buffer with pH of 7, 7.5, 8, 9, 10 and 11 is monitored for 24 h (Fig. 24.36A), and the DNA release amount increased as pH increased. However, the DNA release amount was only 12% even at pH 11. This suggests that DNA exhibits a strong interaction with imogolite. The amount of released DNA decreased as NaCl concentration increased to 0.1 M (Fig. 24.36B). The results are the opposite of the salt-induced desorption of DNA complexes with polycations (Dubas and Schienoff, 2001). The adsorption of DNA on imogolite protected DNA from attack by salt ions. The amount of released DNA decreased at lower temperatures (Fig. 24.36C). As temperature decreased, thermal fluctuations continuously reduced; therefore,

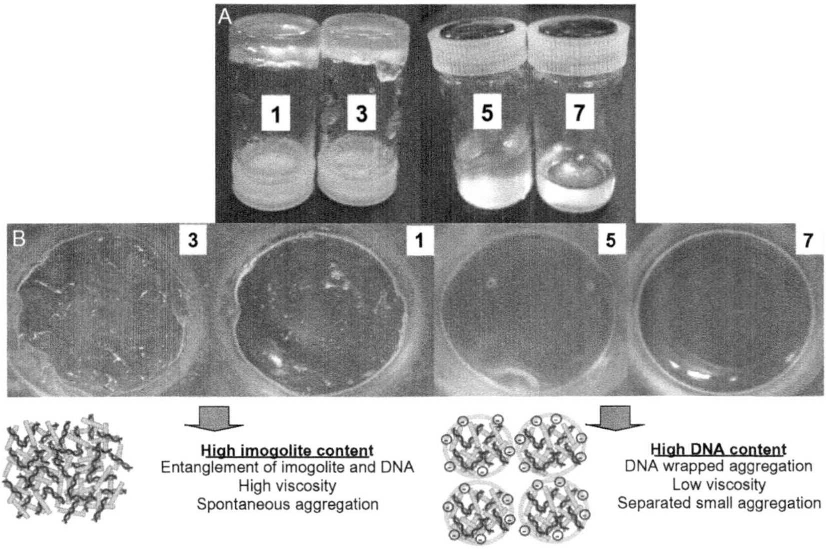

FIG. 24.35 Photographs of imogolite–DNA hybrid gels 1, 3, 5 and 7: (A) *front view* and (B) *up view* and their model structures. DNA/imogolite feed ratio for each gel is shown in Table 24.7. *Reprinted with permission from Jiravanichanun et al. (2012). © 2012, American Chemical Society.*

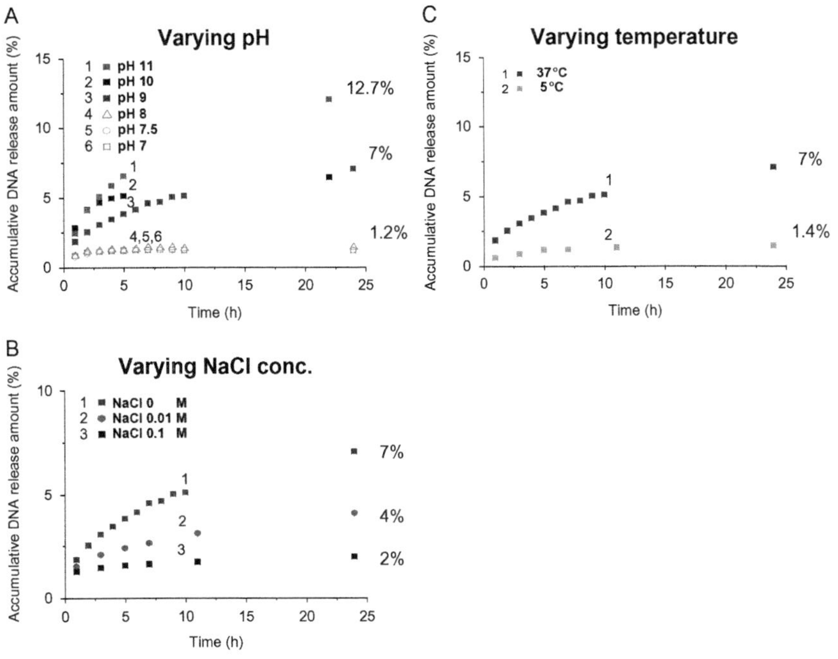

FIG. 24.36 DNA-releasing behaviour from imogolite–DNA hybrid gel at various (A) pH values, (B) NaCl concentrations and (C) temperatures.

DNA motion decreased, leading to lower amounts of DNA release. Under severe conditions, such as strong alkaline solutions or NaCl solutions, the hybridization with imogolite provided enough protection to DNA molecules. This is probably due to the very strong interactions, such as electrostatic interaction between aluminol groups of imogolite and phosphate groups of DNA. The other reason for this is that an entanglement of DNA molecules and imogolite fibres provided physical protection from the added salt or alkaline solution.

24.6 CONCLUDING REMARKS

Their unique structure and positively charged external surfaces are the driving forces for imogolite clay nanotubes emerging as one of the most promising building blocks for various organic–inorganic nanohybrids. Several approaches for dispersing imogolite nanotubes into synthetic matrices, including both hydrophobic and hydrophilic polymers, to prepare nanocomposites were introduced in this chapter. Imogolite polymer nanocomposites based on various composite designs are illustrated in Fig. 24.37. Due to their good dispersion, as well as the transparency feature of imogolite itself, transparent imogolite–polymer nanocomposites were successfully prepared. Mechanical properties of the original polymers were improved by interaction with imogolite. The capability of imogolite for the gel formation of biomolecules was demonstrated by fabricating imogolite-based enzymes and DNA hybrid gels. The amount of biomolecules in the hybrid gels was found to be very high because of the large surface areas of the imogolite nanotubes and the strong interactions between the biomolecules and the imogolite surfaces.

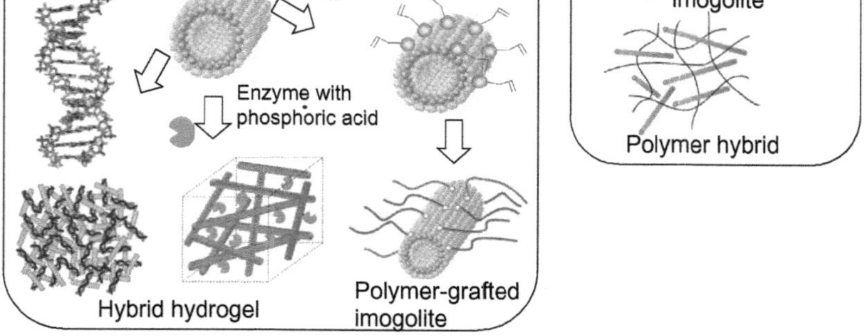

FIG. 24.37 Summary of imogolite–polymer nanocomposites based on various CPN designs.

REFERENCES

Alvarez, A.J., Khanna, M., Toranzos, G.A., Stotzky, G., 1998. Amplification of DNA bound on clay minerals. Mol. Ecol. 7, 775–778.

Ando, H., Nakahama, M., Yamamoto, M., Itoh, K., Suzuki, M., 1996. Infrared reflection absorption spectroscopic study on self-assembling processes of (methacryloyloxy)alkyl dihydrogen phosphates on evaporated silver films. Langmuir 12, 6399–6403.

Barbarella, G., Pudova, O., Arbizzani, C., Mastragostino, M., Bongini, A., 1998a. Oligothiophene-S,S-dioxides: a new class of thiophene-based materials. J. Org. Chem. 63, 1742–1745.

Barbarella, G., Favaretto, L., Sotgiu, G., Zambianchi, M., Antolini, L., Pudova, O., Bongini, A., 1998b. Oligothiophene S,S-dioxides. Synthesis and electronic properties in relation to the parent oligothiophenes. J. Org. Chem. 63, 5497–5506.

Barbarella, G., Zambianchi, M., Pudova, O., Paladini, V., Ventola, A., Cipriani, F., Gigli, G., Cingolani, R., Citro, G., 2001. Oligothiophene isothiocyanates as a new class of fluorescent markers for biopolymers. J. Am. Chem. Soc. 123, 11600–11607.

Cradwick, P.D., Wada, K., Russell, J.D., Yoshinaga, N., Masson, C.R., Farmer, V.C., 1972. Imogolite, a hydrated aluminium silicate of tubular structure. Nat. Phys. Sci. 240, 187–189.

Donkai, N., Miyamoto, T., Kubota, T., Tanei, H., 1992. Preparation of transparent mullitesilica film by heat-treatment of imogolite. J. Mater. Sci. 27, 6193–6196.

Dubas, S.T., Schienoff, J.B., 2001. Polyelectrolyte multilayers containing a weak polyacid: construction and deconstruction. Macromolecules 34, 3736–3740.

Farmer, V.C., Fraser, A.R., Tait, J.M., 1977. Synthesis of imogolite: a tubular aluminium silicate polymer. J. Chem. Soc., Chem. Commun. (13), 462–463.

Farmer, V.C., Adams, M.J., Fraser, A.R., Palmieri, F., 1983. Synthetic imogolite; properties, synthesis, and possible applications. Clay Miner. 18, 459–472.

Forato, L.A., Bernardes-Filho, R., Colnago, L.A., 1998. Protein structure in KBr pellets by infrared spectroscopy. Anal. Biochem. 259, 136–141.

Guimares, A., Tnyashin, A.E., Frenzel, J., Heine, T., Duarte, H.A., Seifert, G., 2007. Imogolite nanotubes: stability, electronic, and mechanical properties. ACS Nano 1, 362–368.

Gustafsson, J.P., 2001. The surface chemistry of imogolite. Clays Clay Miner. 49, 73–80.

Hoshino, H., Ito, T., Donkai, N., Urakawa, H., Kajiwara, K., 1992. Lyotropic mesophase formation in PVA/imogolite mixture. Polym. Bull. 29, 453–460.

Iijima, S., Ichihashi, T., 1993. Single-shell carbon nanotubes of 1-nm diameter. Nature 363, 603–605.

Inoue, K., Huang, P.M., 1986. Influence of selected organic ligands on the formation of allophane and imogolite. Soil Sci. Soc. Am. J. 50, 1623–1633.

Inoue, N., Otsuka, H., Wada, S.-I., Takahara, A., 2006. (Inorganic nanofiber/enzyme) Hybrid hydrogel: preparation, characterization, and enzyme activity of imogolite/pepsin conjugate. Chem. Lett. 35, 194–195.

Jagur-Grodzinski, J., 2001. Preparation of functionalized polymers using living and controlled polymerizations. React. Funct. Polym. 49, 1–54.

Jakubowski, W., Min, K., Matyjaszewski, K., 2006. Activators regenerated by electron transfer for atom transfer radical polymerization of styrene. Macromolecules 39, 39–45.

Jiravanichanun, N., Yamamoto, K., Yonemura, H., Yamada, S., Otsuka, H., Takahara, A., 2008. Fabrication of conjugated polymer hybrid thin films with radially oriented aluminosilicate nanofibers by spin-assembly. Bull. Chem. Soc. Jpn. 81, 1663–1668.

Jiravanichanun, N., Yamamoto, K., Irie, A., Otsuka, H., Takahara, A., 2009. Preparation of hybrid films of aluminosilicate nanofiber and conjugated polymer. Synth. Met. 159, 885–888.

Jiravanichanun, N., Yamamoto, K., Kato, K., Kim, J., Horiuchi, S., Yah, W.-O., Otsuka, H., Takahara, A., 2012. Preparation and characterization of imogolite/DNA hybrid hydrogels. Biomacromolecules 13, 276–281.

Johnson, L.M., Pinnavaia, T.J., 1990. Silylation of a tubular aluminosilicate polymer (imogolite) by reaction with hydrolyzed (.γ-aminopropyl)triethoxysilane. Langmuir 6, 307–311.

Johnson, L.M., Pinnavaia, T.J., 1991. Hydrolysis of (γ-aminopropyl)triethoxysilane—silylated, imogolite and formation of a silylated tubular silicate-layered silicate nanocomposite. Langmuir 7, 2636–2641.

Kim, B.H., Park, T.H., Baek, S.J., Lee, D.S., Park, S.J., Kim, J.S., Park, Y.W., 2008. N-type carbon nanotube by alkaline-earth metal Sr doping. J. Appl. Phys. 103, 096103.

Koenderink, G.H., Kluijtmans, S.G.M.K., Philipse, A.P., 1999. On the synthesis of colloidal imogolite fibers. J. Colloid Interf. Sci. 216, 429–431.

Kong, J., Zhou, C., Yenilmez, E., Dai, H., 2000. Alkaline metal-doped n-type semiconducting nanotubes as quantum dots. Appl. Phys. Lett. 77, 3977–3979.

Krimm, S., Liang, C.Y., Sutherland, G.B.B.M., 1956. Infrared spectra of high polymers. V. Polyvinyl alcohol. J. Polym. Sci. 22, 227–247.

Kuroda, Y., Kuroda, K., 2008. Layer-by-layer assembly of imogolite nanotubes and polyelectrolytes into core-shell particles and their conversion to hierarchically porous spheres. Sci. Technol. Adv. Mater. 9, 025018.

Kuroda, Y., Kuroda, K., 2011. Expansion of intertubular mesopores of imogolite nanotubes by thermal decomposition of an imogolite poly(sodium 4-styrenesulfonate) composite. Chem. Lett. 40, 46–48.

Lee, R.S., Kim, H.J., Fischer, J.E., Thess, A., Smalley, R.E., 1997. Conductivity enhancement in single-walled carbon nanotube bundles doped with K and Br. Nature 388, 255–257.

Levard, C., Rose, J., Masion, A., Doelsch, E., Borschneck, D., Olivi, L., Dominici, C., Grauby, O., Woicik, J.C., Bottero, J.-Y., 2008. Synthesis of large quantities of single-walled aluminogermanate nanotube. J. Am. Chem. Soc. 130, 5862–5863.

Levard, C., Masion, A., Rose, J., Doelsch, E., Borschneck, D., Dominici, C., Ziarelli, F., Bottero, J.-Y., 2009. Synthesis of imogolite fibers from decimolar concentration at low temperature and ambient pressure: A promising route for inexpensive nanotubes. J. Am. Chem. Soc. 131, 17080–17081.

Lvov, Y.M., Shchukin, D.G., Mohwald, H., Price, R.R., 2008. Halloysite clay nanotubes for controlled release of protective agents. ACS Nano 2, 814–820.

Ma, W., Otsuka, H., Takahara, A., 2011a. Preparation and properties of PVC/PMMA-g-imogolite nanohybrid via surface-initiated radical polymerization. Polymer 52, 5543–5550.

Ma, W., Kim, J., Otsuka, H., Takahara, A., 2011b. Surface modification of individual imogolite nanotubes with alkyl phosphate from an aqueous solution. Chem. Lett. 40, 159–161.

Ma, W., Otsuka, H., Takahara, A., 2011c. Poly(methyl methacrylate) grafted imogolite nanotubes prepared through surface-initiated ARGET ATRP. Chem. Commun. 47, 5813–5815.

Ma, W., Yah, W.O., Otsuka, H., Takahara, A., 2012a. Surface functionalization of aluminosilicate nanotubes with organic molecules. Beilstein J. Nanotechnol. 3, 82–100.

Ma, W., Yah, W.O., Otsuka, H., Takahara, A., 2012b. Application of imogolite clay nanotubes in organic–inorganic nanohybrid materials. J. Mater. Chem. 22, 11887–11892.

Mackenzie, J.D., Chung, Y.J., Hu, Y., 1992. Rubbery ormosils and their applications. J. Non-Cryst. Solids 147–148, 271–279.

McCrum, N.G., Read, B.E., Williams, G., 1967. Anelastic and Dielectric Effects in Polymeric Solids. John Wiley & Sons, New York, p. 327.

Maniwa, Y., Fujiwara, R., Kira, H., Tou, H., Kataura, H., Suzuki, S., Achiba, Y., Nishibori, E., Takata, M., Sakata, M., Fujiwara, A., Suematsu, H., 2001. Thermal expansion of single-walled carbon nanotube (SWNT) bundles: X-ray diffraction studies. Phys. Rev. B: Condens. Matter Mater. Phys. 64. 241402/1.

Matsuno, R., Yamamoto, K., Otsuka, H., Takahara, A., 2004. Polystyrene- and poly(3-vinylpyridine)-grafted magnetite nanoparticles prepared through surface-initiated nitroxide-mediated radical polymerization. Macromolecules 37, 2203–2209.

Matyjaszewski, K., Jakubowski, W., Min, K., Tang, W., Huang, J., Braunecker, W.A., Tsarevsky, N.V., 2006. Diminishing catalyst concentration in atom transfer radical polymerization with reducing agents. Proc. Natl. Acad. Sci. U. S. A. 103, 15309–15314.

Mordkovich, V.Z., Baxendale, M., Yoshimura, S., Chang, R.P.H., 1996. Intercalation into carbon nanotubes. Carbon 34, 1301–1303.

Mukherjee, S., Bartlow, V.M., Nair, S., 2005. Phenomenology of the growth of single-walled aluminosilicate and aluminogermanate nanotubes of precise dimensions. Chem. Mater. 17, 4900–4909.

Nakano, A., Teramoto, N., Chen, G., Miura, Y., Shibata, M., 2010. Preparation and characterization of complex gel of type I collagen and aluminosilicate containing imogolite nanofibers. J. Appl. Polym. Sci. 118, 2284–2290.

Nicole, L., Rozes, L., Sanchez, C., 2010. Integrative approaches to hybrid multifunctional materials: from multidisciplinary research to applied technologies. Adv. Mater. 22, 3208–3214.

Oh, J., Chang, S., Jang, J., Roh, S., Park, J., Lee, J., Sohn, D., Yi, W., Jung, Y., Kim, S.-J., 2007. Imogolite as an electron emitter and a water sensor. J. Mater. Sci. Mater. Electron. 18, 893–897.

Ohashi, F., Tomura, S., Akaku, K., Hayashi, S., Wada, S.-I., 2004. Characterization of synthetic imogolite nanotubes as gas storage. J. Mater. Sci. 39, 1799–1801.

Okada, A., Usuki, A., 2006. Twenty-year of polymer clay nanocomposites. Macromol. Mater. Eng. 291, 1449–1476.

Okada, A., Kawasumi, M., Kurauchi, T., Kamigaito, O., 1987. Synthesis and characterization of nylon 6-clay hybrid. ACS Polym. Prepr. 28, 447–448.

Park, J., Lee, J., Chang, S., Park, T., Han, B., Han, J.W., Yi, W., 2008. Current–voltage characteristics of water-adsorbed imogolite film. Bull. Korean Chem. Soc. 29, 1048–1050.

Perruchot, C., Khan, M.A., Kamitsi, A., Armes, S.P., Werne, T.V., Patten, T.E., 2001. Synthesis of well-defined, polymer-grafted silica particles by aqueous ATRP. Langmuir 17, 4479–4481.

Philips, A.P., Wierenga, A.M., 1998. On the density and structure formation in gels and clusters of colloidal rods and fibers. Langmuir 14, 49–54.

Russell, J.D., Mchardy, W.J., Fraser, A.R., 1969. Imogolite: a unique aluminosilicate. Clay Miner. 8, 87–99.

Sanchez, C., Julian, B., Belleville, P., Popall, M., 2005. Applications of hybrid organic–inorganic nanocomposites. J. Mater. Chem. 15, 3559–3592.

Shchukin, D.G., Sukhorukov, G.B., Price, R.R., Lvov, Y.M., 2005. Halloysite nanotubes as biomimetic nanoreactors. Small 1, 510–513.

Shelley, S., Mather, P.T., DeVries, K.L., 2001. Reinforcement and environmental degradation of nylon-6/clay nanocomposites. Polymer 42, 5849–5858.

Shikinaka, K., Koizumi, Y., Osada, Y., Shigehara, K., 2011. Reinforcement of hydrogel by addition of fiber-like nanofiller. Polym. Adv. Technol. 22, 1212–1215.

Shikinaka, K., Abe, A., Shigehara, K., 2015. Nanohybrid film consisted of hydropobidized imogolite and various aliphatic polyesters. Polymer 68, 1279–1283.

Sielecki, A.R., Fedrov, A.A., Boodhoo, A., Andreeva, N.S., James, M.N.G., 1990. Molecular and crystal structures of monoclinic porcine pepsin refined at 1–8 A resolution. J. Mol. Biol. 214, 143–170.

Spitalsky, Z., Tasis, D., Papagelis, K., Galiotis, C., 2010. Carbon nanotube–polymer composites: chemistry, processing, mechanical and electrical properties. Prog. Polym. Sci. 35, 357–401.

Suzuki, M., Fujishima, A., Miyazaki, T., Hisamitsu, H., Ando, H., Nakahara, M., Yamamoto, M., Itoh, K., 1997. A study on adsorption structures of methacryloyloxyalkyl dihydrogen phosphates on silver substrates by infrared reflection absorption spectroscopy. J. Biomed. Mater. Res. 37, 252–260.

Tang, J., Hartley, B.S., 1970. Amino acid sequences around the disulphide bridges and methionine residues of porcine pepsin. Biochem. J. 118, 611–623.

Tanigami, T., Hanatani, H., Yamaura, K., Matsuzawa, S., 1999. Melting of the blends between syndiotacticity-rich and atactic poly(vinyl alcohol)s. Eur. Polym. J. 35, 1165–1171.

Tiselius, A., Henschen, G.E., Svensson, H., 1938. Electrophoresis of pepsin. Biochem. J. 32, 1814–1818.

Tsuchida, H., Ooi, S., Nakaishi, K., Adachi, Y., 2005. Effects of pH and ionic strength on electrokinetic properties of imogolite. Colloids Surf., A 265, 131–134.

Usuki, A., Kojima, Y., Kawasumi, M., Okada, A., Fukushima, Y., Kurauchi, T., Kamigaito, O., 1993. Synthesis of nylon 6-clay hybrid. J. Mater. Res. 8, 1179–1184.

Usuki, A., Koiwai, A., Kojima, Y., Kawasumi, M., Okada, A., Kurauchi, T., Kamigaito, O., 1995. Interaction of nylon 6-clay surface and mechanical properties of nylon 6-clay hybrid. J. Appl. Polym. Sci. 55, 119–123.

Vestal, C.R., Zhang, Z.J., 2002. Atom transfer radical polymerization synthesis and magnetic characterization of MnFe2O4/polystyrene core/shell nanoparticles. J. Am. Chem. Soc. 124, 14312–14313.

Wada, S.-I., Eto, A., Wada, K., 1979. Synthetic allophane and imogolite. J. Soil Sci. 30, 347–355.

Wang, C., Guo, Z.-X., Fu, S., Wu, W., Zhu, D., 2004. Polymers containing fullerene or carbon nanotube structures. Prog. Polym. Sci. 29, 1079–1141.

Yah, W.O., Yamamoto, K., Jiravanichanun, N., Otsuka, H., Takahara, A., 2010. Imogolite reinforced nanocomposites: multifaceted green materials. Materials 3, 1709–1745.

Yah, W.O., Irie, A., Jiravanichanun, N., Otsuka, H., Takahara, A., 2011a. Molecular aggregation state and electrical properties of terthiophenes/imogolite nanohybrids. Bull. Chem. Soc. Jpn. 84, 893–902.

Yah, W.O., Irie, A., Otsuka, H., Sasaki, S., Yagi, N., Sato, M., Koganezawa, T., Takahara, A., 2011b. Molecular aggregation states of imogolite/P3HT nanofiber hybrid. J. Phys.: Conf. Ser. 272, 012021.

Yamamoto, K., Otsuka, H., Wada, S.-I., Takahara, A., 2001. Surface modification of aluminosilicate nanofiber "imogolite" Chem. Lett. 30, 1162–1163.

Yamamoto, K., Otsuka, H., Takahara, A., 2002. Preparation of a novel (polymer/inorganic nanofiber) composite through surface modification of natural aluminosilicate nanofiber. J. Adhes. 78, 591–602.

Yamamoto, K., Otsuka, H., Wada, S.-I., Sohn, D., Takahara, A., 2005a. Preparation and properties of [poly (methyl methacrylate)/imogolite] hybrid via surface modification using phosphoric acid ester. Polymer 46, 12386–12392.

Yamamoto, K., Otsuka, H., Wada, S.-I., Sohn, D., Takahara, A., 2005b. Transparent polymer nanohybrid prepared by in situ synthesis of aluminosilicate nanofibers in poly(vinyl alcohol) solution. Soft Matter 1, 372–377.

Yamamoto, K., Otsuka, H., Takahara, A., 2007. Preparation of novel polymer hybrids from imogolite nanofiber. Polym. J. 39, 1–15.
Yang, H., Chen, Y., Su, Z., 2007. Microtubes via assembly of imogolite with polyelectrolyte. Chem. Mater. 19, 3087–3089.
Yang, H., Wang, C., Su, Z., 2008. Growth mechanism of synthetic imogolite nanotubes. Chem. Mater. 20, 4484–4488.
Yoshinaga, N., Aomine, S., 1962. Imogolite in some Ando soils. Soil Sci. Plant Nutr. 8, 114–121.

Chapter 25

Imogolite for Catalysis and Adsorption

E. Garrone and B. Bonelli*
INSTM Unit of Torino Politecnico, Politecnico di Torino, Turin, Italy
*Corresponding author: e-mail: barbara.bonelli@polito.it

25.1 INTRODUCTION

The composition of its inner and outer surfaces and its unique porous structure render imogolite a very interesting material for applications requiring the presence of cavities with different dimensions, different chemical properties and large specific surface areas (SSA), like catalysis, gas adsorption and storage, removal of ions from aqueous solution, etc.

The chemical composition of imogolite, $Al_2SiO_3(OH)_4$, can be also written as $(OH)_3Al_2O_3SiOH$, going from the outer to the inner surface of nanotubes (Cradwick et al., 1972). It results that the inner surface of imogolite is lined by silanol (SiOH) groups and its outer surface by both Al–O–Al and Al–OH–Al groups, which confer it an amphoteric character (Bonelli et al., 2009). The inner and outer surfaces of imogolite have different acid/base properties (inner SiOH are more acidic than outer Al(OH)Al; Gustafsson, 2001; Guimaraes et al., 2007) and ion exchange capacity.

The behaviour of the material in water [see Eqs (12.1) and (12.2) in Chapter 12] is crucial, and indeed scientific interest for imogolite initially arose from its possible use for the removal of cations and anions from polluted water (Parfitt et al., 1974; Clark and McBride, 1984a,b; Harsh et al., 1992; Denaix et al., 1999; Arai et al., 2006).

Several studies have shown that the interaction of ions with imogolite in water is not merely electrostatic, in that specific ions adsorption may occur, by ligand exchange mechanism and/or ions intercalation within bundles (Clark and McBride, 1984a; Wada, 1984; Su et al., 1992).

Imogolite nanotubes are highly dispersed in water but tend to organize in a hexagonal packing in powder samples. They have an inner diameter of ca. 1.0 nm, corresponding to nanotube inner pore A; internanotubes pore B, about

Developments in Clay Science, Vol. 7. http://dx.doi.org/10.1016/B978-0-08-100293-3.00025-X

0.3–0.4 nm wide, corresponds to cavities among three nanotubes aligned in a bundle and not accessible to small molecules like water, and larger C pores are disordered slit pores (mainly mesopores) among bundles (see Fig. 12.1 in Chapter 12).

Such pores have different dimensions and chemical composition, so that three kinds of surface are distinguished: (i) A surface, very hydrophilic and lined by SiOH groups accessible to gas molecules only after dehydration; (ii) B surface, not accessible to small molecules like water, but becomes accessible in bundles of some chemically modified imogolites and (iii) C surface, related to slit mesopores among bundles, has the same composition of B surface, and an amphoteric character due to the presence of both Al–O–Al and Al–OH–Al groups, with Al in octahedral coordination (Al(VI)).

Besides ion adsorption, the presence of pores with different chemical composition and dimensions, along with the occurrence of aligned nanotubes, render both natural and modified imogolite good candidates for (selective) adsorption of gases and heterogeneous catalysts. Concerning the latter application, several tests reaction have been used in the literature to probe the acid/base properties of the material, as well as the actual accessibility to imogolite pores, in particular pores A.

25.2 CATALYTIC PROPERTIES OF IMOGOLITE

Imamura et al. (1993) tested the catalytic properties of natural imogolite in the isomerization of 1-butene at 50°C. Double bond isomerization to 2-butene and skeletal isomerization to isobutene may occur on both acidic and basic catalysts. With acidic catalysts, the products distribution depends on the strength of acidic sites (Wichterlova et al., 1999). In the presence of mildly acidic Brønsted sites, secondary carbocations form as short-lived intermediates, leading to a *cis/trans*-2-butene ratio close to 1. Stronger acidic sites lead, instead, to the preferential formation of isobutene, due to isomerization of secondary carbocations to tertiary ones. With strong basic catalysts, only 2-butenes are obtained, in a *cis/trans* ratio larger than 10 (Baba et al., 1994; Zhu et al., 1999). For this reason, isomerization of 1-butene can be used as a test reaction to probe acidic/basic properties of solids (Bonelli et al., 2002).

A soil containing natural imogolite was obtained by pumice beds of volcanic origin. Imamura et al. (1993) extracted and purified this raw imogolite in order to get rid of impurities like iron, alumina, silica and carbonaceous material. The purified imogolite was then calcined at different temperatures in the 80°C–950°C range, and the acidic properties of the obtained materials were tested by adsorption of ammonia. The largest amount of NH_3 was adsorbed on imogolite calcined at 400°C and 500°C, then this amount decreased on samples calcined at higher temperature, in concomitance with a drop of SSA and formation of other crystalline phases. The higher acidity of samples calcined at 400°C–500°C was likely due to the formation of a thermally collapsed phase (MacKenzie et al., 1989; Zanzottera et al., 2012b), which has more acidic sites than proper imogolite (Bonelli et al., 2009). In agreement with NH_3 adsorption results, the highest reaction rates of 1-butene isomerization were obtained with imogolite calcined at 400 and 500°C, with a *cis/-trans*-2-butene ratio in the range 0.86–0.98. Similar results were obtained on mesoporous amorphous aluminium-silicates bearing mildly acidic sites, further confirming the medium acidic strength of the studied materials, as compared to other heterogeneous catalysts with stronger Brønsted sites, like protonic zeolites (Bonelli et al., 2002).

If follows that not only proper imogolite, but also imogolite-related phases may have some catalytic activity. On the basis of such results, the gas-phase methylation of phenol (Phe) with methanol was used as a test reaction to probe not only the surface properties of a synthetic imogolite, but also of its precursor (proto-imogolite) and of the collapsed phase (Imo-c) stemming from imogolite thermal decomposition above 300°C (Bonelli et al., 2009).

Proto-imogolite is an amorphous phase occurring as particles with hollow spheres morphology (Fig. 25.1), with a SSA of ca. 180 $m^2 g^{-1}$ and both Brøntsed and Lewis sites due to the presence of accessible Al^{3+} ions (Bonelli et al., 2009).

Imo-c has a residual SSA of ca. 200 $m^2 g^{-1}$, due to the presence of micropores (Zanzottera et al., 2012b), and new surface acidic sites formed after collapsing (Bonelli et al., 2009), related to the presence of Al in coordination different from the octahedral one, namely Al(V) and Al(IV), penta- and tetra-coordinated aluminium (MacKenzie et al., 1989).

The gas-phase alkylation of Phe with methanol (MeOH) was chosen as test reaction, since it may occur on either acidic or basic sites, even in the presence of weaker sites (Ballarini et al., 2007, 2008), and the nature of the

FIG. 25.1 SEM (scanning electron microscopy) image of proto-imogolite. *Adapted with permission from Bonelli et al. (2009). Copyright 2009 Elsevier.*

formed products is affected by the strength of active sites. Weaker sites generally lead to the preferential formation of O-alkylation products (anisole), whereas stronger sites lead preferentially to C-alkylation, with production of cresols and poly-alkylated phenols. In the presence of strong acidic sites, MeOH may undergo side reactions, with formation of dimethylether first, and then of both aliphatic and aromatic hydrocarbons, precursors of coke (Dahl and Kolboe, 1994). The latter reactions are catalysed by strong acidic sites rather than by weak ones.

Preliminary measurements, carried out by following by infrared (IR) spectroscopy adsorption of both Phe and MeOH on the catalysts, showed that both molecules may interact with inner SiOH groups of imogolite. Inner SiOH are enough acidic to activate the considered reactants, but access to A pores is hindered, especially to Phe molecules that also interact with the amphoteric outer surface of imogolite, by forming adsorbed phenolate ions (Bonelli et al., 2009). The two reactants indeed have different molecular sizes and were supposed to have very different diffusivity properties. Therefore, Phe and MeOH were used to selectively access external Al(OH)Al and internal Si–OH, respectively.

The effect of temperature on the catalytic performance of proto-imogolite (Fig. 25.2) showed a relevant catalytic activity, with 27% Phe conversion at 300°C, and the formation of *o*-cresol, poly-alkylated phenols and anisole as

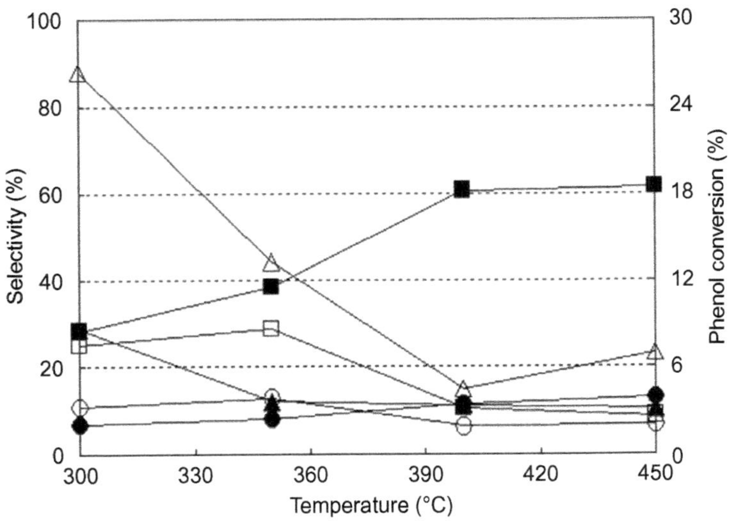

FIG. 25.2 Effect of reaction temperature on the catalytic performance of proto-imogolite. *Symbols*: Phe conversion (△), selectivity to *o*-cresol (■), selectivity to anisole (▲), selectivity to *p*-cresol (●), selectivity to 2,6-xylenol (○) and selectivity to other poly-alkylated aromatics (□). *Adapted with permission from Bonelli et al. (2009). Copyright 2009 Elsevier.*

main products, whereas *p*-cresol and poly-alkylated anisoles formed in a lower amount. The observed products distribution was similar to that obtained with H-mordenite and H-β zeolites (Bregolato et al., 2007). However, at higher reaction temperatures, deactivation took place with a remarkable decline of both Phe conversion and selectivity to poly-alkylated phenols, and an increase of selectivity to *o*- and *p*-cresol, due to the decreased contribution of consecutive reactions. The deactivation behaviour of proto-imogolite was similar to that observed for zeolites, due to the accumulation of coke formed by the hardening of aliphatic and aromatic hydrocarbons originated by MeOH transformation (Mikkelsen and Kolboe, 1999; Haw et al., 2003). Further experiments carried out on the fresh catalyst showed that the progressive decrease of Phe conversion led to an increase of selectivity to *o*-cresol and to a decrease of that to poly-alkylated phenols. Surprisingly, the selectivity to anisole was not affected by the change of Phe conversion, and that to *p*-cresol decreased and became nil after 2 h reaction time. This meant that the formation of poly-alkylated aromatics occurred by transformation of *o*-cresol, whereas the *para* isomer did not undergo consecutive reactions. The spent proto-imogolite catalyst was fully coked, confirming its acidic properties: catalytic tests demonstrated that proto-imogolite has medium-strength acid sites, fully accessible to both Phe and MeOH molecules, in agreement with IR results (Bonelli et al., 2009).

Imogolite was instead inactive at 300°C, with 0.03% Phe conversion, the only two detected products being *o*-cresol and anisole, each one with approximately 50% selectivity. This result was in agreement with IR results showing that Phe mainly interacts with the outer surface of imogolite, by forming chemisorbed phenolate species, and therefore cannot react with MeOH molecules within A pores. On the other hand, the conversion of MeOH was approximately 20%, with formation of dimethylether (and water) and 2,2-dimethoxypropane. The latter forms by the acid-catalysed reaction between MeOH and acetone, which, in turn, formed by dehydrogenation (via H-transfer) of isopropanol. Isopropanol is the product of the acid-catalysed hydration of propene, the olefin typically obtained by transformation of MeOH over strongly acid sites. Therefore, the acidity properties of imogolite were able to catalyse MeOH transformations. The amount of coke formed on the spent imogolite catalyst was much lower than that formed with proto-imogolite. Since the formation of coke was unlikely to occur inside imogolite, this indicates that MeOH reacted with both inner SiOH, yielding lighter compounds (ie, dimethylether and 2,2-dimethoxypropane), and with external Al(OH)Al sites, giving rise to the formation of heavier compounds and, finally of coke.

The catalytic behaviour of Imo-c (Fig. 25.3) was much more active than imogolite, in spite of the low conversion at 300°C (ca. 3%). Comparison among Imo-c and proto-imogolite catalytic behaviours showed no evident deactivation phenomenon with the former catalyst, at variance with

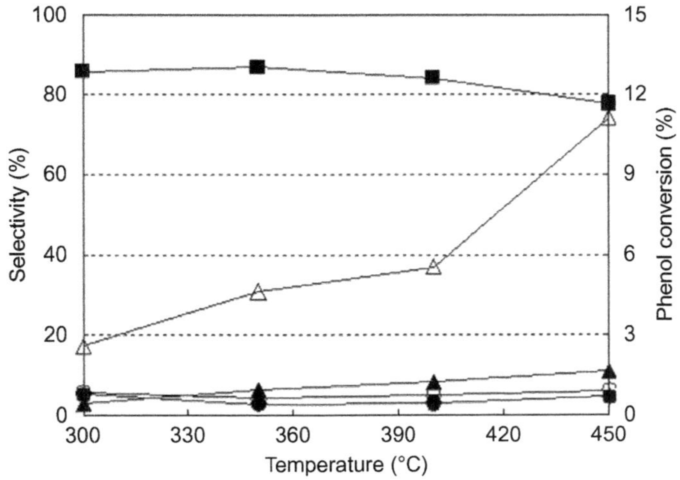

FIG. 25.3 Effect of temperature on catalytic performance of Imo-c. *Symbols*: phenol conversion (△), selectivity to *o*-cresol (■), selectivity to anisole (▲), selectivity to *p*-cresol (●) and selectivity to 2,6-xylenol (○). *Adapted with permission from Bonelli et al. (2009). Copyright 2009 Elsevier.*

proto-imogolite. The products distribution was also very different: with Imo-c, the main product was *o*-cresol, with a selectivity close to 80% regardless of both temperature and conversion. At 450°C, the selectivity to anisole was 10% and that to *p*-cresol 4%, the only poly-alkylated compound in the products being 2,6-xylenol, formed with selectivity of 3%. With proto-imogolite, at the same temperature and comparable Phe conversion, selectivity to 2,6-xylenol, poly-alkylated phenols and *p*-cresol was 8%, 10% and 13%, respectively. This difference indicated that the type and strength of the active sites in the two materials was different.

In some aspects, the behaviour of Imo-c was similar to what observed with basic materials (Ballarini et al., 2008), showing high chemo- (C-alkylation largely preferred over O-methylation) and regio- (ortho-C-alkylation largely preferred over para-C-alkylation) selectivity in the studied reaction. This result is in agreement with the experimental finding that Imo-c gives the strongest interaction with Phe and generates phenolate species. Due to the high nucleophilicity of the O atom, the phenolate ion adopts the orthogonal orientation typically observed with basic oxides, which explains the regioselectivity experimentally observed. On the other hand, the presence of medium-strength acidic sites is responsible for the small, but not negligible, formation of anisole. Therefore, Imo-c appeared to have an amphoteric behaviour, resulting from the combined reactivity of the acidic and basic sites present on its surface. Strongly nucleophilic O^{2-} ions were likely produced by the high-temperature dehydration of imogolite, leading to strained Al–O–Al(Si) bonds which easily open by abstraction of the acidic phenolic proton and coordination of the phenolate to the Al^{3+} cation. The same occurs in strongly dehydrated alumina, where the nucleophilicity of the O^{2-} is responsible for the 'basic' behaviour of the solid towards Phe (Klemm et al., 1968; Porchet et al., 1996; Fu et al., 1998).

Another peculiarity of the Imo-c material was the absence of relevant deactivation phenomena, as confirmed by the relatively low amount of coke accumulated on the catalyst, a behaviour also typical of basic catalysts (Ballarini et al., 2007, 2008). However, a contribution deriving from the lamellar-type morphology (MacKenzie et al., 1989) of the sample, which may favour the counter-diffusion of products limiting the generation of the hydrocarbon-pool precursor of coke formation, could not be excluded.

25.3 CATALYTIC PROPERTIES OF MODIFIED IMOGOLITE

The catalytic properties of the following modified imogolites have been studied, so far: (i) copper exchanged imogolite (Cu-loaded imogolite); (ii) 3-aminopropyltriethoxysilane (3-APTES)-grafted imogolite on which OsO_4 was supported (Qi et al., 2008) and (iii) Fe-doped imogolite (Shafia et al., 2015a).

The performance of Cu-loaded imogolite as a shape-selective catalyst in the decomposition, in liquid phase, of *tert*-butylhydroperoxide and 1,1-bis(*tert*-butyldioxy)cyclododecane with chlorobenzene as a solvent has been studied by Imamura et al. (1996). Copper exchanged imogolite was obtained by dispersing preformed imogolite (with a surface area of ca. 350 $m^2\ g^{-1}$) in a solution of $Cu(NO_3)_2$, stirring overnight, filtering, washing and then calcining at 500°C in air for 3 h. The total Cu content was 2.59%. A sample, denoted as $Cu–SiO_2$, with the same Cu content was obtained by dispersing silica gel (SSA of ca. 500 $m^2\ g^{-1}$) in aqueous $Cu(NO_3)_2$, followed by evaporation to dryness at 60°C and calcination in air at 500°C. Extended X-ray absorption fine structure (EXAFS) analysis of Cu-loaded imogolite showed that copper (II) ions were in exchange position within A pores, since Cu ions electrostatically interact with $Si–O^-$ groups. Conversions achieved with the two reactions are compared in Fig. 25.4 for imogolite, Cu-loaded imogolite and $Cu–SiO_2$.

Cu-loaded imogolite showed high conversion in the case of *tert*-butylhydroperoxide, such behaviour being assigned to a facile diffusion of the reactant, which has the smallest molecular diameter of ca. 5 Å. With 1,1-bis (*tert*-butyldioxy)cyclododecane, having the smallest diameter of ca. 7 Å, the catalytic performance of Cu-loaded imogolite was worst as compared to $Cu–SiO_2$, because of the larger molecule size, so that $Cu–SiO_2$ gave higher conversions due to a higher accessibility of Cu sites at the surface of amorphous silica. It should be noticed, however, that calcination at 500°C, adopted for the preparation of Cu-loaded imogolite, could have brought about some loss of nanotubular structure (MacKenzie et al., 1989).

The use of modified imogolite as a support of a catalytically active phase was reported by Qi et al. (2008), who used a 3-APTES-grafted (Imo-APTES),

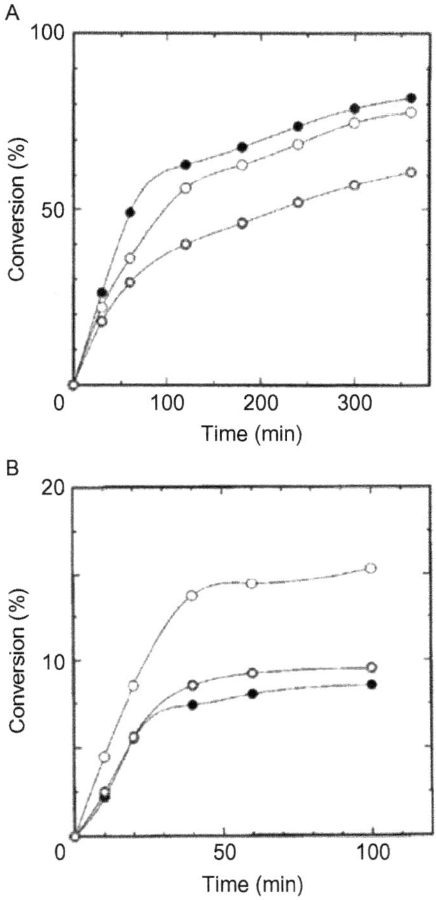

FIG. 25.4 (A) Decomposition of *tert*-butyl hydroperoxide at 90°C on imogolite (**O**), Cu-loaded imogolite (●) and Cu–SiO_2 (○). (B) decomposition of 1,1-bis(*tert*-butyldioxy)cyclododecane at 70°C on imogolite (**O**), Cu-loaded imogolite (●) and Cu–SiO_2 (○). *Reprinted with permission from Imamura et al. (1996). Copyright 1996 Elsevier.*

on which amino groups at the outer surface of nanotubes were able to coordinate osmium tetroxide (OsO_4, Fig. 25.5). OsO_4 is used as a catalyst in dihydroxylation (DH) of olefins, including asymmetric dihydroxylation (ADH) in which olefins react with chiral ligands. Osmium catalysed DH and ADH reactions could be useful for the synthesis of drugs and fine chemicals, but the high cost of OsO_4 along with its toxicity, which poses serious contamination issues, reduce its use in industry. For this reason, Imo-APTES was used with the attempt to immobilize the catalytically active phase, so obtaining a catalyst denoted as Imo-APTES-OsO_4.

X-ray photoelectron spectroscopy (XPS) analysis of the material showed the presence of both Os(VI) and Os(IV) species, with an overall osmium

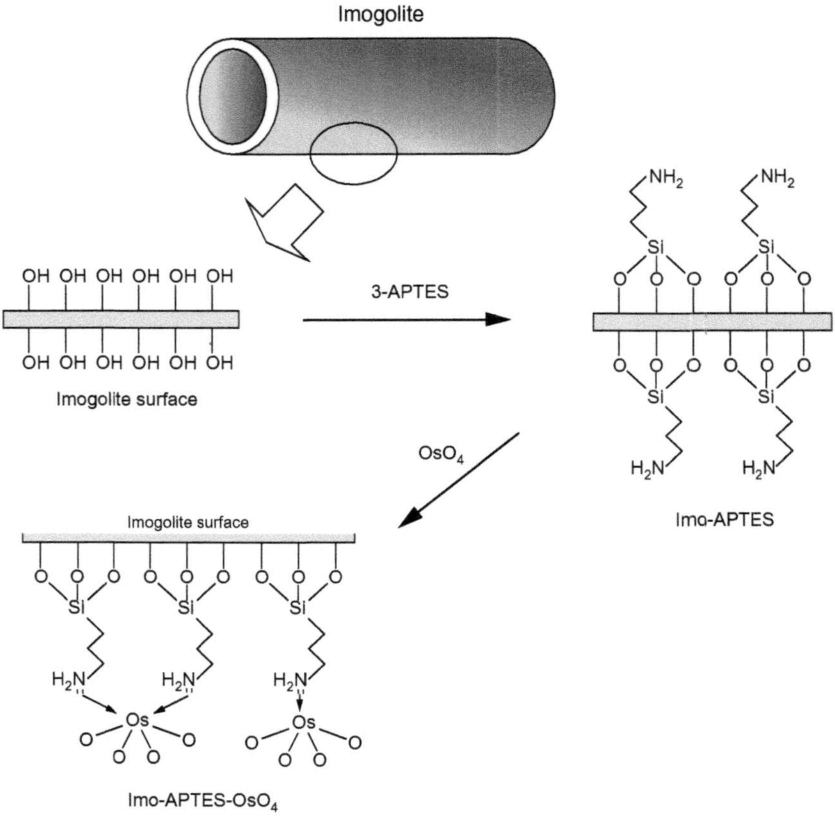

FIG. 25.5 Immobilization of OsO_4 on the surface of 3-APTES-grafted-imogolite (Imo-APTES) leads to Imo-APTES-OsO_4 catalysts. *Reprinted with permission from Qi et al. (2008). Copyright 2008 Elsevier.*

loading of 0.5 mmol g^{-1} of Imo. The chosen reaction was the AD of olefins with NMO (*N*-methylmorfoline-*N*-oxide) carried out at room temperature (r.t.) in acetone/water mixture as a solvent (Table 25.1).

Several olefins were tested (as detailed in Table 25.1). The reaction was completed within the first several hours with all the adopted substrates, with the exception of electron deficient olefins (cinnamonitrile and cinnamide, respectively) that gave no conversion even after 24 h. The catalyst was reused three times, but the disappointing result was that the reaction yield decreased a good deal, likely due to the partial leaching of osmium into the reaction mixture.

Fe-doped imogolite (Ookawa et al., 2006; Ookawa, 2012; Shafia et al., 2015a,b) can be obtained by isomorphically substituting Fe^{3+} for octahedral Al^{3+} at the outer surface of imogolite. Isomorphic substitution is obtained

TABLE 25.1 Catalytic Results Obtained in the AD of Olefins with Imo-APTES-OsO_4 as the Catalyst

Olefin	Reaction time (h)	Isolated yield (%)
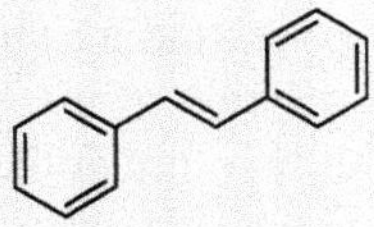	2	86
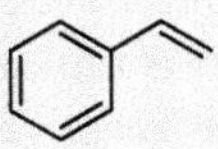	2	83
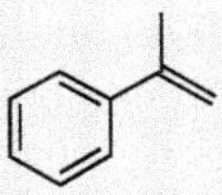	2	89
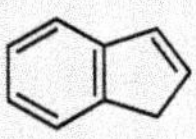	2	68
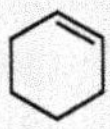	2	75
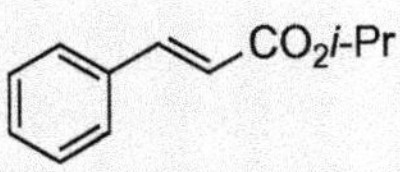	2	85
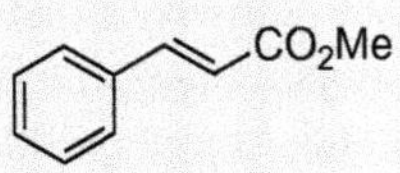	2	79
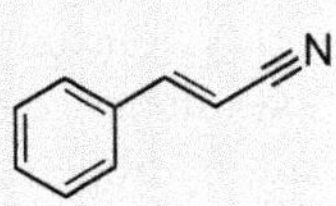	24	No reaction
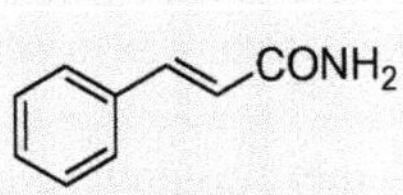	24	No reaction

for a Fe content up to 1.0% by mass, then formation of Fe_2O_3 clusters occurs (Shafia et al., 2015b). Fe-doped imogolite was tested as a catalyst for several oxidation reactions of organic molecules by H_2O_2 in different solvents.

The oxidation of either cyclohexene or benzene (Bz) with hydrogen peroxide solution (30%) in acetonitrile was carried out in the temperature range 50–60°C (Ookawa et al., 2006; Ookawa, 2012). Bare imogolite and a sample on which $FeCl_3$ was adsorbed on imogolite from aqueous solution of the iron (III) salt (Fe-imogolite) were studied for comparison. With bare imogolite, a low conversion was measured during cyclohexene oxidation, with very low yields of both 2-cyclohexene-1-ol (1-ol) and 1,2-epoxycyclohexene (EP) (0.5% and 0.8%, respectively). Notwithstanding the low conversions attained, the results indicated that the outer surface of bare imogolite is reactive per se in the presence of H_2O_2. Comparison with results obtained in the presence of gibbsite, boehmite or Al_2O_3 as catalysts (Mandelli et al., 2001) showed the formation of EP in the presence of Al_2O_3 catalyst in anhydrous conditions. Similarly, in a preliminary study, the outer surface of imogolite was found to be able to activate H_2O_2 molecules in the catalytic degradation of azo-dyes in aqueous environment (Shafia et al., 2015a).

The conversion of cyclohexene was observed to increase with Fe-doped imogolite catalyst, and a mixture of products was obtained, including both *trans*- and *cis*-1,2-cyclohexanediol, besides EP and 1-ol, 2-cyclohexen-1-one. The first two products did not form with Fe-imogolite, which also gave lower conversions, indicating that the nature of iron sites in the two samples was different.

Oxidation of Bz on Fe-doped imogolite gave a 10.6% molar yield of Phe, in acetonitrile solution: the direct oxidation of Bz to Phe is not easily achievable, so showing that Fe-doped imogolite may be an interesting catalyst for this reaction. Moreover, Bz oxidation on Fe^{3+}-loaded Al_2O_3 showed a different products distribution, indicating that the state of Fe in doped imogolite differs from that of a Fe^{3+}-loaded Al_2O_3 catalyst (Monfared and Amouei, 2004).

Oxidation of cyclohexane on Fe-doped imogolite (Ookawa, 2012) was carried out in acetonitrile at 60°C: an overall conversion of cyclohexane of ca. 25% was obtained, with production of a mixture of cyclohexanone, cyclohexanol and a third compound, most probably cyclohexyl hydroperoxide (C_6H_{11}–OOH). The proposed mechanism of reaction involved firstly the formation of the C_6H_{11}–OOH, as an intermediate peroxide that was further oxidized to cyclohexanone and cyclohexanol.

With Fe-doped imogolite, the catalytic oxidation of other aromatic compounds such as Phe and chlorobenzene was also attempted (Ookawa et al., 2008): in that case, the aromatic ring was oxidized. When the side-chain of the aromatic ring was a hydrocarbon group such as methyl, both Bz ring and the side-chain group reacted, but the side-chain group was preferentially oxidized, so that oxidation of toluene lead to benzaldehyde, *p*-cresol and *o*-cresol with yields of 5.8%, 2.2% and 1.3%, respectively.

Finally, it should be noted that the photocatalytic properties of Fe-doped imogolite have been unexplored thus far, although both theoretical (Alvarez-Ramírez, 2009) and experimental studies (Shafia et al., 2015b) showed that upon isomorphic substitution of Al^{3+} by Fe^{3+}, a change in the band gap of imogolite occurs, which makes Fe-doped imogolite a promising candidate for photocatalytic reactions.

25.4 ADSORPTION PROPERTIES OF NATURAL AND MODIFIED IMOGOLITE

Concerning gas/vapour adsorption, the available studies on the subject mainly deal with the adsorption of methane (CH_4), carbon dioxide (CO_2) on either bare or modified imogolite, as well as with the behaviour of water molecules within very hydrophilic A pores.

Adsorption of CH_4 and CO_2 was first studied by Ackerman et al. (1993) on both a synthetic imogolite sample and a modified sample with composition $(OH)_3Al_2Si_{0.5}Ge_{0.5}OH$, in which 50% Si atoms were replaced by Ge atoms. The two samples outgassed at the same temperature (275°C) showed very similar values of SSA (around 400 $m^2\ g^{-1}$), but slightly larger pores occurred in Ge-modified imogolite, in agreement with the larger size of germanium with respect to silicon. Adsorption of CH_4 and CO_2 was measured at 0°C in the 0–1.07 bar range: at subatmospheric pressure, proper imogolite adsorbed ca. 1.6 times more CO_2 and CH_4 with respect to Ge-modified imogolite, notwithstanding their similar SSA. This behaviour was attributed to the effect of the slightly smaller diameter of imogolite with respect to Ge-modified imogolite, since the two samples had similar SSA and no large differences were expected for the interaction strength of gases with either SiOH or GeOH groups (Ackerman et al., 1993).

Adsorption of CH_4 and H_2O on both natural and synthetic imogolite was measured by Ohashi et al. (2004) at 21°C and high pressure (up to 80 bar). Langmuir-type adsorption isotherms of CH_4 were obtained, indicating that monolayer adsorption took place. The reported CH_4 storage capacity was 42.5 $mg\ mL^{-1}$ at 40.9 bar for natural imogolite and 50.6 $mg\ mL^{-1}$ at 40.5 bar for synthetic imogolite. The latter adsorbed amount was higher than the value of compressed natural gas storage (28 $mg\ mL^{-1}$ at 40 bar).

Adsorption isotherms of H_2O on the same materials showed the presence of hysteresis loops in both cases: natural and synthetic imogolite adsorbed up to 60 and 80 mass% H_2O, respectively (Ohashi et al., 2004). Such high hydrophilicity of imogolite was in agreement with the increase of SSA, with the outgassing temperature reported by several researchers (Ackerman et al., 1993; Bottero et al., 2011), due to dehydration occurring up to 300°C without thermal collapse of the nanotubes.

The high adsorption capacity towards water has been the subject not only of experimental studies, but also of theoretical ones. Molecular dynamics study of hydrated imogolite (Creton et al., 2008) showed that the inner surface of imogolite is more hydrophilic than the outer one.

Such behaviour could be ascribed to different factors, including the higher ionicity of SiOH with respect to Al(OH)Al groups (Guimaraes et al., 2007). The electric field within imogolite results stronger than at their outer surface, therefore causing a preferential adsorption of water within nanotubes. According to Creton et al. (2008), a single water molecule likely interacts by H-bonding through its O atom with two SiOH, and only with one external Al–OH–Al group, due to a curvature of the surface, which does not allow the formation of two H-bonds with external hydroxyls.

The high SiOH density within A pores also plays a role. Bonelli et al. (2009) calculated a SiOH density of 9.1 OH nm^{-2}, ca. twice as much as the average SiOH density at the surface of hydrated amorphous silicas, ca. 5 OH nm^{-2} (Iller, 1979). According to Creton et al. (2008) the interaction of the first adsorbed layer with the inner surface of imogolite is so strong that water molecules do not diffuse, but just perform shutting motion along the circumference of the nanotube. Nonetheless, empty nanotubes have probably an ellipsoidal section (Fig. 25.6) and resemble a cylindrical one only when they are filled by water.

Another theoretical work concerns the flexibility of hydroxyl groups of inner SiOH in imogolite (Zang et al., 2010): by means of grand canonical Monte Carlo (GCMC) simulations, adsorption of water, methanol, CO_2 and CH_4 was studied, and the results compared with experimental adsorption isotherms. For H-bonded molecules, the flexibility of inner hydroxyls groups had to be taken into account in order to obtain a good agreement with experimental adsorption data. The comparison between experimental isotherms of water

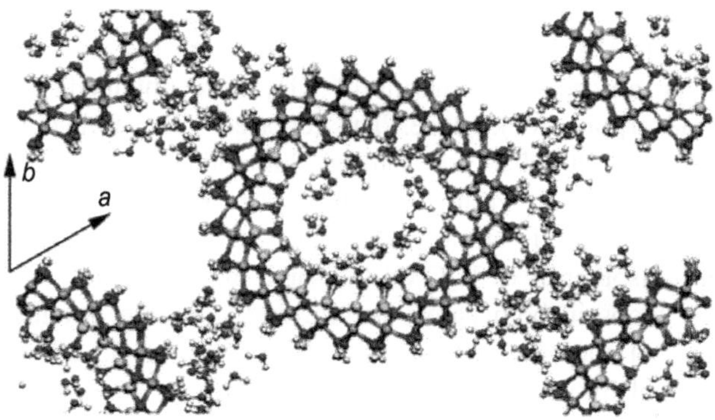

FIG. 25.6 Snapshot of hydrated imogolite as obtained by molecular dynamics simulation. The *b*- and *a*-axes coincide with the *y* and *z* Cartesian axes, respectively. *Reprinted with permission from Creton et al. (2008). Copyright 2008 the PCCP Owner Societies.*

adsorption and isotherms calculated by assuming either rigid or flexible surface hydroxyls (Fig. 25.7) shows that a good agreement was reached in the latter case. Such flexibility leads to the possibility for a water molecule to form up to three H-bonds with water molecules (Fig. 25.8).

The adsorption selectivity of water towards methanol was predicted to be larger than 100 times, which make imogolite a promising candidate for alcohol dehydration. Such flexibility effects were less significant with CH_4 and, to a minor extent, with CO_2; ie, two molecules that cannot give H-bonding.

Controlling the microporosity and mesoporosity of the material, as well as hydrophilic/hydrophobic properties, may lead to the production of adsorbents

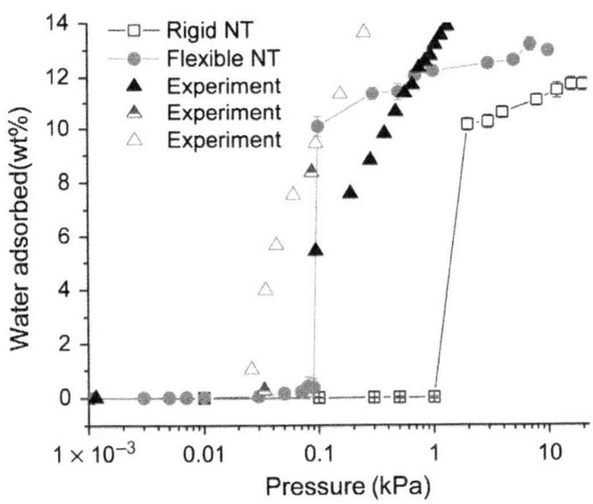

FIG. 25.7 Comparison of experimental (*triangles*) and calculated adsorption isotherms of water with either rigid (*squares*) or flexible (*circles*) hydroxyls on the inner surface of imogolite. *Reprinted with permission from Zang et al. (2010). Copyright 2010 American Chemical Society.*

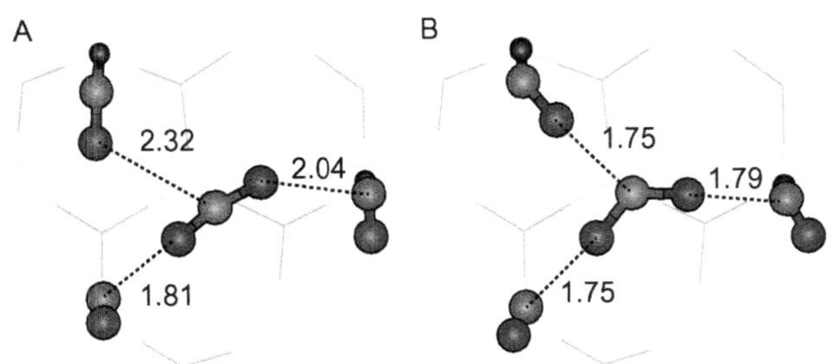

FIG. 25.8 Energy minimized configuration of a single water molecule in rigid (A) and flexible (B) imogolite. *Blue* (*dark grey* in the print version): H atoms; *red* (*grey* in the print version): O atoms; *grey*: Si atoms. Selected O···H distances are shown in Å. *Reprinted with permission from Zang et al. (2010). Copyright 2010 American Chemical Society.*

with tailored adsorption properties. On the ground of these findings, a material with chemical composition $(OH)_3Al_2O_3Si–CH_3$ (Me-imogolite), having an inner hydrophobic surface due to the substitution of SiOH by $Si–CH_3$ groups, was prepared by Bottero et al. (2011). With respect to imogolite, Me-imogolite has larger A pores (ca. 2.0 nm in diameter), being a mesoporous material with an average SSA of 650 $m^2\ g^{-1}$. The presence of a hydrophobic inner surface along with an outer hydrophilic one renders Me-imogolite a material that in principle could be employed in adsorption processes aimed at gas separation (eg, hydrocarbon recovery from wet gaseous streams) or storage. For these reasons, Me-imogolite was tested in the adsorption of CH_4 at 30°C in the high-pressure range (5–35 bar CH_4 partial pressure). The same was done for a synthetic Imo: both samples were previously dehydrated at 300°C and, for each sample, the first adsorption run was followed by a desorption step and a second adsorption run was carried out after reoutgassing samples at 300°C, to check both reversibility of the process and materials stability (Fig. 25.9).

Both imogolite and Me-imogolite gave reversible CH_4 adsorption in the explored pressure range (Fig. 25.9). Furthermore, second adsorption run values (stars) were not affected by the additional outgassing treatment, confirming thermal stability of the adsorbents. Curve-fits obtained by assuming a Langmuir model of adsorption showed good agreement with experimental data: both the maximum adsorbed volumes (76.81 and 29.65 $cm^3\ g^{-1}$ for Me-imogolite and imogolite, respectively) and equilibrium constants (0.03 and 0.06 bar^{-1} for

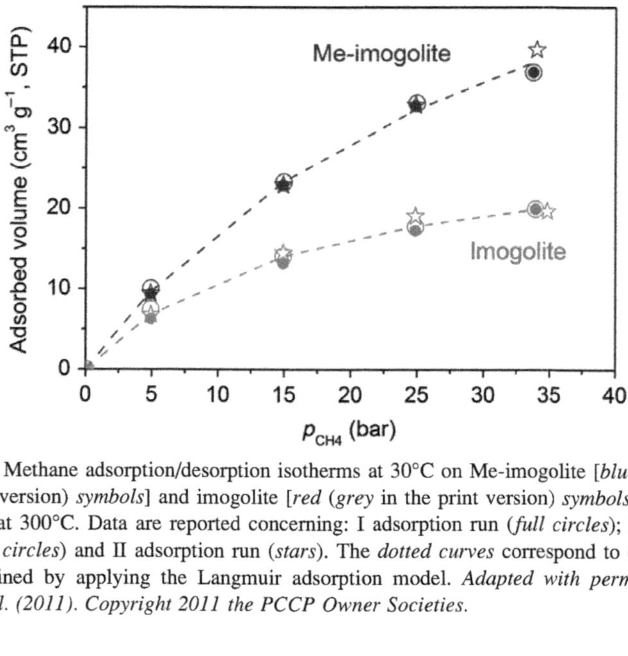

FIG. 25.9 Methane adsorption/desorption isotherms at 30°C on Me-imogolite [*blue* (*dark grey* in the print version) *symbols*] and imogolite [*red* (*grey* in the print version) *symbols*] previously dehydrated at 300°C. Data are reported concerning: I adsorption run (*full circles*); I desorption run (*hollow circles*) and II adsorption run (*stars*). The *dotted curves* correspond to curve-fitting results obtained by applying the Langmuir adsorption model. *Adapted with permission from Bottero et al. (2011). Copyright 2011 the PCCP Owner Societies.*

Me-imogolite and imogolite, respectively) were calculated. Though the obtained average specific capacities were below those reported by other materials proposed for an efficient methane storage (Menon and Komarneni, 1998; Senskova and Kaskel, 2008), Me-imogolite adsorption capacity at 35 bar was 2.7 mass%, ie, about 2.5 times larger than that of imogolite (1.4 mass%). This result is in agreement with both the larger pore volume of Me-imogolite and the presence of a methylated surface, indicating that the material has interesting properties towards selective adsorption of methane.

Kang et al. (2011) produce imogolite with modified interior surface by postsynthesis grafting organic functionalities on dehydrated imogolite: materials were obtained with up to 35% inner SiOH substituted by organic functionalities by reaction with the following reactants: R = acetyl chloride (imogolite functionalized with acetyl chloride (Imo-A) sample), trimethylmethoxysilane (Imo-M sample) or trichlorosilane (Imo-T sample). Adsorption isotherms of H_2O were measured at 25°C in the 0–0.027 bar range (Fig. 25.10) and compared to that of bare imogolite.

After surface modification with the organic moieties, the water uptake capacity of the materials decreased to about 60%–75% with respect to bare imogolite. Simultaneously, lower pore volumes were measured by nitrogen isotherms on modified imogolite. GCMC simulations showed that water adsorption in imogolite occurs by formation of multiple layers. The first layer is due to the formation of H-bonding between water molecules and inner

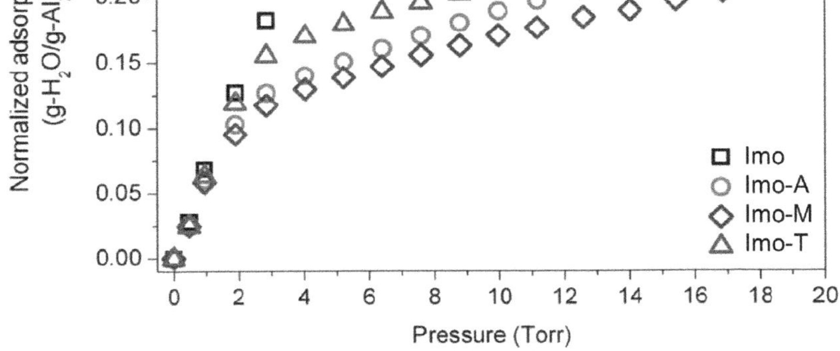

FIG. 25.10 Water adsorption isotherms at 25°C on bare imogolite (Imo) and modified imogolite (Imo-A, Imo-M and Imo-T). *Reprinted with permission from Kang et al. (2011). Copyright 2011 American Chemical Society.*

SiOH and then multi-layer adsorption occurs, in that successive layers form by H-bonding between water molecules.

The model proposed for water adsorption in Fig. 25.11 included (i) the formation of a first adsorbed layer, in which the water molecules only interact with residual SiOH (ie, non substituted by the organic moieties) and (ii) subsequent formation of a water layer, adjacent to the first water layer, by H-bonding between two adjacent water molecules.

The mechanism of water adsorption was modelled according to the Brunauer–Emmett–Teller (BET) method. The BET plot was applied in the $0.1 < P/P^0 < 0.35$ region, so obtaining linear plots (Fig. 25.13).

The fitted slope of the simplified BET equation

$$\frac{P/P_0}{n(1-P/P_0)} = \frac{1}{n_m}(P/P_0) + \frac{1}{n_m C} \tag{25.1}$$

was used to evaluate the monolayer water coverage, n_m (n_m = g-H_2O/g-$Al_2O_3SiO_2$). The following values of n_m were obtained: 0.174 (bare imogolite), 0.118 (Imo-A), 0.107 (Imo-M) and 0.131 (Imo-T). The decrease of n_m values for modified imogolite further confirmed that a fraction of inner SiOH was replaced by more hydrophobic organic functionalities (Fig. 25.12).

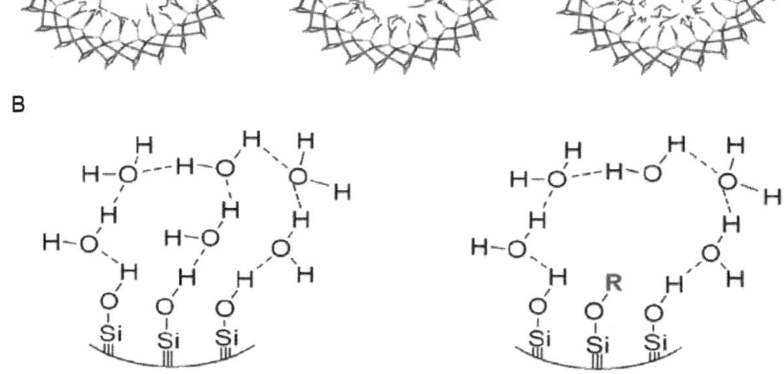

FIG. 25.11 (A) Simulated bare imogolite-water models at different water loadings as obtained by GCMC simulations. (B) Proposed water adsorption mechanisms in the channels of bare and modified imogolite. *Reprinted with permission from Kang et al. (2011). Copyright 2011 American Chemical Society.*

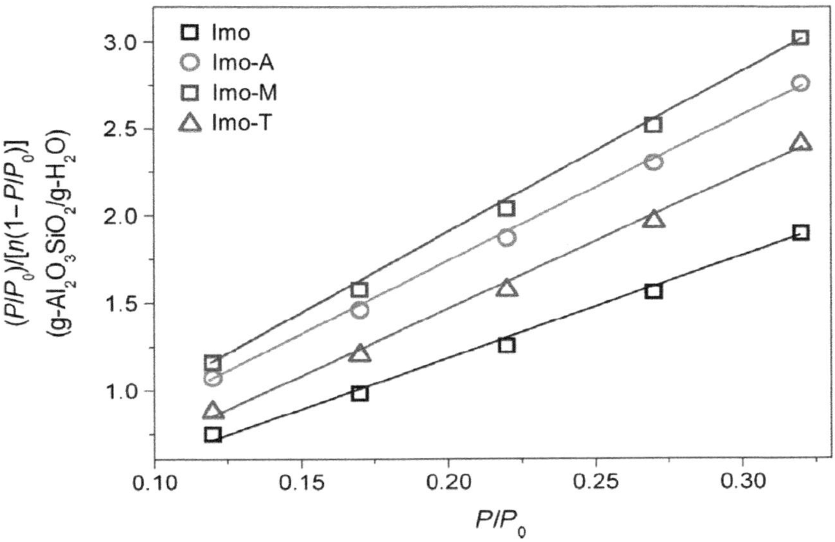

FIG. 25.12 BET plots derived from water adsorption isotherms for bare imogolite (Imo) and modified imogolite (Imo-A, Imo-M and Imo-T). *Reprinted with permission from Kang et al. (2011). Copyright 2011 American Chemical Society.*

The presence of water within A pores may inhibit the adsorption of other species, as shown by a study on Bz displacement by Wilson et al. (2002) who performed a nuclear magnetic resonance (NMR) study of Bz adsorption on both hydrated and dehydrated imogolite. They concluded that on hydrated imogolite, Bz only adsorbs at the C surface, where it is tightly bound since it does not evaporate upon adsorption. On dehydrated imogolite, Bz can adsorb on both A and C pores. In the former case, since the Bz size is about half the diameter of A pores (according to Adams (1980) the kinetic diameter of Bz is 0.62 nm), Bz molecules are less mobile than those adsorbed at the C surface and are not displaced by either water or air. Concerning adsorption at C surface, at least two types of adsorbed Bz were detected by NMR: one loosely bound, the other more tightly bound. Accordingly, a further IR study by Bonelli et al. (2013) showed that Bz molecules interact with both A and C surfaces. Bz adsorption at r.t. on dehydrated imogolite gave rise to several bands in the 2300–1300 cm^{-1} range of the corresponding IR spectra (Fig. 25.13). Bands at 1967 and 1825 cm^{-1} were assigned to overtones of out-of-plane C–H deformation of adsorbed Bz, occurring at 1961 cm^{-1} ($\nu_{11}+\nu_{19}$) and 1915 cm^{-1} ($\nu_{18}+\nu_{19}$) in the free molecule. In π-complexes, such bands shift to higher wavenumbers, with respect to the gas phase ($\Delta\nu$ is positive), whereas if end-on complexes form, $\Delta\nu$ is either zero or negative (Ramis et al., 1993). The most intense band at 1479 cm^{-1} is slightly positively shifted with respect to the gas phase (1482 cm^{-1}) like for Bz adsorbed on hydroxyls (Archipov et al., 2009). In the OH stretch region, a negative band

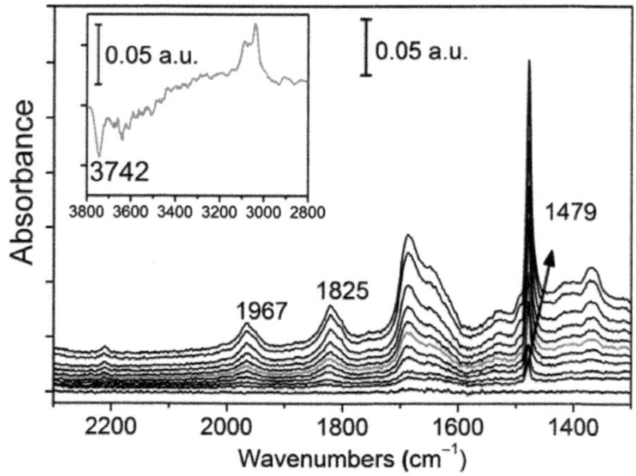

FIG. 25.13 IR spectra, in the 2300–1300 cm^{-1} region, obtained after dosing Bz (equilibrium pressure in the 0.05–40 mbar range) on dehydrated imogolite. *Inset* shows the OH stretch range of the *red* (*grey* in the print version) spectrum (equilibrium pressure of 5.0 mbar). Different spectra are reported, as obtained by subtraction of the spectrum of the bare sample. *Adapted with permission from Bonelli et al. (2013). Copyright 2013 the PCCP Owner Societies.*

was seen at 3742 cm^{-1}, indicating that interaction also occurred with SiOH; ie, Bz molecules may probe both A and C surfaces.

Concerning selective adsorption, one of the most studied subjects is selectivity to CO_2, which is relevant for carbon capture from flue gas and natural gas purification (Tagliabue et al., 2009; Bollini et al., 2011). Adsorptive selectivity of CO_2 from CO_2/CH_4 and CO_2/N_2 mixtures were determined by Kang et al. (2014) on imogolite modified with up to 15% inner Si–OH replaced by Si–CH_2–NH_2 groups (hereafter referred to as 'amine functionalized nanotubes').

Fig. 25.14 reports single gas adsorption isotherms of CO_2, CH_4 and N_2 on both bare imogolite [blue (dark grey in the print version) diamonds] and amine functionalized nanotube [red (grey in the print version) squares], with equilibrium pressures below 10 bar.

Fig. 25.14 shows that bare imogolite has larger adsorption capacities towards the three studied gases. This was explained by the fact that inner Si–CH_2–NH_2 have lower affinities than SiOH for the studied gas. Ideal adsorption selectivity was calculated as the ratio of the single component adsorption uptake of two different adsorbates at the same partial pressure and the obtained values for CO_2 are reported in Fig. 25.15.

Aminomethyl groups greatly suppressed adsorption of both CH_4 and N_2 as compared to CO_2, leading to an increased selectivity towards carbon dioxide, and so amine functionalized nanotube showed a higher selectivity for CO_2 adsorption, due to the presence of inner amino groups.

CO_2 adsorption was also studied by Zanzottera et al. (2012a) on imogolite, Me-imogolite and a material obtained by functionalization of Me-imogolite

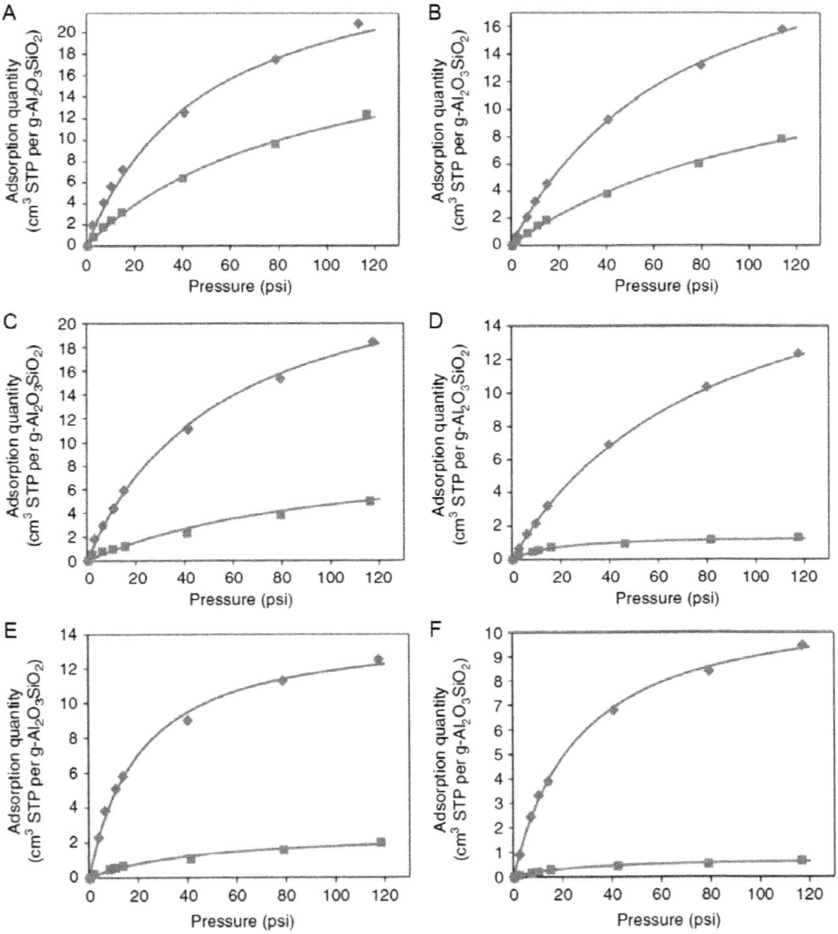

FIG. 25.14 Adsorption isotherms of (A) CO_2 at 25°C, (B) CO_2 at 67°C, (C) CH_4 at 25°C, (D) CH_4 at 67°C, (E) N_2 at 25°C and (F) N_2 at 67°C on imogolite [*blue* (*dark grey* in the print version) *diamonds*] and amine functionalized nanotube [*red* (*grey* in the print version) *squares*]. The *solid lines* correspond to curve-fittings determined by applying a Langmuir adsorption model. *Reprinted with permission from Kang et al. (2014). Copyright 2014 Nature Publishing Group.*

with 3-APTES (Me-Imo-NH_2, Zanzottera et al., 2012c), so bearing amino-functionalities at the outer surface of imogolite (Fig. 25.16).

The three materials exhibited rather high SSA values (355–665 $m^2\ g^{-1}$) and were accessible to CO_2 molecules. Two kinds of measurements were carried out: adsorption of CO_2 at r.t. was followed by IR spectroscopy, to get qualitative information on the type and relative amounts of adsorbed species, and volumetric isotherms were measured at 0°C, to obtain quantitative information on the total adsorbed amounts.

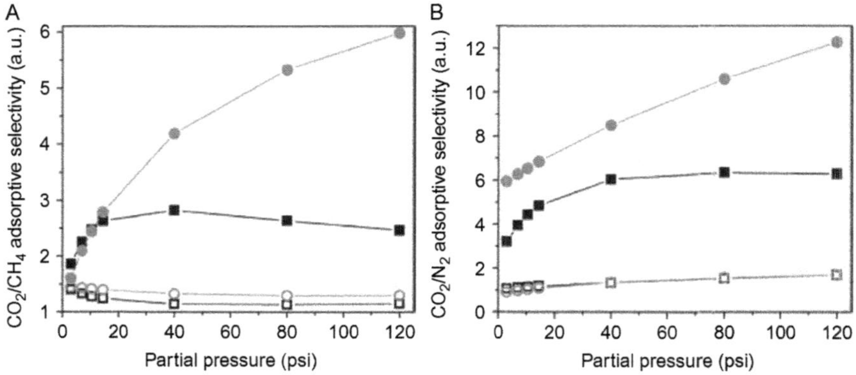

FIG. 25.15 CO_2 adsorption selectivity of imogolite (*open symbols*) and amine functionalized nanotube (*full symbols*) for gas pairs (A) CO_2/CH_4 and (B) CO_2/N_2 at 25°C (*squares*) and 67°C (*circles*). *Reprinted with permission from Kang et al. (2014). Copyright 2014 Nature Publishing Group.*

IR spectra showed that CO_2 may interact in a variety of ways with imogolite and imogolite-like materials:

(i) at the inner surface of imogolite, linear molecular species are reversibly formed by interaction with SiOH, while at the outer surface carbonate-like species are given rise with partial reversible character;
(ii) with Me-imogolite, no interaction took place at the inner surface, whereas linear molecular species formed in the inter-tube nanopores B, as well as carbonate species, as in the case of imogolite;
(iii) with Me-Imo-NH_2, all species present in Me-imogolite were found, as well as reversible carbamate species arising from the reaction with amino groups.

Optical isotherms obtained by IR spectra concerning adsorption of molecular CO_2 are shown in Fig. 25.17. They all showed Langmuir character, whereas those for the reversible formation of carbonates/carbamates species were of the Henry-type.

On the basis of IR results, volumetric isotherms (Fig. 25.18) were interpreted as being due to two independent families of adsorption sites (Langmuir and Henry, respectively). Comparison between optical isotherms (measured at ca. 33°C) and volumetric isotherms (measured at 0°C) allowed a semiquantitative estimate of the adsorption enthalpy for molecular species, corresponding to ca. $-20\ \mathrm{kJ\ mol^{-1}}$, for linear species reversibly formed by interaction with inner SiOH in imogolite and to a relatively high adsorption enthalpy for molecular species formed in the larger inter-tube nanopores of Me-imogolite (ca. $-32\ \mathrm{kJ\ mol^{-1}}$). The former value was compatible with a weak H-bond interaction, and indeed was close to the value reported for CO_2 adsorbed on silica (Ueno and Bennett, 1978). The latter value, which was closer

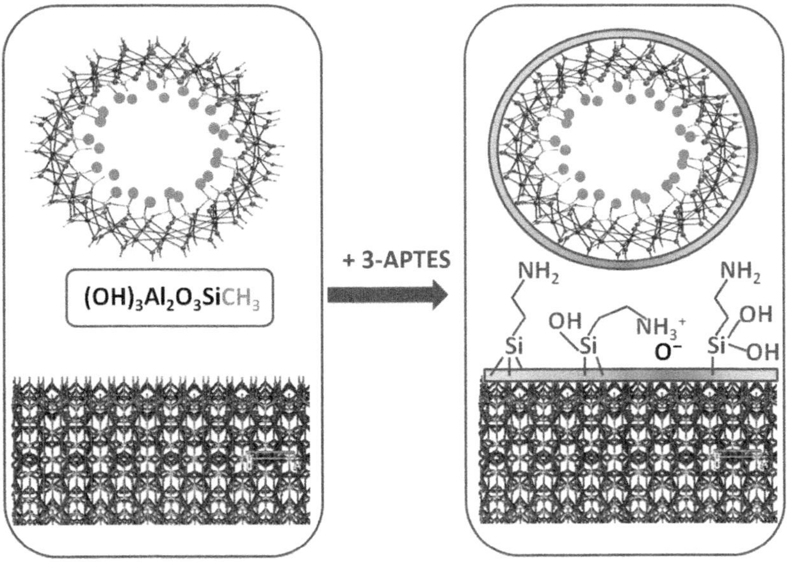

FIG. 25.16 Sections of natural imogolite with an inner diameter of about 1.0 nm; imogolite-like materials [namely, Me-imogolite (Me-Imo, chemical formula $(OH)_3Al_2O_3SiCH_3$), with methyl groups replacing inner SiOH in imogolite] and Me-Imo-NH_2, obtained from Me-imogolite by grafting with 3-APTES. *Reprinted with permission from Zanzottera et al. (2012c). Copyright 2012 American Chemical Society.*

to the values measured for the CO_2 interaction with protonic zeolites (Armandi et al., 2009), was in agreement with a van der Waals interaction with narrow B pores that are accessible in Me-imogolite bundles in virtue of the large diameter of B pores (Zanzottera et al., 2012a; Bonelli et al., 2013).

Another aspect of imogolite adsorptive properties concerns its ability to adsorb charged species from aqueous environment. According to Farmer et al. (1983), it was uncertain whether imogolite has a small ion exchange capacity or simply acts as an adsorbent for salts, acids or bases from solution. The same researchers stated that imogolite salt adsorption capacity lies in the 0.15–0.30 mmol g^{-1} range for both strong and weakly adsorbed salts, which are probably competing for the same adsorption site.

As mentioned in Section 25.1 earlier in this chapter, at neutral pH, inner SiOH are partially dissociated, whereas the outer surface of imogolite is positively charged due to the protonation of Al–OH–Al groups (Gustafsson, 2001). In addition to Al(OH)Al bridges, a few Al–OH defective groups, due to broken bonds, occur at the outer surface, most likely at the nanotube ends. They are more abundant in allophane, and at low pH values, they are protonated, as are Al(OH)Al groups (Parfitt and Henmi, 1980).

A charge density of one positive charge per 700 Å was calculated at low pH, corresponding to about one charge every 40 Al atoms (Theng et al., 1982). Since

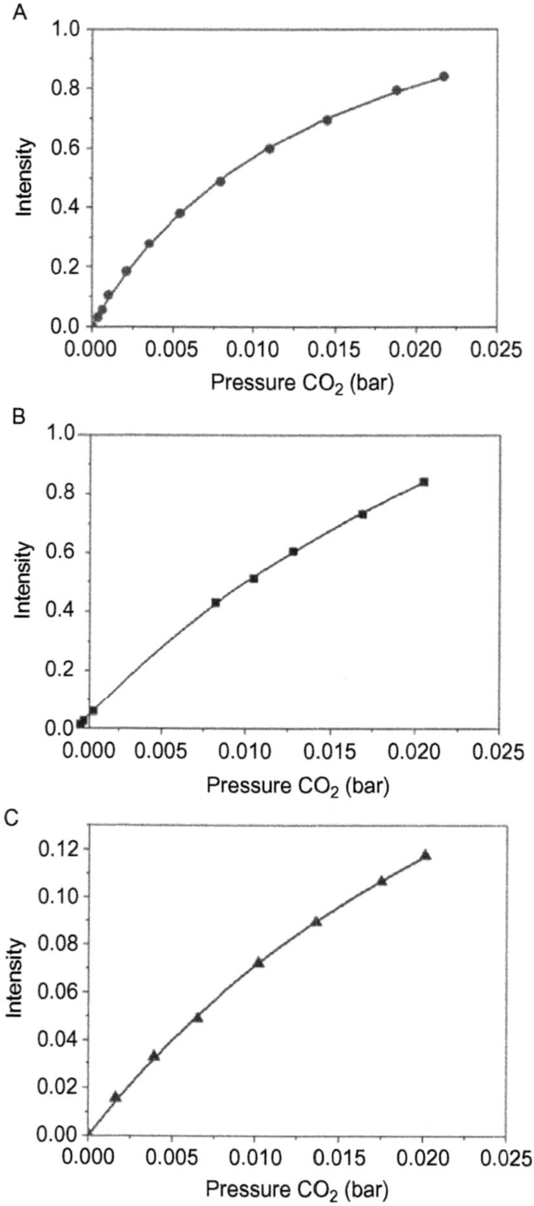

FIG. 25.17 Optical isotherms as obtained by plotting the intensity of the ν_3 CO_2 IR bands as a function of equilibrium pressures for imogolite (A), Me-imogolite (B), and Me-imogolite-NH_2 (C) dehydrated at 300°C. The curve-fittings were obtained by applying the Langmuir adsorption model. *Reprinted with permission from Zanzottera et al. (2012a). Copyright 2012 American Chemical Society.*

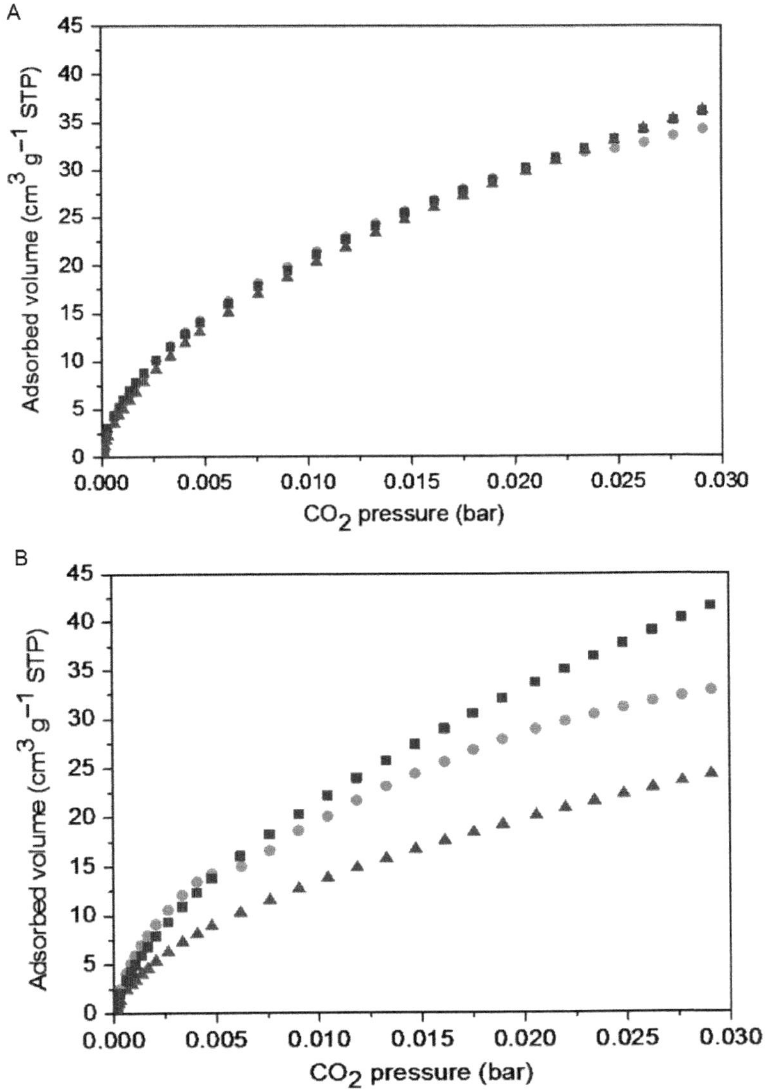

FIG. 25.18 CO_2 adsorption isotherms at 0°C (0.000–0.030 bar) on imogolite (*circles*); Me-imogolite (*squares*) and Me-imogolite-NH_2 (*triangles*) outgassed at r.t. (A) and dehydrated at 300°C (B). *Reprinted with permission from Zanzottera et al. (2012a). Copyright 2012 American Chemical Society.*

such value was obtained by considering a SSA of 1450 $m^2\ g^{-1}$ (Wada, 1977), a greater charge density may be expected for samples with lower SSA.

An important parameter to evaluate the behaviour of imogolite in water is the point of zero charge (PZC), which for imogolite is around 9.0–10, a value very close to that of alumina nanoparticles (Rotoli et al., 2014). The variation of PZC was also used as a method to follow the processes occurring during the formation of nanotubes in synthetic imogolite. Depending on the synthesis procedure, PZC increases from values around 6.6–7.1 in imogolite precursor to values in the 9.2–10.5 range for imogolite (Arancibia-Miranda et al., 2011).

The presence of charged species in the surrounding environment affects the surface charge of imogolite. In Fig. 25.19, ζ-potential measurements (Rotoli et al., 2014) of imogolite in deionized water and in two cell culture media, referred to as dichlorodifluoromethane (F-12) and 2-methoxyethoxymethyl (MEM), in either the presence or absence of foetal bovine serum (FBS) are shown. ζ-potential of imogolite was much less in the two cell culture media supplemented with FBS: very similar curves were obtained with both F-12 (full squares) and MEM (full triangles) supplemented with FBS, since the two media have similar compositions, with the PZC shifting to lower pH values (pH 4.2). Ions present in the two media were likely adsorbed by imogolite, finally lowering the net charge. Alternatively, the loss of positive charges might be due to adsorption of proteins from the serum present in the cell culture media. Measurements run in the two media, in the absence of FBS, gave similar curves with both F-12 (white squares) and MEM (white triangles), but PZC shifted to higher values, indicating a nonnegligible role of FBS on adsorption on imogolite. FBS indeed tends to adsorb at the surface of alumina (Al_2O_3), which has similar composition to imogolite external surface (Rezwan et al., 2004).

Anions are expected to interact with positively charged sites at the external surface of imogolite: these are mainly bridged Al(OH)Al, but also Al-related

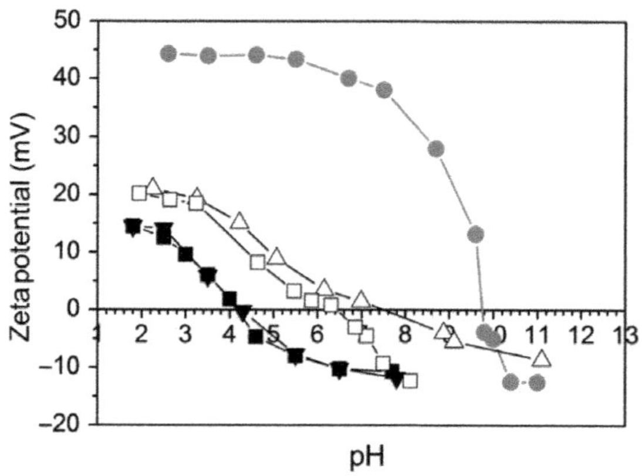

FIG. 25.19 ζ-potential curves of imogolite in: water [*red* (*grey* in the print version) *circles*], F-12 medium (*black triangles*) and MEM medium (*black squares*) both supplemented with 10% FBS, and F-12 (*white triangles*) and MEM (*white squares*), both without FBS. *Reprinted with permission from Rotoli et al. (2014). Copyright 2014 American Chemical Society.*

defects due to broken bonds, Al–OH groups, since both species are protonated. Such Al–OH defects are supposed to be the same species that occur in allophane, where a larger number of broken bonds are exposed (Parfitt and Henmi, 1980), whereas in imogolite, they should occur in smaller amounts, mostly at the ends of the nanotubes.

In allophane, phosphate ions $\left(PO_4^{3-}\right)$ preferentially adsorb on Al–OH defects by ligand exchange mechanism: the adsorption capacity of imogolite towards PO_4^{3-} is less than allophane due to the lower number of Al–OH defects. However, the affinity of imogolite for PO_4^{3-} is greater than for NO_3^-: adsorption of phosphate from 2.7 mM PO_4^{3-} solution was found to inhibit adsorption of NO_3^- from 5-mM solution, and only 0.1 mmol g^{-1} NO_3^- was adsorbed after PO_4^{3-} adsorption (Farmer et al., 1983). Theng et al. (1982) showed that imogolite had a much higher affinity for Cl^- ions, which are expected to adsorb on positively charged (and more abundant) Al–(OH)–Al groups. Retention of small anions (Cl^- and ClO_4^-) was observed at pH levels above the values at which any positive charge should be present (Clark and McBride, 1984a). The same phenomenon did not occur in the case of larger (organic) anions, indicating that only small anions could be intercalated within pores B. This result was confirmed by a TG-mass (Thermogravimetry coupled to Mass spectrometry) study carried out by Zanzottera et al. (2012b), who observed the thermal decomposition of ClO_4^- ions trapped in the B pores of imogolite bundles, synthesized in the presence of $HClO_4$. Accordingly, Wada (1984) explained the excess 'salt absorption' (eg, NaCl) due to intercalation of both cation and anion among nanotubes. Also, Farmer et al. (1983) hypothesized that adsorption of $BaCl_2$ occurred at internanotube pores B. Indeed, recent results have shown that chloride ions may be occluded within B pores in imogolite bundles (Shafia et al., 2015a) when preformed imogolite is put in contact with a solution containing Cl^- ions.

Considering important anion pollutants, removal of arsenate ion [As(V)] is an important environmental challenge, and it has been studied on several adsorbents, including bare imogolite and imogolite magnetite nanocomposites (Arancibia-Miranda et al., 2014). The adsorption of As(V) on the imogolite outer surface is favoured since both Al–(OH)–Al and Al–OH groups are protonated; arsenate is expected to adsorb through a ligand exchange mechanisms rather than by electrostatic interaction:

$$Al-OH+H_2AsO_4^- = Al-O_4AsH_2+OH^- \quad (25.2)$$

$$Al-OH_2^+ +H_2AsO_4^- = Al-O_4AsH_2+H_2O \quad (25.3)$$

The kinetics of As(V) removal obtained for bare imogolite, for imogolite magnetite nanocomposites (Fe_3O_4-Imo) and Fe_3O_4 particles are illustrated in Fig. 25.20.

When using imogolite Fe_3O_4 nanocomposite, up to 94.2% arsenate was removed and equilibrium was attained in the first 20 min. The curves were

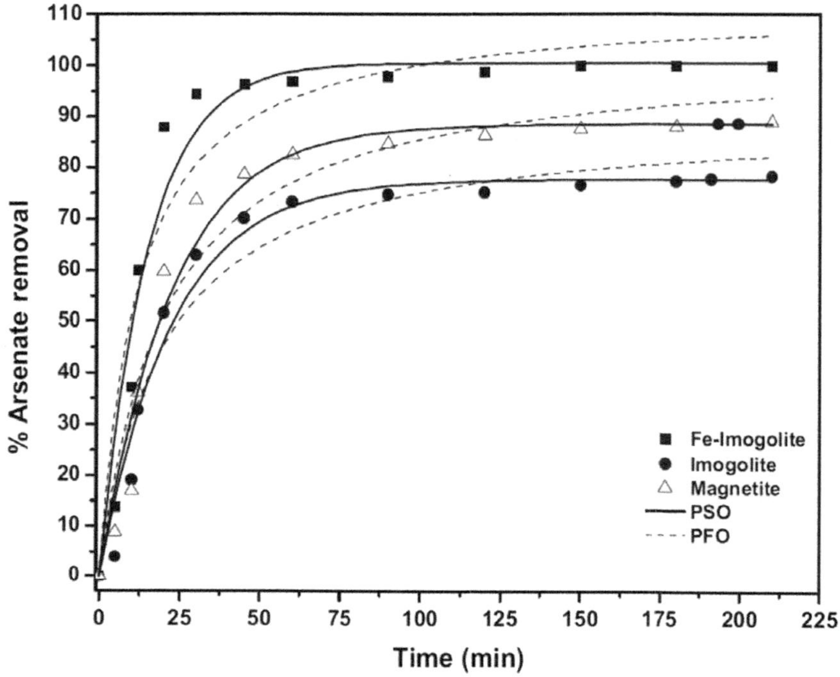

FIG. 25.20 Kinetics of arsenate adsorption on imogolite Fe_3O_4 nanocomposite (Fe-Imogolite, *squares*), Imogolite (*circles*) and magnetite (*triangles*). Curves obtained by either applying a pseudo-first-order or pseudo-second-order kinetic model. *Reprinted with permission from Arancibia-Miranda et al. (2014). Copyright 2014 Elsevier.*

modelled by both pseudo-first-order and pseudo-second-order models, with the latter providing better agreement with experimental results. According to Arancibia-Miranda et al. (2014), with imogolite Fe_3O_4 nanocomposite, the presence of new surface sites related to Fe-oxo/hydroxides should increase the surface affinity for As(V), by forming mono- and bidentate-binuclear complexes through the following ligand exchange mechanisms:

$$Fe-OH+HAsO_4{}^{2-}+H^+=FeHAsO_4{}^{2-}+H_2O \quad (25.4)$$

$$2(FeOH)+HAsO_4{}^{2-}+2H^+=Fe_2HAsO_4{}^{2-}+2H_2O \quad (25.5)$$

Reactions of As(V) ions with Fe-related hydroxyls species end up with arsenate ions coordinated by iron. A higher adsorption rate was obtained for Fe_3O_4-imogolite than for imogolite, indicating a higher affinity of As(V) for the imogolite Fe_3O_4 nanocomposite, although the presence of Fe_3O_4 brought about a reduction of both SSA (206 $m^2\,g^{-1}$ with respect to 303 $m^2\,g^{-1}$ of imogolite) and surface charge, with respect to imogolite, as shown in Fig. 25.21.

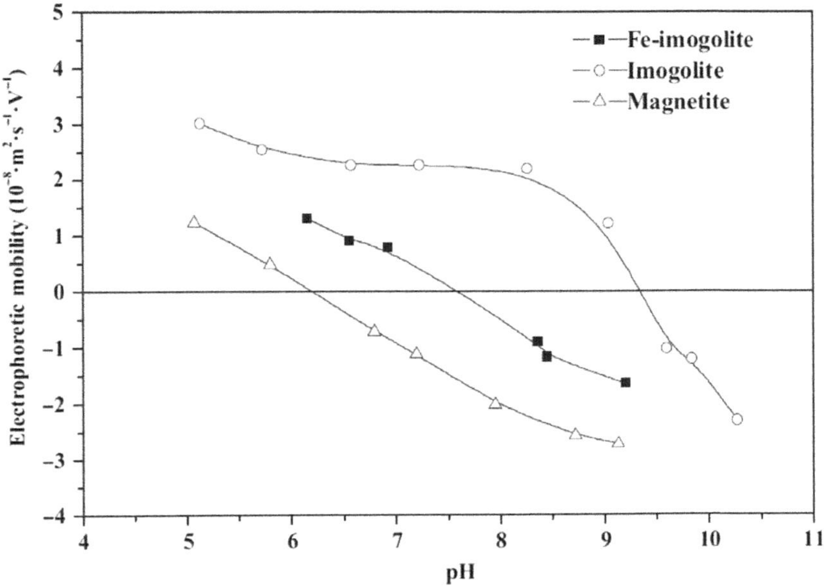

FIG. 25.21 Electrophoretic mobility vs pH plot of magnetite (*triangles*), imogolite Fe_3O_4 nanocomposite (*squares*) and imogolite (*circles*). *Reprinted with permission from Arancibia-Miranda et al. (2014). Copyright 2014 Elsevier.*

Adsorption of (small) cations at the inner surface of imogolite is expected to occur by electrostatic interaction with SiO^- groups (Harsh et al., 1992), and indeed, the adsorption capacity of imogolite towards Na^+ ions from 0.05 M NaCl solution was found to increase with pH (Theng et al., 1982).

Cations seem to compete for adsorbing sites: both ion size and hydration probably play a greater role than the ion charge in competitive adsorption (Farmer et al., 1983). The following cation selectivity was measured by contacting imogolite with 25 mM solutions of the respective chlorides: $Ba^{2+} > Mg^{2+} > K^+$. Low affinity was instead measured for bulky organic cations like cetylpyridinium (Farmer et al., 1983).

Denaix et al. (1999) measured the affinity for Pb^{2+}, Cu^{2+} and Cd^{2+} ions by potentiometric titration and found that adsorption of both Cu^{2+} and Pb^{2+} also occurs when the charge of the solid is positive. Such result was assigned to a specific adsorption of both Cu^{2+} and Pb^{2+} on imogolite, whereas Cd^{2+} gave nonspecific adsorption (ie, electrostatic attraction). Concerning specific adsorption, from adsorption isotherms of Cu^{2+} on both imogolite and allophane, Clark and McBride (1984b) derived that both specific and nonspecific adsorption occurred. These authors showed by electron spin resonance (ESR) that the presence of monomeric Cu^{2+} coordinated to a single Al ion forming a binuclear complex; they concluded that another kind of less energetic adsorption occurred by displacement of a single proton of a SiOH or AlOH group.

Functionalization of imogolite may be adopted to favour the adsorption of dangerous cations at the outer surface of imogolite. The selective adsorption of rubidium has been studied on both bare and functionalized imogolites (Fig. 25.22) with either 2-mercaptothiazoline (Imogolite-MTZ) or 2-mercaptobenzilimidazole (Imogolite-MBI) (Guerra et al., 2011).

The organic linkers were able to act as chelating agent, favouring the adsorption of Rb^+ from drinking water, with respect to another dangerous radionuclide (Th^{4+}) (Fig. 25.23). The competitive adsorption was examined at different pH levels. The distribution coefficient (K_d) was assumed to be a

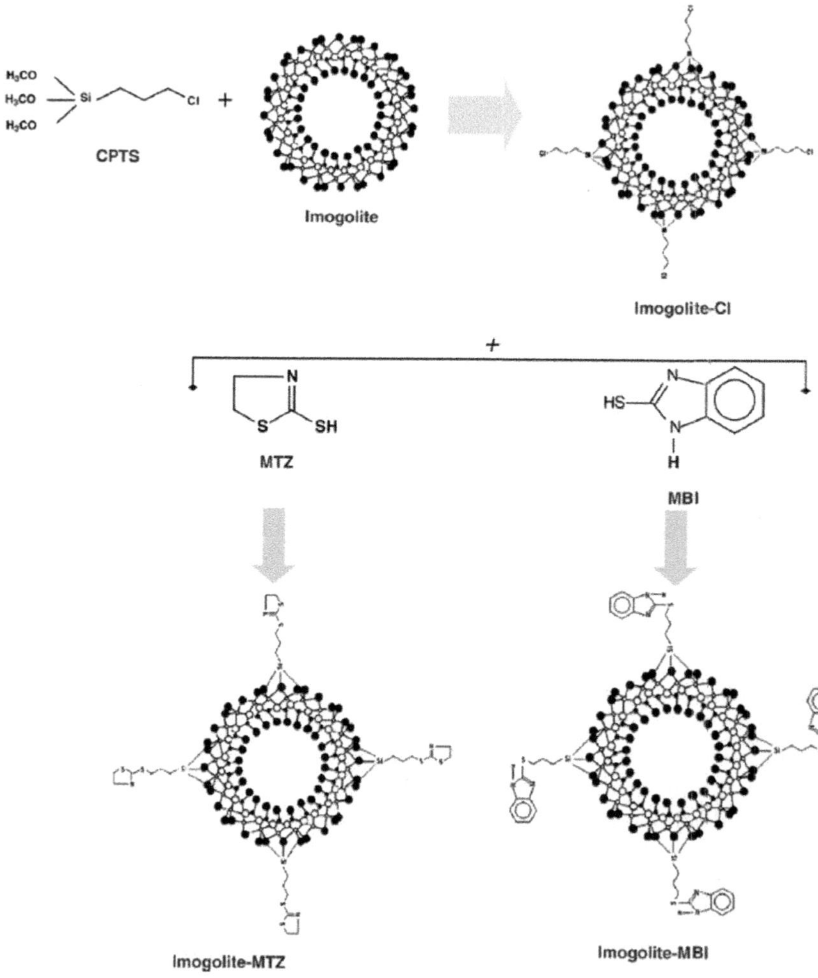

FIG. 25.22 Incorporation of 2-mercaptothiazoline (MTZ) and 2-mercaptobenzilimidazole (MBI) at the outer surface of imogolite. *Reprinted with permission from Guerra et al. (2011). Copyright 2011 Elsevier.*

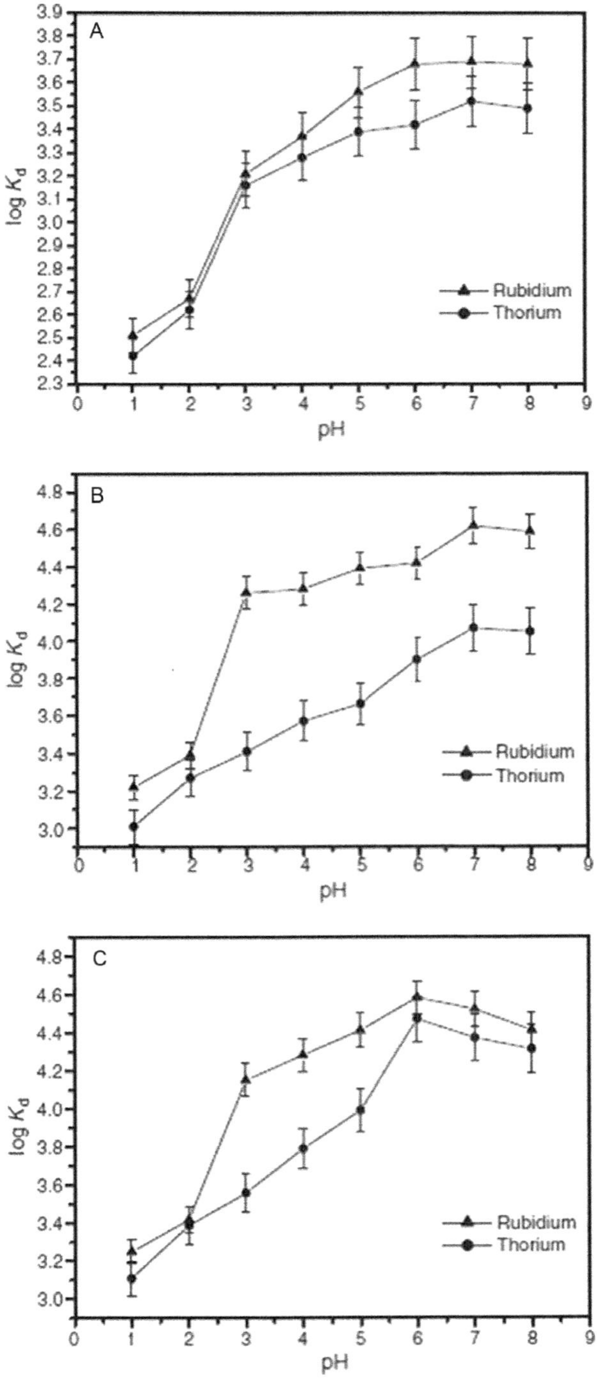

FIG. 25.23 Plot of log K_d (distribution coefficient) vs pH for adsorption of Th^{4+} and Rb^{+} on Imogolite (A), Imogolite-MBI (B) and Imogolite-MTZ (C). *Reprinted with permission from Guerra et al. (2011). Copyright 2011 Elsevier.*

representative value for selectivity measurements if all the tested radionuclide ions compete with each other in binding to the selective functional groups of the chelate-forming adsorbent.

Finally, Arai et al. (2006) observed a complexation of uranyl ions by the outer surface of bare imogolite, through the formation of several complexes by reacting with octahedral Al sites in a wide pH range (5–9).

25.5 CONCLUDING REMARKS

The catalytic applications of imogolite in gas-phase reactions are limited mainly by its thermal stability and by the difficult diffusion of reactants and products within its high-aspect-ratio nanotubes.

Both proto-imogolite and the structure formed upon thermal collapse of imogolite exhibit interesting catalytic properties: those of proto-imogolite are likely due to its open structure and to the presence of stronger acidic sites; those of collapsed imogolite are probably ascribable to the formation of new Al-related sites, with aluminium in a different coordination with respect to Al(VI), along with the presence of residual porosity. The catalytic properties of the latter material, however, still deserve some further study, mainly related to the acidic strength of the newly formed surface.

Another interesting aspect is the intrinsic reactivity of imogolite outer surface with hydrogen peroxide in water, which could be exploited in the future for reactions requiring the activation of H_2O_2. Concerning modified-imogolite catalysts studied so far, Fe-doped imogolite exhibits interesting catalytic properties in (partial) oxidation reactions occurring in liquid phase. To the best of our current knowledge, the photocatalytic behaviour of Fe-doped imogolite deserves further study due to the possibility of having a shift of the band gap by isomorphic substitution of Al^{3+} by Fe^{3+} ions.

The available studies on gases adsorption show some interesting outcomes, such as the fact that modified imogolite may selectively adsorb CO_2 from mixtures. However, further improvements are needed in order to exploit such materials in processes of economical relevance, like carbon capture and storage or methane adsorption. The latter two processes will probably occur with higher performance on modified imogolite, since the high hydrophilicity of bare imogolite may limit such applications due to the presence of water. For other kinds of applications, like gas chromatography, modified imogolite could be useful for producing selective columns.

ABBREVIATIONS

1-ol	2-cyclohexene-1-ol
ADH	asymmetric dihydroxylation
3-APTES	3-aminopropyltriethoxysilane

BET	Brunauer–Emmett–Teller
Bz	benzene
DH	dihydroxylation
EP	1,2-epoxycyclohexene
F-12	dichlorodifluoromethane
FBS	foetal bovine serum
GCMC	grand canonical Monte Carlo
Imo	imogolite
Imo-A	imogolite functionalized with acetyl chloride
Imo-c	imogolite collapsed phase
Imo-M	imogolite functionalized with trimethylmethoxysilane
Imo-T	imogolite functionalized with trichlorosilane
IR spectroscopy	infrared spectroscopy
Me-imogolite	imogolite with Si–CH_3 groups replacing Si–OH groups
Me-Imo-NH_2	Me-imogolite after functionalization of the outer surface with 3-aminopropyltriethoxysilane
MEM	2-methoxyethoxymethyl
MeOH	methanol
MTZ	2-mercaptothiazoline
NMO	*N*-methylmorfoline-*N*-oxide
NMR	nuclear magnetic resonance
Phe	phenol
PZC	point of zero charge
SSA	specific surface area

REFERENCES

Ackerman, W.C., Smith, D.M., Huling, J.C., Kim, Y.W., Bailey, J.K., Brinker, C.J., 1993. Gas/vapor adsorption in imogolite: a microporous tubular aluminosilicate. Langmuir 9, 1051–1057.

Adams, M.J., 1980. Gas chromatographic adsorption studies on synthetic imogolite. J. Chromatogr. 188, 97–106.

Alvarez-Ramírez, F., 2009. First principles studies of Fe-containing aluminosilicate and aluminogermanate nanotubes. J. Chem. Theory Comput. 5, 3224–3231.

Arai, Y., McBeath, M., Bargar, J.R., Joye, J., Davis, J.A., 2006. Uranyl adsorption and surface speciation at the imogolite-water interface: self-consistent spectroscopic and surface complexation models. Geochim. Cosmochim. Acta 70, 2492–2509.

Arancibia-Miranda, N., Escudey, M., Molina, M., García-González, M.T., 2011. Use of isoelectric point and pH to evaluate the synthesis of a nanotubular aluminosilicate. J. Non-Cryst. Solids 357, 1750–1756.

Arancibia-Miranda, N., Escudey, M., Pizarro, C., Denardin, L.C., García-González, M.T., Fabris, J.D., Charlet, L., 2014. Preparation and characterization of a single-walled aluminosilicate nanotube-iron oxide composite: its applications to removal of aqueous arsenate. Mater. Res. Bull. 51, 145–152.

Archipov, T., Santra, S., Ene, A.B., Stoll, H., Rauhut, G., Roduner, E., 2009. Adsorption of benzene to copper in CuHY zeolite. J. Phys. Chem. C 113, 4107–4116.

Armandi, M., Garrone, E., Areán, C.O., Bonelli, B., 2009. Thermodynamics of carbon dioxide adsorption on the protonic zeolite H-ZSM-5. ChemPhysChem 10, 3316–3319.

Baba, T., Handa, Y., Ono, Y., 1994. Superbase character of alumina loaded with potassium by impregnation from ammoniacal solution. J. Chem. Soc. Faraday Trans. 90, 187–191.

Ballarini, N., Cavani, F., Maselli, L., Montaletti, A., Passeri, S., Scagliarini, D., Flego, C., Perego, C., 2007. The transformations involving methanol in the acid- and base-catalyzed gas-phase methylation of phenol. J. Catal. 251, 423–436.

Ballarini, N., Cavani, F., Maselli, L., Passeri, S., Rovinetti, S., 2008. Mechanistic studies of the role of formaldehyde in the gas-phase methylation of phenol. J. Catal. 256, 215–225.

Bollini, P., Choi, S., Drese, J.H., Jones, C.W., 2011. Oxidative degradation of aminosilica adsorbents relevant to postcombustion CO_2 capture. Energy Fuel 25, 2416–2425.

Bonelli, B., Ribeiro, M.F., Antunes, A.P., Valange, S., Gabelica, Z., Garrone, E., 2002. Al-MCM-41 systems exchanged with alkali-metal cations: FT-IR characterization and catalytic activity towards 1-butene isomerization. Microporous Mesoporous Mater. 54, 305–317.

Bonelli, B., Bottero, I., Ballarini, N., Passeri, S., Cavani, F., 2009. IR spectroscopic and catalytic characterization of the acidity of imogolite-based systems. J. Catal. 264, 15–30.

Bonelli, B., Armandi, A., Garrone, E., 2013. Surface properties of alumino-silcate single-walled nanotubes of the imogolite type. Phys. Chem. Chem. Phys. 15, 13381–13390.

Bottero, I., Bonelli, B., Ashbrook, S.E., Wright, P.A., Zhou, W., Tagliabue, M., Armandi, M., Garrone, E., 2011. Synthesis and characterization of hybrid organic/inorganic nanotubes of the imogolite type and their behaviour towards methane adsorption. Phys. Chem. Chem. Phys. 13, 744–750.

Bregolato, M., Bolis, V., Busco, C., Ugliengo, P., Bordiga, S., Cavani, F., Ballarini, N., Maselli, L., Passeri, S., Rossetti, I., Forni, L., 2007. Methylation of phenol over high-silica beta zeolite: effect of zeolite acidity and crystal size on catalyst behavior. J. Catal. 245, 285–300.

Clark, C.J., McBride, M.B., 1984a. Cation and anion retention by natural and synthetic allophane and imogolite. Clay Clay Miner. 32, 291–299.

Clark, C.J., McBride, M.B., 1984b. Chemisorption of Cu(II) and Co(II) on allophane and imogolite. Clay Clay Miner. 32, 300–310.

Cradwick, P.D.G., Farmer, V.C., Russell, J.D., Masson, C.R., Wada, K., Yoshinaga, N., 1972. Imogolite, a hydrated aluminium silicate of tubular structure. Nat. Phys. Sci. 240, 187–189.

Creton, B., Bougeard, D., Smirnov, K.S., Guilment, J., Poncelet, O., 2008. Molecular dynamics study of hydrated imogolite: 2. Structure and dynamics of confined water. Phys. Chem. Chem. Phys. 10, 4879–4888.

Dahl, I.M., Kolboe, S., 1994. On the reaction mechanism for hydrocarbon formation from methanol over SAPO-34: I. Isotopic labeling studies of the co-reaction of ethene. J. Catal. 149, 458–464.

Denaix, L., Lamy, I., Bottero, J.Y., 1999. Structure and affinity towards Cd^{2+}, Cu^{2+}, Pb^{2+} of synthetic colloidal amorphous aluminosilicates and their precursors. Colloid Surf. A 158, 315–325.

Farmer, V.C., Adams, M.J., Fraser, A.R., Palmieri, F., 1983. Synthetic imogolite: properties, synthesis, and possible applications. Clay Miner. 18, 459–472.

Fu, Y., Baba, T., Ono, Y., 1998. Vapor-phase reactions of catechol with dimethyl carbonate. Part IV: synthesis of catechol carbonate over alumina loaded with cesium hydroxide. Appl. Catal. A Gen. 178, 219–223.

Guerra, D.L., Batista, A.C., Viana, R.R., Airoldi, C., 2011. Adsorption of rubidium on raw and MTZ- and MBI-imogolite hybrid surfaces: an evidence of the chelate effect. Desalination 275, 107–117.

Guimaraes, l., Enyashin, A.N., Frenzel, J., Heine, T., Duarte, A., Seifert, G., 2007. Imogolite nanotubes: stability, electronic, and mechanical properties. ACS Nano 1, 362–368.

Gustafsson, J.P., 2001. The surface chemistry of imogolite. Clay Clay Miner. 49, 73–80.

Harsh, J.B., Traina, S.J., Boyle, J., Yang, Y., 1992. Adsorption of cations on imogolite and their effect on surface charge characteristics. Clay Clay Miner. 40, 700–706.

Haw, J.F., Song, W., Marcus, D.M., Nicholas, J.B., 2003. The mechanism of methanol to hydrocarbon catalysis. Acc. Chem. Res. 26, 317–326.

Iler, R.K., 1979. The Chemistry of Silica: Solubility, Polymerization, Colloid and Surface Properties, and Biochemistry of Silica. John Wiley and Sons, New York.

Imamura, S., Hayashi, Y., Kajiwara, K., Hoshino, H., Kaito, C., 1993. Imogolite: a possible new type of shape-selective catalyst. Ind. Eng. Chem. Res. 32, 600–603.

Imamura, S., Kokubu, T., Yamashita, T., Okamoto, Y., Kajiwara, K., Kanai, H., 1996. Shape-selective copper-loaded imogolite catalyst. J. Catal. 160, 137–139.

Kang, D.Y., Zang, J., Jones, C.W., Nair, S., 2011. Single-walled aluminosilicate nanotubes with organic-modified interiors. J. Phys. Chem. C 115, 7676–7685.

Kang, D.Y., Brunelli, N.A., Yucelen, G.I., Venkatasubramanian, A., Zang, J., Leisen, J., Hesketh, P.J., Jones, C.W., Nair, S., 2014. Direct synthesis of single-walled aminoaluminosilicate nanotubes with enhanced molecular adsorption selectivity. Nat. Commun. 5, 3342.

Klemm, L.H., Shabtai, J., Taylor, D.R., 1968. Alumina-catalyzed reactions of hydroxyarenes and hydroaromatic ketones. I. Reactions of 1-naphthol with methanol. J. Org. Chem. 33, 1480–1489.

MacKenzie, K.J., Bowden, M.E., Brown, J.W.M., Meinhold, R.H., 1989. Structural and thermal transformation of imogolite studied by ^{29}Si and ^{27}Al high-resolution solid-stated magnetic nuclear resonance. Clay Clay Miner. 37, 317–324.

Mandelli, D., van Vliet, M.C.A., Sheldon, R.A., Schuchardt, U., 2001. Alumina-catalyzed alkene epoxidation with hydrogen peroxide. Appl. Catal. A Gen. 219, 209–213.

Menon, V.C., Komarneni, S., 1998. Porous adsorbents for vehicular natural gas storage: a review. J. Porous. Mater. 5, 43–58.

Mikkelsen, Ø., Kolboe, S., 1999. The conversion of methanol to hydrocarbons over zeolite H-beta. Microporous Mesoporous Mater. 29, 173–184.

Monfared, H.H., Amouei, Z., 2004. Hydrogen peroxide oxidation of aromatic hydrocarbons by immobilized iron (III). J. Mol. Catal. A Chem. 217, 161–164.

Ohashi, F., Tomura, S., Akaku, K., Hayashi, S., Wada, S.I., 2004. Characterization of synthetic imogolite nanotubes as gas storage. J. Mater. Sci. 39, 1799–1801.

Ookawa, M., 2012. Synthesis and characterization of Fe-imogolite as an oxidation catalysts. In: Marta, V., Grazyna, S. (Eds.), Clay Minerals in Nature—Their Characterization, Modification and Application. InTech, Croatia, pp. 239–258.

Ookawa, M., Inoue, Y., Watanabe, M., Suzuki, M., Yamaguchi, T., 2006. Synthesis and characterization of Fe containing imogolite. Clay Sci. 12, 280–284.

Ookawa, M., Takata, Y., Suzuki, M., Inukai, K., Maekawa, T., Yamaguchi, T., 2008. Oxidation of aromatic hydrocarbons with H_2O_2 catalyzed by a nano-scale tubular aluminosilicate, Fe-containing imogolite. Res. Chem. Intermed. 34, 679–685.

Parfitt, R.L., Henmi, T., 1980. The structure of some allophones from New Zealand. Clay Clay Miner. 28, 285–294.

Parfitt, R.L., Thomas, A.D., Atkinson, R.J., Smart, R.St., 1974. Adsorption of phosphate on imogolite. Clay Clay Miner. 22, 455–456.

Porchet, S., Kiwi-Minsker, L., Doepper, R., Renken, A., 1996. Catalyst development for the selective methylation of catechol. Chem. Eng. Sci. 51, 2933–2938.

Qi, X., Yoon, H., Lee, S.H., Yoon, J., Kim, S.J., 2008. Surface-modified imogolite with 3-APS-OsO_4 complex: synthesis, characterization and its application in the dihydroxylation of olefins. J. Ind. Eng. Chem. 14, 136–141.

Ramis, G., Busca, G., Lorenzelli, V., 1993. Determination of the geometry of adsorbed unsaturated molecules through the analysis of the CH out-of-plane deformation modes. J. Electron Spectrosc. Relat. Phenom. 64/65, 297–305.

Rezwan, K., Meier, L.P., Rezwan, M., Textor, J.M., Gauckler, L.J., 2004. Bovine serum albumin adsorption onto colloidal Al_2O_3 particles: a new model based on zeta potential and UV-vis measurements. Langmuir 20, 10055–10061.

Rotoli, B.M., Guidi, P., Bonelli, B., Bernardeschi, M., Bianchi, M.G., Esposito, S., Frenzilli, G., Lucchesi, P., Nigro, M., Scarcelli, V., Tomatis, M., Zanello, P.P., Fubini, B., Bussolati, O., Bergamaschi, E., 2014. Imogolite: an aluminosilicate nanotube endowed with low cytotoxicity and genotoxicity. Chem. Res. Toxicol. 27, 1142–1154.

Senskova, I., Kaskel, S., 2008. High pressure methane adsorption in the metal-organic frameworks $Cu_3(btc)_2$, $Zn_2(bdc)_2$dabco, and $Cr_3F(H_2O)_2O(bdc)_3$. Microporous Mesoporous Mater. 112, 108–115.

Shafia, E., Esposito, E., Manzoli, M., Chiesa, M., Tiberto, P., Barrera, G., Menard, G., Allia, P., Freyria, F.S., Garrone, E., Bonelli, B., 2015. Al/Fe isomorphic substitution versus Fe_2O_3 clusters formation in Fe-doped aluminosilicate nanotubes (imogolite). J. Nanoparticle Res. 17, 336.

Shafia, E., Esposito, S., Armandi, M., Bahadori, E., Garrone, E., Bonelli, B., 2015. Reactivity of bare and Fe-doped alumino-silicate nanotubes (imogolite) with H_2O_2 and the azo-dye acid orange 7. Catal. Today. http://dx.doi.org/10.1016/j.cattod.2015.10.011. in press.

Su, C., Harsh, J.B., Bertsch, P.M., 1992. Sodium chloride sorption by imogolite allophanes. Clay Clay Miner. 40, 280–286.

Tagliabue, M., Farrusseng, D., Valencia, S., Aguado, S., Ravon, U., Rizzo, C., Corma, C., Mirodatos, C., 2009. Natural gas treating by selective adsorption: material science and chemical engineering interplay. Chem. Eng. J. 155, 553–566.

Theng, B.K.G., Russell, M., Churchman, G.J., Parfitt, R.L., 1982. Surface properties of allophane, halloysite, and imogolite. Clay Clay Miner. 30, 143–149.

Ueno, A., Bennett, C.O., 1978. Infrared study of CO_2 adsorption on SiO_2. J. Catal. 54, 31–44.

Wada, K., 1977. Allophane and imogolite. In: Dixon, J.B., Weed, S.B. (Eds.), Minerals in Soil Environments. Soil Science Society of America, Madison, Wisconsin, America, pp. 603–638.

Wada, K., 1984. Mechanism of apparent salt absorption in ando soils. Soil Sci. Plant Nutr. 30, 77–83.

Wichterlova, B., Zilkova, N., Uvarova, E., Cejka, J., Sarv, P., Paganini, M.C., Lercher, J.A., 1999. Effect of Broensted and Lewis sites in ferrierites on skeletal isomerization of n-butenes. Appl. Catal. A 182, 297–308.

Wilson, M.A., Lee, G.S.H., Taylor, R.C., 2002. Benzene displacement on imogolite. Clay Clay Miner. 50, 348–351.

Zang, J., Chempath, S., Konduri, S., Mair, S., Sholl, D.S., 2010. Flexibility of ordered surface hydroxyls influences the adsorption of molecules in single-walled aluminosilicate nanotubes. J. Phys. Chem. Lett. 1, 1235–1240.

Zanzottera, C., Armandi, M., Esposito, S., Garrone, E., Bonelli, B., 2012a. CO_2 adsorption on aluminosilicate single-walled nanotubes of imogolite type. J. Phys. Chem. C 116, 20417–20425.

Zanzottera, C., Vicente, A., Armandi, M., Fernandez, C., Garrone, E., Bonelli, B., 2012b. Thermal collapse of single-walled alumino-silicate nanotubes: transformation mechanisms and morphology of the resulting lamellar phases. J. Phys. Chem. C 116, 23577–23584.

Zanzottera, C., Vicente, A., Celasco, E., Fernandez, C., Garrone, E., Bonelli, B., 2012c. Physico-chemical properties of imogolite nanotubes functionalized on both external and internal surfaces. J. Phys. Chem. C 116, 7499–7506.

Zhu, J.H., Chun, Y., Wang, Q., Xu, Q., 1999. Attempts to create new shape-selective solid strong base catalysts. Catal. Today 51, 103–111.

Chapter 26

Health and Medical Applications of Tubular Clay Minerals

C. Aguzzi[a], G. Sandri[b], P. Cerezo[a], E. Carazo[a] and C. Viseras[a,*]
[a]*University of Granada, Granada, Spain*
[b]*University of Pavia, Pavia, Italy*
[*]*Corresponding author: e-mail: cviseras@ugr.es*

26.1 INTRODUCTION

This chapter aims to review the possible applications of halloysite and imogolite, two nanosized clay minerals with a hollow tubular structure, in nanopharmaceutics (ie, the use of nanotechnology in pharmaceutics). Pharmaceutics is concerned with the scientific and technological aspects of the design and manufacture of medicinal products and medical devices (Aulton and Taylor, 2013). Medicinal products are defined as "(i) Any substance or combination of substances presented as having properties for treating or preventing disease in human beings; or (ii) Any substance or combination of substances which may be used in or administered to human beings either with a view to restoring, correcting or modifying physiological functions by exerting a pharmacological, immunological or metabolic action, or to making a medical diagnosis" (European Directive, 2001). Medical devices are defined as "Any instrument, apparatus, appliance, software, material or other article, whether used alone or in combination, together with any accessories, ..., for the purpose of diagnosis, prevention, monitoring, treatment or alleviation of disease,..., which does not achieve its principal intended action in or on the human body by pharmacological, immunological or metabolic means,..." (European Directive, 2007).

Medicinal products and medical devices are regulated throughout their entire life cycle, including basic and preclinical research, clinical trials, marketing authorisation and ongoing postauthorisation surveillance. It must be remembered that for every 10,000 new drug candidates studied in basic research, only 250 proceed to the next stage. In preclinical research, the safety of drug molecules and their efficacy expectations are studied before starting trials on humans, and in the end only five molecules reach the clinical trial stage. Clinical trials establish the efficacy of the medicinal product in humans

Developments in Clay Science, Vol. 7. http://dx.doi.org/10.1016/B978-0-08-100293-3.00026-1

(or animals, in the case of veterinary medicinal products) in the treatment of the disease for which it is intended to treat with an acceptable profile of adverse reactions. At the end of the process, only one molecule of the 10,000 initially investigated arrives to marketing authorisation. Similar proportions would be expected with nanotechnology materials and their applications in marketed medicinal products (Fig. 26.1). Academic research plays a key role in this process, driving research and industrial development, as well as through numerous preclinical studies. Still, the market of nanopharmaceuticals could be estimated to be over 25 billion euros in 2020. What are the reasons that explain the success of nanopharmaceuticals in such a hostile environment? First, the scientific push associated with the knowledge of molecular processes linked to diseases, as well as the advances in nanotechnology and subsequent new nanomaterials, increase the number of offerings. On the other hand, there is a strong demand for innovative solutions to early diagnosis of cancer and other pathologies, transport of drugs across the blood–brain barrier, as well as targeting of diseased tissue and organs or development of implant materials with longer lifespans. The strong demand and the increased number of offerings explain the success of nanopharmaceutics.

Both medicinal products and medical devices are materials that, once manipulated and controlled, are used to achieve some health objectives, including therapeutics (drug delivery), regenerative medicine and diagnostics (Fig. 26.2). The development and application of "materials" at the nanoscale plays a relevant role in the design and manufacture of medicines and medical devices, and pharmaceutical nanotechnology (or nanopharmaceutics) is

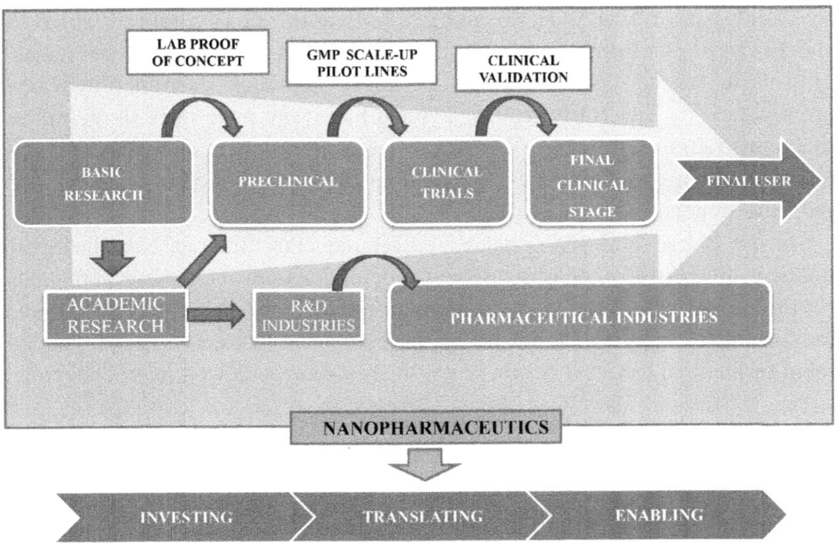

FIG. 26.1 Nanopharmaceuticals value chain, team members and network devoted to the successful fulfilment of nanopharmacy to human health care.

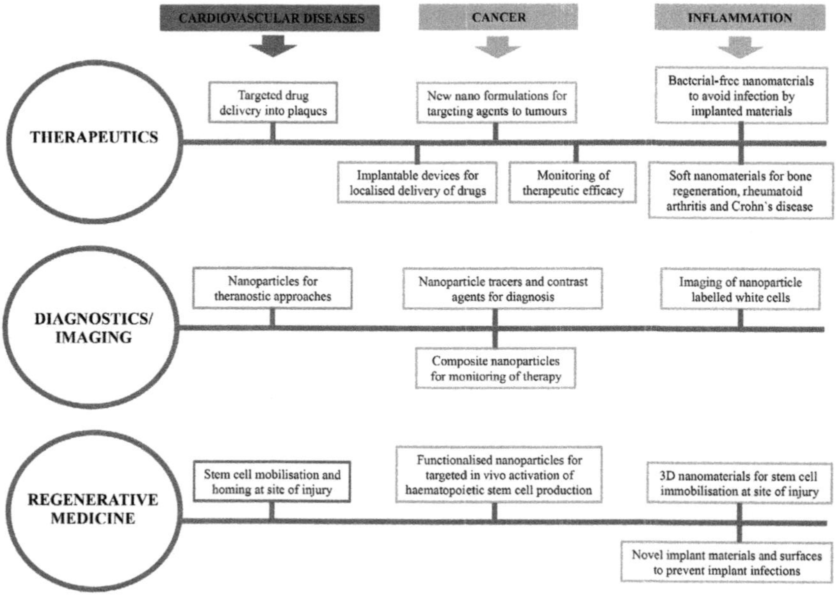

FIG. 26.2 Prospective uses of nanosized tubular clay minerals in nanopharmaceutics.

becoming an emerging academic and industrial area (Wagner et al., 2006). In particular, two nanosized tubular clay minerals, halloysite and imogolite, have been proposed as possible nanovectors for drug encapsulation based on their nanotubular structures, as well as in the restoring, maintaining and repairing of tissues and organ functions, on behalf of their demonstrated biocompatibility (safety) and efficacy. Halloysite is a polymorph of kaolinite in which the sheets curl up to form empty cylinders, and imogolite nanotubes are curved gibbsitelike sheets ($Al(OH)_3$) forming tubes in which the inner hydroxyls are replaced by $SiO_3(OH)$ groups. Typical halloysite nanotubes show internal diameters ranging from 20 to 100 nm, whereas imogolite nanotubes are significantly smaller ($\sim$1 nm). These dimensions are high enough to allow most of the drug molecules to be loaded.

In the last several years, nanosized tubular clay minerals have been proposed in diagnostic scenarios as well, opening a new interesting application for these nanomaterials. Together with these uses in the manufacture of medicinal products, nanosized tubular clay minerals have been used in the development of medical devices. All these applications are reviewed in this chapter.

26.2 USE OF NANOSIZED TUBULAR CLAY MINERALS IN DRUG DELIVERY

Drug delivery (particularly drug targeting) allows the selective attack of organs or cells, while saving healthy organs or tissues from nonspecific drug toxicity. Entrapping of drugs into the lumen of nanosized tubular clay minerals is a good strategy not only to modify drug release, but also to protect drugs against chemical and enzymatic degradation.

26.2.1 Natural Nanosized Tubular Clay Minerals

Natural nanosized tubular clay minerals provide well-documented support for drug encapsulation and delivery (Lvov et al., 2014; Viseras et al., 2015; Yuan et al., 2015). Several drug molecules have been loaded onto halloysite nanotubes (Price et al., 2001; Levis and Deasy, 2003; Kelly et al., 2004; Veerabadran et al., 2007; Lvov and Price, 2008; Lvov et al., 2008; Forsgren et al., 2010; Abdullayev and Lvov, 2011; Zhang et al., 2013). The kind of interactions between the functional groups of the drug and the active sites of halloysite outer and inner surfaces are key factors in kinetics and thermodynamic features of drug/halloysite systems (Viseras et al., 2008a, 2009). In particular, the loading process of halloysite nanotubes with an antiinflammatory drug (5-aminosalicylic acid) was successfully explained as a combination of two separate processes: a rapid adsorption of the drug at the external surface of the clay mineral particles, and a slower adsorption occurring inside the halloysite lumen and aggregates (Fig. 26.3) (Viseras et al., 2008a). Recently, Elumalai et al. (2015) used Monte Carlo simulations to point out the importance of the diameter and length, together with particle charge, of halloysite nanotubes on the loading and release of some model drugs. Drug release from halloysite nanotubes is often characterised by biphasic profiles with an initial burst followed by a slower release phase,

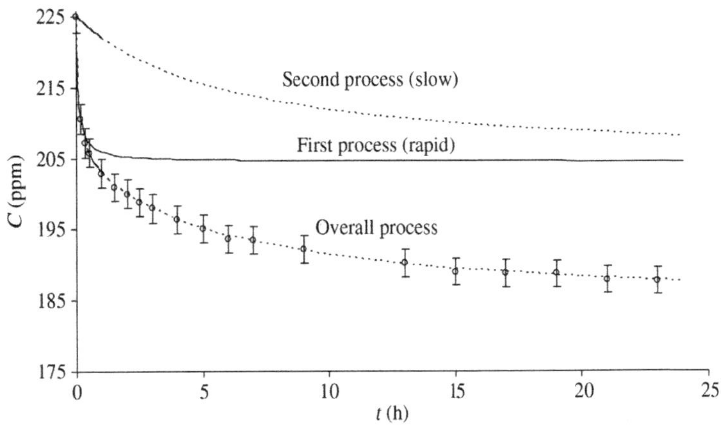

FIG. 26.3 Individual and overall kinetic adsorption isotherms of 5-aminosalicylic acid onto halloysite at 30°C. *Reprinted with permission from Viseras et al. (2008a). Copyright 2008 Elsevier.*

making it difficult to fit the experimental data and to explain the underlying release mechanism using a single mathematical model. A newer model based on kinetic laws able to describe biphasic release curves and provide information about kinetic order and specific release rate from these supports has been successfully postulated by Aguzzi et al. (2013). Briefly, the release of 5-aminosalicylic acid retained onto halloysite nanotubes was explained as an overall desorption process resulting from the contribution of individual simple processes governed by diffusion that included desorption of the drug adsorbed on the external clay mineral surfaces, interparticle spaces or both (fast process) and desorption of drug molecules adsorbed into the halloysite lumen (slow process) (Fig. 26.4). This model provides real hypotheses about mechanisms implicated in drug release from nanosized tubular clay minerals and is able to satisfactorily fit all the experimental data with a simple equation. More examples concerning the use of halloysite as a drug carrier can be found in Chapter 22.

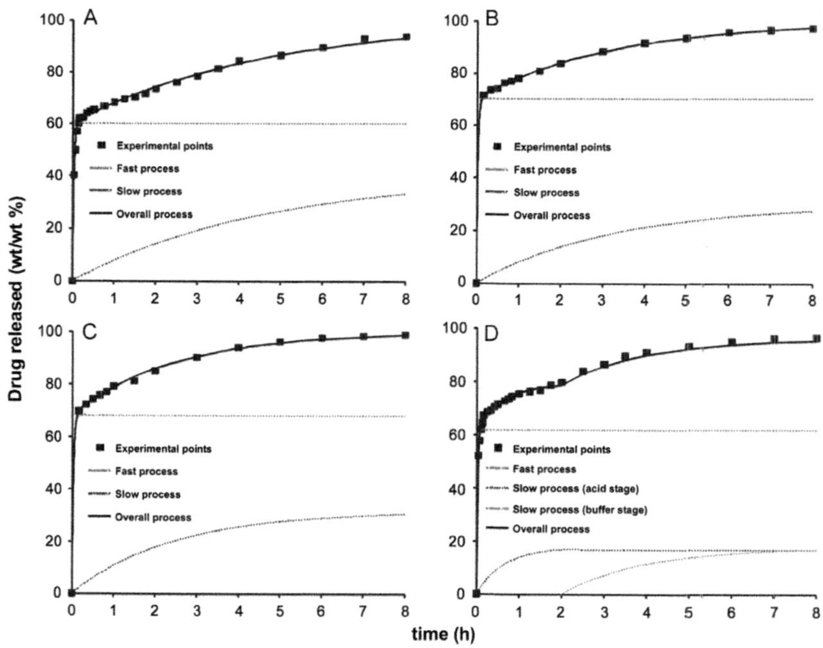

FIG. 26.4 Amount of drug released (mean values ± s.d.; $n=6$) of 5-aminosalycilic acid from halloysite in three dissolution media (A); theoretical dissolution curves (*lines*) of overall and simple release processes in purified water (B), 0.1 M HCl (C) and pH 6.8 buffer (D). *Reprinted with permission from Aguzzi et al. (2013). Copyright 2013 Elsevier.*

26.2.2 Functionalised Nanosized Tubular Clay Minerals

Halloysite nanotubes can be modified by interactions with specific functional groups or thermal processing in order to improve their performance as drug carriers (Yah et al., 2012; Yuan et al., 2012a,b; Joo et al., 2013; Tan et al., 2013). See more details concerning the modification of nanosized tubular clay minerals in Chapters 8 and 12. Functionalised nanotubes find several applications in both conventional and advanced (ie, gene delivery) therapies. Natural halloysite nanotubes modified with 3-aminopropyltriethoxysilane were successfully used as carriers for aspirin (Lun et al., 2014) and ibuprofen (Tan et al., 2014), showing a higher loading capacity and enhanced drug dissolution rate than unmodified halloysite. Functionalisation with carboxyl acid (Zargarian et al., 2015) or poly (*N,N*-dimethylaminoethylmethacrylate) (Hemmatpour et al., 2015) also provided halloysite nanotubes with improved loading capacities for drugs as diphenhydramine and diclofenac. Coating with polyorganosilanes was exploited to change halloysite surface from hydrophilic to superhydrophobic, providing a slow release rate of diclofenac (Fan et al., 2014). Release of antiseptic agents was also greatly reduced by coating halloysite with a benzotriazole-Cu^{2+} chelate film (Wei et al., 2014). The obtained systems showed antibacterial activity against *Staphylococcus aureus* cell cultures for up to 72 h, which means that they look promising to be administered as bandages for wound healing. Halloysite loaded with tetracycline has been coated with polyvinyl alcohol and polymethyl methacrylate films to reduce the initial burst effect that was observed in the release of the drug from the natural nanotubes alone (Ward et al., 2010). Other studies showed that halloysite nanotubes modified with tetrazolium salts were able to enhance the in vitro activity of the antitumour drugs curcumin (Riela et al., 2014) and cardanol (Massaro et al., 2015a) against several tumour cell lines. Curcumin was also loaded on thermoresponsive poly(*N*-isopropylacrylamide)-halloysite nanotubes in order to prevent drug degradation (Cavallaro et al., 2015). Massaro et al. (2014) prepared functionalised halloysite with both cyclodextrins and thiosaccharide units. The obtained modified nanotubes were aimed to act as drug delivery systems and to mimic the binding occurring between cell proteins (lectins) and sugars during cellular recognition events. Layer-by-layer adsorption of oppositely charged polyelectrolytes is another strategy to obtain novel nanovectors for anticancer agents (Guo et al., 2012; Mitchell et al., 2012a,b; Vergaro et al., 2012; Shutava et al., 2014). Cyclodextrin-modified halloysite nanotubes can offer the possibility for simultaneous encapsulation of different drugs in multiagent therapies, as demonstrated with two natural antitumoural agents (silibinin and quercetin) (Massaro et al., 2015b). Advanced therapy based on administration of genetic matter is also greatly attractive as a treatment for cancer disorders. Halloysite nanotubes could efficiently improve intracellular delivery and enhance the antitumour activity of antisense oligodeoxynucleotides (Shi et al., 2011). The clay mineral also has been proposed as a transgene transmission reagent (Campos et al., 2011) because of its cellular uptake and cytocompatibility properties (Kommireddy et al., 2005; Vergaro et al., 2010; Verma et al., 2012).

Imogolite nanotubes present a positive alumina surface and negative inner silica surface, combined with significantly smaller lumen diameter in comparison with halloysite, provoking a decrease in the efficacy of adsorption of biological macromolecules and drugs (Gustafsson, 2001). However, imogolite nanotubes have been successfully exploited for DNA encapsulation in order to protect it from degradation by nucleases but maintain its biological activity (Jiravanichanun et al., 2012).

26.2.3 Nanosized Tubular Clay Minerals/Biopolymer Nanocomposites

In addition with the abovementioned functionalisation strategies, clay mineral/biopolymer nanocomposites are designed to obtain hybrid materials with enhanced properties compared to the pure components (Viseras et al., 2008b, 2010). Natural mineral nanotubes can be assembled with a large variety of polymers yielding nanostructured hybrid materials used to design nanoscale drug delivery systems (Viseras et al., 2008b, 2010; Ghebaur et al., 2012; Abdullayev and Lvov, 2013). Diphenhydramine hydrochloride has been loaded into halloysite/polyvinyl alcohol nanocomposites to control drug delivery (Ghebaur et al., 2012). Biocompatible chitosan and gelatine multilayers were used for the encapsulation of halloysite tubes loaded with dexamethasone by means of layer-by-layer methodology (Veerabadran et al., 2009). Encapsulation of the drug in the mineral nanotubes, accomplished by the assembly of natural polymer shells, resulted in controlled release of the drug molecule. Halloysite/hyaluronate nanocomposites intended for colon target delivery of 5-fluorouracil were recently developed, showing promising pH-dependent drug release profiles (Rao et al., 2014).

26.3 USE OF NANOSIZED TUBULAR CLAY MINERALS IN TISSUE ENGINEERING AND REPARATIVE MEDICINE

Biocompatibility is essential for the potential application of nanosized tubular clay minerals in tissue engineering and regenerative medicine. Halloysite nanotube interactions with cells and the subsequent intracellular uptake have been studied in different cell lines (Kommireddy et al., 2005; Veerabadran et al., 2009; Vergaro et al., 2010; Lai et al., 2013). Cell adhesion and proliferation on biomaterials is a crucial point in tissue engineering and biotechnology. Specific interactions between cells and substrates are important to regulate cell function, tissue homeostasis and matrix remodelling. The presence of divalent cations, such as Ca^{2+} and Mg^{2+}, associated with clay minerals, significantly influences cellular adhesion. Clay mineral/cell interactions depend on the shape, size and surface charge of the particles. It is well known that the cell membrane surface is negatively charged and that positively charged nanoparticles can penetrate deep into cell membranes, while negatively charged particles cannot enter the cell wall. Charge density modulates the biocompatibility and antimicrobial properties of clay minerals, and this is related to the aspect ratio (ie, the relation between structure sizes) (Rawat et al., 2014). Halloysite improved fibroblast cell attachment and spreading and allowed cells to maintain cell phenotypes (Kommireddy et al., 2005). In vitro cell toxicity of halloysite has been studied using human dermal fibroblast and breast cancer cells, and a low inherent toxicity of the pure mineral has been found (Veerabadran et al., 2009). Vergaro et al. (2010) performed in vitro cytotoxicity testing that demonstrated good biocompatibility of halloysite at a concentration of 0.1 mg/mL. It was observed that halloysite was nontoxic towards cervical adenocarcinoma cells (HeLa) and breast cancer cells (MCF-7), as well as the fact that halloysite accumulated in cells without preventing their proliferation. The process of cellular uptake was based on both nanotube concentration around cells and their penetration into cells. Studies with other cellular lines (Caco-2 and HT29-MTX) have also shown a high degree of biocompatibility of halloysite nanotubes below 0.2 mg/mL (Lai et al., 2013). Cell viability and membrane integrity were preserved for middle-term exposure in vitro (6 h) and appeared unlikely to have toxic effects at moderate levels of exposure. At high concentrations (100 mg/l), a significant exposure change in protein expression was observed: this seemed related to the stimulation of the processes of cell growth and proliferation, which is similar to the response to cell infection, irritation and injury and enhanced antioxidant activity, all of which are characteristic of an overall adaptive response to exposure.

With regard to imogolite, its morphology (high length-to-diameter ratio) and expected persistence could suggest potential toxicity; however, the high hydrophilicity and external alumina layer contribute to the relative inertness of this mineral (Rotoli et al., 2014). Ishikawa et al. (2010a,b) declared that the amount of normalised protein per osteoblastlike cell cultured on imogolite was twice as much than when it was cultured on a culture dish, and they also pointed out that the cell-imogolite bindings was stronger than the cell-culture dish ones. These results make evident the good biocompatibility between cells and imogolite.

26.3.1 Tissue Engineering

Tissue engineering offers a means to provide biocompatible replacement tissue with mechanical and functional integrity. A critical step in tissue engineering is the designing of scaffolds with structure, composition, physicochemical, mechanical and biological features analogous to natural tissues. Scaffolds provide a three-dimensional (3D) environment that supports cell attachment, proliferation and differentiation, as well as enabling the transportation of nutrients and cell metabolites in tissue engineering (Patterson et al., 2010; Chrzanowski et al., 2013; Gaharwar et al., 2013; Song et al., 2014). Hydrogel (Gaharwar et al., 2013; Song et al., 2014) has been studied as a scaffolding material, especially for soft tissue regeneration, even if it usually has inferior mechanical properties as well as lacking adequate functionality, which hinders its wide application in regenerative medicine (Biondi et al., 2008). The incorporation of clay mineral nanoparticles improved the mechanical properties and functionalities of these systems, fulfiling the requirements of tissue engineering. The 3D geometry is related to pore size, porosity and interpore connectivity that represent key issues for accelerated tissue engineering application. With these premises, different clay minerals have been proposed as innovative platforms for tissue regeneration and biomaterial design because of their capacity to enhance stem and progenitor cell proliferation and differentiation (Dawson and Oreffo, 2013). Polymer electrospun with halloysite showed potential applications in the fields of tissue engineering, wound dressing and drug delivery (Wang et al., 2012). Halloysite and polyvinyl alcohol were used to prepare clay polymer nanocomposites (CPN) with improved osteoblast adhesion (Zhou et al., 2010). Electrospun CPN with tissue engineering (and drug delivery of tetracycline) applications were prepared with halloysite and polylactic-co-glycolic acid (Qi et al., 2010, 2012, 2013). According to its haemocompatibility, halloysite polymer nanocomposites were considered suitable as artificial tissue/organ substitutes (Zhao et al., 2013). Halloysite polymethylmethacrylate nanocomposites have been used as bone cements doped with gentamicin, providing slow release of the drug without compromising the CPN mechanical strength while enhancing bone adhesiveness (Wei et al., 2012). Halloysite polycaprolactone nanocomposite scaffolds showed improved mechanical properties, substantial protein adsorption, enhanced mineralisation and osteogenic differentiation compared to the polymer scaffold without the clay mineral (Nitya et al., 2012). Polyglycerol sebacate was filled with halloysite nanotubes to obtain elastomeric nanocomposites with optimal combination of compliance and a degradable profile compared with the pure counterpart, showing excellent resilience and satisfactory biocompatibility in vitro to be used as a soft tissue engineering material (Chen et al., 2011). Halloysite chitosan nanotubes nanocomposite scaffolds were prepared by combining solution-mixing and freeze-drying techniques exhibiting a high porous structure and cytocompatibility with great potential for tissue engineering (Liu et al., 2013).

Imogolite is also receiving considerable attention in tissue engineering for its excellent adsorption properties and nanotubular shape. Reinforcement of polymeric materials with imogolite nanotubes has recently been described in the research literature (Yamamoto et al., 2005a,b; Yah et al., 2010; Shikinaka et al., 2011). These findings have been exploited to obtain collagen- (Nakano et al., 2010) and gelatine-based (Teramoto et al., 2012) nanocomposite hydrogels, showing improved properties and promising potential in pharmaceutical and tissue engineering development. Imogolite nanocomposite scaffolds exhibited good biocompatibility and enhancement effects of proliferation over mouse osteoblastlike cells (Ishikawa et al., 2010a). With regard to safety requirements, imogolite nanotubes must demonstrate no potential toxicological effects on humans. The possible cytotoxic and genotoxic effects of synthetic imogolite-like nanotubes have been evaluated, revealing no cytotoxic activity in human fibroblast cells up to 0.1 mg/mL of nanoparticles, whereas DNA damage depended on the cell uptake and aspect ratio of the nanotubes (Liu et al., 2012). Imogolite nanotubes offer a wide range of positive characteristics as scaffolds for osteoblastic proliferation and differentiation and are a both suitable and interesting alternative in bone tissue engineering (Ishikawa et al., 2010b). Imogolite nanotubes were applied to cell cultures to compare their properties as scaffolds with a conventional culture dish and a carbon nanotube scaffold. The scaffold containing more imogolite presented the best roughness, wettability and protein adsorption ability. In addition to these advances, modification and functionalisation of imogolite nanotubes (Kang et al., 2010, 2011) will probably offer new perspectives on biomedical field in the next several years.

26.3.2 Reparative Medicine

A particular interesting use of nanosized tubular clay minerals in regenerative medicine is their use in design of wound-healing treatments. Wound healing involves multiple cell populations, extracellular matrix bioactive molecules as soluble mediators and particularly cytokines and growth factors (Velnar et al., 2009). Chronic wounds are the results of different causes and pathological states, such as burns, vasculitis, arterial and venous insufficiency, malignancies, diabetes and neuropathic and pressure ulcers. In these cases, the normal stages of healing fail, indicating incomplete tissue reparation. Various factors could disturb the normal process, prolonging one or more stages, such as haemostasis, inflammation, proliferation and remodelling. In particular, tissue infections and hypoxia, necrosis and the presence of excessive levels of exudates and inflammatory cytokines could prolong a nonhealing state. A continuous state of wound inflammation causes a negative cascade of tissue responses, and the healing process results in poor functional and anatomical outcomes and frequent relapsing. Generally chronic wounds are often subjected to infections and result in complicated wounds. Different medical approaches and therapeutic choice could affect wound healing and the subsequent steps involved in tissue repairing. The loading and the localisation of bioactive molecules in clay minerals, having a role in cell growth and proliferation, may induce cell differentiation (Salcedo et al., 2012). However a more interesting aspect is the possibility of stimulating cell differentiation per se due to the signals produced by charge density and ion concentration.

Moreover, it was reported that the elasticity or the structure of the matrix environment regulated stem cell differentiation that depends not only on chemical differentiation factors in the culture medium, but also on characteristics of the matrix environment, such as the materials, the surface type and the mechanical properties of the culture substrates (Engler et al., 2006; Miyahara et al., 2006; Reilly and Engler, 2010; Kotobuki et al., 2013). The presence of silicon and magnesium elements has been described as being able to promote the osteogenic differentiation of mesenchymal stem cells. A biomimetic scaffold should be an artificially designed scaffold that enhance neotissue genesis via cell recruiting, adhesion and proliferation (Ma, 2008). The potential application of halloysite nanotubes for wound-healing purposes was pointed out by Lvov and Abdullayev (2013), showing that halloysite polycaprolactone nanocomposite scaffolds loaded with antiseptic agents were effective materials for skin regeneration. Halloysite has been shown to enhance the haemostatic and wound-healing properties of chitosan sponges in rats (Liu et al., 2014). The inclusion of the nanosized tubular clay minerals promotes in vivo epithelialisation and collagen deposition after 1 week, in comparison with pure chitosan. As for future trends, mineral coating of biological cells can be used to obtain surface-functionalised 3D structures (whole cells) with biomimetic properties. Yeast cells have been used to obtain hollow halloysite-based microcapsules (so-called cyborg cells) by layer-by-layer assembly of nanosized tubular clay minerals and subsequent thermal decomposition of the cells (Konnova et al., 2013). These new assemblies may find applications in controllable cell growth.

26.4 USE OF NANOSIZED TUBULAR CLAY MINERALS IN DIAGNOSTIC AND MEDICAL DEVICES

At the moment, the use of nanosized tubular clay minerals in diagnosis has not been explicitly raised by any research group. However, some recent articles allow to venture that this will be one of the most interesting research areas of application of nanosized tubular clay minerals in the near future. In cancer therapy, the count of circulating tumour cancer cells in blood can be a diagnostic tool to monitor the progression of the disease and detect possible metastasis. It was demonstrated that halloysite is effective to enhance cell adhesion of circulating cancer cells (Hughes and King, 2010; Hughes et al., 2012). Similar hybrid systems have been also used to detect DNA damage, and these systems could have many biological and medical applications (Rawtani et al., 2013).

The frontier between medicinal products (European Directive, 2001) and medical devices (European Directive, 2007) is not always clear, particularly with those systems based in nanotechnology. For example, an active implantable medical device may be intended to administer a medicinal product. Obviously, the consideration of this system as a medical device or medicinal product will greatly affect the possibilities of marketing. In any case, without going into a final consideration, some authors have proposed the use of halloysite specifically in the design of such devices. Recently, modified halloysite nanotubes have been used for capturing circulating tumour cells from the bloodstream (Mitchell et al., 2015). Functionalisation with surfactants alters the charge of naturally occurring halloysite nanotubes and induces differential adhesion of tumour cells and blood cells to nanotube-coated surfaces under flow. It is not clear at this point whether these systems are medical devices, but the significance of the basic research in this area is obvious.

26.5 CONCLUDING REMARKS

Nanosized tubular clay minerals have attracted a great deal interest in recent years as materials for new, advanced biomedical applications. They are used as they naturally occur or after functionalisation, including clay mineral biopolymer nanohybrids, both in drug and gene delivery. Together with their possibilities in drug delivery, nanosized tubular clay minerals show good features in the development of scaffolds for tissue engineering, medical devices and diagnostics. Clay minerals and clay biopolymer nanocomposites have been used as carriers of matrix proteins, growth factors and genes in tissue regeneration. The interaction of nanosized tubular clay minerals with biological structures open up opportunities for tissue engineering and offer a range of possibility to develop systems able to facilitate cell adhesion, proliferation and neotissue formation. CPN are characterised by porous network structure, swelling/deswelling behaviour and mechanical properties, which provide physical support for cell adhesion and proliferation as well as excellent biocompatibility. Moreover, cells could interact with clay minerals, suggesting new opportunities in the functionalisation of surfaces for enhanced reparative response. An important aspect is the capability of clay minerals to induce cell differentiation without the employment of other inductors avoiding problems related to uncontrolled cell proliferation in other regions far from the application site. Nanosized tubular clay minerals could be also potentially used in the design of new noninvasive systems to diagnostic of disorders such as cancer and degenerative diseases.

Of the challenges for the medical application of nanosized tubular clay minerals, a particularly important one is the very moderate interest of the pharmaceutical and medical device industry in nanotechnology. Even though recent research includes many innovative studies based on the application of clay minerals in drug delivery and tissue regeneration/reparation, hard work will be needed to transfer the acquired knowledge to clinical practice, requiring high implications of the academic research. The first step to accomplish will be to create spinoffs that can develop current ideas and results to find major pharmaceutical or medical device companies that licence their technologies or partner with the spinoffs to bring their novel approaches through the regulatory approval process. The second challenge will be to develop nanopharmaceutical products with adequate cost-effectiveness in comparison to conventional alternatives, as health-care systems increasingly face cost pressures.

Nanosized tubular clay minerals have demonstrated their potential in both target delivery and diagnostics, and it is expected that in the near future, they will be used in theranostic platforms that can comprehensively carry out in situ diagnostics, then deliver drugs, genes or both to the diseased tissue or cell and monitor the resultant therapeutic response. These are the new frontiers of using natural mineral nanotubes in health-care applications.

REFERENCES

Abdullayev, E., Lvov, Y., 2011. Halloysite clay nanotubes for controlled release of protective agents. J. Nanosci. Nanotechnol. 11, 10007–10026.

Abdullayev, E., Lvov, Y., 2013. Halloysite clay nanotubes as a ceramic "skeleton" for functional biopolymer composites with sustained drug release. J. Mater. Chem. B 1, 2894–2903.

Aguzzi, C., Viseras, C., Cerezo, P., Salcedo, I., Sánchez-Espejo, R., Valenzuela, C., 2013. Release kinetics of 5-aminosalicylic acid from halloysite. Colloids Surf. B 105, 75–80.

Aulton, M.E., Taylor, K.M.G., 2013. Aulton's Pharmaceutics. The Design and Manufacture of Medicines, fourth ed. Churchill Livingstone-Elsevier, Edinburgh.

Biondi, M., Ungaro, F., Quaglia, F., Netti, P.A., 2008. Controlled drug delivery in tissue engineering. Adv. Drug Deliv. Rev. 60, 229–242.

Campos, V.F., de Leon, P.M.M., Komninou, E.R., Dellagostin, O.A., Deschamps, J.C., Seixas, F.K., Collares, T., 2011. NanoSMGT: transgene transmission into bovine embryos using halloysite clay nanotubes or nanopolymer to improve transfection efficiency. Theriogenology 76, 1552–1560.

Cavallaro, G., Lazzara, G., Massaro, M., Milioto, S., Noto, R., Parisi, F., Riela, S., 2015. Biocompatible poly(N-isopropylacrylamide)-halloysite nanotubes for thermoresponsive curcumin release. J. Phys. Chem. C 119, 8944–8951.

Chen, Q.Z., Liang, S.L., Wang, J., Simon, G.P.J., 2011. Manipulation of mechanical compliance of elastomeric PGS by incorporation of halloysite nanotubes for soft tissue engineering applications. J. Mech. Behav. Biomed. Mater. 4, 1805–1818.

Chrzanowski, W., Kim, S.Y., Neel, E.A.A., 2013. Biomedical applications of clay. Aust. J. Chem. 66, 1315–1322.

Dawson, J.I., Oreffo, R.O.C., 2013. Clay: new opportunities for tissue regeneration and biomaterial design. Adv. Mater. 25, 4069–4086.

Elumalai, D.N., Lvov, Y., Derosa, P., 2015. Implementation of a simulation model of the controlled release of molecular species from halloysite nanotubes. J. Encapsulation Adsorpt. Sci. 5, 74–92.

Engler, A.J., Sen, S., Sweeney, H.L., Discher, D.E., 2006. Matrix elasticity directs stem cell lineage specification. Cell 126, 677–689.

European Directive, 2001. Directive 2001/83/EC of the European Parliament and of the Council of 6 November 2001 on the Community code relating to medicinal products for human use. Off. J. Eur. Union 311, 67. 28.11.2001.

European Directive, 2007. Directive 2007/47/EC of the European Parliament and of the Council of 5 September 2007 amending Council Directive 90/385/EEC on the approximation of the laws of the Member States relating to active implantable medical devices, Council Directive 93/42/EEC concerning medical devices and Directive 98/8/EC concerning the placing of biocidal products on the market. Off. J. Eur. Union 247, 21–55. 21.9.2007.

Fan, L., Li, B., Wang, Q., Wang, A., Zhang, J., 2014. Superhydrophobic gated polyorganosilanes/halloysite nanocontainers for sustained drug release. Adv. Mater. Inter. 1, 1300136. 7 pp.

Forsgren, J., Jamstorp, E., Bredenberg, S., Engqvist, H., Stromme, M., 2010. A ceramic drug delivery vehicle for oral administration of highly potent opioids. J. Pharm. Sci. U.S. 99 (1), 219–226.

Gaharwar, A.K., Peppas, N.A., Khademhosseini, A., 2013. Nanocomposite hydrogels for biomedical applications. Biotechnol. Bioeng. 111, 441–453.

Ghebaur, A., Garea, S.A., Iovu, H., 2012. New polymer–halloysite hybrid materials-potential controlled drug release system. Int. J. Pharm. 436, 568–573.

Guo, M., Wang, A., Muhammad, F., Qi, W., Ren, H., Guo, Y., Zhu, G., 2012. Halloysite nanotubes, a multifunctional nanovehicle for anticancer drug delivery. Chin. J. Chem. 30, 2115–2120.

Gustafsson, J.P., 2001. The surface chemistry of imogolite. Clay Clay Miner. 49, 73–80.

Hemmatpour, H., Haddadi-Asl, V., Roghani-Mamaqani, H., 2015. Synthesis of pH-sensitive poly (N,N-dimethylaminoethyl methacrylate)-grafted halloysite nanotubes for adsorption and controlled release of DPH and DS drugs. Polymer 65, 143–153.

Hughes, A.D., King, M.R., 2010. Use of naturally occurring halloysite nanotubes for enhanced capture of flowing cells. Langmuir 26, 12155–12164.

Hughes, A.D., Mattison, J., Western, L.T., Powderly, J.D., Greene, B.T., King, M.R., 2012. Microtube device for selectin-mediated capture of viable circulating tumour cells from blood. Clin. Chem. 58, 846–853.

Ishikawa, K., Akasaka, T., Yawaka, Y., Watari, F., 2010a. High functional expression of osteoblasts on imogolite, aluminosilicate nanotubes. J. Biomed. Nanotechnol. 6, 59–65.

Ishikawa, K., Akasaka, T., Abe, S., Yawaka, Y., Suzuki, M., Watari, F., 2010b. Application of imogolite, almino-silicate nanotube, as scaffold for the mineralization of osteoblasts. Bioceram. Dev. Appl. 1, 1–3.

Jiravanichanun, N., Yamamoto, K., Kato, K., Kim, J., Horiuchi, S., Yah, W.-O., Otsuka, H., Takahara, A., 2012. Preparation and characterization of imogolite/DNA hybrid hydrogels. Biomacromolecules 13, 276–281.

Joo, Y., Sim, J.H., Jeon, Y., Lee, S.U., Sohn, D., 2013. Opening and blocking the inner-pores of halloysite. Chem. Commun. 49, 4519–4521.

Kang, D.Y., Zang, J., Wright, E.R., McCanna, A.L., Jones, C.W., Nair, S., 2010. Dehydration, dehydroxylation, and rehydroxylation of single-walled aluminosilicate nanotubes. ACS Nano 4, 4897–4907.

Kang, D.Y., Zang, J., Jones, C.W., Nair, S., 2011. Single-walled aluminosilicate nanotubes with organic-modified interiors. J. Phys. Chem. C 115, 7676–7685.

Kelly, H.M., Deasy, P.B., Ziaka, E., Claffey, N., 2004. Formulation and preliminary in vivo dog studies of a novel drug delivery system for the treatment of periodontitis. Int. J. Pharm. 274, 167–183.

Kommireddy, D.S., Ichinose, I., Lvov, Y.M., Mills, D.K., 2005. Nanoparticle multilayers: surface modification for cell attachment and growth. J. Biomed. Nanotechnol. 1, 1–5.

Konnova, S.A., Sharipova, I.I., Demina, T.A., Osin, Y.N., Yarullina, D.R., Ilinskaya, O.N., Lvov, Y.M., Fakhrullin, R.F., 2013. Biomimetic cell-mediated three-dimensional assembly of halloysite nanotubes. Chem. Commun. 49, 4208–4210.

Kotobuki, N., Murata, K., Haraguchi, K., 2013. Proliferation and harvest of human mesenchymal stem cells using new thermoresponsive nanocomposite gels. J. Biomed. Mater. Res. 101, 537–546.

Lai, X., Agarwal, M., Lvov, Y.M., Pachpande, C., Varahramyan, K., Witzmann, F.A., 2013. Proteomic profiling of halloysite clay nanotube exposure in intestinal cell co-culture. J. Appl. Toxicol. 33, 1316–1329.

Levis, S.R., Deasy, P.B., 2003. Use of coated microtubular halloysite for the sustained release of diltiazem hydrochloride and propranolol hydrochloride. Int. J. Pharm. 253, 145–157.

Liu, W., Chaurand, P., Di Giorgio, C., De Méo, M., Thill, A., Auffan, M., Masion, A., Borschneck, D., Chaspoul, F., Gallice, P., Botta, A., Bottero, J.Y., Rose, J., 2012. Influence of the length of imogolite-like nanotubes on their cytotoxicity and genotoxicity toward human dermal cells. Chem. Res. Toxicol. 25, 2513–2522.

Liu, M., Wu, C., Jiao, Y., Xiong, S., Zhou, C., 2013. Chitosan–halloysite nanotubes nanocomposite scaffolds for tissue engineering. J. Mater. Chem. B 1, 2078–2089.

Liu, M., Shen, Y., Ao, P., Dai, L., Liu, Z., Zhou, C., 2014. The improvement of hemostatic and wound healing property of chitosan by halloysite nanotubes. RSC Adv. 4, 23540–23553.

Lun, H., Ouyang, J., Yang, H., 2014. Natural halloysite nanotubes modified as an aspirin carrier. RSC Adv. 4, 44197–44202.

Lvov, Y., Abdullayev, E., 2013. Functional polymer clay nanotube composites with sustained release of chemical agents. Prog. Polym. Sci. 38, 1690–1719.

Lvov, Y., Price, R., 2008. Halloysite nanotubules a noble substrate for the controlled delivery of bioactive molecules. In: Ruiz-Hitzky, E., Ariga, K., Lvov, Y. (Eds.), Bio-Inorganic Hybrid Nanomaterials. Wiley, Berlin, London, pp. 440–478.

Lvov, Y.M., Shchukin, D.G., Mohwald, H., Price, R.R., 2008. Halloysite clay nanotubes for controlled release of protective agents. ACS Nano 2, 814–820.

Lvov, Y., Aerov, A., Fakhrullin, R., 2014. Clay nanotube encapsulation for functional biocomposites. Adv. Colloid Interf. Sci. 207, 189–198.

Ma, P.X., 2008. Biomimetic materials for tissue engineering. Adv. Drug Deliv. Rev. 60, 184–198.

Massaro, M., Riela, S., Lo Meo, P., Noto, R., Cavallaro, G., Milioto, S., Lazzara, G., 2014. Functionalized halloysite multivalent glycocluster as a new drug delivery system. J. Mater. Chem. B 2, 7732–7738.

Massaro, M., Colletti, C.G., Noto, R., Riela, S., Poma, P., Guernelli, S., Parisi, F., Milioto, S., Lazzara, G., 2015a. Pharmaceutical properties of supramolecular assembly of co-loaded cardanol/triazole-halloysite systems. Int. J. Pharm. 478, 476–485.

Massaro, M., Piana, S., Colletti, C.G., Noto, R., Riela, S., Baiamonte, C., Giordano, C., Pizzolanti, G., Cavallaro, G., Milioto, S., Lazzara, G., 2015b. Multicavity halloysite–amphiphilic cyclodextrin hybrids for co-delivery of natural drugs into thyroid cancer cells. J. Mater. Chem. B 3, 4074–4081.

Mitchell, M.J., Chen, C.S., Ponmudi, V., Hughes, A.D., King, M.R., 2012a. E-selectin liposomal and nanotube-targeted delivery of doxorubicin to circulating tumour cells. J. Control. Release 160, 609–617.

Mitchell, M.J., Castellanos, C.A., King, M.R., 2012b. Nanostructured surfaces to target and kill circulating tumour cells while repelling leukocytes. J. Nanomater. 831263 (10 pp).

Mitchell, M.J., Castellanos, C.A., King, M.R., 2015. Surfactant functionalization induces robust, differential adhesion of tumour cells and blood cells to charged nanotube-coated biomaterials under flow. Biomaterials 56, 179–186.

Miyahara, Y., Nagaya, N., Kataoka, M., Yanagawa, B., Tanaka, K., Hao, H., Ishino, K., Ishida, H., Shimizu, T., Kangawa, K., Sano, S., Okano, T., Kitamura, S., Mori, H., 2006.

Monolayered mesenchymal stem cells repair scarred myocardium after myocardial infarction. Nat. Med. 12, 459–465.

Nakano, A., Teramoto, N., Chen, G., Miura, Y., Shibata, M., 2010. Preparation and characterization of complex gel of Type I collagen and aluminosilicate containing imogolite nanofibers. J. Appl. Polym. Sci. 118, 2284–2290.

Nitya, G., Nair, G.T., Mony, U., Chennazhi, K.P., Nair, S.V., 2012. "In vitro" evaluation of electrospun PCL/nanoclay composite scaffold for bone tissue engineering. J. Mater. Sci. Mater. Med. 23, 1749–1761.

Patterson, J., Martino, M.M., Hubbell, J.A., 2010. Electrospun poly(lactic-co-glycolic acid)/halloysite nanotube composite nanofibers for drug encapsulation and sustained release biomimetic materials in tissue engineering. Mater. Today 13, 15–22.

Price, R.R., Gaber, B.P., Lvov, Y., 2001. In-vitro release characteristics of tetracycline HCl, khellin and nicotinamide adenine dineculeotide from halloysite; a cylindrical mineral. J. Microencapsul. 18, 713–722.

Qi, R., Guo, R., Shen, M., Cao, X., Zhang, L., Xu, J., Yu, J., Shi, X., 2010. Electrospun poly(lactic-co-glycolic acid)/halloysite nanotube composite nanofibers for drug encapsulation and sustained release. J. Mater. Chem. 20, 10622–10629.

Qi, R., Cao, X., Shen, M., Guo, R., Yu, J., Shi, X., 2012. Biocompatibility of electrospun halloysite nanotube-doped poly(lactic-co-glycolic acid) composite nanofibers. J. Biomater. Sci. Polym. Ed. 23, 299–313.

Qi, R., Guo, R., Zheng, F., Liu, H., Yu, J., Shi, X., 2013. Controlled release and antibacterial activity of antibiotic-loaded electrospun halloysite/poly(lactic-co-glycolic acid) composite nanofibers. Colloids Surf. B Biointerfaces 110, 148–155.

Rao, K.M., Nagappan, S., Seo, D.J., Ha, C.S., 2014. pH sensitive halloysite-sodium hyaluronate/poly(hydroxyethyl methacrylate) nanocomposites for colon cancer drug delivery. Appl. Clay Sci. 97–98, 33–42.

Rawat, K., Agarwal, S., Tyagi, A., Verma, A.K., Bohidar, H.B., 2014. Aspect ratio dependent cytotoxicity and antimicrobial properties of nanoclay. Appl. Biochem. Biotechnol. 174, 936–944.

Rawtani, D., Agrawal, Y.K., Prajapati, P., 2013. Interaction behavior of DNA with halloysite nanotube-silver nanoparticle-based composite. BioNanoScience 3, 73–78.

Reilly, G.C., Engler, A.J., 2010. Intrinsic extracellular matrix properties regulate stem cell differentiation. J. Biomech. 43, 55–62.

Riela, S., Massaro, M., Colletti, C.G., Bommarito, A., Giordano, C., Milioto, S., Noto, R., Poma, P., Lazzara, G., 2014. Development and characterization of co-loaded curcumin/triazole-halloysite systems and evaluation of their potential anticancer activity. Int. J. Pharm. 475, 613–623.

Rotoli, B.M., Guidi, P., Bonelli, B., Bernardeschi, M., Bianchi, M.G., Esposito, S., Frenzilli, G., Lucchesi, P., Nigro, M., Scarcelli, V., Tomatis, M., Zanello, P.P., Fubini, B., Bussolati, O., Bergamaschi, E., 2014. Imogolite: an aluminosilicate nanotube endowed with low cytotoxicity and genotoxicity. Chem. Res. Toxicol. 27, 1142–1154.

Salcedo, I., Aguzzi, C., Sandri, G., Bonferoni, M.C., Mori, M., Cerezo, P., Sánchez, R., Viseras, C., Caramella, C., 2012. "In vitro" biocompatibility and mucoadhesion of montmorillonite chitosan nanocomposite: a new drug delivery. Appl. Clay Sci. 55, 131–137.

Shi, Y.F., Tian, Z., Zhang, Y., Shen, H.B., Jia, N.Q., 2011. Functionalized halloysite nanotube-based carrier for intracellular delivery of antisense oligonucleotides. Nanoscale Res. Lett. 6, 1–7.

Shikinaka, K., Koizumi, Y., Osada, Y., Shigehara, K., 2011. Reinforcement of hydrogel by addition of fiber-like nanofiller. Polym. Adv. Technol. 22, 1212–1215.

Shutava, T.G., Fakhrullin, R.F., Lvov, Y.M., 2014. Spherical and tubule nanocarriers for sustained drug release. Curr. Opin. Pharmacol. 18, 141–148.

Song, F., Li, X., Wang, Q., Liao, L., Zhang, C., 2014. Nanocomposite hydrogels and their applications in drug delivery and tissue engineering. J. Biomed. Nanotechnol. 10, 1–13.

Tan, D., Yuan, P., Bergaya, F., Yu, H., Liu, D., Liu, H., He, H., 2013. Natural halloysite nanotubes as mesoporous carriers for the loading of ibuprofen. Microporous Mesoporous Mater. 179, 89–98.

Tan, D., Yuan, P., Bergaya, F., Liu, D., Wang, L., Liu, H., He, H., 2014. Loading and in vitro release of ibuprofen in tubular halloysite. Appl. Clay Sci. 96, 50–55.

Teramoto, N., Hayashi, A., Yamanaka, K., Sakiyama, A., Nakano, A., Shibata, M., 2012. Preparation and mechanical properties of photo-crosslinked fish gelatin/imogolite nanofiber composite hydrogel. Materials 5, 2573–2585.

Veerabadran, N.G., Price, R.R., Lvov, Y.M., 2007. Clay nanotubes for encapsulation and sustained release of drugs. Nano 2, 115–120.

Veerabadran, N.G., Mongayt, D., Torchilin, V., Price, R.R., Lvov, Y.M., 2009. Organized shells on clay nanotubes for controlled release of macromolecules. Macromol. Rapid Commun. 30, 99–103.

Velnar, T., Bailey, T., Smrkolj, V., 2009. The wound healing process: an overview of the cellular and molecular mechanisms. J. Int. Med. Res. 37, 1528–1542.

Vergaro, V., Abdullayev, E., Lvov, Y.M., Zeitoun, A., Cingolani, R., Rinaldi, R., Leporatti, S., 2010. Cytocompatibility and uptake of halloysite clay nanotubes. Biomacromolecules 11, 820–826.

Vergaro, V., Lvov, Y.M., Leporatti, S., 2012. Halloysite clay nanotubes for resveratroldelivery to cancer cells. Macromol. Biosci. 12, 1265–1271.

Verma, N.K., Moore, E., Blau, W., Volkov, Y., Babu, P.R., 2012. Cytotoxicity evaluation of nanoclays in human epithelial cell line A549 using high content screening and real-time impedance analysis. J. Nanopart. Res. 14, 1137–1147.

Viseras, M.T., Aguzzi, C., Cerezo, P., Viseras, C., Valenzuela, C., 2008a. Equilibrium and kinetics of 5-aminosalicylic acid adsorption by halloysite. Microporous Mesoporous Mater. 108, 112–116.

Viseras, C., Aguzzi, C., Cerezo, P., Bedmar, M.C., 2008b. Biopolymer–clay nanocomposites for controlled drug delivery. Mater. Sci. Technol. 24, 1020–1026.

Viseras, M.T., Aguzzi, C., Cerezo, P., Cultrone, G., Viseras, C., 2009. Supramolecularstructure of 5-aminosalycilic acid/halloysite composites. J. Microencapsul. 26, 279–286.

Viseras, C., Cerezo, P., Sanchez, R., Salcedo, I., Aguzzi, C., 2010. Current challenges in clay minerals for drug delivery. Appl. Clay Sci. 48, 291–295.

Viseras, C., Aguzzi, C., Cerezo, P., 2015. Medical and health applications of natural mineral nanotubes. In: Pasbakhsh, P., Churchman, G.-J. (Eds.), Natural Mineral Nanotubes: Properties and Applications. Apple Academic Press, Oakville, Canada; Waretown NJ, pp. 437–448.

Wagner, V., Dullaart, A., Bock, A.K., Zweck, A., 2006. The emerging nanomedicine landscape. Nat. Biotechnol. 24, 1211–1217.

Wang, S., Zhao, Y., Shen, M., Shi, X., 2012. Electrospun hybrid nanofibers doped with nanoparticles or nanotubes for biomedical applications. Ther. Deliv. 3, 1155–1169.

Ward, C.J., Song, S., Davis, E.W., 2010. Controlled release of tetracycline–HCl from halloysite–polymer composite films. J. Nanosci. Nanotechnol. 10, 6641–6649.

Wei, W., Abdullayev, E., Hollister, A., Mills, D., Lvov, Y.M., 2012. Clay nanotube/poly(methyl methacrylate) bone cement composites with sustained antibiotic release. Macromol. Mater. Eng. 297, 645–653.

Wei, W., Minullina, R., Abdullayev, E., Fakhrullin, R., Mills, D., Lvov, Y., 2014. Enhanced efficiency of antiseptics with sustained release from clay nanotubes. RSC Adv. 4, 488–494.

Yah, W.O., Yamamoto, K., Jiravanichanun, N., Otsuka, H., Takahara, A., 2010. Imogolite reinforced nanocomposites: multifaceted green materials. Materials 3, 1709–1745.

Yah, W.O., Takahara, A., Lvov, Y.M., 2012. Selective modification of halloysite lumen with octadecylphosphonic acid: new inorganic tubular micelle. J. Am. Chem. Soc. 134, 1853–1859.

Yamamoto, K., Otsuka, H., Wada, S.I., Sohn, D., Takahara, A., 2005a. Preparation and properties of [poly(methylmethacrylate)/imogolite] hybrid via surface modification using phosphoric acid ester. Polymer 46, 12386–12392.

Yamamoto, K., Otsuka, H., Wada, S.I., Sohn, D., Takahara, A., 2005b. Transparent polymer nanohybrid prepared by in situ synthesis of aluminosilicate nanofibers in poly(vinilalcohol) solution. Soft Matter 1, 372–377.

Yuan, P., Tan, D.Y., Bergaya, F., Yan, W.C., Fan, M.D., Liu, D., He, H.P., 2012a. Changes in structure, morphology, porosity, and surface activity of mesoporous halloysite nanotubes under heating. Clay Clay Miner. 60, 561–573.

Yuan, P., Southon, P.D., Liu, Z.W., Kepert, C.J., 2012b. Organosilane functionalization of halloysite nanotubes for enhanced loading and controlled release. Nanotechnology 23, 375705 (5pp).

Yuan, P., Tan, D., Bergaya, F., 2015. Properties and applications of halloysite nanotubes: recent research advances and future prospects. Appl. Clay Sci. 112–113, 75–93.

Zargarian, S., Haddadi-Asl, V., Hematpour, H., 2015. Carboxylic acid functionalization of halloysite nanotubes for sustained release of diphenhydramine hydrochloride. J. Nanopart. Res. 17, 218–231.

Zhang, Y., Chen, Y., Zhang, H., Zhang, B., Liu, J., 2013. Potent antibacterial activity of a novel silver nanoparticle-halloysite nanotube nanocomposite powder. J. Inorg. Biochem. 118, 59–64.

Zhao, Y., Wang, S., Guo, Q., Shen, M., Shi, X.J., 2013. Hemocompatibility of electrospun halloysite nanotube- and carbon nanotube-doped composite poly(lactic-co-glycolic acid) nanofibers. J. Appl. Polym. Sci. 127, 4825–4832.

Zhou, W.Y., Guo, B., Liu, M., Liao, R., Rabie, A.B.M., Jia, D.J., 2010. Poly(vinyl alcohol)/halloysite nanotubes bionanocomposite films: properties and in vitro osteoblasts and fibroblasts response. J. Biomed. Mater. Res. A 93, 1574–1587.

Chapter 27

Industrial Implications in the Uses of Tubular Clay Minerals

O. Poncelet* and J. Skrzypski
CEA DRT/LITEN/DTMN, Grenoble Cedex 9, France
**Corresponding author: e-mail: olivier.poncelet@cea.fr*

27.1 MARKETING PRODUCTS CONTAINING HIGH-ASPECT-RATIO NANOPARTICLES REASONABLY

Imogolite and halloysite both have to be considered as nanoparticles; moreover, these nanoparticles exhibit a high-aspect-ratio morphology. This fact will have to be taken into account before an industrial company can put them in products. Environmental agencies in the United States and European Union are implementing new regulations that are becoming increasingly restrictive. It can be expected that eventually, the U.S. and EU standards concerning environmental legislation will become the basic worldwide standards for environmental material and chemical regulations. Registration, Evaluation, Authorization and Restriction of Chemicals (REACH) regulation in Europe, for example, will ultimately require safety data sheets provided to product users to list all the chemical components contained in the product, even if this component is only used at trace concentrations. Under this regulation, the import of products containing forbidden chemical substances will be first limited and finally stopped using international trade treaties. The ultimate goal is to definitively eliminate some dangerous chemicals in the EU community. Over the years, the list of forbidden chemicals has been growing; while intense industrial lobbying can slow down this phenomenon, it cannot stop it. Consequently, commercial enterprises that need a wide range of chemicals to manufacture their products are obliged to continuously modify their technological road maps in anticipation of the new regulations, and also sometimes to halt the manufacture of products containing potentially dangerous chemicals (eg, asbestos and fluorosurfactants). That also means that these companies will have to financially support the cost (US $200 K) of all the analytical testing that must be carried out in order to meet the governmental requirements (eg, material safety data sheets, life cycle of products, waste

Developments in Clay Science, Vol. 7. http://dx.doi.org/10.1016/B978-0-08-100293-3.00027-3

management, and CO_2 footprint) before receiving a market agreement from a governmental agency.

Finally, the value of the brand of a large enterprise is also a critical parameter that acts directly on the value of stock options. Large companies are very sensitive to this parameter. Building a great brand is a very long process that requires the continuous attention of management. Large companies (and their scientists) aspire to develop safe materials. All the participants in this process (suppliers, workers, end users and finally the people who will be in charge of the recycling) are included in the safety policy. Nobody wants to see a scandal that could endanger an enterprise (as has happened in the past with the chemical company Eternit, which has faced asbestos lawsuits in Europe and the United States).

So, whatever "magical" physical properties are in imogolite or halloysite, the first major difficulty is to convince the company's management team to put these substances in a product and launch this product on the market.

Fortunately, people around the world have been using naturally sourced imogolite (paper, biogrowth control, water cleaning) and halloysite (high-value porcelain) for centuries. That fact implies that toxicity of imogolite or halloysite would have been identified after all this time if it were an issue. This was the case for asbestos ores: in ancient times, people did not understand the nature of the danger, but they avoided living in bad fields (such as Corsica). However, with modern standards, studies indicating that an eventual toxicity of these peculiar nanocharges should be detected and estimated cannot be ignored. These same studies, of course, have to take into account the particular fields of use (eg, health, including veterinary purposes; food and cosmetics; organic-based nanocomposites; or inorganic nanocomposites), and the level of risk will be drastically different depending on the situation (for instance, a nanofilament embedded in a thermoplastic matrix probably will be less problematic than a nanofilament in a colloidal dispersion or in a spray). Some research advances regarding the safety of imogolite or halloysite were introduced in Chapter 20. For industrial purposes, the fact that there is the possibility of supply by exploiting mines is a good thing once it is possible to secure the supply chain in quantity, which may be complicated by political difficulties in the mining areas.

27.2 SYNTHETIC OR NATURAL MATERIALS?

In some high-value technical fields such as health and pharmaceuticals, the nature of chemical by-products, including the presence of other aluminosilicate-type materials, which are often associated with naturally sourced materials, can be a concern. So in this last case, synthetic materials are always better and can compete in terms of cost with highly purified, natural materials. However, for other purposes that do not require the substances to be colourless and of high purity (eg, iron-free materials), natural materials must always be considered to have a relevant competitive cost advantage. This is particularly true for halloysite materials (eg, natural halloysite can compete with silica for use in tire manufacturing).

27.3 TECHNICAL ADVANTAGES OF IMOGOLITE AND HALLOYSITE

The addition of high-aspect-ratio nanocharges to a material (such as polymers and other spherical particles) always improves the mechanical performance of the obtained nanocomposite matrices (ie, flexibility, control of cracks). These advantages allow lightweight high-tech products to be designed without needing to compromise in terms of mechanical performance, using less inorganic charges (silica, TiO2, Al2O3, clays...). Lightweight materials are and will remain a tremendous and a strategic field of investigation (for application in the car, truck and plane industries, for example). However, due to their differences in the size and level of crystallinity, imogolite and halloysite are not suited to the same markets.

Imogolite nanofilaments, which are nanocrystalline materials, exhibit a very high-aspect ratio (2 nm in diameter for several micrometres of length) greater than 2 microns. In the case of synthetic imogolite, the final aspect ratio is controlled by the choice of starting salts and the length of the digestion. Imogolite nanotubes can be formulated in water-soluble binders and used to coat various substrates (glass, plastics), becoming totally transparent in the visible range and crack-free films or monoliths. If the surface of the imogolite nanotubes is modified by the grafting of an organic moiety using silane coupling agents, the imogolite nanotubes can be dispersed in a solvent-soluble binder and coated. In this case, too, it is possible to obtain colourless and crack-free films or bulk materials. However, the inner part of the imogolite nanotubes will remain filled with water molecules, which are virtually impossible to eliminate without destroying the tubular structures.

Halloysites are crystalline products with a particle length ranging from 200 nm to a few microns (depending on the ore) and a lumen size of 30–40 nm. Due to these size parameters, it is very difficult to obtain transparent films or coatings even if the refractive indices between the halloysites and the binders are suitably matched. Halloysites can also be modified by grafting, to be dispersed in a large range of solvents. The lumen volume of halloysite can be filled with organic materials to design controlled releasing devices. This unique property of halloysite nanotubes opens many areas of application. The Exxon company created synthetic halloysite using hydrothermal treatment in monel vessels. This synthetic halloysite exhibits the same crystallinity as natural materials but a different morphology (flakes instead of nanotubes). It is possible to integrate some cations into the halloysite structure in a controlled way (Exxon Research and Engineering and Robson, 1978).

Both imogolite and halloysite are flame retardant; the sources of water in these substances are the delivery of entrapped water molecules and water created by the oxolation of aluminol and silanol groups. Imogolites are more attractive in terms of performance (Commissariat à l'Energie Atomique et aux Energies Alternatives and Poncelet, 2012). The remaining thermal by-products (mainly alumina and silica) are able to trap toxic organic materials from the smoke.

These two aluminosilicate-based materials exhibit chemical stability in a large pH range (3–11). This property is very important for ensuring long-term stability of sensitive entrapped organics (dyes, biocides, drugs, plasticizers, monomers for self-healing purposes).

To be used safely, high-aspect-ratio nanoparticles must be embedded in a matrix. The idea is to limit the creation of dangerous dust. This basic safety rule protects the workers at industrial plants (very often, nanoparticles are used in slurry form in a solvent, and major companies (Kodak, large paper industries...) are clearly ready to pay for that) once the nanoparticles are embedded into the final matrix (whether organic, inorganic, or having a mix of materials), forming a nanocomposite. This second matrix will constitute a second, efficient, long-term barrier. However, due to new environmental legislation being passed in countries with high levels of technological sophistication, the used products containing nanoparticles eventually must be identified, collected and then eliminated or recycled if possible. To collect and to put industrial waste in suitable landfills is always expensive; however, companies can greatly improve the perception of their brand by communicating their environmental policy to the public.

27.4 IMOGOLITE AND IMOGOLITE-LIKE MATERIALS AT EASTMAN KODAK

27.4.1 Antistatic Coating

The coating of silver halide–based emulsions on a plastic substrate (nitrocellulose, cellulose triacetate, polyethylene terephthalate, etc.) at high speeds (over 300 m/min) required the antistatic properties of coatings to be improved. This was the case for two reasons: (i) to avoid the emission of luminescent electrostatic discharges that can be caught by light-sensitive silver halide–based emulsions; and (ii) for safety reasons in the particular case of solvent-based silver halide emulsions (silver behenate), wherein electrostatic discharges can cause explosions in alcohol- and ketone-saturated atmospheres.

Antistatic materials for silver halide–based photographic film must exhibit two other important physical properties, which are to be totally transparent in the visible range and can survive the developing ($pH > 11$) and fixing ($pH < 4$) processes. Antistatic materials can be ionic conductors (mixtures of hydrated salts and ammonium alkyls). In this case, of course, the conductivity is sensitive to hygrometry, which could be problematic in very dry environments (eg, industrial X-ray films being used in the Arctic); or antistatic materials could be semiconductors that are intrinsically insensitive to hygrometry. Of course, these materials are brightly coloured, and the transparency of the coatings is directly correlated to their coverage rate.

Eastman Kodak developed and patented its own family of antistatic materials based upon V_2O_5 gels (doped with Li or Ag). V_2O_5 gels are both an electronic conductor and a protonic conductor due to their ribbonlike perovskite structure, wherein some water molecules are entrapped (Eastman Kodak Company and Guestaux, 1980; Eastman Kodak Company et al., 2000a). The unit component of V_2O_5 gel is a filament (2–3 nm of the cross section and several micrometres of length) greater than 2 microns. These filaments, on which electrostatic charges can easily percolate, brought a competitive advantage in terms of the metal coverage required to manage the accumulation of electrostatic charges. Classically, only 10 mg/m^2 of vanadium were necessary, as opposed to 120 mg of tin to dissipate the electrostatic charges (Tin Oxide in Fuji silver halide technology).

However, V_2O_5 gels also caused some problems. Their syntheses required the use of toxic raw materials and a long ageing step, wherein the V_2O_5 filaments grow. V_2O_5 gels are strongly acidic and oxidize the other chemical components of the silver halide–based emulsions (the total formulation of a Kodacolor emulsion requires hundreds of chemicals), and finally, part of the antistatic properties does not survive to developing and fixing baths. Some vanadium can be found in aged developing baths. Eastman Kodak Research made a continuous effort to identify other proprietary antistatic materials. Clearly, the great interest in using very high-aspect-ratio particles to obtain very transparent coating and very efficient antistatic coating by optimizing electrostatic charge percolation has been demonstrated. This is why Kodak researchers concentrated their efforts on identifying new high-aspect-ratio nanoparticles or filamentous gels. Imogolite nanotubes were evaluated in 1992. Natural imogolite products made in Japan showed unexpectedly suitable antistatic properties (Eastman Kodak Company et al., 1998a, 1999b, 2001, 2004b; Eastman Kodak Company and Poncelet, 2002). However, the presence of metal impurities that can interact with silver halide leads quickly to evaluating the value of using synthetic imogolite instead. At this time, there was no industrial supplier, so Eastman Kodak worked on the synthesis of pure imogolite filaments. In 1994, synthetic imogolites were routinely synthesized in a 6-m^3 reactor in Kodak facilities in Canada and Europe. Synthetic imogolites were supplied to Kodak laboratories under powder (freeze-dried) or slurry in water (20 g/dm^3). This is a crucial step for the development of a new material in a large company because the new material will receive an internal registry number, so all the Kodak researchers were able to order the new material and could introduce it into their products. That also means that the company started to address the relevant safety concerns and contacted the Government Environmental Agencies collecting the required data to put products containing imogolite on the market (to file the U.S. environmental protection agency docket concerning imogolite, 200 K$ were spent). This was the beginning of internal scientific communications leading to the evaluation of imogolites within all the business units of Eastman Kodak.

At the same time, internal fundamental research demonstrated that the imogolite filaments must be considered as a protonic conductor. It also has been demonstrated that due to their peculiar structure (hollow but filled with water), imogolite filaments were able to maintain a high level of humidity in the coatings over long periods of time, even in very dry environments. Therefore, pragmatically, imogolite tubes have been formulated with ionic conductors, leading to a low-cost, efficient family of antistatic materials.

As soon as imogolite filaments are made useable by the Kodak Research Department, it has been necessary to address the question of manufacturing costs. This has led to major efforts in the area of manufacturing. The initial synthetic pathway developed by Eastman Kodak Company allowed very pure imogolite nanotubes to be obtained by working in a very diluted way (10^{-3}mol.dm^{-3}) and by controlling hydrolysis followed by 4 days of digestion. The industrial synthetic cost of imogolite gels was around US $20/kg. Quickly, some business units asked for lower-cost, synthetic imogolite. This goal was achieved by synthesizing a mixture of allophane and imogolite rather than a pure phase of imogolite. Imogolite nanotubes were separated from allophane nanospheres by tangential filtration. The synthetic cost of imogolite gels moved to US $8/kg. In 1999, Eastman Kodak externalized the synthesis of imogolite gels to two European chemical suppliers; however, synthetic imogolite batches were reserved for Kodak products (Eastman Kodak Company et al., 1999a, 2000b, 2001; Eastman Kodak Company and Poncelet, 2002, 2004).

27.4.2 Cleaning of Water and Trapping of Metals

In the mid-1990s, in France, the Netherlands, Belgium, Luxemburg and Italy, legislation concerning the water consumption of silver halide–based health X-ray photoprocessors changed drastically. The total annual water consumption of photoprocessor was reduced from 120 to 20 m^3 for the same processed surface of silver halide films. Very quickly, two main concerns were raised by Kodak's customers. Because silver halide–based materials contained a great amount of gelatin (the historic binder of silver halide technology), biofilms appeared very quickly on health X-ray photoprocessors, leading to time and money being wasted on removing it. Moreover, the concentration of silver salts in washed water very quickly exceeded the legal limits on what could be disposed of in sewers. To preserve the European market of health X-ray films, Eastman Kodak helped its customers comply with the new European regulations. A device called 'multifilter' was designed and installed on the health X-ray photoprocessors, allowing for keeping the washing tanks clean and trapping the metal traces before the water went into the sewer

(Eastman Kodak Company et al., 1997, 1998b, 2003, 2004a; Eastman Kodak Company 2001). The first active part of the filter was made of imogolite gels that entrapped a biocide cocktail and acted as a slow delivery device (taking over 6 months), and the second part was made of an imogolite gel, wherein the external aluminol surface was grafted with a thiol moiety for silver trapping. Quickly, the multifilter device also was installed on industrial X-ray photoprocessors throughout the world—not for environmental reasons, but because water consumption is a concern in some areas. The manufacturing of the two cartridges was outsourced to a French company that used 20 tonnes of synthetic imogolite gels per year (2006).

27.4.3 Trapping of Acetic Acid

Very early on, the long-term keeping of silver halide–based negatives (which contain the analogical data of a movie) had been a major concern requiring negative films to be stored in a cold and dry place. Paradoxically, the most sensitive part of these processed silver halide negatives is not the dyes and the gelatin coatings, but the plastic support. The first generation of plastic was cellulose nitrate, but it was quickly replaced by cellulose triacetate for safety considerations (aged cellulose nitrate is unstable and highly explosive). But during ageing, cellulose triacetate generated acetic acid molecules that could interact with dyes and modify the mechanical properties of the negative films (ie, brittleness). In the professional motion picture industry, this phenomenon is called 'vinegar syndrome', and it requires very expensive solutions in order to keep and maintain film negatives. A postprocessing coating of imogolite gels allowed acetic acid to be trapped, and thus to preserve negative films over time (Eastman Kodak Company et al., 1998c). Some water-based imogolite sprays also have been used for the curative treatment of very old colour negatives of films.

27.4.4 Inkjet Receiver

Within Eastman Kodak, the last industrial use of imogolite and imogolite-like materials (allophane) was the design of ink-jet receivers. Quality photo ink-jet prints need to be laminated in order to make them easy to use. Unfortunately, the humectants of inks (mainly poorly volatile ether alcohols) that remained trapped under the surface-laminated plastic film slowly interacted with the plastic, creating bubbles. Adding a small amount of imogolite filaments to the receivers allowed to keep entrapped definitively the humectants, avoiding the formation of bubbles. However, imogolite and imogolite-like materials found an unexpected application in porous ink-jet receiver technology as dye stabilizers (ie, colorlast technology). Aluminol surface groups of imogolite surfaces were able to prevent dyes from fading in response to time, light and atmospheric oxidizing gases (Ilford Imaging Switzerland Gmbh et al., 2004;

Eastman Kodak Company, 2006; Eastman Kodak Company et al., 2006a,b,c, 2007). Due to the required volume for this application, Eastman Kodak tried to identify a natural source, but without success. Both imogolite and allophane were synthesized in Kodak chemical facilities in the United States.

27.5 CONCLUDING REMARKS

The properties of nanotubular clay minerals are unique, and academic research has uncovered many potential applications for nanotubular clay minerals usually largely described in patents. Some of these applications find some niche markets. However, the main industrial applications remain mainly in pottery for halloysite (Imerys) and in antistatic material coating for imogolite (Eastman Kodak). The example of applied research on imogolite conducted by Eastman Kodak has been described in detail.

The emergence of a whole imogolite family and the development of modified halloysite materials open many new application possibilities. If new mainframe industrial applications will emerge is a complex question, but the potential is evident (low toxicity, low cost, and no critical elements, among others).

REFERENCES

Commissariat à l'Energie Atomique et aux Energies Alternatives, Poncelet, O., 2012. Aluminosilicate polymer as fire retardant. US patent 8,287,779.

Eastman Kodak Company, 2006. Ink jet recording element and printing method. US patent 7,083,836.

Eastman Kodak Company, Guestaux, C., 1980. Radiation-sensitive elements having an antistatic layer containing amorphous vanadium pentoxide. US patent 4,203,769.

Eastman Kodak Company, Poncelet, O., 2002. Method to prepare an aluminosilicate polymer. US patent 6,468,492.

Eastman Kodak Company, Poncelet, O., 2004. Method for separating a mixture of colloidal aluminosilicate particles. US patent 6,685,836.

Eastman Kodak Company, Wettling, D., 2001. Aluminosilicate organic–inorganic polymer. US patent 6,179,898.

Eastman Kodak Company, Poncelet, O., Wettling, D., Rigola, J., 1997. Organic/inorganic gels for delivering controlled quantities of active compounds in aqueous solutions. US patent 5,683,826.

Eastman Kodak Company, Poncelet, O., Rigola, J., 1998a. Anti-static composition and photographic material containing a layer of that composition. US patent 5,714,309.

Eastman Kodak Company, Poncelet, O., Wettling, D., Rigola, J., 1998b. Organic/inorganic gels for delivering controlled quantities of active compounds in aqueous solutions. US patent 5,846,555.

Eastman Kodak Company, Poncelet, O., Rigola, J., Boualem, M., 1998c. Method for improving the conservation of a photographic product with a cellulose ester type support. US patent 5,853,970.

Eastman Kodak Company, Poncelet, O., Rigola, J., 1999a. New polymeric conductive aluminosilicate material, element comprising said material, and process for preparing it. US patent 5,888,711.

Eastman Kodak Company, Poncelet, O., Rigola, J., 1999b. Antistatic composition and photographic element containing a layer of this composition. US patent 5,910,400.

Eastman Kodak Company, Eichorst, D., Gardner, S., Apai II, G., Duong L., 2000a. Imaging element comprising an electrically-conductive layer containing intercalated vanadium oxide. US patent 6,013,427.

Eastman Kodak Company, Poncelet, O., Rigola, J., 2000b. Aluminum-based polymer material and use of this material in a photographic product. US patent 6,027,702.

Eastman Kodak Company, Rigola, J., Poncelet, O., Martin, D., 2001. Inorganic polymer based on aluminum and silicon. US patent 6,296,825.

Eastman Kodak Company, Poncelet, O., Roussilhe, J., Wettling, D., 2003. Method to extract silver from a photographic developer. US patent 6,555,008.

Eastman Kodak Company, Poncelet, O., Wettling, D., 2004a. Improved composite material for treating photographic effluents. US patent 6,680,066.

Eastman Kodak Company, Rigola, J., Poncelet, O., Martin, D., 2004b. Inorganic polymer based on aluminium and silicon. US patent 6,699,451.

Eastman Kodak Company, Kapusniak, R., Romano, C., Shaw-Klein, L., Schultz, T., Ghyzel, P., 2006a. Mordanted inkjet recording element and printing method. US patent 7,052,748.

Eastman Kodak Company, Kapusniak, R., Romano, C., Shaw-Klein, L., Schultz, T., Ghyzel, P., 2006b. Inkjet recording element comprising subbing layer. US patent 7,052,749.

Eastman Kodak Company, Kapusniak, R., Romano, C., Shaw-Klein, L., Schultz, T., Ghyzel, P., 2006c. Non-porous inkjet recording element and printing method. US patent 7,056,562.

Eastman Kodak Company, Kapusniak, R., Romano, C., Shaw-Klein, L., Schultz, T., Ghyzel, P., 2007. Ink jet recording element with core shell particles. US patent 7,223,454.

Exxon Research and Engineering, Robson, H., 1978. Synthetic halloysites as hydrocarbon conversion catalysts. US patent 4,150,099.

Ilford Imaging Switzerland Gmbh, Steiger, R., Brugger, P.-A., 2004. Recording sheets for ink jet printing. US patent 6,780,478 B2.

Rawtani, D., Agrawal, Y.K., 2012. Multifarious applications of halloysite nanotubes: a review. Rev. Adv. Mater. Sci. 30, 282–295.

Chapter 28

Epilogue

F. Bergaya[a], A. Thill[b,*] and P. Yuan[c]
[a]*CNRS—ICMN (Interfaces, Confinement, Matériaux et Nanostructures), Orléans Cedex 2, France*
[b]*Laboratoire Interdisciplinaire sur l'Organisation Nanométrique et Supramoléculaire, NIMBE, CEA, CNRS, University Paris-Saclay, Paris, France*
[c]*CAS Key Laboratory of Mineralogy and Metallogeny / Guangdong Provincial Key Laboratory of Mineral Physics and Materials, Guangzhou Institute of Geochemistry, Chinese Academy of Sciences (CAS), Guangzhou, China*
**Corresponding author: e-mail: antoine.thill@cea.fr*

The focus of this book is to bring some attention to the study of two interesting nanotubular clay minerals that up to now have been treated separately by different scientific communities.

Since the discovery in 1950 and 1962 of halloysite and imoglite, respectively, these two clay minerals have seen great progress in the understanding of their formation mechanisms and physicochemical properties, which are presented in the various chapters of this book.

What is evident from the whole content of this book is that pluridisciplinary approaches have been particularly effective in bringing about these recent developments. An interesting feature of this book is that it offers a clear view of the potential and actual applications of imogolite and halloysite. Like many other clay minerals, these anisotropic nanoparticles are useful as reinforcing fillers for polymer nanocomposites, ceramics, catalyst supports, liquid crystals and other products. But unlike other layered clay minerals, halloysite and imogolite have a nanotubular shape, which enables a number of very specific applications.

28.1 TWO SIMILAR CHARACTERISTICS OF THESE POROUS CLAY MINERALS

Halloysite and imogolite share two interesting characteristics.

First, their atomic composition is based on silicon (Si) and aluminum (Al), the two most abundant elements on Earth. This is a great advantage, as sustainable development considerations are growing in importance in the choice of relevant materials to use in the future.

Developments in Clay Science, Vol. 7. http://dx.doi.org/10.1016/B978-0-08-100293-3.00028-5

Second, they both offer nanotubular morphologies, giving rise to an internal confined lumen (pore) and two distinct internal and external surfaces.

Adsorption inside the nanotubes, on the internal surfaces or in the confined lumen is uniquely useful for environmental and medical applications. These applications require the encapsulation and controlled release of drugs and preventing release of trapped pollutants. In addition, the external surfaces of halloysite and imogolite, which are easier to modify than other materials, has been the subject of intensive research in order to adapt the surface properties of nanotubes composed of these substances to many different uses. Finally, selective modifications of internal and external surfaces of these nanotubular clay minerals provide new possibilities for many versatile applications.

28.2 THREE MAIN DIFFERENCES BETWEEN HALLOYSITE AND IMOGOLITE

28.2.1 Tubular Morphology Deriving from Curvatures in Opposite Ways

The tubular shapes of both halloysite and imogolite derive from the curvature in opposite ways of a dioctahedral aluminum layer (gibbsite-like), induced by the action of silicate tetrahedral, leading to open-ended nanotubes.

Indeed, the gibbsite-like surface is exposed on the internal surface of halloysite but on the external surface of imogolite (as illustrated by the yellow color sheet of the 3D schematic structures on the cover of this book). The external surface of imogolite and the internal surface of halloysite, based on Al–OH groups, are prone to easy and useful functionalization.

The internal surface of imogolite is covered by Si–OH groups, whereas the external surface of halloysite is composed of polymerized tetrahedra with siloxane groups (as illustrated by the blue color sheet of the 3D schematic structures on the cover of this book). What is interesting to highlight is the original internal surface chemistry of imogolite, exclusively based on a high density of Si–OH groups, which is a unique type of surface chemistry in clay science.

28.2.2 Spiraled Multi-Walled Halloysite vs Single-Walled or Double-Walled Imogolite

Nanotubes made of halloysite can have different lengths, varying from 50 to 5000 nm, and the size of the spiral outer diameter ranges from 20 to 200 nm. The inner diameter of the lumen pores vary from 5 to 70 nm. However, imogolite is generally a unique, single-walled nanotube with a diameter of about 2 nm, 10 times smaller than the smallest diameter of halloysite. Imogolite is also generally monodisperse, as opposed to halloysite, which is rather polydisperse. This different intriguing property is not yet understood.

28.2.3 Easy Synthesis of Imogolite vs Inexistent Synthesis of Halloysite

The last important difference between the two nanotubular clay minerals comes from the possibility of synthesising these two materials.

Imogolite can be relatively easily synthesised up to industrial scale. However, until now, a commercial supply of imogolite has not existed. Research has been limited to a rather small number of specialised laboratories that have developed the required ability and equipment to synthesise imogolite. Recently, a laboratory-scale production dedicated for research purposes has emerged at a CEA laboratory of Paris-Saclay University (where one of the co-editors works). However, as imogolite can be easily synthesised, impressive recent discoveries have been made about their formation mechanism and on the refined control of their composition and shapes. Nowadays, imogolite can be considered as an emerging family of tunable, monodisperse and microporous nanoparticles.

The situation is almost the opposite for halloysite. Surprisingly, nanotubular halloysite has not yet been successfully synthesised at any scale via direct precipitation from SiO_2- and Al_2O_3-containing solutions under hydrothermal conditions, but it is exclusively available through purification of raw ores from natural deposits. The commercial availability of raw halloysite has enhanced the research into many of its potential applications as opposed to imogolite. However, as halloysite cannot be synthesised through a solution chemistry method, it has not been possible to work on the control of their composition and shape using a bottom-up route. Nevertheless, an indirect synthetic route is available, as the derivation of halloysite from delamination of natural kaolinite has been achieved. The control of the morphology and porosity of halloysite is also possible using a vast variety of posttreatment methods.

28.3 PERSPECTIVES

Two major challenges have emerged from the body of research on halloysite and imogolite:

- Discovering a synthesis process through the low-cost (less than US $10/Kg) production of imogolite on a large scale to make competitive products and to boost the research on its use as a new material.
- Inventing a fast synthesis process to obtain well-crystallised nanotubular halloysite from their chemical precursors to explore the tunability of halloysite composition and structure.

Other important aspects not fully considered, which should attract the attention of the scientific community, are to ensure the low toxicity of all new recipes and compositions, as well as developing 'safe-by-design', environmentally friendly, high-tech nanomaterials based on natural clay minerals. Halloysite and imogolite have many advantages to successfully handle this challenge: namely, the fact that they feature no critical atoms in their composition, have mostly water-based processing and offer biocompatibility of their raw materials.

Low-cost, high-tech, and sustainable new substances are expected to help solve the tremendous challenges facing the future of scientific advances, such as: (i) replacement of critical resources, (ii) water supply, (iii) energy saving and (iv) new disease treatments. Thanks to their exceptional properties, the nanosized tubular clay minerals halloysite and imogolite may play a vital role in the resolution of these challenges.

Index

Note: Page numbers followed by "*f*" indicate figures, and "*t*" indicate tables.

D

E

H

J

K

L

M

N

O

P

R

T

U

V

W

X

Z

Edwards Brothers Malloy
Ann Arbor MI. USA
July 25, 2016